Critical Issues in Systems Theory and Practice

Critical Issues in Systems Theory and Practice

Edited by

Keith Ellis and Amanda Gregory

University of Humberside
Hull, England

Bridget R. Mears-Young and Gillian Ragsdell

University of Hull
Hull, England

Plenum Press • New York and London

Library of Congress Cataloging-in-Publication Data

On file

Proceedings of the Fourth International Conference of the United Kingdom Systems Society, on Critical Issues in Systems Theory and Practice, held July 10–14, 1995, in Hull, England

ISBN 0-306-45100-X

A Division of Plenum Publishing Corporation
233 Spring Street, New York, N. Y. 10013

10 9 8 7 6 5 4 3 2 1

Printed in the United States of America

PREFACE

The systems movement, now 40 years old, is made up of many associations of systems thinkers from different disciplines all over the world. The United Kingdom Systems Society (UKSS) was formed in 1978. Today it has over 300 members and is committed to the development and promotion of "systems" philosophy, theory, concepts and methodologies for improving decision making for the benefit of organizations and wider society.

The first UKSS International Conference was held at the University of Hull in July 1989. Since then we have held International Conferences at the Universities of Huddersfield (1991) and Paisley (1993). The UKSS International Conferences are now an established biannual event and this, our fourth international conference, will be jointly hosted by the Universities of Hull and Humberside.

Systems science is considered to be a trans-discipline which promotes critical and effective intervention in complex organisational and social problem situations. As such it traverses "hard", through "soft" to "critical" systems thinking and methodologies. Yet, despite the currently robust state of the UKSS the systems movement cannot be described as an international movement: different subdisciplines are at different stages of development and are often engaged in pursuing their own particular interests and themes with little "conversation" between the subdisciplines despite their common interest in systems.

Consequently, it may be argued that the future of the systems movement depends upon our ability to address the question: "How can the various disciplines of the systems movement work together in addressing the problems which face the human race?" We feel that the systems movement is facing a crisis. The age of the different disciplines working independently is gone. The role of the trans-disciplines, such as the systems movement, as the means by which communication across the disciplines may be facilitated, is becoming increasingly important. Our concern for this crisis, and the important role being bestowed on the systems movement, is reflected in our choice of the conference theme "Critical Issues in Systems Theory and Practice".

This volume contains many important papers which have been selected for the contribution they are likely to make to the debate on how the subdisciplines might best come together and on the critical issues which will effect this process. We urge that this is not a sterile, inward-looking debate but one which will help us, the members of the systems movement, to take the steps necessary to make a creative and useful contribution to the difficult age in which we live.

Keith Ellis
Amanda Gregory
Bridget R. Mears-Young
Gillian Ragsdell

PROGRAMME COMMITTEE

Keith Ellis,[1] Amanda Gregory,[1] Bridget Mears-Young,[2] and Gillian Ragsdell[2]

[1] University of Humberside
[2] University of Hull

ACKNOWLEDGMENTS

The Programme Committee would like, on behalf of the United Kingdom Systems Society, to thank a variety of people and institutions for their help in the organisation of this conference.

We are grateful to Professor Roger King, Vice-Chancellor of the University of Humberside; Professor Bob Flood, Head of the Department of Management Systems and Sciences; University of Hull, and Professor Mike Jackson, Dean of the School of Computing and Information Systems, University of Humberside, for their encouragement and generosity.

Our thanks go to Mr. Peter Wood, Secretary to the Lord Mayor of Hull, and to Councillor Paddy Doyle, Leader of the Hull City Council, for their giving their support to the conference.

Our gratitude also goes to Gillian Purdey, Hull Conference Bureau; Yootha Edenbrow, University of Humberside; and Melissa Ward, University of Hull, for their assistance and guidance on the organisation and promotion of the conference.

We owe a special debt to Doreen Gibbs for her commitment to the project as a whole and for her work in putting this volume together. Finally, we could not have managed the compilation of these proceedings without the help of our managing editor, Joanna Lawrence of Plenum Publishing Company.

We should also like to extend our thanks to those organisations and institutions which have sponsored this conference:

Ernst and Young
Hull City Council
University of Hull
University of Humberside

Co-sponsoring societies:

Dutch Systems Society
European Systems Union
Hellenic Systems Society
International Society for the Systems Sciences
Polish Systems Society
Spanish Systems Society

International Advisory Committee:

Dr. Evelyne Andreeswsky (France)
Prof. Rodriguez Delgado (Spain)
Prof. Wojceck Gasparski (Poland)
Dr. David B. Keever (USA)

Prof. Alfonso Rivas (Mexico)
Dr. Norma Romm (UK with special responsibility for South Africa)
Dr. Julie Travis (Australia)
Dr. Ioanna Tsivacou (Greece)
Prof. Gerard de Zeeuw (Holland)

CONTENTS

Plenary Papers

Student Papers

Systems Methodology and Practice

Information Systems Theory and Practice

Critical Issues in Information Management

Critical Systems Thinking

Critical Issues in Business Systems

Medical Information and Health Care Systems

Educational Systems

Community and Environmental Systems

PLENARY PAPERS

1

SYSTEMS THEORY AND MANAGEMENT THINKING

Peter Checkland

University of Lancaster
Lancaster, United Kingdom

ABSTRACT

Two inquiring systems developed since the 1960s - Vickers's concept of the appreciative system and the soft systems methodology, are highly relevant to the problems of the 21st century. Both assume that organizations are more than rational goal-seeking machines and address the relationship-maintaining and Gemeinschaft aspects of organisations, characteristically obscured by functionalist and goal-seeking models of organization and management. Appreciative systems theory and soft systems methodology enrich rather than replace these approaches.

INTRODUCTION

Two rich metaphors provide a useful frame within which any consideration of the problems facing us in the late 20th century can, with advantage, be placed. As a result of the first industrial revolution, based on energy, and the current second one, based on information, the world is increasingly Marshall McLuhan's 'global village.' More and more problems need to be examined in a global rather than a local context and, as we do so, we need to remember that we are all of us, in Buckminster Fuller's great phrase, 'the crew of Spaceship Earth.' Thanks to the material successes of the two industrial revolutions we are a crew with rising expectations of high living standards. But we are increasingly aware that the wealth-generating machine may not be able to meet those expectations without doing unacceptable damage to Spaceship Earth, which together with the free supply of energy from our sun, is the only given resource we have.

This triangle - of expectations, wealth generation, and protection of the planet - will have to be managed with great care at many different levels as we enter the 21st century if major disasters are to be avoided. Unfortunately, our current ideas on *management* are rather primitive and are probably not up to the task. They stem from the technologically oriented thinking of the 1960s, and they now need to be enlarged and enriched. This may well be possible from the systems thinking of the 1970s and 1980s, which has placed that body of thought more firmly within the arena of human affairs.

Critical Issues in Systems Theory and Practice, Edited by K. Ellis et al.
Plenum Press, New York, 1995

This article will examine the legacy of thinking about management and organizations that we get from the 1960s and develop a richer view that stems from more recent systems thinking, especially Vickers's work on the theory of *appreciative systems* and work on soft systems methodology, which can be seen as a way of making practical use of Vickers's concepts. This, it is argued, is more relevant than the current conventional wisdom to managing the problems of the new century.

MANAGEMENT AND ORGANIZATION

In spite of a huge literature - some of it serious, much of it at the level of airport paperbacks - and courses in colleges and universities world-wide, the role of the manager and the nature of the process of managing remain problematic, whether we are concerned with trying to manage global, institutional, or personal affairs. Anyone who has been a professional manager in an organization knows that it is a complex role, one that engages the whole person. It requires not only the ability to analyze problems and work out rational responses but also, if the mysterious quality of leadership is to be provided, the ability to respond to situations on the basis of feelings and emotion.

One of the reasons the manager's role remains obstinately problematic stems from our less-than-adequate thinking about the context in which managers perform, namely the organization. Some basic systems thinking indicates that if we adopt a limited view of organization then the conceptualization of the manager's role will inevitably also be rather threadbare. Thus a manager at any level occupies a role within a structure of roles that constitutes an organization. The activity undertaken by managers can be seen as a system of activity that serves and supports and makes its contribution to the overall aims of the organization as a purposeful whole. Now, if one system serves another, it is a basic tenet of systems thinking that the system that serves can be conceptualized only after prior conceptualization of the system served (Checkland, 1981, p.237). This is so because the form of a serving system, if it is truly to serve, will be dictated by the nature of the system served: That will dictate the necessary form of any system that aspires to serve and support it.

Now there is a conventional wisdom about the nature of organization that persists in spite of the fact that anyone who has worked within an organization knows that this image conveys only part of the story. The conventional model is that an organization is a social collectivity that arranges itself so that it can pursue declared aims and objectives that individuals could not achieve on their own. Given this view of organization, the manager's role is to help achieve the corporate goals, and it follows that the manager's activity is essentially rational decision making in pursuit of declared aims. This is the conventional wisdom even though intuitively we all have a rich sense that organizations in which we have worked are more than rational goal-seeking machines. The experienced day-to-day reality of organizations is that they have some of the characteristics of the tribe and the family as well as the characteristics necessary if they are to order what they do rationally so as to achieve desired objectives such as, in the case of industrial companies, survival and growth. In spite of this folk knowledge, the orthodoxy has been very strong, and we can see this both in the literature of organization theory and in that of management science.

Organization Theory

This is not the place to discuss the development of organization theory in any detail, but it is useful for present purposes to mark the general shape of this field as it emerges in such wide-ranging studies as Reed's (1985) *Redirections in Organisational Analysis*. The general shape is that of the establishment of an orthodoxy (the systems/contingency model that held sway from the 1930s to the 1960s) and the challenge to that orthodoxy since then, with no single dominant alternative. Nevertheless, the challenging models do, in general, have in common the fact that they see organizations not as reified objects independent of organizational members, as in the orthodox systems model, but as the continually changing product of a human process in which social reality is socially constructed: the title of Berger and Luckmann's (1966) well-known book - *The Social Construction of Reality* - neatly captures this alternative strand of thinking.

At a broad level of generalization, we can see the two major approaches as reflecting the two main categories of thinking about organizations on which a pioneering sociologist, Ferdinand Tonnies, built his account. In his major work *Gemeinschaft und Gesellschaft* (1887) (translated as *Community and Association* by Loomis, 1955), Tonnies constructed models of two types of society or organization. There is the natural living community into which a person is born, the family or the tribe, (Gemeinschaft), and there are the formally created groupings (Gesellschaft) that a person joins in some contractual sense, as when he or she becomes an employee of a company or joins a climbing club.

In general, the orthodox view of organizations emphasizes their Gesellschaft nature, that they are created to do things collectively (*achieve goals* in the usual language) that would be beyond the reach of individuals. The alternatives emphasize rather that all social groupings take on some flavour of Gemeinschaft: Being in an organization is something like being part of a family. Intuitively, the lived experience of organizations that we all gradually acquire gives us the folk knowledge that organizations exhibit some of the characteristics of both models.

That the orthodox view has been dominant can be seen by perusing college textbooks, which present students with the conventional wisdom. For example, in Khandwalla's (1979) *The Design of Organizations*, the view of organizations as open systems devoted to achieving corporate objectives is described as "the most powerful orientation in organization theory today" (p.251). Much attention is paid to well-known work aimed at correlating an organization's structure with its core tasks carried out in an environment with which it interacts (Lawrence & Lorsch, 1967; Pugh & Hickson, 1976; Woodward, 1965; etc). Reed's (1985) survey argues that "...systems theorists...had dominated organizational analysis since the 1930s" (p.35) but that by the 1960s there was no common history or intellectual heritage. By the 1970s, a systems-derived approach was "...struggling to retain its grip on organizational studies" (p.106). This does not mean that the orthodoxy has lost its adherents, however. In the same year that Reed's book was published, Donaldson (1985) brought out his *In Defence of Organization Theory*, the defense being of the "...relatively accepted contingency-systems paradigm" (p.ix).

Both Reed and Donaldson make much reference to a book that marks as much as any other the challenge to the orthodox systems view: Silverman's (1970) *The Theory of Organizations*. Silverman contrasts the systems view from the 1950s and 1960s with what he calls "the Action frame of reference" in which action results from the meanings that members of organizations attribute to their own and each other's acts. Organizational life becomes a collective process of meaning attribution; attention is displaced away from the apparently impersonal processes by means of which, in the conventional model, a reified organization as an open system responds to a changing environment. Some of Silverman's subheadings convey the nature of his argument: Action not behaviour, Action arises from meanings, Meanings as social facts, Meanings are socially sustained, Meanings are socially changed.

This important work opens the way to various alternatives to the systems orthodoxy. Donaldson's discussion, for example, includes social action theory, the sociology of organizations, and the strategic choice thesis. Just as the orthodoxy draws on a positivist philosophy and a functionalist sociology, the alternatives are underpinned philosophically by phenomenology, and sociologically by an interpretive approach derived from Weber and Schutz.

It has to be said that the orthodox view provides a much clearer model of organization, and hence the manager's role, than is provided by the alternatives. Concentrating on the Gesellschaft aspects of an organization, the conventional view sees it as an open system seeking to achieve corporate objectives in an environment to which it has to adapt. Its tasks are analyzed and assigned to groups within a functionalist structure, and the managers' role is essentially that of decision making in pursuit of corporate aims that also provide the standards against which progress will be judged. No similarly clear picture is provided by the alternatives, beyond the notion that organizations are characterized essentially by discourse that establishes the meanings that will underpin action by individuals and groups.

It is not at all surprising that that section of management literature most concerned with intervening in, in order to influence and shape, real-world situations, namely management science, should itself focus on the orthodox systems model.

Management Science

In examining briefly the state of thinking in management science, it is useful to focus on the work of Herbert Simon. There are two reasons for this. First, it has been a dominating contribution in the field; second, in developing an approach based on the work of Vickers, we find that he explicitly contrasted his approach with that of Simon, drawing attention to the reliance of Simon on a goal-seeking model of human action that he himself was deliberately trying to transcend.

In the period after the Second World War, strenuous efforts were made to apply the lessons from wartime operations research to industrial companies and government agencies. In doing this, a powerful strand of systems thinking was developed - it would now be thought of as 'hard' systems thinking - concerned broadly with engineering a system to achieve its objectives. Systems were here assumed to exist in the world; it was assumed that they could be defined as goal seeking; and ideas of system control were generalized in cybernetics. These ideas mapped the orthodox stance of organization theory discussed in the previous section, and they conceptualized the manager's task as being to solve problems and take decisions in pursuit of declared goals. Indeed, this paradigm is succinctly expressed in Ackoff's (1957) assumption that problems ultimately reduce to the evaluation of the efficiency of alternative means for a designated set of objectives.

This is the field to which Simon has made such a significant and influential contribution, the flavor of which is captured in the title of his 1960 book: *The New Science of Management Decision*.

At a round table devoted to his work, Zannetos (1984) summarized Simon's legacy as "...a theory of problem solving, programs and processes for developing intelligent machines, and approaches to the design of organizational structures for managing complex systems" (p.75).

Overall, Simon sought a science of administrative behaviour and executive decision making. In an intellectually shrewd move that has no doubt helped to make this body of work so influential, Simon wisely abandoned the notion that managers and administrators seek to *optimize*, replacing it with the idea of *satisficing*: the idea that the search is for solutions that are *good enough* in the perceived circumstances, rather than optimal (March & Simon, 1958). Nevertheless, the flavour of hard systems thinking is retained in the claim that the search is "motivated by the existence of problems as indicated by gaps between performance and goals" (p.73).

Similarly in another of Simon's major contributions, the development with Newall of GPS (general problem solver), a heuristic computer program that seeks to simulate human problem solving, the whole work is built on the concept of problem solving as a search for a means to an end that is already declared to be desirable (Newall & Simon, 1972. Simon (1960) stated, "Problem solving proceeds by erecting goals, detecting differences between present situation and goal, finding in memory or by search tools or processes that are relevant to reducing differences of these particular kinds, and applying these tools or processes. Each problem generates sub-problems until we find a sub-problem we can solve - for which we have a program stored in memory. We proceed until, by successive solution of such sub-problems, we eventually achieve our overall goal - or give up (p.27)."

This is an especially clear statement of the thinking, derived from the systems theory of the 1950s, that has dominated management science and that underlies organization theory's orthodox model of what an organization is.

It is the argument here that this goal-seeking model, largely adequate though it was in the management science that contributed to post-Second World War industrial development, is not rich enough to support and sustain the management thinking now needed by the crew of Spaceship Earth, that spaceship having become akin to a global village.

An alternative, richer perspective is provided by the systems thinking of the 1970s and 1980s, and in particular by Vickers's development of appreciative systems theory and by an approach to intervention in human affairs that can be seen as making practical use of that theory, namely, soft systems methodology.

These are discussed in the next section, but it may be useful to point out at once that these are developments in what is now known as 'soft' systems thinking, as opposed to the hard systems thinking of the 1950s and 1960s that permeates both orthodox organization theory and Simonian management science. The usual distinction made between the two is that the hard systems thinking tackles well-defined problems (such as optimizing the output of a chemical plant), whereas the soft approach is more suitable for ill-defined, messy, or wicked problems (such as deciding on health care policy in a resource-constrained situation.) This is not untrue, but it fails to make an intellectual distinction between the two. The real distinction lies in the attribution of systemicity (having the property of system-like characteristics). Hard systems thinking assumes that the world is a set of systems (i.e., is *systemic*) and that these can be *systematically* engineered to achieve objectives. In the soft tradition, the world is assumed to be problematic, but it is also assumed that *the process of inquiry* into the problematic situations that make up the world can be organized as a system. In other words, assumed systemicity is shifted: from taking the world to be systemic to taking the process of inquiry to be systemic (Checkland, 1983, 1985b).

Thus in the following section both appreciative systems theory and soft systems methodology describe inquiring processes - the former with a view to understanding, the latter with a view to taking action to improve real-world problem situations.

Finally, we may note that soft systems thinking can be seen as representing the introduction of systems thinking into Silverman's action frame of reference, although the organization theory literature is apparently at present innocent of any knowledge of post-1960s development in systems thinking (Checkland, 1994).

APPRECIATIVE SYSTEMS THEORY

The Nature of an Appreciative System

The task that Vickers set himself in his 'retirement' after 40 years in the world of affairs was to make sense of that experience. In the books and articles that he then wrote he constructed: "...an epistemology which will account for what we manifestly do when we sit round board tables or in committee rooms (and equally though less explicitly when we try, personally, for example, to decide whether or not to accept the offer of a new job)." (G. Vickers, personal communication, July 1974).

In his thinking as this project developed Vickers first rejected the ubiquitous goal-seeking model of human activity; then he found systems thinking relevant to his task; but he also rejected the cybernetic model of the steersman (whose course is defined from outside the system), replacing it by his more subtle notion of 'appreciation' (Vickers, 1965, is the basic reference). He expressed his intellectual history in the following terms in a letter to the present writer in 1974:

> "It seems to me in retrospect that for the last twenty years I have been contributing to the general debate the following neglected ideas:
>
> 1. In describing human activity, institutional or personal, the goal-seeking paradigm is inadequate. Regulatory activity, in government, management or private life consists in attaining or maintaining desired relationships through time or in changing and eluding undesired ones.
> 2. But the cybernetic paradigm is equally inadequate, because the helmsman has a single course given from outside the system, whilst the human regulator, personal or collective, controls a system which generates multiple and mutually inconsistent courses. The function of the regulator, is to choose and realise one of many possible mixes, none fully attainable. In doing so it also becomes a major influence in the process of generating courses.
> 3. From 1 and 2 flows a body of analysis which examines the 'course-generating' function, distinguishes between 'metabolic' and functional relations, the first being those which serve the stability of the system (e.g. budgeting to preserve solvency and liquidity), the second being those which serve to bring the achievements of the system into line with its multiple and changing standards of success. This leads me to explore the nature and origin of these standards of success and thus to distinguish between norms or standards, usually tacit and known by the mismatch signals which they generate in specific situations, and values, those explicit general concepts of what is humanly good and bad which we invoke in the debate about standards, a debate which changes both." (G. Vickers, personal communication, 1974).

In developing the theory of appreciative systems and relating it to real-world experience, Vickers never expressed the ideas pictorially, in the form of a model, although this seems a desirable form in which to express a system. (His explanation for this lack was disarming: "You must remember," he said, "that I am the product of an English classical education" (G. Vickers, personal communication, 1979). What follows is an account of the model of an appreciative system developed by Checkland and Casar (1986) from the whole corpus of Vickers's writings.

From those writings we may highlight some major themes that recur:

- A rich concept of day-to-day experienced life (compare Schutz's (1967) *Lebenswelt*).
- A separation of judgements about what is the case, *reality judgements*, and judgements about what is humanly good or bad, *value judgements*.
- An insistence on *relationship maintaining* as a richer concept of human action than that popular but poverty-stricken notion of goal seeking.
- A concept of action judgements stemming from reality and value judgements.
- A notion that the cycle of judgements and actions is organized as a system.

The starting point for the model is the Lebenswelt, the interacting flux of events and ideas unfolding through time. This is Vickers's 'two stranded rope,' the strands inseparable and continuously affecting each other. Appreciation is occasioned by our ability to select, to choose. Appreciation perceives (some of) reality, makes judgements about it, contributes to the ideas stream, and leads to actions that become part of the events stream. Thus the basic form of the model is that shown in Figure 1. There is a recursive loop in which the flux of

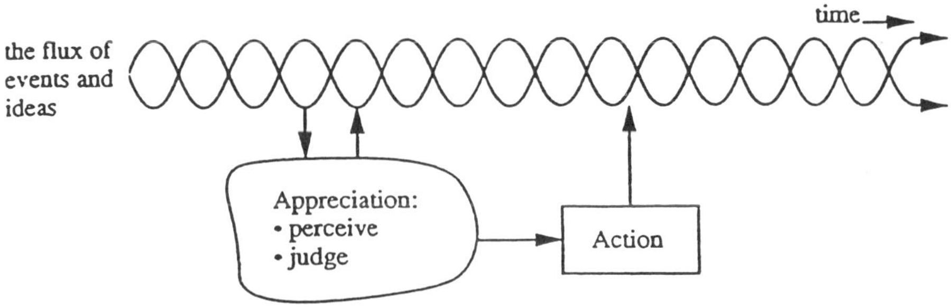

Figure 1. The structure of an appreciative system. Source: Checkland and Casar (1986).

events and ideas generates appreciation, and appreciation itself contributes to the flux. Appreciation also leads to action that itself contributes to the flux.

It is now necessary to unpack the process of appreciation. From Vickers's writings we take the notion of perceiving reality selectively and making judgements about it. The epistemology of the judgement making will be one of relationship managing rather than goal seeking, the latter being an occasional special case of the former. And both reality and value judgements stem from standards of both fact and value: standards of what *is*, and standards of what is good or bad, acceptable or unacceptable. The very act of using the standards may itself modify them.

These activities lead to a view on how to act to maintain, to modify, or to elude certain forms of relevant relationships. Action follows from this, as in Figure 2.

The model also tries to capture Vickers's most important point and greatest insight, namely, that there is normally no ultimate source for the standards by means of which what is noticed is deemed good or bad, important or unimportant, relevant or irrelevant, and so on. The source of the standards is *the previous history of the system itself.* In addition, the present operation of the system may modify its present and future operation through its effect on the standards. These considerations, together with those already discussed, yield Figure 2 as a model of an appreciative system. The most difficult aspect to model is the dynamic one, but it should be clear from Figure 2 that the dynamics of the system will be as shown in Figure 3. The form of the appreciative system remains the same, whereas its contents (its *setting*) continually (but not necessarily continuously) change. An appreciative *system* is a

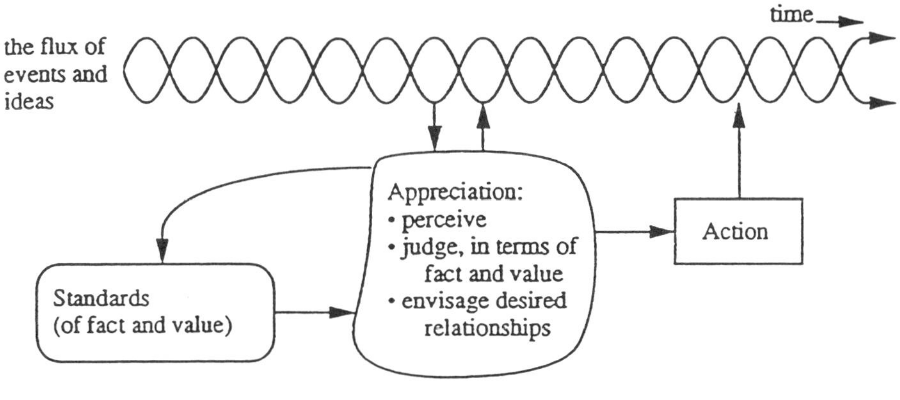

Figure 2. The structure of an appreciative system expanded. Source: Checkland and Casar (1986).

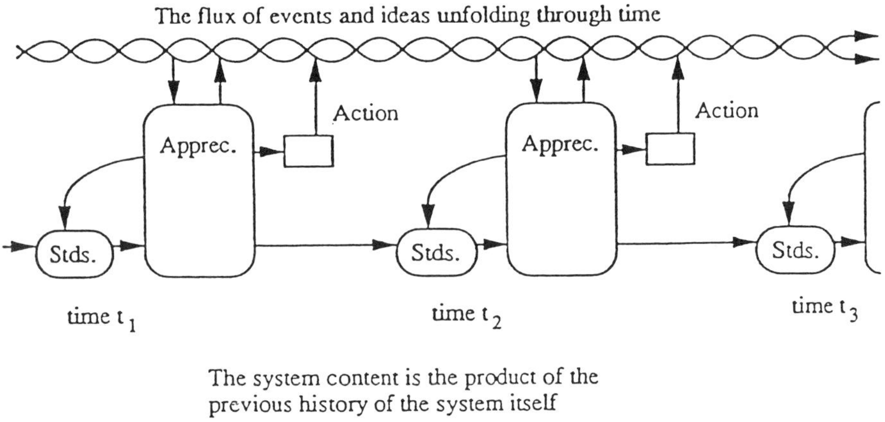

Figure 3. The dynamics of an appreciative system. Source Checkland and Casar (1986).

process whose products - cultural manifestations - condition the process itself. But the system is not operationally closed in a conventional sense. It is operationally closed via a structural component (the flux of events and ideas) that ensures that it does not, through its actions, reproduce exactly itself. It reproduces a continually changed self, by a process that Varela (1984) called the 'natural drift' of 'autopoietic systems' (Maturana and Varela, 1980), systems whose component elements create the system itself. Through its (changing) filters the appreciative system is always open to new inputs from the flux of events and ideas, a characteristic that seems essential if the model is to map our everyday experience of the shifting perceptions, judgements, and structures of the world of culture.

Vickers's claim was that he had constructed an epistemology that can provide convincing accounts of the process by which human beings and human groups deliberate and act. The model in Figure 2 and 3 is a systemic version of the epistemology.

Checkland and Casar (1986) used it to give an account of the learning in a systems study of the Information and Library Services Department of what was then ICI Organics (a manufacturer of fine chemicals within the ICI Group), a study that has been described in detail elsewhere (Checkland, 1985(a); Checkland and Scholes, 1990). This study was carried out by a group of managers in the function with some outside help in the use of soft systems methodology (SSM), which was the methodology used. It is a way of making practical use of the notion of an appreciative system, and it will be discussed briefly in the next section. It entails structuring a debate about change by building models of purposeful activity systems and comparing them against perceptions of the real world as a means of examining what the appreciative settings are in the situation in question and how they and the norms or standards are changing. In the study in question, there were three cycles of this learning process.

In the first cycle, the study team's interest and concern was to rethink the role of their function in a changing situation. They perceived many facts relevant to this, which resulted in 26 relevant systems. They selected and judged these facts in terms of a conception of a particular relationship and standards relevant to it: They accepted the relevance of a simple model that took as given that their function was a support to the wealth-generating operations of their company, and they implicitly made use of standards according to which a good version of this relationship would be to make efficient, effective, and timely provision of information to other parts of the company.

These considerations contributed to the ideas stream of the Lebenswelt and led to the action of exploring several perceptions of the relationship between the function and the rest of the company in greater depth. In this second methodological cycle, the focus was still on the relationship between function and company but the appreciative settings began to change. This can be expressed as a change in standards resulting from the first cycle of appreciation. The shift was in the concept of what would constitute as good relationship:

> "The focus shifted from ILSD (Information and Library Services Department) as a reactive function responding quickly and competently to user requests and having the expertise to do it, to ILSD as a proactive function, one which could on occasion tell actual and potential users what they *ought* to know." (Checkland, 1985(a), p.826).

In the third cycle, the new concept of ILSD was developed and, in the language of Figure 2, several hypothetical forms of relevant relationships were considered. This led to attention being given both to internal relationships within the function (How different would they have to be to sustain a proactive role?) and to the relationship between the function and the company. These considerations led to decisions on actions necessary to broaden the appreciative process. The actions taken were to make both internal (within ILSD) and external presentations of the results of the study. These events entered the company's Lebenswelt and had the effect of starting to bring about the change in the company's appreciative systems, as evidenced by the remark made by the research manager at the external presentation, namely, that I have known and worked with ILSD for 20 years and I came along this morning out of a sense of duty. To my amazement I find I now have a new perception of ILSD" (Checkland 1985(a), p.830).

Finally, the company's subsequent allocation of significant new resources to ILSD can be described as illustrating its implicit adoption of new standards with respect to the Information and Library Services function, standards whose change stems from the recent history of the company's appreciative system, involving input of ideas and events from the systems study itself.

The Appreciative Process in Action: Soft Systems Methodology

It is not appropriate here to give a detailed account of SSM, which is described in numerous books and articles since the early 1970s. (The basic books describing its development are Checkland, 1981; Checkland and Scholes, 1990; and Wilson, 1984; a burgeoning secondary literature may be sampled via, for example, Avison and Wood-Harper, 1990; Davies and Ledington, 1991; Hicks, 1991; Patching, 1990; and Waring, 1989).

SSM was not an attempt to operationalize the concept of an appreciative system; rather, after SSM had emerged in an action research program at Lancaster University, it was discovered that its process mapped to a remarkable degree the ideas Vickers had been developing in his books and articles (Checkland, 1981, chap. 8.)

The Lancaster program began by setting out to explore whether or not, in real-world managerial rather than technical problem situations, it was possible to use the approach of systems engineering. It was found to be too naive in its questions (What is the system? What are its objectives? etc.) to cope with managerial complexity, which, we could now say, was always characterized by conflicting appreciative settings and norms. Systems engineering as developed for technical (well-defined) problem situations had to be abandoned, and SSM emerged in its place.

The development of SSM has been charaterized by four points in time at which what can now be seen, with hindsight, as crucial ideas moved the project forward (Checkland and Haynes, 1994). The first was the realization that all real-world problem situations are characterised by the fact that they reveal human beings seeking or wishing to take *purposeful action*. This led to purposeful action being treated seriously as a systems concept. Ways of building models of human activity systems were developed. Then it was realized that there can never be a single account of purposeful activity, because one observer's terrorism is another's freedom fighting. Models of purposeful activity could only be built on the basis of a declared Weltanschauung. This meant that such models were never models *of* real-world action; they were models *relevant* to discourse and argument about real-world action; they were epistemological devices that could be used in such discourse and debate; they were best thought of as *holons*, using Koestler's (1967) useful neologism, which could structure debate about different ways of seeing the situation. This led to the third crucial idea, that the problem-solving process that was emerging would inevitably consist of a learning cycle in which models of human activity systems could be used to structure a debate about change. The structure was provided by carrying out an organized comparison between models and perceived real situations in which accommodations between conflicting perspectives could be sought, enabling action to be taken that was both arguably desirable - in terms of comparisons between models and perceived situation - and culturally feasible for a particular group of people in a particular situation with its own particular history. (The fourth crucial idea, not relevant here, was the realization that models of human activity systems could be used to explore issues concerning what information systems would best be created to support real-world action - which took SSM into the field of information systems and information strategy).

Given these considerations, SSM emerged as the process summarized in Figure 4. This is a picture of a *learning system* in which the appreciative settings of people in a problem situation - and the standards according to which they make judgements - are teased out and debated. Finally, the influence of Vickers on those who developed SSM means that the action to improve the problem situation is always thought about in terms of managing relationships - of which the simple case of seeking a defined goal is the occasional special case.

CONCLUSION: THE RELEVANCE OF APPRECIATIVE SYSTEMS THEORY AND SSM TO MODERN MANAGEMENT

It is not difficult to envisage the situations in both industry and the public sector in which the thinking about problems and problem solving would be significantly helped by the models underpinned by hard systems thinking , namely the models that see organizations as co-ordinated functional task systems seeking to achieve declared goals and that see the task of management as decision making in support of goal seeking. These models would be useful in situations in which goals and measures of performance were clear-cut, communi-

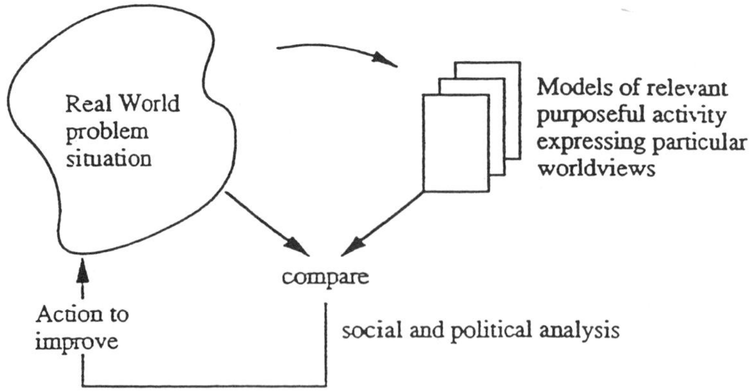

Figure 4. Soft systems methodology as a learning system. Source: Checkland and Scholes (1990).

cations between people were limited and prescribed, and in which the people in question were deferential towards the authority that laid down the goals and the ways in which they were to be achieved. But this image has never accurately described life in most organizations as most people experience it, and it has become less and less true since the end of the Second World War. Since that time the trends have been toward much increased capacity for communication, greater complexity of goals as economic interdependence has increased, much reduced deference toward authority of any kind, and the dismantling of monolithic institutionalized power structures. The dethronement of the mainframe computer by the now ubiquitous personal computer is at once both a metaphor for these changes and one of the catalysts for their occurrence.

In such a situation richer models of organization and management will be helpful, and it has been argued that those based on Vickers's appreciative systems theory and SSM have a role to play here. More important, they do not replace the older models but rather subsume and enhance them. In SSM, focusing on a unitary goal is the occasional special case of debating multiple perceptions and proceeding on the basis of *accommodations* between different interests. For Vickers, managing relationships is the general case of human action, the pursuit of a goal the occasional special case.

Vickers himself has usefully differentiated his stance from that of Simon in remarks that relate to the latter's *Administrative Behaviour* (Simon, 1957).

> "The most interesting differences between the classic analyses of this book and my own seem to be the following:
>
> 1. I adopt a more explicitly dynamic conceptual model of an organisation and of the relations, internal and external, of which it consists, a model which applies equally to all its constituent sub-systems and to the larger systems of which it is itself a part.
> 2. This model enables me to represent its 'policy makers' as regulators, setting and resetting courses or standards, rather than objectives, and thus in my view to simplify some of the difficulties inherent in descriptions in terms of 'means' and 'ends'.
> 3. I lay more emphasis on the necessary mutual inconsistency of the norms seeking realisation in *every* deliberation and at *every* level of organisation and hence on the ubiquitous interaction of priority, value and cost.
> 4. In my psychological analysis linking judgements of fact and value by the concept of appreciation, I stress the importance of the underlying appreciative *system* in determining how situations will be seen and valued. I therefore reject 'weighing' (an energy concept) as an adequate description of the way criteria are compared and insist on the reality of a prior and equally important process of 'matching' (an information concept).
> 5. I am particularly concerned with the reciprocal process by which the setting of the appreciative system is itself changed by every exercise of appreciative judgement." (Vickers, 1965, p./22).

As an example of the relevance of SSM to current problems of managing complexity, we offer recent work done within the National Health Service (NHS) in the United Kingdom. (Some of this is described in Checkland and Holwell, 1993; Checkland and Poulter, 1994; and Checkland and Scholes, 1990, chap. 1).

In recent years the NHS has been subjected to several waves of government imposed change. First there was the imposition of a system of accountable management, replacing the previous consensus management of teams of professionals. This had hardly settled down before it was replaced by an internal quasi-market. In this development the old district health authorities (into which the previous change had introduced district general managers) became purchasers of health care for a defined population, whereas hospitals and some general practitioners became providers of health care, the two being linked by contracts (although not legally binding ones) for particular services at a negotiated price. All these changes have entailed a considerable shift in appreciative settings for health professionals, and the NHS has been experiencing a period of considerable turmoil.

In the study described in Checkland and Scholes (1990), the problem was addressed of how a Department of Community Medicine in what was then a district health authority could evaluate its performance. Clearly the evaluation standards would depend completely on this department's image of itself and its roles within the district. This is not a casual consideration, because concepts of community medicine range from *providing epidemiological data* to *managing the delivery of health care*. In this work, SSM-type models of purposeful activity relating to concepts of community medicine were built, with participation of members of the department, and eventually an evaluation methodology was developed. This was based on a structured set of questions derived from models that members of the department felt expressed their shared appreciative settings with regard to their image of the role of community medicine, which in their case was a very proactive interventionist one.

More recently, much work has been done in NHS hospitals and purchasing authorities as they assimilate and adapt to the purchaser-provider split (Checkland and Holwell, 1993). The new appreciative settings have been explored with participants via models of notional systems to enact the purchaser and provider roles. These have served to structure coherent debate concerning the requirements of the new roles.

In a recent study in a large teaching hospital, the work was part of a project to re-create an information strategy for the hospital suitable to cope with the new arrangements (Checkland and Poulter, 1994). In this work half a dozen teams of hospital workers representing the different professions were set up; members included clinicians, nurses, professionals from the finance and estates office, and so on. Over a period of about 6 months, with a plenary meeting of team leaders every month, the teams discussed their activity and its contribution to meeting the requirements of the contracts for providing particular health care services that the hospital would in future negotiate with purchasers. Activity models were built and then used to structure analysis of required information support. This was related to existing information systems, and the information gaps identified helped in the formulation of the new information strategy.

One incident that occurred during this process may be recounted. It illustrates, in microcosm, the change of appreciative settings that can occur in the process of using SSM. It concerns a working group made up of nurses in the teaching hospital, led by a senior nurse. The group was building activity models relevant to providing nursing care, before using them to examine required information support.

Within SSM, when would-be relevant activity models are built, careful concise accounts of them as transformation processes are formulated (so-called Root Definitions). Various questions are asked in clarifying these definitions, one of which is 'If this notional system were to exist, who would be its victims or beneficiaries?' Nurses asking this question naturally wish to answer, 'The patients.' That is what their whole ethos, education, and professionalism tell them. That illustrates why they are in the profession. It was therefore something of a shock to this group - brought to their attention by the structured requirements of the SSM process - to realize in discussion that under the new arrangements the *technically* correct answer is nearer to being 'The hospital contracts manager.' This is because, under the so-called internal market, each contract for a health care service that involves nursing care ought technically to include the cost of providing a certain level (and quality) of nursing care. The nurses' task is then to provide what the contract calls for. Beyond this, of course, there is a theory according to which the interests of patients will, in fact, best be met by the new purchaser-provider contracts.

But it is not easy for nurses to accept this. The senior nurse who described this incident at one of the plenary discussions said that this question, and the issue it exposed, occupied the team for much of one of their meetings. It gave her insight into the NHS changes and helped her to understand her own misgivings about a supposed internal market in health care. Geoffrey Vickers would have appreciated this story.

Given our self-consciousness and the degree of mental autonomy that we seem to possess as human beings, that part of our thinking that is beyond the unreflecting stream-of-consciousness involvement in everyday life can itself be thought about. This can be done by examining the mental models that we use to make sense of our worlds. It is entirely plausible that our perceptions will be coloured by those mental models. And it follows that they need both to be better than primitive and to change as our human and social world changes.

It has been argued here that the models of organization and management that have been useful since the 1950s need to be enriched. It has been argued that appreciative systems theory and SSM can help to provide such enrichment. They do not replace the earlier functionalist and goal-seeking models: They enclose and enhance them in ways more appropriate to institutional life at the end of the century.

ACKNOWLEDGMENTS

This paper is reprinted from American Behavioural Scientist 38:1, 1994, (Special issue on Vickers' work) by permission of Sage Publications Inc.

REFERENCES

Ackoff, R.L., 1957, Towards a Behavioural Theory of Communication, *in*: W. Buckley ed., "Modern Systems Research for the Behavioural Scientist", (pp. 209-218), Aldine, Chicago, IL.

Avison, D.E. and Wood-Harper, A.T., 1990, "Multiview: An Exploration in Infomation Systems Development", Blackwell, Oxford.

Berger, P., and Luckmann, T., 1966, "The Social Construction of Reality," Penguin, Harmondsworth.

Checkland, P., 1981, "Systems Thinking, Systems Practice," John Wiley & Sons, Chichester.

Checkland, P., 1993, OR and the Systems Movement, *Journal of the Operational Research Society*, 34, 661-675.

Checkland, P., 1985(a), Achieving Desirable and Feasible Change: An application of Softy Systems Methodology, *Journal of the Operational Research Society*, 36, 821-831.

Checkland, P., 1985(b), From Optimizing to Learning: a Development of Systems Thinking for the 1990s, *Journal of the Operational Research Society*, 36, 757-767.

Checkland, P., 1994, Conventional Wisdom and Conventional Ignorance: The Revolution Organization Theory Missed, *Organization*, 1(1), 29-34.

Checkland, P., and Casar, A., 1986, Vickers' Concept of an Appreciative System: A Systemic Account, *Journal of Applied Systems Analysis*, 13, 3-17.

Checkland, P., and Haynes, M.G., 1994, Varieties of Systems Thinking: The Case of Soft Systems Methodology, *System Dynamics Review*, 10, 189-197.

Checkland, P., and Holwell, S., 1993, Information Management and Organizational Processes: An Approach through Soft Systems Methodology, *Journal of Information Systems*, 3, 3-16.

Checkland, P., and Poulter, J. 1994, "Application of Soft Systems Methodology to the Production of a Hosptial Information and Systems Strategy," United Kingdom: HISS Central Team of the MHS Management Executive.

Checkland, P., and Scholes, J., 1990, "Soft Systems Methodology in Action," John Wiley & Sons, Chicheser.

Davies, L., and Ledington, P., 1991, "Information in Action: Soft Systems Methodology," Macmillan, London.

Donaldson, L., 1985, "In defence of organization theory," Cambridge University Press, Cambridge.

Hicks, M.J., 1991, "Problem Solving in Business and Management," Chapman & Hall, London.

Khandwalla, P.N., 1979, "The Design of Organizations," Harcourt Brace, New York.

Koestler, A., 1967, "The Ghost in the Machine," Hutchinson, London.

Lawrence, P.R., and Lorsch, J.W., 1967, "Organization and Environment", Harvard University Press, Cambridge, MA.

March, J.G., and Simon, H.A., 1958, "Organizations", Wiley, New York.

Maturana, H.R., and Varela, F.J., 1980, "Autopoiesis and Cognition", Reidel, Dordrecht.

Newall, A., and Simon, H.A., 1972, "Human Problem Solving", Prentice-Hall, Englewood Cliffs, NJ.

Patching, D., 1990, "Practical Soft Systems Analysis", Pitman, London.

Pugh, D.S., and Hickson, D.J., 1976, "Organization Structure in its Context", Saxon House, Farnborough.
Reed, M.I., 1985, "Redirections in Organisational Analysis", Tavistock, London.
Schutz, A., 1967, "The Phenomenology of the Social World", Northwestern University, Evanston, IL.
Silverman, D., 1970, "The Theory of Organizations", Heinemann, London.
Simon, H.A., 1957, "Administrative Behaviour", 2nd Edition, Macmillan, New York.
Simon, H. A., 1960, "The New Science of Management Decision", Harper and Row, New York.
Tönnies, F., 1955, "Community and Association", (Gemeinschaft und Gesellschaft) (C.P. Loomis, Trans.). Routledge and Kegan Paul, London. (Original work published 1887).
Varela, F.J., 1984, "Two Principles of Self-Organization", *in:* "Self-organisation and Management of Social Systems", H. Ulrich and G.J. Probst, (eds.), Springer-Verlag, Berlin, pp 25-32.
Vickers, G., 1965, "The Art of Judgment", Chapman and Hall, London.
Waring, A., 1989, "Systems Methods for Managers", Blackwell, Oxford.
Wilson, B., 1984, "Systems: Concepts, Methodologies and Application", Wiley, Chichester.
Woodward, J., 1965, "Industrial Organization: Theory and Practice", Oxford University Press, London.
Zannetos, Z.S., 1984, "Decision Sciences and Management Expectations", *in:* "Operational Research '84", J.P. Brans, (ed.), North-Holland, Amsterdam, pp 69-76.

2

TOTAL SYSTEMS INTERVENTION

Critical Success Factors for a Systems Based Problem Solving System

Robert L. Flood

Centre for Systems Studies
School of Management
University of Hull
Hull, East Yorkshire, HU6 7RX
United Kingdom

INTRODUCTION

Well developed problem solving skills are vital possessions that all successful managers must hold. This is because managers essentially are problem solvers. Complementing, expanding and improving problem solving skills will help managers to cope even more effectively with organisational problems that they face and will broaden their perspective on how to face them.

Making improvements to problem solving skills is important. Unfortunately several obstacles get in the way. None of these are the fault of managers. There are perhaps 3 main obstacles.

- Time available for managers to make improvements is limited.
- Help given to managers seeking to make improvements is not good enough.
- Making improvements means change that may seem threatening to managers and not in their interests.

So, what can be done?

It is simple to reason out what can be done and where to begin. Time available for managers can only be increased if they are able to do their work more efficiently and effectively. Being more efficient and effective is only possible if manager's skills are improved. Improving the skills is only possible if the help given is good enough. The help given is not good enough. We need to begin, then, by substantially improving help given to managers and to convince them that it is in their interest to make improvements.

I have come up with 9 Critical Success Factors (CSFs) that indicate where help given to managers can be improved. These CSFs state what must be achieved so that managers are adequately supported in efforts that they make to improve their problem solving skills. The 9 CSFs are catalogued below.

Critical Issues in Systems Theory and Practice, Edited by K. Ellis et al.
Plenum Press, New York, 1995

- There exists in the literature an enormous and diverse range of problem solving methods. Faced with this enormity and diversity, managers are not surprisingly left stunned. It is impossible for them to know where to begin. The barrier has been allowed to rise because no-one has stood back and addressed the broad question that asks 'Which method should be used, when and why?'

CSF1 *A problem solving system is needed that guides managers through the process of choosing methods relevant to the main problems that they face.*

- Systematic investigation of problem solving methods poses huge difficulties for managers. To become familiar with a diversity of methods means coming to terms with their great number of concepts; concepts that differ from method to method. Some of the concepts are obscure and this accentuates the challenge. It is a challenge made unnecessarily difficult. I come across managers, again and again, each doing their best to learn about and understand new methods, but finally having to admit defeat. They are defeated by the volume and obscurity of the concepts.

CSF2 *A problem solving system is needed that cuts back on the number of concepts used to explain methods and translates those that are kept into everyday language and then integrates them into one coherent whole system.*

- Managing organisations is becoming ever more challenging. This statement may sound hackneyed but it is true. Today, managers have to cope with rapidly developing technology for the workplace, escalating amounts of management information moving around at great speed, rising consumer awareness, growing expectations of employees, as well having to handle corporate culture, office politics and so on. Organisations have become very messy things with problems that are hard to understand. Methods that are able to stimulate and provoke creative and perceptive thought about these messes are needed. Methods for creative thinking do exist but are rarely used alongside problem solving methods.

CSF3 *A problem solving system is needed that incorporates powerful tools to stimulate and provoke creative and perceptive thought about organisations and their problems.*

- Methods are frequently presented like recipes. They describe what to do without explaining how the method works. It is essential, however, that an explanation is given so that managers are able to understand the kind of changes that are supposed to occur. This knowledge will improve the likelihood of achieving desired impact for all concerned.

CSF4 *A problem solving system must be supported by an explanation of what is happening when people use the system properly, according to the principles of the system.*

- There are many pitfalls to avoid when problem solving. They rarely are avoided because managers are not worried about their existence nor their intricacies. The pitfalls therefore must be catalogued and organised into a knowledge base. Managers equipped with this knowledge of things to avoid are more likely to achieve desired impact with the methods that they choose.

CSF5 *A problem solving system must be used in the light of knowledge about potential pitfalls to avoid.*

- Shallow and gimmicky books written about problem solving are extremely common. They lead their readers inevitably to failure because the ideas in the books are without foundation. This can cause uncountable damage by disappointing and putting-off managers keen to acquire genuine and worthy problem solving skills.

CSF6 *A problem solving system must not be shallow or gimmicky.*

- Many books on problem solving provide case examples centred on one geographic region, such as the USA, the UK or the Far East. This centricism fails to explore problem solving at an international level. It also means that cases are of limited use to the bulk of managers across the globe. An international perspective on problem solving adds value.

CSF7 *A problem solving system has added value if it is demonstrated to be applicable at an international level.*

- Problem solving is often made to seem gruelling and frustrating. It is true, problem solving can be hard work, but it can often be fun and even funny - especially when a coherent, whole problem solving system is used.

CSF8 *Problem solving should be fun.*

- Problem solving entails being critical. Being critical is usually interpreted as being negative. A cartoon in the Times Higher Education Supplement said it all - a note was pinned on the door of a Professor of Critical Theory that read 'Enter, then start knocking.' Two points given below argue that being critical does not mean being negative.
- Being critical is the most effective way of learning about, understanding and hence making improvements where organisational problems are encountered. As I see it, any improvement to organisations is a positive thing.
- Our work is critical of many current problem solving practices. Criticism is based on experiences of hundreds of managers who are well seasoned when it comes to intervention using problem solving methods. They have evaluated methods and their outcomes, learnt from this, achieved better understanding and with this knowledge helped me to improve the effectiveness of our problem solving system. As I understand it, improving problem solving effectiveness is also a positive thing.

CSF9 *A problem solving system must include procedures that are critical of outcomes it generates and must include procedures which ensure that the system remains critical of itself.*

We have developed a problem solving system that squares-up to the 9 CSFs highlighted above. In doing this, it helps managers to make good use of time available to make improvements to their problem solving skills. It is respectful of managers and the management *modus operandi*. It meets all 9 CSFs in the following way.

- It cuts back on the number of concepts used within problem solving, translates the remaining concepts into everyday language and integrates them into one whole system.
- It incorporates powerful tools to stimulate and provoke creative and perceptive thought and in this way helps to develop an informed understanding of problems faced.
- It steers the problem solver through the process of choosing methods relevant to problems faced.
- It comes with an explanation of what is happening when people use the system properly.
- It comes with the knowledge of potential pitfalls in problem solving that need to be avoided.

- It is not shallow or gimmicky.
- It has added value because it is demonstrated to be applicable at an international level.
- It is critical of outcomes it generates and is critical of itself.
- It encourages fun.
- It is called Total Systems Intervention first reported in Flood and Jackson (1991) and subsequently in Flood (1993).

Total: It draws together and integrates methods for

- *creativity* (to think creatively and perceptively about problems faced)
- *choice* (to choose a method(s) relevant to those problems)
- *implementation* (to apply chosen methods to tackle those problems).

Systems: It guides managers toward a whole or systems view. It takes into account all relevant aspects of a problem solving situation. It deals with whole systems. It is a total problem solving system.

Intervention: It is an activity that brings about progressive improvement where problems are found through intervention - Total Systems Intervention (TSI).

TSI is explained below in terms of its philosophy, principles and process, 'the 3 Ps of TSI'.

THE PHILOSOPHY

The main idea of TSI's philosophy is that we think about an organisation as a whole. It is only when a good understanding of the whole organisation is grasped by taking into account the viewpoints of all concerned that effective management can be achieved. If we fail to develop a whole understanding of the organisation then unexpected things will happen that foil otherwise well thought out plans. When applied to organisations, then, TSI insists that we develop and work with a whole view of them.

With TSI, organisations are conceived of as whole systems that comprise parts which are continually interacting. Parts form a system (the horizontal dimension) that is a subsystem of a larger system and has subsystems itself (the vertical dimension). An organisation is therefore thought about as if it were a vertically and horizontally integrated series of systems. The whole system is also open to its environment, that is it exchanges products and services, money, information, people, etc., with its suppliers, consumers, and other external agencies.

The question remains how we can enrich our understanding of the whole organisation in these systemic terms. The following 4 key dimensions of organisations guide our thinking. We need to have a thorough understanding of the *organisation's processes*, the flows and controls from suppliers right through to consumers. We need to understand the *organisation's design* within which the processes happen. This means identifying the possible functions of the organisation and how these functions (rather than lines of authority) are controlled and co-ordinated.

We must also appreciate *individual and cultural differences and similarities* that exist between people. People are different. People need different things, respond to things in different ways and understand things differently. In this sense, there is no single organisation. There are as many organisations as there are people thinking about the organisation.

Despite this differentiation, cohesion can be attained. This is partly because of the possibility of corporate culture. A strong corporate culture with good cohesion comprises people who share a common history, hold a common understanding and have a common sense of belonging. Corporate culture is where people in an organisation share many ideas, such as rules and practices of the organisation, although there will rarely be a total overlap. In a dynamic corporation differences of opinion or interest inevitably arise. A number of these will be expressed through organisational politics.

Having an appreciation of *political dynamics* helps to round things off, to complete the whole picture. It is important to know who holds power and how this power is used to serve certain interests. It leads us to explore who is in a position to bring resources or biases to bear to get their own way. The question is, 'Who in the scheme of things may be regarded as influential in determining strategy, policy and what people do?'. It also allows us to examine how power may be spread in ways which indeed subvert official positions of power.

It is therefore the argument of TSI that we can only get to know the whole organisation if we develop an understanding of it in terms of control in the following 4 key dimensions.

- *Organisational processes* - flows, and controls over flows.
- *Organisational design* - functions, their organisation, co-ordination and control.
- *Organisational culture* - control in the form of mediation of behaviour in terms of social rules and practices.
- *Organisational politics* - control in terms of power and potency to influence the flow of events.

Unless all 4 key dimensions to organisations are taken into account, i.e., a whole system view is developed, then problem solving is bound to be ineffective. It will be ineffective because crucial influential factors will have been neglected. Leaving influential things hidden gives rise to a predominance of unexpected and usually unwanted changes. Allowing organisations to be run in this way is ineffective management.

The problem solving system TSI must therefore operate a system of methods that is able to deal with problems arising from each of the 4 key dimensions. There are methods available to do this. Methods can be grouped according to which key dimension they are best suited to, i.e., what types of problem they are best at tackling. Each of the key dimensions of organisations is the source of a type of problem, which accordingly means that there must be 4 main types of problem solving method.

In a nutshell, there are 4 key dimensions to organisations and 4 main types of problem solving method that tackle problems which arise respectively in each key dimension. Every problem solving method with its own immediate and given purpose can be categorised in this way.

The image of organisations pieced together above builds a framework on which an ideal whole system view can be constructed. The ideal can be captured as a model of organisational dynamics that helps to achieve effective management. The model is an ideal because it represents the ideal dynamic of an organisation from a TSI perspective. It is an ideal picture of organisational dynamics that we would like to achieve by employing TSI. Of course ideals can never be achieved. The point is that we should try to approximate the ideal as best as possible. The ideal whole system view has 6 stages of construction as set out below.

- *An organisation comprises technical and human activities.* Interactions in organisations are represented by an interactive mixture of technical (organisational processes and organisational design) and human (organisational culture and organisational politics) activities. The whole system framework is, then, a horizontally and vertically integrated set of technical and human activities.
- *Activities of an organisation must be efficiently and effectively controlled (i.e., co-ordinated and mediated) whilst maintaining viability of the organisation.* Activities are co-ordinated and mediated by technical procedures, and socio-cultural and socio-political rules and practices. Procedures, rules and practices must

attune so that viability is achieved. Environmental factors are also influenced and controlled.

- *Activities of an organisation must be directed to achieve some purpose.* An organisation normally will have an officially declared mission to which these activities are directed. Ideally this purpose will allow for interpretation and feedback between the declaration and implementation of purpose.
- *People in organisations appreciate things in different ways.* Individuals and groups naturally make their own interpretations of the interacting activities, the way activities are controlled, and the organisation's purpose. They hold a view of their own role and purpose in the organisation. It is this discord that generates vital organisational dynamism. However it can cause conflict, a lack of cohesion, inefficiency, ineffectiveness, and ultimately non-viability in the organisation. It can lead to polarisation with people rigidly adhering to their own position. Even shades of difference can lead to non-negotiable conflict in the long run. People's views therefore must be taken into account as a matter of course and conflict must be resolved or accommodated fairly.
- *The last two stages must be harmonised through organisational design and management style.* An organisational design and management style must be chosen that balance people's needs with the organisation's needs, remembering that the organisation's needs will also reflect the business or organisational context.
- *Managers and problem solvers must accept responsibility for the impact of their decisions and policies on the physical, biological and social environment.*

A summary sees TSI holding an ideal picture of organisational dynamics that managers should strive to achieve. It is the view of TSI that unless managers do strive toward the ideal then they will be dogged constantly by inefficiency, ineffectiveness, conflict and strife. [You may also share with me the need to strive for the ideal on moral as well as practical grounds.]

Clearly TSI must provide principles, processes and methods for problem solving that help us to intervene in those dimensions of the organisation that do not match up to the ideal. Only then will it be possible to approximate the ideal. Happily, I can say that these components of an ideal problem solving system are provided in *Solving Problem Solving* (Flood, 1995). TSI helps managers approximate the ideal.

When applied to organisational problem solving TSI extends this systemic understanding of organisations. Of great importance here is TSI's awareness of the process of problem solving. For TSI, problem solving actually means getting to know about and then managing interacting issues as opposed to solving identifiable problems. [*Despite adoption of the term problem in this text - for readability - TSI's actual position is that there is no such thing as a problem, only interacting issues to be managed. 'Problem' must be read in this text to mean 'interacting issues'*.] Problems arise from the interaction of technical and human activities, how they are co-ordinated and mediated, interaction of the organisation with its environment, the organisation's mission, organisational design and management style, and people's interpretations of these, and how people choose to exert power that they hold. Problem solving is a particular type of human activity that by definition is also a part of the organisation's activities. An organisation, then, can be understood as a complex of interacting activities and problem solving as a continuous process of managing them. Problem solving is a part of the problem to be dealt with.

Intervention accordingly will go something like this. Problems are surfaced through creative thinking. By undertaking careful and communicative reasoning, an approach most suitable to tackle the problems surfaced is chosen. Change proposals to deal with the problems are worked out using the chosen approach. Implementation of change proposals takes place. Some problems are dealt with purposefully and directly, some purposefully but indirectly (an oblique use of methods), whilst others are surprisingly impacted on as a result of our never having an absolute understanding of the whole organisation. These surprises are counter-intuitive, i.e., things that occur counter to our intuitive understanding where formal analysis has not penetrated. New problems arise as a further consequence of implementation. The organisation as described by its problems changes and reformulated implementation directed at new problems takes place. TSI therefore is a continuous process of dealing with problems throughout the whole organisation.

THE PRINCIPLES

Principles are extremely important. They provide a mechanism to evaluate intervention. A TSI intervention must be evaluated against its own principles. If, and only if, intervention operates according to the principles established in this paper can the intervention be said to be a valid use of TSI. As a minimum, this means in practice endeavouring to achieve the principles.

There are 4 main principles to be extracted from the philosophy - being systemic, achieving meaningful participation, being reflective, and the goal of enhancing human freedom.

The foundational principle tells us to study the organisation as if it were *systemic*. This means 'take into account the whole.' It also means that there is a form of hierarchy. 'The system' under study is a part of a greater whole. It also comprises interacting parts. For TSI the hierarchy comprises technical (organisational processes and organisational design) and human (organisational culture and organisational politics) activities at 3 hierarchical levels - 'the system', 'the subsystems' and 'the suprasystem'. 'The system' is always the hierarchical level that becomes the focus of attention, although the level focused on may change during analysis. All interactions between all parts, of technical and human sorts, at the 3 levels must be taken into account during the process of continuous management of problems. This principle helps to prevent undesirable counter-intuitive consequences from occurring. It therefore leads to more effective management.

The principle of *meaningful participation* follows the systemic principle. If we are to develop an adequate appreciation of all interactions between all parts, of technical and human sorts, at 3 levels at any one time, then the perceptions of all people involved and affected must be drawn into the picture. To a large extent the organisation is what people think it is. It is important therefore that we know what people think. If participation is not achieved then only a limited understanding of the organisation can be developed. This is not a whole system perspective, meaning the systemic principle is violated. This will lead to less than effective management.

The principle of *reflection* follows the previous 2 principles. There are 2 reflective needs with TSI.

- To reflect upon the relationship between different organisational interests - demonstrating that people may be dominated who, as a result, cannot meaningfully participate.
- To reflect upon the dominance of favoured approaches to intervention - demonstrating where the use of one (or a few) method(s) prevails, restricting the capability of managers to effectively tackle the full range of technical and human issues they face.

The first reflective need is to pinpoint where power is exerted and domination over people exists, inside and outside the organisation. Often a dominant view holds sway. Domination prevents meaningful participation of involved and affected people and the inclusion of their viewpoints in decision making. This in turn means that less than a whole system understanding is achieved during analysis.

The second reflective need is to overcome a common mistake made by problem solvers. Normally they use a small number of methods with restricted problem solving capability. They do not know about the limitations in the methods they are using. Each method is limited, however, in the kinds of problem that it is best employed to tackle. A full range of approaches is required, sufficient to tackle all sorts of technical and human problems in the 4 key dimensions of organisations. This can be achieved through critical reflection on the strengths and weaknesses of each method, thus bringing these matters to the fore and so linking methods to the sort of problem they are best directed at. Alternatively, if methods are being used to deal with problems they were not originally designed to address, the consultant or manager should be made aware of this through critical reflection.

The reflective principle can be summarised neatly as follows. The first reflective need is to ensure that a whole system understanding is achieved - surfacing all problems to be dealt with. The second reflective need is to ensure that all problems are dealt with - employing methods as and when relevant. Achieving both needs promotes fair and effective management.

The fourth principle introduces TSI's ideology and indeed its moral justification. TSI argues that an explicit ideology of human freedom must enter management practice. This ideology is a personal commitment for some people, but I am also introducing it in TSI as a part of a coherent theory about effective management. It logically follows the preceding 3 principles. Freedom is enhanced through the process of reflection. Reflection helps to achieve meaningful participation, which in turn promotes being systemic and taking into account the whole. Taking into account the whole is an important step toward better informed management, effective problem solving and avoidance of counter-intuitive consequences.

THE PROCESS

The process of TSI has 3 modes of operation.

- Critical Review Mode.
- Problem Solving Mode
- Critical Reflection Mode.

The Critical Review Mode reviews critically methods bidding to be incorporated in the Problem Solving Mode of TSI. It does this by assessing the ways in which the method under review can be incorporated within and operated by the process of TSI. There are 3 stages to this process.

- Review problem solving methods using TSI to judge which of its 3 phases - Creativity, Choice and/or Implementation - the method contributes to.
- Further review problem solving methods by assessing which of the 4 key dimensions of organisations the method is capable of engaging.
- Ultimately, ask whether TSI can learn anything from methods reviewed in terms of its own philosophy, principles, process, and its own methods such as the method to operationalise the Critical Review Mode.

The Problem Solving Mode of TSI employs a system of methods for problem solving brought together through the Critical Review Mode. It operates through the 3 phases of TSI as follows.

- Think creatively about problems faced and identify the core ones.
- Choose the right method(s) in the circumstances to deal with the core problems.

- Use the chosen method(s) to develop and implement innovative change proposals that tackle the core problems.

The term circumstances is used above in a systemic sense, to mean the whole situation. The whole situation means everyone and everything. The notion of correctness therefore is subject to those circumstances. It is not the case that there is literally a right method independent of the problem solvers. If circumstances change such that there is a different mix of people, the decision on correctness may change too. It all depends on circumstances and the involved people are part of those circumstances. The Critical Reflection Mode helps to judge correctness in the circumstances - a point that must be remembered throughout this book and whenever problem solving.

The Critical Reflection Mode uses the 3 phases of TSI to help problem solvers to reflect upon the adequacy of the output of the Problem Solving Mode. It asks the following questions about each phase. In the circumstances

- Is the method(s) in use the right one?
- Is the output of the method(s) right?

(Procedures that operationalise the 3 modes are detailed in *Solving Problem Solving* (1995)).

The 3 Ps of TSI have each been briefly discussed. It remains to conclude.

CONCLUSION

The problem solving system TSI has been developed to provide managers with a practically useful systems-based approach to problem solving. It provides procedures to integrate all methods for problem solving in a process which ensures that they are employed only to tackle the issues they are best suited to. TSI achieves this through 3 modes of its use; Critical Review Mode, Problem Solving Mode and Critical Reflection Mode. Ultimately this builds up a system of methods for creative thinking, choice of method for implementation, and methods for implementation, within a reflective process that has 3 main phases. The phases are Creativity, Choice and Implementation. Each of the 3 phases has three sub-phases, creativity, choice and implementation, that follow in a recursive structure.

The process of TSI remains true to 4 principles. These principles are, being systemic, achieving meaningful participation, being reflective, and achieving human freedom. The principles support each other. A valid use of TSI would be able to demonstrate to those involved and affected that each of these principles has been met as far as is reasonably possible in practice.

The process of TSI has matured over about 8 years of practice and reflection upon this and its original theoretical foundations. TSI has been used by many consultants and managers in many countries. Its use has occurred in two ways - in consultancy and in the process of management. The process of TSI arguably operates well in practice as seen in the case studies in *Solving Problem Solving* (1995), although there is always scope for further evaluation such as checking across the board whether all organisational members agree about TSI's relevance. I can see no reason at this time to want to change the process as it is expressed by its 3 phases (although further reflection on TSI may change this?). The task of each activity and the expected outcome have become well established through application. There are a number of points that I wish to make, however, about the tools of each activity of TSI in the problem solving mode.

There is enormous opportunity to further develop the tools of the Creativity phase of TSI. Additions and/or replacements for Idea and Image Generation and evaluation need to be explored. It would be wise to continue to apply the fundamental idea of creative thinking to the Creativity phase itself.

The Choice phase has presented the most exacting challenge of the 3 phases. The idea and principles of choice of method are clear enough. Difficulties have arisen when translating these into an agreeable procedure. The latest development is to organise methods into groups. The groups are categories that comply with TSI's 4 key dimensions of organisations. Every type of problem has a set of relevant methods that can be chosen. Therefore a basis exists for aligning methods to problems and choosing the right one in the circumstances. This categorisation and the procedures used to operationalise it might be improved following further investigation.

The Implementation phase has required the most extensive work to be done. A number of extant problem solving methods have been translated into lines of questioning and embedded in the process of TSI in a system of methods. This provides a satisfactory way of including them in the process. It overcomes difficulties that otherwise occur because the original versions of the methods were designed outside of the philosophy, principles and process of TSI. Further translation work is likely to be beneficial.

Much has been achieved using TSI in practice. Much more can be achieved if adequate efforts are invested in the further work to be done. I hope that I and the rest of the team at the Centre for Systems Studies at the University of Hull, succeed in persuading many more managers (with a small m!) to contribute to the exciting programme of further research that we are engaged in.

In this paper I have covered the 3 Ps of TSI that match up to a number of the CSFs documented at the outset, in particular CSFs 1,2,3,6 and 9. The book *Solving Problem Solving* demonstrates this achievement in detail whilst at the same time providing further material that shows TSI meets the remaining CSFs.

And finally, as already stated, there is much more work yet to be done to develop our knowledge about intervention. Whatever developments occur with TSI, however, I believe that there will be one common thread. The common thread is that problem solving will be based on lines of questioning like the version of TSI in this book. In this way it will remain critical. It must be just so. But this kind of notion is not new. As long ago as 1902, Rudyard Kipling established the importance of questioning, questioning and questioning again to promote everyday thinking (although perhaps not expressed in today's politically correct terms hence my addition in italics).

> 'I keep six honest serving men *(sic)*
> They taught me all I knew;
> Their names are What and Why and When
> And How and Where and Who.'
> Rudyard Kipling, 1902,
> *Just So Stories,*
> Macmillan, London.

REFERENCES

Flood, R.L., 1993, "Beyond TQM," John Wiley and Sons, Chichester.

Flood, R.L., 1995, "Solving Problem Solving," John Wiley and Sons, Chichester.

Flood, R.L., and Jackson, M.C., 1991, "Creative Problem Solving: Total Systems Intervention," John Wiley and Sons, Chichester.

3

THE WILL TO SYSTEMS

From Making Sense to Enframing

Ramsés Fuenmayor

Departamento de Sistemología Interpretativa
Universidad de Los Andes
Mérida 5101
Venezuela

INTRODUCTION

"Taking stock historically" was the first subtitle of Professor Peter Checkland's presidential address to the International Society for General Systems Research in June 1987. The title of the address is suggestive: "Images of Systems and the Systems Image" (1988a). By 'taking stock historically' Checkland means to make a chronological narrative of different attempts at systems thinking and check what we have 'stocked' in what seems to be a common project comprising those different attempts. This teleological written chronicle leading to a 'stock' intends to give historical meaning to 'soft systems thinking.' A cursory inspection of the last 25 years of systems thinking in UK is enough to accept that critical systems thinking is historically indebted to Checkland's soft systems thinking. Hence, the problem about the historical meaning of soft systems thinking is at the root of any other type of systems thinking derived from it; even though such a derivation be by means of logical opposition.

I would like to make another attempt to understand the historical meaning of soft systems thinking, but under a different notion of *historical inquiry* from the one guiding Checkland's 'historical stock.' You will later see the difference.

'FROM OPTIMIZING TO LEARNING'

The first thing to notice is Checkland's attempt to open a breach in the solid and homogeneous instrumental rationality (and underlying instrumental interest) of would-be 'systems-thinking' at present age. 'Soft' systems thinking shows itself as a revolution against the reductionism involved in 'hard' (instrumental) systems thinking.

Checkland's revolution can be represented by his phrase "From Optimizing to Learning" (Checkland, 1985). "From optimizing to learning" means a shift *from* a mere will to control and efficiency *to learning about* ".. the relevance of our limitations and to question those assumptions about ourselves which are most inept to the activity and the experience of being human now" (Vickers, 1970, p.15). I am using Vickers's words to explain Check-

Critical Issues in Systems Theory and Practice, Edited by K. Ellis et al.
Plenum Press, New York, 1995

land's meaning of *'learning'* because Checkland himself states that he is adopting Vickers' 'intellectual framework' as the conceptual platform of his 'Soft Systems Methodology' (SSM). SSM is thus seen by his author as a 'way of applying' Vickers' intellectual framework (1985).

In view of that shift *from optimizing to learning*, the new way of 'systems thinking', called 'soft' systems thinking, unlike the 'hard' (instrumentalist) tradition in systems thinking, "regards systems models as models relevant to *arguing* about the world, not models of the world; this leads to 'learning' *replacing* 'optimizing' or 'satisficing'." (Checkland, 1985, p. 765, my emphasis). But why is 'learning' to *replace* 'optimizing and satisficing'? We need to understand better the meaning of 'learning.' For that purpose, let us take a look at the paragraph containing the former quote from Vickers.

> "A trap is a trap only for creatures which cannot solve the problems that it sets. Man-traps are dangerous only in relation to the limitations on what men can see and value and do. The nature of the trap is a function of the nature of the trapped. To describe either is to imply the other....We the trapped tend to take our own state of mind for granted . we the trapped tend to take our own state of mind for granted-which is partly why we are trapped. *With the shape of the trap in our minds, we shall be better able to see the relevance of our limitations and to question those assumptions about ourselves which are most inept to the activity and the experience of being human now."* (Vickers, 1970, p. 15) (My emphasis)

The 'shape of the trap' represents a holistic view of a situation X under study. Systems thinking, by which I mean *holistic* thinking,* *is impelled by the will to providing a holistic account of a sit'ation X. Vickers' metaphor suggests that if we do not try to see the whole 'shape of the trap', we cannot see our limitations and hidden* assumptions. This means that, at least in Vickers quote, *systems thinking* must in-form an authentic process of *learning* about our limitations and hidden assumptions.

Now, according to an old tradition in philosophy, this business of attempting to see *holistically* 'the shape of the trap', so that we can see our limits and hidden assumptions concerning a situation X, is called a *critique* of X. This is what Immanuel Kant meant by *critique* in his three 'Critiques' (of 'Pure Reason', of 'Practical Reason' and of 'Judgment.'). Indeed, according to Kant, *transcendental critique* of something is the examination of the possibilities and limits of experiencing that something (1781/87, pp. B25-B26). Under this definition, it seems that Checkland's notion of (holistic or systemic) *learning* (about our limitations and hidden assumptions) coincides with Kant's notion of *critique*. This can be seen even more clearly if we observe the holistic character of Kant's critical thinking (Reason). Indeed, the whole work of Kant pivots around the systemic nature of Reason. "Reason —writes Kant— is impelled by a tendency of its nature to [...] the completion of its course in a self-subsistent systematic whole" (Kant, 1781/87, B825). This 'tendency' is nothing more than the 'will to systems' of reason. The highest level of (the system of) human reason, namely, holistic reason, is thus directly driven by the will to systems.

It seems, so far, that Vickers' 'intellectual framework' and, in consequence, Checkland's methodology (and the sort of systems thinking in which it is embedded) belong to the Kantian tradition of *critique*. There is, however, something that disturbs this hypothesis: When Checkland 'takes stock historically' about the systems tradition, he writes that the creation of the notion of *systems thinking* ".. is recent, occurring only in the late 40's, and its prime creator is Bertalanffy." (1988(a), p37). Checkland seems to forget that systems thinking was the hallmark of the type of Modern thinking represented by the Enlightenment and German Idealism. To be sure, the whole philosophical project of German Idealism can

---

* Checkland has also proposes the use of 'holistic' in order to avoid confusion with instrumental approaches (1988(b))

be defined —as the French philosopher Luc Ferry does (Ferry, 1991, p. 79)— as a "systems project". So, is Checkland forgetting history or is there really a mismatch between Enlightenment critical systems thinking and Checkland's shift to a systems thinking based on his notion of 'learning'? It seems as if this question takes us to probe deeper into both Enlightenment critical systems thinking and Checkland's soft systems thinking.

In terms of Vickers' metaphor, this 'probing deeper' would mean to try to find out the 'shape of the trap' of both *historical* ways of thinking; and that means to dig into their corresponding limits and hidden assumptions. *The 'shape of the trap' would, in this case, be the historical situation which cannot easily be seen as a whole, from beyond, because we are trapped in it.* The very notion of 'trap' indicates its hidden character. We cannot thus expect to see the shape of the trap of any epochal 'systems thinking' just making a historical chronicle. It is necessary to dig into the *epochal foundations*, which due to their very essence of *foundation (fundamentum),* must be hidden.* Here a comment is due about the notion of 'epoch' and of 'historical inquiry.'

The historiographic conception of history —that of the written chronicle embedded in Checkland's 'taking stock historically'— conceives history as a chain of happenings that takes place in a fixed stage. On the contrary, what I am calling 'ontological history' (inspired in Heidegger's work after his 'turn') supposes that the stage is rather a whole 'space, with its rules for the constitution of the 'objects' contained in it; a 'space' that changes along history. It is a 'space' (in the algebraic sense of the word) from which the shape of the 'fact' depends. To give a simplistic example: the shape of the 'facts' in an Euclidean and in a Riemannian space are quite different (e.g. the shape of straight lines in both spaces are different). Those 'changing spaces' are *epochs*. An *epoch* is a temporal ground from which *what-appears* depends.

What are then the *epochal* conditions in which both Enlightenment critical systems thinking and current soft systems thinking inhere? Before attempting to provide a possible reply to these questions it is necessary to uncover the 'ontologic-historical' relationship between our notion of '*epoch*' and 'systems thinking.'

SYSTEMS THINKING AND METAPHYSICS: MAKING SENSE

Before modernity and still during the first 'wave' of modernity (from Descartes to Kant), the practical question, 'What I ought to do?' (decision making on moral grounds) was a question addressed to the totality, to the *ground of beings* (*Theos*, among presocratic ancient Greeks —see Heidegger, 1985), in search for a sign that tells man how to act so that his action harmonizes with that totality. This way of posing the practical question and the sort of thinking derived from it is systems thinking in its more original form within Western history (and, maybe, also in other non-western cultures). However, this practical question is not a 'moral' question in the sense we give to the term 'moral' today. Today, by 'moral', we mean a sort of 'varnish' or 'glaze' added once in a while to our judgment about human actions, which are primarily (before adding varnish) regarded as 'facts' within a naturalistic and dualistic purview (Fuenmayor, 1994). On the contrary, before modernity and still during the first part of modernity, making holistic sense and acting practically (meaningfully) were

* At stake here is an attempt to see the 'shape of the trap' of 'soft'systems thinking. Soft systems thinking, if it is to be cogent with its principle of 'learning', should not avoid the task of a holistic view of systems thinking (i.e. do some systems thinking about systems thinking) This attempt to see the shape of that trap will suggest that Checkland's 'from optimizing to learning' (and other types of 'soft' systems thinking) is not really very far from what is *essential* to 'hard' systems thinking.

non-separable. There was not such thing as a realm of thinking separated from a realm of facts where human action belongs. The idea of *knowledge applied to action,* which lies at the essence of the notion of 'methodology' as we understand it today (e.g. SSM, Critical Heuristics, a System of systems methodologies, etc.), would have been meaningless. That non-separability in which holistic sense and action merge is precisely the principle of a systemic *ethos* before our current epoch.

However, there is a major problem in posing the practical question to the *ground of beings*: The *ground of beings, theos,* is not a manifested being; it is concealment from which the unconcealed is brought forth. The *ground of beings* is mysterious, or, rather, it is *mystery* itself. So, how can the practical question be posed to a non-being, to mystery itself?

According to Heidegger,

> "The ground is that from which beings as such are what they are in their becoming, perishing, and persisting as something that can be known, handled, and worked upon .... [The ground] brings beings in each case to presencing. The ground shows itself as presence. The present of presence consists in the fact that it brings what is present each in its own way to presence." (Heidegger, 1969, p. 432).

Notice that, although the *ground of beings* is concealment, pure mystery, it has a way of showing itself (or, as we will see later, of completely hiding itself) as *presence*. The *ground* cannot be shown (be present) as *that-which-becomes-present*; it is shown as *presence*. This way of showing itself of the *ground*, *presence*, changes through history. An *epoch* is *the-present* of *presence.*[*] In different *epochs, presencing* of *that-which-becomes-present* occurs in different ways. Each of those ways is associated with a *type of presence*, a way of showing itself (or completely hiding) of the *ground*. When the *ground* shows itself somehow, there is something suprasensory (*presence*, the way the *ground* shows itself) apart from the sensory (apart from that-which-becomes-present). In this case, the sensory depends upon the suprasensory. When this is the case we can talk of a *metaphysical epoch.*

According to Heidegger, Occidental history, the history of metaphysics, from Heraclitus and Parmenides to Nietzsche, is the succession of *epochs* that began with the separation between the sensory and the suprasensory, maintaining the supremacy of the suprasensory over the sensory. *History* is the *destining* (Heidegger, 1962) of the ways of showing itself of the *ground of beings* (Being). In this way, "in accordance to the given *type of presence*, the *ground* has" different characters: grounding as "the ontic causation of the actual" (ancient Greek thinking after Parmenides), grounding as "the transcendental making possible of the objectivity of the object" (from Descartes to Kant), grounding as "the dialectical mediation of the movement of absolute spirit and of the historical process of production" (the last Kant, Hegel and the German Idealism to Marx), and grounding as "the will to power positing values" (Nietzsche). (Heidegger, 1969, p. 432). These different 'characters' of the *ground* as grounding are types of *presence*, i.e. modes of showing itself of the *ground*. Those *types of presence* define metaphysical *epochs*.[†] The last three represent what we are calling (following Ferry, 1991) the three 'waves' of Modernity.

Now, metaphysical thinking, "thinks beings as a whole with respect to Being [*ground of beings*]." (Heidegger, 1969, p. 432). Hence, *metaphysical thinking is systems thinking, thinking in terms of the ground of beings so that sense, holistic sense, be brought forth.* For that purpose, "metaphysical thinking, starting from what is present, represents it in its

---

[*] Notice the difference between *'presence'* (as a noun), 'to presence,' 'that-which-becomes-present,' 'presencing' and 'the-present.'

[†] If one thinks of time in linear (Cartesian) terms, the end of an *epoch* does not mark the beginning of the next. Each epoch slips beneath the other so that, although discourses keep similar shapes, their underlying meanings begin to change

presence and thus exhibits it as grounded by its ground." (Heidegger, 1969, p. 432). Notice that posing the practical question and making holistic sense are the same.

Since the practical question cannot be asked directly to the *ground of beings*, it is posed to a *representative* of the epochal *type of presence* (epochal way of showing itself of the *ground*). The representative may, thus, be 'nature' as a whole (not the set of natural things). In this case, one tries to find *signs* in nature in order to know how to behave. But it can also be the Church as the 'house' of God in earth.

Now we can return to the question concerning the epochal conditions of Enlightenment systems thinking and of current soft systems thinking. Let us start with the first.*

THE EPOCHAL CONDITIONS OF MODERN SYSTEMS THINKING

The will to systems of the Enlightenment is propelled by a will to emancipation from hidden assumptions so that human decision making can be autonomous. Remember that, according to Kant, "Reason is impelled by a tendency of its nature to [...] the completion of its course in a self-subsistent systematic whole." (Kant, 1781/87, B825). That 'endeavor' of Reason, writes Kant, has "its source exclusively in the practical interest of Reason". The practical interest of reason is for Kant the will to freedom. "By 'the practical', I mean everything that is possible through freedom." (Kant, 1781/87, B828). The will to freedom means the will to take decisions on *moral* grounds according to *universal* reason. This, for Kant, means to ask the question 'What I ought to do?' to our faculty for reasoning and not to internal desires or to external influences.

In an extraordinary short piece of writing called *"Was ist Aufklärung?"* (Kant, 1784, p. 85), Kant defines the Enlightenment, the heart of Modernity, as "man's release from his self incurred tutelage." In turn, 'tutelage' is defined as *"man's inability to make use of his reason without direction from another."* The 'resolution and courage' to use reason is the will to autonomy. Enlightenment is thus clearly defined in terms of a 'will' to fight against the tutelage of some other *order* that does not arise from man's own use of reason. At stake here is the need to change the representative (the Church) of the medieval *type of presence* (the Christian God), so that human action can be autonomous. This would lead to a change of the *type of presence*.

The emancipatory process in which Kant was involved concerns the transformation of an *epochal ontological order* into another. The whole work of the main philosophers of modernity, from Descartes to Hegel, can be conceived as contributing to this deep revolutionary process. Such process required the design of new *type of presence*, i.e. a new way by means of which the *ground of beings* could be manifested, so that the practical question could be posed directly to it, without intermediary. Only in that way, would it be possible to preserve the holistic character of the practical question *and* strive towards human autonomy.

The revolutionary project was started by Descartes with the fundamental and first axiom of modern thinking: *cogito ergo sum*. Such an axiom, endowed with self-certainty, would constitute the basis for a new source of truth which could substitute 'revelational truth.' It declares Reason as a new *type of presence*. Indeed, Reason, which could be directly consulted by rational beings, would be, without intermediary, the new *type of presence* of the new *epoch*.† It is important to notice that, until Kant and still with Kant, Reason is not a

* For a more detailed explanation of this topic see Fuenmayor (1993 and 1994).

† A more detailed account on the conditions imposed to the new *type of presence* and how Reason could comply with those conditions can be seen in Fuenmayor, 1994.

property of the human mind. Reason is a *type of presence* before which human beings have limited (finite) access.

The Cartesian *cogito ergo sum* is, in this way, the birth-cry of a new *epoch* (in the ontological-historical sense explained above), a new way of disposing, classifying and giving meaning to what *presences*. New distinctions were made (new beings were created), new meanings were imposed to old names. It was, metaphorically speaking, the birth of a new '*constellation*' of the realm of beings that pivoted around two new centers of gravity: the 'subject of knowledge' that became the new essence of man, and the 'object of knowledge' that became the new essence of what-ever-is-the-case (which now is declared as opposed to the essence of man).* In this way, *that-which-becomes-present* begins to be presented as *object* and slowly ceases to be *created thing* (as in the Middle ages). The *type of presence* that allows the new way of *presencing* passes from *the Creator* to *Reason*; and Reason becomes the *domain of the subjectum*. The *subjectum* is thus the "transcendental making possible of the objectivity of the object." The work of Kant was to design the *system of Reason* so that this "transcendental making possible of the objectivity of the object" could be explained. That work had to be done according to the main rule of metaphysical thinking (systems thinking), i.e., "to represent what is present in its presence and thus exhibit it as grounded by its ground." (Heidegger, 1969, p. 432)

Now we are in a position to better understand the fundamental role of systems thinking in modernity.

The first wave of Modernity represents a new epoch in the history of metaphysics. A new *type of presence* is constituted; but the basic general feature of metaphysical thinking, systems thinking, remains. It still "thinks beings as a whole with respect to Being [*ground of beings*]." In this way, the first wave of modernity preserves the sort of systems thinking that belongs to any form of metaphysical thinking. However, over this basis, the revolutionary process directed to change the former *type of presence* is impelled by a new *will to systems* beyond traditional systems thinking. In this way, *modern systems thinking* (the will to systems and the constructed systems) *represents both a preservation of the sort of metaphysical thinking that had to be kept so that the modern revolution could take place and the propellant of such a revolution.*

This is how systems thinking, contrary to what is common belief in our present systems community, was the hallmark of modern thinking. Now we can comeback to our original question about the meaning of 'soft' systems thinking and its principle of *learning* in order to compare it with modern systems thinking.

THE EPOCHAL CONDITIONS OF POSTMODERN SYSTEMS THINKING

Our first examination of the meaning of the principle of *learning* animating soft systems thinking revealed that it could be conceived as a renaissance of the systems project of modernity (the only doubt that weakened such hypothesis was the oblivion of this modern origin in Checkland's 'historical stock'). Now we know that the systems project of modernity was a deep *revolutionary* project intended to change an *epoch*, but preserving the basic systems thinking of metaphysics. Is current soft systems thinking also propelled by a deep revolutionary project? If so, which *order* is it attempting to change?

* The new modern constellation is an onto-epistemology that could bear the generic name of 'dualism.' Modern philosophy, modern science and modern technology are originally built on the basis provided by the new constellation.

Although Checkland's discourse is not emphasizing revolutionary terms like 'courage, 'liberation', 'emancipation', 'autonomy', one could think that the shift *from optimizing to learning* is a call for emancipation. One thus would think that Checkland is trying to make a revolutionary break with some sort of rational order: instrumental rationality. However, there is something that hinders this hypothesis:

At the beginning of this article I quoted a piece extracted from "From Optimizing to Learning" in which Checkland writes about "'learning' replacing 'optimizing' or 'satisficing'" in soft systems thinking. Immediately after that, we can read: "...this tradition talks the language of 'issues' and '*accommodations*' rather than 'solutions'." (Checkland, 1985, p. 765, my emphasis). In the previous page, Checkland categorically states that "SSM thus seeks *accommodations* among conflicting interests." (p. 764). The shift *from optimizing to learning* is propelled by the interest in seeking *accommodations* (p. 766).

The notion of 'accommodation' as used by Checkland means finding one's place within a given order. *Accommodation* requires a deal between conflicting parties representing conflicting interests so that, although the conflict does not disappear, each party can be *accommodated* within the present order. The conflicts, writes Checkland, "are subsumed in an accommodation which different parties are prepared to 'go along with'" (Checkland and Scholes, 1990, p. 30). *Accommodation* is thus a conservatory call for accommodating everybody into *a given order* so that the stability of such an order is maintained. Checkland's notion of *accommodation* is not altered in its essence if the circle of the affected is widen to all the inhabitants of the planet (which is very unlikely to happen) through the *use* of Ulrich's 'Critical Heuristics' (1983). Neither it is altered if coercive contexts are considered *using* Jackson's 'System of Systems Methodologies' (Flood and Jackson, 1991). In any case, the final purpose of these forms of current systems thinking is to *accommodate* a few or many (normally a few) within a given ontological order. This is, obviously, quite opposed to that deep revolutionary will animating the systems project of modernity, whose basis was the transformation of an *epochal order*. So, while the latter requires a deep *critique* of the present order, *accommodation* among conflicting interests does not require to see the *holistic 'shape of the trap'*; it is enough to detect conflicts in order to *accommodate* conflicting parties.

Then, what is that *epochal order* that the interest in *accommodation* of current systems thinking is so eagerly striving to maintain? What is the *type of presence* behind this *order*? How does *that-which-becomes-present presences* in that *epochal order*?

Enframing

One hundred years ago, Nietzsche caught sight of the exhaustion of the modern *constellation*, much before the signs were as clear as they are now. What Nietzsche caught sight of, he summarized in a simple and largely misunderstood sentence: 'God is dead.' As Heidegger explains it, 'God is dead' means much more than people do not believe in God any more. 'God is dead' means the 'overturning of metaphysics', the end of the supremacy of the suprasensory over the sensory. Through the overturning of metaphysics, "there remains for metaphysics nothing but a turning aside into its own inessentiality and disarray." (Heidegger, 1952, pp. 53-54). With the overturning of metaphysics, we are at the threshold of a new order different from any traditional order before Modernity and different from the *constellation* of Modernity. But, how is such an order constituted?

According to Heidegger (1962), the postmodern epoch* is characterized by '*enframing*', the technological way of 'revealing.' *Enframing* is a way of *presencing* in which the *type of presence* disappears (the 'suprasensory becomes void'), i.e. the *ground of beings* does

* Heidegger does not use the term 'postmodern.'

not have any way to show itself. As a consequence, the two realms of being that characterize metaphysics are reduced to one: *that-which-becomes-present*. But, since there is not such thing as a *ground of beings*, *that-which-becomes-present* does not appear as depending on anything; it just *stands in itself*. It appears as '*standing-reserve*.' *Standing-reserve* is the way of *presencing* of the technological instrument: ready to process its stored energy. Everything: technological devices, birds, mountains, songs, knowledge, persons become present as *standing-reserve*, ready to be *used*. There is a strong will to *accommodate* everything within the frame of *enframing*, as *standing-reserve*. "The suprasensory is transformed into an unstable product of the sensory. And with such a debasement of its antithesis, the sensory denies its own essence." (Heidegger, 1952, pp. 53-54). This is how the moral, understood a 'moral values' from the 19th Century, adopt the varnish quality referred earlier. "The deposing of the suprasensory culminates in a "neither-nor" in relation to the distinction between the sensory and the non-sensory. It culminates in meaninglessness." (Heidegger, 1952, pp. 53-54).

Enframing is a 'trap' in the fullest sense of Vickers' metaphor. Enframing precludes the will to making holistic sense; hence systems thinking becomes senseless. It is a very sophisticated trap because it disguises itself with discourses about freedom and about other key terms of modernity.

Indeed, we certainly keep talking with the words (that rapidly become no more than ruins), not only of a moral discourse and order, as MacIntyre suggests (1985), but of the whole *constellation* of the realm of beings that Modernity reshaped from other traditional orders. However, we lost sight of the order of such constellation from which those words stem.

> "The suprasensory world, the Ideas, God, the moral law, the authority of reason, progress, the happiness of the greatest number, culture, civilisation, suffer the loss of their constructive force and become void." (Heidegger, 1952, p. 65).

Clue words like liberalism, democracy, justice, freedom, rationality mean today something quite different from what they meant during the Enlightenment. Furthermore, other words, fundamental to the former notions and to the whole realm of beings, have also changed their meaning considerably. Indeed, what we mean by very basic notions like 'man', 'world', 'thing', 'another person' is quite different from the corresponding eighteenth-century meanings. Modern man becomes, as Foucault suggests, the murderer of God. But, as the murderer of God and hence, of the very essence of the *subjectum*, modern man becomes his own murderer (Foucault, 1966). Philosophy disappears or becomes associated with the 'occult sciences'; Modern Science succumbs to technology and, beneath all that, the will to systems and, hence, modern systems thinking, becomes meaningless.

Furthermore, if Heidegger (and many other contemporary philosophers) is right in his idea of the overturning of metaphysics, not only *modern* systems thinking becomes baseless, but any other form of systems thinking of which we can think historically is meaningless in the current epoch. Indeed, since, "metaphysics *thinks beings as a whole* with respect to Being [ontological ground], with respect of the belonging together of beings in Being," the overturning of metaphysics implies the oblivion of the metaphysical totality (*ground of beings*) and its ways of showing itself (types of *presence*). The notion of a practical question addressed to the totality (*ground of beings*) proper of systems thinking becomes thus meaningless.*

---

* If the epoch we are living were another metaphysical epoch defined by a *type of presence*, we could think that the interest in *accommodation* of soft systems thinking represents a renaissance, not of modern systems thinking, but of pre-modern systems thinking. *Accommodation* could thus be understood as a call to harmonize with the totality. But that, as already argued, seems very far from being the case.

Nevertheless we are here today still talking about systems thinking, we even associate the notions of *learning and critique* to such a way of thinking. What is then the (historical) meaning of our present systems thinking? How can the notions of holistic critical learning suggested by Vickers' metaphor of the 'trap' match with that of 'accommodation'?

As far as I can see there are only two possibilities for our epochal systems thinking: 1) To be totally absorbed within what Dávila (1994) and Suárez (1994) call the 'managerial technologies' (Total Quality Management, Re-engineering, the Fifth Discipline), so fashionable today, which represent sophisticated strategies to ensure *accommodation* (of those which are liable to do so) within *enframing*, or, 2) to think carefully, under an ontologic-historical purview, about the shape of our trap.*

The first possibility has almost all the chances. The second has almost no chance. Can we choose?

REFERENCES

Checkland, P.B., 1985, From optimizing to learning a development of systems thinking for the 1990s, *J. Opl. Res. Soc.* 9:757-767.

Checkland, P.B., 1988(a), Images of systems and the systems image, *J. Appl. Syst. Anal.* 15:37-42.

Checkland P.B., 1988(b), The case for 'Holon'. *Syst. Prac.* 1: 235-238.

Checkland, P.B., and Scholes, J., 1990, "Soft Systems Methodology in Action," John Wiley & Sons, Chichester.

Dávila, J., 1994, Crítica Parcial de la 'Calidad Total', Proceedings of "VII Congreso Latino Iberoamericano de Investigacion de Operaciones e Ingenieria de Sistemas.", Santiago de Chile.

Ferry, L., 1991, "Filosofía Política," Fondo de Cultura Económica, Mexico.

Flood, R.L., and Jackson, M.C., 1991, "Creative Problem Solving," John Wiley & Sons, Chichester.

Foucault, M., 1966, "The Order of Things," Pantheon, New York.

Foucault, M., 1991, What is Enlightenment?, *in:* "The Foucault Reader," P. Rabinow, ed., Penguin, Harmondsworth.

Fuenmayor, R.L., 1993, The trap of evolutionary organicism. *Syst. Prac.* 6: 469-499.

Fuenmayor, R., 1994, 'Systems science: addressing global issues'—the death rattle of dying era?, *Systemist*, 16:110-157.

Heidegger, M., 1952, The word of Nietzsche: 'God is dead', *in:* "The Question Concerning Technology and Other Essays," (1977), Harper & Row, New York.

Heidegger, M., 1962, The Question Concerning Technology, *in:* "The Question Concerning Technology and Other Essays," (1977), Harper and Row, New York.

Heidegger, M., 1969, The End of Philosophy and the Task of Thinking, *in*: "Basic Writings," (1993), Routledge, London.

Heidegger, M., 1985, "Schelling's Treatise on the Essence of Human Freedom," Ohio University Press, Athens.

Kant, I., 1784, What is enlightenment?, *in:* "Foundations of the Metaphysics of Morals," Bobbs-Merrill, New York.

Kant, I., 1781/87, "Critique of Pure Reason," transl. by Norman Kemp Smith, 1929, Macmillan, London.

MacIntyre, A., 1985, "After Virtue: A Study in Moral Theory," Duckworth, London.

Suarez, T., 1994, Pequeña reflexion sobre la Quinta Disciplina, Working Paper, Department of Interpretive Systemology, Universidad de Los Andes, Venezuela.

Ulrich, W., 1983, "Critical Heuristics of Social Planning," Haupt, Berne.

Vickers G., 1970, "Freedom in a Rocking Boat," Penguin, Harmondsworth.

* The latter would be what Foucault has called an "ontology of the present" or a "critical ontology of ourselves" (Foucault, 1991).

4

THE HOLOGRAPHIC ORGANIZATION

An Introduction to A Values Based Relational System for the Future (VBRS)

Brian P. Hall

Values Technology
740 Front Street
Santa Cruz, California 95060

INTRODUCTION

This paper is the conclusion of approximately 20 years of research and practical application of a values based relational systems approach to organizational development named the 'Holographic Organization'. An organization is holographic to the degree to which it is operating on a healthy and congruent set of values that all employees hold minimally in common. The consequence is that it will run more efficiently, with internal congruence and improved customer alignment. Hall and research associate Benjamin Tonna of the University Malta developed the initial conceptual framework for human values development between 1974 - 82 (1).

The development and validation of several values measurement instruments occurred between 1981-87. Instruments were developed for individual and organizational settings. The initial work was done in social agencies and institutions for higher learning and in global religious organizations (700 plus members). In the last 6 years a California based team has applied the process as a Values Based Systems methodology for congruent organizational transformation in several global for-profit organizations. This paper will draw on recent experience with banking, telecommunications and high-tech silicon industries.

When Hall and Tonna came together in the mid 70's they had both had extensive experience as change agents in major organizations. Hall from the perspective of organizational psychology, and Tonna from that of organizational sociology. *It was concluded that people make decisions from their own mental models of reality, and the values that underpin them, even if the facts advise otherwise.* At this point (1971) they decided to attempt to construct a theory of values, and to develop instruments for their measurement in human organizational settings, based on the following two assumptions:

- That there is an elemental set of values common to all human beings that underpin human relationships. This would be the basis for comparison and measurement.

Critical Issues in Systems Theory and Practice, Edited by K. Ellis et al.
Plenum Press, New York, 1995

- That by developing an instrument that could objectively measure the values of individuals, of groups and written documents, the values priorities and mindsets of the components of a system could be measured, which could provide the basis for a practical methodology for permanent organizational change.

In the rest of the paper I am going to cover three summations to date, or summaries based on the research team's collective experience to date. Summation I will look at values development and measurement, Summation II will give an overview of the present resulting values based systems model. Finally, Summation III will outline the intervention methodology and process.

SUMMATION I. VALUES DEVELOPMENT

Origins

Peters and Waterman, *In Search of Excellence: Lessons from America's Best Run Companies*, (1982) conclude their study by saying "Every excellent company we studied is clear on what it stands for, and takes the process of value shaping seriously. In fact, we wonder whether it is possible to be an excellent company without clarity on values, and without having the right sort of values." (Peters and Waterman, 1982) The use of the word Values in current organizational literature (2), particularly in the popular press by authors like Covey and Peters implies that values are behavioral, stand behind human and organizational motivation and represent qualities of excellence that humans somehow ought to have.

This concept is deeply rooted in the Western traditions going back to pre-Socratic Greece.

The nouns ARETH (Arete) and KALOS (Kalos) in ancient Greece gave the combined sense of the modern word value. For Socrates and Aristotle qualities of human excellence were reflected in the values priorities of a person, and led to all knowledge and action. Aristotle suggested they were manifest through our habits and skills (2). They named about half a dozen of them, including Courage, Truth and Justice. For the ancients, values were not ends in themselves, but rather they were values precisely because they contributed to the good of the larger community or the good of society.

When we come into the first century (AD) rather than an emphasis on Arete in the Greek, the new testament turns to the concept of signs and ethical lists. Paul in Galatians (Galatians, 5:22-26). and Ephesians (Ephesians, 4:11-16) lists behaviors that he feels are essential to the development of the good person or good life. But again as with the Greeks they were not ends but means to the greater wholeness of the community at large. When Aristotle was rediscovered in the West in the 12th Century understandably these two traditions were put together by Thomas Aquinas. The discussion went on in the 19th century with many participants such as Locke, Hobbes and Kant.

The Modern Period

The flowering of the sciences in the eighteen hundreds brought discovery, mechanization, opportunity and wealth at a global level such as had never been known previously. Science became side-tracked and, based on a narrow Newtonian model, believed itself to be values-free substituting in their place objectivity and measurement. In the 1880's Friedrich Nietzsche declared God dead and in the 1930's Hitler's mindless science machine drove across Europe. Not surprisingly, the end of World War II ushered in a new era and a new interest in values. A work of monumental significance that affected my own perspective was

Toward a General Theory of Action written in 1951 by Parsons and Shills. Their work introduced a sociological perspective into the discussion.

In this work, they clearly established that human behavior and motivation is underpinned by values. Values are viewed as desired ends, or ideals that individuals aspire to (Talcott Parsons and Shills). Parsons also went on to establish that values are global (Habermas) and are common underpinnings of all culture and organizational systems. Parsons also noted that values motivate individuals not groups. At the same time he claimed that the values of the collective such as an institution, were an essential expression of its organizational culture. This work becomes the foundation for the current discussion of values in the corporate setting. This then was a third strain in our understanding of values.

The final strain came again for the field of psychology and psychoanalysis, demonstrating how values underpinned all human behavior and motivation in the manner they were prioritized. Rokeach and Maslow came out with values lists, reminiscent of the traditional virtues of Aristotle and the ethical lists of the first century A.D.. But they went much further illustrating how values priorities explain moral and ethical behavior, social conflict, and racial and political differences (Rokeach, 1979).

Within the field of educational psychology Hall was trained in and influenced by psychoanalysis and the work of Brazilian psychologist Paulo Freire (3) who noted the relationship of values to language. He suggested that values were encased in emotionally laden words in the language itself. Hall noted this had foundations also in the religious traditions and termed this phenomenon the Genesis Effect (4).

The Work of Hall and Tonna

The author and associates, building on the previous contributions, scanned literature from all the major religious and philosophical traditions, psychoanalysis, moral and educational psychology, sociology, anthropology, history and the arts. They also scanned current information in the pure sciences, particularly physics and chemistry. They posited the hypothesis that there are 125 elemental values common to all human beings (Hall, 1986). These 125 values divide into two classes: goal values that describe the ideals and desired ends of an individual, and means values which are skill and capability based. A cross-disciplinary team tested out the concept in cross-cultural settings for 4 years. The final list of 125 values was ratified in 1979.

This then led to a standardized definition of values based on the historical discussion: *Values are ideals that give significance to our lives through the priorities we live by.* The concept is values are consciously chosen priorities, reflected in human behavior, that are the matter of what gives meaning to a person: drives and motivates them. A combination of any five or six values in a variation of priority clusters will be able to describe any human behavior. By standardizing the meaning of these values we were then able to develop objective measurement instruments for use in personal and organizational assessment and development. (Hall, Harrari, Ledig and Tondow, 1986.)

Values Patterns

Following the discovery of the 125 values it became evident that values form patterns or clusters that explain a lot about human behavior. Subsequently we posited four phases of development, each with its own distinct world view, mindset, leadership and management style. The concept is that all these patterns of behavior are underpinned by distinct clusters of values in each of the phases, as Figure 1 illustrates (5).

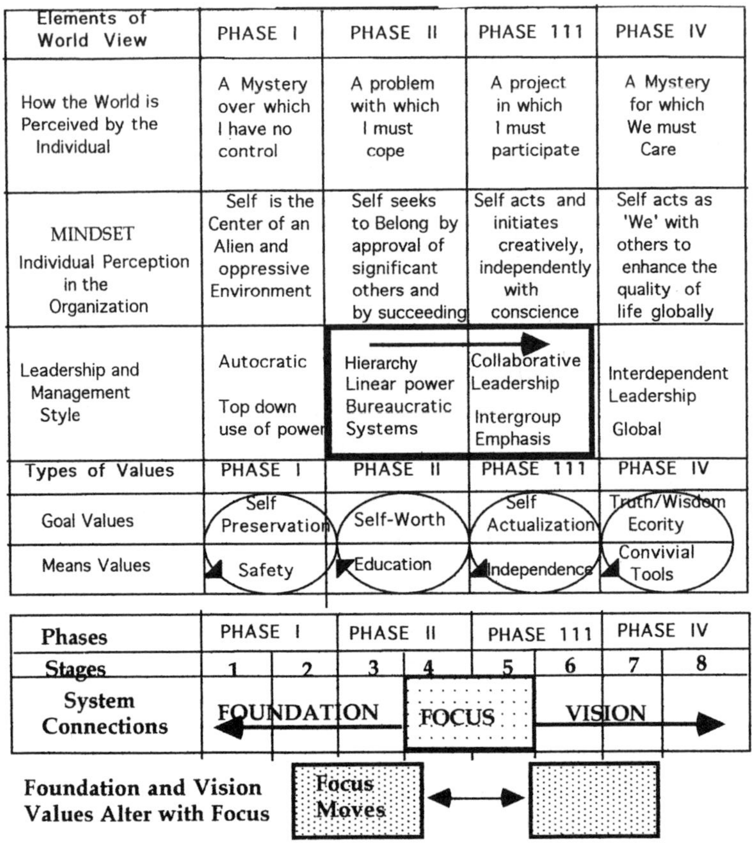

Figure 1. Values and management systems.

SUMMATION II. VALUES AND ORGANIZATIONAL SYSTEMS ANALYSIS

In this section we will see how values analysis of groups within an organizational setting led to the emergence of a values-based systems model. What makes this model unique is that it is both cybernetic and relational in form through the measurement of the values that underpin both factors. Since it is relational, it is therefore also developmental; as most other systems models are not developmental they are missing a critical factor. Problems and issues unexplainable in other models such as the Soft Systems Methodology or VSM are easily handled within a developmental framework.

For example, returning to Figure 1 note the black heavy line around phases II and III in the Leadership and Management Style row. This indicates the common paradigm shift evident in modern management from a formal, task-oriented hierarchical model, to a less formal relational-based collaborative model. Moving up and down the column we realize that individuals sitting in these two camps are going to have different values, mindsets and leadership styles, making communication difficult and sometimes impossible. This communication breakdown, easily explained in terms of differences of values, mindset and world view, would be difficult to explain in terms of a socio-technical or VSM solution.

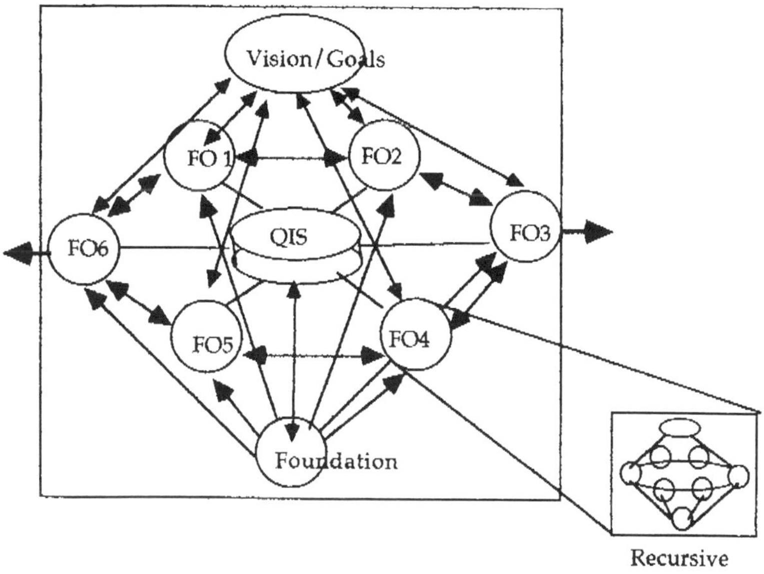

Figure 2. The values-based relational system.

In the last row named 'System Connections' we note that the phases each have two further stages, giving us a scale of eight stages. A measurement instrument in the form of a questionnaire was developed and has now been administered to over 3000 people, including at least 800 executives, in global organizations. It was found that for the majority of people tested the values that reflected their present behavior fell into two adjacent stages somewhere between stages 2 and 7, as indicated in Figure 1. The values priorities before this (Stage 1 up) described their foundational needs and behavior under stress. The values to the right described the underpinnings of their vision for themselves and/or for their organization.

More importantly a methodology was developed and tested (described in the next section) whereby the collective or group of person could come to consensus on the values that they felt should underpin the Foundation, Focus and Vision components of the organization.

'Should' here implies values that would maximize the effectiveness of a given organization in its own particular environment. Traditionally, this has been considered mostly in terms of Financial Capital. In today's climate 'Intellectual Capital' is taking an equal place. The concept of Intellectual Capital acknowledges that relationships precede task in maximizing the creative output of an organization. Effective relationships with human dignity outcomes relative to quality of life, gender and difference equality. The Intellectual Capital components also measures values that are healthy for the environment and in line with customer values. These values patterns have been transferred into a recursive systems design diagrammed above as "The Values-Based Relational System" (VBRS) (7)

This is a three dimensional model that has been applied successful in three global organizations. See Figure 2. The Focus of operations (e.g. Beer's VSM) is depicted by six possibilities as FO 1 - FO 6. The management frame for each of these is recursive in design as the side picture from FO 4 illustrates. QIS or quality information system is a computer data base of all value based and all other relevant information needed and shared equally among the relational partners FO 1 - FO 6.

The concept is that in order to develop a system that is primarily human and therefore relational, the FO partners need to be free to create and not tied down by tasks better executed by the foundational component, so as to maximize creativity. Therefore, the ball titled Foundation is the unit, also recursive that supplies all the support function for the system. Traditionally this would be Human Resources, Education, Finance, Benefits and Legal Departments. The Vision Goals Ellipse are the vision of the organization through a consensus set of values chosen by the organization. Both the Vision component and Foundation feed into and contribute to the QIS system. Next we look at the methodology as it has been piloted in the three global organizations in the last two years.

SUMMATION III. THE PRACTICE

The actual methodology will take a minimum of three years in most organizations, but is, in fact, a recursive or ongoing process. The following processes are all based on a very important educational assumption: ***That no one in an organization will commit themselves to the values of the organization unless they can connect them to their own values. Each person must therefore be minimally aware of what their own values are*** (6)

The method overall has two components:

1. Values Clarification and Formation.
2. Values Alignment and Integration.

1. Values Clarification and Fromation

A. Executive Envisioning. This is where the process begins. The purpose is to enable the Chief Executives of a corporation through a values measurement and coaching process to come to clarity about their vision for the firm. Figure 3 illustrates the process.

The process is initiated when the executives fill out a 125 question questionnaire the results of which are returned at session 1 of the Values Mapping and Envisioning process. This is a 2 hour process. In session 2, also 2 hours, the individual examines the skills behind his or her values and is then coached to come up with a one, three and five year plan for the Firm through the filter of the chosen values.

B. Group Consensus. The group is now brought together for Session 3 which is a meeting of all the executives together for one day. There are three goals for the day:

1. To come to a group values consensus of the Core Values of the organization and a philosophical statement for the group. This will have been anticipated to some

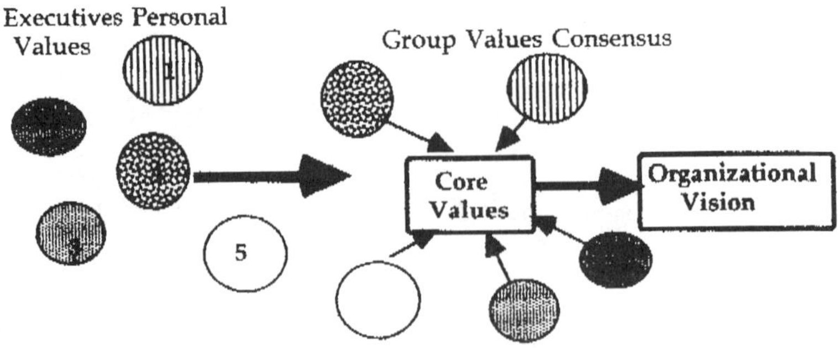

Figure 3. Core values and vision for the organisation.

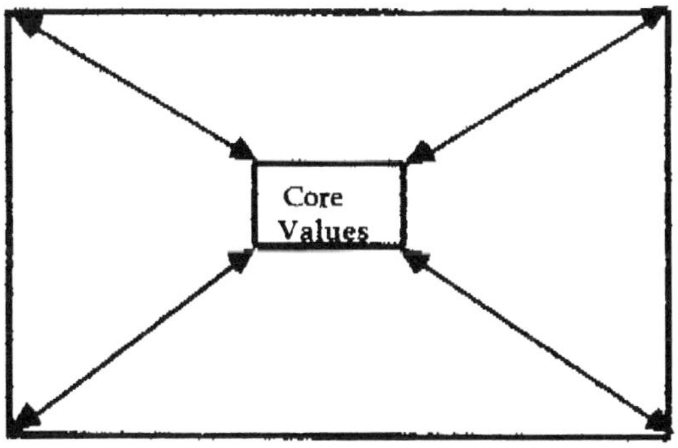

Figure 4. Boundary of minimal values, behaviours and policies.

extent by the coaching process and a computer-generated group and document analysis reports.

2. To discover the immediate blocks to the vision over the next year.
3. To complete an initial action plan based on a prioritization of the blocks to the vision.

Figure 4 illustrates the goals and meaning of this initial process of Values Clarification and Formation

The core values identified at this level now become a partial description and measurement tool of the organizational culture. This is the first values boundary for the overall VBRS diagrammed earlier. Repetition of the values questionnaire on a bi-annual basis can become a measure of the group's progress and therefore of the underpinning of the 'Intellectual Capital' of the system. The core values are now converted into 360 degree performance evaluations and hiring practice tools as mechanisms for early detection of misalignment.

The core values are then the minimal boundary for all employees, and usually include such basics as Courtesy, Making a Profit and Loyalty on the one hand, and long term goals such as Excellence in the Industry, or being an efficient Collaborative team on the other.

The employees now have a minimal set of guidelines on which to build working relationships, team tasks, inter-team relationships and customer or client service. The process is however only partially completed.

2. Values Alignment and Integration

This is a process that emphasizes the values orientation of the local group while retaining and integrating the global values set referred to above as the core values in which the local executives had participated. In this way the local cultural values are maintained with appropriate dignity in each area of operation, while at the same time preserving the company culture that is a part of the institution's intellectual capital.

Figure 5 illustrates the process. This could be an international organization where groups 1-8 are national offices. It could also be a university with the groups being various faculties. In this case, the model is that of a global communications company.

This process now becomes a values integration process whereby individual teams in the local arena can enter into a similar process to test values alignment as a way of bonding the organization throughout. The process here does not require coaching but simple education conducted locally by trained persons.

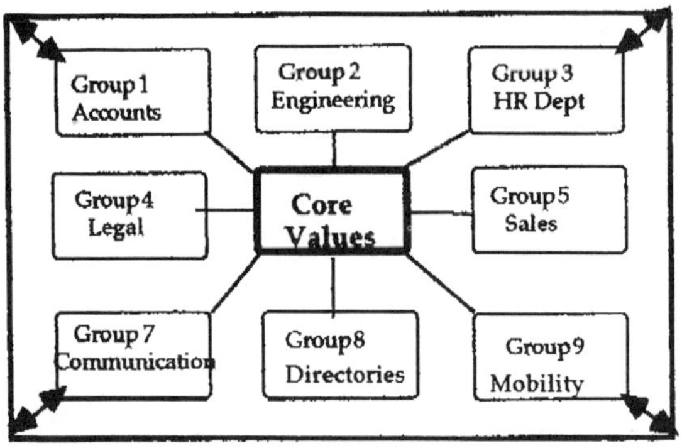

Figure 5. Values alignment.

What is taking place is a values consensus process as described above under 'Values Clarification and Formation' with all autonomous groups in the organization.

There is a difference however, in that each group, such as Accounting or Engineering is going to have a unique set of values determined by its unique tasks, roles and norms. The job then is to come up with a consensus set of values that integrates in the original core set as the minimal boundary for the whole organization.

Finally the levels will go all the way down to the point of customer interface. At this point a special values-based interview program can also measure customer values to maximize the integration between the companies values and those of the customer. Values measurement at this level becomes a way of maximizing team efficiency by aligning individuals in a way to minimize unproductive conflict, and by choosing leadership conducive to maximum team development. The final part of the process is to convert the values at each level into behaviors, by role, policy and goal alignment.

CONCLUSION

The basic purpose of this paper has been to share information about a new systems approach to human organizational development, that integrates systems thinking to certain degree, but which much more forcefully introduces the values component. As such it is fairly new. The values research team has been much more thorough in the values arena. In the systems arena, we are new but have had, at face validity anyway, significant success. Therefore we expect holes to appear.

The central purpose of the paper is to introduce a new systems measurement technology and invite collaboration from people in the field so that something significant may be built in the future.

NOTES AND REFERENCES

(1) Early Development

Elexpuru, Itziar., Portilla, Micaela., Alvarez, Juan Calzon., Torralba, Luis., Izaguirre, Mercedes., Lopez, Javier. 1993, "Los Valores En La Ley Organica de Ordenacion General del Sistema Educativo: Un Analisis Documentos A Traves de La Metodologia de Hall-Tonna" I.C.E., Universidad De Deausto, Bilbao, Spain.

Hall, Brian P., 1976, "The Development of Consciousness: A Confluent Theory of Values", Paulist Press, New York.
Hall, Brian P., 1979, "Developing Leadership by Stages: A Value-Based Approach to Executive Management", Manohar Publications, London and New Delhi.
Hall, Brian., Harari, Oren., Ledig, Barbara., Tondow, Murray., 1986, "Manual for the Hall-Tonna Inventory of Values", Paulist press, New York.
Hall, Brian P., 1987, "The Genesis Effect: Human and Organizational Transformation", Paulist Press, New York.
Hall, Brian P., 1994, "Values Shift : personal and Organizational Development", Twin Lights Publications, New York.
Tonna, Benjamin., 1982, "Gospel for the Cities: A Socio-Theology of Urban Ministry", Translated by William E. Jerman, Ah. Maryknoll, Orbis Books, New York.
Tonna, Benjamin., 1993 and 1994, "Malta Trends 1993 and 1994", Institute for the signs of the times. Media centre, Malta.

(2) The following reference some of the current thinking relative to values: popular version

Aristotle. The *Nicomachean Ethics,* Book II, : The Works of Aristotle, Volume II, The Great Book of the Western World, Encyclopaedia Britannica, Inc, NY, 1952.

References on the diverse conception of virtues can be found in Plato's dialogues of Socrates: *Meno* 174-190, *Gorgias* 257- 284 and the *Republic*, Book VII, 390. In Aristotle: Categories chapter 8, 26-34. Ethics, Books I and II. Thomas Aquinus: *Summa Theologica*, Part I and II Q 54-71. John Locke in *Human Understanding,* Book chapter 11. Thomas Hobbes, *Leviathan*, Part I.. Immanuel Kant, see *Critique of Pure Reason* and *Fundamental Principles of the Metaphysic of Morals..*

Covey, Stephen R., 1990, "The Seven Habits of Highly Effective People", Simon and Schuster, New York.
DePree Max., 1992, "Leadership Jazz", Dell, New York.

These references give some of the basic writing that relates values to virtues historically

Peters, Thomas J., and Waterman, Robert H. Jr., 1982, "In Search of Excellence", Wamer Books, New York.
Schein, Edgar R., 1992, "Organizational Culture and Leadership", , Jossey-Bass, San Francisco.

(3) His best known works are

Freire, Paulo., 1978, "Pedagogy in Process", Seabury Press, New York.
Freire, Paulo., 1972, "Pedagogy of the Oppressed", Herder and Herder, New York.

(4) The following works reference recent values research in the modern period

Parsons, Talcott., Shills, Edward., Editors, 1951, "Toward a Theory of Social Action: theoretical Foundation for the Social Sciences", Harper Torchbook, New York.

The concept of the Genesis Effect takes it's title from publication: in (1) above

Works of special; reference are

Habermas, Jürgen., 1990, "On the Logic of the Social Sciences", translated by Shierry Weber Nicholsen and Jerry A. Stark. Polity Press, Boston.
Jackson, Michael C., 1991, "Systems Methodology for the Management Sciences", Plenum Press, New York.
Maslow, Abraham., ed., 1959, "New Knowledge in Human Values", Penguin Books, New York.
McCarthy, Thomas., 1991, "The Critical Theory of Jürgen Habermas", MIT Press, Cambridge.
Raths, Louis E., Harmin, Merrill, and Simon, Sidney B., 1966, "Values and Teaching: Working with Values in the Classroom", Charles E. Merrill Publishing Company, Columbus, Ohio.

Rokeach, Milton, ed., 1979, "Understanding Human Values: Individual and Societal", Free Press, New York.
Toffler, Alvin., 1990, "Power Shift", Bantam Books, New York.

(5) This figure is dealt with in detail in the book: *Values Shift*, (Hall, 1986)

(6) Value Basics: These issues are covered by three papers available from the Author: *Values Basics I, II and III*

(7) Intellectual Capital

For a detailed review of the concept of "Intellectual Capital" see the comment by Hubert Saint Onge and Leif Edvinsson, both collaborators with author. Fortune Magazine *Your Company's Most Valuable Asset: Intellectual Capital: New Ways to build it and measure it. page 68 . October 3 1994.*

5

WHY IS SYSTEMIC THINKING 'DIFFICULT TO SELL'?

Enrique G. Herrscher*

School of Business Administration - IDEA
Moreno 1850
1094 Buenos Aires
Argentina

INTRODUCTION

Systemic thinking has, as **input**; **(a)** a "Weltanschauung" or world view; **(b)** General Systems Theory or GST; and **(c)** the experiences and problems of the real world. As **output**, it has **(a)** human attitudes and behaviour; and **(b)** solutions to organizational or socio-political problems or issues. This input-output analysis of a discipline is inspired by Van Gigch and Pipino (1986).

Those of us concerned primarily with the output, often wonder why attitudes, behaviour and attempted solutions around us are **'not sufficiently systemic'**. In other words, why systemic thinking is 'difficult to sell'.

Generalizations are dangerous, and no applicable statistics have come to our knowledge. However, from a purely empirical viewpoint, we may say that **(a)** organizational culture is still based more on **fragmentation** (each defending his or her territory) than on shared basic goals; **(b) conflict** is still seen more as a menace to power than as normal interplay of different actors; **(c)** each sector of society is still viewed more as an **isolated system** than as a subsystem integrated to a broader whole; **(d)** and even science and professional profficiency is taught and applied more in **little pieces** than as interrelated elements.

It would therefore seem appropriate to analyse the reasons why, after four decades of theoretical development, the intellectual paradigm and human behaviour is still mainly pre-systemic. Without in the least 'taking sides with the opposition', this article tries to find some answers.

1. CHANGE

From a Kuhnian viewpoint, it should not surprise us that change is more difficult to achieve than maintaining the statu quo. Adapting somewhat an idea expressed in a broader

*Tel: 541 49 0287; Email: herrsche@idea.satlink.net.

Critical Issues in Systems Theory and Practice, Edited by K. Ellis et al.
Plenum Press, New York, 1995

context (Daniel Quinn's breathtaking book 'Ishmael', 1993), we may say that 'it was bad enough' to realize that the earth is not the center of the universe. Systemic thinking pushes us a step further: **'our own world', our sector or organizational niche, is not the 'center' either,** but simply a part of a greater whole. Quite a shock!

2. TERRITORY

Realizing that one's area of action or influence, one's 'territory' is not just a system standing by itself, but also a subsystem of a greater system, does not only mean a change of mentality: it implies (or, at least, is perceived as) **a loss of power.** Whether such loss is real or erroneously perceived as such is irrelevant here, because the issue is precisely what someone in charge of an area **thinks** is happening to him/her, when he/she realizes the **conceptual hierarchy,** the systemic ends-means interrelation, not just the organizational structure, in which his/her sector is embedded. There is a vast literature on the 'territory' issue, curiously more abundant in developing than in developed countries. Some reasons may be found in Yates (1985).

Few persons relinquish power 'for nothing'. The fact that thinking/acting in terms of a greater whole will often improve efficiency, increase output and may not reduce power at all, but even reinforce it, will not easily be accepted. Two cases may therefore be distinguished: **(a)** when greater interrelation will **really** reduce power; and **(b)** when this reduction is only **apparent,** i.e. a greater interdependence 'benefits all'. The distinction is important, because in the second case the issue is **information:** the more said fact carries conviction, the sooner the 'misunderstanding' will be overcome; whereas the first case is not one of 'information' at all (except for 'manipulation'), but of **power:** some stakeholders will gain and others will lose.

3. ORGANIZATION

A systemic attitude organization-wise will not only mean changes in mentality, territorial identification and power. It will almost always affect organizational structure and culture as well. The argument here is that, in most cases, **our organizations are built according to a 'fragmented hierarchical model'** in which true 'systems practice' is out of place.

An example will illustrate our point. In Argentina and in other places, proposals are often made to Universities to include Systems Theory or Thinking in the curriculum. In all cases we know, the University's President was 100% in favour of the project. But when it came to implementation, he had to rely on one of his Faculty or Department Deans. There the trouble began. Deans are **'area oriented':** there is no Dean in charge of 'overall wisdom' or 'general knowledge'. In most cases, the proposal already died at the first step, when deciding which **'specialized area'** should handle this 'unspecialized' teaching and research. In other cases, one area of the existing structure was selected, but this generally meant losing much of the 'overall outlook' of Systems: Business Schools would lean on managerial applications, Engineering on hard systems, and so on. This was one of the conclusions of Michael C. Jackson's diverse presentations in Buenos Aires, Argentina, in July 1994.

This same problem has been abundantly identified in the corporate world. Who is there, below the CEO level, to think and act systemically, if everyone else is committed to a specific area or responsibility? So much so, that the proposal has been made to create a new function: the **connector**. The proposal is contained in several seminars and papers of ex IBM executive Sigfrido Lichtenthal of Columbia University. By not being attached to a single sector, this 'connector' would be the ideal person to assure **efficient interrelation among all parts of the whole,** which seems to be the No.1 problem in many complex organizations.

4. SERIOUSNESS

We will move now from 'acting' to 'thinking'. In other words, we will focus on the intellectual rather than the social web, close interrelation between both notwithstanding.

Our first observation is that many people consider that, **when you abandon specialized knowledge, you lose scientific depth:** you are 'less serious'.

There is a sort of 'quantitative truth' in this argument: if a research fellow dedicates all his/her time to **one field**, the time spent on **that** field will be more (all other things being equal) than a colleague who also wants to cover other connected areas of knowledge.

Here the **difference between depth and broadness** should be pointed out. A 'vertical specialization' will often be the most useful way -if not the only one -to do research and find solutions, when facing well defined problems with clear objectives. But, as shown by Flood and Jackson (1991), we know that in many complex situations, particularly involving human elements, unidirectional research will get us nowhere, whereas a broader approach, even sacrificing depth, will be more 'solution-efficient'. However, the fact remains that, for many 'hard-science-liners', this very broadness will be critiziced as 'lack of depth'.

5. RATIONALITY

A slightly different argument may also point to 'lack of seriousness'. Since the times of Descartes, rationality has been asociated with serious scientific research. This is, in our opinion, still the prevailing paradigm, undisputed until Simon (1947).

We do not believe that there is anyone who thinks that the systems approach is non-rational or - even less - irrational. However, the point can be made that the systems approach comprises **both rational and non-rational elements.** Particularly since soft systems thinking took the lead from the hard systems approach, many social, political and psychological issues became more relevant, weakening the rational part of the rational/non-rational mix or, at least, introducing (in the words of Charles Francois) a psychical and sociological rationality, complementary to the hard systems approach.

Few social scientists would object, but some quantitative or closed-model-oriented scientists might.

6. OBJECTIVITY

In a similar vein, the subjective contents of the systems approach may find critics. The importance of the **observer**, the concept that different observers will see different things, that there is not **'one single reality out there',** but that the observed and the observer are subsystems of a system comprising both (as shown in an unforgettable presentation by von Voerster in Buenos Aires some years ago), are aspects that will not fail to alienate those who believe only in objectivity, in neutral observers and in one single reality equally perceived by everybody.

The fact, put forward by Prigogine and Stangers (1984), that even the traditionally 'objective' sciences, such as physics or chemistry, have lost a significant portion of their original objectivity, should have diminished the number of such critics. Curiously, in this case it is more the humanistic sector where some 'objectivity criticism' may still come from, as shown by Capra (1982).

7. MUTUAL CAUSALITY

Very structured persons, of the 'H' or hierarchical mindscape (in Maruyama's terminology) (1992), may find fault with one specific feature of systems approach: that while A causes or increases B, B may in turn cause or increase A (either directly or through C, D, etc.).

This is the basis of feedback and feed forward processes, of self-correcting servo-mechanisms and thermostats, of the idea of recursiveness and -broadly speaking- of the whole **cybernetic contents** of the systems approach.

It seems almost unthinkable that anyone could deny today such processes or ideas. However, we often see rigid-minded people thinking in simplistic terms along a linear, non recursive causality mode, without being aware of doing so.

One typical example can be found in widespread attitudes regarding the decades-long inflationary process that used to take place in Argentina. Time and again the solutions and even the diagnosis failed, because the majority of people, from the man of the street up to the Minister of Economy, thought that inflation **must have one cause,** and that once that one cause has been identified and corrected, inflation would disappear. But not only does inflation have multiple causes, but many of them are recursive, constituting one of the frequent **vicious circles** that are typical features of developing countries (Herrscher, 1988). In this instance, we have even a case of 'A causes A' or a partly self- generating process: one of the multiple causes of inflation is inflation itself, i.e. the awareness of inflation makes inflationary attitudes almost inevitable (Herrscher, 1992).

This example shows that only a full understanding of multiple causes and mutual causality can cut the vicious circles of this type of complex processes: monetary, fiscal, structural, legal, symbolic, behavioral, socio-economic and political aspects must all be attacked at once in order to achieve results. Not understanding this for more than 40 years shows how embedded the **'single cause - single effect' model** was in the Argentine society.

8. VALUES

All preceding 'arguments' have been presented without any doubt, from our view-point, as having 'missed the point'. In other words, all seven items have nothing to do with our own opinions as systemists. This last item, however, represents an argument that cannot be put aside as easily.

The subjectivity and relativistic features commented earlier may lead many people to view the systems approach as devoid of **'absolute values'** or at least to be extremely **permissive** in this respect. In fact: how can values be anything else than absolute? And how can such 'absolutes' be reconciled with the all-comprising globality of the systems approach?

One case of interest is the religious aspect. In our view, systemic thinking is one of the various ways to perceive God: if everything can be thought of as a system and, as such, as a subsystem of a greater system, which in turn is a subsystem, and so on, such logical chains will lead us to a sort of very last suprasystem, that easily can be called God. This should make the systems approach very dear to anyone involved in a religious creed and/or a religious institution (which is not the same). In spite of this, the systems approach is often rejected by those persons, on the ground of alleged 'lack of absolute principles' derived from said subjectivism.

We do believe that some absolute principles are necessary, even if they arise from within each of us. As shown by Quinn (1993), without a universal value system comprising all living creatures, our planet is headed for disaster. How do we reconcile this with the 'otherwise' systemic relativity? Very simple: by admitting that there is here a contradiction, and by pointing out that **the possibility of contradictions is one of the typical features of systemic thinking.**

Nevertheless, it must be admitted that a 'systemically' all- comprising globality would envision 'good' and 'bad' as subsystems or 'halves' of a world view. From a poetic viewpoint, this dichotomy has been described by Hesse (1925) and commented by Herrscher (1989). Thus ethics, however vital for our survival, would arise from individual judgement. This would alienate those who consider ethics as a 'collective' imperative, and do not accept above possibility of an **'acceptable dilemma'.**

CONCLUSION

It should be clear (we hope) that this list of possible objections to systemic thinking does not imply that we share them. We certainly do not. On the other hand, if we are truly interested in widespread use and comprehension of systemic thinking and attitudes, ignoring such objections would amount to something akin to **'marketing myopia'.**

Therefore, the message is to **analyse strategically** these objections and, by better understanding them, overcome them both in thought and action. This should then be one major challenge for any entity involved in the systems approach.

REFERENCES

Capra, F., 1982, "The Turning Point", Simon & Shuster, New York.
Flood, R.L,. and Jackson, M.C., 1991, "Creative Problem Solving - Total Systems Intervention", John Wiley and Sons, Chichester.
Herrscher, E.G., 1988 (i), Cause-effect loops as explanation of underdevelopment, *Human Systems Mgmt.7:88.*
Herrscher, E.G, 1988 (ii), Mutual causality in developing countries, *Systems Practice.1:3.*
Herrscher, E.G., 1989, Design of the modern inquiring system - Hermann Hesse, *Systems Research.6:1.*
Herrscher, E.G., and others, 1992, Hyperinflation, culture and morphogenetic economics in Argentina, Chapter 2 in: "Context and Complexity - Cultivating Contextual Understanding", *in*: M. Maruyama (Ed), Springer, New York.
Hesse, H., 1989, "Demian", Suhrkamp, Berlin.
Maruyama, M., 1992, "Context and Complexity - Cultivating Contextual Understanding", Springer, New York.
Prigogine, I. and Stangers, I., 1984, "Order out of Chaos - Man's New Dialogue with Nature", Bantam, New York.
Quinn, D., 1993, "Ishmael", Bantam, New York.
Simon, H.A., 1947, "Administrative Behaviour".
Van Gigch, J.P and Pipino, L.L., 1986, In search of a paradigm for the discipline of information systems, *Future Computing Systems*.71:97.
Yates, D., 1985, "The Politics of Management", Jossey - Bass, San Francisco.

6

INTERACTIVE MANAGEMENT

Framework, Practice and Complexity

F. R. Janes

Department of Systems Science
City University
Northampton Square
London, EC1V OHB
United Kingdom

1. INTRODUCTION

This paper begins by outlining a framework for Interactive Management (IM). Here, IM is described as a methodology that enables the work of participant groups to become more productive when attempting to tackle complex issues of management and planning. This is followed by a consideration of IM in practice viewed from three perspectives. This includes an analysis of the activities involved in a complete IM intervention. Finally, the way in which IM attempts to tackle the situational, cognitive and pluralistic aspects of complexity is examined.

2. THE INTERACTIVE MANAGEMENT FRAMEWORK

In using the expression Interactive Management (IM), the term 'interactive' refers to the way that IM both promotes communication and understanding among interacting participants, and enables the interactions among problem components to be examined. The term 'management' refers to the way that IM is concerned both with the management of issues and problems, and with the management of the process of group work.

The framework for IM referred to here involves a number of important aspects. In particular, the notion that work with groups can be made more productive through due attention to five components: the group, the facilitator, the methods, the computer support, and the decision support room. Also, the notion that appropriate methods may be selected for the intelligence, design and choice phases of management decision making. The framework owes its origins to the work of Warfield (1976, 1982, 1984). Initial implementation of the framework was undertaken by Warfield together with Christakis and others (Wood and Christakis, 1984; Christakis, 1985). This work has steadily expanded

Critical Issues in Systems Theory and Practice, Edited by K. Ellis et al.
Plenum Press, New York, 1995

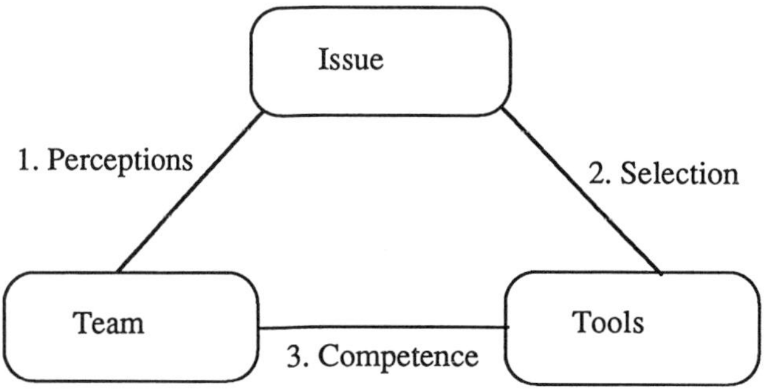

Figure 1. Connections between issue, team and tools (adapted from Warfield, 1976).

(Warfield, 1990; Warfield and Cárdenas, 1994) with, latterly, increasing emphasis on the design phase.

2.1. Issue, Team and Tools

In order to investigate a complex issue, it is often both necessary and desirable to assemble people of diverse backgrounds who can work together as a team. As shown in Figure 1 the connections between the issue, team and methodological tools themselves give rise to a complex situation that needs careful management. Warfield (1976:26) has described these connections as "..the fundamental triangle of societal problem solving". Connection 1, between the team and the issue, indicates that a group has to be assembled with appropriate involvement of stakeholders who have knowledge specific to the issue in order to explore it properly and recognises that the participants will have different perceptions of the situation. Connection 2, between the tools and the issue, indicates that a large range of methodological tools may be available to the group and the appropriate ones for the issue at hand have to be selected. Connection 3, between the team and tools, concerns the fact that even if the appropriate methods exist for the issue, the group may not be aware of them or may not be competent in their use. IM provides a framework for exploring complex issues in which these connections are explicitly recognised and managed.

2.2. Participative Principle

In his writings on planning, Ackoff (1981) emphasises three operating principles of interactive planning: the participative, continuity and holistic principles. These all have a direct bearing on the framework of IM (Christakis, 1983). In particular, the participative principle emphasises the involvement in the planning process of managers who will be responsible for implementing a plan. It stresses moving away from the idea of an organisation employing external or internal consultants whose function is to do the planning for an organisation. By participating in the planning process, members of an organisation acquire a better understanding of the objectives and ideals of the organisation and learn how other managers see their own roles and problems. They also come equipped with a knowledge of the structure, operation and culture of the organisation. Participation enables managers both to contribute to the development of the plans themselves and to learn about the reasons, logic and arguments that have led to particular decisions. It encourages a sense of ownership of the plans. Managers are thus more likely to be committed to implementing the plans than

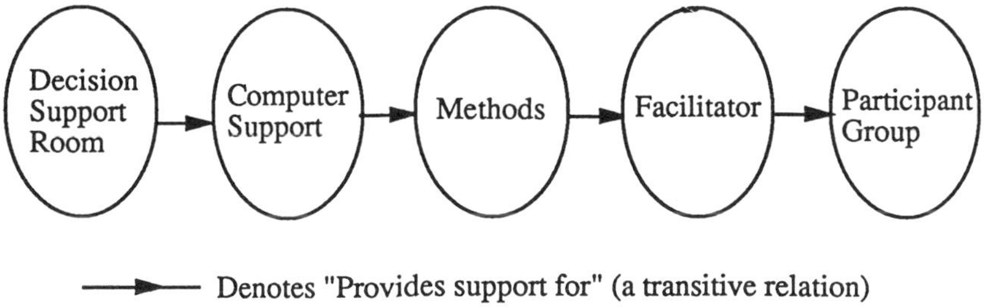

Figure 2. The five components shown as a transitive digraph.

they are to plans which have been imposed upon them without their involvement. For the above reasons it may be argued that participation in the planning process is as important as the products of planning, i.e., the plans themselves.

2.3. Making Group Work More Productive

The need to bring together individuals of diverse backgrounds when tackling complex issues such as planning often results in group work involving meetings. Handy (1985:154) suggests that, on average, 50% of the time of managers is spent working in one kind of group or another, whilst for senior managers this may be as high as 80%. Christakis (1984) reported that a study by the EXXON and XEROX Corporations in 1982 had shown that 17.5 million meetings per day took place in the USA. This is approximately four billion meetings per year. However, all too often dissatisfaction with meetings is widespread. Wainwright (1987:5) lists 16 sources of dissatisfaction including: badly run meetings, avoidance of decisions, attendance by the wrong people, wandering from the point, and lack of participation. Poor preparation and poor chairmanship are two primary sources of such dissatisfaction.

In the context of meetings of groups for complex problem solving, the IM framework incorporates five components which help to ensure that the resulting meetings are productive and overcome many of the common sources of dissatisfaction. Warfield (1984) lists these components as: the participant group, the facilitator, the computer and peripherals, the situation room, and the consensus methodologies. In adapting the framework for use in the UK, the author has found it helpful to define the five components in the following ways.

- **Participant Group.** Individuals, including appropriate representation of stakeholders, with content knowledge relevant to the issue.
- **Facilitator.** To manage the work of the group and guide participants through the steps of the methods used.
- **Methods.** To assist the group in intelligence, design and choice activities.
- **Computer Support.** To run software associated with the methods and to provide display facilities and word processing capability.
- **Decision Support Room.** To provide an environment equipped for efficient group work.

These five components for productive group work support one another as shown in Figure 2. The arrows in the figure represent the transitive relation 'provides support for': thus the facilitator provides support for the group whilst the methods support both the group and the facilitator; the decision support room supports all four other components.

2.4. Methods

The methods form a core part of the IM framework. Each provides a process which, under the guidance of a facilitator, can be used to structure the participants' activities and focus their thinking and effort. Warfield (1990: 199) states that ".... because they [the methods] embody attributes that support consensus or near-consensus, they were designated as 'Consensus Methodologies' ...". In practice, the word 'consensus' is used to signify a number of important characteristics concerning the participative nature of the methods. These are: that the meetings involved are conducted in an open fashion, whereby all participants have an opportunity to contribute through structured discussion of the issue being explored; that the ideas of participants are acknowledged by the group so that members feel that their views have been listened to and that they have had a fair chance of influencing decisions; that individuals can formally indicate their judgements, e.g., by voting; and that participants accept that whilst everyone may not be in complete agreement with all aspects of the products resulting from the use of the methods, they accept the outcome as a way forward. Thus it should not be inferred that the methods necessarily achieve consensus in the literal sense of unanimity, viz., being of one mind. It should be noted that the term 'method' is used in this paper rather than 'methodology' to describe the individual methods, in the sense that each method provides a way of doing something, or a process for attaining an objective. The IM framework as a whole may then be considered as offering a flexible process of inquiry constituting a 'methodology', in the sense that it provides a set of procedures and, as a field of study, it employs a body of methods, rules and postulates (Webster's New Collegiate Dictionary).

Six methods selected by Warfield (1982, 1990) are as follows:

- **Ideawriting.** A quick and efficient idea-generation method for use by groups.
- **Nominal Group Technique.** An efficient idea-generation method for use by groups which allows for clarifying, editing and obtaining an initial ranking of the generated ideas.
- **Interpretive Structural Modelling.** A computer-assisted method for mapping the elements of a complex issue into an agreed diagrammatic structure by an informed group.
- **Options Field**. A method for conceptual design by groups which shows in one diagram all the conceived dimensions of the prospective designs and all the conceived options within each dimension.
- **Options Profile.** A method for selecting specific alternative conceptual designs from an options field.
- **Trade-off Analysis Method.** A method for evaluating and choosing between alternative designs by comparing the differences between the alternatives, and for clearly documenting the basis of decisions made.

3. INTERACTIVE MANAGEMENT IN PRACTICE

The experience of running IM workshops enables some conclusions to be drawn about the methodology of IM. Three perspectives on the methodology are shown in Figure 3. The first embodies the wider sequence of activities in a complete IM intervention, from

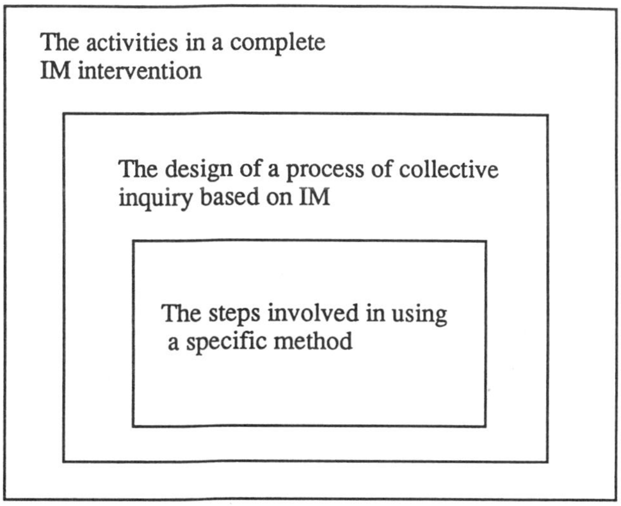

Figure 3. Three perspectives on the methodology of IM.

initial contact with a potential client organisation, through the planning workshop itself, to the implementation of the results. The second perspective, at a higher level of resolution, considers the design of a process of collective inquiry into an issue, based on IM. This includes, for example, how the various methods are to be sequenced during a particular planning workshop. The third perspective, at a still higher level of resolution, considers the steps involved in using a specific method; this is not discussed further here; it is dealt with in depth by Warfield (1990: Chap. 7 and App. 3).

3.1. The Activities in a Complete IM Intervention

Figure 4 , illustrating the first perspective, shows the activities and roles involved in an IM intervention. It is not intended to be an exhaustive list but to pick out important activities and to illustrate a typical pattern of events. In practice, in particular cases, the activities, their sequence and the associated roles may show deviations from the figure.

3.2. Designing a Process of Collective Enquiry

In preparing for an IM workshop, it is essential that careful thought is given to the process of collective inquiry that is to be undertaken during the workshop. The different methods may be combined together as an effective process of collective inquiry in a variety of ways. Designing such a process will be aided by developing with a client a clear understanding of the aims of the workshop. Once these have been agreed, consideration can be given to the way in which the methods may be used to accomplish those aims. Judgement, based upon knowledge of the principles and methods of IM, is important here as well as experience with the use of the methods.

Figure 5. shows the activities involved in an IM planning workshop, the purpose of which was to help a group of participants design a proposed centre for enterprise management. First, Ideawriting was used to identify the requisite activities to achieve a

Stages	Activities	Roles
	Receive request to investigate a complex issue	Potential client and one of IM team
Situation assessment	Examine issue with client organisation	IM consultant with client
	Decide whether IM could provide an appropriate approach	IM consultant
Workshop preparation	Agree aims of workshop	IM consultant with client
	Design process of collective inquiry into the issue, based on IM	IM consultant with IM team and client
	Select participant group	Client with IM consultant
	Prepare detailed workshop plan	IM consultant with IM team and client
	Inform participants about planning workshop and provide relevant documentation	Client with IM consultant
Group work	Conduct planning workshop using methods and principles of IM	Facilitator and support staff with participant group
	Distribute, immediately, the products of the group work	Support staff
Report writing	Prepare report on planning workshop with recommendations for client organisation	IM consultant with facilitator and support staff
Implementation	Implement results and recommendations	Client
	Follow up	IM team

Figure 4. Important activities and primary roles in an IM intervention.

viable centre. Nominal Group Technique was then employed to generate possible objectives for the centre. This was followed by two Interpretive Structural Modelling sessions, first to build an intent structure and then to set priorities for the centre. Finally, a mission statement was prepared. The eight activities represent a complex sequence of group work and illustrate the careful thought needed in designing such a process of inquiry.

Activity / method		Product
Ideawriting	→	Requisite activities
Nominal Group Technique	→	Objectives
Interpretive Structural Modelling	→	Intent structure
Group discussion	→	Sub-groupings in intent structure
Idea categorisation	→	Categorised objectives
Interpretive Structural Modelling	→	Composite priority structure
Group discussion	→	Linear priority structure
Group work	→	Mission statement

Figure 5. Sequence of group activities and resulting products.

4. TACKLING COMPLEXITY

4.1. Aspects of Complexity Addressed by IM

In attempting to explore the meaning of complexity, Flood (1987) distinguishes between two different aspects of it. The first aspect is associated with the system or situation under study. Here, complexity is considered to arise because of the presence of many parts or elements in the system with numerous interrelationships between them; this is labelled by Flood as the 'systems dimension' of complexity. Within the field of systems, this aspect relates to what might be thought of as a traditional view of complexity. The second aspect of complexity distinguished by Flood is associated with the interpretation of a system or situation by observers and is considered to depend upon their capabilities, interests, notions and perceptions, and values and beliefs; this is denoted as the 'people dimension' of complexity. Warfield (1990:133) draws a distinction between 'situational complexity' and 'cognitive complexity'. The former may be interpreted as those aspects of the situation under study that the mind is able to intercept. Cognitive complexity relates to those aspects of our understanding of the situation that make interpretation difficult. Both Flood's and Warfield's explanations of the meaning of complexity have something in common. They are of interest in the context of IM because of the way in which they quite explicitly consider complexity to be a function of both the situation or system being explored and the way that the situation is interpreted by an observer(s).

In considering those aspects of complexity that are addressed by IM, it will be helpful to draw upon the ideas of both Flood and Warfield referred to above. This is done in Figure 6, where Warfield's term 'situational complexity' is used to refer to the complexity of the issue, situation or problem being explored. This is preferred to Flood's term 'systems dimension' in that it may or may not be appropriate to describe the issue, situation or problem being explored as a 'system'. Furthermore, as the focus of IM is on work with groups of participants, it is desirable when considering the 'people dimension' of complexity to distinguish explicitly between the 'cognitive complexity' relating to an individual's understanding and interpretation of a situation, and the 'pluralistic complexity' resulting from the fact that individuals in a group will interpret issues differently.

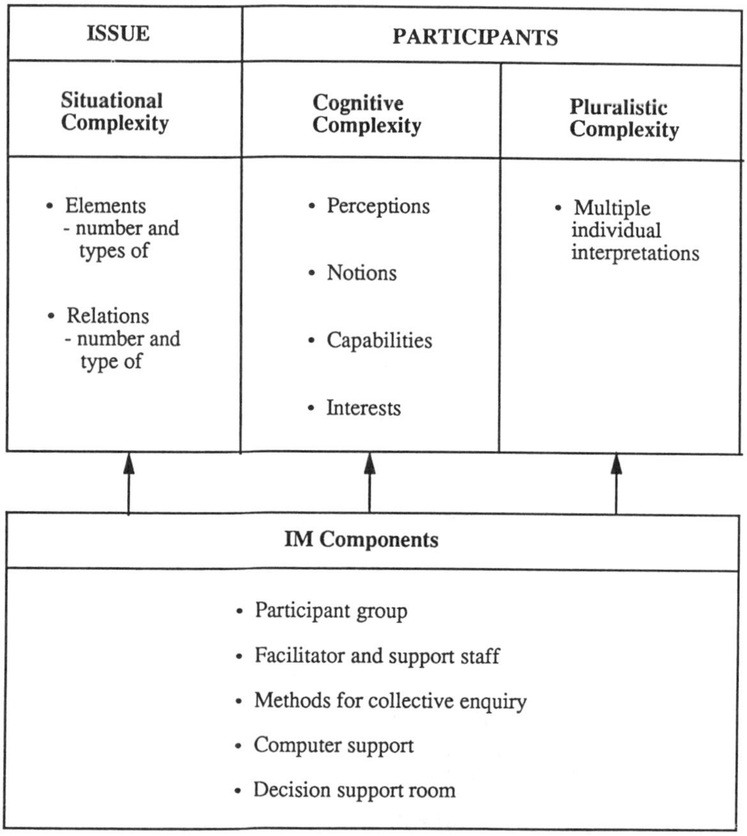

Figure 6. Aspects of complexity addressed by IM.

Figure 6 indicates that IM attempts to address all three areas of complexity: situational, cognitive and pluralistic.

Clearly, there can be little doubt that IM attempts to address **situational complexity.** The identification of elements and explicit examination of the relations between them are a major focus of the methods, e.g., Interpretive Structural Modelling, and one of the things that they were designed to do. With regard to **cognitive complexity**, it will be helpful to make use of some of the terms suggested by Flood in connection with his ‘people dimension’, i.e., perceptions, notions, capabilities and interests. The participative nature of an IM workshop, the structured debate which forms part of the methods used, the clear diagrammatic modes of representation, and the subtle pressure on people to contribute, all help to tackle this aspect of complexity. The third column in Figure 6., which is labelled **pluralistic complexity**, explicitly recognises the multiple individual interpretations of a situation that will arise as a result of individual cognitive complexity. The participative nature of IM should help the individuals involved to appreciate one another’s perception of a situation and to learn more about it. In so doing they may, of course, find that the situation is more complex than they had previously appreciated. Nonetheless, the improved understanding should enable those involved to address the situation in a more informed manner.

4.2. Complexity of the IM Approach

Figure 6. also serves to illustrate the complexity involved in running an IM planning workshop. The workshop designer has available various resources which need to be carefully integrated to make the workshop productive. These include: personnel including a facilitator, computer operator and other support staff; a variety of methods which may be combined to provide a process of collective inquiry; computers, software and peripheral displays; and the equipment and layout of the decision support room. The design of the workshop requires that these be combined into a complex system which enables the three types of complexity associated with the issue and the participants to be addressed simultaneously. The implementation of the workshop by the facilitator(s) is particularly demanding in that he or she has to ensure that the resources are coordinated so as to make effective and efficient use of the participants' time throughout the duration of the workshop; there will probably not be a second opportunity to get it right!

5. CONCLUSIONS

The work of participant groups attempting to deal with complex issues may be made more productive through due attention to five interrelated components : the group, the facilitator, the methods, the computer support, and the decision support room. These components are central to the framework of IM. The practice of IM has been viewed from a number of perspectives. An analysis of the activities constituting the widest perspective, that of a complete IM intervention, illustrates the steps involved in the planning and conduct of such an intervention. At a higher level of resolution, the design of a process of collective inquiry needs careful thought; such a process enables a facilitator to guide a group in the application of a sequence of methods in order to explore a complex issue of concern. The aspects of complexity that IM addresses have been examined. These are described as situational, cognitive and pluralistic complexity; it is considered that IM attempts to address all three aspects. However, it is noted that the complexity of the IM approach itself means that a professional IM intervention should not be embarked upon lightly without due attention to the resources required and many activities involved.

REFERENCES

Ackoff, R.L., 1981, "Creating the Corporate Future", John Wiley & Sons, New York.

Christakis, A.N., 1983, "Planning, design, and implementation : the sigma five strategy", paper presented at the Cairo Conference on Planning and Implementation Strategies for Alleviating Congestion, Cairo, April 17, 1983.

Christakis, A.N., 1984, Management of complexity : the interactive approach, in Continuing the Process, Newsletter, Graduate School, US Department of Agriculture, Washington, DC, 1(4), 1-2.

Christakis, A.N., 1985, The national forum on nonindustrial private forest lands, *Systems Research*. 2(3), pp189-199.

Flood, R.L., 1987, Complexity : a definition by construction of a conceptual framework, *Systems Research*. 4(3), pp177-185.

Handy, C.B., 1985, On the workings of groups, *in:* "Understanding Organizations", Penguin, London.

Wainwright, G.R., 1987, "Meetings and Committee Procedure", Hodder and Stoughton, London.

Warfield, J.N., 1976, "Societal Systems : Planning, Policy and Complexity", John Wiley & Sons, New York (Reprinted, Salinas, California : Intersystems, 1989).

Warfield, J.N., 1982, Organisations and systems learning, *General Systems*. 27, pp5-74.

Warfield, J.N., 1984, Principles of interactive management, *in:* "Proceedings International Conference on Cybernetics and Society", IEEE, January, 1984, New York, pp746-750.

Warfield, J.N., 1990, "A Science of Generic Design : Managing Complexity Through Systems Design", Intersystems Publications, Salinas, California (2nd edition, Iowa State University Press, 1994).

Warfield, J.N., and Cárdenas, A.R., 1994, "A Handbook of Interactive Management", Iowa State University Press, Ames, Iowa.

Wood, W.C., and Christakis, A.N., 1984, A methodology for conducting futures-oriented workshops, *Technological Forecasting and Social Change.* 26, pp281-297.

7

WHAT IS THIS THING CALLED CRITICAL SYSTEMS THINKING?

Gerald Midgley

Centre for Systems Studies
Department of Management Systems and Sciences
University of Hull
Hull, HU6 7RX
United Kingdom

INTRODUCTION

This paper is about Critical Systems Thinking (CST), a research perspective which is said to embrace three fundamental commitments. These are commitments to:

- *critical awareness*—examining and re-examining taken-for-granted assumptions, along with the conditions which give rise to them;
- *emancipation*—ensuring that research is focused upon 'improvement', defined temporarily and locally, taking issues of power (which may affect the definition) into account; and
- *methodological pluralism*—using a variety of research methods in a theoretically coherent manner, becoming aware of their strengths and weaknesses, to address a corresponding variety of issues.

The idea that there are three commitments in CST was introduced into the literature by Flood and Jackson (1991(a)), but the definitions given above are my own. They are inevitably an over-simplification of the range of issues considered important by critical systems thinkers, but are useful for indicating the general interests pursued by proponents of the perspective.

Now, the title of this paper asks a question: 'what is this thing called CST?' My immediate response is to reply, 'there is no one such thing'. Many different perspectives on CST are offered in the literature. What they all have in common, however, is an interest in questions about, and surrounding, the nature of the above commitments. But there is no consensus on their definition. Indeed, different writers have evolved very different understandings, and continue to develop their ideas in communication with others. CST can therefore be seen as an evolving discourse around a set of themes that are considered important by a significant number of systems practitioners. The term 'discourse' is central here as it emphasises dynamism and continued development rather than the stasis of a final definition.

Critical Issues in Systems Theory and Practice, Edited by K. Ellis et al.
Plenum Press, New York, 1995

In consequence, the view I will present in this paper of CST should be seen as a further contribution to the discourse rather than as a statement of a final position. And I will be treating the work of other writers in a similar manner. I intend to focus upon one particular vision of CST (presented in a number of places by Flood and Jackson, but most notably in Flood and Jackson, 1991(a)), and will highlight some problems with their argument that I believe require further attention. I will then draw for inspiration upon earlier work by Churchman (e.g., 1979) and Ulrich (e.g., 1983) in order to argue that it is possible to address these problems. A different vision of CST will emerge as a result.

However, before presenting Flood and Jackson's views, I need to say why I have decided to focus upon their work rather than that of the many other authors who have written about CST. My answer is that, in the last decade, Flood and Jackson have produced a substantial body of literature communicating their own particular vision of CST, and this has been quite influential. Two particularly notable contributions are Jackson (1991) and Flood and Jackson (1991(a)). The first seeks to provide concise definitions of the commitments (although Jackson identifies five rather than three). The second is a book of edited readings by a variety of authors, and in my view the commentary surrounding the chapters is exceptionally clearly written. In fact, both of these works are *so* clearly written that I believe there is a danger that Flood and Jackson's vision could be regarded as definitive, and this could threaten the continued development of CST.

So, what is CST from Flood and Jackson's point of view? I will answer this question by taking each of the three commitments in turn, clarifying the meanings the two authors give them. In producing this review, I do not wish to imply that the authors share the same perspective on every single issue, but there is a substantial amount of common ground. I should also make it clear that the accounts I shall give will inevitably be over-simplified. Interested readers should consult the original literature as referenced.

METHODOLOGICAL PLURALISM

Let us start with the commitment to methodological pluralism, or 'complementarism' as Flood and Jackson also sometimes call it. According to Jackson (1987(a)) and Flood (1989), methodological pluralism makes explicit use of a meta-theory to identify the strengths and weaknesses of different methodologies, which are thereby viewed as complementary.

Jackson and Keys (1984) and Jackson (1987(b)) have developed just such a meta-theory which classifies systems methodologies according to the assumptions they make about social reality. For example, quantitative, modelling methods assume that there is agreement on what the research problem is (if there isn't, then it is difficult for the researcher to know what to model), while qualitative, debating methods work on the assumption that there is disagreement that needs to be discussed (if there is full agreement then debate becomes redundant). The authors argue that, because different methodologies make different assumptions, it is appropriate to use them in a complementary fashion to deal with a variety of different contexts. Their meta-theory, 'the system of systems methodologies', aligns various methodologies with the contexts that, in an ideal world, they should be most capable of dealing with. Flood (1990) describes this treatment of methodologies as 'meta-paradigmatic': the system of systems methodologies sits above the more limited methodological paradigms that are governed by it. It can therefore be used in practice to aid critical reflection on methodology choice.

This is the instrumental aspect of Flood and Jackson's work on methodological pluralism, where the focus is on how methodologies can most appropriately be used in practice. However, it has also been underpinned by an epistemological theory (a theory about the nature of knowledge) originally proposed by Habermas (1972). Habermas calls this the 'theory of knowledge-constitutive interests', and it was first introduced into Flood and Jackson's vision of CST by Jackson (1985). While Habermas' work is immensely broad, and cannot be summarised adequately in a few paragraphs, it is nevertheless possible to sketch his core idea. Habermas claims that all human beings have three fundamental interests: a 'technical interest' in predicting and controlling our natural and social environment, a 'practical interest' in pursuing mutual understanding, and an 'emancipatory interest' in freeing ourselves from constraints imposed by power relations.

It is this theory that Flood and Jackson claim can be used to underpin the system of systems methodologies. However, the two authors are not in total agreement on how this underpinning should be achieved, and a review of their individual positions can be found in Midgley (1992(a)). To keep matters simple, I will concentrate on the position that has been discussed most widely in the literature—that proposed jointly by Flood and Jackson (1991(b)). In short, they suggest that "...'hard' and cybernetic systems approaches can support the technical interest, soft methodologies the practical interest, and critical systems heuristics can aid the emancipatory interest".

To explain in more detail, 'hard' and cybernetic systems approaches are those which have modelling as their central activity. In relation to the theory of knowledge-constitutive interests, these methodologies are viewed as supporting our technical interest in predicting and controlling our environment. In contrast, 'soft' methodologies involve managing debate between people so that learning may be facilitated, ideas evaluated, and plans for action developed. In relation to the theory of knowledge-constitutive interests, these support our practical interest in achieving mutual understanding. Finally, critical systems heuristics is concerned with subjecting assumptions in planning to ethical critique. It asks both the researcher and participants in dialogue to address a number of questions concerning the issue of whose views should enter into the planning process, and how this should be achieved. According to Flood and Jackson (1991(b)), this can support our emancipatory interest in freeing ourselves from restrictive power relations.

We therefore see that Flood and Jackson have a particular understanding of methodological pluralism which critically aligns methodologies with contexts for use, and which supports this alignment with an epistemological theory of universal human interests in prediction and control, mutual understanding and freedom from oppressive power relations. We can now move on to examine Flood and Jackson's understanding of the commitment to emancipation.

EMANCIPATION

According to Jackson (1991), "...critical systems thinking is *dedicated to human emancipation* and seeks to achieve for all individuals the maximum development of their potential. This is to be achieved by raising the quality of work and life in the organizations and societies in which they participate" (emphasis in the original). Jackson (1991) argues that emancipation, in these terms, can only be achieved by addressing all three of Habermas' human interests (in prediction and control, mutual understanding and freedom from oppressive power relations) as and when issues concerning them arise in systems practice. This view is also promoted by Flood and Jackson in their (1991(a)) joint work.

Having clarified the meaning for Flood and Jackson of the commitment to emancipation, we can now look at their work on critical awareness.

CRITICAL AWARENESS

Between them, the two authors identify three interlinked forms of critical awareness: "...understanding the strengths and weaknesses and the theoretical underpinnings of available systems methods, techniques and methodologies" (Jackson, 1991), understanding both the context of application (assessing alternative interpretations of it) and the possible consequences of using various methodologies once the context has been defined (Flood, 1990; Jackson, 1990), and "...closely examining the assumptions and values entering into actually existing systems designs or any proposals for a systems design" (Jackson, 1991).

The first two forms of critical awareness clearly support the commitment to methodological pluralism. The first involves critical thinking *about* methodology, so that effective meta-theories can be developed, and the second involves the critical *use* of methodology, where the focus of critique is the context of application. The third form of critical awareness primarily supports the commitment to emancipation. It is through ethical critique that power relations can be understood and 'improvement' defined.

CRITICAL ISSUES

Having given a brief description of CST as interpreted through the writings of Flood and Jackson, I wish to highlight six problems that I see in their position. I will list these briefly below, and will then move on to address them through the production of an alternative vision of CST.

The first problem is in Flood and Jackson's understanding of methodological pluralism. Flood (1990) claims that their work is 'meta-paradigmatic'—above and beyond all paradigms. I would argue that this cannot be the case given that Flood and Jackson make assumptions about human knowledge drawn from Habermas' (1972) theory of knowledge-constitutive interests. These assumptions are alien to, and incommensurable with, assumptions made by the proponents of other systems paradigms. Far from being meta-paradigmatic, CST embodies its own unique assumptions, meaning that it is trying to establish the foundations for a *new* paradigm. For further details of this argument, see Midgley (1989(a)(b)).

It is also possible to identify a significant problem with Habermas' (1972) theory of knowledge-constitutive interests. By suggesting that human beings have an interest in 'predicting and controlling' the natural and social worlds, Habermas risks perpetuating the myth of the human domination of nature. This myth leads people to regard natural phenomena as 'resources' for control and consumption, with often unpredictable side effects. As a number of authors have pointed out, many of our current ecological problems stem from this myth. It would be preferable to talk in terms of human beings having an interest in building and preserving a sustainable, interactive relationship with their non-human environment (Eckersley, 1992; Midgley, 1992(a)).

There are also difficulties with Flood and Jackson's vision of the commitment to emancipation, the first of which follows on from my previous point concerning ecological sustainability. Jackson (1991) is quite specific that his concern is with *human* emancipation. While I would argue that human well-being is served by building and preserving a sustainable, interactive relationship with our non-human environment, the ramifications in terms of the need to change organisational, societal and global priorities are not well appreciated in many societies. Even when they *are* appreciated, the enforced dependency of people, especially in the West, on unsustainable forms of social life (traditional forms of employment, high energy consumption, the frequent use of cars, etc.) makes it very difficult for them to change their lifestyles, even if they want to (Gregory, 1992). It is for this reason that the term 'human emancipation' will usually be interpreted as the promotion of human well-being *separate from* consideration of the 'environment'. I place inverted commas around the word 'environment' because the separation I am talking about is even present in the language I am using. In my view, talking about a commitment to human emancipation, rather than a more general commitment to 'improvement', does more harm than good: it does not encourage people to challenge the automatic prioritisation of a human boundary in systems practice. For further thoughts on the problem of uncritically prioritised human boundaries, see Midgley (1994).

Another difficulty that arises in connection with the commitment to emancipation is the extent to which we can talk about improvement of human well-being in absolute, or near absolute, terms. Flood and Jackson only make very general comments on the commitment to emancipation, but their use of Habermas' (1972) theory of knowledge-constitutive interests raises a question about whether they agree with Habermas that human society is in a process of social evolution. If they do, this would mean that the term 'emancipation' could be tied in with the idea of Mankind's supposed 'march of progress'. Given the problems of deciding what is or is not 'progress' (progress for some may be a setback for others), many authors (e.g., Rorty, 1989) have argued that theories of social evolution have little credibility. I do not want to put words into the mouths of Flood and Jackson, as there is no positive indication that this is what they actually do believe, but there is a need for further clarification here.

The fifth problem I need to raise relates to the commitment of CST to critical awareness. We need to ask, 'what form does critical awareness take?'. Flood and Jackson talk about studying the strengths and weaknesses of methodologies, asking questions about the nature of the context when intervening, and revealing the ethical assumptions that flow into systems practice. But *how*? While Flood (1990) proposes a general theory of critical awareness—there is a need both to "liberate knowledges" (alternative views of the situation) and then choose between these in practice—this is not translated into methodology. Interestingly, the problem would appear to stem from Flood and Jackson's vision of methodological pluralism. A writer who has considered the need for a *practice* of critical awareness is Ulrich (1983), and he proposes the methodology of critical systems heuristics. In Flood and Jackson's vision, however, critical systems heuristics is just one methodology amongst many, and is specifically aligned with contexts characterised by coercion. This means that, in practice, it will only be used on an occasional basis. Now, I am not suggesting that critical systems heuristics gives us all we need to enact critical awareness (several criticisms can be raised against it, as Romm, 1995, and others have pointed out), but the question remains, 'how is the commitment to critical awareness enacted in situations where coercion is *not* initially identified?'. For further reading around this question, see Gregory (1992), Tsoukas (1993) and Ulrich (1993).

The sixth and final problem follows on from this. Because a lot of systems practice is commissioned by organisations wishing to deal with problematic issues, the lack of a methodological aspect to critical awareness (other than that offered when coercion is identified) could have important implications. There could be a tendency to take the commissioning organisation's agenda as given, and there could also be a failure to take sufficient account of the effects of that agenda on the organisation's wider environment (both people and planet). This can be described as taking an organisational boundary for granted. To see whether this is indeed a problem for Flood and Jackson's vision of CST, we need to look at some examples of their practice. Flood and Jackson (1991(b)) contains several reports of the authors' systems practice, but it is beyond the scope of this short paper to study these in detail. Nevertheless, it is worth highlighting the fact that the reports contain little discussion of the boundaries defining who determined the remit of the work. Also, all the reports except one (a report of the use of critical systems heuristics) concern problems of design or planning *within* an organisation. There is no indication in these reports, apart from the one describing the use of critical systems heuristics, that any boundary other than the organisational one was considered. I understand, however, that this issue is being addressed in a future work (Flood, 1995).

THE ETHICAL CRITIQUE OF BOUNDARY JUDGEMENTS

Having highlighted some problems with Flood and Jackson's vision of CST, it is now time to look at how we can address them. In the process, a different vision of CST will emerge. My starting point for creating this new vision is further reflection upon the problem of critical awareness mentioned towards the end of the previous section. My intention is to address the problem of critical awareness by seeking inspiration from the work of Churchman and Ulrich, both of whom have explored the concept of boundary critique. I will first summarise the main strands of their work before going on to look at its implications for CST. Let us start with Churchman.

Churchman has made a major contribution to systems thinking (e.g., 1968(a)(b), 1971, 1979), and only a flavour of his writings can be given here. For Churchman, the business of defining improvement is a systems problem. It is a systems problem because the *boundary* of analysis is crucial. As he makes clear, what is to be included in, or excluded from, the analysis is a vital consideration: what may appear to be improvement within a narrowly defined boundary may not be improvement at all if the boundaries are pushed out. This way of thinking involves a fundamental shift in our understanding of the nature of 'system'. Prior to the work of Churchman, many people assumed that the boundaries of a system are 'given' by the structure of reality. In contrast, Churchman made it clear that boundaries are constructs that define the limits of the knowledge that is to be taken as pertinent. There is also another important element of Churchman's understanding of 'system'. When it comes to human systems, pushing out the boundaries of analysis may also involve pushing out the boundaries of who may legitimately be considered a decision maker (Churchman, 1970). Thus, the business of setting boundaries defines both the knowledge to be considered pertinent *and* the people who generate that knowledge (and who also have a stake in the results of any attempts to improve the system).

Not only did Churchman introduce this fundamental change in our understanding of 'system', but he also discussed critique. In defining 'improvement', Churchman (e.g., 1979) followed Hegel, who stressed the need to expose our most cherished assumptions to the possibility of overthrow. To be as sure as we can that we are defining improvement adequately, we should seek out the strongest possible 'enemies' of our ideas and enter into a process of rational argumentation. Only if we listen closely to their views and our arguments survive should we pursue the improvement.

Churchman produced a great deal of highly influential work in the 1960s and 1970s, and in the 1980s several other authors began to build upon it in significant new ways. One of these authors was Ulrich, who has acknowledged his considerable intellectual debt to Churchman in several 'tribute' papers (Ulrich, 1988(a)(b)). Once again, only a flavour of Ulrich's work can be given here, so I recommend consulting his original writings, particularly Ulrich (1983).

Ulrich was actually the first person to use the terms 'critical' and 'systems' together. This was when he created the methodology of critical systems heuristics. For Ulrich, the term 'critical' has its roots in the work of Habermas, but mostly his later writings (Flood and Jackson tend to draw upon Habermas' earlier work). Habermas (e.g., 1976) argues that critique is a dialogical process emerging from the inherent potential of language to allow us to question. However, dialogue may be distorted through the effects of power either directly, when one participant coerces another, or indirectly, when participants make unquestioned assumptions about the absolute necessity for, or inevitable future existence of, particular social systems. To overcome these effects of power, we need to move towards what Habermas calls an 'ideal speech situation': a situation where any assumption can be questioned and all viewpoints can be heard.

However, while Ulrich (1983) accepts the *principle* of Habermas' understanding of critique, he nevertheless criticises him for being utopian. For all viewpoints to be heard, the ideal speech situation would have to extend debate to every citizen of the world, both present and future. This is quite simply impossible. Ulrich sees his task as the *pragmatisation* of the ideal speech situation, and the marriage between critical and systems thinking is the means by which this can be achieved. Truly rational inquiry is said to be *critical*, in that no assumption held by participants in inquiry should be beyond question. It is also *systemic*, however, in that boundaries always have to be established within which critique can be conducted. Indeed, Ulrich claims that both ideas are inadequate without the other. Critical thinking without system boundaries will inevitably fall into the trap of continual expansion and eventual loss of meaning (as everything can be seen to have a context with which it interacts, questioning becomes infinite). However, systems thinking without the critical idea may result in a 'hardening of the boundaries' where destructive assumptions remain unquestioned because the boundaries are regarded as absolute.

A priority for Ulrich is to evolve practical guidelines that can help people steer the process of critical reflection on the ethics of drawing system boundaries. For this purpose, Ulrich (1983) developed a list of twelve questions which can be used heuristically to define what the system currently is and what it ought to be. It is important to note that some of these questions relate to who should be involved in discussing boundary judgements in the first place, meaning that there is always the possibility for people to enter or leave discussions. Furthermore, Ulrich claims that his twelve questions can be employed to enhance critical thinking about boundary judgements *regardless of the systems method that is being used*: Whether the intention is to engage in quantitative modelling or qualitative debate, reflection on boundary judgements is still important.

Now, I do not want to suggest that the work of Churchman and Ulrich alone will answer all the issues I have raised in relation to Flood and Jackson's vision of CST. Neither do I want to claim that they have said all there is to be said about critical awareness. On the contrary, significant criticisms have been levelled at both Churchman (e.g., Jackson, 1982) and Ulrich (e.g., Flood and Jackson, 1991(b); Babüroglu, 1992; Romm, 1995). What is important for this paper, however, is that the *principle* of boundary critique—being critical about systems—is sound, and that it is possible to translate it into methodology. While some of the background theory may require further work, the basic principle and the possibility of methodology is not damaged by the theoretical difficulties so far identified.

THE IMPLICATIONS FOR CRITICAL SYSTEMS THINKING

I will now use the work of Churchman and Ulrich on critical boundary judgements as a starting point for addressing the issues I raised earlier concerning Flood and Jackson's vision of CST. I will take the issues one by one, starting with the problem of operationalising the commitment to critical awareness. A different vision of CST will emerge through this process.

It will be remembered that operationalising the commitment to critical awareness becomes a problem in Flood and Jackson's vision because critical systems heuristics, the only methodology specifically designed to do this, is reserved for situations where coercion is identified. My proposal for a first step to correct this problem is to support Ulrich's (1993) argument that critical systems heuristics should be used to enhance critical thinking up-front—both when we enter into interventions, and periodically after that. I say that this is a 'first step' because there is still a crying need for further research, both to enhance critical systems heuristics (in theory and practice) and to develop other approaches to making critical boundary judgements. In particular, we must begin to move away from 'mechanical' applications of critical systems heuristics, where participants in inquiry simply answer the twelve questions in the form of a list, towards a situation where they become an integral part of the whole inquiry process, interwoven (where appropriate) with other systems methods. Encouragingly, research along these lines is already well under way, (see, for example, Cohen and Midgley, 1994; Gregory, Romm and Walsh, 1994; Flood and Romm, 1995). What I am proposing, then, is to define the commitment to critical awareness in terms of the ethical critique of boundary judgements, and to continue to conduct research to enhance both the theory and practice of boundary critique.

Now, it is important for me to give further support to this position, especially as it challenges Flood and Jackson's understanding of methodological pluralism, where all the different systems approaches are aligned with their most appropriate contexts of application. Flood and Jackson may well reply by saying that boundary critique is redundant in situations where coercion has not been identified. However, the most immediate questions that spring to mind are, 'how do we identify coercion?' and, 'whose views do we take into account in the identification process?'. Answering these questions will involve the researcher and other interested parties in *making critical boundary judgements*. In other words, making up-front boundary judgements cannot be avoided in *any* research situation. Failure to realise the full implications of this will inevitably result in some of the most important boundary judgements—those which determine who the researcher will talk to and how the initial remit of the work will be defined—being made uncritically.

Having dealt with the first problem of critical awareness in Flood and Jackson's vision of CST, we can now move on to the second: the tendency to give uncritical priority in interventions to an organisational boundary. This is directly addressed by bringing to the fore the notion of making critical boundary judgements. If the researcher and/or participants engage in a process of questioning boundary judgements, then the proposed use of an organisational boundary would have to be justified. While some researchers might fear that commissioning organisations will be put off by this kind of questioning, my own experience is that this rarely happens: many people welcome the chance to look at how the problem they have identified interfaces with others, and appreciate systemic logic. People often understand that their own activities have a wider impact that needs to be taken into account if unanticipated, negative reactions from others are to be avoided. Examples of studies conducted by critical systems thinkers that use boundaries other than the organisational one are now appearing regularly in the literature: these are mostly multi-agency interventions (e.g., Cohen and Midgley, 1994; Gregory and Midgley, 1994; Gregory, Romm and Walsh, 1994; Flood, 1995; Midgley and Milne, 1995) and systems studies involving community participation in service planning and evaluation (e.g., Midgley, Kadiri and Vahl, 1995).

Bringing critical boundary judgements to the fore in CST also gives us the means to address both of the issues I raised concerning Flood and Jackson's understanding of the commitment to emancipation. The first of these was a concern that the term 'human emancipation' will often be interpreted as the promotion of human well-being *separate from* consideration of the 'environment'. Certainly, the ethical critique of boundary judgements will, to an extent, address this, but I would also suggest a change of terminology. Instead of talking about a commitment to human emancipation, why not call it a commitment to improvement? This de-emphasises the 'human', and brings CST more in line with Churchman (1970), for whom 'sustainable improvement' is a concept of central importance.

The second issue concerning Flood and Jackson's understanding of the commitment to emancipation relates to the possibility of interpreting it in absolute, or near absolute, terms. Habermas (1972) argues that human society is in a process of social evolution. If Flood and Jackson agree with this, then their understanding of 'emancipation' could be tied in with the idea of Mankind's supposed 'march of progress'. In contrast, I would argue that 'progress' is difficult to define, except locally and temporarily from the viewpoints of stakeholders. Now, if we bring the notion of the ethical critique of boundary judgements to the fore, the 'march of progress' interpretation is no longer possible. Any vision of 'improvement' or 'emancipation' that is employed in systems practice would inevitably be a function of the boundary judgements defining who is involved in the process of generating that vision. Of course, this has profound implications for systems practice, and is an issue that requires further research.

We may now move on to the difficulties with Flood and Jackson's understanding of methodological pluralism. The first one I will tackle is their acceptance of Habermas' view (1972) that human beings have a fundamental interest in 'predicting and controlling' their non-human environment rather than building and preserving a sustainable, interactive relationship with it. They thereby run the risk of perpetuating the myth of the human domination of nature. Elsewhere (Midgley, 1992(a)), I have argued that we should abandon Habermas' theory of knowledge-constitutive interests because of this problem and turn to Habermas' later work on validity claims (e.g., 1976, 1984). This can be used to underpin methodological pluralism with a philosophical theory (see Midgley, 1992b, for details) without making the assumption that human beings have an inherent interest in 'predicting and controlling' the natural world.

The final problem with Flood and Jackson's vision of CST is the claim that it is 'meta-paradigmatic' (Flood, 1990). I have argued that this cannot be the case given that Flood and Jackson make assumptions about human knowledge that are alien to, and incommensurable with, assumptions made by the proponents of other systems paradigms. Far from being meta-paradigmatic, CST is trying to establish the foundations for a *new* paradigm. Interestingly, Flood (1990) accepts the incommensurability argument: he suggests that it is possible to have methodological *commensurability* while acknowledging theoretical *in*commensurability. However, it seems to me that this stands in opposition to his claim that CST is 'meta-paradigmatic': it means that, if CST is indeed theoretical in nature, we must recognise that it is incommensurable with other perspectives based on different theoretical assumptions. This does not mean that critical systems thinkers cannot learn from other paradigms, or that the proponents of other paradigms cannot learn from CST (Gregory, 1992), but it does make the claim to a meta-paradigmatic status problematic. Now, my understanding is that this claim came about through the wish to promote a form of methodological pluralism where no one methodology was seen to dominate. However, if we take on board the idea that the critique of boundary judgements is of central importance, then there is no longer a need to describe methodological pluralism in this 'neutral' manner. We can accept that there may be many different visions of methodological pluralism, including one which rests on the assumptions of CST outlined here, and our task is then to justify why any one vision is preferable (Midgley, 1992(b)). We can therefore drop the claim that CST is 'meta-paradigmatic'.

I have now addressed all the issues I raised earlier in response to Flood and Jackson's understanding of CST, and the outline of a new vision has emerged. My hope is that this can be used as the basis for further theoretical and practical research. In the spirit of the (revised) commitment of CST to improvement, it will be important to continue to improve CST itself.

CONCLUSION

In this paper I have described CST as an evolving discourse around a set of themes that are considered important by a significant number of systems practitioners. In the spirit of discourse, I have focused upon Flood and Jackson's vision of CST, which has been particularly influential in recent years. I have raised some issues which concern me, and in addressing them I have returned to the earlier work of Churchman and Ulrich for inspiration. As a result, the outline of a new vision of CST has emerged. Whether or not you agree with what I have proposed, I hope you are stimulated by it and wish to engage in dialogue. It is only through further discourse that CST will continue to develop and change.

ACKNOWLEDGEMENTS

I would like to acknowledge the excellent feedback I received from Bob Flood, Wendy Gregory and Norma Romm on the first draft of this paper. Also, my apologies to Alan Chalmers for appropriating the title of his (1982) book, "What is this thing called Science?", and twisting it to my own devices.

REFERENCES

Babüroglu, O., 1992, Tracking the development of the Emery-Trist systems paradigm, *Syst. Pract.* 5:263-290.

Chalmers, A.F., 1982, "What is this thing called Science? An Assessment of the Nature and Status of Science and its Methods," 2nd ed., Open University Press, Milton Keynes.

Churchman, C.W., 1968(a), "Challenge to Reason," McGraw-Hill, New York.

Churchman, C.W., 1968(b), "The Systems Approach," Dell, New York.

Churchman, C.W., 1970, Operations research as a profession. *Man. Sci.* 17:B37-B53.

Churchman, C.W., 1971, "The Design of Inquiring Systems," Basic Books, New York.

Churchman, C.W., 1979, "The Systems Approach and its Enemies," Basic Books, New York.

Cohen, C. and Midgley, G., 1994, "The North Humberside Diversion from Custody Project for Mentally Disordered Offenders: Research Report," Centre for Systems Studies, University of Hull, Hull.

Eckersley, R., 1992, "Environmentalism and Political Theory: Toward an Ecocentric Approach," UCL Press, London.

Flood, R.L., 1989, Six scenarios for the future of systems 'problem solving', *Syst. Pract.* 2:75-99.

Flood, R.L., 1990, "Liberating Systems Theory," Plenum, New York.

Flood, R.L., 1995, "Solving Problem Solving," John Wiley & Sons, Chichester.

Flood, R.L. and Jackson, M.C., 1991(a), "Critical Systems Thinking: Directed Readings," John Wiley & Sons, Chichester.

Flood, R.L. and Jackson, M.C., 1991(b), "Creative Problem Solving: Total Systems Intervention," John Wiley & Sons, Chichester.

Flood, R.L. and Romm, N.R.A., 1995, Enhancing the process of choice in TSI, *Syst. Pract.* 8:in press.

Gregory, W.J., 1992, "Critical Systems Thinking and Pluralism: A New Constellation," Ph.D. Thesis, City University, London.

Gregory, W.J. and Midgley, G., 1994, Planning for disaster: developing a multi-agency, post-disaster counselling service, *in*: "Community Works: 26 Case studies showing Community Operational Research in Action," J. Bryant, C. Ritchie and A. Taket, eds., Pavic Press, Sheffield.

Gregory, W.J., Romm, N.R.A. and Walsh, M.P., 1994, "The Trent Quality Initiative: A Multi-Agency Evaluation of Quality Standards in the National Health Service," Centre for Systems Studies, University of Hull, Hull.

Habermas, J., 1972, "Knowledge and Human Interests," Heinemann, London.

Habermas, J., 1976, "Communication and the Evolution of Society," English edition, 1979, Heinemann, London.

Habermas, J., 1984, "The Theory of Communicative Action, Volumes I and II" Polity Press, Cambridge.

Jackson, M.C., 1982, The nature of soft systems thinking: the work of Churchman, Ackoff and Checkland. *J. Appl. Syst. Anal.* 9:17-29.
Jackson, M.C., 1985, Systems inquiring competence and organisational analysis, *in*: "Proceedings of the 1985 Meeting of the Society for General Systems Research," 522-530.
Jackson, M.C., 1987(a), Present positions and future prospects in management science, *Omega* 15:455-466.
Jackson, M.C., 1987(b), New directions in management science, *in*: "New Directions in Management Science," M.C. Jackson and P. Keys, eds., Gower, Aldershot.
Jackson, M.C., 1990, Beyond a system of systems methodologies, *J. Op. Res. Soc.* 41:657-668.
Jackson, M.C., 1991, The origins and nature of critical systems thinking, *Syst. Pract.* 4:131-149.
Jackson, M.C. and Keys, P., 1984, Towards a system of systems methodologies, *J. Op. Res. Soc.* 35:473-486.
Midgley, G., 1989(a), Critical systems and the problem of pluralism, *Cyb. & Syst.* 20:219-231.
Midgley, G., 1989(b), Critical systems: the theory and practice of partitioning methodologies, *in* "Proceedings of the 33rd Annual Meeting of the International Society for the Systems Sciences," held in Edinburgh, Scotland, on 2-7 July 1989.
Midgley, G., 1992(a), "Unity and Pluralism," Ph.D. Thesis, City University, London.
Midgley, G., 1992(b), Pluralism and the legitimation of systems science, *Syst. Pract.* 5:147-172.
Midgley, G., 1994, Ecology and the poverty of humanism: a critical systems perspective, *Syst. Res.* 11:67-76.
Midgley, G., Kadiri, Y. and Vahl, M., 1995, Managing stories about quality, *Int. J. Technol. Man.* 10:in press.
Midgley, G. and Milne, A., 1995, Creating employment opportunities for people with mental health problems: a feasibility study for new initiatives, *J. Op. Res. Soc.* 46:35-42.
Romm, N.R.A., 1995, Some anomalies in Ulrich's critical inquiry and problem-solving approach, *in:* "Critical Issues in Systems Theory and Practice," K. Ellis, A. Gregory, B. Mears-Young and G. Ragsdell, eds., Plenum, New York.
Rorty, R., 1989, "Contingency, Irony, and Solidarity," Cambridge University Press, Cambridge.
Tsoukas, H., 1993, The road to emancipation is through organizational development: a critical evaluation of total systems intervention, *Syst. Pract.* 6:53-70.
Ulrich, W., 1983, "Critical Heuristics of Social Planning: A New Approach to Practical Philosophy," Haupt, Berne.
Ulrich, W., 1988(a), C. West Churchman—75 years, *Syst. Pract.* 1:341-350.
Ulrich, W., 1988(b), Churchman's "process of unfolding"—its significance for policy analysis and evaluation, Syst. Pract. 1:415-428.
Ulrich, W., 1993, Some difficulties of ecological thinking, considered from a critical systems perspective: a plea for critical holism, *Syst. Pract.* 6:583-611.

8

DOES 'INFORMATION SYSTEMS' NEED SYSTEMS?

Frank Stowell

Department of Computing and Information Systems
University of Paisley
Paisley
Scotland
United Kingdom

INTRODUCTION

At the 3rd UKSS biannual International Conference at the University of Paisley it was argued that the systems epistemology could provide the intellectual basis of the emerging discipline of Information Systems (Stowell, 1993). As if to support this argument the largest stream of the ten streams that made up the 1993 conference was the Information Systems (IS) stream in which a variety of systems based ideas about IS were expressed. Since the 1993 conference the single most important concern for IS practitioners and academics alike has been their active attempts to get IS recognised as a domain of education in its own right (Jones, 1994). Their actions lift IS as a discipline out of a purely intellectual debate into the higher education funding arena. However, a potential change in 'status' for IS has important ramifications. If IS is to be treated as a discipline, a 'department of knowledge', this will mean important ramifications in terms of funding, assessment, professional accreditation and the location of the subject domain into one of the accepted domains of academic pursuit. Some of the issues are relevant to Systems. This paper will address some of those issues and consider what effect the potential changes in IS might have upon Systems and what, if any, contribution systems might make to the developing IS field.

CAN INFORMATION SYSTEMS BE CONSIDERED AS A DISCIPLINE?

There have been many attempts to define Information and Information Systems (e.g. Yourdon and Constantine, 1979; Galliers, 1987; Amadio, 1989; Avison and Wood-Harper, 1990; Checkland and Scholes, 1990; NCC, 1993; Stowell and West, 1994) but as yet there is no universally accepted definition. What definitions of IS there are seem to fall into three categories. The first reflects the technical view in which attempts are made to take into account the social impact of the use of IT. In the technical definitions often the words

Critical Issues in Systems Theory and Practice, Edited by K. Ellis et al.
Plenum Press, New York, 1995

Information Technology and Information System and Data and Information are interchangeable. The second category of definition seems to come from cognate domains outside computing. For example, there are a number of definitions of IS that can be found within the management science literature. Here emphasis is given to the organisation and to those that manage it. Definitions of IS include reference to company mission and strategic planning and is often based upon a notion of organisation in which goals and objectives are prime. The third camp belongs to those who seek to establish a philosophy for the domain arguing that without such a philosophical basis a discipline of IS cannot exist. Mistakes of the recent past are cast up as evidence of the need for a rethinking of Information Systems in the light of technological developments.

Although the foregoing account summarises some of the fundamental differences between these three intellectual 'camps' perhaps the more important point to make is not the differences but their similarity, namely, the deliberate attempt to treat Information Systems as a distinct cognate domain rather than an adjunct to another.

WHAT IS A DISCIPLINE?

The Shorter Oxford Dictionary defines a discipline as " (2) a branch of instruction; a department of knowledge and (3) the training of scholars and subordinates to proper conduct and action by instructing and exercising in the same; mental and moral training" Perhaps this provides us with a starting point. What is the branch of instruction that we claim for IS? If we look at the literature we see papers on a variety of topics. These include the social impact of information technology, e.g. employment, personal skills, papers that are concerned with the use of technology to improve business performance e.g. strategic planning, decision support systems, papers which are concerned with the definition, provision and development of IT support e.g. methodologies, application generators and finally papers concerned with the improvement of technical design e.g. database, object orientation. In addition there are a few papers which are concerned with the future of IS and those which suggest a decline in the popularity of computing as an area of study. The range of disciplines from which these papers come and the noticeable epistemologies evident from the papers seems to belie the hope of demonstrating strong unifying principals and practices that are normally associated with a discipline.

The reservoir of literature claiming to relate to IS is a rich one but paradoxically it also highlights the difficulty of defining the domain. We can say that the literature demonstrates the interdisciplinary nature of IS, we can also, rather cynically, say that IS is a 'sink' for papers that will not fit any of their 'home' journals. The latter can be dismissed fairly quickly since a significant number of the IS papers are published in journals rated to be of International standing and unless we are prepared to re-examine the whole learned journal accreditation process then we must accept that the majority of papers are of intellectual significance. Moreover, the establishment of Professorial posts suggests that IS receives recognition from other areas of learning. This leaves IS with the rich but wide range of subjects and consequently the attendant problems that that creates for recognising IS as a discipline.

SYSTEMS DEPARTMENTS

What can be learnt from the systems field itself? Systems as a fundable area of academic study has enjoyed a bitter-sweet relationship within education and industry. There has been on one hand the acceptance of Systems as an area of study but on the other hand there is a failure to get Systems funded as a distinct field of interest. The teaching of Systems as a subject or methods which are based upon Systems ideas is widespread. Many academic institutions include Systems as a core element in undergraduate and postgraduate courses. In many business enterprises Systems methods are taught to managers as part of management training. Paradoxically few, if any, independent Departments of Systems survive. Even the University of Lancaster the 'home' of Systems for many years no longer has an independent Department of Systems.

Perhaps IS can learn from the problems that the lack of formal recognition brings. The lack of formal recognition from the educational establishment means no allocation of funding the consequence of which may impair the development of the domain. A lack of formal recognition denies the possibility of professional accreditation for practitioners and hence little or no quality control mechanisms for them. In one important area though, Systems has formal recognition and that is in the literature. Systems as a field of interest does not suffer from the lack of a philosophical base and as evidence of this there is a considerable literature available relating to the Systems epistemology and the associated ideas. This literary recognition does not, however, carry enough weight to provide the domain with a secure independence. Putting to one side the interdisciplinary nature of Systems, and maybe the apparent intellectual desirability of the subject not being seen as a separate department, the reality of the situation is that few, if any, departments of Systems exist. One consequence of this is that there is no focal point for resource allocation a matter which may provide IS academics with food for thought.

LACK OF THEORETICAL UNDERPINNING

Paradoxically, the success of Systems in establishing a wealth of literature in which philosophical and practical issues are dealated has not succeeded in obtaining formal recognition whilst other domains with perhaps less philosophical credibility are resourced. One criticism that is sometimes heard is that some formally funded cognate domains have no theories of their own. An argument is made along the lines that theories commonly used have been taken straight from another intellectual area. If the IS literature is a guide then the theories drawn upon by its authors come from engineering, social science and science itself. Whilst this fact can be used as an example of the 'woolliness' of the subject and a reason why IS cannot be considered as a separate domain of intellectual study it also provides an example of the interdisciplinarity of IS.

In my presentation at the last conference I proposed that Systems may provide a suitable intellectual framework upon which to build IS. The notion of 'Wholeness' is a fundamental part of the Systems epistemology and seems to be appropriate to the thinking that is necessary for IS.

Systems ideas are appropriate to IS because its practitioners are concerned about the relationship of technology to perceived human information systems and the consequent skill base that that task implies. IS academics who teach the growing number of topics said to be relative to IS have themselves come from a wide variety of first degree disciplines. As a discipline IS includes all areas of knowledge which are appropriate to human decision making. From that assertion we might be justified is claiming that IS is not a discipline but it is multidisciplinary. However, such an assertion does little for those who purport to be an IS professional since by inference no one individual can be. Even if we accept IS as multidisciplinary it would seem to require an intellectual base and a defined set of skill requirements which together provide a clear department of knowledge. One useful task for the IS community would be to agree an educational agenda for IS.

WHERE IS INFORMATION SYSTEMS PRACTISED?

IS academics can be found in many University departments and some even located in departments which do not offer IS subjects but allocate 'service' lecturers to other departments where an IS module is offered. As things stand presently the most common location for IS as a body of academics are found in management and in computing departments. It is argued that as a discipline IS is as much out of place in Management Science as it is in Computing Science. The latter subject areas can provide examples of the impact of technology upon human information systems but neither exclusively represent it.

There are an increasing number of courses both at degree and postgraduate level in which the words Information Systems appear in the title. In industry itself it is rare to read references to 'Data Processing Manager' or 'Computer Systems Manager' and commonplace to read 'Information Systems Department' or 'Information Systems Manager'. A recent initiative by the Jupiter Consortium (1993) draws attention to the needs of the modern manager to both manage Information and the business through the TEAM scheme of postgraduate education. In this education programme the emphasis is very much upon information systems and the management of information technology and little upon the technology itself.

In the past four years there has been the establishment of several academic chairs in Information Systems and more recently the setting up of a committee of Information Systems Professors. These examples provide a clear practical illustration of IS as a domain of expertise, at least as far as the academic world is concerned. Ironically Information Systems Professors are 'intellectual cuckoo's' because they exist within a host domain which is not their parent. If the majority of the Professors of Information Systems are to be found in a particular cognate domain is it not possible that IS is in danger of becoming a sub-set of that domain? Many practitioners may remember the problems for computing which arise from the legacy of early business computing departments which were set up under the department of finance. If we are to avoid this kind of thing from happening in IS we need to ensure that the representations being made currently for a separate discipline of Information Systems succeed.

Despite the evidence that IS exists as a field of academic pursuit in a formal sense as far as the academic world is concerned there is no cognate domain of Information Systems. Most of those who work within information systems do so from Computing Studies, Business Studies or Management. Academics are mindful of the importance of the many assessment exercises that Higher Education faces these days, e.g. research and quality, and of the 'rules' which surround these exercises. To this end every attempt is made to contribute to the success of the department and therefore the publications which attract the most effort also relate primarily to the cognate domain of the department. This has two potentially harmful effects. The first of these is that the intellectual content of the publications may, necessarily, be biased towards the 'host' discipline and, secondly, the degrees of freedom allowable to those who seek to investigate Information Systems as a discipline are constrained.

HOMELESS

Intellectually, where will IS as an emerging discipline be located? Let us consider Computing Science as one possibility. Unfortunately Computing itself is not recognised by many as a discipline as such although it is accepted as a cognate domain of expertise for the purposes of assessment. Computing like Information Systems did not emerge as a distinct area of study until relatively recently. Computing as a subject began its existence in departments of mathematics and electrical engineering. For many people computing is a subset of engineering. Can information systems be considered as an engineering discipline? Many have emphasised the social concerns of information systems which mark it out from computing. Does this mean that IS is a social science?

Should information systems have as its intellectual home management science? This too creates difficulties. For example, is Information Systems Strategy or Business Information Systems really what we mean when we talk of IS as a discipline? Many writers highlight the deficiency of management education in its failure to recognise information technology as a potentially revolutionary technology which too often concentrates upon the use of spreadsheets as a token towards management use of technology. To many it is unthinkable that IS should be separated from technology.

Perhaps one immediate difficulty that we must face is where does IS fit into current educational categories e.g. is IS science, engineering or social science. There is little to be gained from denying the existence of these categories since education is funded through their 'existence' Information Systems does not readily fit under any of these headings and the danger is that if current attempts to gain recognition succeed it may be made to fit. There would seem to be as much an urgency to address the practical problems of IS as its philosophical underpinning.

SYSTEMS IN INFORMATION OR INFORMATION SYSTEMS?

What has created this upsurge of interest in IS? There are many reasons but one important element seems to be the exploitation of Information Technology. Modern IT has enabled many forms of human communication and methods of enquiry which seem limited only by the imagination that creates them. The processing power of modern IT means that the speed and execution of the application itself appears to be almost instantaneous. The actuality of these many developing technologies are less important than the possibilities that each development itself inspires.

If we consider the potential impact of one area of development, Virtual Reality. Virtual reality currently accounts for about one twenty fifth of a \$116m market (Houlder, 1994) and is expected to rise. The technology is being used in a wide area of applications but particulary in training. The media enables tasks to be undertaken as if the operator were performing the real activities. This development is particularly useful in areas which are complex, potentially dangerous and demand a high level of safety, e.g the nuclear energy industry, avaiation. For example, the refitting of engines in passenger aircraft is an area in which virtual reality is providing an excellent training method. Virtual reality systems are being used to provide architects with the opportunity to walk through their buildings before they are built, engineers the chance to test complex structures before they are assembled and surgeons to perform operations without any risk to patients.

Consider the development of a complex computer-based model which combines hard facts and the softer socio-historical context in which the 'model' exists. For example, a model of the building of a medieval cathedral using multi-media or a virtual reality presentation . It is possible to reproduce in a three dimensional model of any of the great cathedrals. The facts for the building itself can be derived from the architectural knowledge available today and from the archive material relating to the architecture of the time. The package could also provide many other details including the architect, the builders and details of the times in which the cathedral was proposed and a history of its construction right through to completion. Clearly the production of such a model includes the collection of a combination of hard and soft facts and contextual information before the development of the software can begin.

It is not too hard to imagine the complexity of the task and of the range of skills that the task will demand.

The problem for the designer of sophisticated computer-based models is a complex one. For example, who has the responsibility for managing the project? who is responsible for the production of the final design? There are a number of potential candidates including the historian, the architect, the social scientist, the software engineer, the systems analyst? Does it matter? It can be argued that this provides an example of the multi-disciplinary nature of IS but who will ensure that the package will represent the whole story in an unbiased way? The end result may well be an important source of reference which will influence countless scholars and any individual interested in medieval architecture. Consideration of such problems raise issues about the whole process of the design process including: knowledge elicitation, the production of the technical specification, the validation of the model and the ethics involved in the presentation of such a powerful and influential knowledge base.

The information that such developments as the one described above can provide is significant and such computer based models will become influential over many fields of academic work and at many levels of education. Such information systems require methods of development that are as advanced as the ideas which the technologist seeks to represent

THE CONTRIBUTION OF SYSTEMS

At the last International Conference the value of systems ideas to IS was advocated by many of the delegates. The developments in technology has emphasised the emergence of a domain of expertise which suggests such a range of skills as to take the domain out of its existing location. Notwithstanding the lack of a universal definition of information systems and a somewhat contradictory set of literature about IS as a discipline many definitions suggest a wide range of skills probably outside the conventional education and training of a systems analyst.

For example, if we take an information system as "The notional whole through which the provision, manipulation, and use of appropriate data is managed to enable appropriate decision making." (Stowell and West, 1994). This is one definition which close examination suggests a number of areas of knowledge. The list will include addressing such issues as - in order to know about the notional whole requires a set of skills, both practical and intellectual, which will' enable that information to be established. The provision and manipulation of data requires a different set of skills both technical and non-technical. Decision making requires a knowledge not just about human decision making but of the context in which the decisions take place The IS practitioner needs to facilitate discussion between a number of experts which will enable them to communicate with and learn from each other and in such a way that he/she can define the information system to enable the technology to be determined and the clients to get from the technology what they require.

THE END OF THE METHODOLOGY ERA?

Contributions from the Systems field have grown steadily in recent times including the incorporation of SSM and associated interpretivist systems thinking into the IS development process (Stowell, 1994). The development of such ideas continues but what is required is not more methodologies but frameworks through which a variety of ideas can be used to achieve a satisfactory outcome. If the framework is based upon systems thinking so much the better since the overarching systems idea of wholeness seems to relate closely to the demands technological development. One such framework if the notion of Client Led Design (Stowell and West, 1994).

CLIENT-LED DESIGN AS A FRAMEWORK FOR THINKING ABOUT IS DEVELOPMENT

Central to Client-Led Design (CLD)is the idea that the client or clients should be enabled to appreciate, define and specify their technology based information system requirement . In CLD the organisation is considered as a relationship maintaining system in which the system attempts to maintain a relationship with its environment by learning about it and adapting to it. Importantly this perspective attempts to represent the system of interest and its environment from the perspective of those involved within the problem situation (Stowell and West, 1994).

The interpretivist perspective which CLD purports to subscribe means that action is taken within the recognition that the most important task for the ISA is to create a learning environment in which client and ISA can appreciate each others views. The analyst acting as teacher and facilitator enables the client, through the processes of CLD, to control the design.

CLD is easiest described as having five phases of development although the five phases should not be thought of as discrete entities but a series of interrelated activities through which a process of learning takes place resulting in a client led information design. The most important part of the development of an IS project is at the beginning of the project. It is at this stage when CLD and client(s) try to communicate what is required and what can be provided. This important activity takes place in Phases I and II of CLD. In Phase I the client and ISA attempt to 'Appreciate' the situation. Appreciation here is used in the Vickerian (1978) sense in which both judgements of value and judgements of reality are exercised. At the start of the project CLD makes explicit the process of learning through the employment of appropriate methods through which both ISA and Client can 'appreciate' the situation. Each participants preconceptions of the situation will be modified by the knowledge gained from the encounter which should be obvious to all parties. Once the participants are satisfied that a level of understanding has been reached which will enable them to represent their ideas about the information system then a means of representing this is selected. The method of representation relies upon the advice of the ISA but an important consideration is that the client should be capable of understanding the representation sufficiently to enable alterations to the model to be made by them if necessary. In Phase three the ISA and client are concerned with the production of a technical specification that will support the information systems defined in Phase two. At this stage the ISA will provide the technical expertise which will enable the production of a technical specification. Phases four and five are concerned with implementation and re-evaluation of the installed information system.

CLD is not intended to be offered as a methodology (it is not) for IS but by way of illustrating how Systems thinking itself, rather than the production of another methodology, might help in the identification and development of the complex information systems that technology can support.

CONCLUSIONS

In this paper I have attempted to address some of the issues relating to Information Systems and the contribution of Systems Thinking. My conclusion is that the Systems epistemology seems suited to the IS domain since both are seeking to address a similar set of problems. However, Systems itself is not a good example to follow since it has no home of its own in the resource allocation sense. In a practical sense, and in some respects this the most urgent problem to confront, IS should be seen as a separate area of intellectual pursuit notwithstanding the interdisciplinarity of the domain. IS has, it is argued, outgrown the various intellectual homes it presently inhabits. If research and the general maintenance of this area of knowledge is to be fulfilled then it needs be recognised and subsequently funded. Economically, well documented computer system failures add weight to the argument of 'a better way' being needed to design computer support, in the business world it is improbable to imagine a successful company not having extensive computer support. Socially, the impact of modern technology is apparent at every level of society . Common sense suggests that a cognate domain of information systems is necessary since common understanding by both practitioner and academic of their area of concern seems to support the need. A common vocabulary and the definition of good practice are two of the important concerns. Moreover, the growing level of wasted investment, both in financial and social terms, provide 'hard' facts to support the assertion.

REFERENCES

Amadio, W., 1989, "Systems Development. A Practical Approach", McGraw-Hill, Singapore.

Avison, D.E., and Wood-Harper, A.T., 1990, "Multiview: An Exploration in Information Systems Development", Blackwell Scientific Publishers, Oxford.

Checkland, P.B., and Scholes, J., 1990, "Soft Systems Methodology in Action", John Wiley and Sons, Chichester.

Duderstadt, J.J., Knoll, G.F. and Springer, G.S., 1982, "Principles of Engineering", John Wiley and Sons, New York.

Jupiter Consortium, 1988, "Management of Technology: The Invisible Advantage", Booklet, Statement by the Jupiter Consortium Technology Management Group, Twickenham.

Jones, M.,1994, What is the Distinctive Nature and Value of Information Systems as a Discipline?:*4th UKSS Information Systems seminar held at the University of Warwick.* November 1994.

Houlder,V., 1994, Virtually in Business, *Financial Times*, 18th November p14, London.

NCC, 1993, (2nd edition),"Handbook of Data Communications", NCC Blackwell, Manchester.

Shorter Oxford Dictionary, 1977, 3rd edition, Revised and Edited Onions, C.T., Clarendon Press, Oxford.

Stowell, F.A., West, D, and Howell, J,G., 1993, Information Systems and Systems Science, *in*, "Systems Science, Addressing Global Issues" ed. Stowell, F.A., West, D., and Howell, J.G., Plenum Press, New York.

Stowell, F.A., and West, D., 1994, "Client-led Design: A Systemic Approach to Information Systems Design", McGraw-Hill, Maidenhead.

Stowell, F.A., (ed), 1994, "Information Systems Provision: The Contributions of SSM", McGraw-Hill, Maidenhead.

Vickers, G, 1978, Rationality and Intuition, *in* , "On Aesthetics in Science", ed. Jarechslep,P., MIT Press, Cambridge, Mass.

Yourdon, E., and Constantine, L.L.,1979, Structured Design, Prentice Hall, Eaglescliffe,New Jersey.

9

DEMANDS IMPOSED ON SYSTEMS SCIENCE BY COMPLEXITY

John N. Warfield

George Mason University
Mail Stop 1B2
Fairfax, Virginia 22030-4444

1. PRELIMINARIES

1.1 Assumptions

All human knowledge is constructed by human beings as collections of models. Models are of three types: formal, informal, and hybrid (a mixture of formal and informal). Formal models are of three types: numerant, structural, and hybrid (a mixture of numerant and structural). Structural models can be linear or nonlinear. A linear structural model is a drawing that is isomorphic to a directed graph, such that a directed line can be drawn passing through all vertexes and lines without touching any of them more than once.

Spaces are mathematical (i.e., they are heavily symbolic and programmable), and are of three major types: root, intermediate, and application-oriented. A root space is any mathematical space that forms a framework for developing formal models, thus formal models can be positioned in root spaces.

Amelioration of undesired consequences of complexity involves the study of situations and systems. A situation is a triad consisting of (a) a human component (an individual or an aggregation of individuals), (b) systems contained in the situation, and (c) their respective environments. A universe is a set of all situations relevant to a chosen investigation. A system (following J. Willard Gibbs) is 'any portion of the material universe which we choose to separate in thought from the rest of the universe for the purpose of considering and discussing the various changes which may occur within it under various conditions' (Rukeyser, 1988:445).

A science is a body of evolving knowledge consisting of three variously-integrated components: foundations, theory, and methodology. Foundations inform the theory and the theory informs the methodology. The volume of knowledge is smallest in the foundations and largest in the methodology. The domain of a science consists of the science and its applications. All science is evolutionary. Evolution typically occurs by comparing the congruence between the science and results observed in its applications.

Critical Issues in Systems Theory and Practice, Edited by K. Ellis et al.
Plenum Press, New York, 1995

0 ∞

ORDINARY	TRAN-SITION RANGE	COMPLEX
Situations on this part of the scal are those commonly dealt with b individuals. Also these situations are overwhelmingly the type that is dealt with in higher education.		Situations on this part of the scale are those with which individuals cannot deal. They require teams working together in a designed environment with designed processes. Typically they are not discussed in any depth in higher education.

Figure 1. A postulated scale of situational complexity.

1.2 Scale of Complexity

In addition to the Assumptions appearing in 1.1, it is assumed that there exists a 'Scale of Complexity' such that a given situation can be positioned somewhere on the scale, and that the scale can be represented as shown in Figure 1. Situations positioned in the leftmost region of the scale are called 'ordinary situations'. Situations positioned in the rightmost region of the scale are called 'complex situations'. It is further assumed that the number of interesting ordinary situations is much greater than the number of interesting complex situations.

1.3 Higher Education and Complexity

The time spent on particular components of higher education is too small to admit careful analysis and design of complex situations. Instead, higher education thrives on incorporating large numbers of components in its student menu, each of which represents an ordinary situation.

Of the many students who are able to comprehend large numbers of ordinary situations, some gradually acquire the confident belief that this prepares and qualifies them to work individually and comprehensively with complex situations. Unfortunately this belief is mistaken, because complex situations inherently do not lend themselves to use of the short-time, superficially-constructed models that characterize ordinary situations.

2. LAWS OF COMPLEXITY: INTERRELATIONS AND CONTEXTS

Seventeen Laws of Complexity have been discovered. It is assumed that the reader either (a) has already studied these individual Laws or (b) is willing to suspend judgment on these Laws in order to allow the discussion to proceed. The justification for this situation is twofold. First, a document is available to the reader who wishes to obtain it from the author that discusses and interprets each of these Laws in considerable detail (Warfield, 1994). Second, the study of the Laws as they interrelate and, as they imply demands on various

		OUTCOMES			
		DESCRIPTION	DIAGNOSIS	PRESCRIPTION (DESIGN)	IMPLEMENTATION
BEHAVIOR	INDIVIDUAL	• Limits • Triadic Compatibility		• Requisite Parsimony • Requisite Saliency	
	GROUP	• Limits • Uncorrelated Extremes	• Inherent Conflict • Structural Underconceptualization • Diverse Beliefs	• Requisite Variety	
	ORGANIZATION	• Limits • Org. Linguistics	• Forced Substitution • Precluded Resolution		
	PROCESS	• Limits • Triadic Necessity & Sufficiency • Universal Priors	• Success & Failure • Universal Priors		• Gradation • Validation

Figure 2. Behavior-outcomes matrix.

domains of behavior and inquiry as suggested in this writing, may provide an incentive to study these Laws without which they might be ignored. Figure 2 shows a Behavior-Outcomes matrix, such that the Laws whose titles appear in a particular matrix cell are particularly relevant to the type of situation represented by that cell.

The behavioral (vertical) indexing deals with the individual, the group, the organization, and process; all aspects of human behavior. The Outcomes (horizontal) indexing refers to components of a generic kind of work program that is broadly relevant to human activity.

By aggregating all of the Laws that occur in a particular row, one can see which of the 17 Laws relate, for example, to the individual human being. By aggregating all of the Laws that occur in a particular column, one can see which of the 17 Laws relate, for example, to the description of a desired outcome.

In studying these Laws, it has been found helpful to recognize that some Laws can be most easily understood if other Laws have been studied first. Figure 3 shows a structural model conveying the pattern of study that recognizes these learning interactions. Submodels can be constructed from the structural model in Figure 3 to apply to any of the matrix entries, for purposes of developing learning sequences. For example, the learning sequence for the

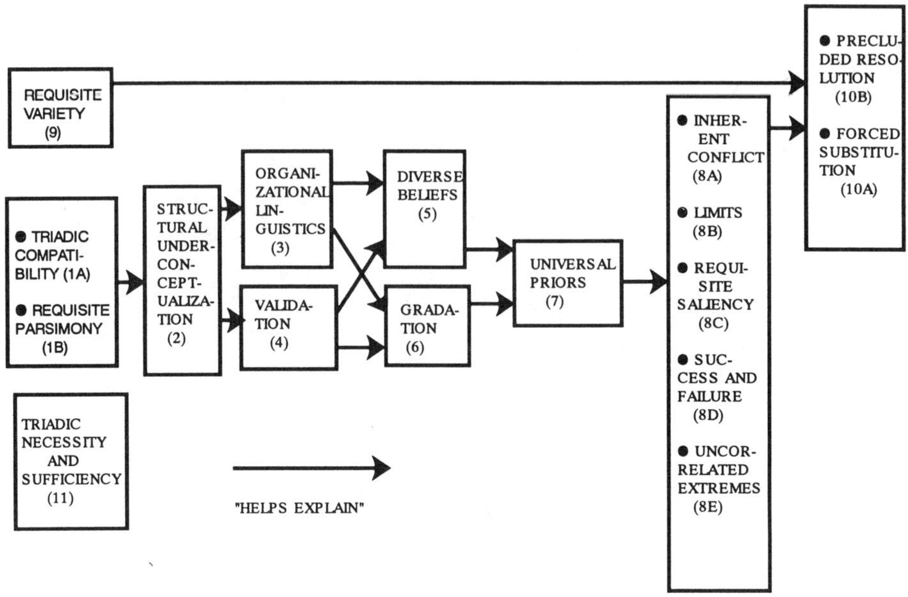

Figure 3. Interactions in interpreting laws of complexity.

individual would be comprised of four Laws studied in the following sequence: Triadic Compatibility (1A), Requisite Parsimony (1B), Limits (8B) and Requisite Saliency (8C). The learning sequence for process could be comprised of six Laws studied in the following sequence: Validation (4), Gradation (6), Universal Priors (7), Success and Failure (8D), Limits (8B), and Triadic Necessity and Sufficiency (11).

3. PEOPLE AND COMPLEXITY

Study of the behavioral components in Figure 3 has revealed that all three of the human components are seriously deficient when they attempt to apply what has been learned about ordinary situations to complex situations.

3.1 The Individual

The individual is subject to the restriction of scope reflected in the Law of Triadic Compatibility. This means that the individual cannot comprehensively deal with, in short term memory, interactions among four or more situational variables (Warfield, 1988). Processes designed to assist the individual to think through progressively the interactions among a variety of situational variables must reflect the contents of the Law of Requisite Parsimony. Each of these Laws contributes to understanding the limits on the individual. The Law of Requisite Saliency is amply illustrated by data taken from many Interactive Management projects. (Warfield and Cardenas, 1994).

The type of analysis outlined in section 2. can be repeated for each of the vertical and horizontal indexes, and then aggregated into an overall set of conclusions. This effort has been contributory to developing the practice called Interactive Management (Warfield et al., 1994). It incorporates process designs that are intended to minimize or annul undesirable impacts found when people try to apply approaches appropriate for ordinary situations to complex situations.

3.2 Groups and Organizations

The corresponding difficulties with groups and organizations are, respectively, Spreadthink (Warfield, 1995) and inadequate shared linguistic domains, illustrated by Laws pertaining to groups and organizations respectively. Spreadthink refers to the empirically found attribute of groups: members of virtually all groups that attempt to work with complex situations involving methods applicable to ordinary situations hold very different viewpoints on the relative importance of situational factors and, moreover, there is almost never a majority point of view on the importance of any particular factor.

Spreadthink is, therefore, an immobilizing attribute that is common to groups attempting to work with complex situations. (Part of the basis for this attribute is that the members do not share a linguistic domain and must, accordingly, engage in a process or dialog that creates such a domain.) When this attribute is extended to the entire organization, the organizational linguistics are seen to be totally inadequate and the organization seems unable to do anything to correct this situation, in the absence of what would probably be seen at present to be a radical change in management style.

4. OUTCOMES AND COMPLEXITY

Following the approach illustrated in 3.1 and 3.2, the impact of the Laws on Outcomes of human activity can be determined. For example, Prescription (Design) is impacted by the Laws of Requisite Parsimony and Requisite Saliency, as they apply to the individual; and the Law of Requisite Variety as it applies to groups. These three Laws relate to the inability of individuals mentally to interrelate more than three things at a time and to the well-established failure of individuals to assess properly the relative saliency of a set of concepts using ordinary practices; and to the requirement that groups strive to discover the dimensionality of the situations considered, and match the dimensionality of the system proposed to alleviate problems in the situation to the dimensionality of the situation.

5. SYSTEMS SCIENCE AND COMPLEXITY

The study of the impact of the Laws on Behavior and Outcomes enables the identification of the impact of complexity on systems science. In turn, this provides a basis for recommending various practices that ought to discipline the restructuring and evolution of systems science.

5.1 Organizing Systems Science

Generally speaking, systems science should be organized into its foundations, its theory, and its methodology, with some sense of consistency among these three components. It seems clear that the archival components of systems science today are very heavily oriented towards methodology, and that most methodologies proposed to date are not supported by any articulated foundations and theory. While the science of generic design put forth by Warfield (1994) may or may not be viewed by the systems community as well-conceived, at least a very strong effort has been made to organize this science along the lines mentioned, not leaving methodology to float on a choppy sea removed from its foundations.

5.2 Relating to Behavior and Outcomes

In the domain of complexity, systems science should be involved with both the horizontal and vertical index sets in Figure 2, not as components to be dealt with in isolation, but as topics to be integratively coupled.

5.3 Domains of Methodologies

One form that such involvement could entail is to review all of the methodologies proposed for use in systems studies, and relegate them to ordinary and complex systems respectively, by demonstrating either that they do or do not comply with the Laws of Complexity.

5.4 Testing Systems Science in Applications

A second form that such involvement could entail is to recognize once and for all the need to couple proposed contributions to systems science with valid testing in applications, and to begin to close the circle in the best tradition of scientific development. A serious review of the extent to which this has taken place so far in the evolution of systems science should be carried out to remind the systems community of the importance of this tradition. In the absence of such testing, all of the difficulties described by Jackson in his carefully crafted paper on fads (Jackson, 1993) will prevent systems science from attaining the kind of credibility it requires to assume an often-sought position of prominence in human affairs.

5.5 Emphasizing Process in Organizations

Systems scientists need feedback from applications in order to improve systems science. Experience suggests that organizations will have to change their management style in order to provide this feedback. One approach that could be taken is to appoint a high-level executive to oversee organizational processes just like, for example, having a high-level executive to oversee organizational finance. The high-level process executive could oversee the specialized training of a set of appointed mid-level process managers. Data from process applications could then be collected and made available to improve systems science.

5.6 Improving Systems Education

Some educational institutions might recognize the need to reorganize in order to provide the specialized infrastructure required to support investigations involving complex situations (Warfield, 1990). If graduates could emerge that understand the impact of complexity upon human life, conceivably they might not only take the positive actions needed to design effective systems, but also they might stifle the extensive and ill-founded attempts to impose poor politically-designed systems upon long-suffering populations.

5.7 Improving Modeling

Finally, perhaps the greatest positive impact on systems science that could be envisaged would be the improvement of system modeling. Far too much emphasis is presently placed on numerant modeling unsupported by the relevant structural models. One of the most malevolent practices currently applied involves the attempt to force nonlinear structural models into inherently linear prose. This practice has to be broadly recognized as destructive to systems science. So it is necessary to place much more emphasis upon structural models than in the past. But that, in itself, will not be appropriate unless the systems community recognizes that ad hoc structural models do not communicate, and are simply another form of inappropriate communication. The use of hybrid models must be subjected to strong intellectual discipline of the type that a strong systems science would provide. Inherent in this practice will be the basing of structural models in root spaces. An especially significant root space is that founded in the Theory of Relations initiated by Augustus De Morgan in London, England, in the first half of the nineteenth century.

6. SUMMARY

Systems scientists should adopt a disciplined approach to systems science, in the interests of having a science that enables it to be responsive to complex situations. This approach should enhance significantly the access to this science of practitioners in industry and government, by organizing systems science so that it becomes much less opaque. Methodology will probably always make up the largest part of systems science, but methodology that is not archivally connected to foundations and theory, but rather is often dependent on unarticulated rationale (and therefore, inaccessible for validation or improvement), and which is subject to highly ambiguous interpretation, must begin to receive official frowns from the leadership in the systems community, no matter how much what they are doing may involve such methodology.

The potential benefit to society from a strong systems science cannot be realized unless the universities redesign themselves to provide appropriate infrastructure for systems science that can be applied to complex situations. Government and industry organizations will continue to work with poorly-designed and publicly-mistrusted large systems until such time as they begin to administer systems-based processes in a manner at least as prominent as that which they apply to managing organizational finance and marketing.

The seventeen Laws of Complexity discovered so far, and those likely to be discovered as time passes can, with further probing and validation, provide a major part of the systems science of tomorrow. Until more candidates are proposed, the already documented science of generic design and its accompanying practice through Interactive Management can serve as a way of managing complexity; since these advances are responsive to the Laws of Complexity.

The bad news is that at the present time few of the demands of complexity on systems science are being met. The good news is that because of the bad news there is a wide arena of opportunity for substantial improvement in systems science and for significant increases in its application.

ACKNOWLEDGMENTS

While the work of many contributors has provided valuable background for the work reported here, it seems especially appropriate at a meeting of the United Kingdom Systems Society to acknowledge relevant contributions of persons identified with the United Kingdom, all of whom are deceased. These persons are: Ross Ashby (1903-1972), George Boole (1815-1864), Lewis Carroll (1832-1898), Arthur Cayley (1821-1895), Augustus De Morgan (1806-1871), Geoffrey Vickers (1894-1982), and Alfred North Whitehead (1861-1947).

REFERENCES

Jackson, M. C., 1993, Beyond the Fads: Systems Thinking for Managers, Working Paper Number 3, The Centre for Systems Studies, The University of Hull, UK.

Rukeyser, M., 1988, "Willard Gibbs", Ox Bow Press, Westbridge, CN.

Warfield, J.N., 1988, The Magical Number Three—Plus or Minus Zero, *Cybernetics and Systems.* 19: 339-358.

Warfield, J.N., 1990, Cybernetics, Systems Science, and the Great University, *Systems Research.* 7(4): 287-294.

Warfield, J.N., 1994, "Laws of Complexity: Origins, Statements, and Interpretations," Fairfax, VA: IASIS.

Warfield, J.N., 1994, "A Science of Generic Design", 2nd Edition, The Iowa State University Press, Ames, IA.

Warfield, J.N., 1995, Spreadthink: Explaining Ineffective Groups, *Systems Research.* 12(1).

Warfield, J.N., and Cárdenas, A.R., 1994, "A Handbook of Interactive Management," 2nd Edition, The Iowa State University Press, Ames, IA.

STUDENT PAPERS

10

APPLICATION OF MULTICRITERIA ANALYSIS TO RANKING AND EVALUATION OF WATER DEVELOPMENT PROJECTS

The Case of Jordan

B. Al-Kloub

Staffordshire University
School of Engineering
Beaconside
Stafford
ST18 0AD
United Kingdom

INTRODUCTION

Water shortage in Jordan is becoming acute, with limited available water and financial resources. To balance current needs and demands, taking into account the possibility of limited investment in the future, the current Jordanian investment programme (1992-1998) contains various national objectives and constraints to be met. The requirement for this study arises from the need for a reliable system methodology to select and rank water development projects. Multicriteria analysis is a realistic approach to this complex problem, in particular, the PROMETHEE method is well adapted to it. Its flexibility enables the Decision-Maker to express his preferences more precisely. In this paper the problem is formulated, criteria are presented and quantified, and a partial and complete ranking of projects is obtained by applying the PROMETHEE method. The computational results are given taking into account the full range of different constraints, i.e., financial (initial cost, operation and maintenance) and regional development issues.

BACKGROUND

Jordan is located in an arid to semi-arid region. Water availability is becoming a major concern and currently demand exceeds supply. In addition, current practices of over-exploiting and over-pumping have resulted in reducing ground water levels in most wells in the country (Al-Kloub, 1994). A more comprehensive system of planning to address the multiple objective nature of water projects is needed (World Bank, 1993). This research is part of this

Critical Issues in Systems Theory and Practice, Edited by K. Ellis et al.
Plenum Press, New York, 1995

planning approach. Ranking water development projects in the current Jordanian investment plan subject to different constraints is carried out in this study.

MULTICRITERIA ANALYSIS

Although there have been many substantial developments in multicriteria methodology, single criterion analysis is still common in many real-world decision making tasks, especially where financial or economic criteria are thought, rightly or wrongly, to be dominant. However, the use of multicriteria techniques and decision-making methods is a must where there is limited investment resources and where an effective investment decision must satisfy the broad objectives defined by socio-political conditions which are sometimes non-commensurable and conflicting in nature. In a multi-objective programming problem a set of "non-dominated" solutions is sought compared to one optimal solution in a single objective programming (Goicoechea et al, 1982). The need for introducing value judgements into the solution process to order the option solutions in the non-dominated set is central. To aid the decision maker in solving problems and to introduce the value judgements and trade-offs, three basic types of method have been developed (Al-Kloub, 1994):

- Outranking methods.
- Trade-off methods which utilise utility functions.
- Various interactive methods.

A tendency has been observed such that the outranking methods are most successful because of their adaptability to real problems and the fact that they are more easily comprehended by decision makers (Mladineo et al., 1992). A very promising method is the PROMETHEE method (Brans et al., 1986).

Figure 1. shows a general overview of the adapted model used to solve the problem utilising PROMETHEE method.

THE PROMETHEE METHOD (PREFERENCE RANKING ORGANISATION METHOD FOR ENRICHMENT EVALUATION)

The PROMETHEE method includes the following three major steps :

- Enrichment of the preference structure by introducing generalised criteria to remove scaling effect.
- Enrichment of the dominance relation by building :
    i. A multicriteria preference index to express to which degree an option is preferred to another.
    ii. An associated outranking graph and outranking flow to express how each option is facing the other options together (strength and weakness of the option).
- Exploitation for decision aid. PROMETHEE I provides a partial ranking, including possible incomparabilities, PROMETHEE II shows complete ranking of options. And PROMETHEE V extends the application of PROMETHEE II method to the problem of selection of several options given a set of constraints.

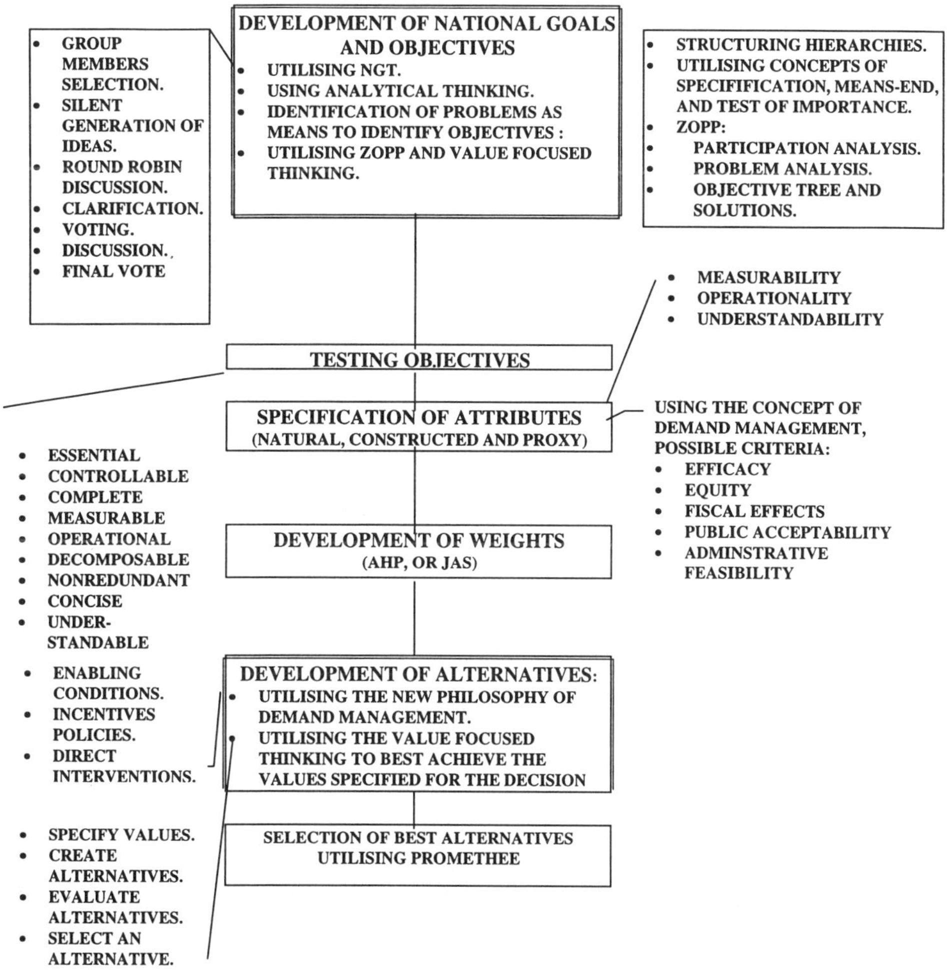

Figure 1. General overview of the adapted model utilizing PROMETHEE method.

THE GAIA VISUAL METHOD

The GAIA program (Geometrical Analysis for Interactive Aid) provides a geometrical presentation of results obtained by PROMETHEE and followed by the PROMCALC method. GAIA is based on the reduction of multi-dimensional problems to two-dimensional plane to allow direct visual presentation. It gives an effective presentation of results and prediction of (what if?) situations.

IDENTIFICATION OF NATIONAL OBJECTIVES

An organised brainstorming workshop was carried out with the help of a few decision makers and a facilitator in Jordan to identify the objectives, the hierarchy, weights, and fundamental objectives. As a result 24 fundamental objectives were elicited utilising the following ideas:

- Nominal Group Technique as a creative process to identify objectives (Delbecq et al., 1976).
- Value focused thinking to establish fundamental objectives (Keeney, 1992).
- Demand management philosophy to overcome the shortcomings of the existing supply augmentation approach. (Winpenny, 1994).
- Management by objectives to create options with the aim to best achieve the values specified for the decision situation (Keeney, 1992).

The set of fundamental objectives and their relative importance, measurement scale, and the relevant threshold parameters were identified (Al-Kloub, 1994). In order to set priorities the JAS software was used to reduce subjectivity and to validate consistency (Islei and Lockett, 1991).

IDENTIFICATION OF WATER RESOURCES ACTIONS, AND THE EVALUATION MATRIX

A comprehensive survey regarding possible feasible options reflecting above fundamental values were developed. The options were taken from different available documents (including the new investment plan). The spreadsheet of PROMCALC and GAIA (evaluation matrix) was defined. It describes the different generalised criteria, associated weights, options, types of preference functions for each criteria and their defined parameters, input data for each option, the problem type (min. or max.), constraints (initial cost, operation and maintenance cost, and regional development) (Al-Kloub, 1994).

RESULTS OF ANALYSIS

A complete analysis for major water actions was carried out, partial and complete ranking was demonstrated and sensitivity analysis was undertaken by changing the weights of the criteria and observing the changes in the ranking of the projects. Results obtained included descriptive statistics for the criteria. Partial and complete ranking of the selected options were presented graphically. The main results of above analysis are presented (Al-Kloub, 1994) considering the following cases :

Case 1: ranking of all water resources actions without introducing constraints (technical, regional, managerial, pricing and regulatory).

Case 2: ranking of all water resources actions after introducing capital, operational and maintenance costs constraints.

Case 3: ranking of each group of actions independently after introducing capital, operational and maintenance costs constraints.

Case 4: ranking of all possible current projects (this is the current situation case where execution of some of the projects is not possible due to different reasons (political and others)) after introducing capital, operational and maintenance costs constraints.

Case 5: ranking of current technical options based on the current situation (possible projects) and introducing regional development, and costs constraints (it is assumed that the technical options must accrue in the approximate proportion to the population of the four regions in the country while other categories of options mostly are beneficial to all regions).

CONCLUSION

In the Middle East, especially in Jordan, the socio-economic environmental dimension and the new values related to water sector are so important and without fundamental changes regarding conservation, efficiency, demand management and to operate on a multi-objective basis the existing problems will escalate.

The PROMETHEE method as a multi-objective technique was utilised after establishing the objectives and weights in a systematic procedure to rank and choice options. Alternative choice investment options that best satisfy the national objectives relevant to the water sector from the point of view of Jordan experts and decision makers were carried out. This process was formulated into a multi-objective decision problem from which an optimal options based on evaluations, constraints and relative importance was carried out. With the application of the methodology to Middle Eastern water resources management and planning, policy makers are given the opportunity to evaluate different points of view, different priorities and carry out sensitivity analysis, consequently, modification can be made to water policies to make it more responsive to the critical situation at hand.

From the presented information it can be concluded that the PROMETHEE method has potentially a decisive positive contribution both to the process of decision making for selection of projects and ranking of these projects. The geometrical representation of the multicriteria problem represents powerful tools in the hands of the system analysts and a valuable help in solving problems which include conflicting criteria.

REFERENCES

Al-Kloub, B., 1994, "Optimisation and Ranking of Water Development Projects Utilising the Multicriteria Analysis (the Case of Jordan)", Internal Report No/TAS/BK/15 School of Engineering, Staffordshire University, England.

Brans, J.P., Vincke P., and Mareschal, B., 1986, "How to select and how to rank projects: the PROMETHEE method", *Eur Jrnl of OR*, 24: 228-238.

Delbecq, A., and Van de ven, A., 1976, "Group Techniques for Program Planning, A Guide to Nominal Group and Delphi Process", Scott Publisher, London.

Goicoechea, A., Hansen, Don R., and Duckstein, L., 1982, "Multi-objective Decision Analysis with Engineering and Business Applications", John Wiley and Sons, Chichester.

Islei, G., and Lockett, G., "Modelling strategic decision making and performance measurement at ICCI pharmaceuticals", *Interfaces*, 6:4-22.

Keeney, R., 1992, "Value Focused Thinking", Harvard University Press, Harvard.

Mladineo, N., Lozic I., Stosic, S., Mlinaric D., and Radica, T., 1992, An evaluation of multicriteria analysis for DSS in public policy decision, *Eur Jrnl of OR*, 61: 219-229.

Winpenny, J.,1994 , "Managing Water as an Economic Resource", Development Policy Studies (ODE), London.

World Bank, 1993, "Comprehensive Water Resources Management", ISBN 0-8213-2636-8, Washington DC.

11

HAVE YOU BEEN TALKING TO YOURSELF, RENÉ?

Tony Brauer

Systems Department
The Open University
Milton Keynes MK7 6AA
United Kingdom

INTRODUCTION

Even if we are sceptical, we make assumptions. Within a given discourse, however, there are conventions for the validation of assumptions. It is argued here that

a. the characterisation of emancipation as an *a priori* assumption is intrinsically invalid within any sceptical systemic discourse, and that

b. the conventions of validation are in any case inadequately theorised. An alternative analysis is offered, using a Cartesian methodology in a conversational (as opposed to introspective) context.

A PRIORI AUTONOMY AND ALTERNATIVES

Lots of people have discussed autonomy. Robert Flood (1993), for example, seems very keen on the idea, though he tends to use terms like *liberation,* and *emancipation* instead. [We could use other terms too; *freedom, enfranchisement,* and *empowerment*, for example.] My dissatisfaction is not with the terminology; it arises from the assumption apparently made by Flood, and others (e.g.: Tsivacou (1992), Ulrich (1991)) that liberation is self-evidently a Good Thing.

This is a shaky assumption. Firstly, unless it can be shown how we can transcend the limits of experience, *a priori* assertions appear to be no more than linguistic devices informing us of conventions in particular language games. This issue will be examined further a little later.

Secondly, this particular *a priori* proposition - that autonomy is necessarily a Good Thing - is empirically dubious. There seem to be people who prefer not to question the *status quo*, who are content to live within the role that has been ascribed to them; and to describe this as *false consciousness* is no help. We know there may be such a phenomenon, but we cannot identify specific instances of it with any certainty whatsoever.

Critical Issues in Systems Theory and Practice, Edited by K. Ellis et al.
Plenum Press, New York, 1995

Thirdly, if moral arguments are to be presented as *a priori* assertions, it is necessary to believe that morality is noumenal rather than a human construct. This seems to be intuitively implausible, to the irreligious, at least. What would it mean to say that morality exists independently of consciousness?

In addition, how can there be a moral argument which doesn't presuppose some set of circumstances? Moral commitments are necessarily preceded by ontological and aesthetic propositions, which are themselves, if not counterfactual, by definition *a posteriori.*

These comments are not intended to deny the beauty and importance of autonomy. It is, however, quite possible to praise autonomy without claiming that it is a universal moral truth - to tell stories about the value of autonomy in the hope that others might inherit the beliefs contained within the story. Hammond (1994), for example, argues for holistic self-organisation not on the ground that it is self-evidently desirable, but as a dynamic which counteracts the pernicious effects of entropy.

I find her story more persuasive than any simple assertion could be, even though it seems incomplete: but what would constitute completion? How do we know when to suspend our scepticism, and reschedule our beliefs as commitments?

AN ASYMMETRICAL CONVERSATION

If no-one is listening, it doesn't matter what I say. If someone else is participating in this asymmetrical conversation, welcome. What can we infer from the assumption that these words, selected by me, are triggering some response in you?

Firstly, my consciousness is implicit in the assumption: and if my talking to you is to make sense, I must suppose that you have a consciousness too. [Let's call it experience for the sake of euphony.]

Secondly, if we had free access to each others' experience, why would we have to talk like this? It seems likely that our experiences are not directly interconnected.

Thirdly, all that I have to work from is that which I experience; and since we do not have free access to each others' experience, we cannot corroborate any ideas we have, and so must tend towards doubt.

Fourthly, we have chosen to work together in this strange disjointed project of scriptural communication, so we evidently have preferences; and it is from such preferences that my sense of what is good is derived.

Fifthly, since we have attributed experience and preferences to each other, we accept each others' moral relevance.

Sixthly, since habitual lack of confirmation prevents us accepting propositions as truths, we will have to use the criterion of usefulness rather than truth for any practical propositions of which we become aware. [Given the commitments made here, usefulness will clearly have to be assessed in terms of the preferences of those who are morally relevant.]

This is excellent. Merely from our participation in a conversation, we have assembled an impressive array of ontological, aesthetic, epistemological and moral commitments that, I hope, we can share. The ontology describes discrete continuities of experience, with an aesthetic element evident in our preferences for some experiences over others. These are, of course, ideas that we would expect many others to share, rather than necessary generalisations; but if the epistemological commitments derived from this ontological ground are accepted, a shared explanation is as far as we can go in the direction of descriptive certainty. Empirical, sceptical pragmatists don't expect anything more than something useful to be going on with.

Having, then, committed ourselves to beliefs about what we are, and to beliefs about the rational grounds of belief, we have grounds on which to construct a morality. Our commitment must surely be concerned with how all those who are morally relevant can have a fair chance of their preferences being satisfied: that is, our moral commitment is to trans-personal consequentialism, and the set of commitments is complete.

WHAT'S GOING ON HERE, THEN?

There are, of course, all sorts of things we could do with this set of ideas, but what I started out to do was to discuss autonomy and boundaries.

Boundaries seem to be an inescapable issue. The boundary of any theory is defined by its implicit assumptions, and, as sceptics, we will challenge those assumptions. Scepticism doesn't, however, suggest that we shouldn't have beliefs, or that we shouldn't operationalise them as commitments, only that we shouldn't parade them as certainties. How, then, do we decide when to stop challenging boundaries?

It could certainly be argued that the sceptical process is sometimes curtailed rather abruptly, and without much explanation. Ulrich (1991), for example, seems to set up the idea of a boundary that should be open to critical challenge, without explaining why the ensuing boundary should not be subject to the same process. Similarly, Midgley (1992) validates pluralism by reference to a meta-theoretical assumption of context. Why, though do they only take one step beyond? Why limit one's ambitions to a critique of the original boundary, or to the acceptance of a meta-theory?

Going beyond is, in itself, clearly unsatisfactory to anyone who is habitually sceptical. Beyond an illusion is only another illusion. *Beyond* is a direction away from an illusion, with no intrinsic explanation as to why it should be in the direction of truth, or of anything else that might be useful. *Beyond* becomes useful only if it is directional; but fortunately there is an aspect of the construction of knowledge which allows us to analyse the *beyondness* of theories. This is the continuum between scope and specificity.

ON THE BORDER

There appears to be a trade off between the scope of a proposition and its efficacy in specific instances. An epistemological commitment doesn't tell us how to make a nut roast with mashed potatoes, peas and gravy, for instance.

On the other hand, if we wish to assess the potential usefulness of a specific proposition, we will tend to refer it to a more general one, to see whether or not it falls into the set of generally useful ideas. Thus, while we might not dismiss out of hand a design for a tunnel to Australia based on flat earth theory, we would probably devote resources to some other plan.

Scope and specificity are thus complementary, and the immediate problem is the identification of the limit of *beyondness* in the direction of generality or scope. Within the domain of conversation, the commitments described here have been presented as lying at the limit of utility; but why has this particular limit been chosen?

The answer is that this set is at the boundary of experience. As Descartes (1912) argued, if we are sceptical, we can only go back and see if there is anything in experience of which we can be certain. He found, not surprisingly, that one can be certain only of experience. [The false imputation of a discrete self does not invalidate his essential methodology.]

The domain of knowledge in which Descartes was operating, however, was not that in which we are involved. For Descartes the boundary of the domain was personal, and the ontological commitment he derived, introspectively, was that there is experience. Here, the boundary of the domain is social, and the ontological commitment derived from the possibility of conversation is that there are segregated, aesthetically ranked, experiences.

The case is not closed, however. Although it seems not unreasonable to assert, [on the grounds of commonality and within the conversational domain], that private experience, preferences, and uncertainty are given in experience, it must be conceded that the derivation of epistemological and moral commitments is theory-laden. On the one hand, this could be described as consistency; the theories which permit the derivation are those which are derived. On the other hand, this could be described as circularity: empiricism has been derived from an assumption that [what is taken to be] unmediated experience defines the limit of scepticism.

Certainly, circularity is not adequate; one cannot lift oneself up by one's toes. At the same time, consistency is necessary. Paradox is simply a quality of language that draws attention to a disjunction between a description and that which it is intended to describe. For this reason, the absence of a consistent alternative to empiricism may be persuasive.

VALIDATING THE BOUNDARY

If experience is not to form the boundary of theory, it would seem we can either seek to transcend the boundary through the use of our analytic powers, which we can describe as rationalism, or ignore theory and rely on intuition.

Intuition is certainly to be valued. We know more than we can tell, as Polanyi (1969) pointed out. We are concerned here, however, with the relationship between scope and specificity, with distinguishing between theories by reference to other theories; and intuition is independent of the articulation of theory. Intuition must, therefore, be treated as a discrete though complementary approach, and though rationalism produces theories that distinguish between other theories, it is, unfortunately, self-refuting.

There are two main routes to the transcendent propositional knowledge of rationalism; propositional logic and the analysis of the necessary categories of experience. Logic fails immediately, in this context. Is the set of all sets which are not members of themselves a member of itself or not? Whether you answer yes or no, you confound the fundamental principle of logic; that something cannot simultaneously have and not have a given property.

The alternative route is typified by Kant's (1787) bid for transcendent knowledge. What, he asked, is necessary to experience? Of all the categories he suggested, all that remains is the possibility that space and time are necessary to experience. Even if this is the case, it is an inadequate ground on which to transcend the limits of our experience. Space and time are qualities of our conception of experience, and tell us nothing of what lies beyond it. Furthermore, both our collective and subjective conceptions of space and time are under continual review. Kant's root commitments are no more transcendent than the empirical assumptions made here.

Through rational argument it is simple to discount rationalism [See also Brauer 1993]; and what reason is there to give credence to a theory that denies itself? This is a particularly pertinent question for the sceptic, since those who are unwilling to dismiss ideas have a special affection for parsimony, whenever a credible opportunity to exercise it arises.

Of course, this dismissal of rationalism reflects also on scepticism, but the qualities of the sceptic's assertions are substantially different from those of the rationalist. The sceptic is merely seeking to present a story that is persuasive, not asserting transcendent truths.

To this sceptical principle of offering stories rather than asserting truths, pragmatism adds the criterion of usefulness, while trans-subjective consequentialism adds the criterion of the collective assessment of usefulness. This, it is suggested, is as great a scope [that is, degree of generality] as we can reasonably achieve in the social construction of knowledge.

Together with the other commitments derived from conversation, it forms the initial stage of distinguishing between the likely usefulness of other, more specific, theories.

CHOOSING STORIES AND THE VALIDATION OF EMANCIPATION

We are now in a position to go back to our original question: how to choose between explanations of the value of autonomy.

Within this discourse, there are three dimensions:

a. whether the evidence comes from experience or from the analysis of language;

b. the extent to which a belief about autonomy is presented as a certainty; and

c. the effects of holding these sets of belief.

In the first and second dimensions, the *a priori* is necessarily out of court. In the third dimension, however, it is necessary to imagine the consequences of holding any given belief as to how autonomy can be validated.

If autonomy is taken as given, we have the unintelligible duty to impose freedom. The means conflict with the ends. On the other hand, if autonomy is a hypothesis exemplified by a story, we can simply offer it to others as a hypothesis. This, I would suggest, is more likely to be effective than the approach implied by the *a priori* assumption, simply because it does not involve a contradiction between the ends and the means. In the instance of autonomy, at least, such a contradiction is difficult to defend, for the following reason.

As Vahl (1992) suggests, stories ascribe actorship, and by so doing change the social environment. The idea that autonomy is an *a priori* moral certainty seems to ascribe actorship to those who are already in power, allowing them to impose their ideas of emancipation or empowerment or liberation on others.

I prefer another story.

GRANNY AND THE CHAIR

Gran used to tell everyone how to suck eggs. She had status and influence, until people started to say that there is no definitive way of sucking eggs, [or of breaking them to make an omelette].

It occurred to Gran to suggest that while it is up to others how they make an omelette, this mob didn't really understand how to go about making that decision. She thought, for example, of the children who never were listened to at all; of the plans frustrated because of unshared information, and she recalled persistent instances of alienation and apathy. Here was an opportunity to relocate the source of her privilege and power; and superficially, at least, she would be doing so for altruistic reasons. Instead of being an expert in eggs, she could become an authority on the social processes of decision-making.

Luckily, she is familiar with some of Gregory's (1994) arguments, and is self-reflective. "How", she said to herself, "can I accept the claim that no-one is in a position to say what is best for others, and then go and tell them how they must reach decisions about what to cook? Instead, I shall tell them the story of *The Goat Who Wouldn't Listen,* and of *The Monkey that Knew It All,* and then everybody can make up their own minds."

ACKNOWLEDGMENTS

I would like to offer thanks to all those people with whom I think I've held conversations during my life, and, in particular to Alan Thomas and Jim Harris; and to Paul Theakston, who brews Black Sheep Ale, two pints of which always persuades me that we can reach each other.

REFERENCES

Brauer, A., 1993, "Why do they try to nail jelly to the sky?", MA dissertation, University of East Anglia.

Descartes, R., (translated by J.Veitch), 1912, "A Discourse on Method", Dent, London.

Flood, R.L., 1993, Practising freedom; designing, debating, and disemprisoning, *International Journal of Management Sciences*. Vol. 21. No1:7.

Gregory, W., 1994, Critical appreciation: thinking, speaking and acting critically, *in*: " New Systems Thinking and Action for a New Century", Proceedings of the 38th annual meeting of ISSS, B.Brady and L.Peeno. eds., Louisville.

Hammond, D., 1994, "Evaluating the Ethical Implications of Systems Thinking", internal paper, University of California at Berkeley, Berkeley.

Kant, I., (1787), "Critique of Pure Reason", 2nd edition, (Translated by Smith, N.K., published in 1933), MacMillan, Bassingstoke.

Midgley, G., 1992, Pluralism and the legitimation of systems science, *Systems Practice*. Vol. 5 No.2.

Polanyi, M., 1969, "Knowing and Being", Routledge and Kegan Paul, London.

Tsivacou, I., 1992, Emancipation and truth in critical systems methodologies, *Systems Practice*. Volume 5 No 2:181.

Ulrich, W., 1991, Critical heuristics of social system design *in*: "Critical Systems Thinking", R.L.Flood, and M.C.Jackson, eds., John Wiley & Sons, Chichester.

Vahl, M., 1992, The use of stories in systemic enquiry, subject to local culture, *in*: "Libro de la Conferencia Internacional de Trabajo", Insitituto Andino de Sistemas, Lima.

12

THE AMBIGUITIES IN OUR UNDERSTANDING OF PARTICIPATION AND THE IMPLICATIONS OF THIS FOR METHODOLOGY

Mandy Brown

Centre for Systems Studies
University of Hull
Hull HU6 7RX
United Kingdom

1. INTRODUCTION

This paper discusses the nature of the ambiguities in our understanding of participation, present in the participation and systems thinking literature. There is a tendency to confuse the two main justifications for participation; namely the arguments for participation on the grounds of efficiency, or on the grounds of a moral right to be included in decision making.

The justification via efficiency, emphasises participants' knowledge as a recognised valuable resource requiring appropriation. This can be described as participation as an 'input' or '*means*', indicating the passive and limited nature of the interaction, which concentrates on issues of structure and information extraction. The second justification concerns the moral and ethical argument which states that people have a right to be involved in decisions which affect their lives. This can be described as participation as a 'process' or as an '*end*', indicating the proactive nature of the interaction, thereby incorporating issues of learning and empowerment. This paper argues that the 'level' of participation, i.e. how it is physically operationalised, is not what determines whether a *means* or *ends* process is being undertaken, but rather that these are only recognisable by revealing the purpose or objectives of understanding the participation and the evaluation procedure.

This has implications for the use of systems thinking methodologies, which are primarily aimed at problem solving. When considering some of even the 'soft' uses of systems thinking, and viewing them from the participation-maximisation stance, we can see that the pressure to intervene and improve the situation is often greater than the time needed for the participants' social skills such as that of negotiation to be learnt. I will discuss whether such constraints bias us towards the *means* approach to participation and therefore move us towards a mainly information extraction form of participation, where skills of participative learning are not learnt, because the facilitator wants to see the problem addressed, and

Critical Issues in Systems Theory and Practice, Edited by K. Ellis et al.
Plenum Press, New York, 1995

evaluates this in terms of whether people rate their problem situation as improved, rather than whether they have learnt dialogical skills.

2. PARTICIPATION

Participation is generally considered an important issue in much of the systems literature, yet the writing on the subject appears ambiguous as to whether it is relating to participation with the objective of efficiency, or the need for inclusion on moral grounds. In this paper I will seek to address this ambiguity and propose to expand our understanding of the participative process and apply this knowledge to systems methodologies. The volume of literature on participation is sizable and in this paper I have attempted to collate theories from numerous articles (Bhasin,1985; Cernea,1985; Chambers,1983; Freire,1973; Friedmann, 1992; Ghai,1988; Goulet,1989; Marsden & Oakley,1990; Oakley,1991; Rahman,1993). These usually have several features. Firstly within the literature the term participation is recognised as an umbrella expression, by which people label numerous forms of interaction, perspectives and modes of operation. Secondly these modes are often divided into several 'levels' of participation, described as a continuum, an example of which is given from literature on participatory rural development in Table 1. Finally there is often present an ambiguity as to the justification or reasons for participation, but implicit in the continuum is the belief that an efficiency justification is implied by levels A to C, whilst the moral justification by rights is implied through levels D to F. This implication I believe to be ambiguous because the levels of A to F could be justified by efficiency gains and conversely all could be justified as moral inclusion. What is important in terms of the argument in this paper is not so much the 'level' of participation but *the purpose of the participation* (whether it is means or ends oriented) and who is setting the overall evaluative objectives. I will seek to explore this further by considering the component parts of a participatory intervention, namely the objectives, structure, process, facilitation and evaluation. (for further expansion see Brown, 1994).

2.1 Participation as a Means to an End

Participation viewed as a *means* of achieving an objective starts with the idea that the point of participation is to generate an involved way of solving a problem situation. Thus participation can be viewed as another input to the decision making process, having no priority over other forms of input. The focus on the 'problem' and problem solving as the preset objective is often defined prior to interaction with participants. I consider there to be two levels of objectives, firstly the overall objective of problem solving, and secondly, a set of sub-objectives such as improved efficiency, efficacy, or effectiveness. The objective of problem solving however is seldom up for negotiation.

The participative structure is conceived as fairly static and one dimensional, often being fixed in membership and composed of formal meetings, with preset agendas and limited flexibility. This also raises the issue of legitimate representation, and the possibility of externally appointed leaders causing the process to be rapidly undermined, as issues of fairness are raised. The process is aimed at meeting some projected overall objective, thereby often constraining participation to an input in the reorganisation of resources, participants may gain greater access to resources, but within the boundary of other competing agency factors.

The facilitation is regulatory in that it gives no greater weight to peoples' views than other factors, whilst being unable to question or define the overall objectives. Reinforcement of the regulatory nature occurs via limited information or knowledge being disclosed, which

Table 1. A participation continuum

A.	Voluntary **contribution** to project, where the people are not expected to take part in the shaping of the project or criticise its content
B.	To sensitise people via increased receptivity, to enhance the ability to **respond** to development projects and encourage local initiatives
C.	To harness peoples' **involvment** in decision making processes, implementing projects, sharing the benefits, and being involved in the evaluation
D.	**Active involvement** of the people in the decision making processes in so far as they affect them (wider remit)
E.	Active process requiring initiative, and the **assertion** of their autonomy to do so
F.	Organised effort to **increase control** over resources and regulatory institutions in a given social situation on the part of groups and movements hitherto excluded.

Based on Oakley and Marsden.

constrains the discussion to focusing on the initial sub-objectives. The facilitator here remains throughout the process external to the participant group, and is usually undertaken by a person considered to be technically competent in the field. This can result in the externally appointed facilitator, who may have an inherent tendency to emphasise the 'expert' nature of knowledge and therefore bias toward scientific, objective and materialist knowledge forms (Feyerabend, 1978). Consequently expert knowledge can come to dominate experiential knowledge, becoming the 'norms' of relevant knowledge.

Finally this approach evaluates and measures the success of the intervention in terms of quantifiable variables, often focused around financial and budgetary factors, reinforcing participation as an efficiency exercise. Participation in this mode has therefore been externally initiated, managed, and controlled, and consequently the forms of quantifiable measurement and evaluation actually measure this process rather than the nature of the participative contributions. We can align such a form of assessment with the hard systems approaches to intervention. However I would argue that even a soft systems intervention can use participation as a means to an end, as the focus is on the problem situation, and although participants evaluate the change in the situation for improvement, evaluation of their learning and dialogical skills is not made, for Checkland (1981) says "In dealing with human activities as perceived problems, the best we can hope for is that in the eyes of the concerned people former problems are now 'solved' or that problem situations are rated as 'improved'". Such a statement is even ambiguous as to who 'the concerned' are, for this could be enacted simply as the problem owner, and not the participants. Thus I would argue soft systems methodology (SSM) as it now stands takes a means approach to participation.

2.2 Participation as an End

Participation as an *end* is linked to the moral justification for inclusion in decision making, as a process of learning modes via dialogical co-generation of options, which enskill the participants. The learned ability to participate is not built on the classical education system of knowledge transfer, but in co-learning workshop environments based on interaction (Freire, 1973). The emphasis is on a co-learning with initiative, control of the content, and process remaining with the participative group, with the systems practitioner incorporated into the group. The significance of this approach is the primacy of people and their knowledge, it focuses on individual human development, potentially leading to synergism in the development of their local environment. This requires recognition that the participants firstly have the ability to deal with the content, and secondly, that having processed their recommendations they have the ability to take action, reflect on the outcomes and adjust their objectives in the light of increased understanding.

Such a participative approach aims to be dynamic and holistic in process, and although initially requiring an external 'contributor', subsequent structural development should be participant directed and continuously evolving. The workshop format incorporates a flexible approach to content, with limited preset areas for discussion. Although interaction may centre around a core membership, people are encouraged to join the process provided they have some level of commitment to it. Emphasis here is given to the emergence of leadership not based on organisational positioning, but skill in the representative process. The process aims at the strengthening of the participants' capabilities, which is considered to be more important than the decisions made. This process obviously has to be centered around some content, but the emphasis is on individual and group capacity building. Consequently all aspects within the potential control of the participants are available for negotiation. This is a learning process which aims to transform a person (object) who is a passive recipient of 'expert' knowledge, into a subject, being an active creator of knowledge (Ulrich, 1983).

Facilitation of this process should recognise that although initially an external contributor for participation may have been required, their evolving role is primarily in supporting the group to acquire strength and understanding by stimulating awareness of the processes undertaken, which strengthens their organisational base. Consequently the primary skills required are social skills, character is of equal, if not greater importance than technical knowledge. The key element is for the facilitator to encourage the participants to develop autonomously, by increasing their own ability, and skills, whilst reflecting upon their actions individually and corporately.

Evaluation is in relation to the development of the participative learning process, focusing on the participants' increased ability to function in an enskilled manner. The crucial factor is assessing not the level of participation in relation to Table 1, but rather the nature and purpose of the participative process. Evaluation is undertaken by the group, focusing on assessing their increased abilities and skills within the designated field, here problem solving and decision making. Has the learning of skills within dialogical/dialectical process occurred? The facilitator may assess for example if the dialogical modes of communication are enhanced, firstly in their role as part of the participant group, and secondly as an individual researcher. SSM currently has the potential to move towards an ends approach to participation, but it would require the refocusing from solely the problem area to include the participants, and the form of intervention undertaken. Such a change would require additional guiding principles within the methodology.

3. IMPLICATIONS FOR OUR ROLE AS FACILITATORS

Within the social development field various arguments have emerged around the issue of addressing development problems through external intervention. Initially the procedure was *prescriptive (means)*, where an expert-driven intervention appeared to have been made, however over time many argued that supposed gains evaporated (Chambers, 1983). These practitioners began to reason that people are not 'developed' when you tell them what to do, or when you limit the extent to which they are involved in the development process, for both of these lead to short-term benefits. Consequently within the development field many practitioners changed their paradigm to an *enabling* approach. It was recognised that if participation is an *end* purpose, then you also gain more effective use of resources, as defined through critical confrontation with alternatives, in that development is more sustainable due to the participants motivation and knowledge of their priorities. This led to the idea that sustainable development occurs where people not only sustain and enhance the initial project, but are also enskilled to tackle new problems as they are identified or emerge. The old adage of teaching a person to fish rather than giving them a fish, would prevail and external facilitation support would gradually decrease.

Is there a potential model here for systems thinkers? Is it not possible for us to address the issue of moral right to inclusion by seeing it as an issue of human development? This would imply moving towards an enabling approach, with participation as an '*end*' purpose, focusing on sustainable improvements in a situation, beyond the period of our direct contact. If we consider our interventions as facilitators do we operate out of a *means* or *ends* approach? I have read few case studies that I would consider to be as committed to enskilling as an *ends* approach. A large number of soft systems interventions appear to cluster around a participative approach which seems to incorporate the objective of participants affecting the system design, however under a means/ends analysis, many predominantly extract information for design, and are thereby actually a *means* approach. The lack of explicit discussion about participative learning within these interventions compounds the confusing messages about the role of participation. Belief appears to be that the methodology has assessed participative issues, however as argued above, most of the assessment of participation, even within soft systems, relates to a means approach. Evaluation of such case studies highlights firstly, the seldom discussed issues or purposes of participation, and secondly, there is often a failure to reassess the interventions 2, 3, or 5 years on for sustainability (a process which is often mandatory in development).

3.1 Belief in our Role as Facilitators

The key issue is what do we as systems facilitators believe our role to be? It could be countered that within systems thinking we do incorporate participants in many of our approaches, however participants rarely share the focus of our attention, for this is narrowly defined in the problem situation and our intervention to improve this situation. For systems thinking has unfortunately developed a tendency to view 'problem' situations through the telescope of our methodologies, often defining the problem in a way that our methodology can improve it. As a profession it is difficult to view 'problems' as learning opportunities, not simply for participants to learn our tools, and our methodologies, but also in terms of enskilling in dialogical debate (Gregory & Romm, 1994). Our priority is to improve the situation, and in so doing we often constrain both the acquisition of social skills on the part of the participants, and their opportunity to learn and use the methodology in a semi-autonomous manner, on an on-going basis. The question is does our intervention remain just an isolated change, which for additional enhancement requires the recalling of the expert? In such an expert-based scenario, improvement would not be endogenous and continual, but exogenous and disconnected. Therefore I would argue that if we believe our role to be that of expert, then we limit the potential impact of intervention.

3.2 Criterion of Participation

We should ask why are systems facilitators ambiguous about participation? I believe this to be threefold; firstly, due to the lack of understanding of participation, secondly due to the lack of principles guiding participation in the methodology, and thirdly due to the threat to our perceived role. Interventions tend to move towards a highly structured framework, with a problem focus and often unintentionally sideline the ethical issues of participation. Similarly our underlying objectives for participation are rarely discussed, either between facilitators, with problems solvers or participants. If they are discussed they are rarely explicitly referred to in case study literature. As seen above even the soft systems methodologies can be used mainly in an information extraction format, due to the under theorising of participation, being often a means approach, using efficiency ultimately as the justification. For an ends approach to be reached the incorporation of people into the context to be negotiated implies a boundary extension and possible problem relocation, for instance locating to the area of enskilling in viewing options (Checkland would do this but the focus is still on problem solving). We do have tools that incorporate participation, however the learning principle within these tools is seldom maximised. The purpose is rarely participation to build learning and capacity into participants, thereby benefiting individuals and synergising development. Learning is considered secondary and due to under-emphasis might rarely occur, for inherently we often exude the role of expert, which precludes a nurturing enabling role and environment, which moves toward a paternalistic prescriptive role.

3.3 Evaluation

Evaluation within systems is firmly rooted in the problem situation. For example soft systems methodology which incorporates the principles of learning and participation assesses success or improvement in terms of whether the participants believe the intervention has improved the situation (with or without them). However my argument is that the problem situation should not be the sole criterion of success, we should include evaluation of whether the participants themselves have developed in their understanding and abilities, as assessed by themselves. Do they believe with the facilitator that through dialectic debate they have an increased capacity to tackle difficult 'messy' situations and design the required systems? This criterion points toward sustainable development having occurred, by individually, and organisationally, having increased their capacities, and therefore moving them towards the goal of more sustainable process.

3.4 Implications

Finally in order to maximise participation, the enabling or learning potential, notably requires a change of emphasis in our approach to *problem owners*. I am suggesting that the initial terms of reference require re-negotiation in order to maximise participation in social change and thereby gain the benefits of effectiveness, moral justification, and sustainability. Organisations are willing to allow staff to attend theoretical training courses on issues they *may* encounter in their positions. However when a real live problem (case-study) occurs in their work environment an external expert is often drafted in to define the problem with the objective of improving the situation. If we could move away from this expert intervention model, and toward an on-the-job-learning approach, we could deliver not only a more sustainable improvement, but also a change of perception towards problem solving, that would hopefully synergise, creating dynamic development. External facilitation is not excluded, but the role would be perceived differently.

4. CONCLUSION

Assessment of participation must be in terms of two criteria? First, what is the objective or purpose underlying the participation, which requires transparency within the participative process, second, what are the evaluation criteria. It must also be asked who is evaluating the process; e.g. the problem owner, the facilitator, or the participants. And what are they evaluating the process in terms of:

i. solely an improved situation;

ii. the evaluation of the situation for improvement by the participants - as argued by Checkland,(1981) which aligns to a means approach to participation;
iii. the human development issue, whether the participants consider themselves to have been enabled and enskilled in the dialogical processor; or
iv. as I would argue (ii) and (iii).

We have identified the constraints on such a process to be threefold. Firstly, our lack of guidance in systems methodologies, secondly the concept we have of our role, which directly impacts the third constraint of poorly negotiated terms of reference for our interventions. We often persist in our perceived role of 'systems expert' at the expense of focusing on the development of people within the organisation. For if we focused on their development with the aim of enhancing their abilities to co-generate understanding of the situation, identify and debate the various options and then design the required systems, with evaluation in terms of the situation and their enskilling, we would be moving towards an 'ends' approach to participation. Therefore we need to understand the issues involved in participation and be open and explicit about our use, and at times our abuse of it.

REFERENCES

Brown, M., 1994, From exclusion to empowerment, some practical and political issues of participation, *Systems Thinking & Progressive Social Change Congress,* Cape Town, South Africa

Cernea, M., 1985, "Putting People First", Oxford University Press, Oxford.

Chambers, R., 1983, "Rural Development: Putting the Last First", Longman, London.

Checkland, P.B., 1981, "Systems Thinking, Systems Practice", John Wiley & Sons, Chichester.

Feyerabend, P., 1978, "Science in a Free Society", New Left Books, London

Freire, P., 1973, "Education: The Practice of Freedom", Writers & Readers Coop, London

Friedmann, J., 1992, "Empowerment: The Politics of Alternative Development", Blackwell, Oxford.

Ghia, D., 1988, Participation in Development : Some Perspectives from Grassroots Development, UNRISD, Geneva.

Goulet, D., 1989, Participation in development: new avenues, *World Development* 17 (2):165.

Gregory, W., & Romm, N., 1994, Developing multi-agency dialogue: the role(s) of facilitation, *Working paper 6,* Centre for Systems Studies, University of Hull, Hull

Marsden, D., & Oakley, P., 1990, "Evaluating Social Development Projects", Oxfam, Oxford.

Oakley, P., 1991, "Projects with People: The Practice of Participation in Rural Development", International Labour Organisation, Geneva.

Rahman, A., 1993, "People's Self Development: Perspectives on Participatory Action Research", Zed Books, London.

Ulrich, W., 1983, "Critical Heuristics of Social Planning: A New Approach to Practical Philosophy", HAUPT, Bern.

13

A SOFT SYSTEMS APPROACH TO INTEGRATING INFORMATION SYSTEMS PLANNING AND DEVELOPMENT

Sharon Dingley

Department of Computer Science and Applied Mathematics
Aston University
Birmingham B4 7ET
Great Britain

INTRODUCTION

Soft systems theory has invoked much debate as to how it can contribute to information systems development practices, which are founded in reductionist techniques. This paper addresses the application of soft systems theory, to the critical issue of integrating the planning and development of information systems.

Information systems are a powerful organisational resource. The use of all resources must be aligned with the strategic direction of the company. Integrating the development of information systems with business strategy is a key concern of Information Systems Managers.

Strategy formulation requires a balanced appreciation of many diverse, internal and external factors. Many analytical tools have been developed to assist strategic planning. However, each tool contributes a specific, limited view of the situation.

The soft systems approach provides a means with which to explore a number of different views of a problem situation. The application of soft systems theory to strategic planning addresses the need to explore different perceptions of the corporate position provided by various analysis tools. The development of a rich strategic picture can be used to assist in the planning of information systems aligned with corporate strategy. The strategic picture can then be used as the basis for actioning the strategy by developing the information systems identified.

This paper presents a framework for aligning information systems development with corporate strategy using techniques from soft systems.

1. BACKGROUND

Information systems (IS) are a powerful strategic resource. Sustained alignment of IS with corporate strategy is a critical issue of concern for IS managers (Wysocki & Young,

Critical Issues in Systems Theory and Practice, Edited by K. Ellis et al.
Plenum Press, New York, 1995

1990; Galliers, 1992, 1993; Galliers *et al.*, 1994). The following section (2.) outlines current approaches to address this concern.

Strategy formulation requires an appreciation of different views of the corporate situation. Soft systems theory provides a means with which to explore differing views of a problem situation. Section three describes previous work applying a soft systems approach to strategic IS planning. A soft systems approach is adopted by the framework proposed in section four to integrate the planning and development of IS aligned with corporate strategy.

2. ALIGNING 'IS' WITH CORPORATE STRATEGY

Strategic IS planning outlines the development of IS to provide continued support to the attainment of corporate objectives. Wysocki & Young (1990) identify an IS planning spectrum: *Stand-alone Planning* (a corporate plan or an IS plan is prepared); *Reactive Planning* (IS plan is prepared to support the corporate plan); *Linked Planning* (IS are matched to organisational needs); *Integrated Planning* (corporate and IS plans prepared interactively).

Integrated planning recognises that in addition to supporting corporate objectives, IS can also influence objectives. Planning IS in isolation from organisational strategy, can lead to the development of systems which do not fully serve the needs of the organisation. Major technological opportunities and threats may be overlooked if IS are not considered during formation of corporate strategy (Martin, 1986).

Current methods to align IS with corporate strategy adopt a reactive or linked approach to IS planning, for example Business Systems Planning (BSP) and Information Engineering (IE). BSP is described as a structured methodology to translate business objectives into information requirements (IBM Corporation, 1984). It is criticised for neglecting analysis of the external environment (Wysocki & Young, 1990). IE (Martin, 1986) is an integrated methodology that aims to link the development of IS to business objectives. It is a data driven approach which can neglect the mission and objectives of the organisation.

Integrated planning requires analysing corporate strategy in the context of IS (Premkumar & King, 1991) and analysing IS in the context of corporate strategy (Piercy, 1991). Strategic IS planning methods can no longer neglect tools such as the TOWS matrix (Weihrich, 1982) used in the formation of corporate strategy.

3. A SOFT APPROACH TO STRATEGIC PLANNING

Strategic planning models have adopted a mechanistic approach to strategy formulation (Ciborra, 1994). The underlying premises of this approach include prescriptive process and structured models (Mintzberg, 1990). This reduces strategic planning to an analytical process of decomposing goals into actionable steps (Mintzberg, 1994). Each model contributes a specific, limited view of the strategic situation. Strategic analysis requires many diverse views of the situation to be considered. Mintzberg (1990) argues strategy formation is not an arbitrary sequence of steps from which a fully formulated strategy emerges but an integrated system of continuous learning. This involves strategic thinking (Mintzberg, 1994) concerned with synthesis and creativity.

The Soft Systems Methodology (Checkland, 1981) offers an alternative, softer approach, to the traditional, hard approach to IS development. An alternative softer approach is now needed to support the following requirements of strategic thinking:

- Richer representation of knowledge releasing strategy from the restrictive 2 x 2 matrix of the 'Boston Box' (Doyle, 1991).
- Synthesis of different views which need to be considered.
- Attainment of a shared understanding of the strategic situation.
- Clarification of the activities needed to attain the objectives identified.

The Soft Systems Methodology (SSM) is considered a suitable basis for strategic planning (Galliers, 1992, 1993).

The application of SSM for strategic IS planning has been explored. Wilson (1990) extends the basic approach of SSM to determine the information requirements needed to support the organisational activities. The first stage of Wilson's approach is to analyse the current situation. This involves the analysis of many complex and diverse factors in both the organisation and the external environment. Although structured strategic planning tools offer a restricted view of the situation, they do provide limited guidance in analysing complexity. This issue is recognised in Galliers scenario-based approach (Galliers, 1992, 1993).

Galliers divides the stage 'analyse current situation' in Wilson's model into two phases to address both internal and external organisational analysis. The use of strategic IS planning tools and techniques is required to assist this analysis. Galliers (1992) states that the aim is to synthesise the range of views provided by such techniques. A number of different scenarios for the future of the organisation may then be identified. Existing and required information needs are compared to form recommendations for the development of an information architecture.

Both approaches propose the complete application of SSM to strategic IS planning. Phases of SSM are extended as necessary to address the specific aspects of strategic planning. The following section presents a framework which adopts a soft approach to IS planning without the complete adoption of SSM.

4. INTEGRATING STRATEGIC 'IS' PLANNING AND DEVELOPMENT

The framework in figure 1 adopts a two phase approach to integrate IS planning and development. The planning phase is supported by a Situation Summary Diagram; the development phase is represented by a Justification Alignment Contour.

4.1. Situation Summary Diagram

The Situation Summary Diagram (SSD) supports strategic analysis by providing both:

- Structured tools with which to analyse specific areas of complexity.
- A soft approach to support the creative art of strategic planning, acknowledging differing views of the strategic situation.

The SSD is a physical diagram which summarises the strategic situation explored using a rich notation. A formal framework underlies the construction of the SSD. The SSD is developed through iteration of structured strategic planning techniques. Organisations demand flexible planning methods (Premkumar & King, 1991); the SSD provides this flexibility, supporting the use of proprietary methods.

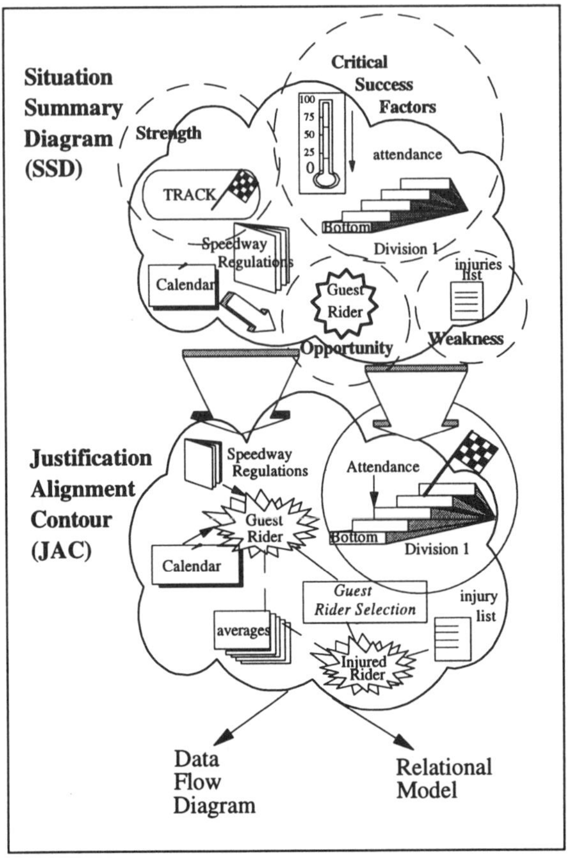

Figure 1. Example Application of Conceptual Framework.

The SSD summarises the corporate strategy by synthesising the information captured from structured strategic planning techniques. Figure 1 illustrates the SSD derived from the use of critical success factor and TOWS analysis for a speedway stadium.

4.2. Justification Alignment Contour

A Justification Alignment Contour (JAC) is prepared for each IS identified from the SSD needed to support or facilitate corporate objectives. A JAC is a sub-set of the SSD relevant to the system under development.

The purpose of the Justification Alignment Contour is to *Justify* the system to be developed, thereby *Aligning* development with the corporate strategy. It provides the starting point for development, using the information collected during strategic planning. A C*ontour* is defined as an outline or a line on a map joining points of the same altitude. JACs outline the IS to be developed. Strategy may be defined at various degrees of altitude within the company, for example at corporate or divisional levels. A JAC aligns IS planning and development with the strategic direction of the company at all levels.

Figure 1. illustrates a JAC identified to improve the strategic situation summarised. A potential IS has been proposed to assist in the selection of guest riders. The JAC provides information regarding the strategic role of the development to all participants in the development. This increases the developers' understanding of the system required in addition to increasing morale and interest in the development. The initial JAC identifies the main objects associated with the IS extracted from the SSD. As the model progresses, more detail concerning these objects is obtained and further objects are identified. Process and relational data models can then be derived from the JAC. Arrows on the JAC indicate specific relationships between components, identified from further analysis. For example the fixture calendar imposes restrictions on the selection of guest riders.

4.3. Summary of Framework

The framework proposed provides an iterative mechanism for the synthesis of strategic planning techniques. Key concepts from the planning phase (SSD) are analysed during development (JAC) closely integrating the two phases. Improved integration between planning and development requires transferring some of the richness of strategy formulation to IS development (Galliers, 1992); this is the aim of the JAC.

Integration of IS planning and development is essential to support approaches to strategic IS planning which advocate continual learning (Mintzberg, 1990) and radical learning (Ciborra, 1994). Such approaches require automated tool support for iterative learning using prototypes. The framework provides the foundation for automated tools which sustain direct alignment between strategy and IS development.

5. CONCLUSION

Strategic IS planning is complex and a range of structured planning methods have been developed to assist this task. IS researchers have recognised the potential contribution of the soft systems approach to strategic IS planning. Previous work has concentrated on the complete application of SSM for strategic IS planning. A framework is proposed adopting a soft approach to address the critical issue of maintaining alignment of IS with corporate strategy.

REFERENCES

Checkland, P., 1981, "Systems Thinking, Systems Practice," John Wiley & Sons, Chichester.

Ciborra, C., 1994, The grassroots of IT and strategy, *in*: "Strategic Information Systems A European Perspective," C. Ciborra and T. Jelassi, eds., John Wiley & Sons, Chichester.

Doyle, J.R., 1991, Problems with strategic information systems frameworks, *Eur. J. Inf. Systs.* 1:273-280.

Galliers, R.D., 1992, Soft systems, scenarios, and the planning and development of information systems, *Systemist* 14:146-159.

Galliers, R.D., 1993, Towards a flexible information architecture : integrating business strategies, information system strategies and business process redesign, *J. Inf. Syst.* 3:199-213.

Galliers, R.D., Merali, Y., and Spearing, L., 1994, Coping with information technology? How British executives perceive the key information systems management issues in the mid-1990s, *J. Inf. Tech.* 9:223-238.

IBM Corporation, 1984, "Business Systems Planning : Information Systems Planning Guide," Fourth Edition, No.GE20-0527-4.

Martin, J., 1986, "Information Engineering," Vol.1, Savant, Great Britain.

Mintzberg, H., 1990, The design school : reconsidering the basic premises of strategic management, *Strat. Mgt. J.* 11:171-195.

Mintzberg, H., 1994, The fall and rise of strategic planning, *Harvard Bus. Rev.* 72:107-114.

Piercy, N., 1991, "Market-led Strategic Change : Making Marketing Happen in YOUR Organization," Thorsons, Great Britain.

Premkumar, G., and King, W.R., 1991, Assessing strategic IS planning, *Long Range Plan.* 24:41-58.

Weihrich, H., 1982, The TOWS matrix - a tool for situational analysis, *Long Range Plan.* 15:54-66.

Wilson, B., 1990, "Systems : Concepts, Methodologies and Applications," Second Edition, John Wiley & Sons, Chichester.

Wysocki, R.K., and Young, J., 1990, "Information Systems : Management Principles in Action," John Wiley & Sons, Singapore.

14

A SYSTEMS STUDY OF THE SCOPE AND SIGNIFICANCE OF INFORMATION SYSTEMS IN THE MANAGEMENT OF THE COLOMBIAN COFFEE INDUSTRY

L. F. García-Montoya

University of Humberside
School of Computing and Information Systems
Hull
HU6 7RT
United Kingdom

INTRODUCTION

The purpose of this paper is to point out different aspects which gave rise to the research project : 'A systems study of the scope and significance of information systems in the management of the Colombian Coffee Industry.'

Organisational intervention concerns resolving problems which involve people. As I see it, understanding people* and helping people to understand themselves better, is a way of resolving problems. To facilitate the dissemination of knowledge about problems and their resolution it is necessary for theory and practice to be united.

With respect to the aims of the research: information systems are seen as a mean for effective organisational intervention, and, the project focuses on the questions which arose from the researcher's different experiences of working with people.

THREE STORIES, THREE QUESTIONS

Since every question has arisen from practice, I will summarise the main aspects of each experience. It is presumed that setting out the facts will make sense of the questions.

* To understand people is a concept which will be used throughout this work. It refers to a phenomenologist understanding of people's actions without prejudices, and assuming not only rational coherent content but also non-rational processes of human interactions.

Story 1: Human Resources and Human Beings

My first degree was in Industrial Engineering. While other engineering disciplines cope with well-structured problem situations, industrial engineering is concerned with human work. In an economic context, human work must assist organisational survival; that means human activity is aimed at producing goods or services. In that context, people are a resource which have to be managed in order to optimize profits or results. Furthermore, it is possible to talk about 'human resource optimization' as well as machines, computers, finances, etc.

However, as Checkland (1981) has shown human activity systems (and, for that matter, designed systems) are fundamentally different in kind from natural systems; as a consequence of self-consciousness, human beings have genuine freedom of choice in selecting their actions, hence it becomes necessary to change from an optimizing to a learning paradigm.

In this context, the question which arises is, 'how can different methodologies be designed to tackle human work cope with human expectations? In other words, how can we change the instrumental view of people, i.e. people as an instrument to make profits, for another view which emphasises learning as the main objective of work?'

Story 2: Information and Participation

My first job was in a social research institute. The aim of the work was to provide organisational support to information systems design and implementation in the Justice area. To illustrate the nature and difficulty of the problem, I will discuss one project in which I was involved - the design and implementation of the National Jurisdiction of Public Order. Since its creation in 1991, this institution has operated efficiently in the fight against drug traffic crimes.*

Despite pressure by the United States to approve an extradition agreement against a supposed Colombian permissive drug traffic policy, Colombia formulated its own strategy; the main aim of which was to curb threats to, and murders of judges overseeing the investigation of those crimes.

This strategy was based on two fundamental actions. First, judgement activities were relocated to the five major cities. The purpose of this relocation was to create a safe infrastructure for the judges, instead of having small courts in little towns without safe work conditions. Secondly, an information system was implemented to maintain a secret link between the judge and the process of judgement. In this way, the judges' deliberations do not belong to a particular individual but to all judges as a group, reducing the previous danger levels. Currently, this strategy has not only stopped the killings of the judges but also maintains a unique and effective system in the fight against drug traffic crimes.

The information system so designed implied a big change in everybody's way of work. The judges had to work by themselves, isolated from their secretaries and personnel support, and the personnel support were not allowed to see the judges. Furthermore, the new working methodology required personnel to follow a step by step procedure in order to maintain judges' security.

The strategy was designed by the head of the institution and there was resistance to accept it at the lower organisational levels because people felt it was an imposition. The

* There is a distinction between drug traffic and drug traffic crimes. Drug traffic is an international problem encompassing demand and supply, with concern for both affected sides. Drug traffic crimes are crimes caused by drug traffic cartels in a single country.

question in this case would be, 'what kind of mechanisms can be used to enable people to be involved in the design and implementation of an information system in order to accommodate everybody's needs?'.

Story 3: Power, Interests and Trust

Next, a policy story with strategy implications. At the beginning of each governmental administration, the President announces the government's programmes which are tied together in a four-year national development plan. The investment budget is managed by the National Planning Department (DNP). This means that the DNP evaluates all projects which need public finance and approves only those that are considered to be in the interests of the national development plan. This process follows a particular pattern using technical criteria in order to warrant an adequate management of public resources.

Lastly, the DNP set an economic open-market policy which takes into account world-wide economic phenomena. This policy refers to internationalisation of the economy, neo-liberalism, economic liberalisation, etc. For Colombian industry it meant the end of protectionism and the need to recognise raising performance standards in order to compete in the international markets. Technological development, for the division of the DNP in which I was employed, implied that the criteria to approve a scientific and technological project was based on its link with the competitiveness problem of the industry. In order to demonstrate project viability the company must have a share of fifty percent of the total cost of the project.

The worst problem was that there was lack of knowledge of the two Industrial and Academic sectors which worked separately, having no link between each other. Industry, having no national scientific and technological research, was not convinced that they would find any solution to their technical problems. It was cheaper just to pay international experts to solve problems rather than finance national research. On the other hand, the academic sector, i.e. universities and research centres, did not find any contact point to obtain industrial resources and co-operation. Also, the academic sector was unaware of industrial matters and concerns and it was more interested in international recognition than in fulfilment of local expectations. In this case, both sectors viewed each other as a threat against its own interests. They could not formulate a strategy to present to DNP in order to obtain research resources because of poor inter-sector communication.

As a matter of fact, the methods used to evaluate investment projects offer efficient macroeconomic mechanisms to control public expenses; however, as a cost of that benefit, economic criteria tend to take precedence over any other considerations. In such cases, the methodologies do not consider important aspects of human communication such as power, interests and trust.

The questions which arise here are, 'whether it is possible to include human communication contents not rationally expressed in the organisational intervention processes and what is the cost of considering them or letting them aside ?'

FACTS FOR A RESEARCH PROJECT

Up to now, I have considered information systems in an organisational intervention manner and a communication problem in the design of public policies. Now, I will consider the information system for the Colombian Coffee Industry. This is an information system which is expected to cope with the most important variables associated with an industry which has long been present in the history of Colombia.

Coffee has been for more than a century the main product in the Colombian economy and, therefore, the one with a major inference on its economic and social development. During 1991, coffee exports represented 18.7% of the total value of all exports. The aggregate value for the bean production reached a total contribution of 6% of the real total Gross Domestic Product. Employment in coffee activity, exclusively agricultural, reached 750 thousand people in 1988, that is 5.2% of the country's economically active population. Moreover, it is possible to assert that the integration of 'coffee' with national development has made improvements in the domestic transportation system, social welfare, public finances, industrial development and the structure of the internal market.

Another important aspect in the history of the coffee industry is related to the role played by the National Federation of Coffee Growers which was created in 1927. At the beginning of the 20th century, the instability of the external market was immediately transmitted to producers. There was no minimum price nor guarantee of the crop. Storage facilities were reduced and there was no technical assistance at all. External market and overseas transportation were dominated by foreign companies. These circumstances were changed by the Federation as it realised that in order to defend the country's interests it was necessary to create an infrastructure to manage every aspect of the coffee business.

That infrastructure includes: regional and local committees for assisting each coffee farm not only in production matters but also in social matters; a national committee with a strong national and international negotiation power; the Fondo Nacional del Café (Coffee National Fund), ALMACAFE (Management Storage Company), CENICAFE (Technical Research Centre), the Flota Mercante Grancolombiana (Merchant Fleet), the Banco Cafetero (Coffee's Bank) and the Compañia Agrícola de Seguros (Insurance Agricultural Company).

As an example of the functional existence of one element of the infrastructure, the National Coffee Fund is a tool to maintain the coffee policy for the benefit of the producers and the country. The existence of this Fund has facilitated the creation of a generalised minimum price. This way, the Federation always purchases in cash, all production available at a price which is announced in advance of harvesting.

Contextual Factors

There is general agreement that the world coffee industry is now facing a serious crisis. The coffee bean prices in the world market have been steadily falling. At the same time, world production has produced high surpluses since the '89/'90 coffee year and Colombia is one of these centres contributing to the surplus. Additionally, the economic clauses of the World Coffee Agreement, which regulated the market, were suspended in July 1989, and from that moment, producing and consuming countries have failed in their different attempts to reach agreement.

There are obvious impacts of this crisis: coffee export incomes fall; nominal purchase prices decrease; domestic prices fall; National Coffee Fund changes; expenses grow rapidly and productive structure is damaged; etc. The most burning question has been the one of National Federation policies and strategies. Some social sectors have argued that there is a case for dissolution of the institution and for the design of other 'modern mechanisms' to manage the coffee business.

Practical Factors

In the middle of this storm, the National Federation of Coffee Growers, through the Economic Research Division, leads the creation of the Coffee Information System SICA.

The SICA's initiation will enable basic knowledge of and a systematic follow-up of, the status, distribution, and characteristics of the national coffee growing activity and of the working and living conditions of the people residing in the coffee areas. An added advantage is that the information will be geo-referenced. This means that the different users of the SICA will be able to know, among other things, the precise physical location of the coffee areas, their expansion or reduction and their state of technological evolution. It will also be possible to associate the socio-economic information to the geo-referenced areas. In this way, SICA will additionally provide support for the program's design as well as local projects.

The SICA will provide the basis for the planning, investigation and execution processes in support of the National Federation of Coffee Growers, and consequently, it shall perform as a management information system; an information network and a research programme. As a management information system, it will attempt to provide controlling tools and indicators for making decisions. As an information network the system will provide mechanisms for data interchange between the different functions of the Federation and its subordinates, in such a way that the feed processes constitute a common current practice of data information enrichment, helping at the same time in the follow-up of the programmes related to coffee growing, producers and their families. As a research programme, SICA will contribute to a better knowledge and analysis of the coffee situation, and this will enable the definition of programmes and policies priorities.

Questions for the Research

Currently, the National Federation is building a database through a National Coffee Survey. This process will take three years. As an independent researcher, my purpose is to formulate a methodology based on the questions that I highlighted before.

In the context of this research, there are specific situations referring to each question formulated in every story of this paper. Referring to the question 'how different methodologies to tackle human work can cope with human expectations', it is necessary to analyse the production structure in the Colombian Coffee Industry. It seems to be that small farms contribute at least 55% of the bean production. For that reason, it seems appropriate to revise the uses of the information system for a sector not marginal to, but as the major client of the system.

Besides the production structure, the Colombian Coffee Industry involves issues of a : technical-; financial-; storage-; transportation- and international marketing support-nature. In light of this arises the question 'what kind of mechanisms can be used to get people involved in the design and implementation of an information system in order to resolve everybody's necessities?'

Moreover, noticing that the Colombian Coffee Industry is as powerful as the national government, means that it is necessary to consider interests and power games concerned with the question 'is it possible to include human communication contents not rationally expressed in the organisational intervention processes?'

It is important to point out that if it is taken for granted that Coffee Industry has developed a successful strategy for survival, a question implied by that characteristic is: 'could it be possible to apply success elements of the strategy to other agricultural products?'

CONCLUSION

The coffee industry is vital to Colombia's balance of exports. A study of the scope and significance of information systems in the manager of the coffee industry was outlined in this paper. Given its systems focus, the study discussed offers a unique opportunity to explain the key factors which contribute to the successes and failures, within this industry.

The potential for this methodology to be used in the analysis of other agricultural products which are critical to the Colombian economy has yet to be assessed.

REFERENCES

Checkland, P.B., 1981, "Systems Thinking, Systems Practice", John Wiley & Sons, Chichester.

15

THE CRITICAL SYSTEMS THINKING (CST) AS A PATH FOR ACKNOWLEDGING HUMAN DIGNITY IN THE ORGANIZATIONS

R. García and S. Motta

ITESM-CEM
DACS-Humanidades
Carr. Lago de Gpe. Km 3.5
Atizapan, Edo. de México
México

INTRODUCTION

Modern organizations of the so called 'modernity' era have been characterised by a high degree of Systematisation in their operations which involves a high degree of specialisation. This has resulted in an almost unbreakable hierarchical order and a strict control on the people who execute the task related operations, who, for this reason, turn out to resemble the pieces of a chess game. "The different pieces of chess are only defined by the functions given by the game rules...these can be substituted by any other without losing their identity. As long as these are moved according to the game rules...its properties have no effect on the system. (Llano, 1994:50)

THEORETICAL BACKGROUND

A person is not, and neither can he be, a (chess) piece determined solely by system rules and properties for he possesses free will which gives him freedom to determine self-direction and determination towards higher ends (than those of the system) by a personal acceptance of its actions. He is capable of searching for his own sense and personnel existence, giving an exclusive nature which is irreducible and therefore incapable of being substituted. In this way a person enjoys a dignity higher than that experienced by lower order 'things', that is every one has something to give with his existence.

Where it is not the system which has the last word of the being of the person, but the person himself, then it is self worth and personal properties which will determine the system. This we interpret in terms of people being the foundation, the determinants of the organizational ends and also the organizational ends.

From this it directly follows that human work should be a personal experience because a person, when doing work and exercising personal power, creates himself and his being. Habermas (1982) would argue that, in every persons work is summarized in the form of personal identity. The work becomes 'his' work through the commitment of his being, that is he is linked with the world, with other persons and with his own ends. However, when a persons' self worth is low his function and being become suspended, "...the person stops being and is invaded by the anguish of his own nothing" (Llano, 1994:51).

It is for this reason that all systems and methodologies should be governed by agreed criteria and ethical principles in order not to lose sight of the dignity of other persons. Ethics tell us what must do, then the essential purposes of our human nature can be established and through the guidance of moral principles our behaviour can be directed towards being perfection.

Moreover, ethics are not necessarily seen as a phenomenon, process or instrument, but as coming from within the individual, from his existential anthropological entity, for the ethic is an expression of the human being in an integral way. And if it is not acknowledged - because human nature is subject to habits - we start to abandon our opportunity for expression of free will becoming objects which can be determined, manipulated and used by others. This can result in an individual becoming irresponsible and, like a building subject to erosion, sooner or later he falls, causing the corruption of the system. As Plato discovered, "...the ethical education is the soul of the society, it is the link uniting the individual with a commonwealth" (Vilchis, 1992:18).

EXPERIENCE

From this basis, we realize, through some experience with the methodology of Total Systems Intervention (TSI) (Flood and Jackson, 1991) in a modern Mexican Company, that TSI is a resource through which to acknowledge human dignity within organizations. The intervening analyst needs to be aware of human dignity so that those who are involved in the problem situation are acknowledged in terms of the human ethic and dignity. Even though TSI permits an analyst to obtain a total vision of an organization, and the context within which people develop, the intervention needs to be governed by an ethical/philosophical approach which will avoid the trap of human beings as 'cogs in the machine'. During our intervention, using TSI, we noticed this tendency when we attempted to determine who were the relevant organizational participants because we encountered the problem of how to determine involved participants without offending the dignity of others. This dilemma raised several questions:

- If only some organisational members were 'classified' as relevant, would the non inclusive members be regarded as of lesser or no importance?
- Would non inclusive members be regarded as having less dignity?
- Would inclusive members be regarded as those who direct the system and, consequentially, those who determine the roles and behaviour of non inclusive members?
- From the previous question is the system to be regarded as being of greater importance than non inclusive members?
- Are non inclusive members to be regarded as expendable and exchangeable parts of a machine? (where the machine is seen to be a System).

For us these were important issues because if a person felt he was being considered as irrelevant, with respect to the problem situation then that person may see himself as of lesser importance than the system of which he were supposed to be a part. And also we can think, as analyists, that his personal value depends on his system functionality and not of his dignity. Because of this he would get into Nazism principles.

On the other hand, the human capacity to approach 'reality' is limited, and an analyst can not always embrace or involve all organizational members in a problem situation. How then could we get closer to the 'total reality' of the organization in which we were intervening?; and how does TSI provide guidelines through which human dignity is acknowledged. This , we feel, is where the need of an ethical approach becomes important.

At the start of our intervention it was evident that the criteria for deciding relevance (for inclusion) or otherwise was based on the organizational hierarchy. Clearly the use of levels of responsibility, power and authority was an impersonal approach. However the selection process should not end here as we would fail to approach total reality. In order for us to understand the operating characteristics of the organization (as a system) we needed to take a phenomenological stance to gain personal involvement with organizational members. In this way we, and organizational members, would gain knowledge of operations and the relationship which exist between people in the organisation.

The context for relationships is constituted by the intersubjective relation between individual consciences. Conscience is an important factor through which an individual determines his purpose, his world view, his duty of being and his sense of freedom, "The sense of existence not only has to be, but can be found, and the conscience guides the man to his search. In a word, the conscience is a sense organ" (Viktor, 1990:31). In other words, not recognising human conscience, meant that we would not be taking into account the personal characteristics associated with organizational context. We would simply be considering organizational functions as determining context which is incomplete and not real. Our argument here is that in failing to acknowledge human dignity places us in a situations of apparent reality.

It is in this sense that TSI with its commitment to human emancipation can reveal that all people in the context of an organizational problem situation, possess dignity. That is because it is a critical methodology which permits the inclusion of all points of view in a group of participants who are attempting to achieve personal goals through their contribution to the development of the organizational purpose. However, we argue that, in order to fully achieve human emancipation, an analyst needs to be, not only phenomenological, but also ethical in his approach in the analysis of an organization. Any other approach can result in failure to address the issue of human emancipation in an organizational intervention.

In the intervention in question the central issue was that of frustration. Without ethical criteria (ie., recognising only the functions relationship) we would have failed to achieve 'total reality' and understand the problem and been able to relate it to appropriate phenomenon. That is, we could have confused alienation, that is inherent to the person, with inefficiency that is only a symptom.

CONCLUSION

We must take into consideration the importance of human beings in a system which should be expressed through their dignity and not just their performance. Although it may not be possible to interview every organisation member, accepting their dignity as a relevant factor will help.

We need to acknowledge the inalienable rights of human beings, amongst these are individual freedom to determine their existence. The organizational context is contingent upon human freedom and any intervention methodology, including TSI needs to take this into account.

This shows that there is a need to intrinsically link the philosophic and ethical approach to the development of any systems based intervention methodology.. Therefore any person who designs and implements a system (who will be subject to personal values, virtues and habits) and will exercise personal freedom in that endeavour, must be willing to change his model in order to make it realizable. In consequence, any person who could be ultimately involved in building this system, through human conscience relations that show, TSI, that they accept the dignity of others working on the building of such a systems. To achieve such a state, human beings need a sound personal conscience, based on ethical vision through which they can face the possibility of different points of view and in so doing achieve a final system acceptance to all.

This 'reconquest' of the person depends on our personal conscience, and the conscience we may be able to arouse in our contextual surroundings. We argue that TSI is a way to awaken the 'sleeping consciences' with respect to human dignity, for we cannot force anyone to change, but we can create a realisation of the consequences of failure.

> "It is possible to have the good luck of doing a lot for the others, the good luck conducting the others were I desire...But this result is not in my hand, it depends on a lot of things, and, above all, depends on the fact that the person wants it or not. For all the eternity is impossible to force a person to accept an opinion, a conviction, a belief. But I can do one thing: I can force him to realise" (Kierkegaard, 1985:63-64).

ACKNOWLEDGMENTS

We thank Dr. Javier Vilchis and Dr. Joel Rodriguez for their kindness, assessment, ideas and reviews of this work.

REFERENCES

Flood, R.L., and Jackson, M.C., 1991, "Creative Problem Solving : Total Systems Intervention", John Wiley & Sons, Chichester.

Kierkegaard, S., 1985, "Mi Punto de Vista", SARPE, Madrid.

Llano, C., 1994, "El Postmodernismo en la Empresa", McGraw-Hill, Mexico.

Vilchis, J., 1992, "Análisis Histórico-Filosófico de la Naturaleza Social del Hombre", Tesis Doctoral, UCIME, Mexico.

Viktor, E.F., 1990, "Ante el Vacio Existencial", Herder, Barcelona.

16

THE MANAGEMENT OF CHANGE IN PRISONS

Toward a Process of Postmodern Systems Thinking

Catherine Gaskell

School of Computing and Information Systems
University of Humberside
Cottingham Road
Hull
HU6 7RT
United Kingdom

INTRODUCTION

Her Majesty's Prison Service is an integral part of the wider criminal justice system, striving to "serve the public by keeping in custody those committed to the courts" (Home Office, 1993). Whilst the relationship to this wider system must always be kept in mind, it should be acknowledged that the Prison Service functions as a system in its own right, consisting of more than 130 establishments. In turn, these individual prisons function systemically both internally and in relation to the wider Prison Service. It is becoming more and more important to ensure that the Prison Service as a whole is structured in terms of systems, one reason being that communication is the key to the realisation of major objectives. Communication links have to be extended not only at establishment level but throughout the whole of society in England and Wales; this is partly due to the degree of accountability which the Service has to ensure to the government and the public. The organisational values of many prisons today stress the importance of systems thinking in assisting communication and ultimately ensuring that Prison Service objectives are met. In practice of course, it is often difficult for the Service and its establishments to operate in a truly systemic way. This is thought to be due to the nature of its business and 'clientele', problems of security and the general sense of secrecy which surrounds the concept of 'the prison'.

Although for some time many Prison Service establishments have been aware that their organisation should function as a system both internally and externally, the adoption of systems thinking as a managerial methodology is a fairly new venture. It is anticipated that a systems methodology could assist prisons in bringing about various types of organisational change, and enable initiatives such as Total Quality Management and accreditation with local Training and Enterprise Councils (TECs) as 'Investors in People' to be achieved.

Critical Issues in Systems Theory and Practice, Edited by K. Ellis et al.
Plenum Press, New York, 1995

If these objectives can be met in the future, then it could be possible that public perception and confidence towards the Prison Service may change and be improved.

BARRIERS TO CHANGE IN THE PRISON SERVICE

Victorian Values

Since its rebirth in the mid - late nineteenth century, the prison system in England and Wales has been associated with a certain image; one of rigid structure and definite inflexibility. Notions of hierarchical power structures are constantly accompanied by perceptions of the prison system as being stiffly opposed to change and reform. Although at the present time the Prison Service is attempting to bring about greater de-centralisation and enhanced devolution, it has been pointed out from within the prison system itself, that a strain of the 'nineteenth century clockwork running of the system' is still in place today. It seems therefore, that even though the Prison Service is attempting to be innovative in its new organisational methodologies, there is still an inherent belief amongst some prison officials, in a 'traditional' prison system with Victorian values. It is likely that the values of many prison officers and their reluctance to 'change with the times' will prove a major obstacle to much organisational change. It must be pointed out however, that Victorian values are not universal within the prison system. On the contrary, there are many people within the service who are willing to strive for and participate in the pursuit of change.

The public demand for more austere and harsh regimes in prisons does however, provide some evidence that Victorian values do exist. The government's recent 'Back To Basics' campaign, only adds more fuel to the argument, and seems to legitimise the authoritarian and hierarchical power structures still inherent as a strain within the Service. It seems to be becoming clear that the Prison Service is in a contradictory position, on the one hand attempting to introduce new methodologies whilst on the other bowing to public demand and appearing to adhere to an archaic value structure. This argument, along with the whole concept of 'the prison' and the custodial function it essentially has to perform, raises the question as to whether any systems methodology can be successfully applied within the Prison Service. Could an organisation, even today so shrouded with secrecy, function efficiently as an open system?

It can be argued then, that because the Prison Service and its establishments have always been so highly structured in terms of authority, power and security, it could prove extremely difficult to introduce new managerial and organisational philosophies and methodologies. It seems relevant at this point to consider postmodern discourse as a means of trying to explain how a new systems methodology could be accepted into the Prison Service, and whether any semblance of current thinking on any aspect of the Service can exist in light of revelations about the postmodern condition.

POSTMODERN DISCOURSE AND THE PRISON SERVICE

Jean-Francois Lyotard in his book, The Postmodern Condition (1984), defines (what has come to be termed) postmodernism as 'the search for instabilities', and because postmodern society, and indeed organisations, are constantly changing, it seems sensible for any methodology to seek out these instabilities or structural inadequacies in order to initiate change. It is only through highlighting what is wrong within an organisation such as the Prison Service, that any improvements to the way establishments are run can be achieved; or at least the limitations of the system recognised. It is essential within the prison environment to constantly question all activity and to highlight any problems which are apparent. For example, one major instability within the prison system is that of secrecy and the failure to operate more fully in relation to the wider environment. If this, and other instabilities are constantly subjected to questioning, then maybe the prison system will begin to realise how to deal with them. If more decision making is devolved at establishment level, with all employees fully understanding the nature of the problems of the organisation, then maybe some new methodologies can be introduced to deal with them. It is essential that staff awareness of issues and instabilities is promoted within the organisation, if any new methodologies are to be accepted.

It may also be necessary for the Prison Service as a system to reverse its view of employees and the organisation. Instead of concentrating on how the *system* functions and how individuals can fit into it, perhaps it should be thought about how the *individuals* can make the system work. Instead of viewing the Prison Service and each establishment as a *totalizing* system, why not look at it more in terms of a *fragmented* system? Postmodernism, according to Cooper and Burrell (1988), argues that it is only through a rejection of the totalizing system or meta-narrative, that new methodologies can be developed. If a systems methodology is to be fully incorporated into the prison system, then the individuals within the organisation must understand and be prepared to question any instabilities; they must see themselves as being the people that make the system function, instead of believing that they are merely small parts of a highly authoritarian, totalizing organisation.

Of course, it can be argued that this notion of individuality over the totalizing system could be nothing more than a manipulative tool. It has been argued at great length, particularly in the work of Michel Foucault, that the notion of the individual is merely a construct; to believe in individuality gives credence to actions that may be dictated to by others. In his collected work, "Ecrits", Jaques Lacan (1977) argues that through the 'Mirror Stage' we exist in the reflection of others, as merely surfaces and that any individual action is reliant on the actions of others. The present structure and organisation of the Prison Service could, to an extent, be an illustration of this point. At the present time, the Prison Service is attempting to introduce more devolution into the system by handing more power over resources and decision making to the governors of individual establishments. On the surface, this concentration of power at a local level, seems to be moving a step away from the idea of the *totalizing system*, more towards that of a *fragmented system*, and as a result, ultimately towards a kind of 'postmodern systems thought'. However, although this perception can be made, it is necessary to look deeper into the reasoning behind the decision to devolve and to promote individuality within the system. Using Lacanian theory, it can be argued that the whole enlightenment notion of such a thing as objective reality, is merely a construct.

The whole concept of appearing to break down the totalizing system must be taken with a great deal of trepidation. Just like all other aspects of the organisation, postmodernism suggests that this change must be questioned and its instabilities exposed. Could this increased devolution and the suggestion of more freedom, merely be a manipulative tool by which the Home Office can shroud their actions in the rhetoric of freedom and individuality? Much of Postmodern thinking argues that contemporary society is a product of what Frederic Jameson (1991) terms 'The Cultural Logics of Late Capitalism'. If this is to be accepted, then the Prison Service along with individuality has entered into the third stage of capitalist production and has become a commodity to be exploited. It can only be a commodity if it is accepted as existing in reality. The prison service, invariably tied to the criminal justice system is, in postmodern terms, a meta-narrative, a totalizing system and therefore should be treated with caution and indeed questioned. It is only by taking a diachronic, linear historical view that the system is seen as existing and having existed with any authority and in 'reality'. By taking the diachronic view, the Service itself can never be wholly isolated from the nineteenth century, and more than that, can never exist in wholly contemporary terms. The 'modern' Prison Service exists more in hyper-reality, where any decision made rests on prior received information from a variety of conflicting sources. It is in Jean Baudrillard's (1988) terms, a simulacrum, a copy of something that has no origin. With the contemporary service being based on an ideal that never existed, it is a notion that is more a social, political and media construct than a physical reality.

The questioning of what constitutes abstract reality is a question that is far removed from the notion that the prison in its most basic form is a physical institution, and for those employed or incarcerated within its walls, it exists in no uncertain terms. The question remains as to how to reform a system which has its ideological basis in effectively nothing at all. The question still remains as to whether it is possible to do anything more than draw attention to the instabilities inherent in the system. Even when the Prison Service does appear to be introducing innovative programmes of change, these must still be questioned and any instabilities uncovered. It can be argued that the only way the notion of a new 'contemporary prison system' can be achieved, is to completely wipe out the existing system and start again. Obviously, this cannot be done, because the only option is to draw attention to problems within the present system and build upon this.

It is becoming clear that the Prison Service's desire to change and to do so using a systems management methodology, will be laden with problems. Existing attitudes amongst employees, and archaic value structures inherent within the present system highlight this. The diachronic view of the Prison Service as never being completely isolated from its history, tells us that it is impossible to change completely. Looking at the Service from a postmodern perspective, maybe it is time to accept that all they can do is highlight current instabilities, and attempt to put them right, to a small degree at least.

Poor communication within prison establishments seems to be a major cause of concern at the present time, and it is partly on this premise that a systems approach to management may have been suggested as a means of bringing about change. Implementing any new methodology within prison establishments however, is going to be fraught with difficulty. Postmodern thought stresses that instabilities should be highlighted, so before new management methodologies are implemented, problems with the actual implementation need to be brought to attention. It cannot be accepted that one system will automatically replace another while meeting no resistance.

CONCLUSION

Toward Postmodern Systems Thought?

There now arises the question as to whether a process of 'postmodern systems thought' could ever exist. The answer in short is no, because any complete school of thought can be seen as totalizing. Postmodernism is merely a way of looking at things, of explaining the contemporary and highlighting instabilities - it is not a totalizing school of thought. We can, however, move *towards* a process of postmodern systems thought. If it can be accepted that systems methodologies may only bring about *temporary* and *contested* improvements, and that constant questioning and the exposure of instabilities is a necessary feature of any managerial method, then postmodernism can play a part in systems thinking. Surely a systems methodology which is tempered with postmodern thinking can do nothing but good within organisations such as the Prison Service. An organisation which is so accountable to the government and the general public, may need to be constantly questioned; if prison establishments can question themselves and their own instabilities, then perhaps as well as running effectively, even public perceptions of the Service will improve. The notion that an essential part of the postmodern condition is the sense of no explanations for anything and no structure, may not necessarily be negative. Human nature and/or inherited modes of thought dictate that we live by rules and therefore give our lives and society some kind of structure; it is unlikely that this will ever change. Postmodern theory can be liberating as opposed to apocalyptic - instead of an end to order, it is an end to totalitarianism, to dogma and to regime. People still may be ideologically created animals, but at least we are aware of it.

REFERENCES

Baudrillard, J., 1988, "Selected Writings", Polity Press, Oxford.

Cooper, R., and Burrell, G., 1988, Modernism, postmodernism and organisational analysis: an introduction, *Organisation Studies*. Vol.9(1), p91-112.

Home Office., 1993, "Doing Time or Using Time: Report of a Review", HMSO, Cm2128, London.

Jameson, F., 1991, "Postmodernism, Or, The Cultural Logic of Late Capitalism", Verso, London.

Lacan, J., 1977, (translated by Sheridan, A.), "Ecrits: A Selection", Tavistock, London.

Lyotard, J-F., 1984, "The Postmodern Condition: A Report on Knowledge", Manchester University Press, Manchester.

17

MERGING THE WORLD VIEWS OF SYSTEMS SCIENCE AND HUMAN VALUES

The Organizational Perspective

Martin L.W. Hall*

School of Computing and Information Systems
University of Humberside
Cottingham Road
Hull, HU6 7RT
United Kingdom

INTRODUCTION

It could be argued that the concept of 'world view' is one that is integral to systems science. In other words, it can be understood that the observer's perspective on the world affects how it is viewed or perceived. It may also be argued that human values help to shape the observer's perspective. Values theories such as those proposed by Graves (1974), Motloch & Woodfin (1993) and Hall (1986, 1993) introduce the idea of world views into their developmental investigations. Jackson (1991) and others have identified (through the use of metaphors) the world views of various systems methodologies.

By merging or integrating the world views of methodologies in systems science with the world views of values theories, one can gain a holistic organizational perspective that merges all levels of a system from the individual view through to that of the organizational (or systemic) view. This paper will attempt to demonstrate and outline a potential outcome of this merger which may become a practical, yet powerful organizational intervention methodology with further research. The linkage between systems and values will be critically examined in an effort to show how intervention in organizational problem situations may be more effectively guided.

Ever since the importance of observer bias has moved into the discussion of systems science, the concept of world view has played an important role. It is a particularly important concept that Checkland (1981) and others have used with Soft Systems Methodology (SSM) to explain the perspective of the stakeholders in the system. Schein (1992) has used the concept of mindsets and culture which relate to how one perceives the world (or in the case of organizational culture, is a context for decision making), and Morgan (1986) has used the

* E-Mail: martin@netcom.com.

Critical Issues in Systems Theory and Practice, Edited by K. Ellis et al.
Plenum Press, New York, 1995

concept of metaphors to describe organizations. Jackson (1991) has extended this idea by using the concept of metaphors to relate systems methodologies to problem context.

DEFINING AND PLACING WORLD VIEWS IN CONTEXT

Human Values and World View

World view is also important in the study of values, in that values help usform our world view. World view gives the context of showing how a person is going to make a decision around core values. In the values field, world view is the perception of the world as built by the values of the individual.

World view, or as Checkland (1981) calls it, *Weltanschauung*, carries with it the idea that an individual's (or group's) decisions will be far from random. There is a context for which decisions (or at least an individuals interpretations of the world) are understood. This author discusses how values are the filters through which we make decisions (Ellis & Hall, 1994). The concept of world view is formed by a clustering of values along a developmental scale of behaviour. This is backed by research in the area of human values (Motloch & Woodfin, 1993; Hall, 1986,1993).

Values cluster in a manner in which creates a core set of values. This permits individuals or organizations to understand the world view they are operating within (Hall, 1993; Motloch and Woodfin, 1993). Schein (1992) would call this culture. Core values will help determine the decision-making and world view of the organization (Senge, 1991). Core values of individuals lead to core values for the organization (Rokeach, 1979; Hall, 1986).

Values lead to a descriptive idea of what the organization's decision making culture is (Schein, 1992). Once the values of the organization have been identified, then the organizational culture can be understood. It is these clusters of values in a core set that form the totality of which we call 'world view'. In other words, values help form our world view. By establishing this we are closer to connecting values and systems thinking.

Systems Thinking and World View

Jackson (1991) borrowing from Morgan (1986) discusses organizations in terms of metaphors. Both Jackson and Morgan discuss metaphors in terms of descriptions for viewing the world, or at least a context for doing so. This further strengthens the discussion of world views in the context of systems thinking.

Through the concept of metaphors (Morgan, 1986; Jackson, 1991) a connection between metaphor and world views can be made. From this a strong connection can be made for matching the world views of values with that of systems science. Jackson's use of metaphors helps to look at various systems methodologies in terms of certain world views.

Metaphors are one way in which humans approach a problem; putting the problem in a describable context. World views, on the other hand, are more encompassing. World view attempts to be about all that one perceives about the world. Metaphors are situational in that the perception is about a specific problem or issue. World view is a general description of the context of a person's perception (within a particular cultural framework). These are related concepts. They deal with context for which one perceives reality. Culture, world view, mindsets, metaphors, and values are related concepts (in descending order from most general to the most fundamental) (Schein, 1992).

The ability to solve problems relies on the ability to reliably create a model of the problem (Robertshaw, Mecca & Rerick, 1978); and then selecting a solution. World views create this context. This is where the power of systems methodologies comes from. Systems methodologies, particularly Soft Systems Methodology, by trying to look at issues holistically, use world view as a context for interpreting what goes on in the world. The metaphors that are often used to relate systems methodologies to problem situations are simply an attempt to put a description on this. Metaphors are therefore integrally related to the concept of world views.

Values and world view are connected quite solidly in the literature on human values (Hall, 1986). But, maybe more importantly, values also connect to the metaphors through the concept of 'storytelling' (Hall, 1993), which is a way of framing the meaning of a cluster of values. Morgan (1986) calls metaphors: "frames of meaning". Thus, an even stronger connection between human values and systems is made.

LINKING SYSTEMS THINKING AND VALUES THROUGH WORLD VIEWS

Why make connections between systems and values? In a previous paper (Ellis & Hall, 1994), thIS author discusses how decision making can be a connection between human values and systems methodologies. World views are connected in a very similar, yet potentially more powerful manner. Decision making created the connection between values and systems methods because both values and systems use decision making. The use of world views allows for a connection to be made to decision making because it is the context for the decision making environment. If the decision making environment of the individual (and the organization) can be linked to specific problem solving methodologies through the same, or similar, world views, a more efficient process can be created.

Both (systems and values) are either, affected by, or affect decision making. Decision-making can be used to tie values to specific systems methodologies. Jackson uses a modified version of Burrell and Morgan's matrix of social paradigms to critique and judge systems methodologies (Burrell and Morgan, 1979). Burrell and Morgan (1979), and Hall (1986) (with Tonna) were considerably influenced by Parsons (see Parsons & Shils, 1951), using his (Parsons') basic theory relative to how to view the world (world view/metaphors). Decision making lays the foundational linkage from which the concept of world view works. The systems science concept of world view and the values concept of world view has this as a core linkage. It is this linkage which gives them another strong, foundational connection.

The idea that the decision-maker (or observer of a system) has their own agenda or bias is not new. This is, of course, a notion that is common to the whole idea of observer-dependency (bias) in systems models (Robertshaw, Mecca, & Rerick, 1978). What is new is that there is a way to measure this bias, and make reasonably scientific judgments about how these biases (values) will impact the system or decisions being made in the system.

The connection is straightforward. World view provides the context for decision making. Values help provide the world view. Systems methodologies exist within these world views.

If the decision-making process is corrupted or clouded by something, it can adversely affect the smooth execution of a given systems methodology. Clarification of values helps to improve the clarity of decision making within an organization. Understanding decision-making or problem-solving is an underlying element to the sucess of many systems methodologies. World view provides this context and connection.

Values can be tied to individuals and organizations. Individuals make decisions in organizations that have a culture that supports those values (Schein, 1992). In essence, the organizations take on the values of the individuals of the organization. A clustering of these individual values allows an organizational world view to be identified.

Flood and Jackson with Total Systems Intervention (TSI) (1991) have tried to show us a method for viewing organizational behavior and then linking it to an appropriate systems methodology. If a linking can be made at the point of compatible or congruent world views then the methodology of selection becomes simpler and more accurate if quantifiable measurement is used (Hall, 1993; Ellis & Hall, 1994). This allows the power and practicality of a methodology to show through because it potentially reduces work and error for the interventionist by using quantifiable data to help select an appropriate intervention methodology, rather than relying on the world view of the interventionist. Therefore it is possible to use measurable data from organization (such as values data) and defined information on systems structure to determine best tool for helping the organization. This presupposes that values can be measured (Ellis & Hall, 1994).

Values link decisions to individuals in the organization. From this, the values and decision-making style of the organization can be measured (Hall, 1986). The process flows from the values of the individuals to the values of the organization to the decision making style of the organization. It is at the level of identification of the decision making style where the world view also is important.

The connection between world view and systems therefore becomes more evident. If world view and problem-solving are key elements of systems methodologies, and values are key to people's decision-making and world view, then it is possible to hypothesize that there is a direct connection between systems methodologies and human values at the point of world view. It then follows that if one understands the world view they are working from, one would get a better understanding of how a given systems methodology might work in a given situation.

World view (through values) and specific skills (actualized with systems methodologies) are key to the human element of organizations (Rokeach, 1979; Colins and Chippendale, 1991), if one can organize these elements effectively then this will be the first step in a useful, practical methodology. Connecting the organization style with its decision-making style can help the organization to become more efficient reducing the number of conflicts due to unclear organizational direction.

CONCLUSION

Why is this merging of world views a critical issue for systems science? First of all, it creates a stronger link between systems and values. This is important because this helps to link the individuals in the organization to overall goals and purpose of the organization. Second, it adds a stronger connection to the human relations element of organizations (through the use of values). Systems methodologies have historically been weak with respect to the human relations element in organizations (Ellis & Hall, 1994). This can create conflict and chaos in the organization if the individuals and organization as a whole are not working together congruently with respect to goals.

Using information on world view with systems intervention methodologies can potentially determine which of the core systems methodologies should be used for a given organization at a given point in time. Values theory can help organizational consultants to look at systems methodologies in terms of certain decision-making cultures and world views.

Values measurement can help identify systems methodologies through the connection of world view. Systems methodologies can aid in measurement of organizational values and actualization of the identified values. In effect, values and systems can work together in a type of recursive rhythm that complements each others strengths.

Using measurable data from an organization (values information), the best tool for helping the organization can be determined. Decision-making (or leadership) style and metaphors/world views can help determine the appropriate core system methodology to be used in a given situation (Jackson, 1991).

Systems and values can use the organizational strengths of each other to diminish each others weaknesses (Senge, 1991). Together they can become a methodological approach that is useful in virtually any situation with virtually any organizational system. Systems and values together can become useful for practical organizational interventions.

Values measurement can allow one to create a quantifiable description of the world view. Measurable values effectively translate to measureable world view. This is the basis of a practical methodology for selecting systems methdologies. And consequently is the cornerstone for a practical systems intervention methodology for organizations.

This paper is part of on-going research in the area of systems and values in organizations.

REFERENCES

Burrell, G., and Morgan, G., 1979, "Sociological Paradigms and Organisational Analysis," Heinemann, London.

Checkland, P.B., 1981, "Systems Thinking, Systems Practice," John Wiley & Sons, Chichester.

Colins, C.J., and Chippendale, P.J., 1991, "New Wisdom: The Nature of Social Reality," Acorn Publications, Sunnybank Hills, Australia.

Ellis, R.K., and Hall, M.L.W., 1994, Systems and values: an approach for practical organizational intervention, *in*: ISSS 94 Conference Proceedings, ISSS, Louisville, KY.

Flood, R. L., and Jackson, M. C., 1991, "Creative Problem Solving: Total Systems Intervention," John Wiley & Sons, Chichester.

Graves, C.W., 1974, Human nature prepares for a momentous leap, *The Futurist*, 8 : 72-87pp.

Hall, B.P., 1986, "The Genesis Effect: Personal and Organizational Transformations," Paulist Press, New York.

Hall, B.P., 1993, "Values Power", Values Technology, Santa Cruz.

Jackson, M. C., 1991, "Systems Methodology for the Management Sciences," Plenum, New York.

Morgan, G., 1986, "Images of Organization", Sage Publications, Newbury Park, CA.

Motloch, J.L., and Woodfin, T., 1993, General systems theory, cultural change, and a human science foundation for planning and design, *Systems Research,* 10, 2 : 1-16.

Parsons, T., and Shils, E.A., Eds., 1951, "Toward a General Theory of Action: Theoretical Foundations for the Social Sciences", Harper Torchbooks, New York.

Robertshaw, J. E. , Mecca, S.J., and Rerick, M.N., 1978, "Problem Solving: A Systems Approach," Petrocelli Books, New York.

Rokeach, M., 1979, "Understanding Human Values: Individual and Societal," Free Press, New York.

Schein, E.H. , 1992, "Organizational Culture and Leadership," Jossey-Bass, San Francisco.

Senge, P.M., 1991, "The Fifth Discipline: The Art & Practice of The Learning Organization," Doubleday Currency, New York.

18

COMMUNICATION IN DECISION SUPPORT

Prospects and Limitations

H. A. Higgo

University of Humberside
School of Computing and Information Systems
Cottingham Road
Hull HU6 7RT
United Kingdom

INTRODUCTION

The complexity usually embedded with understanding organisations and their decision making processes, necessitates a theoretical framework where knowledge can be represented and communicated. It is the systems thinking approach as a way of intervening to improve problem situations, which is equipped to embark itself in this trans-disciplinary type of study. This paper is enshrined upon communication as the basic pre-requisite in dealing with decision problems in organisational contexts. Getting people together to negotiate their differences, in a problem context, necessitates the existence of a facilitator equipped with his or her skills and qualifications, to influence the situation and to understand the initial problem language or the access language depending on the context of the problem. Understanding of such a language may facilitate developing an extended language which may contribute and enhance the problem solution. This paper is meant to contribute to the paradigm shift in the process of support needed and the potential role of the computer. It is also meant to emphasize the possibility of embedding a computer based-system in this negotiation process, to facilitate an extended language and to increase the communicative competence between the participants.

NEGOTIATION AND CONFLICTING GOALS

The importance of communication embraces construction of a mechanism or a structural framework whereby negotiation processes can be undertaken. Introduction and institutionalisation of these processes may be important in discussing and bargaining differences in organisational contexts. The literature on communication and negotiation tells us about situations where it is undesirable to act in a certain way; something which may jeopardise the negotiation process (Raifa, 1982). A case in point is when the mediator, whose existence is often of vital importance, sometimes may introduce a scenario which is worse

Critical Issues in Systems Theory and Practice, Edited by K. Ellis et al.
Plenum Press, New York, 1995

than the worse case scenario. Humphreys and Wisudaha, (1989) show us how to be cautious in cases where special techniques are needed to convince the negotiators to play the game. They write:

> "extending the terms of reference of what can be considered in a social decision making situation to admit material drawn from an enlarged "Small world" encompassing the small worlds that each participant would like to explore may well be an unrealistic solution to the problem of handling differences between problem owners with different interest in the decision. This would undermine the background of safety which participants need if they are to negotiate the knowledge structure within which all the various participants' judgements about options and consequences may be represented." (Humphreys & Wisudaha, 1989)

Nevertheless, the articulation of the mechanism through which these processes can be carried out must be sought. Communication then, in this manner, is the basic pre-requisite for interaction, co-ordination and action (Habermas, 1987). Embedding a computer-based decision support system (DSS) with structuring capabilities, e.g. Multi-attribute Utility Structure, to direct the negotiation process may be possible. The emphasis is upon influencing the individuals' ideas and perspectives. At this stage, the automation may facilitate the articulation of the extended language which may bring consensus (Phillips, 1986) around the ideas and render a shared vision of action. Using the computer in this sense implies a philosophical shift from its traditional utilisation; where it dictates the language and the solution scope. It is here where the computer can increase the communicative competence which is usually necessary in the negotiation scope.

Potential Advantages of Communication

- Communication in a wide scale can foster genuine democratic behaviour through emancipation of intrinsic oppressed views. This may lead to a new atmosphere respecting others ideas and behaviours. This is particularly important in situations where freedom and justice are the missing rigours.
- It may also foster and emphasize attention to more complex situations where more thorough and knowledgeable communication is needed. Thus it may provide a new culture in contexts and domains where communicative competence is needed.
- The importance of communication can be perceived in its role as a learning process in evolutionary context. This occurs as a result of transference of knowledge and skills and ultimately through exchange of ideas, messages, and experiences as well as behaviours.
- Communicating in such a way to share knowledge and experience is a notion which is supported by advance information technology. Securing such communication can support future complicated implementations; i.e. to be used as one of the important ingredients of pre-requisite structure when support is needed from the vast technological base.

The complexity of life and the multiplicity of practices needed in various situations, makes it inevitable to face a range of complex decision problems. Intuitively, we need to deal with them, try to overcome their difficulties; and even to come to terms with their outcomes. This means that not only our education i.e. skills, qualifications and experiences, but also our organisational structures are the basic resources to both individuals and groups necessary to deal with survival problems in one form or another and ultimately to govern our rationality. This means whatever complicated real life situations are, we have to live with them and try our best to explore and invent the most appropriate mechanisms to organise our action and behaviour. This is of particular relevance to strategic decision problems where representation and communication of knowledge and experiences is vital.

DECISION PROBLEMS AND SUPPORT

Generally speaking, any change process in one way or another may be associated with a decision of some kind. A change may be needed in a modernisation-driven situation; or adaptation or even survival in turbulent environments. This may implicate that, any interaction in real life contexts, to bring about change involves an action situation. However, it is the complexity and diversity of decision problems coupled with our cognitive limitations which bring the matter forward to seek support. It is the limited information processing capacity which lead decision makers and problem solvers to relegate their desires and aspirations and of course their actions, to fast information processing machines with unprecedented power of calculations. The basic flaw here is that, human cognitive processes cannot be encaptured and matched by machine processes. Need arises where the computer can be used to extend knowledge rather than distort knowledge. Many critiques chased the use of artificial intelligence and expert systems which nevertheless used to address readily packaged type of problems i.e. well structured type of problems (Paprika, et al., 1985; Humphreys, 1989; and Hawgood et. al, 1987). This indicates, the existence of a gap in trying to find a viable procedure in this support with regard to the potential role of the computer. While the need for support is pretty obvious, however, still, how, where and when to provide support alludes us. The answer to this question entails dealing with representation and communication of knowledge in the social context; which is beyond the scope of this paper. Nevertheless, does the support necessarily be an automated one? and if that is needed, then what is the rational behind it?

Decision Support and Uncertainty

Decision makers, in their search for support, usually suffice rather than maximize (Jannis and Mann, 1977). This follows Simon's revolutionary ideas of bounded rationality (Simon, 1972). In doing so, they also do not want to leave a great margin of error to chance. However, it is inevitable also to confront greater complexities with vested ambiguities and uncertainties. In such situations we must struggle to communicate our knowledge and discuss our mutual/conflicting interests to bring about successful change which does not come cheap. However, given our cognitive limitations coupled with our desire to achieve the best possible outcome, necessitates support. Nevertheless, support via negotiation process can sometimes be time consuming and can also be fraught with irreconcilable goals, interests and perspectives. In such situations, additional, somewhat, objective resource, is needed.

Such help may be provided by information technology in the following sense. When decisional structures support communication, this implies a restriction of variety in what is communicated. Unpredictablilty increases this variety. Information technology may help to anticipate such an increase by simulation. This makes it possible to explore what additional decisions to make, and thus what to communicate to other decision makers. This makes for more competent forms of communication.

The importance of organisational decision support is implicated in the design of its structure (Jackson, 1992); as a means of rationality (Simon, 1986). Information systems utilisation to handle information processing activities, in general, enhances interaction between the structure and process in organisational contexts. Something which illustrates the integration of technology to support human decision processes at various levels in the organisational hierarchy. This is, of course, a pattern of communicating knowledge and experience in social contexts. This may also indicate that the information technology when it is used in the appropriate manner can extend the communicative performance within and between groups; and hence enhance the capabilities to achieve better and undeniable results.

This is in direct contrast, when it is imposed to do jobs which alienate the individuals, by betraying the initial problem language, rather than increase the communicative competence between them.

The effort in supporting organisational decision making processes, which exit in a social reality environment, has culminated in the need for articulation of a soft language which may encompass the subjectivities and subtleties of the situation. The search for this language, depends upon the individuals engaged in the negotiation process on one hand, and the problem and its context on the other. The communication between these ingredients determines the scope of a successfully and agreed upon decision; or otherwise power may operate. In certain contexts, a facilitator, possibly from within the organisation, can provide support to extend the language and foster consensus around their ideas and aspirations (de Zeeuw, 1992).

In organisational research literature, we observe the coin between successful implementation of strategic decisions and organisational survivability or otherwise transformability. These problems need articulation of a problem language and increase in communicative competence.

LIMITATIONS AND IMPORTANT NOTIONS

- Decision theory in its current stage of development, does not provide us with a way of dealing with side-effects; something which is vital for any study which needs to emphasize the behaviour and implications of a social context. Moreover, it does not tackle the socio-cultural aspects; which is necessary in the process of representing and communicating knowledge about organisational phenomena. Solution to this dilemma may be sought by representing and communicating knowledge about organisational systems in a comprehensive yet recursive way. With clear attention to the socio-cultural and behaviourial aspects we may police coherence between pre-and-post decision processes.
- Decision Support in its automated form is fraught with grave problems. A lot of authors emphasise the need for a paradigm shift to combat the 'rallying cry' associated with respect to the methodology as well as approach for developing DSS. The need for shift is particularly needed to provide a subjective language to substitute the objective one which contrasts the original problem language i.e. the soft language. This objective language constrains rather than enhances the problem solution in real life problem situations.
- It is often that strategic decision problems do not have clear answers; i.e. the scope and the nature of the solution depends upon the nature of the problem and its domain; and the values, goals and interests of those concerned as well. This illustrates that problem solutions are context-dependent; i.e. decision solutions can not be transferred among contexts.
- Decision problems are usually inherently concealed with conflicts and clashes. Rationalisation in such circumstances could be an extremely difficult job.
- Sometimes reconciliation between ideas, goals, views and perception will be hard to achieve. A creative approach may be necessary to motivate people and to increase communicative performance. Sometimes also people may find it difficult to communicate as a result of their conflicting goals. For example, an actor in a certain role in a certain context, may also be influenced by various other roles. This makes his or her actions coloured by virtue of satisfying other goals which might not be related to the contextual issue at stake.

- Sometimes the problem domain can lead to other adverse problems, making the original problem very complex to solve. For instance, in the Sudan, the turbulence in the economic environment makes it difficult to use statistical information - if it is available - as it is not enjoying the relative stability information used to have in other societies. This by definition makes it difficult, particularly for strategic decisions with broader time horizon.
- The automated decision support with modelling capabilities, can introduce a germ of determinism in structuring decision problems (Angell & Smithon, 1991). By doing that, it introduces a problem of its own creation which may obscure our understanding of the problem's uncertainties. This renders it difficult to simulate social reality problem situations to receive controlled expectations.
- The solution of a problem through negotiation process usually needs additional resources; i.e. an analyst to drive the wagon. As the qualification and the skill of the analyst varies from one specialist to another; something which may be needed in carrying out a successful strategy in politically-sensitive situations.
- Implementation of complex decisions, which always affects various interest groups, is usually confronted with conflicting loyalties, resistance to change and resentment. Overcoming such problems in real life problem situations without resorting to power and influence is very difficult. This may get in conflict with the idea of increasing communicative competence in situations where there are losers and gainers.
- Often real life organisational structures mask a completely different structure (Jaques, 1989). In this situation, power and influence may distort communicative competence. Handy, (1993), calls this a power culture.
- Decision analysis tools and techniques which are useful in various contexts, probably because they match our hierarchical organisational structures, nevertheless - given their manipulative power - they may empower the decision maker more and more; something which may constrain the negotiation process and hence distort communicative competence.
- In real life organisational problem situations, there are usually problems owners, sub-owners, proposers, experts and consultants. The engagement of these people in the decision making process, renders the rationality, which governs the decision, to be organisationally-driven; however, the reconciliation of such views is something difficult in non-unitary type of organisations. This is why shared assumptions are highly valued in organisational success (Jackson, 1992).

CONCLUSION

This paper emphasized the decision making process in social contexts as a process of negotiation in which communication of knowledge and experience is the core of this interaction. Negotiating decision problems in such a manner, necessitates the emergence of a language which may be developed by the help of a facilitator or an analyst; from within or outside the organisation. In this context, what is necessary then is the communicative competence between the negotiators. It is important to note that I have defined communicative competence as the situation whereby different stakeholders, e.g. groups of people or interested parties or even committee members, find it difficult to proceed in negotiating their differences; and when power and influence is likely to dominate. The use of a computer-based system in this process, to provide an extended language and to increase the communicative competence is highly appreciated. The application of the computer in such a manner indicates a philosophical shift from its traditional role as a problem solver. This may illustrate clear implication on the methodological and design aspects for such systems. In this respect, it is important to emphasize the role played by the information technology when it is used to enhance and extend human contextual knowledge and interaction and ultimately to lead to co-ordination and action. This is in direct contrast when it is used to constrain rather than enhance communication in social contexts.

REFERENCES

Angell, I., and Smithson, S.,1991, "Information Systems Management: Opportunities and Risks"; Macmillan, London.

de Zeeuw, G., 1992, Soft language accumulation and the use of competence, *Systems Practice*, 5, 192-223.

Habermas, J., 1987, "The Theory of Communicative Actions: The Critique of Functionalist Reason", Volume II, Polity Press, Cambridge.

Handy, C., 1993, "Understanding Organisations" (4th edition), Penguin, London.

Hawgood, J., and Humphreys, P., 1987, "Effective Decision Support Systems", Technical Press, London.

Humphreys, P.C., 1989, Intelligence in decision support: a process model, *in*: "Knowledge Based Management Support Systems", G.I. Doukidis, F. Land, and E. Miller (eds.) Ellis Horwood, Chichester.

Humphreys, P.C., and Wisudaha, A., "Techniques and Tools providing Strategic Support: A Framework, Review and Guidelines". Decision Analysis Unit Technical Report 89-1, LSE, 89, London.

Jackson, M.C., 1992, "Systems Methodology for the Management Sciences", Plenum, New York.

Jannis, I.L., and Mann, L., 1977, "Decision Making: A Psychological Analysis of Conflict, Choice, and Commitment", Free Press, New York.

Jaques, E., 1989, "Requisite Organisation: The CEO's Guide", Casson Hall, Arlington, Virginia.

Paprika, Z., and Kiss, I., 1985, Interactions in decision support systems: division of labour in DSSs, *Engineering Costs and Production Economics*, 8, 281 - 289.

Phillips, L.D, 1986, Conferencing to consensus, *Datamation,* April, 26-29.

Raifa, H., 1982, "Communication Between Negotiators and the Role of the Mediator", Addison Wesley, Reading, Mass.

Simon, H.A., 1972, Theories of bounded rationality, *in*: "Decision and Organisation", C.B. Mcguire and R. Rander, ed., Amsterdam, Holland.

Simon, H.A., 1986, Introduction, *in*: "Empirical Research on Organisational Decision Making", E.Witte and J.H.Zimmerman, eds., Amsterdam, Holland.

19

INFORMATION SYSTEM DEVELOPMENT

Selection of Development Methodologies

M. E. James

University of Humberside
School of Computing and Information Systems
Hull HU6 7RT
United Kingdom

1. INTRODUCTION

My objective is, 'to research the selection of information system development methodologies in order improve information systems.'

This paper is a statement of the need I perceive for my research. In it I attempt to explain the need for a meta method to be implemented at the beginning of the process of developing an information system. This meta method will have three uses. The first is as a basis for the understanding of an organisation, its need for and use of information systems. The second is as a tool for the selection of an appropriate system development method. The third is as a tool for the understanding of development methodologies, and as an aid for research into the nature of the numerous methodologies available.

This paper will consider the relationship between an information system development method and the resultant system. It also looks at the need for a meta-method to aid the selection of the method used to structure the development of an information system.

Connor (1985) says, "If the science of information system specification and design were to be compared to a large country, it should be taken for granted that visitors could obtain and use a road map which would enable them to plot a route across the country by road and by following this route, reach their destination. ... It is truly unfortunate that the science of information system specification, design, and development has no such road map." Connor suggests that there is no existing method for the selection of the correct method for, or route to, the development of an information system. I propose a need for this map, and an approach for its development.

2. A RELATIONSHIP BETWEEN DEVELOPMENT METHOD AND RESULTANT SYSTEM?

My research is based upon three initial assumptions.

Critical Issues in Systems Theory and Practice, Edited by K. Ellis et al.
Plenum Press, New York, 1995

1. All information system development methods are based upon 'world views'. They are therefore based upon a unique set of assumptions about what information systems should be.
2. Assumptions made at the beginning of the development of an information system will affect the structure and use of the resultant information system and organisation.
3. Different organisations use information in different ways. Therefore information systems should be specific to each organisation.

In order to support these assumptions I will use the example of an engineering company in Yorkshire. This company has traditionally sent out representatives to seek tender enquiries all over the world. When an enquiry is made technical staff develop a process unique to that enquiry that is designed to a level of detail where a tender can be made for the contract. This process is resource and time intensive.

A couple of years ago a new managing director was appointed. He decided to change the way that this company operates. His dream is that tenders will be prepared for enquiries by assembling processes from a set of standard design components. The main advantages of this way of operating are reductions in enquiry response time, and a reduction in resource requirements.

As part of the implementation of this change one division within this company has implemented a project to develop an engineering information system. This system will enable the rapid compilation of tenders using standard components.

2.1. Methodologies and World Views

Methlie (1979) suggests that there are two perspectives (world views) upon which information system development methods are based. These he called a datalogical perspective and an infological perspective. The development approaches based upon the datalogical perspective regards the existing data flows as satisfactory. Therefore they commonly simply computerise the existing processes. A development method based on the infological perspective addresses information and its use within the organisation. It therefore tends to attempt the development of changed processes.

Thus we have two different world views, transfer information systems from one medium to another or develop altered information systems. These are two perspectives or world views on the development of information systems, and will lead to distinctively different methods.

Using the engineering firm outlined above, using a datalogical approach will result in a system to support the way the organisation currently operates, whereas the company has actually taken an infological approach by redesigning the way it operates. Two different perspectives will lead to two different information system development methods. So we see that the world view of the method developer has an effect on the method itself.

2.2. Assumptions and Organisations

Senior managers in the company above find it difficult to appreciate that their assumptions will change the way the organisation operates. If the infological approach is followed to the final implementation of the engineering information system it may leave the organisation in a difficult operational position. There are a number of competitors who operate by offering 'off the peg' processes and they have a number of years lead over this company in this type of business. However undertaking work for customers who require made to measure, 'bespoke', processes with unusual requirements will be either extremely difficult, or impossible after this system is implemented. Therefore the engineering information system developed from the infological perspective, along with the assumptions that this is the best way to operate, will allow the company to work in only one way.

Assumptions made at the beginning of the development process will affect the resulting information system, and consequently the way the organisation operates.

2.3. Organisations and Information

As I argue above, information systems affect the way organisations operate. It therefore follows that since organisations all operate in different ways they should all have significantly different information systems.

In the example above if the engineering information system were developed from a datalogical perspective the resulting system would have supported the bespoke approach to process development. This, however, would have made it difficult for the company to compete with the response times of companies that assemble processes from pre designed components.

It is true therefore that the information system should be specific to the organisation.

3. INFORMATION SYSTEM DEVELOPMENT

I have not yet shown that a meta method is required. Even if the assumptions above are correct, is there a method that aids the understanding of information systems development methods by analyzing the basic assumptions implicit in them? In addition is there a method that aids the selection of a suitable method for a specific situation? If these methods exist are they widely applied? In this section I will attempt to answer these questions.

3.1. Practice

A survey of UK companies undertaken by KPMG Peat Marwick McLintock (KPMG, 1990) found that "30% of the UK's biggest computer projects were massively over budget, overtime, and if ever completed, failed to do the job they were meant to". Additionally Willcocks (1992) states that "By 1992 UK Expenditure on IT was exceeding £10 billion per year", this figure excludes public sector spending. These figures together suggest that there is a need for a meta method if only to reduce some of this approximately £3 billion waste.

Examples of this waste due to failure in development are numerous, I will highlight two.

The first, the TAURUS project, is well documented in the press, after having read a small amount of the available material I will summarize my understanding. The project became bogged down in power and organizational politics, eventually becoming so complex and difficult to manage, with regular changes in functionality and no final specification that it tripped over itself and was abandoned. I do understand that a lot of the conceptual work undertaken during the TAURUS project has been utilized in the system now used, but by basic criteria it is a failure.

The second is a downsizing exercise documented and analysed by Mansell (1993). The original specification for the project was the transposition of the mainframe system to a PC network with no change in functionality. The method chosen for the transfer was SSADM. This is an infological system development method that should be used for the development of new information systems. It includes analysis of user requirements, when the requirement, already specified, was the downsizing, from mainframe to PC based network, with no change in functionality. This contradiction between infological approach that raised user expectations and the required datalogical approach, where users should not have been asked for their input, caused conflict. The conflict caused the prolonging of the project which resulted in increases in size and cost to such a level where Mansell (1993) calls it a failure.

Are these examples unusual? Do managers normally select information system development methodologies by taking into account the requirements of the firm, and matching these to a compatible method? A CFM Group report (1994) summarizes research into executive decision making criteria for outsourcing. Findings for IT functions are that the top three perceived benefits of IT outsourcing are; predictability of costs, reduction of costs, and flexibility of resources. These were the three least important considerations in outsourcing of non IT functions such as legal and accounting roles. This suggests that decisions in the IT sectors, unlike decisions made in many other critical areas, are predominantly based upon cost. When selecting a contractor, or consultant to manage or develop an organizations information system, a decision based mainly upon cost, giving a low priority to development method, will, I believe, be more likely to result in an unsuitable system.

3.2. Theory

In this section I address two works on the selection or categorisation of development methodologies. Both methods are related to the development of computer based systems. But it is my belief that lessons learnt here may be extrapolated to the development of a total information system, whether based on computers, paperwork, human interaction, or a combination of these or other information communication or storage media. In order to keep this paper within the specified length I have simply skimmed the work of Connor (1985) and Olle(et. al.) (1991) in order to highlight what I believe to be their strengths and weaknesses.

3.2.1. A Road Map. Connor (1985) suggests that system developers need an understanding of different types of development method in order to be able to select the correct route between an awareness of the need for a designed information system and that systems implementation. Connor then goes on to describe a number of different methodologies and their theoretical application to the same problem situation. Although the descriptions are concise, and the applications provide insight, there is no attempt to define the methodologies in terms that allow comparison between each one. Selection is still left to the system designer making his selection by comparing each with his needs. Equally important is a recognition that Connor writes his book for system designers, and expects them to choose impartially. In practice humans do not make impartial judgements, I believe that it is more likely that selection of a development method will be based upon the question, 'Which method do I know best?' and 'Can I make this familiar method fit the problem situation?'.

3.2.2. A Framework For Understanding. It has been suggested "There are still many questions on the topic of information systems methodologies, the more significant of which include:

1. Are there really so many substantially different ways to design a computerised information system.?
2. If not, how are these ways similar? If so how are they different?
3. Should we be using one method for one kind of information system and a different method for another?
4. If so, which method should we use for which kind of system?" (Olle, et. al., 1992:2)

Olle et. al. (1992) then goes on to address a number of points related to these questions throughout the book. They do not however explicitly answer any of them, and it is left mainly to the reader to derive answers from the information presented.

3.3. What Is Presently Available

The section above, practice, seems to suggest that methods that aid selection of information system development methods in relation to the organisation or need are not commonly used. In addition although a number of authors have attempted to aid system developers in their search for the best method for a particular application. Most have simply taken what they believe to be a representative group, and have described their application and in some cases their strengths and weaknesses. I have not yet found an attempt to classify methodologies by either their strengths and weaknesses, or by the assumptions about the world that they are based upon in order to allow consistent selection and to minimise the subjectivity involved in that selection.

4. THE BEGINNING

I believe that a meta method is necessary in order to aid the consistent development of good information systems. If method selection decisions are based upon criteria specific to the chooser's perception then they are unlikely to take into account the 'world view' of the method developer They will therefore be unlikely to match this 'world view' to the organisation's need. The result of development based on choices made in this way is likely to be a system that poorly matches the organisations need.

It is true that a number of individuals and groups do develop information systems applying tools and methodologies to specific needs. I believe that these choices are based upon cumulative experience, and individual skill. There are two main problems in relying on these people or groups. The first is that their choice is prejudiced by their own personal perception. So a failure due to inappropriate application in the past may rule out a specific tool in a situation where it's application would be valuable. The second is that their choice is based upon a set of random criteria which may not even be understood by the chooser. This makes the knowledge volatile, you will lose it if you lose its human origin.

Returning to Olle (et. al.) (1991) questions one and three respectively. Because each information system development method is based upon a unique world view they are likely to be significantly different. If this is true then we need to match these world views to the position and needs of the organisation. I therefore need a method for the selection and understanding of methods.

4.1. A Meta Method

I believe that the answer to Olle (et. al.) (1992) question two will lead me to one dimension of a tool within this selection methodology. This will allow me to analyse and define development methods based upon assumptions made about the nature of information systems.

This criteria on its own is useful if all organisations are similar, unfortunately this is not true. I now need to answer the fourth question. I therefore need to analyse and define each method based upon assumptions made about the nature of organisations. Presently it is my belief that this may be achieved by using Jackson's (1994) three sets of assumptions about relationships within organisations, defining them as unitary, pluralist, or coercive.

I therefore have two criteria for analysis and selection, an as yet unchosen set based upon assumptions made about information systems, and a tentative set of assumptions about the nature of the organisation.

This however is only the beginning of a tool for the selection or analysis of information system development methodologies. I need to develop methods for using this proposed tool in order to make it useful and to fulfil my objective. "To research the selection of information system development methodologies in order improve information systems".

REFERENCES

CFM Group Limited., 1994, "Outsourcing: Fitting into the business equation", CFM Group, Marlow.

Connnor, D., 1985, "Information System Specification & Design Road Map," Prentice-Hall, New Jersey.

Jackson, M.C., and Keys, P., 1984, Towards a system of systems methodologies, , JORS, 35: 473-86.

KPMG Peat Marwick. 1990, "Runaway Computer Systems", KPMG, London.

Mansell, G.J., 1993, The failures method and soft systems methodology, *Systemist* Vol 15: 4.

Methlie, L.B., 1979, Systems requirements analysis - methods and models, *In*: Proceedings of International Federation of Information Processing conference, Bonn, June 1979.

Olle, T.W., Hagelstein, J., Macdonald, I.G., Rolland, C., Sol, H.Q., Van Assche, F.J.M., Verrijn-Stuart, A.A., 1991, "Information Systems Methodologies: A Framework for Understanding," Addison-Wesley, Amsterdam.

Willcocks, L., 1992, Evaluating information technology investments: research findings and reappraisal, *Journal of Information Systems* 2: 1-26.

20

A KNOWLEDGE-BASED SYSTEM FOR THE SYNTHESIS OF PARTS IN A DESIGN PRODUCT

The Wind

Manolya Kavakli*

Faculty of Architecture
Istanbul Technical University
Taskisla, Istanbul
80191, Turkey

INTRODUCTION

Building knowledge-based design systems for a class of well-known products or systems is an important topic of interest for researchers in artificial intelligence. Systems which are able to assist in analysing complicated cases and situations, in applying relevant regulations, in pointing to necessary actions, in giving aid in calculating can be of value for a lot of specific tasks in production. This relates to tasks like designing a new machine or tool, scheduling a smooth production process from design to the end product, checking how parts of different material will go together, etc. For this purpose, a knowledge-based system called WIND (*Synthesis of WINDow parts*), is developed for the synthesis of parts in a design product. In this study, the design product is a dwelling window holding assemblies of profiles like a machine.

Knowledge sources of the WIND are the documents related to the subject and the experts working at the associated firm. As a knowledge-based system, WIND needs to be fed with the specific know-how of the firms, for internal use. Besides other knowledge sources, design research, as Archer (1981) defines, investigates the designers' approach to design while designing. In this study, the design process in the firms is analysed in detail to code the domain knowledge and its phases are formulated step by step to simulate the designer's behaviour. An empirical design research approach is applied to investigate the knowledge bases of domain experts. In this paper our concern will be primarily with knowledge structuring rather than its acquisition. Information flow and operation on the artifact in the WIND are automated. We shall concentrate, therefore, on the data structures. Automated design systems require two main kinds of structured data (Latombe, 1976):

* e-mail: Mikavak @ tritu.bitnet.

Critical Issues in Systems Theory and Practice, Edited by K. Ellis et al.
Plenum Press, New York, 1995

- The state of the design process, i.e. sequences of tasks, choices made by the system, failure situations etc.
- The designed object itself, or the world of the design, at physical and conceptual levels, i.e. geometric data, technological components in use, stresses and strains in various situations etc. The first one refers to the phases of design and the second one involves knowledge representation.

PHASES OF DESIGN

'In traditional approaches', the first phase is the *Analysis* that is to diagnose, define, prepare, understand the problem and produce an explicit statement of goals according to Asimow (1962). Jones (1970) defines this phase as *Divergence* that is concerned with breaking the problem into pieces and defining the boundaries of a space in which a fruitful search for a solution can take place. Bachman et al (1993) calls this phase *Clarification of the task* that involves the collection of information about the requirements to be met by the solution and also about the constraints. In manufacturing firms, this phase includes getting specifications from the architect of the project, analysing sections, details of the system and plans. Static calculations are generated. General characteristics of elevations such as axes, opening casements, position of horizontal and vertical bars are searched.

Second phase is the *Synthesis* (Asimow, 1962) that involves finding plausible solutions. Jones (1970) calls this phase *Transformation* that involves putting the pieces together in a new way. A cycle is implied in which the solution is revised and improved by reexamining the analysis. In the first cycle of the human thinking process, this phase can be defined as *Conceptual design* (Bachman et al, 1993; Tomiyama and Yoshikawa, 1985; Hatvany et al, 1993). Conceptual design involves the establishment of function structures, the search for suitable solution principles, and their combination into concept variants. It commences with the specification of a product requirement from market considerations, involves creative innovation and decision making processes leading to the evolution of a product outline, and corresponds to setting up the first design model, i.e. to fixing the temporary design solutions or goals. For the WIND, this cycle is preferred to be named as *Initial* or *Fundamental design*. In the second cycle, it becomes *Basic design* (Tomiyama and Yoshikawa, 1985) that is an evolving process of first model to the assembly drawing through several intermediate models or *Embodiment design* (Bachman et al, 1993) in which the designer determines the layout and shape for all the parts and develops a technical product or system accordance with technical and economic considerations. In the third cycle, it turns into *Detail design* (Bachman et al, 1993; Tomiyama and Yoshikawa, 1985; Hatvany et al, 1993) in that the minute data of the product required for production are settled. Detail design is the design of all the component assemblies and units for the manufacture of the product. In detail design, the arrangement, form, dimensions, and surface properties of all the individual parts are laid down, the materials are specified. In manufacturing firms, this phase includes choosing the system part corresponding to the performance requirements in the specifications. The accessory that is standard for each system is chosen. The nodes are solved such as attachment to the suspended ceiling and the parapet, joints of the special corners and fastening to the finishings. The product parts are synthesized. Synthesis of product parts requires criteria in relation to the location and the combination of parts.

The third phase is *Evaluation* (Asimow, 1962) that concerns judging the validity of solutions relative to the goals and selecting among alternatives. Jones defines it as *Convergence* (Jones, 1970) that is testing to discover the consequences of putting the new arrangement into practice. The output of this phase is *Final design* in the WIND. In manufacturing firm, architect's approval on the detail design is applied in this phase. Design product can be modified by the architect of the project. Then the cost of the whole system is calculated.

KNOWLEDGE REPRESENTATION

Data groups in each design phase represent the domain knowledge of human experts. Representation of knowledge for this design task can be classified as follows (Kavakli and Bayazit, 1994):

1. Representation of Product Parts

Product parts are objects that are units of information. The code *Lc* consists of complete list of the product parts. Graphical representation of objects produced by the firms are coded as *Pf*. Each part has some properties represented in terms of different attributes coded as *Att* and their values. Values of these attributes are represented as *Db-c* -Data base of the parts. *Rough* is the rough graphical representation of parts, refers to the instances. *Rough(s)* are the rough models of the parts in a system, in evolutionary design process that are replaced with final models later.

2. Representation of the Problem Environment

Different codes are used to represent facts & conditions, design descriptions, and goals & objectives of the problem environment. *Dc* represents internal and external conditions of the problem environment. *Df* represents facts. Design descriptions, as well as the physical and the general characteristics of a system part, are represented by the elevation of the window named *We*. The section of the window is called *Ws*. The plan of the window called *Wp*. Goals and objectives of an artifact are defined at the initial stages of the design process. Design is a process of producing a set of descriptions for an artifact that satisfy a set of given performance requirements and constraints. Parameters of performance requirements are called *Ppr*. Performance requirements are named *Pr* that should be defined by the customer.

3. Representation of Domain Knowledge

Domain knowledge can be characterized with the statements about mapping of different facts. These mappings enable us to derive new facts from existing ones. There are many ways of representing these mappings. One of them is to define rules that can be classified as follows:

3.1. Technical Rules (Tech). Knowledge, that matches the attributes of the parts to the parameters of performance requirements, is called technical rules.

3.2. Design Rules (Des). These rules contain strategies that follow problem solving procedure. In the professional practice, a number of primary design rules are defined to restrict the solution space.

3.3. Relational Rules (Rlt). Relational rules relate to the attachment of parts together. Relations, which are labels attached to the groups of information units (design objects), may also map onto what we regard as relationships between physical objects such as '*next to*', '*inside*', '*longer than*', etc.

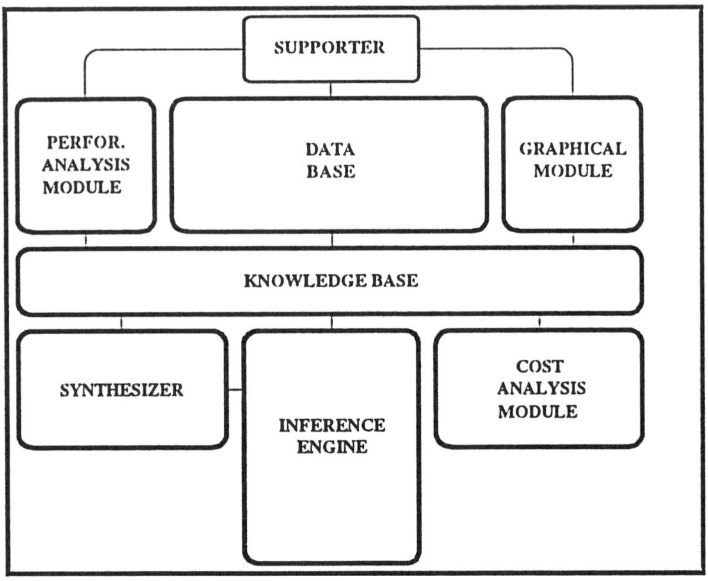

Figure 1. Modules of the WIND.

3.4. Hierarchical Rules (Hrc). Rules about the hierarchical structure of parts (rules used to define classes, subclasses, and instances) are called hierarchical rules. Frames of parts used in design (*Frm*) refers to the classes, list of the parts (*Ls-c*) matching their attributes with the parameters of performance requirements, refers to the subclasses. *Rough* and *Pf* refers to the instances.

3.5. Rules about Elimination and Configuration. Rules in relation to the elimination of alternatives (*Elm*) define strategies of elimination, while rules of fitting (*Fit*) defines appropriate actions to perform for producer configurations.

KNOWLEDGE STRUCTURING

WIND is developed by combining several modules such as a supporter, a data base, a knowledge base, an inference engine (SYNTHESIZER), a performance analysis module, a graphical module (PEN), and a cost analysis module.

Each module has a different task to support designer during the design process. The design process executed by the human expert is simulated by a software. Knowledge can be structured in the following phases (Kavakli and Bayazit, 1994):

1. Input

Customer defines performance requirements (*Pr*) by giving specifications to the firm. Expert investigates internal and external conditions of the problem domain (*Dc*) and defines facts (*Df*). Drawings represent the list of parts (*Lc*), and physical description of the window such as the elevation (*We*), the section (*Ws*), and the plan of the window (*Wp*).

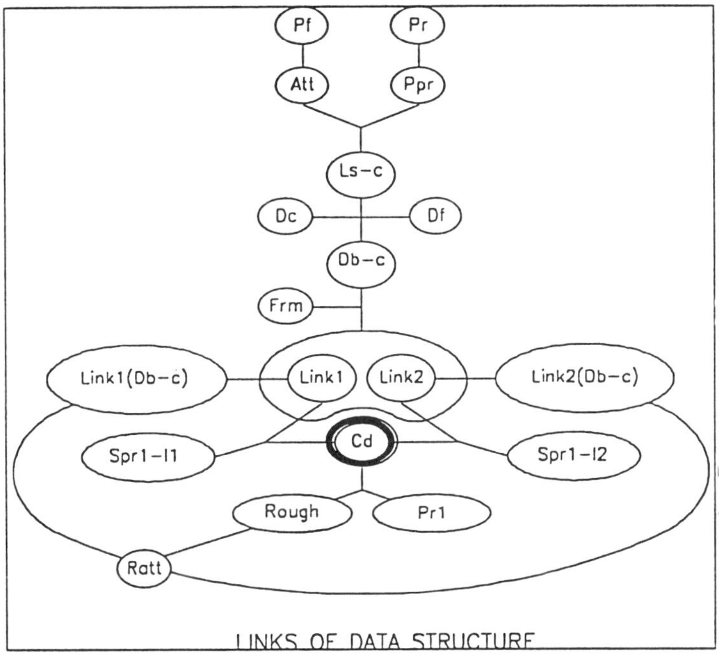

Figure 2. Links of data structure and the information flow.

2. Defining Parameters of Performance Requirements

Getting the input, concerning to the performance requirements (*Pr*) from the customer or specifying necessary conditions for problem environment, expert calculates parameters of performance requirements (*Ppr*).

3. Matching the Attributes of Window Profiles to the Parameters of Performance Requirements

The attributes of window profiles (*Att*) match with the parameters of performance requirements (*Ppr*). The lists of parts (*Ls-c*) whose attributes matching up to the parameters of performance requirements are defined by expert.

4. Choosing the Parts of a System

The expert sorts out the data base of the parts (*Db-c*) from the list of parts (*Ls-c*) whose attributes matching up to the parameters of performance requirements, by evaluating internal and external conditions of the problem domain (*Dc*), and the facts (*Df*). Thus, the real parts (*Db-c*) are extracted.

5. Synthesis of Product Parts

In the WIND, the initial design is generated by the SYNTHESIZER that enumerates performance requirements by asking end-user for their priorities. Firstly, the design rules are searched in relation to the primary performance requirement. Thus, approximate graphical representation of the parts (*Rough*) corresponding to the primary performance requirement

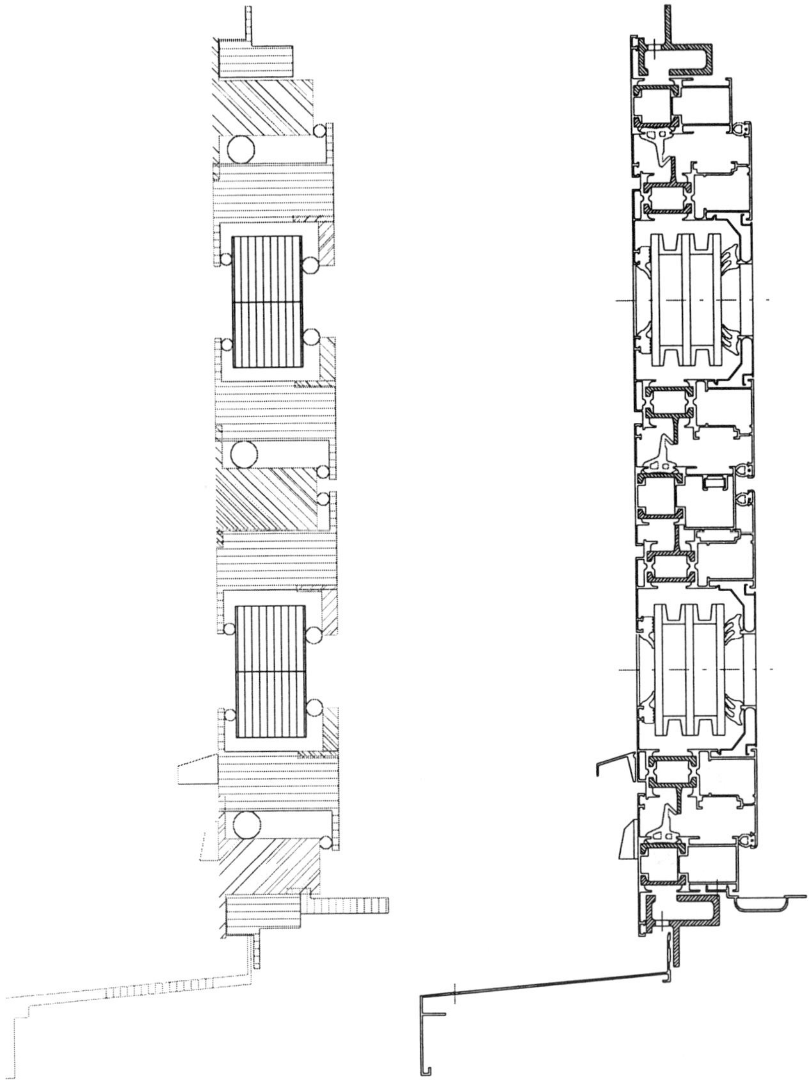

Figure 3. Synthesized product parts.

(*Pr-1*) are placed in the graphical representation of the initial design (*Cd*) by using hierarchical (*Hrc*) and relational rules (*Rlt*). Approximate graphical representation of the parts (*Rough*) has the same attributes (*Ratt*) matching up to the attributes in the data base of the parts (*Db-c*). User defines which firm's products to be used. Then, product parts are classified and structured hierarchically. The links of parts that can be used together are generated in hierarchical structures such as *Link1*, *Link2*, etc. Possible links are enumerated. Parts (*Link1*(*Db-c*)) of the first link (*Link1*) match with the attributes (*Ratt*), in the rough representations (*Rough*). Rough representations of the parts (*Rough*) are replaced by *Link1* of one firm's profiles (*Pf*). This procedure generates the solution for the primary performance requirement (*Spr1-11*), and is an output monitoring on graphical interface called PEN. Operation is repeated for every possible link and then other solutions are generated such as

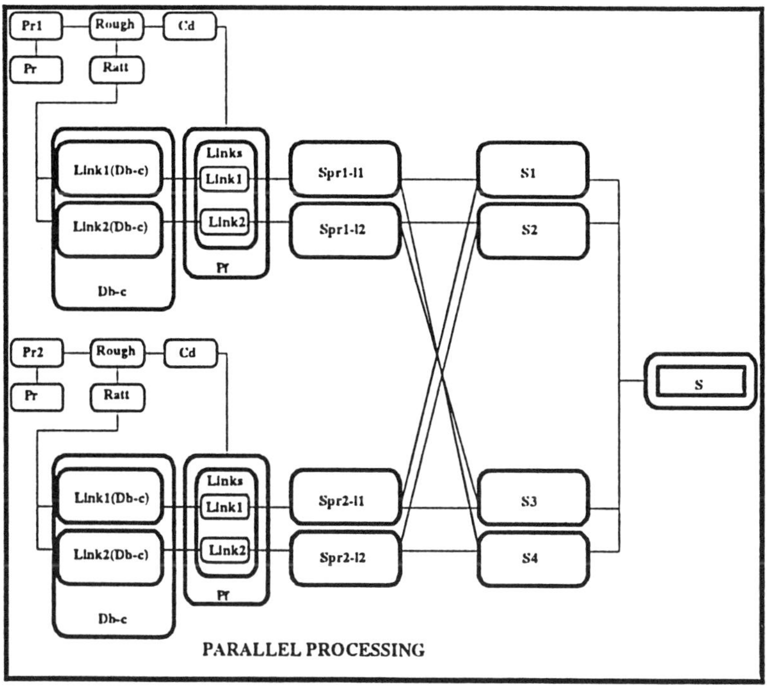

Figure 4. Parallel processing.

Spr1-l2, *Spr1-l3*, *Spr1-l4*, etc. Solutions for all the performance requirements can be find following the same procedure (*Spr2-l1*, *Spr2-l2,...,Sprn-ln*).

6. Decision Making

In the WIND, alternative solutions are arranged in relation to the levels of satisfying performance requirements such as *Spr1-l1* better than *Spr1-l3,* and so on. But the most important one is the decision of the architect who designs window facade. Therefore, a cooperated work or the approval of the architect is needed. In this operation, rules are used to eliminate the alternatives (*Elm*). Then, the rules of fitting alternative solutions (*Fit*) are fired.

There are two kinds of synthesis method to extract final design:

1. In parallel processing, enumerated but not evaluated alternative solutions for each performance requirement, combined with each other (for example *Spr1-l1* + *Spr2-l1*, *Spr1-l2* + *Spr2-l2*, etc.) by the rules of fitting selected alternative solutions (*Fit*). A morphological matrix, is generated representing all the possible solutions, such as *S1, S2, S3, S4,...,Sn. S* is chosen as final design.
2. In serial processing, the most effective solution for primary performance requirement (for example, *Spr1-l1*) is tried to combine with the alternative solutions for secondary performance requirements (*Spr2-l1, Spr2-l2 or Spr2-ln*). If it is combined, primary solution for final design (*Sn*) is obtained, and the alternative solutions for the third performance requirement are evaluated. If it does not fit, *Spr1-l1* is modified and tried to fit *Spr2-l1* by using rules of fitting selected alternative solutions (*Fit*). Then, *Sn* is obtained. Following this procedure, secon-

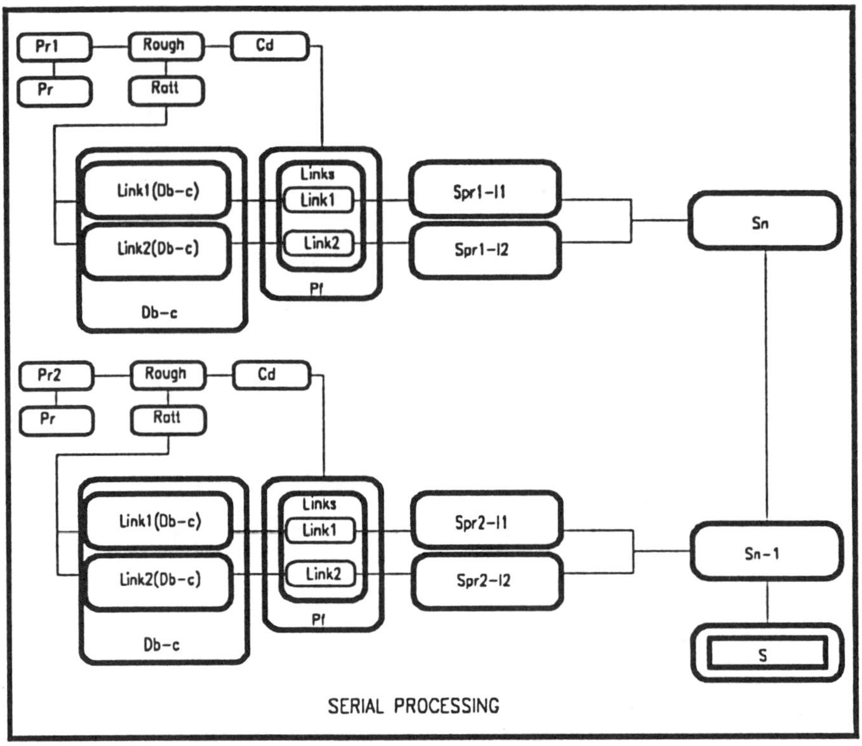

Figure 5. Serial processing.

dary solution for final design (*Sn-1*) is derived from *Sn*. At the end, final design (*S*) is derived from these temporary solutions. This is an evolutionary process.

To obtain approval of the architect, the elevation of the window with respect to final design is drawn by the graphical module, and it is evaluated whether the elevation of final solution and the sketch of the architect (*We*) is alike or unlike. If they are unlike, another solution is evaluated in the same way.

7. Calculating the Cost of Whole System

The values of the attributes in relation to cost in the parts of the final solution (*S*) are used to define real cost of the window. Some of the firms have already been using a computer program to make calculations of the cost, getting input from data base of the product parts.

CONCLUSION

The synthesis of the parts (profiles) of a window system in professional practice is simulated by using presented knowledge structuring approach in the inference engine of the WIND. WIND is developed as a system that analyzes the design specifications and fixes the approximate layout of the design solution. This knowledge-based system can synthesize product parts, then, generates alternative solutions for each performance requirement to obtain final solution. Although, theoretically, it is possible to produce many links for each performance requirement, practically, there is a few links of profiles in the manufacturing firm and solution space is not so large. At last, it checks the feasibility of the candidate solution. Thus, the actions of a human expert in each design phase corresponds to a computer program and, finally, the system is completed by combining those programs that overcome different design tasks.

ACKNOWLEDGMENT

This knowledge-based system is the subject of a Ph.D. thesis in the Faculty of Architecture, Istanbul Technical University. This thesis has done under the guidance of Prof. Dr. Nigan Bayazit, whose advice, support, and valuable ideas played an important role throughout the design process of the system.

REFERENCES

Archer, L.B., 1981, A view of the nature of design research, *in*: "Design : Science : Method" , R. Jacques and J.A. Powell, eds., IPC Business Press Ltd, London.

Asimow, W., 1962, "Introduction to Design", Prentice-Hall, New Jersey.

Bachman, T., Daniel, M., Pahl, G., and Rix, J., 1993, Feature-based modeling in support of embodiment design, *Computers & Graphics*, 17(3): 285-294.

Hatvany, J., Newman, W.M., and Sabin, M.A., 1993, World survey of computer-aided design, *Computer-Aided Design*, 25(12): 776-798.

Jones, J.C., 1970, "Design Methods", John Wiley & Sons, New York.

Kavakli, M., Bayazit, N., 1994, Knowledge structuring of an expert system developed for the synthesis of parts in industrial design, *in*: "Cybernetics and Systems' 94, Proceedings of the Twelfth European Meeting on Cybernetics and Systems Research", R.Tappl, ed., World Scientific Pub.Co., 1: 539-546, Vienna.

Latombe, J.C., 1976, AI in CAD; the TROPIC system, *in*: "CAD Systems", J.J.Allan, ed.,North Holland, Amsterdam.

Tomiyama, T., and Yoshikawa, H., 1985, Knowledge engineering and CAD, *Future Generations Computer Systems*, 1(4): 237-243.

21

STABILITY INDICATORS

A Tool for Organizational Planning and Control in Carbocol

Clemencia Morales Montejo

School of Computing and Information Systems
University of Humberside
Cottingham Road
Hull HU6 7RT
United Kingdom

ABSTRACT

This paper is a summary of my MPhil Thesis (Morales, 1992) and discusses a methodology that helps to identify and measure Stability Indicators (SI's). SI's are used as a tool for organizational planning and control. A SI is a measure of those organizational variables that are characteristic of the overall state of the organization. The organization is assumed to be stable when the organization is controlled and management plans are realized in practice.

The factors that motivated this research were the need for integrating the functions of planning and control and the need for flexible evaluation methods in organizations. The work uses systems approach, cybernetic, and strategic planning approaches as a theoretical framework.

The methodology was used in a project with Carbocol (The Colombian National Coal Company). Managerial ability has been strongly applied in order to learn from the company's past and present problems, and also as means of solving them. Arriving at the set of necessary SI's for control is the result of a learning process. Permanent feedback from the whole system helps to define strategies because it is characterized by a progressive knowledge of institutional learning programme.

1. BACKGROUND OF CARBOCOL

The Colombian Government undertook the task of creating Carbocol in 1976. Carbocol's major objective is to finance a variety of activities in the coal industry, to give financial and technical assistance to miners, to improve their social and economical conditions and also to control environmental pollution. Carbocol could not be considered to be an isolated entity, but, rather, it is a dynamic organization taking an important part in the Colombian economy. Today it has had great social and economic effects on the country. Therefore, it is related to the other organizations which make up the country's economy; it is thus affected by the world economy.

Critical Issues in Systems Theory and Practice, Edited by K. Ellis et al.
Plenum Press, New York, 1995

Due to the last series of oil crises, the conversion from oil to coal created an enormous demand for the importation of steam coal. The crises helped new coal producing countries such as Colombia to obtain a foothold in the market. Indeed, the crises represented a great opportunity for Colombia as studies revealed coal deposits which account for more than 40% of the Latin American Reserves.

The methodology for the identification and measurement of stability indicators used in a pilot-prototype for the Central Cerrejon project was chosen. As an opencast coal project, it produces steam coal both cheaply and profitably. Carbocol managed the mine with a mine operator and commercialized the coal directly.

2. PROBLEMS IN PLANNING AND CONTROL AND THE NEED FOR STABILITY INDICATORS

There are some issues which make organizing, planning, and controlling a firm such as Carbocol very difficult. These difficulties are essentially technical, financial and managerial. The technical problems we find include, for example, the inexperience of Colombian executives in negotiating major technical projects, and the inexperience of Colombian engineering companies in dealing with major technical projects. Some of the important financial problems are mining projects representing long-term investments that make capital return rather slow, and the difficulty of predicting coal prices in the international market in the face of hard competition. As regards management problems, the national political instability jeopardized the continuity of project.

SI's alert management to changes in the state of the organization. There is stability in the organization if the indicators fall close to standard value. Fluctuation in the value of the SI's points to the organization's being unstable. Thus, SI's increase management's awareness of the state of the organization and, consequently, management's ability to control it.

3. THE METHODOLOGY

The techniques of systems, cybernetics and strategic planning offer ways of improving, understanding and learning about the reality of organizations. The systems approach gives an overview of the organization as a whole and identifies the different systems, sub-systems and processes present in the organization. Cybernetics increases the capacity to interpret and analyze extremely complex situations. Strategic planning offers ways of designing or generating strategies as well as ways of analyzing and evaluating them.

The techniques of all three fields were used in the design of a methodology for the identification and measurement of SI's.

The methodology used to identify and measure SI's in the Carbocol study is shown in Figure 1. According to the figure, in order to attain an efficient Indicator System, it is necessary to:

A. look for the origin of problems in planning and control in the organization;
B. create the best organizational conditions for the methodology to be operating all of the time;
C. structure correctly the organization under study;
D. identify and handle Stability Indicators;
E. measure indicators; and
F. handle efficiently the implementation process.

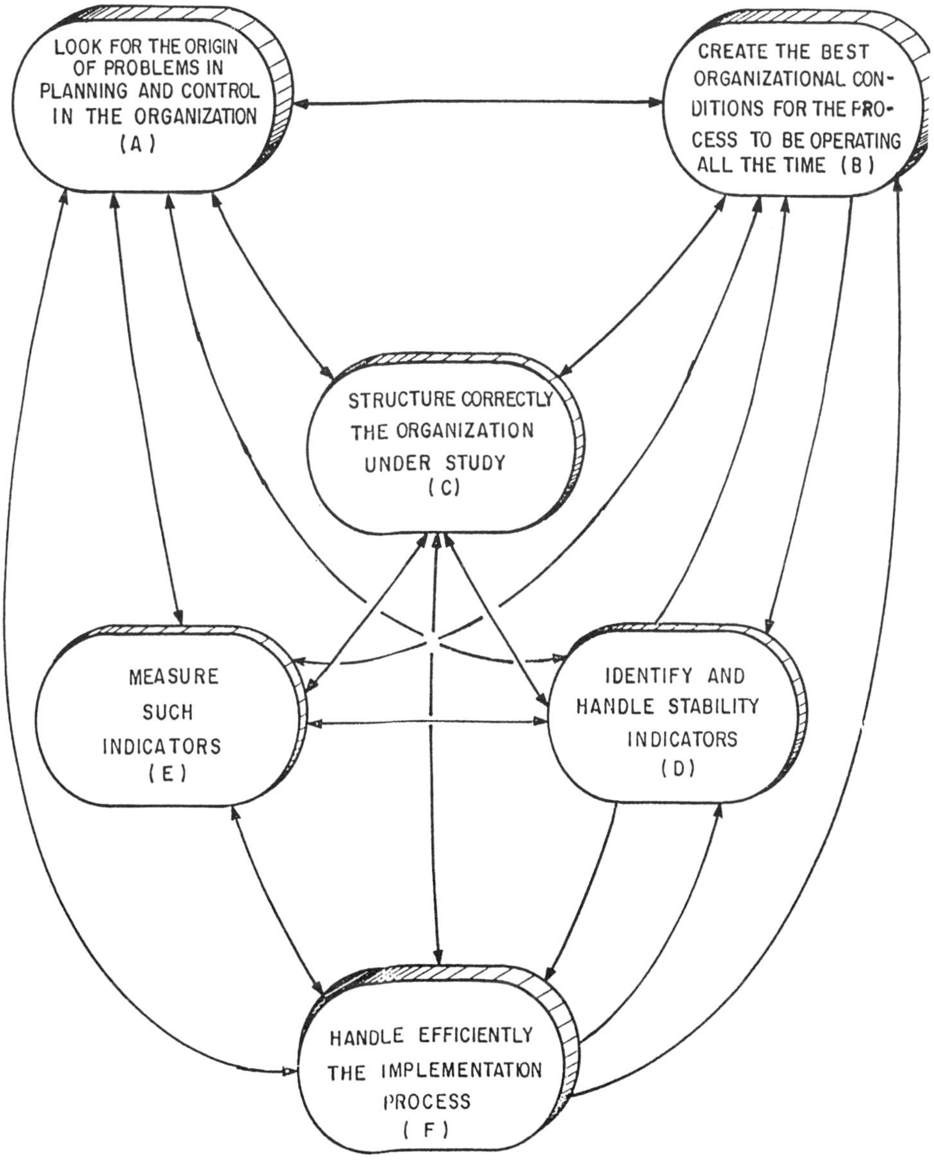

Figure 1. Methodology.

As shown in Figure 1, there is interaction between the different stages of the methodology. The explanation here of each of the stages of the methodology is based upon the Carbocol study:

3.1. Looking for the Origin of Problems in Planning and Control

The intention in this stage was to perform a diagnosis of the possible causes of problems in Carbocol planning and control. Some of the difficulties that were detected included:

i. a lack of commitment to strategic planning;
ii. poor integration of the planning and control function,;
iii. weakness in plan structuring;
iv. weakness in the planning process;
v. weakness in measure in the plan;
vi. lack of internal and external coordination and participation,
vii. poor communication in Carbocol.

3.2. Creating Conditions for Effectively Solving the Problem of Planning and Control in Carbocol

In conducting the Carbocol study it was realized that employees' opinions are very helpful to the evaluation of the organization. Further, it may be said that a successful indicator system is viable only with the cooperation of executives involved and their understanding of the organization through their participation in observations, interviews and continuous debates comparing the 'real world' and the 'abstract model'.

3.3. Structuring the Organization Under Study

The most appropriate stability indicators should be easily identifiable after having structured the organization. Then, it is important to find the best structure for the organization in order for it to cover the basic activities that should be measured. This step is grouped in three principal phases: organizational modelling, operational modelling and strategic planning.

Organizational modelling defines Carbocol's recursion structure. It segregated the organization into strategic business units (SBU). For each SBU, a Viable System Model (VSM) (Beer, 1979; 1985) was defined according to each level of recursion.

Operational modelling should be the output of rationalizing different organizational processes according to 'what should be done' at each level of recursion. Carbocol received the coal, on trucks, in the mine, and then had to take care of the land transport to the railway. A second operator was in charge of the land transport and a third operator handled the coal in port until it was shipped.

The strategic plan defines and discusses Carbocol's strategic planning. The focus was firstly centred on an external and then an internal analysis of the organization as a complete entity. In this way, it guarantees that the decisions to be taken at each level are appropriate to its level of recursion.

3.4. Identify and Handle Stability Indicators

In the process of definition, the important thing is to select the essential indicators that result in the optimization of the cost/ benefit ratio. A constant evaluation of Indicators enables the identification of critical areas, activities or operations. In the Carbocol study Indicators were classified, according to their degree of importance, in terms of efficiency, efficacy, and effectiveness. Importance may be based on the cost to the organization which might be incurred were the indicator left out of the analysis.

3.5. Measure Stability Indicators

For measuring purposes, the indicator is considered as a reason or quotient, to which events or numbers, depending on a set of pre-established rules, are assigned. Managers may effect producing processes, either by increasing or reducing the input resources, varying the efficiency of the transformation process, or varying the efficacy of results according to what is expected from the organization. The criteria of measurement are defined in terms of indices of productivity, latency, and performance as the reason for measuring actuality, capability, and potentiality. A sophisticated performance measure derives from a careful analysis of indicators, beginning with the whole organizational system and segregating each level until it has throughly measured the performance of all lower levels.

3.6. Handle Efficiently the Implementation Process

Once SI's have been established and measured, the implementation of the Stability Indicator System should be guaranteed. This system must, essentially, permit the identification, measuring and handling of Stability Indicators to be adjusted and actualized. The importance of continuously generating debates among managers, plus a systematized information system, may contribute to the use of the indicator system.

The permanent feedback from each of these steps must be a continuous learning process for the managers; their information, derived from the behaviour of each indicator, could lead either to organizational adjustments, or to re-defining the indicator, the measuring patterns and the patterns of behaviour. An alternative is, simply, to adjust the methodology.

The absolute value of indices provides not only a measurement of the evolution of each indicator in time, but also helps executives to adjust to the organizational complexity in accordance with the inherent dynamics; thus, it becomes a powerful tool in decision making. The complete Stability Indicators System makes up an Information System which has the main objective of providing the organization only with the necessary reliable and timely information, in the best place, at the best moment, with the right quality and cost, to develop organizational activities. This information system will support fulfilment, follow up, and monitoring of objectives and strategies for the organizational mission to be duly attained.

4. CONCLUSIONS

The methodology makes it feasible to improve the organizational performance and to include corrective actions. Based on the indicators' behaviour, it would be possible to decide on the best action for improving Carbocol's performance. The Stability Indicators System has a monitoring mechanism to be used as permanent feedback which should serve to improve the obtained results of every managerial situation. Thus, it is also indispensable that, from the policies of mine operators, commercial or any other kind of contract, Carbocol could obtain all the information needed to feed the Indicator System.

The Stability Indicator System should improve management's understanding of the organization's behaviour, because it offers them the opportunity to develop a faster adapting and learning capacity. The system enables management to evaluate, follow up, restate and forecast the best strategies to facilitate the feedback process of the strategic plan in the organization.

The methodology brings a system that is used to support every decision. It also helps managers to be prepared for organizational deficiencies even before they take place. In implementation it is important to have the institutional support (human, financial, and technical resources) necessary for the normal development of its operation. The success in implementing the methodology depends on a successful combination of organizational culture with changes in the structure, systems and people. The existing political atmosphere in the organization may deeply affect the implementation of the methodology.

5. RECOMMENDATIONS

The analysis of SI's must not be a cold calculation or an interpretation of figures isolated from reality. On the contrary, it should encompass information about the environment in which the enterprise operates. The mechanical usage of indicators can lead to a series of wrong practices. However, sometimes it may seem to be more important to obtain mathematical accuracy rather than to choose the indicators that provide the most valid information.

It is almost impossible to have guidelines that enable the simplification of reality without detrimental effects on the accuracy and knowledge of the variables relevant to identified indicators. It is difficult to find which criteria serve to define the correct indicator; for example, to define indicators which categorize and compare organizations, sometimes it is necessary to use arbitrary criteria that, when applied as standards for the sector where the enterprises operate, would favour the image of certain organizations and harm others.

It is very difficult to define and measure quantifying indicators of every kind of organizations and situations, in particular in the public sector where organization evaluation is not concerned with quantitative indicators; for example, the evaluation of improvements in social and economic conditions affecting miners working for Carbocol. Also, there are problems associated with measuring intangible indicators, e.g., communication, technology, know-how, managerial action, loyalty, etc. It is difficult to identify and measure intangible indicators as by definition they cannot be expressed quantitatively. For example, it is difficult to identify significant behavioural measurements for managers, since their job generates no end products, either tangible or calculable.

It is important to recognize the need for other methodologies and tools to be applied in the public sector. Resultants from combining the systems approach, cybernetics and strategic planning need to be correlated. There should be emphasis on the complementarism and integration of systemic, cybernetic, and strategic planning approaches for improving the methodology for identification of Stability Indicators and developing a set of their measurements.

REFERENCES

Beer, Stafford, 1979, "The Heart of the Enterprise," John Wiley and Sons, Chichester.

Beer, Stafford, 1985, "Diagnosing the System for Organizations," John Wiley and Sons, Chichester.

Morales, Clemencia , 1992, Stability Indicators, A Tool for Organizational Planning and Control, *Master of Philosophy Thesis,* The University of Aston in Birmingham.

22

THE CASE FOR CRITICAL CREATIVITY IN TSI

A Necessity, A Practical Possibility, or A Contradiction of Terms?

Gillian Ragsdell

Department of Management Systems and Sciences
University of Hull
Hull HU6 7RX
United Kingdom

INTRODUCTION

Creativity is desirable in many activities. No longer is it restricted to Fine Arts and poetry; and no longer is it a pursuit for a chosen few. Creative thinking is an increasingly acknowledged practice in organisational activities - problem solving being one such activity. Interaction between diverse departmental interests, wide ranges of technology and a changing operational environment continuously generates unique issues for management. Originality is essential for successful management of complexity; creative thinking is the key.

This paper explores creativity, initially in a general context and then in its relationship with the problem solving framework of Total Systems Intervention. Finally, 'Critical Creativity' is introduced.

CREATIVITY

What is Creativity?

Considering 'creativity' is such a well-used term, it is difficult to find a precise, succinct definition. Words such as 'innovation', 'imagination' and 'ingenuity' spring to mind, yet one is aware of their inadequacy. Many authoritative writers, however, have contributed informed expressions. Koestler (1981), for instance, speaks of three forms of creativity - "artistic originality, scientific discovery and comic inspiration" - and believes that all creativity falls into one of, or a combination of, these categories. Hausman (1981, p189) puts forward four criteria of creativity: "(1) created outcomes have intelligible structures that are irreducible; (2) the structures of created outcomes are unpredictable; (3) the structures of created outcomes are inherently and usually instrumentally valuable; (4)

Critical Issues in Systems Theory and Practice, Edited by K. Ellis et al.
Plenum Press, New York, 1995

and the acts that lead to created outcomes include an element of spontaneity so that although they are directed and are controlled, they are discontinuous."

The Image of Creativity

The perception and role of creativity in society have changed over time. Much of the mystery which once shrouded creativity has been lifted and its measurement and assessment has even become feasible, for instance Vergara-Cabrera (1981) and Metcalfe (1983). Anderson, in his introduction, (1959, p x) writes of this transition in the regard for creativity. "In the Old Testament man was not creative; only God was creative. Until two or three centuries ago the notion that man - any man - could be creative in thought and in action was regarded as blasphemy. This transition in human thinking which is taking so long - so many centuries - is itself a lesson in human creativity." More recently, Lamb (1991) tells of three schools of thought which have surrounded creativity, referring to them as 'myths about creativity'. He mentions the heroic theory of creativity and its close link with the romantic tradition and the philosophical tradition of Plato. Both of these deny that anything informative can be said about creativity. In current writings, Cohen (1994) chooses to use the term 'magic' in preference to 'creativity' when relating specifically to corporations and organisations, because she feels that creativity is personalised. She writes "'Magic' is creativity on the large scale, and it has implications for individual self-actualization, contentment and a corporate flexibility." (Cohen 1994, p 255) Just from these few examples it is clear that 'creativity', in various guises, has been a topic of discussion for a long time.

Value and Creativity

Briskman (1981), writing a couple of decades after Anderson, seems to tell of the popularisation of creativity with a certain cynicism. Yet he also reinforces the high esteem which he feels the concept warrants - " there seems to me to be one aspect of creativity that will remain constant, and that is that creativity is something *valuable*, and that the notion of creativity is permeated with *evaluation*. To adjudge something to be "creative", in others words, is to bestow upon it an honorific title, to claim that it deserves to be highly valued for one reason or another, Hence, without standards and values, creativity ceases to exist, just as morality ceases to exist. But as with morality, how high (or low) we set the standards is partially a matter for our decision." (Briskman 1981, p 129) The idea of value judgments is fundamental to the development of critical creativity. Creativity is valued - valued by particular parties in particular contexts at particular times.

TSI AND CREATIVITY

TSI - Its Process and Foundations

TSI is a framework for problem solving which has recently come to the fore in management and systems arenas, (see for example, Flood and Jackson (1991(a)), Flood (1995)). Flood (1995) summarises it quite neatly: "TSI has been developed to provide practitioners with a usable approach that sticks to the original intent of systems thinking, to be holistic. It integrates all approaches to 'problem solving'.... in a process that ensures they are employed only to tackle the issues they are best suited to. TSI achieves this by embedding 'problem solving' approaches in an interactive 3 phase process." In a drastic simplification, the cyclic process of TSI incorporates three phases - Creativity, Choice and Implementation.

The context of the problem situation is explored from varying view points and described, using metaphors or other means of free expression - Creativity. Translating the metaphor into a location in a complementarist framework, the problem manager is guided in the selection of a systems methodology(ies) - Choice. Application of the methodology follows - Implementation. Each phase has a recursive structure - there is movement between Creativity, Choice and Implementation, and there is reliance on reflection and enquiry being active integral parts.

The philosophical foundation of TSI is critical systems thinking. In fact, TSI has been said to "offer a practical face of critical systems thinking" (Flood and Jackson (1991(b) p 336). Thus, 'complementarism', 'sociological awareness' and the promotion of 'human well-being and emancipation' are intrinsic to the process.

'Creativity' Phase of TSI

For any systems thinking to be effective, creativity is necessary. An open - minded state needs to be upheld, preventing obstruction to new ideas. Creativity in TSI was originally discussed with respect to selecting metaphor(s) from Morgan's (1986) range developed to characterise organisational life, which aptly described the problem situation. One might have suggested that creativity was a little restricted since a problem manager's thinking was being constrained and limited into metaphorical terms. The reconstituted version of TSI (Flood 1995) takes a wider perspective of Creativity and in fact is now keen to encourage all forms of 'brain storming' techniques. However, methodological guidelines are not yet provided.

Metaphors

The potential of using metaphors can be seen from this summary of Barrett and Cooperrider (1990):

1. Metaphor is an invitation to see the world anew.
2. Metaphor facilitates the learning of new knowledge.
3. Metaphor provides a steering function for future action and perceptions.
4. Metaphor invites active experimentation in areas of rigidity and helps others overcome self-defeating defenses.

However, as with other creativity enhancing techniques, the purpose of the exercise, along with the method of introduction and facilitation needs careful consideration. Yeoman and Peters (1994) encountered practical difficulties in their experience of using metaphors and suggested that "Participants perceived metaphors to be very subjective and unreliable".

Merali and Martin (1994) have usefully employed metaphors in the teaching of information systems, but express concern over five areas in which metaphors may show limitations - reification, distraction, distortion, manipulation and disfunctionality. Metaphors have proved useful in various scenarios - as an agent for group development and organisational change (Barrett and Cooperrider, 1990) and as an aid to knowledge elicitation (Behrooz and Marshall, 1993) to give just two examples.

METHODS OF ENHANCING CREATIVITY

Although I have intimated that everyone is capable of being creative, it could be suggested that young children are masters of its practice. (I appreciate, however, that there are arguments to the contrary e.g. Pickard (1979) asserts the argument that children cannot be creative, where creativity is defined as the 'ability to reconstruct reality'.) One only has to watch infants at play to realise that their environment - physical, mental and emotional - is near-perfect for overt exercise of the imagination. They have not yet experienced the pressure to conform to institutional expectations or to peer pressure. Nor do they yet know what it means to respond to personal desires of ambition and for self-esteem. Their innocence of such barriers to creativity is perhaps something for 'life battered' adults to attempt to reproduce.

There have been studies which have concentrated on identifying the personality traits found in so-called creative people. Stein (1974), Whitfield (1975) and Hare (1982) mention many. It is suggested that the creative individual is generally widely informed, has wide ranging interests and is versatile. S/he is also open to feelings and emotions, is aesthetic in his judgment and value orientation but low in economic values. The creative person is observed to possess freer expression of what have been described as feminine interests and lack of masculine aggression; has little interest in interpersonal relationships; is emotionally unstable but capable of using his instability effectively. Hare (1982, p 157) concludes that: "....the creative person is essentially a non-conformist with the capacity to pursue nonconforming and creative ideas in the face of societal pressures to see things as others have seen them and leave things as others have found them."

Even if one put absolute faith in the psychometric approach and managed to select a team of individuals with all the traits of creativity, a creative process would not be guaranteed. Sometimes modification, possibly to the physical setting or to the emotional atmosphere, is needed to increase the level of creativity. Stein (1974; 1975) provides useful sources of creativity enhancing methods.

Role Playing

Role playing is probably one of the oldest and the most widely practised techniques for stimulating creativity, given the fact that it is an activity most children will quite naturally engage in. The process of assuming another perspective by putting oneself in 'someone else's shoes' and 'acting out the part' to change one's self-concept can be useful for triggering new ideas.

Brainstorming

Brainstorming was developed by Osborn (1963), although the term is so popular that it is often used to refer to all creativity stimulating techniques. Participants are invited to suggest ideas to solve the given problem, following four basic rules -

1. Criticism is ruled out
2. Free wheeling (building on a previous idea) is welcomed
3. Quantity is desirable
4. Combination and improvement are sought

Attribute Listing

Attribute listing calls for the isolation and selection of the major characteristics of an idea, object or product. The next step is to modify each of these attributes in turn, in every conceivable way; there is to be no evaluation of the changes at this stage. Evaluation and selection of the change options follows only after the flow of modifications has completely dried up.

Synectics

Synectics is a Greek word which refers to the joining together of different and apparently irrelevant elements. The synectics process, originally developed by Gordon (1961), relies on the use of metaphors and analogies to restate and review a problem situation. During the synectics process an 'excursion', a journey of the mind, is taken which often results in the attainment of creative solutions.

There are numerous examples of other creativity enhancing methods in the literature e.g. Input-output technique, morphological analysis, use of the ridiculous, checklists and forced relationships to name a few. More recent methods include the use of computer technology. It is perhaps worth stressing that it is 'methods' and not 'methodologies' that are being discussed here. Hence, their application will be more prescriptive and systematic than the enquiring mode of application one expects to adopt for methodologies.

CRITICAL CREATIVITY

From Two Elements to a Compound

From the philosophy of CST, we have an understanding of the term 'critical'. Being 'critical' involves encouraging complementarism, sociological awareness, and human well-being and emancipation. To look at the other element, it could be said that 'creativity' is a spontaneous process which results in an output that is both original and valuable to the parties concerned. When we combine the two elements we produce a metaprocess of creativity where reflection and evaluation are intrinsic; without which the creative process is corrupt.

The complementarist stand point would be displayed in several ways. There would be an appreciation that creativity can arise from a number of different origins, but no one particular source is superior. Also, that varying approaches to creativity can yield different outcomes and different forms of output. There would be acknowledgement that particular methods for stimulating creativity have emerged from varying schools of thought and consequently possess varying strengths and weaknesses for different organisational contexts.

The sociological awareness of critical creativity would help the interventionist become aware of the social consequences of using particular creativity enhancing methods. A sensitivity towards the nature of the problem to be tackled and towards ongoing matters in the participants' environment, perhaps.

In a similar vein, critical creativity would seek "to achieve for all individuals, working through organisations and in society, the maximum development of their potential", (Flood and Jackson 1991(a) p 49), thus disclosing an emancipatory intent. A rather ambitious proposal, one might say, but one which could be made real by considering the participants and the group dynamics, in particular.

Necessity, Practical Possibility or a Contradiction in Terms?

Creativity is open to many interpretations. Thus, it is not sufficient to talk of 'creativity' per se; for example, its interpretation within TSI must be more precisely directed based on reflection in the context of the intervention. Creativity must be in a form which is appropriate for the organisation and its players. Critical creativity is called for. Being critical and being creative could be interpreted as two ends of a spectrum. However, it has already been mentioned that there is a value judgment associated with creativity. An appropriate value system will be a balance between the value judgments of the interventionist and the clients. Challenging each others value basis and interacting outside the boundary of the organisation is important for discussion and adjustment of views. The explicit introduction of critical creativity into TSI is to ensure that the 'Creativity' phase and the subsequent recursions will be seen as 'good' in their context.

In practical terms, attention must be given to the creative process - both in terms of its outcomes and outputs. For example, whereas as a design manager may feel relaxed at the prospect of venturing on a synectics excursion, a production manager may be more inclined to participate in a brainstorming exercise. Similarly, a manager, whose hobby is amateur dramatics, may revel in role-playing, while another may be horrified at the prospect. We also need to consider the environment for the exercise. Availability of time, space and facilities need to be taken into account. Scheduling a brain storming session with the accounts department at the end of a factory audit may not be conducive to the process. An appropriate style of facilitation also needs to be adopted, with forethought given to aspects such as leadership style and the method of communication.

SUMMARY

The airing of different views of creativity, eventually focusing on creativity in TSI and presenting examples of creativity enhancing methods, provided a background for developing the idea of critical creativity. I have suggested that a critical form of creativity is appropriate for use in TSI and put forward various practical measures for its pursuit. Finally I have shown that although, at first sight, critical creativity may seem to be a contradiction in terms, it is something which is necessary and practically possible in TSI.

REFERENCES

Anderson, H. H., (ed.), 1959, "Creativity and its Cultivation," Harper and Row Publishers, New York.

Barrett, F. J., and Cooperrider, D. L., 1990, Generative metaphor intervention: a new approach for working with systems divided by conflict and caught in defensive perception. *J. Appl. Behavioral Science*, 26:2, pp 219-239.

Behrooz, A., and Marshall, G., 1993, Metaphors as seen by a knowledge elicitor, *in*: "Systems Science: Addressing Global Issues," F.A.Stowell, D. West, and J.G.Howell, eds., Plenum Press, New York, pp 649-652.

Briskman, L., 1981, Creative product and creative process in science and art, *in*: "The Concept of Creativity in Science and Art," D. Dutton and M. Krausz, eds., Martinus Nijhoff Publishers, The Hague, pp 129-155.

Cohen, C., 1994, 'Magic' and a voluntary organization: the implications for community OR, *J. Opl. Res. Soc.* 45:3, pp 255-260.

Flood, R.L., 1995, Total systems intervention (TSI): a reconstitution , *J. Opl. Res. Soc.* (in print)

Flood, R.L., and Jackson, M.C., (1991 (a)), "Creative Problem Solving: Total Systems Intervention," John Wiley and Sons, Chichester.

Flood, R.L., and Jackson, M.C., (1991(b)), "Critical Systems Thinking: Directed Readings," John Wiley and Sons, Chichester.

Gordon, W.J.J., 1961, "Synectics", Harper and Row Publishers, New York.

Hare, A.P. 1982, "Creativity in Small Groups," Sage Publications, California.

Hausman, C.R., 1981, Criteria of creativity, *in*: "The Concept of Creativity in Science and Art," D. Dutton and M. Krausz, eds., Martinus Nijhoff Publishers, The Hague. pp 75-89.

Koestler, A., 1981, The three domains of creativity, *in*: "The Concept of Creativity in Science and Art," D. Dutton and M. Krausz, eds., Martinus Nijhoff Publishers, The Hague, pp 1-17.

Lamb, D., 1991, "Discovery, Creativity and Problem-Solving," Avebury Academic Publishing Group, Aldershot.

Merali, Y., and Martin, A., 1994, Metaphor in information systems teaching and research, *Systemist* 16:1, pp 12-19.

Metcalfe, R.J., 1983, "Assessment of Creativity - Rediguide 17," Nottingham University, School of Education.

Morgan, G., 1986, "Images of Organization," Sage Publications, London.
Osborn, A.F., 1963, "Applied Imagination," Scribners, New York.
Pickard, E., 1979, "The Development of Creative Abillity," NFER Publishing Company Limited, Windsor, Berks.
Stein, M.I., 1974, "Stimulating Creativity -Vol.1, Individual Procedures," Academic Press Inc., New York.
Stein, M.I., 1975, "Stimulating Creativity - Vol. 2, Group Procedures," Academic Press Inc., New York.
Vergara-Cabrera, E., 1981, "Creativity in Strategic Planning," PhD Thesis, Wharton School, University of Pennsylvania.
Whitfield, P.R., 1975, "Creativity in Industry," Penguin Books, Harmondsworth.
Yeoman, I., and Peters, S., 1994, Soft operations research methods, facilitation and the group process, Research Centre, Birmingham College of Food, Tourism and Creative Studies. (unpublished)

23

A THEORETICAL FRAMEWORK FOR THE DEVELOPMENT OF A SOCIAL ACCOUNTING SYSTEM

Alfonso Reyes

Aston Business School
Aston University Aston Triangle
Birmingham
B4 7ET
United Kingdom

"As a net is made up of a series of ties, so everything in this world is connected by a series of ties. If anyone thinks that the mesh of a net is an independent, isolated thing, he is mistaken. It is called a net because it is made up of a series of interconnected meshes, and each mesh has its place and responsibility in relation to the other meshes" (Buddha).

INTRODUCTION

The aim of this paper is to present a theoretical framework for the study of the internal coherence of social systems. Maintaining this internal coherence, in spite of continuous perturbations from its environment, is what we understand as the viability of the system. On the other hand, an effective organisation is one that preserves its viability and a system that permits the continuous measurement of this effectiveness is what we are naming as a social accounting system.

BASIC CONCEPTS

Building a sound theoretical framework requires precise definitions of the concepts underpinning it, therefore we will begin with the establishment of some definitions that will set our particular viewpoint from which we are looking at social systems.

A (social) organisation is defined as a closed network of interpersonal recurrent interactions (Espejo, 1994, p205). Its identity is the set of relationships defining the organisation as a whole, independently of the particular actors involved (Maturana and Varela, 1987, p47), and the actual resources and interactions constituting these relationships in a particular time and context define the structure of the organisation.

Critical Issues in Systems Theory and Practice, Edited by K. Ellis et al.
Plenum Press, New York, 1995

These definitions have a straightforward consequence: they imply that a social organisation is an autonomous entity as indeed is any system exhibiting organisational closure (Varela, 1979, p58). Therefore, changes in social organisations are structure (internally) determined because the environmental perturbations affecting the system are compensated by mutual adjustments of the recurrent relations of its constituting actors. In other words, external perturbations trigger but do not determine structural changes in social organisations. As long as these changes preserve the organisation's identity the system will be viable within its environment.

It should be noted that while there is a one-to-one relationship between a social organisation and its identity, there is a one-to-many relationship between a social organisation and its structure. The domain of this many-to-one mapping defines the domain of the organisation's viability, e.g., any structural change that falls outside this domain will imply the system losing its identity. This leads to the question of what are the 'rules' for structural change that maintain viability, or in other words, to preserve the organisation internal coherence? To answer this question we have to appeal to the notion of complexity.

COMPLEXITY

The complexity of a situation is the number of distinctions we can make on it. However, in a social organisation there are, at least, two different ways of doing these distinctions: either we are observing the interactions of actors constituting the organisation or we are participating in these interactions as actors ourselves. These two different but complementary ways of perceiving the world constitutes what have been called the informational and the operational domain respectively (Espejo, 1993, p77; 1994, p206).

These domains are closely related to Giddens' model of the social actor as proposed in his Structuration Theory. Here he differentiates between two non-disjunctive levels of consciousness: an individual's practical consciousness and an individual's discursive consciousness. Practical consciousness "consists of all the things actors know tacitly about how to 'go on' in the contexts of social life without being able to give them direct discursive expressions" (Giddens 1984, p.xxiii). Therefore the distinctions we make in our operational domain are distinctions made at the level of our practical consciousness. We call these distinctions our individual complexity. On the other hand, the distinctions we make in our informational domain are distinctions made at the level of our discursive consciousness.

From our discussion above about structural change in an organisation it should be clear that the individuals' complexities are the ones we have to consider in order to understand the internal coherence of a social organisation. To see this point let us return to our initial definition of a social organisation.

A FORMALISM

Let us consider a closed network of recurrent interactions constituted by a set $A = \{a_1, a_2, a_3, \ldots a_n\}$ of n actors. Let us define the set $C = \{c_1, c_2, c_3, \ldots\ldots c_n\}$ where each c_i represents the individual complexity of actor a_i at any given time. Let us classify these interactions according to the (average) number of individuals involved on them. Now, after engaging in a particular recurrent interaction, each actor will have the same or a different individual complexity depending on how his/her individual distinctions were affected by that interaction. That is, the same interaction could lead to different effects upon the individual complexities of its participants. We can formally represent this by defining the operator $R_{i,j}(a_1, a_2, \ldots a_i, \ldots a_j)$ as a recurrent interaction of j actors considered from the point of view of actor a_i. Let us agree that the 'value' of applying this operator is the individual complexity c_i, (e.g. the individual complexity of actor a_i) after the interaction.

Of course each particular actor could be engaged in many different recurrent interactions with many different actors during the constitution of the social organisation. Therefore, at any given moment the network of recurrent interactions can be formally represented by a set of simultaneous equations similar to the following:

$$c_1 = R_{1,3}(c_1, c_2, c_5) + R_{1,2}(c_1, c_4) + R_{1,2}(c_1, c_5) + \ldots\ldots$$

$$c_2 = R_{2,31}(c_1, c_2, c_5) + R_{2,21}(c_2, c_4) + \ldots\ldots.$$

$$\ldots. = \ldots\ldots\ldots\ldots\ldots\ldots\ldots\ldots\ldots\ldots\ldots\ldots\ldots\ldots$$

$$c_n = R_{n,31}(c_1, c_2, c_5) + R_{n,2}(c_1, c_5) + R_{n,2}(c_3, c_5) + \ldots\ldots. \quad (1)$$

where the + symbol stands for the common union operator of set theory. In general this can be represented as :

$$c_1 = S_1(c_1 , c_2, \ldots., c_n)$$

$$c_2 = S_2(c_1 , c_2, \ldots., c_n)$$

$$\ldots. = \ldots\ldots\ldots\ldots\ldots\ldots\ldots\ldots\ldots$$

$$c_n = S_n(c_1 , c_2, \ldots., c_n) \quad (2)$$

where S_i represents all the recurrent interactions in which actor a_i is engaged while participating in the constitution of the organisation.

Note that each individual complexity is a function of all the recurrent interactions in which a particular actor participates. Therefore if at any moment an individual complexity (say c_i) changes, this may change the individual complexity of other actors - as suggested by (2) - and this, in turn, will affect back the individual complexity of c_i and so on in an ongoing process.

Now, it can be proven that the formalism we have just presented constitutes what is known as a continuous algebra and therefore a system of equations represented by (2) has a solution of the form: $\underline{c} = S(\underline{c})$ where $\mathbf{\underline{c}}$ is the n-tuple $(\underline{c}_1, \underline{c}_2, \ldots.., \underline{c}_n)$ and S is the vector representing $(S_1, S_2, \ldots.., S_n)$. This value is known as a fixed point of the system of equations and in our present discussion is interpreted as a set of individual complexities that gives the organisation its stability, or in other words, its internal coherence. We call this set of distinctions 'eigen-distinctions' and proposed it to be the operational definition of the organisation's complexity. An important consequence of this definition is that organisational complexity is not the sum of the individual complexities of the actors constituting the organisation but the organisational closure of these individual complexities as represented by their eigen-distinctions.

Therefore, the history of structural changes in an organisation preserving its viability can be characterised by a sequence of eigen-distinctions $\underline{C} = \{\underline{c}_a, \underline{c}_b, \underline{c}_c, \ldots..\}$ each one measuring its corresponding organisational complexity. But before exploring in some detail this structural coupling (Maturana and Varela, 1987, p75) of a social organisation with its environment, let us consider some of the implications of the formalism just presented.

First, if we identify S as the structure of the organisation, in total agreement with our previous definition of structure, we note that the pair $(\underline{c}, S)$ represents a complementary pair in the sense that they specify each other. In other words, the set of distinctions $\underline{c}$ is both constituted by and constituting the structure of the organisation in an ongoing manner. In particular, if the structure of the organisation changes (for example as a result of an external perturbation) the organisational complexity will probably change as a result of this structural adjustment. But, at the same time, the way the organisation will react to an external perturbation will depend on its organisational complexity, that is, on its internal coherence.

Secondly, we see that while $\underline{c}$ represents a discrete value, S is a continuous process. In this way the formalism developed makes explicit the idea of time in the constitution of the organisation's complexity. In particular it reinforces Giddens' concept of reversible time in social institutions (Giddens, 1984, p35). In fact this is not the only connection between this formalism and Giddens' Structuration Theory as we will see in what follows.

GIDDENS' STRUCTURATION THEORY AND COMPLEXITY

Let us consider again the solution $\underline{c}$ of the set of equations (2). Each component of this vector can be written as $\underline{c}_i = S_i(\underline{c}_i)$. Now, $\underline{c}_i$ is a subset of the individual complexity of actor a_i, so it is a set of distinctions made by him/her at the level of his/her practical consciousness while participating in the constitution of the organisation. This set of distinctions is a sort of 'practical knowledge' used by actor a_i when engaging in interactions and so it coincides with Giddens' definition of 'rules' as "tacit knowledge" (Giddens, 1984, p21).

Therefore, each component of the solution of equation (2) may be interpreted as representing each actor applying a set of rules (e.g., bringing them forth in the moment of his/her interactions) and getting as a result of its application the same set of rules. In other words, these rules are both the outcome and the medium for the constitution of the organisation. But this is exactly Giddens' principle of the "Duality of Structure" (Giddens, 1984, p25) that is the corner stone of his Structuration Theory! Note, however, that what Giddens calls structure is what here we call eigen-distinctions.

Summarising so far, we have presented a set of concepts such as organisation, identity, structure, operational domain, informational domain, individual complexity and organisational complexity. These concepts are the basis of a new language of complexity rested upon the paradigm of the so called 'second order cybernetics' or cybernetics of the observing system in contradistinction to the cybernetics of the observed system (Von Foerster, 1979, p8; Espejo, 1993, p70). We believe that social cybernetics belongs to this higher order category as we have tried to make apparent by making an operational interpretation of part of a current sociological theory (e.g. Giddens' Structuration Theory). A complete interpretation will be developed elsewhere.

A SOCIAL ACCOUNTING SYSTEM

Let us now go back to our original question and study with more detail the internal coherence of a social organisation. Let suppose we concentrate in a particular organisation and consider the network of recurrent interactions that constitute it. Actors participating in these processes ascribe tacit purposes to their actions by doing certain things and no others, this is their theory in use (Argyris and Schön, 1978). When a set of these recurrent interactions achieves organisational closure an autonomous unit will emerge. We may say that these actors by maintaining these recurrent interactions are ascribing a 'common tacit purpose' to this autonomous unit. In a similar way different sets of autonomous units will emerge. This is a characterisation of a self-organising process. If we consider now each of these autonomous units as a 'black box' then a similar process, but now considering recurrent interactions between autonomous units, will result in a set of autonomous units at a higher level of abstraction - or to use Beer's terminology, at a higher level of recursion (Beer, 1979, p314). By continuing with this process an autonomous unit corresponding to the organisation as a whole will emerge at a particular recursion level. We call this process, carried out in the operational domain, the 'synthesis of organisational complexity' as a contradistinction of Espejo's (1989, p90) "unfolding of complexity" that is carried out in the informational domain.

Let us now look at two consecutive recursion levels L_i and L_{i+1}. An autonomous unit belonging to the level L_{i+1} will have $\underline{c}_{i+1}$ as its organisational complexity. This complexity emerges as an eigen-distinction of the network of interactions of autonomous units at the level of recursion L_i. Let us suppose there are n of such units each one with its respective organisational complexity. Let us identify these complexities by $\{\underline{c}_{i,1}, \underline{c}_{i,2}, \underline{c}_{i,3}, \ldots, \underline{c}_{i,n}\}$. Of course each $\underline{c}_{i,j}$ is, in turn, an eigen-distinction of a network or recurrent interactions occurring at a still lower level of recursion.

Now, the relations keeping the stability of the unit al level L_{i+1} are restricting the organisational complexity of the autonomous units at the level L_i. This is because from all the possible values that their complexities can take, only those corresponding with eigen-distinctions will keep the stability of the unit at the level L_{i+1}. In other words we can see the unit at level L_{i+1} as 'controlling' the units of the level immediately below. Here we understand 'control' as autonomy restriction. Again, the same argument applies for each unit at the level L_i with its autonomous units at a level L_{i-1} and so on until we reach the lowest relevant recursion level. Therefore, we claim that one of the rules for structural change that maintain viability is this balance between control and autonomy at each pair of recursion levels in the structure of an organisation. In other words as long as a structural adjustment resulting from an external perturbation preserves this all-levels balance between autonomy and control, the system will keep its internal coherence as a whole.

The practical criterion for monitoring this balance is, of course, the Law of Requisite Variety (Ashby 1964) as applied between two consecutive recursion levels. That is, as long as the residual variety (the one that is left unattended by internal self-regulation) of the units at the level of recursion L_{i-1} is matched by the variety at the level of recursion L_i by means of designed attenuators and amplifiers, the balance will be maintained.

Therefore, a social accounting system continuously measures the organisational complexity of autonomous units at each level of recursion and applies the Law of Requisite Variety to two consecutive levels as a criterion of balance.

Summing up, we have presented some concepts that are the basis for a new language of complexity. A language we believe is suitable to the study of social organisations specially because it makes apparent the distinction between our operational domain and our domain as observers. We think that a failure in doing this distinction is probable one of the reasons some paradoxical problems arise from the study of social organisations, (e.g. Geyer and Van der Zouwen, 1986).

Central to our discussion was the idea of autonomy and its companion of internal coherence that led us to our proposal of a social accounting system. The theoretical framework presented in the paper is, however, an initial step in developing a more robust approach including a sound methodology. We are aware that some methodological questions were left unattended as, for example, how can we properly identify 'common tacit purposes' that are fundamental for defining the boundaries of autonomous unit. However, trying to answer them is what motivates our current efforts of research.

REFERENCES

Argyris, C., and Schön, D., 1978, "Organizational Learning: A Theory of Action Perspective," Addison-Wesley, Reading, Mass.

Ashby, R., 1964, "An Introduction to Cybernetics," Methuen, London.
Beer, S., 1979, "The Heart of Enterprise," John Wiley & Sons, Chichester.
Espejo, R., 1989, The viable system model revisited, *in:* "The Viable System Model: Interpretations and Applications of Stafford Beer's VSM," R. Espejo and R. Harnden, ed., John Wiley & Sons, Chichester.
Espejo, R., 1993, Management of complexity in problem solving, *in* "Organisational Fitness," R. Espejo and M. Schwaninger, eds., Campus Verlag, Frankfurt.
Espejo, R., 1994, What is systemic thinking?, *System Dynamics Review*. 10:199.
Geyer, F. and Van der Zouwen, J., eds., 1986, "Sociocybernetic Paradoxes," Sage, London.
Giddens, A., 1984, "The Constitution of Society," Polity Press, Cambridge.
Maturana, H., and Varela, F., 1987, "The Tree of Knowledge: The Biological Roots of Human Understanding," New Science Library, Boston.
Varela, F., 1979, "Principles of Biological Autonomy," Elsevier, New York.
Von Foerster, H., 1979, Cybernetics of cybernetics, *in:* "Communications and Control in Society," K. Krippendorff, ed., Gordon and Breach, New York.

24

EXPLORING THE NOTION OF CHANGE IN BUSINESS AND ORGANIZATIONAL SYSTEMS

Francis Stickland

Engineering Management Centre
Department of Systems Science
City University
Northampton Square
London, EC1V 0HB
United Kingdom

ABSTRACT

The paper seeks to open up a debate on the nature of change at a conceptual and theoretical level. It argues that the abundance of methodologies and strategies for managing change within business and organisational systems belie a desperate lack of any clear understanding of the very nature and essence of change itself. This deficiency is seen as a critical issue which is currently constraining further theoretical development of intervention approaches for organisational systems, and must be addressed. A greater understanding of the fundamental dynamics of change is called for. A cross discipline approach is proposed and the paper concludes with some suggestions for work.

INTRODUCTION

Change is a concept which has increasingly become a focus of much interest and study. The pace of change seems to be accelerating away in many spheres of human activity, and is now reaching unparalleled levels within our technological and industrial systems, as Callan (1993) has noted. Organisation theory and management science in particular have focused closely on developing effective strategies for managing innovation and change within business and organisational systems. Managers are now confronted with a bewildering array of methodologies, techniques and strategies to assist them in changing some aspect of their business for the better.

However the main thrust of much of this work has been to answer the practical 'how' questions - how change can be effectively dealt with. Arguably, the more fundamental 'what' questions have yet to be seriously asked at a generic level: What is the *nature* of change? What characteristics and attributes does change possess? What structural features, interactions and processes define change? What basic principles govern change, if any? What marks

Critical Issues in Systems Theory and Practice, Edited by K. Ellis et al.
Plenum Press, New York, 1995

the beginning and end of change? What perception and measurement issues are associated with change? It is the firm belief of the author that these and other such probing questions need to be asked in an attempt to reveal the nature of change at a fundamental level. Despite the plethora of approaches and tools for managing change within business systems, there has yet to emerge any solid theoretical framework for exploring and describing its nature and dynamics. Practical approaches to change management continue to multiply in number, often as 'one off' solutions to specific organisational change problems. But as one commentator has observed: "Practice is static. It does well what it knows. It has however no principle for dealing with what it doesn't know.... Practice is not well adapted for rapid adjustment to a changing environment. Theory is light footed. It can adjust itself to changing circumstances, think out fresh combinations and possibly, peer into the future" (Litwick 1987).

Organisational change practice - for better or worse - is well established but far from fleet of foot. A little theory and conceptual insight would make it far more responsive to the capricious demands being placed upon it. By necessity what is written here can only be a somewhat incomplete and inadequate initial foray into the subject, and will raise more questions than it answers. The purpose is not to advocate some grand theory of change, but rather to encourage the exploration of change as a concept and hopefully to stimulate more debate in the literature. In time, it is hoped that the investigative approach proposed will lead to a deeper and more complete understanding of the dynamics of change within organisations.

THE PROBLEM HIGHLIGHTED

Vickers (1980) alluded to the need for such a basic understanding of the nature of change when he said: "The view of entities as both systems and constituents of systems raises intriguing questions about identity and continuity. When does something, or somebody, retain its identity and continuity through change? When by contrast does it cease to be its old self and either vanish or become something new or different? The question is not frivolous or merely metaphysical but may be of great practical concern" (Vickers, 1980).

Some practitioners may point to the apparent success of certain well established applied research techniques and strategies for dealing with organisational change, and query why such fundamental questions need to be asked at all. However, as Golembiewski et al (1976) have observed: "There is a truism about applied research that an inadequate concept of change leads to diminished or misguided applied research."

While exploring the concept of change within organisations in the late 1970's, Berg (1979) came to the conclusion that "...this has been a much neglected area of research...[and].... that much of the present confusion and ambiguity in the field stems from the fact that we actually know very little about the nature of change". March (1981) has also highlighted the weak theoretical base of organisational change literature. In his discussion of broad theoretical notions about organisation action, he states that "...a comprehensive development of managerial strategies...requires a more thorough understanding of change in organisations, *not a theory of how to introduce any arbitrary change*." (italics added). Here again we see concern expressed at the apparent over emphasis on answering the *how* question. More recently Morgan (1986) has identified the need for a deeper comprehension of the fundamentals of change. He argues that "...we need to try and understand how the discrete events that make up our experience of change are generated by a logic unfolded in the process of change itself". Writing on organisational change back in the late 1960's, Bennis et al (1970) stated: "We seem, quite often, to become lost at the crossroads of a false dichotomy; the purity and virginity of theory on the one hand and the anti-intellectualism of some knowledge-for-what adherents on the other". It would seem that in the intervening years, we have passed this crossroads, having chosen to tread with vigour the practical path - down which we are now making great strides. Sadly, relatively few have ventured out upon the theoretical road. As Wilson (1992) recently lamented, "...much of the current vogue in management theory is for delineating the steps through which successful change can occur...". He goes on to highlight the explosion of *change recipes* "...which have little empirical or theoretical foundation [and] should give cause for grave concern amongst all who work in organisations of every type." (Wilson, 1992: 120).

Various calls then have been made over the years right up to recent times, for greater attention to be given to understanding the *nature* of change. These appeals have largely been made in the conviction that organisational change practice would benefit considerably from such research. It would however, be unfair to say that they have gone totally unheeded. Initial attempts have been made to explore the phenomenon of change as it is manifested in organisational and business systems. Some of these will now be examined briefly.

SOME TENTATIVE BEGINNINGS

Morgan (1986) realised the need for a more profound understanding of the concept of change. He has examined the organisation through what he terms the metaphor of flux and transformation, considering how organisations may exist within deep structures and processes which possess their own logics of change. He proposes three such 'logics': self organisation principles drawn from biology; concepts of mutual causality taken from cybernetics; and dialectics and the notion of opposites taking ideas from Tao philosophy and Marxian analysis.

In discussing processual analysis of organisational change strategies, Van de Ven (1987) defines change as "...an empirical observation of differences in time on one or more dimensions of an entity." Here the concept of change is closely associated with the passage of time and physical observation. He argues that the *process* of change can only be indirectly perceived as "...conceptual inferences about the temporal ordering of relationships among observed changes." (Van de Ven, 1987: 331). Building on the work of Dahrendorf (1959) and Hernes (1976), Van de Ven suggests that a good, robust theory of change in social structures like organisations should satisfy four basic requirements:

- It should explain how change, behaviour and structure are interconnected at both macro and micro levels of analysis;
- It should describe how change is a function of internal and external factors;
- It should account for both change and stability;
- The theory should incorporate time as the *key historical metric*.

Pettigrew (1990) argues there has been an over emphasis with prescriptive writing in the literature, leading to under concern with descriptive analysis and conceptualisation. He highlights the contextual nature of change within organisation theory, emphasising:

> "...embeddedness, studying change in the context of interconnected levels of analysis";
>
> "...temporal interconnectedness, locating change in past, present and future time"
>
> "...the need to explore...how context is a product of action and vice versa..."
>
> "...causation of change is neither linear nor singular."

Pettigrew's research identifies the need to study change across different levels of analysis and different time periods, and goes a considerable way towards refuting the simplistic, one dimensional and discontinuous view of change within management thinking.

Smith (1982) has attempted to analyse some of the philosophical problems that cloud our thinking about organisational change. He concludes that an organisation consists essentially of "...relations among parts and relations among relations..." (Smith, 1982: 318). Therefore any concept of organisational change must be founded upon changing *relationships*. These he argues, can only be altered by changing the metaphors, analogies and metonymies we use to describe them. He discusses the notion of boundary as the place where change occurs, and emphasises that it too is a relation, and not part of the structure of the organisation. He draws heavily on the language of biology and other life sciences to explore the concept of change, making a clear distinction between the rules which govern internal structure changes, and those which determine changes of order at the level of the whole.

Lauer (1971) argues that research into the nature of change within social systems has been constrained and 'neutralized' by certain misleading common sense assumptions. These he calls *fallacies* and suggests that they have gained such scientific legitimacy as to impede further study into the nature of change:

- Change represents some deviance from the norm;
- Change is associated with trauma and crisis, to be regarded as foreign and unwanted;
- Change is considered to be unidirectional and deterministic, converging inevitably towards some predestined end;
- The difference between theories of change and theories of stability constitutes a semantic illusion.

Lauer concludes that recognition and understanding of these fallacies would serve to inspire further theoretical exploration into the nature of change. His analysis represents a genuine attempt to explore the nature of change in a social context, by examining some of the common misconceptions which often distort our understanding of the concept.

Clearly then, some efforts have been made to define and explore the nature of organisational and social change. The foregoing is by no means a comprehensive review and others have attempted to survey the literature: for example Goodman (1982); Zaltman & Duncan (1977). Nonetheless, it does capture some of main themes which have been examined to date. However, to take this initial work forward a multi-disciplinary approach is called for. Change is a phenomenon which occurs across the disciplines. Specialists from different knowledge domains have their own language, concepts and theories for describing and explaining it. Exploring what other subject areas have to say about the nature of change and abstracting any potentially useful theoretical insights they may offer at a generic level, could lead to a greater understanding of change within business and organisational systems. One such approach will now be described.

A MULTIDISCIPLINARY APPROACH

In his discussion of General Systems Theory (GST), the economist Boulding (1956) outlined a strategy for dealing with what he termed phenomena of 'universal significance'. This was "...to look over the empirical universe and pick out certain general phenomena which are found in many different disciplines, and to seek to build up general theoretical models relevant to these phenomena" (Boulding, 1956) Arguably, change is just such a phenomenon. According to von Bertalanffy (1968) one of the original stated aims of GST was to "...investigate the isomorphy of concepts, laws, and models in various fields, and to help in useful transfers from one field to another...". This noble objective is perhaps far more embracing and ambitious than what is being proposed here with regard to change. Nonetheless, comparing and pooling concepts of change from across the disciplines does serve as a useful starting point. As a first stage then, well documented change phenomena from the physical and natural sciences are identified and examined. This will involve an analysis of the theories and notions employed by various scientific communities to describe and explain change phenomenon. Seemingly disparate change phenomenon may reveal many similarities in form and process when examined in this manner. Such analysis would serve to identify:

- Fundamental principles or themes of change;
- General characteristics and attributes of change;
- Problem issues associated with system intervention to measure or manage change.

If successful, the results of this analysis could be used to form the beginnings of a general framework which embodies the fundamental dynamics and characteristics of change at a generic level. The initial utility of such a framework would be its descriptive and explanatory ability - uncovering some of the underlying processes and dynamics of change. It would also serve as a useful aid for identifying and classifying different types of change according to attribute and characteristic. Figure 1 below summarises the overall approach.

Fruitful metaphors and analogies thrown up by the phenomena under study which describe some aspect of change are of particular interest, however obscure. While the flux metaphor of Morgan (1986) discussed earlier, provides a basis for considering different concepts of change, it is very general. Of more interest here is the identification of specific, well defined metaphors *for* change rather than the collective metaphor of change itself. Metaphorical thinking is becoming well established as an explicit tool with which to study various facets of system behaviour, particularly within organisation theory: Tsoukas (1991); creative management, Henry (1991); GST, Rappaport (1988) and Critical Systems Thinking, Flood & Jackson (1991). The derivation of new and original change metaphors using this approach could do much to revitalise our conception of the nature of change within organisations. It is acknowledged that exploring the concept of change in this manner may not in the short term yield the comprehensive and robust methodologies which change management practitioners desire. The initial thrust of the approach is somewhat abstract and

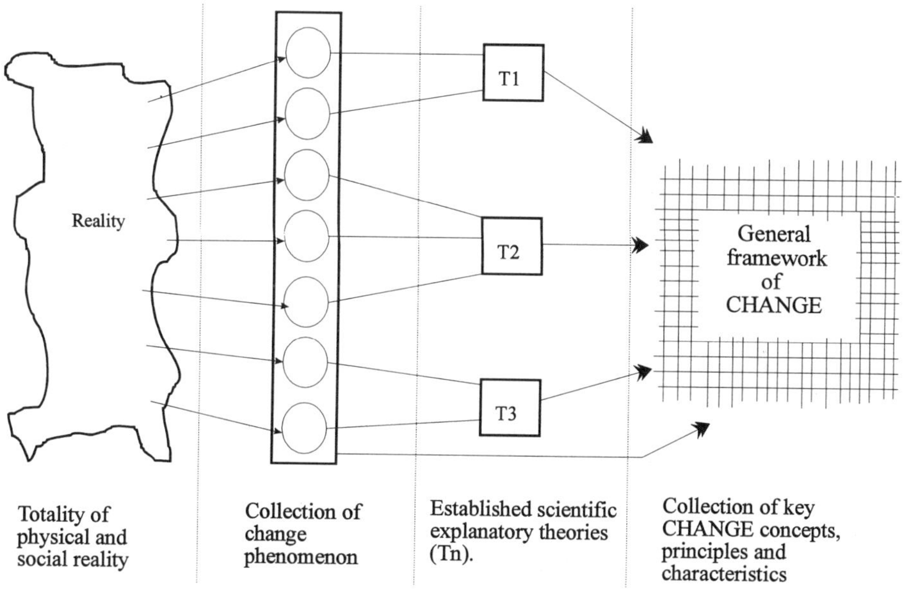

Figure 1. Exploring the concept of change.

conjectural in nature. However, as Lundberg (1984) observed while discussing the notion of change within organisations, "...even speculative reasoning which is carefully done and which probes the pragmatic dimensions of a major, increasingly crucial phenomenon, has utility, for it begins to inform and guide practice and to stimulate inquiry". Pursuing the approach outlined here should open up an abundance of change concepts from across the disciplines, making them available to specialists within organisation theory and management science. It is hoped that in the long run, these will provide organisational change theorists and practitioners with a fresh source of potentially fruitful concepts, metaphors and analogies from which to draw inspiration and enhance their language, theories and methods.

SCOPE FOR FURTHER WORK

There are many manifestations of change phenomena well documented in the natural and physical sciences, some of which have been explored elsewhere (Stickland & Reavill, 1994). Chaos theory and non-linear dynamics contain many ideas relevant to the study of organisational change, including for example the concepts of strange and chaotic attractors. Work is already well under way in applying chaos theory to organisational behaviour: see Zimmerman & Hurst (1993) but much remains to be done in exploring the concept of change itself. Work being done in the various branches of complexity science, such as genetic algorithms, neural networks and adaptation, cellular automata and artificial life is all concerned with modelling and understanding the process of change over time in dynamic complex systems, and is likely to contain many change concepts worth exploring.

CONCLUSIONS

It is the author's belief that lying hidden among the multitude of specialist disciplines of natural and physical science are a wealth of rich and useful change concepts suitable for abstraction at a general level, and appropriate for application as metaphors and structural analogies at a conceptual level into other fields of study, particularly organisation theory and management science. Theories of change in the social sciences also have much to offer, and some are currently being explored by the author. In time, organising all these ideas into some semblance of order could lead to the development of an initial conceptual framework describing the nature of change. Such a framework would form a much needed theoretical foundation for the organisational change practitioners, providing them with a sound foundation from which to develop new strategies and tools for change management. It may also serve as a useful basis from which to assess the theoretical footing and ontological validity of the many and varied methodologies for organisational change already in existence.

REFERENCES

Bennis, W.G., Benne, K.D., and Chin, R., eds., 1970, "The planning of change", Holt Rinehart and Winston, London.

Berg, P.O., 1979, "Emotional Structures in Organisations: A Study of the Process of Change in a Swedish Company", University of Lund Student Literature, Sweden.

Bertalanffy, L. von. 1968, "General Systems Theory", George Braziller, New York.

Boulding, K.E., 1956, General systems theory, *Mgmt Sci.* 2 (3):199.

Callan, V.J., 1993, Individual and organisational strategies for coping with organizational change, *Work and Stress*. 7 (1):63-75.

Dahrendorf, R., 1959, Social theory, social research and a theory of action, *Am. J. Soc.* 16:1309-1335.

Flood, R.L. and Jackson, M. 1991, Total systems intervention, *Syst. Pract.* 4(3): 197-213.
Golembiewski, R.T., Billingsley, K. and Yeager, S., 1976, Measuring change and persistence in human affairs, *J. App. Behav. Sci.* 12:133.
Goodman, P.S., and Associates, Eds. 1982, "Change in Organizations", Jossey-Bass, San Francisco.
Henry, J., 1991, "Creative Management", Sage Publications, London.
Hernes, G., 1976, Structural change in social processes, *Am. J. Soc.* 82(3):513-545.
Lauer, R., 1971, The scientific legitimation of fallacy, *Am. Soc. Rev.* 36:881-889.
Litwick, R., 1987, quoted *in*: "Encylopedia of Organisational Change Methods," A. Huczynski, Gower Publications, Aldershot.
Lundberg, C.C., 1984, Strategies for organisational transitioning, *in*: "New Futures", J.R. Kimberly and R.E. Quinn, eds., Dow Jones-Irwin, Homewood, IL.
March, J., 1981, Footnotes to organisational change, *Admin. Sci. Quart.* 26: 575.
Morgan, G., 1986. "Images of organisation", Sage Publications, London.
Pettigrew, A.M., 1990, Longitundinal field research on change, *Org. Sci.* 1(3):269.
Rappaport, A., 1988, The relevance of the systemic outlook to our present predicament, *in*: "Yearbook of the International Society for the Systems Sciences," 31:5-12, New York.
Smith, K.K., 1982, Philosophical problems in thinking about organizational change, *in*: "Change in Organisations", P.S. Goodman & Associates, eds., Jossey-Bass, San Fransisco.
Stickland, F. and Reavill, L.R., 1994, Understanding the nature of system change, *in*: Procs. 38th Annual Meeting of the Int. Soc. Sys. Sci. 2:1509-1520, Pacific Grove, CA.
Tsoukas, H., 1991, The missing link: A transformational view of metaphors in organizational science, *Acad. Mgmt Rev.* 16(3):566-585.
Van de Ven, A.H., 1987, Review essay: Four requirements for processual analysis, *in:* "The management of strategic change," A. Pettigrew, ed., Basil Blackwell, Oxford.
Vickers, G., 1980, Some implications of systems thinking, *in*: "Systems Behaviour", J. Beishon & G. Peters, eds., Harper and Row, Milton Keynes.
Wilson, D.C., 1992, "A Strategy of Change: Concepts and Controversies in the Management of Change," Routledge, London.
Zaltman, G., and Duncan, R., 1977, "Strategies for Planned Change", John Wiley and Sons, Chichester.
Zimmerman, B.J. and Hurst, D.K., 1993, Breaking the boundaries: The fractal organization, *J. Mgmt Inquiry.* 2 (4):334-355.

25

THE PHILOSOPHICAL FOUNDATIONS OF CRITICAL SYSTEMS THINKING

Beyond Habermas, toward Foucault

Néstor Valero-Silva

Centre for Systems Studies
Department of Management Systems and Sciences
University of Hull
Hull
United Kingdom

INTRODUCTION

Michel Foucault and Jürgen Habermas represent two of the most influential contemporary philosophers of the post-war era. Their studies of modern society have contributed to continuous debate and development in moral and legal philosophy, sociology, gender studies, and more recently, in systems science - specially within the paradigm of Critical Systems Thinking. However, it is important to highlight that most of the Critical Systems literature (e.g., Ulrich, 1983; Jackson, 1991) has concentrated almost exclusively on the work of Jürgen Habermas. Only Flood (1990) has considered Foucault in any depth, but he has tended to focus on Foucault's earlier work. As far as I am aware, his later work has not yet been assessed by Critical Systems thinkers at all. There may be possible reasons for the dominance of Habermasian thinking. One is Habermas's well established reputation as part of the Frankfurt School's efforts to challenge traditional conceptions of science and social theory, as initially proposed by Horkheimer and Adorno. Another is Foucault's refusal to systematically construct and argue for a theory, a feature that he shares with other contemporary thinkers. A third is that Foucault's philosophy does not provide a normative framework to guide and evaluate social action.

A further crucial problem, for some people even more controversial, arises when trying to track Foucault's 'background'. In this sense, he has followed the difficult path initiated by Nietzsche. Along with Sartre, Foucault was a militant during the student protests in the sixties. Most of his biographers dedicate entire chapters to his homosexuality and to his frequent visits to sadomasochistic bars in San Francisco, trying to unearth 'special connections' between his private life and his ideas. He has been associated with the controversial postmodernists and poststructuralists, he has been called nihilist, and neo-conservative. No wonder that the systems literature has avoided the work of Foucault, which is at first sight incoherent, nihilistic, full of interesting stories, but frightening when he pictures

Critical Issues in Systems Theory and Practice, Edited by K. Ellis et al.
Plenum Press, New York, 1995

the development of biopower. At the end, he leaves readers with a feeling of uncertainty, and a pessimistic view of the future.

There are a number of questions that is beyond the scope of this paper to tackle. Questions like whether Habermas's (1984) description of the colonisation of the lifeworld by the system, or his evolutionary model, or his theory of communicative competence can bring hope for the future, or provide a more feasible model for the assessment of the present state of affairs, or even a normative framework to guide social action. In any case, Foucault's refusal of 'improvement', at least in universal terms, does not appeal to the systems community, which is to a large extent confronted in its daily work with issues related to the 'improvement' of the efficiency of organisations.

However, it is important to highlight here that I do not assume that Habermas's thought is less critical about modern society than Foucault's, but rather, that they both represent alternative ways of continuing with the tradition of critical thought initially proposed by Horkheimer and Adorno. In this context, I will highlight some of the differences between Foucault's and Habermas's ideas, which make them theoretically incommensurable, and finally, I will reflect briefly on what Foucault can offer to the Critical Systems project.

FOUCAULT AND THE FRANKFURT SCHOOL

A major problem when studying both Habermas's and Foucault's work has always been to establish their connections (Flood, 1990). However, I think that one must approach their ideas assuming, as Couzens-Hoy (1994) does, that "...whereas Foucault is interested in the *historicity* of reason, Habermas is interested in the *theory* of reason. Each sees his question as the only possible way out of what Horkheimer and Adorno called the dialectic of enlightenment. That is, the modern search for knowledge that promised enlightenment and freedom but has produced domination and barbarism as well." (page 146).

However, we must also highlight that Foucault acknowledges the relevance of Habermas's work in the development of contemporary philosophy, while regretting the lack of communication that existed between them. In this sense, in an interview with Raulet (1983) he says,

> "If I had been familiar with the Frankfurt School, if I had been aware of it at the time, I would not have said a number of stupid things that I did say and I would have avoided many of the detours which I made while trying to pursue my own humble path - when, meanwhile, avenues had been opened up by the Frankfurt School....And when I was a student, I can assure you that I never once heard the name of the Frankfurt School mentioned by any of my professors" (page 200).

From the other side, Habermas, in a separate interview, praised Foucault's "...masterly description of the moment reason bifurcated. This bifurcation was unique. It happened once. At a certain point, reason took a turn which led it to instrumental rationality, an auto-reduction, a self-limitation. This bifurcation, if it is also a division, happened once and once only in history, separating the two realms with which we have been acquainted since Kant." (Raulet, 1983, p. 201).

While the two thinkers clearly appreciated each others' work, the quotation above illustrates a key difference between the two philosophers. While Habermas acknowledges a bifurcation of reason, that since it only occurred once can be rectified, Foucault assumes that reason is in a constant process of bifurcation in some way or another. In this sense Foucault remarks,

> "I would not speak about *one* bifurcation of reason but more about an endless, multiple bifurcation - a kind of abundant ramification. I do not address the point at which reason became instrumental. At present, for example, I am studying the problem of techniques of the self in the Greek and Roman antiquity; how man, human life and the self were all objects of a certain number of *technai*

> which, with their exacting rationality, could well be compared to any technique of production...this does not constitute *the* bifurcation of reason....it was an important event, or episode; it had considerable consequences, but it was not a *unique* phenomenon." (Raulet, 1983, p. 201).

Let us also review some other differences of paramount importance in the debate between Habermas and Foucault. Firstly, whereas Habermas assumes that the rational critique of rationality requires a previous theory of 'what rationality really is' in order to assess current versions of rationality, Foucault thinks that 'reason is self-created', which means that humans develop different conceptions of rationality given specific historical conditions. As a consequence, in Foucault's terms, the assessment of different forms of rationality according to some universal standard is meaningless.

In this context, Foucault (1984) praises Kant's reflections about the present which do not give any privilege to the present over the past and, in his interview with Raulet, he attacks those who give a special status to the present over the past, in either positive or negative terms.

> "The solemnity with which everyone who engages in philosophical discourse reflects on his own time strikes me as a flaw...(we should not)...allow ourselves the facile, rather theatrical declaration that this moment in which we exist is one of total perdition, in the abyss of darkness, or a triumphant daybreak, etc. It is a time like any other, or rather, a time which is never quite any other" (p. 206).

In sum, Foucault does not assume, as Habermas (e.g. 1984) seems to do, the superiority of present rationality over other practices of rationality, i.e. that previous forms of rationality can be understood as more primitive in an evolutionary sense. Neither does he assume the position claimed by other contemporary thinkers that we are witnessing the 'collapse of reason'. In this respect, he emphatically says,

> "For me, no given form of rationality is actually reason. So I do not see how we can say that the forms of rationality which have been dominant in the three sectors I have mentioned (types of knowledge, forms of technique, and modalities of government or domination) are in the process of collapsing or disappearing. I cannot see why we should call this transformation a collapse of reason. Other forms of rationality are created endlessly. So there is no sense at all to the proposition that reason is a long narrative which is now finished, and that another narrative is under way" (Raulet, 1983, p. 205).

WHAT CAN FOUCAULT OFFER?

Before considering what Foucault can actually offer, it is necessary to discuss another issue that, in my opinion, is not generally appealing to the 'systems community'. That is, Foucault's refusal to provide universal standards to guide and evaluate social action.

As mentioned before, Foucault's refusal to construct a theory of rationality, together with his studies about the development and expansion of an increasingly disciplinary society, provide a pessimistic view of the future. In this sense, he denies the possibility for improvement, in any other than historically contingent terms, making it meaningless to aim for the restoration of reason as it was constructed at any previous historical moment.

This does not mean, however, that his ideas necessarily involve a fatalistic view about the future, nor that the future of our society is already determined. On the contrary, if we consider for a moment what his main project, 'the critical ontology of ourselves', assumes (see Valero-Silva, 1994, for a review), we will find that since our present is the actualisation of one of the possibilities embodied in our past, we could attempt to materialise, in our future, one of the possible alternatives embodied in our present situation, or even create an entirely new one. A key point is that genealogical studies represent particular explanations of the present, yet changes we introduce create new, different, unique, and problematic versions of present, since, as Foucault says, all alternatives are dangerous. It is because of his awareness of danger that, when Foucault was asked whether his work on Greek ethics might offer a basis for an alternative to modernity, he replied:

> "No! I am not looking for an alternative; you can't find the solution of a problem in the solution of another problem raised at another moment by other people. You see, what I want to do is not the history of *solutions*, and that's the reason why I don't accept the word "alternative". I would like to do the genealogy of *problems*, of *problematiques*. My point is not that everything is bad but that everything is *dangerous*, then we always have something to do. So my position leads not to apathy but to a hyper- and pessimistic activism. I think that the ethico-political choice we have to make every day is to determine which is the main danger" (Hiley, 1984).

As a consequence, since different self-understandings are just different interpretations that appear in contingent historical moments, without any logical necessity, we can only adopt a critical position about our present problematics 'from within'. In this sense, we can change what we consider undesirable, being able to 'unlearn', and construct ourselves in a different way. This critique and evaluation of any particular self-understanding 'from within', without universal standards, is possible, according to Nehemas (1988), since

> "We acquire our beliefs, our desires, our values at different times, in different circumstances, from various sources. Each of us belongs to a number of distinct but overlapping communities, and each community has its own but overlapping standards of rationality. We do not need to stand outside everything in order to criticise and evaluate the standards and practices, the "reason" of any particular community. All that we need is a position outside it. This position is provided by the many other communities of which we are all members" (p. 34).

However, since Foucault's project is a critical ontology of 'ourselves' we need to discuss his assumption of social diversity, and the possibility of social interaction and consensus.

According to Couzens-Hoy (1994, p. 178), Foucault assumes that one must not be *for consensuality*, but one must be *against nonconsensuality*. This is because both entail the danger of an intolerance of difference. In this sense, we can highlight how Foucault understands social action, i.e. without the belief that any discussion of social problems would result in consensus about the best solution. However, if consensus is reached, it would be historically contingent, and not the result of evolutionary development; i.e. what Foucault objects to is not consensuality or social solidarity per se, but the elevation of solidarity and consensus to the status of a superior modus operandi.

I believe that Foucault's refusal to provide universal standards should be seen as an opportunity rather than as lack or a deficiency, leaving us with the possibility of no longer being, doing or thinking what we are, do, or think. This process must be based on our contingent understanding of the events that have "...led us to constitute ourselves, and to recognise ourselves as subjects of what we are doing, thinking, saying" (Foucault, 1984).

CONCLUSIONS

Foucault's genealogical project of how problems came to be considered as such, in historically contingent circumstances, is a provocative framework for further research in systems theory, particularly for Critical Systems thinkers. However, it is necessary to analyse its potential for social action from a different perspective to the one generally assumed through the work of Habermas.

Firstly, as Foucault (1984) says, the critical ontology of ourselves aims to address the question of "...how we are constituted as subjects of our own knowledge, as subjects who exercise power, and as moral subjects of our own actions" (p. 49), in order "...to create a history of the different modes by which, in our culture, human beings are made subjects....subjects to someone else by control or dependence, and tied to his own identity by a conscience or self-knowledge. Both meanings suggest a form of power which subjugates and makes subject to" (Foucault, 1982, p. 212).

Secondly, we must understand that the absence of any normative framework, or universal standards, does not mean that genealogy is neutral or unengaged. In this sense Hiley (1984) remarks,

> "Since the genealogy of power is self-consciously written in the present and situated in the danger it seeks to see, it is not value-free, if by value-free one means that it is neutral and unengaged. But also, because it is necessarily caught in the present, it is not normative, if by normative one means having criteria or a meta-story to tell the better from worse, legitimate from illegitimate exercise of power" (p. 197).

Finally, if a particular form of rationality represents just one alternative amongst others, and consequently, if we assume that both ourselves and our problems are constructed, we could aim to deconstruct ourselves, change characteristics that we may consider undesirable, and rebuild ourselves in different ways. As a consequence, we should regard Foucault's ideas, as Hiley (1984) points out, as an invitation for political engagement, but without liberal hope or comfort.

ACKNOWLEDGMENTS

I would like to thank Gerald Midgley for his advice in the preparation of this paper.

REFERENCES

Couzens-Hoy, D., 1994, Critical theory and critical history, *in:* "*Critical Theory*", McCarthy, T. and Couzens-Hoy, D., eds., Blackwell, Oxford.

Dávila, J., 1993, Foucault's interpretive analytics of power, *Systems Practice*. 6: 4.

Dreyfus, H., and Rabinow, P., 1982, "Michel Foucault, Beyond Structuralism and Hermeneutics", University of Chicago Press, Chicago.

Flood, R. L., 1990, "Liberating Systems Theory", Plenum Press, New York.

Foucault, M., 1984, What is Enlightenment ?, *in* "*Foucault Reader*", Rabinow, P., ed., Pantheon, New York.

Foucault, M., 1982, The subject and power, *in:* "*Michel Foucault, Beyond Structuralism and Hermeneutics*", Dreyfus, H., and Rabinow, P., University of Chicago Press, Chicago.

Habermas, J., 1984, "The Theory of Communicative Action", 2 vols., Beacon Press, Boston, Mass.

Hiley, D. R., 1984, Foucault and the analysis of power: political engagement without liberal hope or comfort, *Praxis International*. 4: 2.

Jackson, M. C., 1991, "Systems Methodology for the Management Sciences", Plenum, New York.

Nehemas, A., 1988, Review of Habermas's philosophical discourse of modernity, *The New Republic*. 198 (22).

Raulet, G., 1983, Interview with Gérald Raulet, *Telos* 55. Spring.

Ulrich, W., 1983, "Critical Heuristics of Social Planning: a New Approach to Practical Philosophy", Paul Haupt, Berne, Switzerland.

Valero-Silva, N., 1994, Michel Foucault: power, knowledge and the "critical ontology of ourselves", *Systemist*. 6: 3.

26

OPERATIONALISING TSI

The Critical Review Mode

Jennifer Wilby

Centre for Systems Studies
Department of Management Systems and Sciences
University of Hull
Hull, HU6 7RX
United Kingdom

Introduction

Total Systems Intervention (TSI) (Flood and Jackson, 1991) is a meta-methodology which operates a range of problem solving methodologies. Flood (1995 (a), (b)) views problem solving as a process of intervention where practitioners can learn about and manage complex interacting issues. The majority of these issues are at the present focused within the organisational setting. However, the principles, practice, utility and ideology underlying TSI are equally applicable in wider problem settings. The overall aim of this work is to advance the operationalisation of TSI within this wider context. The aim of this paper is to discuss the structure of TSI's critical review mode within TSI. The critical review mode is one of three modes of operation in the implementation of TSI explained below: the critical review mode, the problem solving mode, and the critical reflection mode.

The critical review mode critically assesses methods that may be used in the problem solving mode of TSI. This is done by drawing upon the 3-phase 'creativity-choice-implementation' cycle: to assess firstly which of these 3 phases benefits most from the addition of this method to TSI; and secondly, to assess how the purpose of the underlying principles, practice, utility and ideology of the method can be categorised for use in the problem solving mode. This analysis ensures that informed use can be made of the method under review in the problem solving mode.

The problem solving mode employs the process, creativity, choice, and implementation in the surfacing of core issues within the problem situation, and then choosing the method(s) best suited to those core issues and implementing the choice of method(s). The third mode is the critical reflection mode, which reflects on the implementation of the process of problem solving asking whether or not the right method was chosen for implementation and whether or not the outcome was of benefit in the situation, and to the individuals involved.

The links between the three modes of TSI are based on the processing of a methodology from it's first review, through its implementation, to its evaluation after implementa-

Critical Issues in Systems Theory and Practice, Edited by K. Ellis et al.
Plenum Press, New York, 1995

tion. For any methodology, it is essential to ensure that the initial critical review fully documents, debates and addresses the methodology's strengths and weaknesses. TSI assesses methodologies in terms of how they address the main principles of TSI (those of being systemic, of promoting a meaningful participation, of being reflective, and of enhancing human freedom, (Flood, 1995(a), (b)). TSI also yields insights about the methodologies as to their ability to contribute to one or more phases, creativity, choice, and/or implementation, and finally, the critical review mode has a responsibility to debate the ethical use of these methodologies in problem solving situations.

In a previous paper (Wilby, 1994) hierarchy theory was assessed using a framework of analysis detailed in *Creative Problem Solving* (Flood and Jackson, 1991). That framework examined hierarchy theory by detailing and discussing its theory, utility, ideology, and methodology. In the light of recent work in both critical systems thinking and TSI, it is now time to move beyond that first framework for critique and incorporate TSI's recent advances as just discussed, and as described in *Solving Problem Solving* (Flood, 1995). However, a more detailed and practical framework for the critical review mode is important for those practitioners and managers who use various systems methodologies in daily problem solving activities.

Now that a workable description of the structure of TSI is in place, it is time to focus on the details of the critical review mode. In this paper, the aim is to: 1) describe the historical and present form of the critical review mode; 2) propose additions to the framework of the critical review mode for analysing methodologies; and 3) illustrate this expanded framework using examples drawn from the previous critique of hierarchy theory.

The Historical Form of the Critical Review Mode

TSI's framework for critique was first presented in *Creative Problem Solving* (Flood and Jackson, 1991). This framework had three main steps: 1) a description of the philosophy, principles and process of the methodology under review; and 2) a critique of the methodology's theory, methodological principles, utility, and ideology in terms of how each of these four areas addressed the technical, practical and emancipatory interests (Habermas, 1972) inherent in problem solving situations; and 3) evaluating the information from steps one and two to determine where this methodology may contribute to the creativity, choice, and/or, implementation phases. It is summarised in the question: what forms of creativity, choice, and implementation are advocated by the methodology? Flood and Jackson stated their aim thus:

> "Our task in each critique is to craft an intertwining of lines of argument, to manufacture a meta-paradigmatic net with which we can assess the legitimacies and limitations of the various systems methodologies at different levels (paradigm being a general term that represents the pulling together of issues relating to ideology, epistemology, ontology, methodology and method). In fact each systems methodology is assessed in terms of its theoretical presuppositions, the methodological guidelines it offers, its ideological biases and its utility." (Flood and Jackson, 1991).

The Present Critical Review Mode

The present Critical Review Mode has developed from this first framework of critique and has evolved into the structure described in *Solving Problem Solving* (Flood, 1995). This new structure, which sees the first formalisation of the critical review mode, seeks information from the first two steps of critique. It also seeks to enrich TSI and the methodology under review by evaluating each in terms of the other.

Step four is designed to gain an understanding of how the methodology may be able to tackle the four key dimensions of organisations as detailed in the following four-part question: How does the methodology tackle the four key dimensions of organisations: a) in the efficient design of processes?; b) in the effective design of the organisation?; c) in the context of organisational culture?; and d) in the evaluation/acceptance of these designs and decisions and their implications to the stakeholders.

Step five investigates how the basic principles of TSI (of being systemic, being reflective, attaining meaningful participation and enhancing human freedom) are met by the critiqued methodology. The following is asked: how does this review further add to the framework and methodology of both the methodology under review and TSI's principles?

The language used to explain the recent changes and improvements to the framework of the critical review mode in TSI has developed through research in the study area of the organisational-enterprise setting. Partly, this has arisen from the biases of the researchers involved in developing the process, but also because the managing of complex situations is an enormous area of concern and will require much more time and effort to move TSI into these expanded areas of research. The next section addresses this issue and attempts to define TSI's use for wider social, and other organisational contexts. In this way the framework itself continues to evolve and widen its applicability and utility.

Proposed Additions to the Critical Review Mode

The additions outlined by this author in this section are proposed changes to the existing framework. The third, fourth and fifth steps of the review process exist between the previous form of the critical review mode and the form proposed in this section. These three steps are discussed in the latest revisions to TSI, but they overlap with the proposed changes in the way that they are applied to each other. In this section the key change discussed is the recursive structure and associated processes of the three modes of TSI and the three phases of TSI. This recursive structure is applied to the methodology under review and TSI itself. Because the three phases and the three modes of TSI are recursive, the phases of creativity, choice, and implementation of the problem solving mode can also be applied within both of the other two major phases of TSI (the critical review mode and the critical reflection mode). In the same way, the need for critique, reflection and problem solving is required in each of the three modes. At every step in the critique, the creativity, choice, and implementation phases, previously found only within the problem solving mode, are applicable.

Specifically, the third step evaluates the information from this basic critical review to determine where the methodology under review contributes to the creativity, choice, and/or implementation phases. In this step, but also threaded through all of the steps, are the three original phases of TSI: creativity, choice, and implementation. Within each step these three phases operate to generate information, make choices about the generated information, and implement the next steps in the critical review mode.

The fourth step deals with the creating of the ideal organisation. The ideal organisation consists of the following six considerations or parts: 1) technical and human activities; 2) efficient and effective control of those activities; 3) a directed purpose to those activities; 4) a recognition of the differences (and similarities) of the people involved; 5) balancing and harmonising 3 and 4 through organisational design and management style; and 6) accepting responsibility for the outcomes and effects of these processes. The fourth step of this operationalised mode is evaluating the methodology to see how well it addresses the process of creating the ideal organisation through utilizing and also satisfying the four key dimensions of organizations (culture, politics, organisation and process). In other words, how does the methodology aid the TSI process so that the four key dimensions are met in the striving for the achievement of the ideal organisation.

The fifth step must have the critique answer how this methodology under review meets the requirements of TSI's four key principles of: 1) being systemic; 2) being reflective; 3) enhancing human freedom; and 4) achieving meaningful participation in the process.

The outcome from the sixth step is a synthesis and evaluation using all of the information and core issues surfaced in the previous five steps. This reflection feeds-back into the critique process and becomes a part of that process itself, reinforcing the underlying principles that TSI is a critical, recursive and reflective process. Some examples of this are discussed in terms of the hierarchy review presented in the next section.

The Previous Review of Hierarchy Theory

The decision of an observer to investigate an interesting phenomenon is the beginning point for any research but is especially relevant in an investigation employing hierarchy theory. Hierarchy theory is a theory of observation. (Allen and Starr, 1982). Researchers in hierarchy theory hold different positions regarding the description and explanation of the relationship between the observer and the observed phenomena. However, the structure, processes, behaviour, and interrelationships within and between a system's parts are the units analyzed in all versions of the theory. Hierarchy theory includes the following tenets: complexity; the importance of the observer; self-reflection; scale and resolution; emergence, connectivity, stability and disturbance, reorganizations of the system, content of the levels; control, constraint and containment; and surfaces and filters. For additional explanation of these tenets see Allen and Starr (1982), O'Neill et al.(1986), Salthe (1985, 1993), and Weiss et al.(1971).

The choice of hierarchy theory for a critical review of its development and application was deemed necessary and appropriate because of its long-term use within general systems theory without undergoing such a critique or other evaluation. This critique was made by using the first and second steps of the critical review mode outlined in section 3. The philosophy, principles and methodology were presented, and the following four sections are summaries of the discussion of the critique done at that time (Wilby, 1994). The full discussion of these items can be found in that article.

Theory

The critique of the theory found that: 1) hierarchy is not a formal theory; it is a collection of principles that have been used for the description of complex systems; 2) the practice of the theory is governed critically by the capabilities and biases of the observer/practitioner of the theory; and 3) the implementation of an hierarchical approach to systems design has to overcome the existing linguistic biases and prior misuse of the concepts of the theory by the social sciences.

Methodology

The critique of the methodology found that: 1) the theory has no practical, well-defined methodology for the design of new hierarchical systems or the illustration of existing complex systems; 2) that there are interesting insights within the theory which are more closely aligned to the interests of soft systems; 3) that the theory needs to develop a capable mathematical structure for hierarchy's middle-number systems; and 4) that in choosing one scale for observation and decomposing the structure for analysis it may be that such a process may alter the system when all of the parts are reassembled.

In answering the last point, the structure is indeed different, but not because there are changes or mistakes made in the re-assembly, but because the observer's presence in any form within the system will alter the system. An equally important reply in terms of the emancipatory or critical interest would be to question the effects of such changes. There are fewer consequences if the observer's effects are noted and the changes to the system are realised. These effects do matter if conflict is produced by the resulting changes being so unmanageable that they threaten system viability.

Ideology

The critique of the ideology found that: 1) the theory belongs with the group of systems approaches designed to address complex situations; 2) hierarchy theory can be viewed as an approach to use in systems where radical change is desired or expected to occur; and 3) hierarchy theory can address practical and emancipatory interests through its commitment to the recognition of observer importance and observer paradigms.

Hierarchy theory belongs in the group of systems approaches described as 'complex' in the 'system of system methodologies' (Jackson and Keys, 1984; Jackson, 1990). As to whether the method is unitary or pluralist, hierarchy theory should not be labeled as unitary since many examples of its systems have many stated purposes or goals, as found in complex social systems, or are not assigned purpose or goals, as in complex natural systems.

With regard to the idea that hierarchy theory is used to preserve existing systems, hierarchy is actually more interested in changes in structure and process. Change is not something that a system must avoid at all costs; in fact, the evolution of systems depends on change.

Utility

The critique of the utility found that: 1) hierarchy theory has the potential for use in concert with other systems approaches if methodological concerns can be addressed in future work; 2) hierarchy theory can address practical and emancipatory interests through its commitment to the recognition of observer importance and observer paradigms; and 3) the theory has an implied rather than stated potential for enhancing self-reflection in systems analysis thereby serving emancipatory interests.

Although not unique to hierarchy theory, its most important insight is the recognition that there is an observer in the system and that person's biases and personal beliefs have to be fully explored prior to any analysis or intervention in a system. Further, the observer must be aware of the effect of his or her presence as it alters the environment and interactions of the system being observed.

Summary

Steps one, two and three are the only steps used in this earliest critique of hierarchy theory. It is obvious that the critique is incomplete if compared to the proposed operationalisation of the critical review mode. It is also apparent from such a comparison of an early critique to the proposed critical review mode of how TSI has developed and changed over time. These apparent difference requires additional work in both the new critical review mode itself and in the previous critiques of methodologies done when the critical review mode was less-developed. Some additional insights about the critical review mode are discussed below but a reworking of this critique of hierarchy must be a topic for further work. Creatively surfacing the core issues has historically belonged only within the problem solving mode, but now, at every step in TSI, creativity must be employed in order to surface information about the methodology, its philosophy, principles, and practice, its theory, method, utility and ideology, and the way in which the researchers treatment of these topics influence the outcome of the critique. Creative methods such as those detailed in *Solving Problem Solving* (Chapter 9) can be used in all these steps to get the necessary information for the critical review mode. In addition, there are other research and investigative methods that may be used to surface considerations and critique.

In using the second phase, choice, you must choose between different means of surfacing information and the use and interpretation of the resulting information about the methodology surfaced in the previous phase. This is dependent on the need to be critically reflective on the underlying biases, values, aims, and philosophy of the method, for example, which are present and unavoidable in the critical review process. There is choice as to where the critiqued methodology will be placed in the meta-paradigmatic net, choice in determining the role of participants in the intervention process, and choice in defining the relationship between the practitioner, the participants, and other affected people. All of these concerns about choice and the practitioner's role call for a careful working of the choice process within all the steps of the critical review mode. All of these concerns about choice and the practitioner's role fall into a specific area in which hierarchy theory's principles may be of use.

In using the third phase of implementation, two levels of implementation are apparent. On one level there is the practice of actually implementing the critical review of the methodology which is a nuts-and-bolts type of consideration, working through the six steps of the review. But on another level, this process also reveals where the methodology fits into the overall framework of TSI for the implementation phases of the problem solving mode and the critical reflection mode. So, one level of implementation is somewhat routine, surfacing and evaluating the methodology, and the other level fits the methodology into the larger picture in terms of its benefits to both the use of the critiqued methodology and the TSI framework as well.

Further work in both the new critical review mode and in the previous critiques of systems methodologies, done when the critical review mode was less-developed, need to be the focus of immediate attention. In the case of hierarchy theory, a complete reworking of this critique is an on-going research topic. Other systems methodologies must be subjected to the same recursive process in order for their contributions to the meta-methodology of TSI to be improved. However, it is not those methodologies themselves that will be improved but rather the meta-methodology of TSI that will benefit from their repeated, recursive review. At this time it is also important to consider that other non-systems methodologies may also be critiqued using this process in the hope that they will also bring diversity to the present and future forms of TSI.

REFERENCES

Allen, T. F. H., and Starr, T. B., 1982, "Hierarchy: Perspectives for Ecological Complexity," The University of Chicago Press, Chicago.

Flood, R. L., 1995(a), Total systems intervention (TSI): A reconstitution, *J. Oper. Res. Soc.* 46: (in press).

Flood, R. L., 1995(b), "Solving Problem Solving," John Wiley and Sons, Chichester.

Flood, R. L., and Jackson, M. C., 1991, "Creative Problem Solving: Total Systems Intervention," John Wiley and Sons, Chichester.

Habermas, J., 1972, "Knowledge and Human Interests," Heinemann, London.

Jackson, M. C., 1990, Beyond a system of systems methodologies, *J. Oper. Res. Soc.* 41:657-68.

Jackson, M. C. and Keys, P., 1984, Towards a system of systems methodologies, *J. Oper. Res. Soc.* 35:473-86.
O'Neill, R. V., DeAngelis, D. L., Waide, J. B., and Allen, T. F. H., 1986, "An Hierarchical Concept of Ecosystems," Princeton University Press, Princeton.
Salthe, S. N., 1985, "Evolving Hierarchical Systems: Their Structure and Representation," Columbia University Press, New York.
Salthe, S. N., 1993, "Development and Evolution: Complexity and Change in Biology," MIT Press, Cambridge, Mass.
Weiss, P., Buechner, H. K., Coleman, J. S., Forrester, J. W., Lorenz, K., McNeill, D., Ozbekhan, H., 1971, "Hierarchically Organized Systems in Theory and Practice," Hafner Publishing Co., New York.
Wilby, J., 1994, A critique of hierarchy theory, *Syst. Pract.* 7:653-670.

SYSTEMS METHODOLOGY AND PRACTICE

27

LEARNING TO LOVE AND LEAVE ACKOFF

Does Interactive Planning Succeed within the Bounds of "Complexity"?

Sam Agyeman

School of Management
University of Humberside
Cottingham Road
Hull, HU6 7RT
United Kingdom

INTRODUCTION

In his seminal article, The Future of Operational Research is Past, Russell Ackoff (1979) outlined his view of organisational problems thus: "Managers are not confronted with problems that are independent of each other, but with dynamic situations that consist of complex systems of changing problems that interact with each other." These he termed *'messes'*[1].

For Ackoff, 'messes' may not be managed via the reductionist, problem-solving approaches so beloved of OR practitioners. They demand holistic treatment, and inputs from all stakeholders associated with the challenge. Such interdisciplinary contributions to mess-management, to him, is the way forward for organisations.

Who would argue with such a well-presented case? I do not. Indeed, Ackoff's arguments and incisive critique of the objectivist approaches of his colleagues in OR have become, for me, a beacon for the critical analyses of problems ranging from those associated with management decision making, through human resource management, to the understanding of organizational behaviour and research methodology.

Ackoff does not leave the argument at this stage, though. To help his audience 'manage' the 'messes' they face, he offers Interactive Planning (IP)[2,3] as his model for addressing various organisational problems. That there has been a plethora of articles and viewpoints questioning aspects of the IP model[4,5], may be taken as the expected cut and thrust of academic discourse on new ideas negotiating the tortuous route towards acceptance or rejection, respectability or opprobrium.

What raised, for me, serious questions about Ackoff's model was his insistence, as I saw it, that in IP is to be found the answer to all the 'messes' any organisation may be called upon to manage[6]. The problem, to him, lay in the negative thinking of decision makers and their advisers, as well as their inability to engage all the stakeholders in a particular problem, in its solution/management. If only the tenets of Interactive Planning were followed the

ORDER ⟹ "COMPLEXITY" ⟹ CHAOS

Figure 1. 'Complexity' within dynamical systems[9].

outcomes would be successful resolution of 'messes' each time. And, he has numerous success stories to back his assertions up!

In this paper, some of the arguments raised against Ackoff's IP model will be discussed within the framework of, and in addition to, aspects of 'Complexity', which has been described by its adherents as "...the emerging science at the edge of order and chaos."[7]

INTERACTIVE PLANNING

Very briefly, in Interactive Planning Ackoff advocates the involvement of:

- all the stakeholders that the planning would involve/affect;
- dissolution (as opposed to solution) of problems by systems redesign;
- searches for explanations of phenomena (as opposed to descriptions of them), and from such explanations;
- testing such hypotheses/explanations in the real world;
- continual evaluations of the implementation of recommendations;
- taking account of the effect of implementation of recommendations on the social and physical environments;
- visualising sets of problems, as well as opportunities in as broad a scope as managers can conceptualise and control; and
- encouraging the learning from experience, and the use of this in adapting to the environment.

'COMPLEXITY' AND ACKOVIAN 'MESSES'

'Complexity', a concept which is at once difficult to specifically define, and also excitingly lucid, has been described as representing that zone "...*at the edge of chaos* (sic)....where the components of a system never quite lock into place, and yet never quite dissolve into turbulence either."[8]

Schematically, this translates as illustrated in Figure 1.

Taking into consideration the turbulent nature of human - and specifically organisational - systems, Figure 2 may seem to provide a more realistic representation of such systems.

Problems, or better still, Ackovian 'messes', would seem, in such a model, to reside within the zones represented by either 'complexity' or chaos. Change is endemic in such a system, with an overall result that may be represented by Figure 3 below.

In such a model, Ackoff's 'messes' would have to be defined and redefined ad infinitum, as the environment changes and the nature of the 'messes' themselves change.

ORDER ⇄ "COMPLEXITY" ⇄ CHAOS

Figure 2. Bi-directional flow of change within dynamical systems

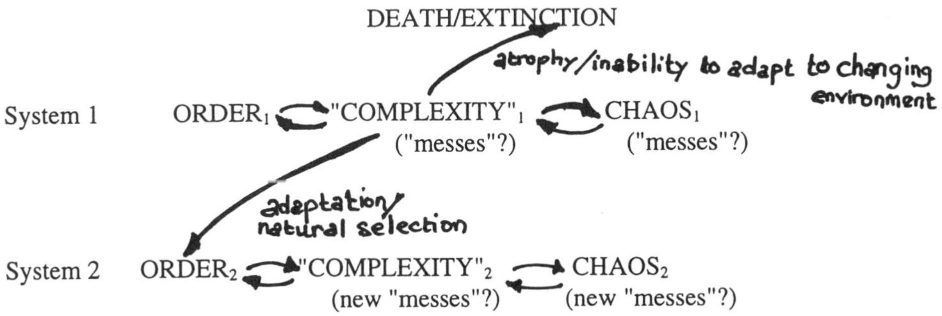

Figure 3. Change within multi-system dynamical systems.

The question, then, is: can any organisational system sustain such flux without one or other of the stakeholders losing commitment to the problem/'mess' and its solution/management?

Even if power considerations (one area in which IP has been criticised)[10] are assumed out of an initial problem situation in say, Figure3,System1, it would not be inconceivable (indeed it would seem to be more than likely) that power struggles would ensue as the new system (Figure 3, System 2) takes hold, with its own peculiar aggregation/requirement of stakeholders. Conflict and disownment of the problem/'mess' would be the result, thus ensuring that a satisfactory outcome is not achieved.

The dynamic nature of any system incorporating 'complexity' would seem to militate against an on-going successful adaptation to an ever-changing environment. Ackoff himself states:

> "Many of the problems we 'solve' do not stay solved because the problems themselves change."[11]

How well would Ackoff's stakeholders fare here? In time, would they not become complacent, apathetic, lose consensus and 'ownership' of the problem, thus ensuring the systems's eventual failure?

For, in 'complexity' those agents that are successful are those who change continually, and seek opportunities to address problems that have yet to be raised. They prepare for enemies yet to manifest in reality; they shore up reserves against future shortages of nutrients, with its inevitable competition for scarce resources. Such alertness requires great energy, commitment, and innovation: characteristics that, over time, most initially successful organisations lose.

SELF-PROPAGATION AND ADAPTATION IN 'COMPLEX' SYSTEMS

In their investigation of the phenomenon of 'complexity' in biological systems, workers at the Santa Fe Institute (in New Mexico) for Studies in the Sciences of Complexity and have argued (and been able to produce viable computer models to back this up) that for any organism to survive, it must do two things. Firstly, it must have a means of appropriating from its environment nutrients to build itself up. It is thus that it can have the energy to compete and adapt to its ever-changing environment. Secondly, such a system must have a means of copying itself[12].

By analogy, for Ackoff's IP model to successfully serve an organisation anytime it is used to solve problems, it would seem logical to argue that it has to have built into it an ability not only to adapt environmental situations to its purposes, but that in addition, it should be able to learn and transmit such learning to its future stakeholders.

In fairness to Ackoff, he does make the ability to learn and adapt to a changing environment one of the operating principles of his IP model. Whether, however, in real life, true learning organisations exist, is another matter.

THE CONCEPT OF INCREASING RETURNS

One of the favourite concepts within the growing science of 'complexity' is that of Increasing Returns. 'To them that have, more shall be added', would seem to be a simple way of describing the basic tenets of the concept of Increasing Returns. The phenomenal success of the VHS video system over its technically superior rival, Betamax, is a case in point. The more customers bought VHS systems, the more the shops stocked the model in preference to Betamax, and therefore the more VHS systems were bought.

When such a process is set in motion, it is difficult to halt. This may explain the sudden fall of otherwise impregnable world empires, stock market crashes, animal and plant population wipe-outs, and so on. A process is set in motion with a momentum of its own that no tampering with via careful selection of stakeholders, definition of means and ends, or assessment of required resources, and so on, seems to be able to stop.

Successful problem-solution in a situation of increasing returns could then be argued to have less to do with whatever problem-solving model was being used, and more to do with the momentum of the process itself. Conversely, it could be argued that if the momentum was towards greater and greater chaos and eventual collapse, no amount of systems redesign, or attempts to dissolve the problem could prevent the system careering towards failure.

As Waldrop (1993) states: "Increasing returns [would imply that]....if small chance events can lock you into any of several possible outcomes, then the outcome that is actually selected may *not* be the best."[13]

THE CULTURE-SPECIFICITY OF ACKOFF'S MODEL

Modelling Reality

Work being done on artificial life at the Santa Fe Institute in the USA has involved attempts to model reality (through computer simulations) in order to understand the behaviour of human and animal populations, the economy, political systems, whole ecosystems, and so on. Joshua Epstein and Robert Axtell's 'Sugarscape' model for example, involves computer agents living in an environment initially dominated by their sole nutrient, sugar[14]. The agents are programmed to have sex, reproduce, fight each other, have differences in their ability to appropriate nutrients from their surroundings, and so on. The 'culture' here involves conflict and inequality. The resulting behaviour of the agents very closely reflect real life systems in terms of success and failure of various agents.

Unlike such models, IP does not take account of the pluralistic nature of many work organisations, with their in-built systems of inequality, intrigue and conflict. Maybe certain Western organisational cultures do reflect consensual interactions. However, within certain world cultures strict taboos concerning who may sit in discussion with whom; and which castes may interact or not with which other castes, do exist. Strict divisions along the lines of gender, ethnic grouping, or age even, may preclude the invitation of certain groupings to contribute to the problem-solving process with others of a different class or caste. Surely, this would make IP virtually impotent? Which change-agent or facilitator would have the skill, time, power, energy, and psychological know-how to create meeting points of hearts and minds, dissolve conflict, and forge consensus in such situations?

Modelling Non-Reality

Not all the computer models of artificial lifers are as realistic as Joshua Epstein and Robert Axtell's. In some cases, economic and other assumptions of say, perfect knowledge or complete equality of agents, are built into the model. For example, model birds are programmed (unrealistically) to have exactly the same characteristics, with the same few instructions governing their *random* movements.

The fascinating result, so far, has been that over time, even such 'unrealistic' models have exhibited some uncanny likenesses to their real-life counterparts. Birds have flocked together and negotiated obstacles *en masse*, economies have boomed and busted, and so on. It would seem that even on computer; and even assuming unrealistic initial agent characteristics, real life will manifest.

Does this phenomenon reflect the reason why Interactive Planning has been so successful? Does IP incorporate all the elements at work in 'complex' systems? Is it another example of a model that would produce successful results in spite of its inherent flaws? Does it provide for adaptation in such a way that 'real life', as defined as successful problem-solution, always manifests?

But surely, no real life system gets it right each time as IP claims to do? What the artificial life models show (and the results have been tested in real life) is not that any system or agent behaves correctly all the time; rather that such systems will behave in certain typical, often predictable, ways whether these are classified as successful behaviour or not. Does Interactive Planning make similar claims?

CONCLUSION

Ackoff's contribution to the Systems movement cannot in any way be denied. In this paper, I have attempted to examine his Interactive Planning model through comparison with aspects of the 'emergent' science of 'Complexity'. Many of the points raised above as criticisms have already been directed against IP[15]. What I do here is attempt a re-focusing of such arguments using some of the basic tenets and implications of 'Complexity'.

The questions that the paper struggles with are these: Does one accept the 'Complexity' model of dynamical systems? If one does, where might one place organisational 'messes'? Does one place them right within the bounds of chaos, or at its peripheries, that area described as exhibiting 'emergent' behaviour, the zone of 'Complexity'? What about Interactive Planning: does it provide the conditions for continual successful dissolution of problems at the edge of chaos, and elsewhere?

Inasmuch as many of the concerns have been raised elsewhere some replies have already been made to them[16]. The aim here has been to provide a *theoretical* critique of the model. That aspects of the model may have inadvertently been misrepresented here may reflect the inevitable consequences of attempting to interpret another person's meaning. That aspects of 'Complexity' may have also been misrepresented is even more likely, as the phenomenon is an on-going hotbed of activity aimed at defining it as a science in its own right, and to understand some of the intriguing issues raised therein.

Hopefully, this paper will re-open discussion on, and interest in, an aspect of the work of a man whose work I still very much respect.

REFERENCES

1. Ackoff, R.L., 1979, The future of operational research is past, J. Operational Res. Soc. 30.2:93-104.
2. Ackoff, R.L., 1980, The art and science of mess management, S^3 *Papers*. Univ. of Pennsylvania, 80-106.
3. Flood, R.L., Jackson, M.C., 1991, "Creative Problem Solving: Total Systems Intervention", John Wiley and Sons, Chichester, UK, 143-165.
4. Flood, R.L., Jackson, M.C., 1991, "Creative Problem Solving: Total Systems Intervention", John Wiley and Sons, Chichester, UK, 143-165.
5. Jackson, M.C., 1982, The nature of soft systems thinking: the work of Churchman, Ackoff and Checkland, *J.Appl.Sys.Analysis*. 9:17-28.
6. Flood, R.L., Jackson, M.C., 1991, "Creative Problem Solving: Total Systems Intervention", John Wiley and Sons, Chichester, *UK, 143-165*.
7. Waldrop, M.M., 1993, "Complexity: the Emerging Science at the Edge of Chaos", Viking, London.
8. Waldrop, M.M., 1993, "Complexity: the Emerging Science at the Edge of Chaos", Viking, London,p12.
9. Waldrop, M.M., 1993, "Complexity: the Emerging Science at the Edge of Chaos", Viking, London, p234.
10. Flood, R.L., Jackson, M.C., 1991, "Creative Problem Solving: Total Systems Intervention", John Wiley and Sons, Chichester, UK, 143-165.
11. Ackoff, R.L., 1981, Beyond prediction and preparation, S^3 *Papers*. Univ. of Pennsylvania, 82-106.
12. Waldrop, M.M., 1993, "Complexity: the Emerging Science at the Edge of Chaos", Viking, London.
13. Waldrop, M.M., 1993, "Complexity: the Emerging Science at the Edge of Chaos", Viking, London, p48.
14. Horgan, J., Nov. 1994, Sex, death and sugar: researchers try to "grow" societies on a computer, *Scientific American*, 271.5: 14-15.
15. Flood, R.L., Jackson, M.C., 1991, "Creative Problem Solving: Total Systems Intervention", John Wiley and Sons, Chichester, UK.
16. A response to one such source of criticism may be found in: Ackoff, R.L., The hard-headedness and softheartedness of M.C. Jackson, *Journal of Management Studies, 20: 56-69.*

28

A FRAMEWORK FOR ANALYSING SSM-BASED STUDIES

A. Barnden,[1] R. Smith,[2] and R.B. Watson[2]

[1] Department of Information Systems
Monash University, Caulfield
Australia 3145
[2] School of Computer Science and Software Engineering
Swinburne University of Technology
Hawthorn
Australia 3122

INTRODUCTION

Many systems practitioners have undertaken studies which purport to be based on the Soft Systems Methodology (SSM) as described by Checkland(1981) and others. In accord with its methodological roots in action research, SSM encourages the involvement of the analyst in reflection on the conduct of the study as it proceeds. A characteristic and strength of SSM is that this reflection on the evolving study should lead to insights that can affect both the particular study and lead to improved future practice. In the present paper a prototype framework is reported which is appropriate to the task of analysing the methodological practice of study teams of which one was not necessarily a member, who have undertaken studies which claim a basis in SSM. The framework seeks to provide a structure and some heuristics for thinking critically about the ways in which a study has been undertaken. The framework seeks to facilitate identification of significant departures from presently documented SSM practice, so supporting debate on the extent to which such departures have been beneficial and might be incorporated in future practice. Initial application of the analysis framework to some studies conducted in Australia is presented in outline.

1. BACKGROUND

Gwilym Jenkins, in establishing the postgraduate Department of Systems Engineering at the University of Lancaster in the mid-1960's, determined that the only way to develop a new subject which sought to investigate the application of the methodology of (hard) systems engineering to the less well defined problems of managers, was to insist upon interaction with real problem situations in the 'action research' mode. This approach has undoubtedly been vindicated, with the publication of a number of seminal works describing

Critical Issues in Systems Theory and Practice, Edited by K. Ellis et al.
Plenum Press, New York, 1995

what is now known as Soft Systems Methodology (SSM) (Checkland, 1981; Wilson, 1984; Checkland and Scholes, 1990; Davies and Ledington, 1991).

With the graduation of students trained in these approaches, and the availability of texts describing SSM, a new class of studies has now appeared - studies conducted within industry by practitioners not interested in subsequent reflection upon possible methodological lessons which might be learned as a result of their work. If methodological lessons are to be learned from such a plethora of work they will have to be extracted by analysts reflecting upon studies in which they did not themselves take part.

The research program introduced in the present paper seeks to lay the groundwork for such critical analysis, developing and applying a framework appropriate to the task of analysing the methodological practice of study teams who have undertaken studies which claim to be based upon SSM, where that analysis is to be undertaken after the event by persons not involved in the study itself. It is anticipated that in the course of the cycle of case studies which it is believed will be needed to refine the framework, insights may well be gained into some significant departures introduced, some of which may well have been beneficial and might be incorporated into future SSM practice.

In Section 2, we build a case for a research program structured as a cycle of action research, with each cycle involving a selection of case studies. In Section 3 we discuss the notion of breaking into that cycle by proposing an initial framework which adapts ideas from the capability maturity modeling approaches which have had some success in Software Engineering. Section 4 then presents some initial observations on attempts made to apply this 'SSM Maturity Assessment' framework to two studies conducted in Australia. Finally, Section 5 draws some tentative conclusions as to the potential of this type of approach.

2. RESEARCH METHOD AND TECHNIQUES

The approach which it is proposed should be adopted follows the 'Action Research' cycle as presented by Checkland (1991), according to which the elements of any piece of research can be regarded as entailing a particular set of linked ideas or Framework (F), used in a Methodology (M) to investigate some Area of interest (A). Viewed in this way, the cycle of action research can be conceived of as dealing in research themes (F) within which lessons can be sought. The researcher joins a real-world problem situation (A), taking part in the deliberations which lead to practical outcomes, and declaring F and M so that reflection can subsequently yield findings of various kinds concerning F and/or M and/or A. In the present case the Area of interest (A) is the practice of analysts applying SSM-based ideas to real-world problems. The Framework of ideas (F) is the existing body of understanding of the philosophy underpinning SSM, and discussion of the process and techniques available (see section 3.1 below). The Methodology (M) however, is somewhat more problematic - indeed the evolution of an appropriate M is a major objective of the present research. In the present case we propose to break into this cycle of research by adapting in the first instance ideas from the body of literature now emerging in the discipline of Process Modeling (Curtis et al., 1992) (see section 3.2 below).

3. AN INITIAL FRAMEWORK

3.1. The Elements of SSM

To break into the cycle of action research discussed above we must define the framework of ideas (F) which constitute the present understanding of SSM. Avison and Fitzgerald (1988) define a methodology as a set of processes supported by techniques and tools to be applied by analysts seeking to address some problem domain, where those processes, techniques and tools are underpinned by a philosophical standpoint which provides coherence to the methodology. In those terms, the definition of SSM provided by Checkland and Scholes (1990, pp. 284-296) might be viewed as:

- Process: The 'system to use SSM' which maps the experiences of using it in terms of 16 human activity systems.
- Techniques: The techniques/concepts to be applied, captured as the SSM epistemology, including: Real world; Systems thinking world; Problem situation; Analyses One/Two/Three; Rich pictures; Root definitions; CATWOE; The 5Es; Conceptual model; Comparison; Desirable and feasible change; Action.
- Philosophy: The five Constitutive Rules of Checkland and Scholes (1990) are an attempt to account for the family resemblances which characterize the whole spectrum of SSM. As such they form, at least, a good proxy for a statement of the philosophy which underpins SSM. In summary, they emphasize:

1. The objective: Improvement in some problematic real-world situation;
2. The approach: Structured thinking based on systems ideas;
3. The boundaries: No automatic assumption that the real world is systemic; A distinction between real world and systems thinking about the real world; Systems thinking involving the construction of holons; Desirable/feasible change defined by comparison of holons with the real world;
4. The adaptation: Any potential use should be characterized by conscious thought about how to adapt it to the particular situation; and
5. The reflection: Methodological lessons can be extracted by conscious reflection on the experience of using SSM.

Note that there has been some tentative research towards establishing tool sets which might support the application of the above techniques, most notably explorations of computer-based support of SSM (Avison et al., 1992). As none have clearly established pre-eminence, it is not sensible to promote any such tool set as a methodological standard.

3.2. Process Maturity

To break into the research cycle a methodology (M) must be proposed by which one might initially attempt to align the framework of SSM above with the practice of real world study teams. The present SSM literature offers little or no help in this choice. Some exploratory work has appeared which attempts to study the behavioural factors in SSM application, focusing in particular on the design problem solving strategies employed (Khushalani et al., 1994). This work requires the collection of concurrent verbal protocols and hence is not applicable to the retrospective analysis of studies proposed herein. There have been a number of case studies published in which participants in SSM studies have reflected on their personal experience (e.g. see Omerod (1992)), however such reports have been largely descriptive and have not seriously questioned the means by which the process of reflection takes place, in particular the danger of post-hoc rationalisation (Mohr, 1985). Mingers and Taylor (1992) have reported the outcomes of surveys of SSM users, however such techniques are of limited value if one seeks to probe the detailed conduct of a study.

In the present research we have chosen to borrow from a different literature as an entry to the cycle of action research; that of the somewhat 'harder' software engineering domain. Therein, a notion called process maturity has been developed to a level of some sophistication (Humphrey, 1990). The process maturity framework characterizes the software development process employed in any organization or in any project into one of five maturity levels, with the intent that by so doing areas where improvement might be achieved can be readily identified. These levels have largely to do with issues of process control. As such, they are applicable only to the activity 'Decide issues concerning mounting and doing the study' in the 'system to use SSM' model, and to extend the idea to analysing the full range of activities in an SSM study we need to recast the maturity level scheme.

In that spirit, we propose initially a three level characterization of maturity of an SSM study, suggested by our own experience in teaching and researching SSM and a reflection on the present SSM literature as indicated below. The levels proposed are:

1. Initial Level: Typical of the studies conducted by students following their initial instruction in SSM. Smith and Watson (1993) describe students formally reporting their studies in the form of a seven-step process, but with the description provided being largely a post-hoc attempt to rationalize essentially ad-hoc recommendations by reference to conceptual models which confuse real world and systems thinking and almost always view the real world as systemic.
2. Functional Level: Typical of studies described by Stowell et al. (1991) as "..limited and functionalist..", wherein the study formally follows the 7-stage or 2-streams of intervention models to investigate some part of the real world, but does not tap many of the more powerful insights that can come by use of the full range of systems concepts and notions such as metaphor (Davies and Ledington, 19991).
3. Internalized Level: Studies wherein the practitioners have so internalized the practice of SSM that their reflections on the everyday flux of events surrounding the study are very naturally viewed using SSM's epistemology and incorporating systems concepts. Such studies may well still be formally cast in the form of a 7-stage analysis, but the reflections which take place are of the type which might be conducted by a practitioner who has reached the level Checkland and Scholes (1990) describe as a Mode 2 thinker.

It must be recognized that the above levels, and indeed the use of the notion of maturity in the SSM context are tentative, and are used as a means of seeding the research process only. The level definitions and granularity may well evolve over the lifetime of the project as indeed may the notion of maturity. In this sense the use of a 'hard' systems methodology such as the capability maturity model of software engineering to break into the research cycle might be seen as paralleling the use of 'hard' systems engineering to break into a research cycle, adopted by the Lancaster University researchers in the early '70's. It should also be noted that a maturity classification is not being viewed as a characterization of correctness of a study, but may be related to it in the following sense. According to the "..system to use SSM.." model (Checkland and Scholes, 1990), the effectiveness of a study is measured by the amount of learning which comes from it (how this is measured is another difficult question). The tools and techniques used in education must be related to the level of knowledge of the learners. So a beginner learning SSM may learn a lot from a level 1 use of it, and nothing from attempting to use level 3 techniques. Similarly, experienced SSM users may not learn much from a level 1 use of it - for them level 1 is incorrect. The maturity level should be seen as a means whereby pointers to possible enhancement in SSM practice and departures from the elements of SSM (see Section 3.1) might be recognized.

3.3. The Maturity of an SSM Application - An Initial Framework

In order to bring together the elements of SSM (Section 3.1) and the concept of maturity (Section 3.2), a matrix of the form outlined in Table 1 is proposed, wherein the factors which characterize a study at each of the three proposed levels are recorded for each SSM element. In Table 1 we tentatively suggest such characteristics for three elements, one from each of the Process, Techniques and Philosophy groupings. The research now in

Table 1. The maturity of an SSM study - Samples from an initial framework

Elements of SSM (Samples only)	Level 1: Initial	Level 2: Functional	Level 3: Internalised
Process: "Select relevant systems in the form of RDs and CATWOE"	Single relevant system only; No checking of the structure of the RD against CATWOE; RDs constructed after development of CMs, only as a rationalisation of that CM	Multiple relevant systems considered; Structure of RD checked against CATWOE; The nexus between RDs and CMs maintained	Multiple relevant systems in the form of RDs and CATWOE used as a basis of powerful debate about issues such as who clients, actors and owners are, and various possible world views
Techniques: Root Definition construction	CATWOE elements missing from RDs; Tendency to only consider primary task definitions; Confusion of what/how distinction	Root Definitions formally in accord with CATWOE; Open to both primary task and issue based definitions; What/how structure present and a hierarchy of RDs developed	Root Definitions formally correct etc., but in addition they are constructed in imaginative ways e.g. by development of multiple, contrasting RDs, possibly introducing the idea of metaphor
Philosophy: Constitutive Rule 3 - A distinction between real world and systems thinking about the real world	No period of disengagement from the real world; RD's and models are simply a representation of the real world, constructed using the language of the real world; No significant use of systems concepts	Practitioners stand back from the real world at some stage; RD's and models are not representations of the real world but systems models which can be usefully compared with the real world	Practitioners have so internalised the practice of SSM that their reflections on the models and the everyday flux of events surrounding the study are very naturally viewed using SSM's epistemology and incorporating systems concepts

progress anticipates that the levels and characteristics of the full set of SSM elements will be refined through a cycle of analysis of various case studies collected from industry practitioners, and that suitable instruments (e.g. structured interviews, document analysis schedules and questionnaires) for collecting and analysing the details of studies will be developed in due course.

In the initial round of studies, essentially unstructured interviews with 10 industry practitioners have been collected together with the available study documentation, and are being examined in the light of the scheme introduced above. Brief comments on two of these follow.

4. APPLICATION - SOME OBSERVATIONS ON TWO STUDIES

In order to refine the methodology, it is being initially applied to two readily available studies. These are:

- Business District Centre Operational Strategy Study: An SSM-based study to develop an operational strategy for the management of Business District Centres in an Electricity Commission; and

- Information and Work Distribution System: An SSM-based investigation of an integrated information automation system for a service organization.

In terms of the maturity level schema above, each would be tentatively classified as functional level studies, adhering to the formal 7 stage model, developing defensible primary task models, whilst not demonstrating in large measure creative use of systems ideas to analyze the wider flux of events surrounding the studies. Some aspects of note, however, include:

- a tendency of some practitioners to work at times totally in terms of the conceptual model, making jumps to formulate significantly different relevant systems without recourse to a reformulation of the root definition;
- no clear indication of formal Analysis One, Two and Three stages, although some unstructured consideration of cultural, social and political dimensions is acknowledged; and
- some evidence of internalization of SSM ideas in one case where one practitioner has synthesized SSM with a technique known as Value Added Management (Technology Transfer Council, 1989).

5. SUMMARY AND OUTLOOK

In the present paper a prototype framework has been outlined which it is believed will be appropriate to the task of analysing the methodological practice of study teams of which one was not necessarily a member, who have undertaken studies which claim a basis in SSM. The framework seeks to provide a structure and some heuristics for thinking critically about the ways in which a study has been undertaken, operationalising the application of ideas such as those presented as the "Constitutive Definition of SSM" (Checkland and Scholes, 1990). The framework seeks also to facilitate identification of significant departures from presently documented SSM practice, so supporting debate on the extent to which such departures have been beneficial and might be incorporated in future practice.

Exploratory application of the analysis framework to studies conducted in Australia is in its infancy, and it is premature to draw conclusions as to the efficacy of the framework. Indeed several cycles of case study evaluation will be required to suitably refine ideas based on the notion of capability maturity which have been used to break into the action research cycle. Some tentative observations on the conduct of two studies have been presented.

REFERENCES

Avison, D.E., and Fitzgerald, G., 1988, "Information Systems Development: Methodologies, Techniques and Tools", Blackwell Scientific Publications, Oxford.

Avison, D.E., Golder, P.A., and Shah, H.U., 1992, Towards an SSM toolkit: rich picture diagramming, *Eur. J. Inf. Systs.* 1:397.

Checkland, P., 1981, "Systems Thinking, Systems Practice", John Wiley & Sons, Chichester.

Checkland, P., 1991, From framework through experience to learning: the essential nature of action research, *in:* "Information Systems Research: Contemporary Approaches and Emergent Traditions", H.E. Nissen, H.K. Klein, and R. Hirschheim, eds., Elsevier, Amsterdam.

Checkland, P., and Scholes, J., 1990, "Soft Systems Methodology in Action", John Wiley & Sons, Chichester.

Curtis, B., Kellner, M.I., and Over, J., 1992, Process modeling, *Communications of the ACM,* 35 No 9:75.

Davies, L., and Ledington, P., 1987, Creativity and metaphor in soft systems methodology, *in:* "Proceedings of 31st annual meeting of the International Society for General Systems Research", June 1987, Vol. 1:176, Budapest.

Davies, L., and Ledington, P., 1991, "Information in Action: Soft Systems Methodology", MacMillan Information Systems Series, Basingstoke, London.

Humphrey, W. S., 1990, "Managing the Software Process", Addison-Wesley, Reading, Mass.

Khushalani, A., Smith, R., and Howard, S., 1994, What happens when designers don't play by the rules: towards a model of opportunistic behaviour in design, *Aust. J. Inf. Syst.* 1 No 2:13.

Mingers, J., and Taylor, S., 1992, The use of soft systems methodology in practice, *JORS.* 43 No 4:321.

Mohr, L.B., 1985, The reliability of the case study as a source of information, *in:* "Advances in Information Processing in Organizations", L.S. Sproull., and P.D. Larkey eds., 2:65, JAT Press, Greenwich, Conn.

Omerod, R.J., 1992, From IS strategy to systems development: combining hard and soft systems practice, *Systemist* 14 No 3:160.

Smith, R., and Watson, R.B., 1993, Systems thinking and the education of the software engineer: reflections on the Swinburne experience, *in*: "Ethical Management of Science as a System", R. Packham, ed., Proc. of the 37th Annual Meeting of the ISSS, University of Western Sydney, 628.

Stowell, F.A., West, D., and Stansfield, M.H., 1991, The application of an expert system shell to an unstructured domain of expertise: using expert system technology to teach SSM, *Eur. J. Inf. Systs.* 1 No 4:281.

Technology Transfer Council, 1989, "VAM in Action", TTC, Melbourne.

Wilson, B., 1984, "Systems: Concepts, Methodologies and Applications", John Wiley & Sons, Chichester.

29

SYSTEM(S) ENGINEERING

Organization versus Practice

J. C. Boarder

Department of Computing and Electronics
Buckinghamshire College
High Wycombe, HP11 2JZ
United Kingdom

INTRODUCTION

System(s) Engineering is a broad-spectrum, multi-disciplined, holistic activity coloured by individual experience. In some organisations, System(s) Engineering drives successful innovation. In other organisations, System(s) Engineering acts as a 'fire brigade' investigating failures. In yet other organisations, System(s) Engineers establish system architecture and project management needs, while at the other end of the spectrum, System(s) Engineering performs progress chasing and production control. Such variation in interpretation is unsatisfactory for industry and engineers. They defocus and inhibit professional involvement. System(s) Engineering is a critical issue linking systems thinking and systems practice.

In this paper, I argue for a separation of concerns in systems thinking. The argument separates a general-purpose, generic organisation, systems engineering, from a special-purpose, goal-oriented process, system engineering. Systems engineering supports a broad spectrum environment, a resource base, for an engineering process. System engineering selects and organises an engineering system developing and supporting a customer's system.

System(s) Engineering, it transpires, is a convolution and an embryonic component of all human activity systems in which processes are exercised over organisations. Systems engineering is an engineering system which creates and supports system engineering. Reciprocally, system engineering creates and supports engineering systems. The convolution is the cause of the multifaceted interpretations of the term System(s) Engineering and needs to be unwound, it is not mere semantics. In the separation, two laws of systems engineering appear, imposed on system engineering. Also, a structured and ordered framework, or platform emerges for system engineering activity enabling it to respond consistently and perhaps completely to the requirements of systems engineering. This platform can be used to guide the system engineer in his practice and systems engineering in process reviews.

CONTEXT

The Draft Guide to the Practice of System Engineering

In the late 1980's, IEE Professional Group M5 (Systems Engineering) initiated a working party that produced the "IEE Draft Guide to the Practice of System Engineering", (Hitchins, Boarder and Moore, 1992). Hitchins (1992), separately, published part of the work as a "Guide to Proper Practice". The guide, drawing on the work reported by Boarder, (1989; 1990), was strongly challenged but, in the opinion of discussion meetings and questionnaire responses, the guide identified the:

- Role of System Engineering
- Aim and Objectives of System Engineering
- Activities needed to satisfy the aim and objectives

The guide attempted to define, in general terms, the necessary and sufficient activities of system engineering so that a graduate engineer with some years engineering experience could recognise and practice the process.

Interestingly, within the IEE establishment, the guide was not accepted but individually among the authors and elsewhere its development continues.

System

The generality of the system concept is typified by Buckminster-Fuller (1975) with the definition:

- System is the first subdivision of Universe

where it is implicit that system is an interpretation of experience. Checkland (1981), similarly, claims system to be a model of a whole entity relative to an observer. When applied to human activity, the model is characterised fundamentally by hierarchy, emergent properties, communication and control but "...the crucial characteristic is that of emergent properties from the whole", Checkland (1981). The working party to the guide took this model of system as a working hypothesis while recognising its incompleteness.

System(s) Engineering develops systems to customer and end-user requirements. Emergent properties of subsystems are integrated in the local environment. Properties of the system of interest mutually and complementarily interact with properties of other sibling systems. These interactions produce emergent properties of the containing system from systems' supplementary properties. Conversely, mutual and complementary demands are placed on subsystems by the system local environment.

System Dynamics and the System Environment

System, containing system, sibling systems, component systems requirements change dynamically. System(s) Engineering is fundamentally concerned with the dynamic of these changes, their influences on emergent properties and the environment.

Engineering System

An evolving process brings a system into being, sustains it and maintains it throughout its life. Ultimately the system is disconnected from the local environment and dispersed. This process is that of an engineering system composed of men, money, material and machines as described by Jenkins (1972). The engineering system causes the required emergent properties to be delivered in the local environment. Unless the engineering system is properly and effectively designed and constructed, the system cannot perform to requirements.

Engineering Management System

Both system and engineering system evolve through a managed process. Processes trigger, monitor and record events in the lifecycle. Many events are expected and planned for including unexpected emergent properties.

System(s) Engineering

Customers' systems, which provide the environments, engineering systems and engineering management systems have requirements, properties and end users. The principles of hierarchy, emergence, communication and control apply. Engineering systems and engineering management systems are embedded in environments over their lifetimes. The development and balance of environments, engineering systems and engineering management systems are proper concerns of System(s) Engineering.

System Engineering

System(s) Engineering is itself conducted within an environment. The convolution arises and is resolved by recognising individual system engineers as working within given engineering systems under given engineering management systems. System engineering is therefore a focused activity. It is concerned with the composition of subsystem properties. It is in this sense that system engineering is defined generically as:

- The structured and ordered modelling of a system, within its environment, with systemic and systematic properties required by customers and end-users

A fundamental problem for the system engineer is highlighted; the differentiation of systemic issues from systematic issues, of supplementary properties from complementary properties, and the needs of the engineering system for their integration. The use of modelling is noted because a distinction is drawn between the development of a system and its application. This distinction is highlighted by defining Project Engineering as:

- The orderly application of well-defined and established procedures for implementing systems

with Project Engineers defined as:

- Project Engineers practice engineering within the defined boundaries of an engineering system

Market and Business Perspectives

The context of an organisation stresses the focused nature of the process. System engineering is conditioned by individual system engineers who bring market knowledge of solutions to customers' problems. Customers' needs are identified by a 'Quality Gap', a gap between customer expectation and engineering achievement. As such, Quality Gap promotes business opportunity. Quality Gap exists at two levels; that of the customer with a requirement and that of multifarious engineering technology suppliers each with partial solutions to perceived markets needs. System engineering sits between the two providing value-added solutions. System engineers respond to mutual expectations integrating suppliers' technology to satisfy customers' needs.

Selection of potential solutions is based on differentiation of requirements, selection and integration of specific tools, methods, procedures and standards that support and manage the system from 'cradle to grave'. System engineering, therefore, manages and attempts to minimise the dynamic Quality Gap over a system lifecycle. These issues are addressed by Boarder (1994), in the context of education systems development and assessment.

Systems Engineering

System engineering is separated from System(s) Engineering but, within an organisation for systems development, there is variety in the systems developed. This makes necessary support for various system-engineering environments and their multifaceted processes. Since this organisation is defocused from the ultimate systems of interest, any process supported cannot be system engineering. This organisation and supporting processes form systems of interest at different organisational levels. Lacy (1992) refers to this system as the "Systems Engineering Management System".

In Summary

An argument has been made for a distinction between System(s) Engineering: Organisation versus Practice. In summary, we state as illustrated in Fig. 1 that System(s) Engineering has Purpose, Product, Process and Position so that through its Position and Process:

- System(s) engineering is a convolution of a managed organisation and a process operating over the organisation.

While,

- Systems Engineering is the managed organisation, a resource, a Position,
- System engineering is the Process practised by the system engineer over the Position

FRAMEWORK

In making a distinction between systems engineering and system engineering, no mere play on words is intended. The distinction facilitates the identification of a framework for system engineering practice. This framework gives clear guidance to the system engineer of necessary and sufficient activities to be invoked. A framework based on a structure of Goal, Aims and Objectives is illustrated in Fig. 1 and defined in Boarder (1994, forthcoming). This framework is being evaluated and reported in an ongoing research programme. The identifiers in the matrix, for example P11, P12, etc., reference specific activity descriptions as given in Boarder (1994, forthcoming).

To establish the framework, answers to four simple questions concerning Purpose, Product, Process and Position are needed. These answers, the Why, the What, the How and the What With of system engineering, are:

- Purpose - System engineering establishes a customer's requirement, its environment; Goal, Aims and Objectives

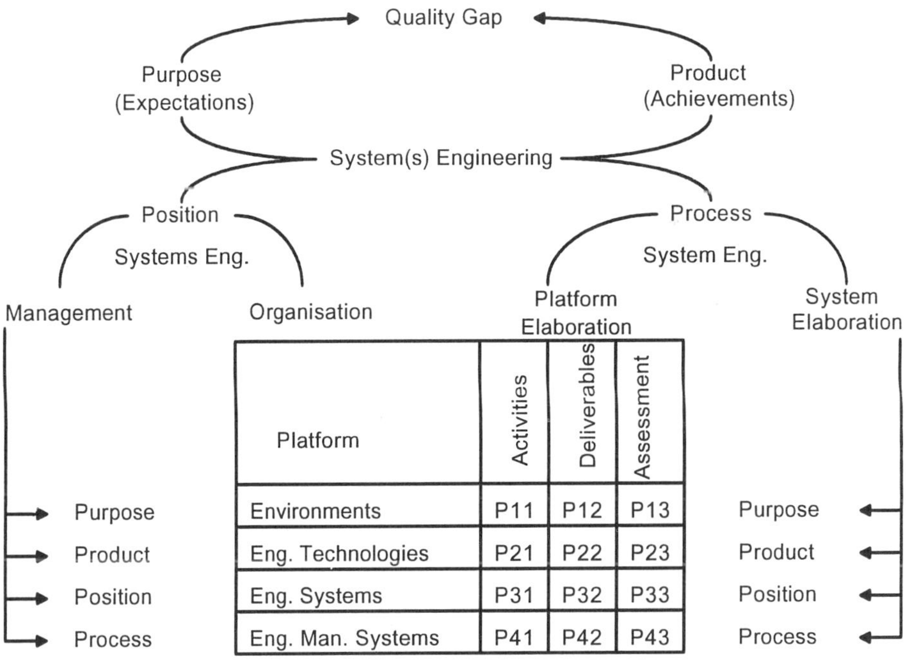

Figure 1. A system engineering framework.

- Product - System engineering delivers system models
- Process - System engineering practises activities leading to well-defined deliverables and assessments
- Position - System engineering selects well-defined materials, methods, tools, practices and principles that realise the models

System engineering, then, models, plans and engineers a response to a requirement. In doing so system engineering elaborates the required system and the Platform for its development from the Position established by systems engineering.

On Systems Engineering

System engineering is conducted over the systems engineering Position. Systems engineering managers will carry overall commercial responsibility for system engineering. Systems engineering managers will establish the training needs for engineers using the outputs of system engineering. In order that the systems engineering position can be sustained, system engineering appears to take place under the guidance of two Laws: namely:

- Law 1: System engineering shall be a traceable process over Systems Engineering
- Law 2: System engineering shall be a tractable process over Systems Engineering

The first law requires decisions taken to be justifiable after the event. An audit trail is essential. The second law requires system engineering to be easily managed, pliant, docile; a teachable and submissive process; teachable to engineers and submissive to the needs of customers. Given these laws, if a process employed in the development of a system is neither traceable nor tractable, then it cannot be system engineering.

CONCLUSION

This paper argues for a separation of systems engineering from system engineering. A context for system engineering is provided and a framework is elaborated. The argument is over the System(s) Engineering convolution concerning Environment, Engineering Technology, Engineering System, Engineering Management System issues which influence systemic properties, systematic processes and subsystems of systems under development.

System engineering, conducted over a systems engineering position, differentiates and integrates these diverse concerns delivering emergent properties while satisfying laws of traceability and tractability. The differentiation and integration are encapsulated in a framework of Goals, Aims, Objectives, Activities, Deliverables and Assessment Criteria. The elaborated framework enables the Quality Gap, to be assessed. The convolution is completed when the customer's system of interest is itself a systems engineering environment.

In the introduction, the confused interpretation of System(s) Engineering was highlighted. Perhaps we should be careful of titles given to engineers, organisations and processes if we intend their function to be traceable and tractable. For, in each of the interpretations, issues of position and process exist. The difference is one of degree, as determined by the observer, which is as it should be.

REFERENCES

Boarder, J.C., 1994, A diagrammatic aid to course development and assessment, *Educational Training and Technology International*, Vol. 31(3).

Boarder, J.C., 1994, (in prep.) "An elaborated system engineering framework".

Boarder, J.C., 1990, Project management and automated support, *in*: "Software Engineering for Technical Managers", ed. R.J. Mitchell, Peter Peregrinus, London.

Boarder, J.C., 1989, ASEE: ATMOSPHERE (ESPRIT II Advanced System Engineering Environment TIP), *in*: "Software Engineering Environments: Research and Practice", ed. K.H.Bennett; Ellis Horwood, Chichester.

Buckminster-Fuller, R., 1975, "Explorations in the Geometry of Thinking: Synergetics", Macmillan, New York.

Checkland, P.B., 1981, "Systems Thinking, Systems Practice", John Wiley & Sons, Chichester.

Jenkins, G.M., 1972, "The Systems Approach", *in*: "Systems Behaviour", edited by Beishon, J., Peters, G., Harper and Row, London.

Hitchins, D.K., Boarder, J.C., Moore,.D.R., 1992, IEE Draft guide to the practice of system engineering, IEE Publications, ref. SP024.

Hitchins, D.K., 1992, "Putting Systems to Work", John Wiley & Sons, Chichester.

Lacy, J.L., 1992, "System Engineering Management: Achieving Total Quality", McGraw-Hill, New York.

30

A SYSTEM ENGINEERING FRAMEWORK AND ITS EVALUATION

J. C. Boarder and J. Laming

Department of Computing and Electronics
Buckinghamshire College
High Wycombe, HP11 2JZ
United Kingdom

INTRODUCTION

An argument has been developed (Boarder, 1995(b)) for resolving the implicit convolution in the terms systems and systems engineering. 'System engineering', it is concluded, is a process conducted over a 'systems engineering' position. 'System engineering' differentiates and integrates, with respect to a specific system requirement, diverse issues while delivering emergent properties into an environment and satisfying laws of traceability and tractability.

It is proposed that the differentiation and integration issues of system engineering can be encapsulated in a generic and parametric framework of goals, aims, objectives, activities, deliverables and acceptance criteria.

In this paper, we set the 'systems engineering' context and further elaborate the 'system engineering' framework. In particular, we summarise three levels of the framework, illustrate the system engineering process and show how generalised workpackage structures emerge. This development sets the context for an evaluation of the framework, currently being undertaken within a configuration management context. The early results of the evaluation will be reported to conference. If the framework has value, that value will lie in the support it provides for complex rather than one-man projects. On the scale proposed, it is hoped that these projects will validate consistency within the framework, identify incompleteness, and advance areas of technology support. On the other hand the framework may be resolutely refuted.

SYSTEMS AND SYSTEMS ENGINEERING

The 'systems' revolution is upon us. Over the years, a constantly changing industrial and engineering environment demonstrates that what once worked simply doesn't now; complexity is the order of the day. 'Systems engineering' is a rational approach to decision-making related to the solution of complex problems in engineering planning, design and

Critical Issues in Systems Theory and Practice, Edited by K. Ellis et al.
Plenum Press, New York, 1995

operation. The evolving 'systems engineering' discipline is a natural successor to the disciplines of physics, engineering, biology, etc., in that it sits above these conventional disciplines, seeking to provide an umbrella and to establish a comprehensive and universal set of principles. There are, it seems, (Hitchins, 1992; Checkland, 1981) two kinds of systems, 'hard' systems and 'soft' systems, and two kinds of 'systems engineering' with varying opinions and thought processes that have lead to a fundamental divide in thinking and approach.

A 'hard' systems approach would be seen as a specific system solution to actual needs. A 'soft' systems approach would explain that, in the real world, a human activity system is by nature messy, and may not exhibit a clear need or purpose. So any solution must evolve in a similar manner to the need. Perhaps there is a middle way that enables messy, 'soft' systems to be captured and 'hard' systems concepts to be applied in their resolution. In this work then, we are concerned with building a bridge between the 'soft' and 'hard' systems approach.

SYSTEMS ENGINEERING AND SYSTEM ENGINEERING

It is argued (Boarder, 1995(b)), that the convolution implicit in the term 'systems' should be resolved. The multifarious interpretations of 'systems engineering', 'system engineering', even 'system's engineering', and 'systems engineering management' need to be separated. These terms should be set in context, as attempted by Lacy (1992), and consistently defined. Alternatively, new terminology is required. Either way, by reducing the complexity of 'systems engineering', by setting its parts in proper context, we may find ways of reducing complexity in systems.

The argument distinguishes 'systems engineering', an organisation, from 'system engineering', a process. The argument shows the relationships between organisation and process and the need for support by a framework which highlights 'system elaboration' and 'platform elaboration'. In system elaboration, the system engineer elaborates the purpose, product, position and process of the system to be delivered. Under platform elaboration, the system engineer establishes a platform, a set of resources, over environmental, technological, engineering systems and management issues, for the system to be delivered. The resources are drawn from the systems engineering organisation.

A SYSTEM ENGINEERING FRAMEWORK

It is shown (Boarder, 1995(a)), that the prime concerns of system engineering are the elaboration of the required system and the elaboration of a platform to develop and support the system. While system engineering is not directly involved in the production and support of a system, it determines what system is produced and supported, how the system is produced and supported and what resources are used in producing and supporting the system. As further developed, this process has four essential phases: monitor, model, define and organise. Thus, system engineering monitors the potential environment of a system, models aspects of the system and its environment, defines proposals for new systems supporting the environment and organises engineering processes for system production and support. A system engineering framework integrates these four phases with respect to a particular system development.

A framework for system engineering, then, is a device that guides, supports and captures system engineering with respect to a specific system development.

- In guiding system engineering, the framework clearly and unambiguously sets out the essential system engineering process and its subprocesses.
- In supporting system engineering, the framework promotes structure and order in information acquisition and system development.
- In capturing system engineering, the framework records activities and their progress for subsequent review.

A system engineering framework therefore provides the link between the rationalisation of 'soft' systems issues, captured in the monitoring and modelling phases of the process, and their resolution using 'hard' system engineering, established by the definition and organisation phases.

In contrast, an aphorism for systems engineering might be: Given a context, we elaborate a framework, within which we practice system engineering with consistency and, hopefully, completeness.

CONTEXT

The definition of system by Checkland (1981) is a model of a whole entity relative to an observer characterised fundamentally by hierarchy, emergence, communication and control. Emergent properties of systems are properties meaningful only when attributed to a whole and not its parts. Hitchins (1992) puts the concept of emergent properties within the context of systems engineering well when stating:

> "The primary task of systems engineering is to identify, realise and maintain the requisite emergent properties of a system to meet customers and end users needs".

Since much emphasis is placed on relations between system and observer, we take it as an axiom that the observer defines the characteristics of system. The system engineer's view must be substantiated, not just by one observer (the system engineer), but consistently among many observers. System engineering, as a goal directed process, structures and orders the system's emergent properties. The definition of system engineering used here is therefore:

- The structured and ordered monitoring, modelling, defining and organising of a system with emergent properties required by a customer, integrated within an end-user's environment

From the foregoing, a framework, as an aid to the system engineer, provides a structured and ordered environment within which system engineering is conducted. Taking one step back in the hierarchy, the framework has a goal which is:

- System engineering with emergent properties required by customers integrated in end-user environments

Here emergent properties include consistency and completeness, the customer is system engineering management and the end user is the system engineer. The framework provides the guidance, support and capture facilities referred to earlier.

FRAMEWORK

The proposed system engineering framework uses a structured interpretation of the terms context, goal, aims, objectives, activities, deliverables and acceptance criteria. This structure was used, in part, in the IEE Draft Guide to the Practice of System Engineering (Hitchins, Boarder and Moore, 1992) which is being thoroughly revised because of its failure as a basis for training and education.

The framework exists at three levels of definition.

First Level Framework

The first level of the framework, or basic structure, defines the terms context, goal, aims, objectives, etc. These definitions are rationalised in Boarder (1995(a)) as a first level framework and stated here as:

- Context: Descriptions of the system, environments, support and management
- Goal: A single definitive statement of the end to be achieved for the customer
- Aims: two statements of intent with regard to the Goal and the basis for its achievement
- Objectives: three statements of lesser goals for Product, Position and Process by which the aims are satisfied
- Activities: Four activities, namely, monitoring, modelling, definition and organisation activities over environmental, technological, engineering and management issues in fulfilment of the Objectives
- Deliverables: Those proposals, definitions, models, structures and processes to be delivered by the activities
- Acceptance Criteria: The acceptance criteria to be used in the assessment of the deliverables

Second Level Framework

The second level framework elaborates system engineering by expanding the first level. In this framework, key words are shown in capital letters. For brevity detail is omitted and the second level framework is as follows:

The system engineering framework has a:

- Goal: SYSTEM ENGINEERING, with emergent properties required by customers integrated in end-user environments

The goal will be achieved by activity directed:

- Aim 1: To elaborate SYSTEM ENGINEERING in end-user environments with emergent properties required by customers
- Aim 2: To elaborate a position from which SYSTEM ENGINEERING can be advanced in end-user environments with emergent properties required by customers

The aims of the framework will be served by a framework which captures and maintains the outputs of activities which have:

- Objective A: Modelled SYSTEM ENGINEERING with emergent properties required by customers and end-user environments
- Objective B: Applied ordered processes in structuring and modelling SYSTEM ENGINEERING emergent properties
- Objective C: Used well-defined practices and established principles in monitoring SYSTEM ENGINEERING emergent properties

The objectives will be achieved by Activity-Deliverable-Acceptance Criteria tasks concerned with:

- Monitoring the SYSTEM ENGINEERING environment, technology, engineering and management
- Modelling the SYSTEM ENGINEERING environment, technology, engineering and management
- Defining the SYSTEM ENGINEERING environment, technology, engineering and management
- Organising the SYSTEM ENGINEERING environment, technology, engineering and management

Third Level Framework

The third level framework applies simple replacement techniques to key words of the second level framework. An example of the framework is being developed to support configuration management. By replacing the key word SYSTEM ENGINEERING at level two by Configuration Management, the third level framework becomes:

The system engineering framework for Configuration Management has a:

- Goal: Configuration Management, with emergent properties required by customers integrated in end-user environments

The goal will be achieved by activity directed:

- Aim 1: To elaborate Configuration Management in end-user environments with emergent properties required by customers
- Aim 2: To elaborate a position from which Configuration Management can be advanced in end-user environments with emergent properties required by customers

The aims of the framework will be served by a framework which captures and maintains the outputs of activities which have:

- Objective A: Modelled Configuration Management with emergent properties required by customers and end-user environments
- Objective B: Applied ordered processes in structuring and modelling Configuration Management emergent properties
- Objective C: Used well-defined practices and established principles in monitoring Configuration Management emergent properties

The objectives will be achieved by Activity-Deliverable-Acceptance Criteria tasks concerned with:

- Monitoring the Configuration Management environment, technology, engineering and management
- Modelling the Configuration Management environment, technology, engineering and management
- Defining the Configuration Management environment, technology, engineering and management
- Organising the Configuration Management environment, technology, engineering and management

In the full model other key words are defined. Fully expanded, the framework comprises activities, deliverables and acceptance criteria for environment, technology, engineering and management issues for application systems to be developed. Space does not permit their display but the framework can be reapplied for lower level systems and expansions of lower level work packages.

PRACTICE

The framework as a basis for the practice of system engineering is described in Figure 1 of Boarder (1995(b)). The practice and the evaluation of the framework is therefore an elaboration of these activities. Their conduct is performed in an orderly and structured manner that, through support for the framework, offers traceability, visibility and tractability of system development. The experiment for configuration management and others will, hopefully, display consistency and completeness in the framework and identify further support resources for industrial and commercial applications. Configuration management of the frameworks, is being investigated as an experiment in system engineering.

CONCLUSIONS

Systems engineering as an convolution of self-similar activities is being resolved into organisation and practice both aspects of which need structured and ordered support. Support can be provided in the form of generic and parametric frameworks that guide and support system engineers in their work and captures their results. One such multi-level framework is proposed and illustrated. The framework is based on generally accepted systems terminology and system engineering activity using a high-level, system engineering process model. The framework is generic and parameter replacement allows specific systems oriented frameworks to be generated. Experiments are in-hand to evaluate completeness and consistency in the framework, among other properties, and more specifically to identify configuration management support requirements. The nature of the framework, the activities, deliverables and acceptance criteria have been outlined. The evaluation results will be reported and areas in need of further technology highlighted.

ACKNOWLEDGMENTS

The authors would like to acknowledge the support of Buckinghamshire College and to thank staff, particularly Sam Groom who was involved in the background work for this paper, for their patience, encouragement and support.

REFERENCES

Boarder, J.C., 1995(a), System engineering: a generic and parametric model; submitted to the First Spring Symposium of the UK Chapter of the National Council On Systems Engineering, London.

Boarder, J.C., 1995(b), System(s) engineering: Organisation versus Practice, *in*: F.A. Stowell, D. West, and J. Howell, (eds.), "Systems Science : Addressing Global Issues", Proceedings of the Fourth International Conference UK Systems Society, Plenum, New York.

Checkland, P., 1981 "Systems Thinking, Systems Practice", John Wiley & Sons, Chichester.

Hitchins, D.K., 1992, "Putting Systems To Work", John Wiley & Sons, Chichester.

Lacy, J.L., 1992, "Systems Engineering Management: Achieving Total Quality", McGraw-Hill, London.

Hitchins, D.K., Boarder, J.C. and Moore, P.D.R., 1992, IEE Draft Guide to the Practice of System Engineering, IEE Publications ref. SP024, London.

31

SYSTEMS FOR PHILOSOPHY

Bill Cameron

University of Glasgow
Glasgow
United Kingdom

INTRODUCTION

The aim of this paper is to explicate the concept 'system' as a working tool for the practice of analytical philosophy. System is defined as a mode of description, as a description of a conceptual holism. Some of the implications of this are considered.

DESCRIPTION

In science or in philosophy, we are concerned to explain. Before we can explain, we and the audience, which may be ourselves, need to recognize what it is that is being explained. Prior to the act of explanation is the act of description. In this paper I wish to describe a particular sort of description. One in which some particular aspect of our world is described not just in terms of its function and form but also in terms of its counterfactuality, how its function and form determine the manifold events that can occur in this aspect of the world. It is in these terms that I wish to explicate the notion of 'system'.

We must distinguish between the act of description and the well formed sentences in the language that represent the knowledge conveyed by the act. It is this abstract entity, which is realised in the text, the marks of the processes of language, that I will refer to by the term 'description'. By locating the description in the text I am not taking a stance in the post-modern, post-structural dispute regarding the determinacy or relativity of the 'meaning' in text, though this could be discussed in system terms. This dispute has been addressed by many writers including Fish (1980) and Midgley (1994). My aim is simply to distinguish the terms of the text from the actual, fictional or hypothetical reality described. The concepts referred to by the terms of the text may differ between author and interpreter. It is this risk of cultural ambiguity that the exponents of techniques of cooperative description such as Cognitive Mapping, Eden (1989), seek to avoid.

By a 'description' I mean more than a denotative expression, although any description in natural language may contain many such phrases. A particular description is a description of something. It is a description of actuality in that the modal operators used in the text are concerned with the factual rather than the counterfactual. It is a description of an actual world rather than of one prescribed by counterfactuals over some range of possible worlds.

Critical Issues in Systems Theory and Practice, Edited by K. Ellis et al.
Plenum Press, New York, 1995

SYSTEM AS A MODE OF DESCRIPTION

The function of a system in language is similar to that of a description. The domain of the entities it describes will range over possible worlds defined by the counterfactual conditionals within the text of the description and it is this generalizing function that characterizes 'system' as a mode of description. This characterization of 'system' is different from that of workers in the system movement such as Checkland. In p 110 of his book Checkland (1981) writes "Let us start with the physical systems that apparently make up the universe". For Checkland, a system is what is described rather than its description. It is real, it is a 'substance'. Klir, (1969) has written, about systems; "In experimental science; an abstraction to be used when nature is examined from a viewpoint. In engineering science; an abstraction when relations are prescribed and we wish to implement them." It is at Klir's level of generality and abstraction that I will use the system mode of description. The narrative mode of description may describe an actual event or an actual structure, it may describe a fictional or hypothetical event or structure. The system mode of description describes how the properties of the entities and the relations between them, prescribe the sorts of structures or events that may occur and 'may' is a modal idiom.

ACTUAL AND POSSIBLE WORLDS

The notion of counterfactual dependence is enmeshed with the notions of causation, theory and law. Everyday statements like 'If you had got home in time we would have got to the party' rely for their interpretation on theories of cause and effect. Lewis, (1973), has given a general analysis of counterfactual conditionals. Lewis contends that our modal idioms are quantifiers over possible worlds. For Lewis, possible worlds are real entities and 'actual' is an indexical like 'I' or 'here' or 'now'. Our interpretation of indexicals depends on the context of utterance and the context of interpretation. Utterance may be in the context of action or of planning. Interpretation may be hearing such utterance, the reading of a narrative description or the reading of an assembly manual in the spatial and temporal contexts of assembly. If, in the text of narrative description, I say "we were then too late to go to the party", I am describing what actually happened. If in the idioms of counterfactual hypothesis, I say, "Had you got home in time we would then have got to the party" then, in Lewis's terms, I am describing a relation between possible worlds. I could have said, "Had you got home in time we would then have got to the party but actually you were late and we did not". It is in this spirit that Lewis (1973) claims that 'actual' is indexical. "It depends for its reference on the circumstances of utterance, to wit on the world where the utterance is located". Lewis does not claim that other possible worlds actually exist. He does claim that, in any text containing counterfactuals, the same logic with the same criteria for the preservation of truth will apply for each alternative text within the combinations of the counterfactuals. It is in this sense that a system may be characterized as a set of possible worlds in which set membership is prescribed by the defined properties of the entities and their relations. Lewis is a realist about possible worlds. For him my definition of system as a mode of description would correspond to abstract, linguistic, ersatz worlds. (Lewis 1986).

Lewis also defines the notion of 'similarity' between sets of possible worlds. If a system description contains a finite number of counterfactual statements then any state of the system is a possible world. Worlds are 'similar' in that the components of the state space vectors that locate them are similar.

THE ROLE OF CAUSATION

Hume, (1748), defined cause in the terms "an object followed by another and where all the objects similar to the first are followed by objects similar to the second. Or in other words where, if the first object had not been the second never had existed." This is two definitions. The first is based on the notion of constant conjunction. It is a recognition of regularity. It implies a diagnosis of cause which we may choose to generalize as a law. The second is a statement of counterfactual implication.

Causal statements may be singular or general. Singular statements are concerned with attributions of cause between specific events, they may be intentional. General causal statements may be law-like or they may be intentional. Material implication may derive from some overarching law or from general causal assumptions that are equivalent to laws within the description. They are explanatory within any alternative outcome.

Counterfactual statements will derive from general causal assumptions. In a system description we are generally concerned to distinguish between counterfactual dependencies that are causal in a law-like fashion and those that are caused by rational intervention. Additionally we may be concerned to include in our explanation the possibility of accident, which may be expressed formally in terms of probability or as recognition of the unexpected. Thus, some of the counterfactual statements within a system will be based on nomological and some on rational, causation. Some will be based on the recognition of accident.

Not all descriptions involving counterfactuality are usefully seen as systems. Any narrative may include clauses as to how things might have turned out. One component of the essence of a system is that it is a description of the way things might be. A system is essentially counterfactual. It is also essentially holistic.

CONTENT, CONCEPT AND HOLISM

The word 'holism' was first coined by Smuts, (1926) and defined as "...the tendency in nature to produce wholes from the ordered grouping of units". When we describe something we may assign identity to that thing within the text of the description. The success of the act of description lies in the changed state of belief in the audience. Belief requires understanding and understanding requires that describer and audience share mastery of those concepts that comprise the description. Peacocke, (1992) has set out a philosophical theory of concepts. For Peacocke the mastery of a concept by a thinker lies in the thinker satisfying formally stated possession conditions and it is these possession conditions that define that concept for that thinker. Concepts are the components of content. It has been suggested, Quine, (1953), that all content is coherent, that "The unit of empirical significance is the whole of science". This global holism is controversial and is too strong for our needs. Peacocke (1992) gives possession conditions for the concepts within a local holism in the terms "Concepts F1.....Fn are those concepts C1.....Cn to possess which a thinker must meet the condition A(C1.....Cn)". Tennant, (1987), distinguishes between molecularism and holism. Following Weir, (1985), <$ F Weir (1985) defines: "A relation 'R' is well founded iff there are no R-cycles nor any infinite descending R-chains." >Tennant shows that if a structure of concept dependency is well founded and is separable then Quine's (1953) 'whole of science', can be seen as "a possible blend of globally separable local holisms" (Tennant, 1987). It is this sort of holism that characterizes what is described in a system description. The concept described by a system will be a logical category as described by Ryle, (1949). Much of Ryle's argument for avoiding category mistakes could be presented in system terms.

For example, a disposition to some particular behaviour could be redescribed as a state of an intentional system.

FUNCTION

We may describe a component within a device as having a function within the device. This is the function of the component. We may describe the device as having a function for us. Function is a property that we attribute to a holism, whether process, event or substance. Evidently, the function is 'of' the holism. What the function is 'for' is less self evident. The notion of function arises naturally in the description of life. With the advent of life came intentional behaviour, behaviour that is for the organism. This behaviour includes the responses of the organism to its environment and the production of artefacts; web, cocoon or space shuttle, that have function for the organism.

The concept 'function' can be a component of the content of a system description or it can be a component of the essence of the holism described. Millikan, (1984), describes the components of language as devices with function. These include the terms of the language, the marks of the language, whether in print or in memory and the hearers and speaker who recognise and interpret these terms. A system description of a natural language would include these as components with function. For Millikan, (1984), "...these natural devices have not literally been 'designed' by someone to serve their functions." Their function has evolved and a 'purpose' is what it 'does'. If the function of language is for intentional communication then, in an intentional functional explanation, we may describe it as for us. Thus we distinguish between the function of a component in an artefact and the function of an intentional act for its instigator. An intentional, functional explanation of language can be founded in the definition of life as establishing function 'for' and without this foundation we cannot bound our explanations. This distinction parallels the distinction between law-like cause, which explains material implication, and intentional cause which initiates change. Function may be one component of the essence of a system description in that it names the function of an artefact that is the realisation of the system and this artefact will have this function for the designers or their customer. The term 'root definition' has been coined by Checkland, (1981), as a functional essence of a system description of a holism, yet to be designed or amended, to address some problem. Not all systems describe the possible worlds of potential artefacts. We may be describing the organisation of health care in a community or the genesis of destructive fires. The ascription of function within a system that describes events in the world must be explanatory. The validity of this explanation will depend on the coherence of the theory that link the concepts of which it is comprised.

VALIDITY AND REALITY

We speak of a model as 'valid' if it is, in some sense, an 'appropriate' or 'correct' representation of the holism that we wish to explain and predict. We speak of the propositions that describe the states of the holism as 'true' in a metalanguage if the conditions set out by Tarski, (1944) are met. Within the meta-language we may state, 'P is T iff p' where 'p' is a statement about the real world and P is a name of a proposition in the natural language. Field, (1972), has reformulated the Tarski truth conditions in terms of denotation. Lewis, (1973), describes how the truth conditions for a set of possible worlds depend on the truth of the counterfactuals that link them. From these three formulations we may define a valid system as one that satisfies the following conditions.

1. The possible worlds described within a system include a description of an actual world.
2. Coherent denotative relations are established between the terms of the description and entities in the world.
3. The counterfactual conditionals are true in the Lewis sense.
4. The propositions in the description of the actual world are true in the Tarski sense.

Any artefact that, by whatever analogues, realizes this system will be a valid model of the real world of possibilities that is described by the system.

SUMMARY OF SYSTEM

A system is the textual realisation of a mode of description. It is essentially counterfactual, it describes a set of possible worlds. A system is essentially holistic, it is a molecular holism of interdependent, grounded concepts. This holism has a place in our world, it has a relevance or a function for us. A system has identity. It is recognized by its unique holism of concept and theory.

APPLICATION

We are in the world and we are of the world. In system terms we may describe our world, including ourselves changing our world and one another. Also, in system terms we may describe our use of language both to communicate in the here and now and to describe how we communicate and describe. By the recursive use of system we may explain ourselves, our language and our world without the need for a transcendental stance. My belief is that self, mind and consciousness can be explained as concepts emergent from system descriptions and this work is ongoing.

ACKNOWLEDGMENT

Acknowledgement is due to Jim Edwards at Glasgow for his support in this work. Errors and prejudices are my own.

REFERENCES

Checkland, P., 1981, "Systems Thinking, Systems Practice", John Wiley & Sons, Chichester.

Eden, C., 1989, Using cognitive mapping for strategic options development and analysis, *in*: "Rational Analysis for a Problematic World", J Rosenhead, ed., J Wiley, Chichester.

Field, H., 1972, Tarski's theory of truth, *Journal of Philosophy*, 69:13.

Fish, S., 1980, "Is There a Text in This Class?", Harvard University Press, Cambridge, Mass.

Hume, D., 1748, "Enquiries Concerning Human Understanding", L.A. Selby - Bigge, and Rev. P.H. Nidditch, eds., Oxford University Press, Oxford. (1975)

Klir, G.J., 1969, "An Approach to General Systems Theory", Van Nostrand Reinhold, New York.

Lewis, D., 1973, "Counterfactuals", Blackwell, Oxford.

Lewis, D., 1986, "On the Plurality of Worlds", Blackwell, Oxford.

Midgley, G., 1994, Five sketches of post-modernism: Implications for systems thinking and operational research, *Systemist* 16:3.

Millikan, R.G., 1984, "Language, Thought, and Other Biological Categories", Bradford Books, MIT Press, Cambridge.

Peacocke, C., 1992, "A Study of Concepts", Bradford Books, MIT, Cambridge.
Quine, W.V.O., 1953, Two dogmas of empiricism, *in*: "From a Logical Point of View", Harper and Row, New York.
Ryle, G., 1949, "The Concept of Mind", Hutchinson, London.
Smuts, J.C., 1926, "Holism and Evolution", Macmillan, London.
Tarski, A., 1944, The semantic conception of truth and the foundations of semantics, *Philosophy and Phenomenological Research*, 4.
Tennant, N., 1987, "Anti-Realism and Logic", Clarendon Press, Oxford.
Weir, A., 1985, Against holism, *Philosophical Quarterly* 35:

32

MEMORIZATION PROCESSES IN HIERARCHICAL ORGANIZATIONS

A Systems Perspective versus the Bureaucracy Pattern

H. Dragomirescu

Academy of Economic Studies
Piata Romana 6
Bucharest 70167
Romania

INTRODUCTION

The amount of knowledge currently available on how organisations memorise could be assessed as being particularly scarce. However, since it was argued that organisations make use of their memory by acting (Nelson and Winter, 1987), there are reasons to assume that getting organisational memorisation process explained could bring about substantially new and meaningful insights as to interpreting the actual organisational functioning patterns. In such a reasoning line, two alternative ways of representing memorisation processes that take place in today's widely spread hierarchical-type organisations are briefly discussed in this paper: The first one is grounded on the traditional, positivist organisational paradigm, the limits of which are highlighted, while the second one pertains to a systems approach. The point to be made is that it is the systems approach that provides a valid interpretation of the memorisation processes taking place within hierarchical organisations thus specific arguments being brought about to support the legitimacy of adopting this approach in both organisational theory and practice.

ORGANISATIONAL MEMORISATION PROCESSES WITHIN THE IDEAL BUREAUCRACY FRAMEWORK

The ideal bureaucracy pattern, as theoretically shaped by Max Weber (1975), is widely recognised as the prototype for hierarchical organisation as that is defined in terms of the positivist paradigm discussed by Ellis (1993); this pattern thus worth being hereby chosen to serve as a referential setting for any inquiry as to how the positivist organisational paradigm would enable to get organisational memorisation processes explained. Max Weber's classical model of the ideal bureaucracy was discussed in rather criticist terms (Makrimiditis, 1993); however, to capture its essential features suffice it referring to the

Critical Issues in Systems Theory and Practice, Edited by K. Ellis et al.
Plenum Press, New York, 1995

opinion that bureaucracies would have been completely cleared of turbulence, ambiguity or redundancy (Nonaka, 1988), as well as to that stating that ideal bureaucracy looks like peopleless organisation (Bergeron et al., 1986). Although Max Weber emphasised the crucial importance of organisational knowledge and job-related technical competence as major sources of the ideal bureaucracy's pretended supremacy in terms of rationality and smooth functioning, no specific evidence as to how bureaucratic organisations are meant to perform their inherent memorisation function are to be found in his model; *this model therefore needs being re-assessed in order to derive its implicit premises that would have led Weber to characterise the ideal bureaucracy as a typically knowledge-based organisation.* A set of assumptions is hereby proposed, that attempt making the Weberian model explicit with respect to the memory-oriented side of the hierarchical organisations functioning under the terms of the bureaucratic pattern. These assumptions could be formulated as follows:

- Bureaucracies are meant to adaptively collect and store information through operating their current formal tasks and procedures rather than pro-actively acquiring knowledge through both individual and group learning.
- Bureaucratic organisations' members are expected to only acquire insights and experience strictly related to their formally assigned task, as defined under the organisational internal division of labour.
- Insights and experience acquisition by bureaucratic organisation's members favours current routines and procedures maintenance and amelioration rather than promotes organisational change from within.
- Formal rules and procedures alteration or even radical change is exclusively assigned to the top management intervention, which in turn is subject to constraints drawing on the amount and accuracy of knowledge available at this hierarchical level.
- As bureaucratic organisations currently function, their overall memory is involved on its instrumental side (files, official records, rulebooks, procedure specifications) rather on its human side drawing on the individual and shared memory resources.
- Within the bureaucracy framework, individual memorisation process is supposed to be top-down, formally regulated and thus constrained to concern itself with supporting required job performance as well as with observing authority-based rules and procedures in force.
- Bureaucratic organisation's environment can be globally scanned and interpreted by the top management and at the organisational level only, where overall objectives are set out, performance assessed and regulatory interventions triggered.
- Assisting in hierarchical control performance is a major reason for organisational, formally regulated action of bureaucracies towards undertaking memorisation tasks.

To summarise the above listed set of assumptions, one can assert that the bureaucratic-type organisational structuring and functioning is to be attributed to the joint, well-balanced application of two principles, that is the *knowledge principle* and the *control principle*, assuming that they are compatible. *The ideal bureaucracy model then holds valid under the strict condition only that a homorphism exists between the intra-organisational distribution of knowledge (including memory) and that of the formal authority.* This particularly tight validity domain reveals the limits of the Weberian model.

Firstly, the *operationality limit* of the ideal bureaucracy pattern should be stressed, which pertains to both feasibility and sustainability of a purely bureaucratic conduct within hierarchical organisations; in practice, there are the mostly impure forms of such a conduct that might be encountered, as a perfect match between the knowledge distribution and that of the authority is unlikely to be maintained. This argument should be provided when raising the question as to whether purely bureaucratic organisations could really exist (Handy, 1988). Organisational practice suggests a rather negative answer to this particular question, even though real-world, contemporary organisations exhibit, in most cases, hierarchically-like patterns and thus adopt some elements of bureaucratic conduct.

On the other hand, mention must be made of the *epistemological limits* of the bureaucratic pattern of organisation as they apply to the memorisation process. From the specific point of view of producing valid insights as to how this process takes place, the ideal bureaucracy model fails due to its intrinsic reductionism; it privileges the instrumentation of an abstract, de-personalised formal memorisation setting, while over-simplifying and thus underestimating the human dimension of organisational memory and knowledge.

A SYSTEMS PERSPECTIVE ON MEMORISATION PROCESSES WITHIN HIERARCHICAL ORGANISATIONS

The limits of the bureaucratic-type view on organisational memorisation implicitly suggest the need for an alternative representation of this process. As Simon (1982) pointed out, an obsolete, contested theory still stands until a more valid one replaces it. In this respect, within the field of contemporary organisation theory, a change of paradigm is taking place, the concept of 'knowledge-based organisation' being thought of as the basic concept of the systemic organisational paradigm as contrasted to the positivist one, whose core concept is that of the 'control-based organisation'. In connection with the systemic organisational paradigm, Le Moigne (1990) developed the concept of 'memory-centred organisation' which he neatly differentiates from both the hierarchical type and the anarchical type of organisation respectively. Without doubting the epistemological validity of the 'memory-centred organisation' concept, it might be asserted that such a pattern could take shape further to a deliberate structuring option rather than by an incremental organisational self-development. Under such terms, it appears that an organisation could function according to the memory-centred pattern in those case only when it is either 'ex-novo' designed as such (Ferrer-Figueras, 1994), or when it comes to embed an organisational information system in the sense defined by Van Gigch and Le Moigne (1990) to serve as a basis for advanced memorisation organisational function.

Capturing the real degree of complexity of the memorisation process taking place within real organisational hierarchical frameworks seems only to be possible by adopting a systems perspective. The subtle character of the respective process, which the ideal bureaucracy model ignores, actually appears to be one of the major hypostasis of this complexity. Thus, *in a real organisational framework, a large part of the memorisation processes cannot be subject of a strict hierarchical control.* In addition to the formally established organisational sub-units and posts, memorisation processes involve informal actors too (groups and individuals); these actors use to memorise according to their own information-utility functions, aiming at using the information they collect to undertake informal, intra-organisational information transactions, as pointed out by Carreaux (1990). With respect to the memorisation processes, *the hierarchical-type organisation actually worth looking at as a complex, multi-actor system, naturally tending to self-organise so that its memorisation capabilities are enhanced.* Consequently, as Morgan (1989) mentioned, redundancy inherently comes to exist due to self - organisation. Actors carrying out memorisation processes have a certain autonomy; none of them could be possibly authoritatively prevented form seeking, collecting and storing information that seems relevant to them.

The conduct of the organisational memorising actors is typically a pro-active one. These actors define and promote - sometimes explicitly, sometimes subtlely - their own memorisation tasks, for whose accomplishment they make use of the support provided by the formal structure; they profit by the positions they hold in the hierarchical framework, by their specialised role in the internal division of informational labour, by the human and technical resources allocated to them. In the same time, they tend to *informational self-sufficiency*, always attempting to acquire that particular information the possession of which could enlarge their autonomy or allow them to get a stronger capacity to both negotiate and influence on major decision-making matters at the overall organisation's level. *The memorising behaviour of organisational actors has therefore a marked selective character* in the sense defined by Breton and Winthrobe (1982); this character becomes manifest by their adopting a contingent approach of the problem of releasing information they possess and which are formally or informally demanded. Thus, releasing a certain information may sometimes be subjected to some informal condition, deliberately untimely conveyed or even disguised.

Under such circumstances, actors' selective informational behaviour comes to condition the capacity of formal leaders to accomplish their official duties; the paradoxical situation described by Crozier (1983) may occur, where those in power lack necessary information, whereas those lacking power were actually holding relevant information, still being unable to use it for decision-making purposes. As a reaction to such a situation, top management tends to develop its own informal communication network; this propensity was mentioned by Sainsaulieu (1987), who restated that an organisation actually needs internal information exchanges more largely and freely developed than formally allowed. Successive alliances or tacit confrontation interactions usually occur between organisational memorising actors, *a contingent information management framework being thus established at the organisation level.* Within such a framework, co-operation between actors could only partly be subjected to formal rules. *Actors are permanently confronted with a trade-off between following their prescribed informational duties or concluding informal transactions involving the information they hold or need.* In the latter situation, either tacitly implementing out-of record information routines or undertaking ad-hoc information exchanges could take place. The information asymmetry, which comes to underlie the propensity for informal transactions, is self-maintained.

An underground information economy develops in parallel with formal information system in place; *the 'market versus hierarchy' issue therefore comes to be relevant to the intra-organisational framework too.* Within the frame of the intra-organisational parallel information economy, a monopolistic competition-like pattern emerges to embed the market-type conducts of the memorising actors involved. It is only when the actors' specific interest cohere towards becoming informationally involved in supporting a certain project that they jointly make use of the information they control.

A pluralistic view on how real-world hierarchical organisations actually work as memorising systems emerges to countervail the reductionist view provided by the bureaucracy pattern. To summarise the above mentioned aspects, one may assert that formally, hierarchically structured, *memorising organisations tend to embed multiple anarchical-type memorisation patterns which, in an information-reach environment, come even to prevail.* In formal terms, the anarchical-type memorisation patterns would appear to be non-rational to the extent that the performance of the prescribed, hierarchical-type organisational type is affected; the occurrence of such patterns cannot be predicted, nor it complies with the formal rules in force. However, adopting by organisational actors of pro-active, selective memorising behaviour enables them profit by their informational autonomy and thus their own specific rationality criteria are met, as long as the memory-centred organisational pattern is still not deliberately adopted.

Concluding Remarks

- The ideal bureaucracy model contains its own abolishment's premises, which are particularly relevant to the memorisation processes area.
- The obvious need for accurately modelling the organisational memorisation processes legitimates the propensity towards systems approach application to get to an in depth understanding of the more general issue of how real organisations work through making use of their knowledge and memorised information resources.

REFERENCES

Bergeron, J.L., Coté-Léger, N., Jacques, J., and Béranger, L., 1986, "Les Aspects Humains de l'Organisation", Gaëtan Morin Editeur, Chicoutini.

Breton, A., and Wintrobe, R., 1982, "The Logic of Bureaucratic Conduct", Cambridge University Press, Cambridge.

Carreaux, G., 1990, La théorie des transactions informelles : une synthèse, *Economie et Sociétés (série Sciences de Gestion)*, 15.

Crozier, M., 1983, De la bureaucratie comme système d'organisation, *in* "L'analyse des Organisations. Une Anthologie Sociologique", F.Seguin-Bernard, and J.F.Chanlat, eds., Editions Préfontaine Inc, Québec.

Ellis, R.K., 1993, Positivism in systems science, *in* "Proceedings of the Second European Congress on Systems Science", published by the European System Union, Prague.

Ferrer-Figueras, L., 1994, Una metodologia para el diseño óptimo y el pilotage de la implantación de un sistema «ex-novo», *in* "Tercera Escuela Europea de Sistemas", L.Ferrer Figueras, A.Casellas Moncho, R.Beneyto Torres, eds, Ajuntament, Valencia.

Handy, C.B., 1988, "Understanding Organisations", Penguin Books, London.

Le Moigne, J.L., 1990, " La Modélisation des Systèmes Complexes", Dunod, Paris.

Le Moigne, J.L., 1993, Sur l'ingénierie de la connaissance organisationnelle, Research paper 93-02, GRASCE, Aix-Marseille III University, Aix-en-Provence.

Makrimiditis, A., 1993, La généralité de la contestation du principe de la hiérarchie dans le cadre de la théorie des organisations, *in* "Proceedings of the Second European Congress on Systems Science", published by the European System Union, Prague.

Morgan, G., 1989, "Images of Organisations" , Sage Publications, Beverly Hills.

Nelson, R., and Winter, S., 1987, An Evolutionary Theory of Economic Change, *in* "The Economic Nature of the Firm", L.Putterman, ed., Cambridge University Press, Cambridge.

Nonaka, I., 1988, Creating organisational order out of chaos, *California Management Review,* XXX, 3.

Sainsaulieu, R., 1987, "Sociologie de l'Organisation et de l'Entreprise", Editions d'Organisation, Paris.

Simon, H., 1982, "Models of Bounded Rationality", The MIT Press, Cambridge, Mass.

Van Gigch, J.P., Le Moigne, J.L., 1990, The design of an organisation information system,. Intelligent artifacts for complex organisation, *Information & Management,* 19, 5

Weber, M., 1975, "The Theory of Social and Economic Organisations", Macmillan, London.

33

THE RESOURCE-BASED THEORY OF THE GROWTH OF THE FIRM

A Systems Perspective

Elizabeth Garnsey

University of Cambridge
Department of Engineering and
The Judge Institute of Management Studies
Mill Lane
Cambridge CB2 1RX
United Kingdom

INTRODUCTION

Not all work with a systems perspective is acknowledged as such. Penrose in her pioneering *Theory of the Growth of the Firm* (1959; 1995) explored the conditions for the growth of a firm, the determinants and limits of growth. She developed an original framework of analysis in exploring these questions, one which has come to be increasingly acclaimed and influential (Best, 1990). In this paper it is argued that Penrose's resource-based theory of the growth of the firm is a form of systems thinking. Penrose did not set out to do a systems study and indeed the influence of her work in economics depended on her use of an accepted economic discourse. However it is argued here that the systemic nature of Penrose's analysis and her awareness of the interpretive filters which mediate between experience and action give special insight to her study. Critical systems thinking aims to provide scope for just such an interpretive approach to the workings of systems in the social world (Flood and Jackson, 1991; Garnsey, 1993). If we recognise Penrose's perspective as an example of systems thinking, then instead of the work appearing to be a one-off analysis, set apart from other approaches, it can be viewed as part of a wider intellectual movement, an implicit application of systems thinking to the problems of the growing firm. If we can identify Penrose's method of conceptual inquiry, we are in a position to apply this method, which has proved so fruitful, in other contexts.

THE FIRM AS A POOL OF RESOURCES

We can begin with Penrose's concept of the firm as a pool of resources. The originality of Penrose's approach consisted above all in substituting a dynamic analysis for the static

Critical Issues in Systems Theory and Practice, Edited by K. Ellis et al.
Plenum Press, New York, 1995

neo-classical theory of price and output determination in the firm. The firm, for Penrose, is essentially a pool of resources the utilization of which is organized in an administrative framework. The firm is conceived as a team of people learning together about how to make the most effective use of resources to achieve their ends in such a way as to effect expansion. Whether or not external opportunities will be perceived and responded to depends on the nature of the productive services available to the firm, and especially on entrepreneurial capacities and managerial competence.

> "Strictly speaking it is never resources themselves that are the inputs in the production process but only the services that the resources can render. The services yielded by resources are a function of the way in which they are used - exactly the same resource when used for different purposes or in different ways and in combination with different types or amounts of other resources provides a different services or set of services. ...resources consist of a bundle of potential services and can for the most part be defined independently of their use, while services cannot be so defined, the very word service implying a function, an *activity*." (Penrose, 1995, p.25, my italics). Thus resources enable action to take place, as in what Checkland calls a 'human activity system' (Checkland, 1981). In studying growing firms, we can conceptualise the enterprise as an input-output system, a 'human activity system' in Checkland's terms, in which resources are converted into products and services, the purchase of which provides the firm with the resources needed to continue and expand activities. Penrose draws attention to this learning process when she emphasises the productive services which resources make possible, the result of the activity that goes on within the firm.

THE FIRM AS A SYSTEM IN WHICH LEARNING OCCURS

Penrose discusses the ways in which, with experience, a unique form of interaction evolves among members of the business, in which their knowledge (of each other, of what works, of opportunities) emerges on an interactive and experiential basis. She points out that when people "... have become used to working in a particular firm or with a particular group ... they become individually and as a group more valuable to the firm in that the services they can render are enhanced by their knowledge of their fellow-workers, of the methods of the firm, and of the best way of doing things in the particular set of circumstances in which they are working. The experience gained is not only of the kind which enables a collection of individuals to become a working unit, but is also of a kind which develops an increasing knowledge of the possibilities for action and the ways in which action can be taken by the group itself, that is, by the firm, This increase in knowledge not only causes the productive opportunity of a firm to change in ways unrelated to changes in the environment, but also contributes to the 'uniqueness' of the opportunity of each individual firm", (Penrose, 1995, p.52).

What Penrose is here referring to can also be conceptualised as the way in which experience and interaction contribute to a unique social system, a whole with emergent properties, that is with attributes beyond those of individual members. These people are working with and on the physical resources to which Penrose refers, the buildings, tools, productive equipment; they take part in a productive process by means of which supplies are converted into saleable output. Penrose anticipates Senge's work (1990) and that of other recent literature in drawing attention to the importance of learning within the organisation.

THE FIRM AS A DYNAMIC SYSTEM

Implicitly, Penrose acknowledges the way in which systems dynamics operate through leads and lags which affect the capacity of the system to alter its internal arrangements and respond to external factors. A major theme in Penrose's work is that unused managerial and other productive services provide an incentive for the expansion of the firm. She also points out that delays inherent in recruitment, training and the inculcation of the ways of the firm limit the rate at which new human resources can be obtained. That is, there are delays in how fast people can be taken on and trained to the point where they have the experience to provide those firm-specific services which are needed for effective expansion. "As management tries to make the best use of the resources available, a truly dynamic interacting process occurs wich encourages continuous growth but limits the rate of growth", (Penrose, 1995, p. 5).

Penrose explores organisational bottlenecks which impede internal synchronisation and adaptability. The limits to growth have been a special concern in the analysis of systems dynamics (Senge, 1990). Penrose (1995) points to three classes of factors which can be used to explain the limits to the growth of firms - product or factor markets, managerial ability and uncertainty and risk (p. 43). The nature of markets is not an unavoidable constraint in her analysis because for her the environment is not a given. Costs of some resources or declining revenue for some products need not in principle prevent a firm from relocating or pursuing alternative openings in other markets, so long as internal resources permit flexibility. The principal constraints Penrose identifies initially lie in the productive services at the disposal of the firm and especially in managerial services. New opportunities cannot necessarily be taken up: "...because the very nature of a firm as an adminstrative and planning organization requires that the existing responsible officials of the firm at least know and approve, even if they do not in detail control, all aspects of the plans and operations of the firm ... The capacities of the existing managerial personnel of the firm necessarily set a limit to the expansion of that firm in any given period of time, for it is self-evident that such management cannot be hired in the marketplace", (Penrose, 1995, p.45). A group of people 'cannot do everything at once,' the existing team cannot oversee expansion beyond a certain point.

By the same reasoning, there are pressures for growth as long as there are unused productive services or capacities in the firm. "Many of the productive services created through an increase in knowledge that occurs as a result of experience gained in the operation of the firm as time passes will remain unused if the firm fails to expand. Thus they provide an internal inducement to expansion as well as new possibilities for it", (Penrose, 1995, p. 54). There is opportunity cost in not pursuing possibilities for expansion so long as these are available and there are resources in the firm capable of exploiting them.

KEY SYSTEMS IDEAS IN PENROSE'S PERSPECTIVES

To summarise: there are three conceptual features of Penrose's analysis in particular which are also found in systems thinking. Penrose stresses the properties of the firm as a whole, as a group capable of joint learning (cf *emergent properties of the system*). She is concerned with the productive activities of the firm which are achieved by turning resources into productive services (the *conversion process of an open system*). She is concerned with response times, with delays in organising to respond to opportunities in the environment and with the potential to carry out activities not yet organised. These resemble the *lags and leads* in response processes in a dynamic system (Senge, 1990).

AN INTERPRETIVE OR 'CRITICAL' SYSTEMS APPROACH

Thus Penrose's (1995) explanatory framework is essentially systemic, but her approach takes into account interpretation as well as behaviour. She does not use a mechanistic systems model. Her firm can be seen as a system of interdependent action in which meaning is attributed to action by the participants. This approach has features in common with the social action perspective used by Weber (1947) and Schutz's conception of the life world (cf Schutz and Luckman, 1974). She develops a sophisticated account of the way in which entrepreneurs have to interpret their world, form perceptions and create expectations which mediate between objective reality and the subjective awareness on the basis of which decisions are made: "...the environment is treated in the first instances as an 'image' in the entrepreneur's mind of the possibilities and restrictions with which he is confronted, for it is after all, such an image which in fact determines a man's behaviour; whether experience later confirms expecations is another story", (p.5).

She goes on to point out that: "Expectations and not objective facts are the immediate determinants of a firm's behaviour ... In the last analysis 'the environment' rejects or confirms the soundness of the judgments about it but the relevant environment is not an objective fact discoverable before the event; economists cannot predict it unless they can predict the ways in which a firm's actions will themselves change the relevant environment in the future", (p 41). Entrepreneurial ideas about what the firm can and cannot do constitute the nature and extent of the " 'subjective' productive opportunity of the firm", (p. 42). In adopting this interpretive perspective, Penrose breaks with classical systems thinking, but moves into the domain of 'critical' systems thinking which makes just such an attempt to acknowledge the importance of the interpretation given to events and actions by participants in a system (Garnsey, 1993).

In a critical systems perspective the environment is not given but enacted in large part by participants of the systems which make it up. Penrose (1995) recognizes that the environment is not externally fixed. Referring to her critique of biological analogies she points out: "Firms not only alter the evironmental conditions necessary for the success of their actions, but, even more important, they know that they can alter them and that the environment is not independent of their own activities", (p. 42).

PENROSE ON AGENCY AND STRUCTURE

In the first part of her book Penrose adopts an voluntarist theory of the entrepreneur, that is she assumes for the sake of argument that entrepreneurs' voluntary or chosen actions shape their experience and that of their firm. For example she explains at some length why the firm's capacity to raise money is ultimately a function of the entrepreneur's ingenuity and subjective outlook on available possibilities (p. 38). However in later chapters she shifts to examining the constraints of structure on the entrepreneur's options for action. She looks into structural restrictions and acknowledges that the conditions on which finance is available to entrepreneurs constitute an objective restriction on the growth of the firm, that small firms do suffer from competitive disadvantages of a structural nature. Now she acknowledges that "There can be no question that for any particular firm the environment 'determines' its opportunities", (Penrose, 1995, p.38).

Here we find a manifestation of the structure-agency problem: are entrepreneurs and others engaged in action as autonomous agents or tied by structural constraints (cf Giddens, 1984)? Again, recent developments in systems thinking have attempted to resolve the impasse which results from adopting one or other perspective exclusively (Garnsey, 1993). Penrose could see the need to shift analytic focus. "If we want to see why ... the environment is different for every firm we must take the 'resources' approach (ie consider the use the firm makes of its resources); if we want to explain why a particular firm or group of firms with specified resources grows in the way it does we must examine the opportunities (in the environment)", (Penrose, 1995, p.217) Interpreting Penrose's analysis from a critical systems perspective, we see that in the first case, an agency approach is appropriate: we can examine the actions open to members of the firm as agents able to make (or failing to make) effective use of their resources. However there are also structural limits to the opportunities available to them; we have also to look at the nature of the environment and the opportunities and constraints it provides to actors. Penrose is implicitly acknowledging that neither approach suffices of itself. Critical systems thinking examines the interlocking of effects the individual agent can have on structural outcomes and of social structures on activity. The concept of social systems as 'enacted' by their participants (Weick, 1979) incorporates this notion of mutual influence.

CONCLUSION

In brief, Penrose presents a resource-based concept of the firm, viewed as the organisational means for converting resources into productive services capable of producing output saleable for a profit. We can readily conceptualise Penrose's firm as a dynamic, open system, drawing in resources from the environment but subject to dynamic interactions internally which both stimulate growth and restrict the firm's ability to expand its organizational capacities to cope with and facilitate growth. Like many systems, the firm is subject to leads and lags which result in instability - in some respects conducive to and in other respects restrictive of growth.

Penrose sees the difficulties of organisational synchronisation required for growth as creating fundamental constraints. Whether these are compensated for and overcome in processes leading to growth depends on the competence which is built up by experience, the group's capacity to learn, to perceive and respond to opportunities. Productive resources provided by staff can be expanded internally through learning and experience. In other words, this kind of system is unique because human perception, learning and competence-building are amongst its crucial features.

Moreover in explaining the capacity for entrepreneurial success, which for her entails growth, Penrose accords central place to the interpretations entrepreneurs form of their opportunities and capacities. She articulated clearly the phenonenon later described as the 'social construction of reality' (Berger and Luckman, 1967). Again Penrose contributes a thoughtful analysis of the agency-structure problem in emphasising the importance of the distinction between objectively ascertainable and subjectively experienced opportunities. A voluntarist analysis attributes responsibility for success on the entrepreneur's drive, ingenuity and vision. A structuralist analysis, in contrast, emphasises the facilitation of and restrictions on opportunity created by overall investment levels and costs. Without labelling the problem in this way, Penrose points out that the (system's) observer in selecting the problem to be analysed determines which set of factors will be highlighted.

Having identified in what the analytical methodology consists, one can go on to apply this approach to other issues. Seen in this light, Penrose's analytic method has multiple applications both to the study of the firm and to other organisations.

Knowledge about social and economic life must be continually renewed but the process of learning can only be cumulative if some degree of conceptual continuity is maintained. Systems thinking provides one of the few consistent and cumulative bodies of understanding available to social scientists. It provides a meta-theoretical approach in terms of which various perspectives can be compared. This can enable us to to build on the understanding provided by past theoretical and empirical work.

ACKNOWLEDGMENTS

I have been fortunate to have helpful discussions with Professor Penrose about the growth of the firm. Her lack of exposure to the systems literature encourages me to feel that conceptual thinking can converge independently, in directions which make sense empirically.

REFERENCES

Berger, P., Luckman T., 1967, "The Social Construction of Reality", Penguin, London.
Best, M., 1990, "The New Competition: Institutions of Industrial Restructuring", Harvard University Press, Cambridge, Mass.
Checkland, P., 1981, "Systems Thinking, Systems Practice", John Wiley & Sons, London.
Flood, R., and Jackson M.C., eds, 1991, "Critical Systems Thinking; Directed Readings", John Wiley & Sons, Chichester.
Garnsey, E., 1993, Exploring a critical systems perspective, *Innovation in Social Sciences Research*, *6*, 2: 229-256.
Giddens, A., 1984, "The Constitution of Society", Policy Press, Oxford.
Penrose, E., 1959, "The Theory of the Growth of the Firm", Blackwells, Oxford. New edition forthcoming, Oxford University Press, 1995.
Senge, P., 1990, "The Fifth Discipline; the art and practice of the learning organization", Doubleday, N.Y.
Schutz, A., and Luckman, T., 1974, "The Structures of the Life-World", Heinemann, London.
Weick, K.,1979, "The Social Psychology of Organizing", second ed., Random House, N.Y.
Weber, M., 1947, "The Theory of Social and Economic Organisation", (T. Parsons ed.), Free Press, 1966.

34

STRUCTURATION AND ENACTED SOCIAL SYSTEMS

Elizabeth W. Garnsey and Séamas B. Kelly

University of Cambridge
Department of Engineering and
The Judge Institute of Management Studies
Mill Lane
Cambridge CB2 1RX
England

INTRODUCTION

Giddens' theory of structuration (1984) represents one of the most influential developments in social theory in recent years. His work focuses on the dialectic interplay between agency and structure within social systems, and has provided the basis for a number of recent empirical studies of organisational life (Walsham, 1993; Roberts, 1990; Barley, 1986). However, the use of this theory in empirical research can be fraught with difficulty. The concepts employed by Giddens are subtle and rich in insight but they are also prone to misinterpretation and distortion. In this paper we make use of the idea of *enactment* which, we argue, an approach to the study of social life, which is conceptually simpler and more intuitive than that of structuration theory. The enactment approach enables us to draw on the insights of structuration theory and couch them within a systems framework.

We begin by discussing the central ideas of structuration theory. This is followed by brief discussion of problems involved in using Giddens' concepts when conducting empirical research. Finally, we explore a simple model of interaction illustrating enactment and discuss the integrative potential of this approach.

STRUCTURATION THEORY

Structuration theory is concerned with the manner in which knowledgeable human agents constantly create and recreate their social world through their actions. Giddens' notions of *structure*, *system* and *the duality of structure* are central to the theory. He defines structures as "...the structuring properties [rules and resources]... the properties which make it possible for discernibly similar social practices to exist across time and space and which lend them 'systemic' form" (1984: 17). Here, rules are defined as "...social conventions and knowledge of them includes knowledge of the contexts of their application" (Giddens, 1981). These are

Critical Issues in Systems Theory and Practice, Edited by K. Ellis et al.
Plenum Press, New York, 1995

taken to be composed of both 'constitutive' (relating to the constitution of meaning) and 'regulative' (relating to the sanctioning of modes of social conduct) components. Resources refer to capabilities of making things happen, and are classified as being either of the 'allocative' (capabilities of generating command over objects, goods or material phenomena) or 'authoritative' (capabilities of generating command over persons or actors) variety. Sets of rules require resources for their actual incorporation into the production and reproduction of social practices.

Structures are ascribed only a virtual existence, i.e. they do not actually exist in time and space. Rather, social phenomena have the capacity to become structured, and it is here that Giddens breaks with the Durkheimian conceptualisation of structures as being external to and coercive over actors. He holds that "...structure is what gives form and shape to social life, but it is not *itself* that form and shape" (1989: 256).

Giddens uses the concept of a social system to refer to a set of reproduced social practices, or "reproduced relations between actors or collectivities, organized as regular social practices" (1984: 25). This definition is idiosyncratic in that it is very close to the conventional sociological use of the term 'structure': "...what most sociologists have thought of as structure, the patterning of relationships between individual or collectivities can be best dealt with by the notion of system" (Giddens, 1981). By distinguishing in this way between these concepts of structure and system, Giddens wishes to emphasise that, as he sees it, social systems do not have structures, but rather, exhibit structural properties. The reproduced practices observable in social systems are manifestations of structure. Structure "...only exists in and through the activities of human agents" (1989: 256). The role of unintended consequences of action in the constitution of these social systems is also emphasised. While some social systems may be the product of intentional human action, many are not. These unintended consequences of action in turn condition further action (i.e. they feed back into action), and it is because of this that human agents can never exert any real control over the social systems in which they participate.

Structures are thus 'instantiated' in social systems. They are also, however, manifest at the individual level where they take the form of "...memory traces orienting the conduct of knowledgeable human agents" (1984: 17). This brings us to Giddens' idea of the duality of structure. He claims that "...[t]he constitution of agents and structures are not two independently given sets of phenomena, a dualism, but represent a duality.... the structural properties of social systems are both medium and outcome of the practices they recursively organise" (1984: 25). Thus, agents draw on structures to guide their action and in so doing reproduce or reinforce those very structures. There is a dialectic relationship between structure and agency, one cannot exist without the other.

PROBLEMS ASSOCIATED WITH THE USE OF STRUCTURATION THEORY AS A BASIS FOR EMPIRICAL RESEARCH

The researcher faces a number of problems in attempting to use structuration theory for the analysis of empirical data. One of these emerges, perversely, from the very complexity and sublety inherent in the formulation of the theory, and the proliferation of concepts, only some of which can be referred to here. Moreover, in his exposition of the theory Giddens makes esoteric use of standard terms. Examples include his usage of the terms 'structure', 'system', 'rules' and 'resources'. The development of a full understanding of these concepts as they are used in structuration theory demands a detailed study of a recondite literature, thus making such knowledge inaccessible to all but the cognoscenti.

The second problem is a corollary of the first. Given that it is difficult to gain a full understanding of the meanings of the terms and concepts employed, simplifications are often used in the application of the theory to empirical work. Such simplifications can lead to the distortion of the theoretical concepts. In particular, structure is sometimes represented in work influenced by Giddens, as simply in the minds of actors, as memory traces. This is the obverse of reifying social structure, and though it is an idea which appears to follow from certain statements in Giddens' work, it overlooks his central notion that knowledge is instantiated in action. Misinterpretations of features of structuration theory, e.g. the 'duality' of structure are not uncommon. An example of a misapplication of the theory is to be found where "...an analytical distinction between social structures within technology and social structures within action" is called for (DeSanctis and Poole, 1994: 125). The subtleties of Giddens' concept of structure have baffled experienced social scientists (Clark, Modgil and Modgil, 1990) so it is not surprising that less experienced researchers should sometimes be confused.

Another issue of concern is the lack of specificity of certain concepts. An example of this is in relation to Giddens' definition of system as "...the patterning of social relations across time-space, understood as reproduced practices" (1984: 377). Cohen identifies a question arising from this when he argues that it "makes systems appear as if they just happened to be patterned and reproduced in certain ways" (Cohen, 1989: 114). A greater degree of specificity in the concept of social system may improve empirical focus and point to relevant causal influences. Thus, in many cases the notion of open systems is useful in highlighting the connections between internal relations of a social system and relations with the wider environment.

Finally, while structuration theory is useful for exploring the sources of stability and continuity in a social system, it is less clearly directed towards the analysis of change. Although the approach illustrates how knowledgeable agents constantly draw on social structures to guide and inform their actions and in so doing produce and reproduce those same structures, it does not provide specific insight into the mechanisms by which these structures are altered over time. This is partly because dynamic processes are not explored in as much depth as in the systems dynamics literature (Senge, 1990).

The difficulties in using structuration theory, as outlined above do not invalidate studies drawing painstakingly on this perspective. However, they lead us to believe that a programme of empirical research would benefit from a conceptualisation which is simpler and easier to use than that provided by Giddens. But much systems thinking is somewhat naive in precisely the areas where Giddens has most to offer. It is therefore important that a simpler conceptualisation draw on Giddens' understanding of the nature of social life as expressed in his idea that knowledge is instantiated in action and that structure is both the medium and outcome of action. These ideas creatively overcome the behavioural-interpretive dichotomy which has long beset the social sciences (Garnsey, 1993). Ideally, the concepts used should be expressed simply and in such a way as to reduce the risk of misinterpretation and distortion. The terminology employed should, wherever possible, be consistent with standard usage. We aim for a conceptualisation that could have the capacity to integrate other approaches which identify processes constitutive of social life (e.g. Havel, 1986; Weick, 1979). We believe that an approach which is congruent with the classic systems ideas of open systems, emergent properties and dynamic feedback processes can draw not only on insight from structuration theory but on the conceptual and empirical richness of systems research.

AN ALTERNATIVE CONCEPTUALISATION : ENACTED SOCIAL SYSTEMS

Here we present the notion of an *enacted social system* in an effort to address the concerns expressed above. This approach draws on and combines two distinctive strands of

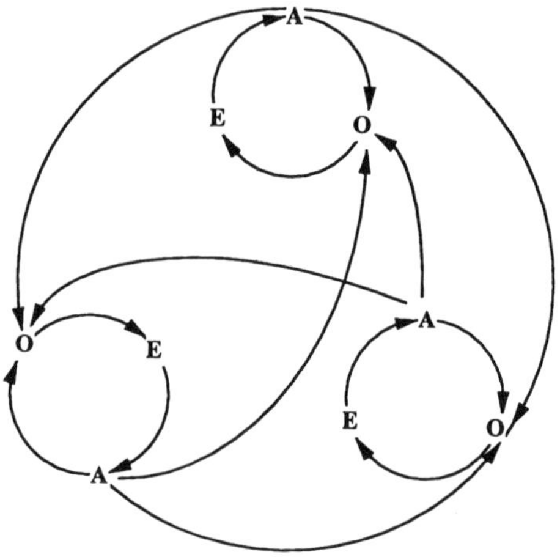

Figure 1. Interaction among three actors; illustrating simple enactment processes.

intellectual thought: the enactment approach, of which structuration theory is the best known version, and systems thinking. Empirical applications have been discussed elsewhere (Garnsey and Rees, 1994). Here we illustrate the approach in its simplest terms in Figure 1.

Figure 1 illustrates enactment processes in the simple case of the interactions between three actors (top, right, left) engaged in cycles of enactment (A-E-O). They each act (A) on the basis of their experience (E) and the expectations (also E) which they create as to how others will act. Their actions have outcomes (O). However, the cycle of action, outcome and experience/expectation is not self contained for each actor. Instead, the outcomes which give rise to actors' experience are the result not only of their own actions but also of those of others. At the same time their own actions give rise to outcomes which impact on the experience and expectations of others. Even among very simple sets of relations of the kind depicted in Figure 1, the outcomes of actions may be unintended. Insofar as these unfolding cycles give rise to recurrent and self-maintaining relations between the actors, stable structures are identifiable. In stable conditions, actors tend to assume roles and follow rules, increasing the predictability of their mutual interactions. With large numbers of people in interaction, non-linear chains of action and reaction are set off with more extensive ramifications than those illustrated in simplified form in Figure 1, and are the basis of dynamic processes often with widespread consequences unintended by any given actor.

Implied in Figure 1 are both patterns of cognition and conduct and the processes that these reinforce. Insofar as expectations are shared among social actors, a common outlook prevails (in its absence, there are divisions). Shared expectations continually renew norms, or what is viewed as acceptable and legitimate conduct. The outcomes of actions shape expectations and there are inevitable interdependencies as actions impact assymetrically on others. These processes - interpretive, normative and reinforcing of interdependencies - resemble the complex features of structuration termed 'signification, legitimation and domination' in structuration theory.

When such interrelated clusters of self maintaining interactions are to be observed, we may identify a social system, an emergent property of which is its capacity for self renewal. Often social systems can be identified as open to a wider environment supplying the resources by means of which they are sustained. Thus in Figure 1, if the actors were drawing on resources which facilitated their actions and using these together to carry out some conversion process which resulted in an output from their activity, they might be identified as forming an open system with a detectable boundary. Clearly such a system would not be an entity in the real world. Systems thinking is a methodology (Checkland and Scholes, 1990). Systems, here, are a way of conceptualising inter-related phenomena.

This conceptualisation incorporates the notion that knowledge and expectations are not in practice separable from action. As Giddens puts it, action instantiates knowledge. By the same token, knowledge of what is expected of actors in day-to-day interaction is inseparable from structure. There are occasions when action is not premeditated but puts into practice or actualises knowledge. Nevertheless, experience and expectations (c.f. mental models) motivate our actions whether or not we are consciously aware of this process. Increasingly, the importance of mental models, those cognitions which mediate between raw experience and motivation to act, is receiving emphasis from a number of writers (Senge, 1990; Hoskings and Morley, 1991).

The idea of enactment traces out the ways in which social systems operate and are subject to dynamic processes, and is thus congruent with many of the classic ideas of systems thinking. Our approach could be taken as an example of paradigm mediation as discussed by critics of the concept of paradigm incommensurability (Aldrich, 1991). The operation of systems dynamics are acknowledged but not developed in Giddens' reference to feedback processes in systems. We can find points in common with Senge's analysis where the latter emphasises the benefits of understanding the wider structures and systems of which actions form part (Senge, 1990). However, Senge does not acknowledge a difference in conceptual order between dynamic systems in the natural world and those of a human activity system; for him these all follow common systems archetypes. We need to make explicit in the idea of enactment the unique properties of human activity systems. In these, systems actors' experience interpreted and expectations held by actors (their Weltanschauung in Checkland's analysis) play a fundamental role, of which there is no precise equivalent in systems in the natural world, though all are subject to dynamic processes.

CONCLUSION

Structuration theory has proved useful in surfacing the 'taken for granted' aspects of interaction, in tracing embedded relations of power, unacknowledged norms and the influence of dominant but implicit interpretations. The approach in its esoteric form is accessible only to those dedicated to mastering the subtleties of this school of thought. But an appeal of the approach is that the basic ideas are simple and intuitive and can provide the basis for a deeper understanding of complex interactions and processes. The enactment approach aims to surface these basic ideas. It aims to trace through interconnections between action and structure, to explore underlying expectations, their sources and outcomes and to uncover relations of power and dependence in similar ways. It is an approach which provides for such an analysis while making it possible to draw on the conceptual and empirical strengths of open and dynamic systems thinking.

REFERENCES

Alrich, H., 1991, Incommensurable Paradigms? Vital Signs from Three Perspectives *in:* M. Reed, M. Hughes, eds, "Rethinking Organization: new directions in organizational theory and analysis", Sage, London.

Barley, S., 1986, Technology as an occasion for structuring: evidence from observation of CT scanners and the social order of radiology departments, *Admin. Sci. Qu.* 31: 78-108.

Checkland, P., and Scholes, J., 1990, "Soft Systems Methodology in Action", John Wiley & Sons, Chichester.

Clark, J., Modgil, C., and Modgil S., 1990, "Anthony Giddens: Consensus and Controversy", The Falmer Press, London.

Cohen, I., 1989, "Structuration Theory; Anthony Giddens and the Constitution of Social Life", Macmillan, London.

DeSanctis, G., and Poole, M.S., 1994, Capturing the complexity in advanced technology use: Adaptive Structuration Theory, *Organization Science* 5, 2, 121-147.

Garnsey E., 1993, Exploring a critical systems perspective, *Innovation in Social Sciences Research* 6, 2: 229-256.

Garnsey E., and Rees, B., 1994, Discourse and Enactment: Gender Inequality in Text and Context, Cambridge Research Papers, *in*: "Management Studies", No. 20, 1993-4, Forthcoming in Human Relations.

Giddens A., 1981, Agency, institution and time-space analysis, *in:* K. Knorr-Cetina, and A. Cicourel, eds., "*Advances in Social Theory and Methodology: toward and integration of micro- and macro-sociologies*", Routledge and Kegan Paul, Boston.

Giddens A., 1984, "The Constitution of Society: an outline of the theory of structuration", Polity Press, Oxford.

Giddens A., 1989, A reply to my critics, *in:* "*Social Theory of Modern Societies: Anthony Giddens and his critics*",*:* D. Held, and J.B.Thompson, eds., Cambridge University Press, Cambridge.

Havel, V., 1986, The power of the powerless, *in*"Living in Truth", *:* J. Vladislav (ed.), Faber, London.

Hosking, D.M., and Morley, I.E., 1991, "A Social Psychology of Organizing: people, processes and contexts", Harvester Wheatsheaf, Hemel Hempstead.

Roberts, J., 1990, Strategy and accounting in a UK conglomerate, *Accounting, Organisations and Society*, 10, 4: 443-456.

Senge, P., 1990, "The Fifth Discipline; the art and practice of the learning organization", Doubleday, N.Y.

Walsham, G., 1993, "Interpreting Information Systems in Organisations", John Wiley & Sons Ltd, Chichester.

Weick, K., 1979, "The Social Psychology of Organizing", second ed., Random House, N.Y.

35

SOFT SYSTEMS METHODOLOGY

Modes of Practice

M. G. Haynes

CSSM
Sarah's Croft
Coniston Cold
Skipton
North Yorks
BD23 4ED
United Kingdom

INTRODUCTION

Systems thinking, in the form of a general theory, emerged during the 1950's and gave birth to what is now known as the 'systems movement'. Within the systems movement a number of different approaches to 'doing' systems thinking appeared. One of these, concerned with the management of human affairs, and which evolved over a 25-year period of intellectual deliberation and organised experiential learning, is Soft Systems Methodology (SSM). The early version of SSM was described in terms of a conventional seven-stage model; this has now matured into a more sophisticated form of structured enquiry involving interacting streams of logic-based and cultural analysis (Checkland and Scholes, 1990). SSM is well known but is still perceived by many to be the preserve of academia rather than a useful tool that can be used to deal routinely with the complex human and information management issues that confront every manager.

The feature that distinguishes SSM from other systems approaches is that SSM is itself a learning system - in other words an organised process of enquiry, the form of which is based on systems ideas. Systemicity is focused on the process of enquiry rather than on the world (Checkland and Haynes, 1994).

This article will set out to reflect on the practical experience of 'doing' systems thinking in the context of an organised process of enquiry, based on SSM, and employed by the author in numerous consulting situations over a six year period. These consulting experiences range from public to private uses of the methodology to deal with a variety of problem situations, and also those where clients expect to acquire SSM skills in compressed time-frames in order to subsequently do their own problem solving. It will describe, briefly, ways in which those in a problem situation, and who are methodologically naive, can become involved in participative problem solving where they use the ideas directly themselves. It

Critical Issues in Systems Theory and Practice, Edited by K. Ellis et al.
Plenum Press, New York, 1995

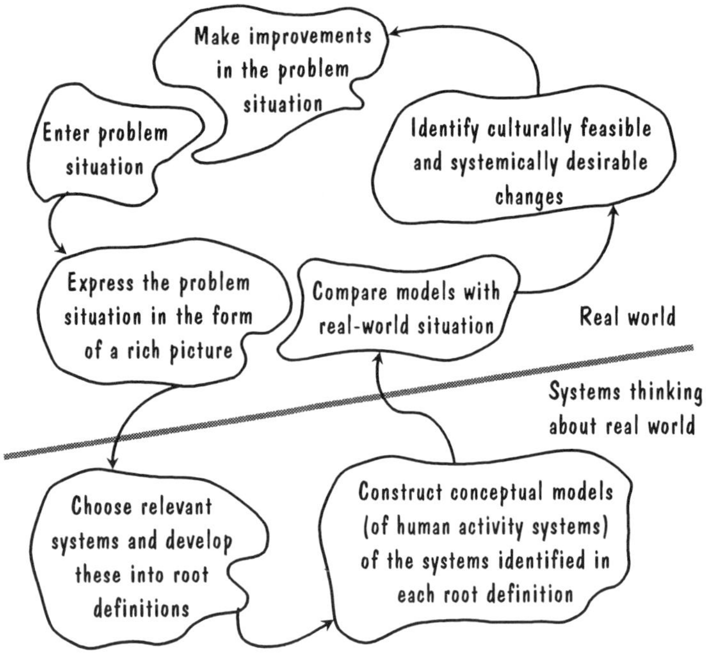

Figure 1. SSM described in the form of a seven-stage model.

will finally draw together the lessons that emerge from this practice and hopefully make a small contribution to the further development of a well-established methodology.

THE FORMS AND MODES OF SSM

The 'popular' version of SSM is that of the conventional seven stages as shown in Figure 1. Although the developers of SSM now regard this as being 'rather bald' (Checkland and Scholes, 1990), first-time users find it easy to assimilate, its 'rule-based' appearance is attractive to many, and it offers a coherent problem-solving process.

Two main weaknesses of this early version can be identified. Firstly, would-be problem solvers tend to slavishly follow through the prescribed seven-stage process, lacking the confidence to adapt it to suit a particular problem situation; as a result problem solving tends to be driven by the needs of the problem solving process rather than the problem situation itself. The second problem is that cultural aspects tend to be overlooked - organised purposeful human activity usually being burdened with hidden agendas, vested interest, power and all the other ingredients that combine to form the rich background tapestry of organisational life. These weaknesses have led to some shallow uses of SSM which sometimes fail to get to the root of a problem and occasionally lead to disillusionment with the methodology.

The learning derived from early experiences led to a developed form, the main elements of which are illustrated in Figure 2 (Checkland and Scholes 1990, Fig 2.6). This is a significant step-change in sophistication for the naive user, requiring a higher level of abstract and conceptual thought. A non-academic user seeking a ready-made process for dealing with messy problem situations is asked to deal with 'two interacting streams'

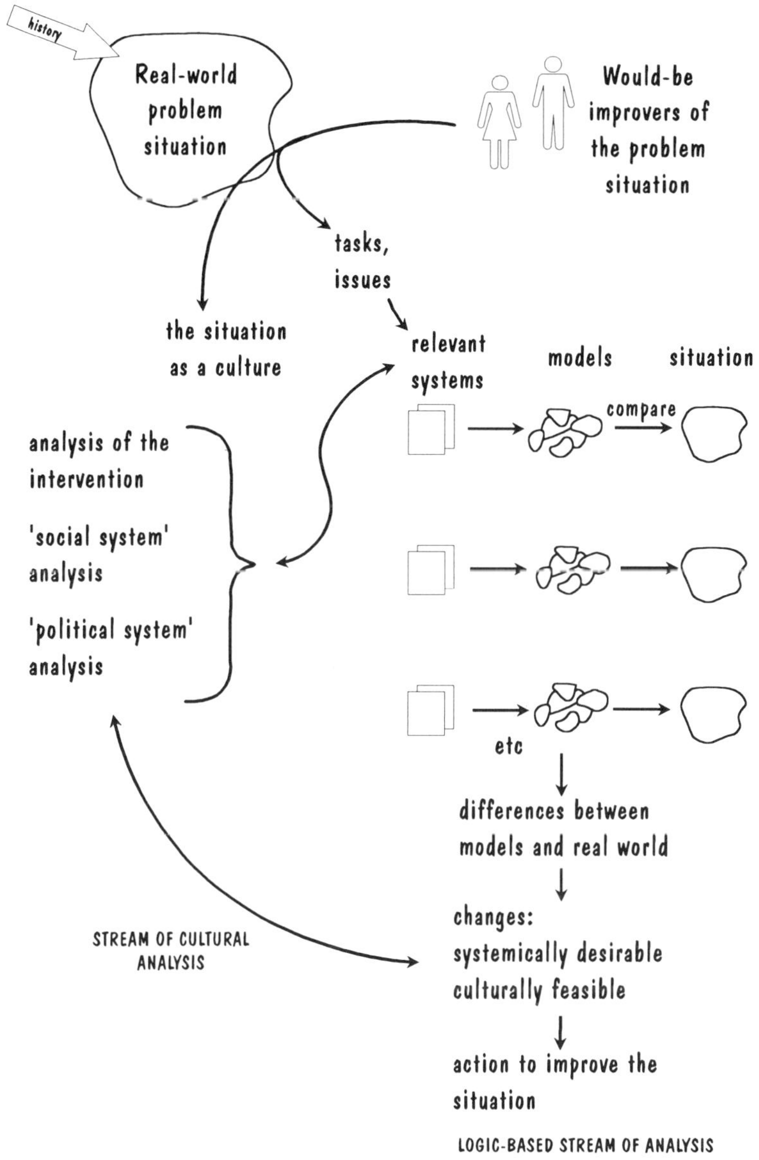

Figure 2. The developed process of SSM. Reproduced by permission John Wiley & Sons, Ltd.

(Checkland and Scholes, 1990) of analysis, one of which, the cultural element, is offered with few 'rules' apart from some general guidance. Primary task and issue-based choices of relevant systems become an explicit requirement - many first-time users find the distinction between the two difficult to understand. In addition the new user is expected to cope with the notion of a holon!

The above paragraphs describe very briefly the development of the SSM process from an algorithmic style seven-stage model to a mature form that expects more of the user. In parallel with this the developers have recognised two ways or 'modes' of use. Mode 1 is described as the 'formal stage-by-stage application of the methodology' in order to 'intervene and conduct an enquiry into a problem situation. Mode 2 use involves the internal mental use of the methodology as a thinking mode to guide interaction with 'the flux of events and ideas' that characterise a particular problem situation and so enable learning towards improvement' (Checkland and Scholes, 1990, pp281-283). For those who are not natural conceptual thinkers Mode 2 is possibly beyond reach in a normal consulting situation.

MANAGERIAL CONSULTING WITH SSM - THE DILEMMA

The consultant employing SSM to guide his or her practice is faced with a dilemma. SSM is well known having achieved considerable penetration into the public domain. Prospective users generally understand that it can be applied usefully to complex, messy problem situations that are essentially human in nature. The seven-stage model described above is clear, easily understood and can be assimilated by most in the limited consulting time usually available. It is the best thing about SSM for the first-time user whose orientation is one of problem solving, rather than methodological appreciation. The main difficulties in application are :

a. persuading people that developing some form of a picture capturing the main elements of the problem situation is a useful thing to do and that considerable artistic talent is not a prerequisite; and

b. ensuring that purposeful activity models constructed are not models of the real world.

The dilemma is that whilst, on the one hand, the seven-stage model offers a practical route to realising the participative philosophy of SSM: "Given its nature as a system of inquiry, the use of SSM has to be participative. The role of the 'expert' in SSM is best thought of as helping the people in the situation carry out their own study" (Checkland, 1988), its shortcomings lie in the lack of an explicit requirement to examine the cultural situation. Thus the analysis can become mechanistic and lack richness with the result that learning is inadequate and a desirable level of consensus between different interest groups is not reached. On the other hand, the alternative is to employ the developed form of SSM from the outset but this involves a greater investment in time; in the commercial world it is unusual to find clients willing to give more than a couple of hours to learning and thinking about a methodology - most wish to press on with the task at hand.

ORGANISING THE PROCESS OF SSM - THE BUILDING BLOCKS

The consultant employing SSM as an approach has at his or her disposal a framework containing a number of ideas. At a meta-level these include the basic building blocks of systems thinking of emergence, hierarchy, communication and control. At a methodological level these include the idea that problem situations that are human in nature can be explored usefully in terms of purposeful activity, that purposeful activity is linked to a particular worldview and that each situation can contain many worldviews (all equally valid), that 'ideal type' models of human activity systems can be constructed and used as devices to learn, to enable accommodations to be reached and improvements made. Finally, the methodology is supported by techniques and rules such as rich pictures, CATWOE, rules for model building (language, dependency, formula, etc.), and comparison tables.

ORGANISING THE PROCESS OF SSM - REFLECTIONS ON PRACTICE

An important consulting objective is to adopt, from the beginning of each assignment, a problem, rather than a methodological, orientation answering the question 'how best can these ideas (wider systems concepts and SSM) be organised to help in this particular situation?'. In other words *giving specific attention to organising the process of enquiry in order for useful learning, for the client, to occur.*

There is considerable variation between problem situations encountered and the way clients wish to proceed. At one end of the spectrum some wish to assimilate (usually in the space of a couple of hours) and then use the methodology themselves with brief subsequent help from the consultant. In the middle there are those who expect a transparent use where they are exposed to the methodology but do not actually become involved in modelling themselves, making use of the models produced by others to learn. At the other extreme are clients, who, having assured themselves that there is some intellectual framework underpinning the work, have no further direct interest in the process - essentially the methodology remains private to the consultant - but they expect learning to emerge from some form of structured debate. Obviously, there are many shades between these examples but they are all connected by a common aim: to learn how to move forwards and make some improvement. The situation with the greatest potential for exploiting the methodology is one where the consultant is given sufficient space and time to fully involve those concerned in participative problem solving with sufficient skill transfer to ensure that iteration continues to occur long after the departure of the so-called 'expert'. Unfortunately the pace at which modern organisations conduct their affairs rarely permits the luxury of this ideal.

The three examples above can be used to illustrate different applications of the ideas. In the case of those wishing to quickly learn and use the methodology there is little alternative but to use the seven-stage model. The problem here is to ensure that some form of cultural analysis is carried out. To a limited extent this shortcoming can be overcome by promoting the use of CATWOE. Originally developed as a means of assisting the formulation of root definitions it has been found that in use it inevitably opens up wider contextual discussion especially the part of the mnemonic concerned with the 'W' - *weltanschauung* or worldview. The consultant is likely to remain operating in mode 1 in this situation and should limit his or her ambitions to transferring an elementary version of the methodology. Hopefully there will be sufficient interest and users will go on to discover the developed version - the author is aware of several instances where this has occurred.

In the second example, of transparent use where an SSM expert is used to facilitate problem solving using SSM ideas, there is much wider scope for employing the developed form of SSM since the expert will guide its use. Entry to the problem situation can occur in any number of ways ranging from group meetings to one-to-one interviewing or a mixture of both. It is usually helpful to start with a cultural analysis collecting the information through face-to-face interviews and so teasing out important social and political aspects of the situation. Once sufficient information is available the construction of logical models can begin and these can be shared with individuals or groups. It is often helpful to share these in 'unfinished' form engaging others in helping to finish them. This seems to stimulate more interest and involvement as people sense they are making a contribution - and they learn by so doing. CATWOE is again a useful tool in both formulating root definitions with a beneficial by-product of provoking wider contextual considerations. Activity models are helpful devices for carrying messages from one interest group to another; although they may be attached to a particular worldview they are perceived to be a neutral representation of that worldview (since they are not models of the real world), non-threatening and therefore open for constructive debate. There are likely to be a series of iterations as the study moves forwards as different lines of enquiry are pursued and these in turn open up new areas. In this example the consultant is guiding an intervention (Mode 1), publicly using SSM ideas to provide a framework for learning, whilst also privately using that same framework to ensure his own interaction (Mode 2) - i.e. maintaining a coherent learning system - remains consistent with the needs of the problem situation as events unfold.

The third example, of totally private use of the methodology where the language of SSM is not used, requires a considerable degree of self-discipline if rigour is to be maintained. This is best achieved by paying close attention to managing the process of enquiry. This means thinking at two levels of iteration. The first is employing SSM ideas in a practical intervention sense - collecting information, doing modelling (privately), and using the learning to structure a debate in ways (not using systems language) that allow those involved to also learn (Mode 1). The second or meta-level being concerned with the learning system itself - reflecting on the intellectual framework being used and its success in interacting with the problem situation and moving things on towards improvement or an 'answer' in the language of the client (Mode 2). In one assignment, involving work for a large corporation, reflecting on the process of enquiry caused the author to halt the work and suggest that the desired outcome was not attainable. The work then took on a new dimension with the client becoming involved in modelling which proved to be helpful.

Finally, experience has shown that access to the methodology and participative problem solving can be facilitated by the use of magnetic media. This is commonly available in the form of hexagon shaped discs which are used in conjunction with metal-backed whiteboards. These can be used to describe the problem situation, recording and mapping out comment, discussion, ideas, etc. These randomly collected points can then be grouped (since the objects tessellate) under theme headings which then can become relevant systems. This approach is particularly useful when working with groups as the map is a record of the verbal contributions of each member and this encourages greater ownership of, and commitment to, the process. In a similar way conceptual model building can be done more easily as a group exercise using the same tools. Whilst lacking the elegance of the traditional freehand drawn rich pictures and models these tools go some way to reducing the inhibitions of first-time users towards these activities.

CONCLUSIONS AND LEARNING

Three main lessons emerge from this reflection on practice. The first concerns the difference between users and developers/experts of the methodology, the second suggests an area for further work on developing the methodology, and the third is about the role of the expert.

- The SSM developers and experts need to remind themselves of the intellectual struggle and effort, often over an extended period of time, to reach the various mileposts that mark the development of SSM. First-time users do not have the benefit of this experience.
- The seven-stage model is widely recognised as a 'hallmark' of SSM. Lessons need to be learnt from this success and used to inform further development and research aimed at making the developed version of SSM more accessible. Users have difficulty understanding how to do the cultural analysis and then integrate this with the logic-based stream.
- The role of the SSM-practising consultant is to design and manage a learning system appropriate to the needs of the client and the problem situation. This entails

gaining a clear understanding of the needs of the client and the organisational context in which learning about improvements will occur. The methodology is not imposed on the situation but adapted to suit each situation.

REFERENCES

Checkland, P.B., 1988, Soft systems methodology: an overview, *Journal of Applied Systems Analysis*. Vol.15, pp27-30.

Checkland, P.B., and Haynes, M.G., 1994 Varieties of systems thinking: the case of soft systems methodology, *System Dynamics Review*. Vol.10,pp189-197.

Checkland, P.B., and Scholes, J., 1990, "Soft Systems Methodology in Action", John Wiley & Sons, Chichester.

SYSTEMICS APPLIED TO THE STUDY OF ORGANIZATIONAL FIELDS

Developing a Systemic Research Strategy for Research in Organizational Fields

Jon-Arild Johannessen

Bodø Graduate School of Business
8002-Bodø
Norway

INTRODUCTION

The question we want to shed light on in this article is: How can we study organisational fields from a systemic angle of incidence?

We will first determine what we mean by a systemic perspective, and then what we mean by an organisational field. Then we design a superior methodological scheme, where the main entities in the systemic approach are clarified. We will then look at the underlying process for each of the main entities in the research scheme.

The purpose of the article is to develop a systemic research strategy for the study of organisational fields.

THE SYSTEMIC APPROACH

The following article is based on knowledge acquired from Bunge (1974(a), 1974(b), 1977, 1979, 1983(a), 1983(b), 1985(a), 1985(b), 1989).

The systemic approach is based on a system-theoretical ontology, where the world is seen as a system consisting of subsystems, and an epistemology combining realism and rationalism.

By emergent is here meant: Let S be a system with composition A, i.e. the various components in addition to the way they are composed. If P is a property of S, P is emergent with regard to A, if and only if no components in A possess P; otherwise P is to be regarded as a resulting property with regards to A (Bunge, 1977:97).

Systemic thinking is based on the premise that society is a concrete system of interrelated individuals, and that some properties are aggregates of individual properties, while others are 'global' and emerge as a result of relations between the individuals. The

Critical Issues in Systems Theory and Practice, Edited by K. Ellis et al.
Plenum Press, New York, 1995

emergent properties must be studied at different levels in a system, and the relations between the levels must also be studied.

The systemic view is that patterns of social behaviour, even if they are created by individuals or groups, are rarely a result of momentary considerations, but are developed in the interaction between social groups in the course of time. Social behaviour is, according to the systemic point of view, limited by laws of nature, ecological circumstances and tradition. The social behavioural pattern fades when the group which they regulate collapses, even if the behavioural pattern can be internalised in the culture to such an extent that it exists long past the point where it has ceased to have a function.

ORGANISATIONAL FIELDS

By organisational fields Barley, Freeman & Hybels (1992:313) mean: "...a community of organisation that have some functional interest in common". They further state (1992:313): "Member in organisational field is not limited to organisational directly involved in developing, producing, or distributing the products or services associated with a functional area; fields also include organisations that provides funding, that have regulatory oversight, and that offer ancillary services."

It is further important to make a distinction between field and network. In an organisational field several networks can exist, but the existence of a network is not a necessary condition for the existence of the field. But if the field is organised in strong networks it becomes a strong basis for power and an instrument for the exercise of control of field interests in relation to those outside the field. Networks on the other side can exist without being coupled to functional interests, i.e. they can to a greater extent be coupled to pragmatic interests, not necessarily functional ones. When organisational fields are coupled as coherent networks; network fields, with great similarity to autopoietic systems, appear, i.e. they are locked into themselves, reproducing their own similarities, as a sort of social variety of the Phoenix myth: The only being having reproduced itself by laying an egg from which it was born itself.

If we start out from the definition of Barley, Freeman & Hybels an organisational field can be determined by three entities:

1. a definite composition
2. a definite environment
3. a definite structure.

Composition is constituted by the co-ordination of organisations with common functional interests. Environment refers to organisations not being part of the field in question, but who have relations to organisations in the field regarding other areas of interest, and who can then make organisational fields with these organisations in other functional interest areas. Structure is here understood as the relations between the persons in the field, as well as between the components and the environment. The relations are further 'filled' with information and/or matter-energy, and are constituted by the processes in progress between the organisational entities of the field.

For an organisational field to be described as existing, certain necessary prerequisites must be present:

1. There must be a certain number of connections between the entities, which are based on functional common interests.
2. There must be a minimum of information regarding this common functional interest, flowing in the connections.

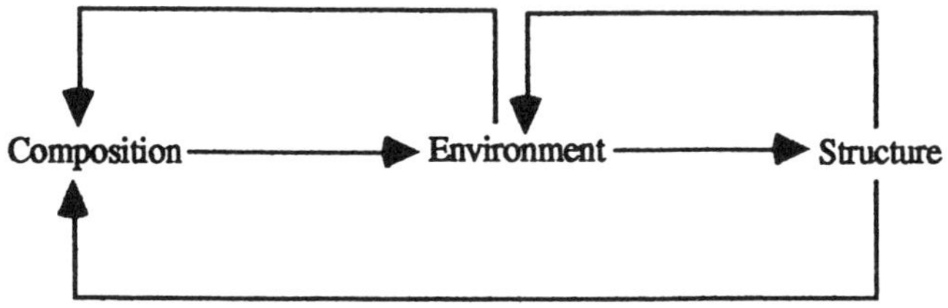

Figure 1. Research elements in the field.

3. There must be a certain amount of expectation of a common result as a result of the connections established and the information flowing through the connections.
4. In every organisational field there will be certain basic common interests and some spontaneously developed interests.

What we here regard as the minimum knowledge in order to gain insight into the dynamics of the organisational field is the composition, environment and structure of the field, and the interaction between them, as shown in Figure 1.

Figure 1. will constitute the minimum model for the systemic approach to an organisational field.

SYSTEMIC RESEARCH STRATEGY FOR THE STUDY OF ORGANISATIONAL FIELDS

The remainder of the article focuses on the three main entities: composition, environment and structure. 'Composition' is meant to answer the question: What keeps the field together? 'Environment' is meant to focus on 'The Boundary Problem'. 'Structure' is meant to answer the question: What mechanisms constitute and reconstitute the social relationships in the field?

Composition

Any phenomenon in the organisational field can be determined by the entities: critical variables, properties, relation, and inter-dependence. The critical variables are chosen by the observer in relation to the motive he/she has with the study, i.e. what they want the model to say something about. What we seek to achieve will also determine the boundaries we draw. This is oriented toward the pragmatic effect, i.e. what pragmatic implications we want to comment on.

The variables being selected must further be interdependent, but distinct, i.e. variable A must have the capacity to be removed without the removal of any of the substance in B, and vice versa. The size of the field is determined by the number of variables on which we choose to focus. The degree of dependence between the variables is determined by 'Connectance' (Ashby, 1970; Johannessen, 1994), which is measured percentage wise as to the influence of one variable on the other variables in the field (the ones selected). Connectance has to do with the interlining of critical variables in a field. If we in an organisational field have selected 7 critical variables (A - G), and a change in A precipitates a change in all other variables, we say that connectance is 100%. If a change in A does not result in a change in the other variables, we say that connectance is 0%. Connectance varies then between 0 and 100 % in a field.

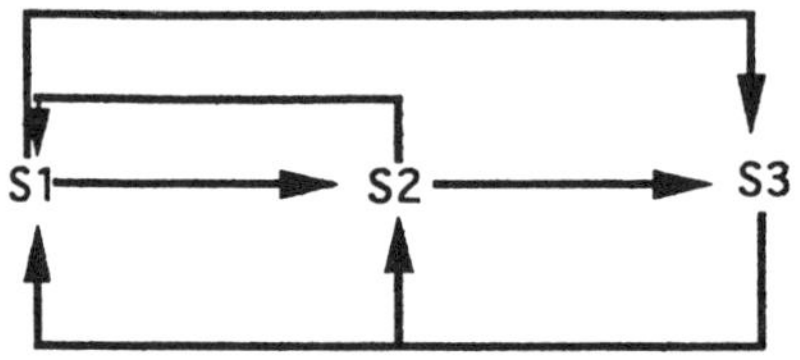

Figure 2. Systems in their environment.

Environment

Social systems, unlike e.g. biological organisms, lack defined boundaries as Habermas (1979) points out. Cetina-Knorr (1981:16) underlines this by arguing: "Social studies of science have long suffered from this difficulty". Indirectly this is an attack on the emphasis on system boundaries on the part of systems theory. We here propose a model of social systems where the dichotomy between a system and its environment is abolished, and where only systems related to each others exist. It is also our opinion that the dichotomy between a system and its environment can be abolished without effecting any change in character of the system theory. Whether there is a relation between a system and its environment, or if you regard the system as elements in an interaction with each other without dealing with the boundary problem, is, according to our understanding, only a matter of extending the system understanding to encompass the total organisational field subject to study.

If we do not operate with a system and its environment, but with a system in its environment, we get the model shown in Figure 2.

The model shows the triviality that S1 influences S2 and S3. S2 and S3 influence each other and S1, i.e. the various 'outputs' from S1 through various interrelated contexts become 'inputs' in S1. In a subtle way 'outputs' from S1 also become 'inputs' in S1. The model can be interpreted in the following manner: A system polluting other systems also pollutes itself.

Another way of exemplifying the model, and at the same time linking it to information and pattern contexts, is to use the following example: The birds and the fish make up a pattern which, as interpreted by the fisherman, implies that there is fish beneath the surface where the birds are hovering. Interpreted by the experienced fisherman this is informative as to the pattern of fish stock, and size in addition to depth of location.

If relevant contexts are cut in patterns, the information generated will be of little service to the observer. In organisational contexts it is also the patterns linking together into entire understanding horizons which will be of interest. Let us assume that in organisational contexts, we as researchers really do not know what we want before we have found it. In such cases an attitude where fragmentation and isolation from relevant contexts will be potentially fatal. We might run the risk of exclusively focusing on the indicators (the birds in the example above), and not what is interesting, i.e. what we are really looking for (the fish in the example above). To cut relevant patterns might also cause us to lose decisive contexts as observers.

What has been described above indicates that patterns only can be perceived through thinking focusing on processes, where part and entirety are embedded in circular loops, and where the patterns are manifested as concrete actions and where events are repeated.

Structure

It is the integrating and disintegrating processes in the field we are focusing on. The extent of integration within a field depends on the couplings between the actors, in addition to the disintegrating processes internally in the field, but also externally.

Integration is only meaningful if it is seen and treated as the opposite of disintegration. These two entities can be summarised as a co-ordination problem in the field, which is also linked to symbolic entities (e.g. ideas, values, codes, etc.), in addition to a political problem linked to decisions and legitimacy.

One of the decisive processes which could have an **integrating** effect in the organisational field is a need for group affiliation. This need expresses the complementary relationships in the field. The need for group belonging might reinforce the need for self-confirmation, which in turn might increase the need for some sort of reward. Both the need for self-confirmation and the need for reward might have a positive effect on the need for group belonging, and also strengthen the need for co-operation. The need for group belonging will also reinforce the need for predictability, which in turn will increase the need for more information. Both the need for predictability and the need for information might have an amplifying effect on the need for group belonging, and also have a reinforcing effect on the need for co-operation.

This entire process with its underlying mechanisms will contribute to an increase in the interdependence in the field. With increasing interdependence in the field common values, ideas, codes, rules and norms towards which the actors orient themselves, will be generated.

One of the main processes which might have a **disintegrating** effect in the organisational field, is a type of relationship based on competition and mistrust (symmetrical relationship). A symmetrical relationship type might develop hierarchical processes with differences in status and rank, which accordingly will bring about a result of variant resource allocation. Symmetrical relationships might also lead to the development of various subgroups in the field, which then can mobilise power with these subgroups. This will then have a reinforcing effect on the symmetrical relationships. A resultant effect of these underlying processes is that a conflict within the field, with a disintegrating effect, is developed.

CONCLUSION

The research strategy argued for in this article is based on three main entities. They are composition, environment, and structure. It is further processes and patterns tying the organisational field together which is our continuous focus of attention. At an organisational level it is relations and causal connections we are trying to uncover. At the behaviour level it is norms and job regulation which is in focus.

Sequentially this research strategy follows the subsequent pattern:

1. We select a phenomenon in an organisational field.
2. We then select the system in focus or the functions which should be in focus for the study. We clarify the selection mechanisms exactly.
3. We select the relevant variables, and clarify explicitly the selection mechanisms.
4. We select the measurement indicators of the variables, and make the selection variables explicit. Variables can be numerical and non-numerical.
5. We investigate the phenomenon by using three elements: a) sensitivity towards the phenomenon and context familiarity, b) personal knowledge of the phenomenon and c) experience and closeness.

Points 1 and 2 are qualitatively different from points 3 and 4. Points 1 and 2 correspond to the ontological level, while points 3 and 4 correspond to the system level, i.e. the description of the ontological field. Points 1 and 2 contain an infinite number of variables, while points 3 and 4 represent the final number. Transformation rules for the variables between the data system (1 and 2) and the object system (3 and 4) must be made explicit.

The distinction between points 1 and 2 and points 3 and 4 are therefore decisive for the research process. It is Ashby (1956) who originally has made this distinction ready, the distinction between the object and its system. Point 5 is meant to ensure the validity in the transformation rules between the data system and the object system.

REFERENCES

Ashby, W.R., 1956, "An Introduction to Cybernetics," Chapman & Hall Ltd, New York.

Ashby, W.R., 1970, Connectance of large dynamic (cybernetic) systems : critical values for stability, *Nature*, 228:5273, Nov. 21.

Barley, S.R., Freeman, J. & Hybels, R.C., 1992, Strategic alliances in commercial biotechnology, *in*: "Networks and Organizations," N. Nohria, & R.G. Eccles, eds., Harvard Business School Press, Cambridge, Mass.

Bunge, M., 1974(a), "Sense and Reference," Reidel, Dordrecht.

Bunge, M., 1974(b), "Interpretation and Truth," Reidel, Dordrecht.

Bunge, M., 1977, "The Furniture of the World," Reidel, Dordrecht.

Bunge, M., 1979, "A World of Systems," Reidel, Dordrecht.

Bunge, M., 1983(a), "Exploring the World," Reidel, Dordrecht.

Bunge, M., 1983(b), "Understanding the World", Reidel, Dordrecht.

Bunge, M., 1985(a), "Philosophy of Science and Technology. Part I," Reidel, Dordrecht.

Bunge, M., 1985(b), "Philosophy of Science and Technology. Part II," Reidel, Dordrecht.

Bunge, M., 1989, "Ethics: The Good and the Right", Reidel, Dordrecht.

Cetina-Knorr, K.D., 1981, "The Manufacture of Knowledge," Pergamon Press, New York.

Habermas, J., 1979, "Communication and the Evolution of Society," Heinemann, London.

Johannessen, J-A., 1994, "Cybernetics and innovation: the hopeless choice between stability and innovation," *Entrepreneurship Innovation and Change* 2:3, pp. 249-263.

TELEOLOGY AND THE END OF SYSTEMS THINKING

Iain Munro

Warwick Business School
University of Warwick
CV4 7AL
United Kingdom

INTRODUCTION

The teleology of the systems approach was first brought into question when Churchman began his critique of the methods of traditional operational research in the sixties (Churchman, 1968, 1971). Since then little interest has been shown in this fundamental aspect of the systems approach, with the exception of Checkland's brief commentary on Churchman's *Anatomy of Systems Teleology*, in which he thoroughly reworked the teleological foundations of the systems approach (Checkland, 1988). However, this article will attempt to show that both of these approaches are grounded in an impoverished view of human nature, where this may be a crucial concern for contemporary systems thinking and practice. The concept of teleology will be analysed by drawing on Alasdair MacIntyre's (1985) recent work on the moral history of liberal society. It is only through this analysis of teleology that we can appreciate the different ways in which values are created within any social practice. In the light of this discussion, the teleology of the systems approach will be criticised in an attempt to enrich the ethical foundations of systems practice.

It should be noted that in such a short space it will be impossible to do justice to MacIntyre's work, but nevertheless, this article will attempt to show how his powerful ideas might be related to contemporary systems theory.

WHAT IS TELEOLOGY?

Teleology may be broadly defined as the study of purposes. The understanding of teleology which underpins the systems approach has been inherited from the Enlightenment tradition and its liberal ideologues. Our understanding of ethics and human nature was radically altered with the advent of the Enlightenment period in the Eighteenth Century, and disastrously so according to MacIntyre (1985). Before the Enlightenment a broadly Aristotelian schema held sway over much of Western thought. Under this ancient framework ethics provided the bridge between two states of being, that is, man in his untutored state and

Critical Issues in Systems Theory and Practice, Edited by K. Ellis et al.
Plenum Press, New York, 1995

man-as-he-could-be-if-he-realised-his-telos (MacIntyre, 1985). Thus an action cannot to be understood as an isolated effect of a person's will, but only as part of a whole pattern of conduct which leads towards an ultimate *telos*. It is only in terms of a *telos* that people can be understood to have some ethical potential on which they can work.

However, these ideas did not arise in a social vacuum and eventually this ancient conception of ethics was abandoned as a consequence of the secularisation of society and the emergence of popular democracy. On the one hand one could no longer appeal to a divine source when searching for a purpose to one's life, and on the other hand, any purpose could not be restricted to a privileged few but must be available to all. In the absence of some substantive *telos* to which people could devote themselves, it became increasingly difficult to understand how ethics might be used to cultivate any kind of ethical potential within a person. In this light, ethics could no longer be understood as the bridge between the human raw material and the good person, and was thus re-conceived to be in direct opposition to human nature.

TELEOLOGY AND THE GOODLIFE

In response to this state of affairs, we need not start over again in search of an ultimate purpose to life true for all people, but rather we must explore the idea of human potentiality and the good. Indeed, it seems almost perverse that today the term 'good' has become a synonym for a mere commodity. To initiate this project MacIntyre has re-examined the idea of the good in terms of his analysis of teleology. This work provides us with an elaborate attempt to reconstruct the idea of ethics as the bridge between the untutored man and man-as-he-could-be-if-he-realised-his-telos.

Therefore, in an attempt to overcome the naive methods of evaluation derived from economics, MacIntyre has discriminated between external goods such as wealth, status and power, and those goods which are internal to a particular social practice. A practice may have internal goods because it is valued as a good way of life in itself, and not merely because it is productive of money or commodities. Thus a painter may consider painting as itself a good way of life, whatever the wealth and status gained through this practice, and likewise for professions such as teaching, nursing, building, and so on. As such, the internal goods of a practice may be properly understood only by those who are themselves pursuing these goods.

This conception of internal goods recognises the profound affect that one's work can have on one's consciousness and one's moral motivations. However, it is important to note that the production of internal goods may well be neglected if the emphasis of work is simply on the pursuit of money or status. For example, the modern notion of efficiency contains no concept of internal goods, where its uncompromising logic may gradually squeeze out the social space and time which are necessary for these internal goods to flourish. Indeed, the distinction between external goods and internal goods is clearly different from the fashionable claims that 'excellence' will make you more money (Peters and Waterman, 1982).

It should be immediately obvious how this critique may be applied to traditional operational research with its implicit assumption that only external goods are valuable, however, the critique also has serious implications for both soft and critical systems thinking, though these must be worked out more fully elsewhere. Indeed, it will be argued below that the existing work in systems thinking and practice contains no such concept of the human telos and as such relies on a singularly impoverished conception of value.

CHURCHMAN'S COMMENTARY ON THE TELEOLOGY OF SYSTEMS

Churchman (1971) has done much to map out a coherent teleology for the *Design of Inquiring Systems*, and his work has been referenced approvingly by many within the soft and critical schools of systems thinking. He has observed that the methodologies of traditional operational research assume that things are merely tools which may be used to achieve predefined ends. Under this scheme of things human beings are also related to in purely instrumental terms, and not as ends-in-themselves (Churchman draws on Kant's rather opaque terminology for his critique). Thus he set about a rigorous moral critique of traditional operational research, but nevertheless, this critique was itself weakened from the outset due to its grounding in an Enlightenment view of teleology.

From this very restricted perspective, teleological behaviour is simply taken to be goal seeking behaviour. Therefore, an action is not understood as part of a whole pattern of conduct, but merely in terms of its fulfilment of the short term goals of the system. Churchman recognizes the existence of 'intangible' values which arise in activities such as poetry and research, but he wrongly describes such activities as ateleological. This would be a complete inversion of the Aristotelian schema, arising from the very restrictive sense in which Churchman considers purposes. In other words, he thinks that an activity is purposeful only in terms of the goals of the immediate task, but this entirely neglects the possibility that creative work may be purposeful in terms of one's overall lifestyle (Aristotle, 1976; MacIntyre, 1985). In this way he fails to give an adequate account of how internal goods are created through work and how they may be built into systems designs. Given this inversion of the concept of teleology, Churchman has also neglected the fact that one may engage in an activity both because of its benefits to the community and for its own sake. As he puts it,

> "The teleologist sees no conceptual embarrassment in saying that a man may pursue an activity for its own sake and not be willing to recognise any benefit that such an activity may have beyond the happening or occasion itself." (1971, p.255)

Through this caricature he has set an ultimate divide between the artist who produces everything for its own sake and the worker who produces nothing for its own sake. In contrast to this worldview MacIntyre's framework opens up possibilities for organising work so as to produce both external goods and internal goods. Even Churchman himself concludes that, based on his own view of teleology, ".. every plan must be partially immoral." (1971, p.256).

Now Checkland's work on the teleology of systems will be examined to see whether it succeeds where Churchman's has failed.

CHECKLAND'S COMMENTARY ON THE TELEOLOGY OF SYSTEMS

The soft systems thinker, Peter Checkland, has criticised Churchman's work for its functionalist ontology (Checkland, 1988). Checkland asserted that goal seeking behaviour is not a characteristic of all systems, since this assumes that systems have an ontological reality independent of our own perceptions of them. In contrast to this Checkland takes a phenomenological stance where "systemicity lies not in the world but in the processes of inquiry into the world. (1988, p.383). However, he still does not escape the Enlightenment view of teleology, and its impoverished conception of value.

Taking a reassuringly liberal view of things, the soft and critical systems thinkers allow for human value judgements to enter into definitions of a problem situation. This derives from their broadly phenomenological insight that reality is a social construction, and thus there may be many views of a problem situation. As Checkland has put it,

> "Man as designer is a teleological being, able to create means of enabling ends to be pursued, and to do so on the basis of conscious selection between alternatives." (1981, p.119)

So one decides the job to be done, ideally in a democratic manner, and then one sets about doing it. Checkland goes on to distinguish between immediate goals and objectives and a more general purpose which is an "end which can be pursued but never finally achieved....e.g. maintaining relationships" (1981, p.316). But again this purpose has been decontextualised. For example, what kind of relationships does a particular practice maintain, and in what ways might they be considered valuable? Thus, despite an explicit concern for bringing values into debate in a problem context, this soft systems methodology remains unconcerned with the ways in which values are created.

CONCLUSION

This paper has attempted to show how important the much neglected concept of teleology is to systems thinking and practice. It is only in terms of a particular practice and its associated telos that the value of that practice can be properly understood. Some examples of contemporary systems thinking have been drawn upon in order to illustrate their weaknesses in this regard. This is particularly noticeable from the lack of consideration given to the internal goods which may be generated within a system. This has implications for enriching the value frameworks underlying the systems approach, not only for traditional operational research, but also for soft and critical systems thinking.

REFERENCES

Aristotle., 1976, "Ethics," Penguin, London.

Checkland, P., 1981, "Systems Thinking, Systems Practice," John Wiley & Sons, Chichester.

Checkland, P., 1988, Churchman's 'Anatomy of Systems Teleology' revisited, *Systems Practice*, 1:377-384.

Churchman, C., 1968, "The Systems Approach," Delta, New York.

Churchman, C., 1971, "The Design of Inquiring Systems," Basic Books, New York.

MacIntyre, A., 1985, "After Virtue: A Study in Moral Theory," Duckworth, London.

Peters, T., and Waterman, R., 1982, "In Search of Excellence: Lessons from America's Best Run Companies," Harper and Row, New York.

38

SOCIAL AUTOPOIESIS

A QUANTITATIVE APPROACH

F. Parra-Luna

Complutense University of Madrid
Spain

INTRODUCTION

Are social systems autopoietic? Bailey's (1992) extensive review includes both authors who have concluded that societies should be considered autopoietic and those that have concluded the contrary, considering the term as originally defined for systemic theory and popularized by Maturana and Varela (1974, 1978, 1980). According to Bailey, Beer (1975), Luhmann (1982, 1984, 1986, 1989) and Robb (1989(a), 1989(b)), for instance, seem to be under the conviction that social systems are autopoietic, while other such as Mingers (1989) and Maturana and Varela themselves, think they are not. Let us begin, then, by discovering what is meant by *Autopoiesis*.

DISCUSSION

There are, then, three major features to the concept:

- 1. It represents the development of the very ability to produce;
- 2. It is due to internal processes so any possible influence brought to bear by the environment should be 'subtracted'; and
- 3. It is a process that takes place over time.

Under these premises and previously developed measuring instruments,(Ref Num1) a first attempt may be made to approach Autopoiesis from a quantitative standpoint. The objective, as in any formulation of this nature, is to express the meaning in concrete terms in order to avoid the proliferation of heterogeneous and incomparable notions and above all to facilitate critique.

This calls for a precise definition of social systems, which consists of the following four fundamental dimensions:

- 1. Inputs (X), consisting in turn of four components:

a. Structural Factors (EF) or initial resources,
b. Needs (N) of individuals in the system,

c. Memory (M) or system's past,
d. The influence of the Environment (E)

- 2. Transformation (T), Organisational dimension of the system that, for the purposes of this paper, acts like a 'Black Box', but which, as we shall see, constitute the fundamental expression of Autopoiesis in so far as it is a subsystem that organises the system's transforming and reproductive capacity.
- 3. Outputs (Y), Dimension that translates what the system does to the benefit and satisfaction of the system population's needs (N). Outputs, which may be Objective Y(O) or Subjective Y(S) mean not what the system says it is (rhetoric), but what the system does (practicality), all of which is expressed as a complex System of Values. Therefore, sociologically speaking, it represents the most important expression of the system. A social system is ultimately evaluated on the basis of its *Outputs* much the same way as a tree is evaluated on the basis of its *fruit*.
- 4. Feedback (R), Broadly speaking, whatever the system puts out comes back into it as Inputs (Y = DX) which are complementary to the initial Inputs (X). But this leads to a final System of Values that may clash with the System of Values initially planned. Hence systems' cybernetic interest in controlling their Outputs to ensure that they have as small an effect as possible on the established axiological programme.

In the first place a system's total *Outputs* (Y), should be considered to be not only the Outputs 'consumed' by its members, but the potential to produce future Outputs as well. In this case the Output level would provide an indication (sight should not be lost of Ashby's law of variety) of the transforming power or systemic production capacity. In order to find Autopoiesis, then, all that need be done is to eliminate the influence the Environment has on those Outputs. Hence, a first approach consists of separating Inputs into components: *internal* (systemic) and *external* (coming from the Environment). We know that system Outputs can be expressed as:

$$Y = T.X \tag{1}$$

where the Outputs (Y), both on the *objective* (statistically measured events) and on the subjective (survey-compiled and quantifiable opinions) level, represent each and every one of the End-Values which, according to the model adopted, any social system should aspire to reach (RefNum2), and where the Inputs (X) can be subdivided into two components: the internal (XI) Inputs, which in accordance with the general definition of Inputs (Parra-Luna, 1983,p.89) are thus limited to:

$$XI = (FE + N + M)/3 \tag{2}$$

and the External Inputs or those coming from the Environment (e); therefore:

$$XE - e \tag{3}$$

Consequently, total Inputs (X) can be represented as follows:

$$X = (XI.3 + XE)/3 + 1 \tag{4}$$

If Outputs (Y) are likewise divided into internal Outputs (YI) and external Outputs (YE), the following expression is valid:

$$YI = T.XI \text{ (Internal or autopoietic Outputs)}$$

and

YE = T.XE (External Outputs or Outputs explained by the Environment)

so that a possible expression for Autopoiesis on the basis of Outputs 'Y' (AU(Y)), stated in a relative and comparable form, would be:

$$AU(Y) = \frac{(Y - YE)100}{Y} \tag{5}$$

which would measure the percentage of total Outputs due to developments involving internal factors in the system only.

This expression does not, however, take the notion of process (times t_0 and t_1) into account, providing a merely static view as it only deals with Outputs at a given point in time.

A second quantitative approach to the Autopoietic process, which accommodates the time factor, would involve centring the question around the explanatory power of the Environment (in terms of variance of Changes in Y) in either of the models represented in figures 18.5 and 18.6 of Parra-Luna (1983) (RefNum3), but preferably the latter. Subtracting the variance explained by the Environment (Ve) from the total variance explained (VTE) would yield the variance of change in Outputs C(Y) that is explained by internal factors during a given period 0-1; here Autopoiesis would be expressed in terms of change AU(C) as:

$$AU(C)_{01} = VTE - Ve \tag{6}$$

There can be no doubt that this expression measures only the 'selected' or *significant* part of total change, the part that is explained solely by the system's internal structures (composed of 'Inputs' and 'organization').It represents the explanation of system progress (or regression) *per se* which includes not only change in the current production of Outputs, but change in the ability to produce future Outputs as well, by virtue of the fact that Outputs are fed back into the system in accordance with the equation Y = DX (RefNum4).

A third way of 'measuring' Autopoiesis, which seems to fit the semantics of the concept better, consists of centring the approach around the notion of 'global systemic productivity' or the system's ability to effect transformation, considering not only the *Outputs* or changes, but the *Inputs* as well, and focusing on 'T' as the factor relating the two. A comparison of T over time would represent changes in production (positive or negative) of production capacity, making 'T' a transforming device that accumulates (or reduces) the power to effectively convert Inputs into Outputs. The mere comparison over time of measurements of 'T'(T_0, T_{11}, ..., T_n) would afford us a more precise view of the autopoietic process. This

third way of perceiving Autopoiesis in terms of transformation capacity (AU(T)) could be formulated for a period 0-1 and expressed as a percentage of the previous period, as follows:

$$AU(T)_{0-1} = \frac{(T_1 - T_0)100}{T_0} \tag{7}$$

Under this expression the influence of the Environment is considered to be simply one of the system Inputs, in accordance with equations (2), (3) and (4):

$$(FE + N + M + e)/4 \tag{8}$$

and its possible effect is implicitly 'deducted' from T in the process of deriving Y. A social system would be Autopoietic to the extent that T (transforming capacity) is able to dynamically maximize the Output/Input ratio.

Practical computation of Autopoiesis can pose serious operational problems in very large or complex systems. Let us consider how to determine the difference in internal Transforming Efficiency between two countries, e.g., France and Spain, during a given period. A knowledge of Outputs, either to explain them as in C(Y) or to relate them to Inputs as in T, should pose no particular problem if a suitable system of empirical indicators is available. But quantification of Inputs, equally necessary in both cases, entails the added difficulty, as it calls for empirical information on the Outputs of the environments surrounding the two countries. An approximation can be obtained from available international statistics, but evaluating these environments with scientific rigour must be a hard task indeed, even where the two countries' environments are practically the same (i.e., the EU countries) such as in this case, and where the statistical information available is relatively acceptable. But there is yet another limitation: in respect of the two countries in question (Spain and France) as well as of the remaining EU countries, we must keep to *objective* Outputs Y(O), since the information on *subjective* Outputs Y(S), which can only be obtained through surveys, is rarely systematic enough to be comparable. In contrast, in micro-social systems (e.g., a classroom) subjective information is easier to obtain than objective information. Comparing the actual degree of knowledge acquired, for instance, in two classrooms or in two different subject matters A and B, by the same group of students (exams are usually read by different graders) is more difficult than surveying these students (common population) to evaluate their subjective perception of the benefit derived from the two systems. This, then, would open the door to yet a fourth way of measuring autopoiesis. Limiting the evaluation of Outputs to Subjective Y(S) Outputs only and considering that the Environment (e) is a constant for the two systems (same physical classroom, comparable timetable, same teaching aids, etc.), the difference in transforming performance (T) should be the result, solely and exclusively, of *internal organisation* of Classroom subject matter (quality of lectures, degree of participation, student freedom, perceived fairness in grading, ongoing grading system, degree of cooperation among students, etc.), which would make Outputs synonymous with Autopoiesis and (necessarily internal) efficiency. The Autopoiesis of Classroom A ($AU(S)_A$) in terms of subjective Outputs will then compare to the Autopoiesis of Classroom B ($AU(S)_B$) in the same way as its subjective Outputs $Y(S)_A$ compare to the subjective Outputs of B $Y(S)_B$. This is yet another approach to the quantitative measurement of Autopoiesis (AU(S)) that in addition to being practical for small social systems (schools, hotels, medical clinics, travel agencies, etc.), is in line with product and service quality standards which should be rated highly, first of all, by consumers or users (RefNum5).

Having reached this point and defined Autopoiesis in operational terms on the basis of the notion of the system's *transforming efficiency*, reference must be made to the full definition of Global Efficiency of the system. Because in the end, a social system is efficient only when: (1) 'T' grows and 2) 'Y' grows. It is recalled that the definition of Systemic Efficiency (ES) was formulated as: ES = T.Y. Autopoiesis may not, then, ignore the two concepts involved: changes in T, and changes in Y, both considered from an intrasystemic or autopoietic standpoint.

Another possible quantitative approach to the concept of Autopoiesis (AU) would be as given by the expression:

$$AU=f[AU(C), AU(T)] \tag{9}$$

that can be averaged by the expression:

$$AU=f[AU(C), AU(T)]/2 \quad (10)$$

or, since percentages are involved, by:

$$AU=[AU(C).AU(T)]^{1/2} \quad (11)$$

which expression also complies with the properties of the concept of Autopoiesis, i.e.:

1. it expresses differences in development by comparing Outputs;
2. this is due to internal factors; and
3. it is representative of a process over time.

Given the lack of a quantitative definition of the concept, I have suggested three approaches to the idea of Autopoiesis that I trust will contribute to a more precise perception thereof. The concept of Autopoiesis is undoubtedly of great importance in social systems, due to the fact that the nature of these systems (invention, creation and imagination which are characteristic of social systems) makes them 'non-zero sum' or self-reproductive systems whose self-development potential would seem to be unlimited, in particular if considered in the long term.

Nonetheless, certain important aspects have yet to be discussed in concrete terms. Firstly, does Autopoiesis imply development? Or, in other words, where a social system is considered to be autopoietic, is it always *developing*? Is it forever growing in the direction its population wishes? And would this be defined by simply measuring the growth in the Progress/Regression Index discussed in (Parra-Luna, 1983, p. 463) (RefNum7) Or, on the contrary, does Autopoiesis call for compulsory development of each and every one of the dimensions (End-Values) of the system? And secondly, if such were the case, would change have to be homogeneous in all systemic variables? Given the intrinsic variability (some progressing, some regressing) of dimensions, objectives or End-Values of social systems, it does not seem difficult to establish, assuming that social systems are Autopoietic, that this concept reflects both positive and negative change, so that in the end it is an expression of both progress and regression and entails a substantial degree of internal variability. In any case, it is on this basis that the possible ways of measuring the concept have been put forward.

The following conclusion may, then, be advanced: if social systems are autopoietic, the definition of Autopoiesis (or, at least social Autopoiesis) would entail the following features:

1. it represents a change in the production capacity of its productive structures *per se*;
2. this change is brought about by internal factors as opposed to environment-provoked change;
3. it is a process that must be defined between two points in time t_0 and t_1;
4. change may be either positive or negative (i.e., may entail an increase or a decrease in production capacity); and
5. the degree of change varies widely. If these five features can not be accommodated by the theoretical concept of Autopoiesis, it may perhaps be affirmed that social systems are not autopoietic.

In any case, what does seem clear is that societies, characterized as they are by the very complex factors that condition the way they function, change both due to internal factors (Autopoiesis) and external factors (Allopoiesis); consequently, if there is a term able to cover the global causes that would explain development or change in such systems under any circumstances, that term, might I suggest, would be, 'Pantopoiesis', from the Greek 'panto' (all) and 'poiesis' (to produce). Since social systems are open to the Environment, as shown in explicative formal systems (Fig. 18.5 and 18.6 in Parra-Luna 1983, p. 453), they are inevitably 'Pantopoietic'.

ENDNOTES

1. A fuller discussion of these instruments can be found in F. Parra-Luna (1974, 1978, 1979, 1980, 1991, 1994), which focuses in particular on operational definitions of Inputs, Outputs, Transformation, Feedback and other concepts deriving therefrom.
2. The model adopted, called Referential Pattern of Values (PRV), proposes the following nine End-Values (Health, Material Wealth, Law and Order, Knowledge, Freedom, Social Justice, Conservation of Nature, Quality of Activities and Prestige). The theoretical grounds for the model can be found in Parra-Luna 1983, chapter 8. Each one of these values becomes operational in accordance with Lazarsfeld's (1965) famous rule, i.e., taking a mental image of the concept, subdividing it into theoretical dimensions and representing each one of these dimensions with reliable and accurate empirical indicators. Despite their heterogeneity (they are expressed in different units), these indicators are statistically integrable into a single global numerical index that represents the social image of the concept. In this manner, '$\underline{Y}$' represents the total Outputs produced by the System, either as 'flows' or as final levels.
3. See F. Parra-Luna (1983, pp. 453-454), where two models for explaining change are discussed. In Fig. 18.5 the model represents a multiple regression equation. In Fig. 18.6 the explanatory model uses a 'Path Analysis' technique, separating direct and indirect effects. The purpose of both consists of computing the percentage of variance in the change produced in Outputs (CY) explained by each one of the model variables.
4. See F. Parra-Luna, 1983, p. 186 *et sequentes*.
5. The Complutense University of Madrid, for example, has been evaluating teaching performance over the past few years on the basis of students' opinion of classroom operation on the basis of the following aspects: teaching style, Content organisation, Clarity of delivery, Academic competence, Motivational skills, Teaching requirements, and Grading fairness, all of which was operationalised into 29 subjective indicators that graded Overall teaching Performance on a scale from 1 to 7. Since:
    1. A distinction can be drawn between professor categories (salaries or costs to the University);
    2. The environment for all other objective conditions (physical space, teaching aids, services, etc.) was the same for all classrooms; and
    3. The students were practically the same; the composite Grade computed (ignoring the fact that it only recorded Subjective Outputs) was in fact an evaluation of Autopoiesis asoccurring in each 'classroom system'.
6. The definition of Systemic Efficiency given in chapter 14 of the book *Is society a System?* (unpublished), expresses the concept in terms of both Outputs (Y) and Transformation (T).
7. See Parra-Luna (1983, pp 463-466), where the concept of *change* is broker down into *positive* change (Progress) and *negative* change (Social Regression), and a Progress Index is defined in terms of the relationship between the former and the latter.

REFERENCES

Bailey, Kenneth. D., ed. 1975, "Cluster Analysis, Sociological Methodology", Jossey-Bass 59-128, San Francisco.

Bailey, Kenneth. D., 1992, "Sociology and the New Systems Theory", State University of New York Press, New York.

Beer, Stafford, ed., 1975, Preface, *in:* "Autopoietic Systems", H.R.Maturana and F.G. Varela, University of Illinois, Urbana.

Luhmann, Niklas., 1982,The world society as a social system, *International Journal of General Systems.* 8: 131-138.

Luhmann, Niklas.,1984, Die wirtschaft der gesellschaft als autopoieteisches system, *Zeitschrift für Soziologie.* 13: 308-327.

Luhmann, Niklas.,1986,The autopoiesis of social systems, *in:* "Sociocybernetic Paradoxes: Observation, Control and Evolution of Self-Steering Systems", R.F. Geyer and J. van der Zouwen, eds., Sage, London, p172-192.

Luhmann, Niklas.,1989, "Ecological Communication", Polity Press, Cambridge.

Maturana, Humberto R, 1974, Cognitive strategies, *in:* "Cybernetics of Cybernetics", H. von Foerster, ed.,.University of Illinois, Urbana.

Maturana, Humberto, 1978, Biology of language: the epistemology of reality, *in:* "Psychology and Biology of Language and Thought: Essays in Honour of Eric Lenneberg", G. Miller and E. Lenneberg, eds., Academic Press, New York, p27-63.

Maturana, Humberto R., and Varela, F.G., 1980, "Autopoiesis and Cognition: The Realization of the Living", Dordrecht, Reidel.

Mingers, John., 1989, An introduction to autopoiesis-implications and applications, *Systems Practic.* 2: 159-180.

Parra-Luna, F., 1974, "Towards Comparing National Social Performances. (Una aplicación de la Teoría de Sistemas)", Universidad de Lausana, Suiza.

Parra-Luna, F., 1978, La mesure du progress dans la perspective systemique, "Comunicación presentada en el Congreso sobre Sociología del Progreso celebrado en Menton", (Francia) del 12 al 17 de Mayo de 1975, Publicado en 'le Progress en Question', ed., Anthropos, Paris.

Parra-Luna, F., 1979, Hacia la medida de la eficacia de los sistemas de seguridad social, *Revista de Seguridad Social*, Abril-Junio, nº 2.

Parra-Luna, F., 1980, "Balance Social y Progreso Empresarial. (La aplicación de la Teoría de Sistemas a través de un ejemplo cifrado de Balance Social)", Editorial CIRDE, Madrid.

Parra-Luna, F.,1983, "Elementos para una Teoría Formal del Sistema Social" ed., de la Universidad Complutense, Madrid.

Parra-Luna, F., 1991, On the cybernetics gearing of social systems, *in:* "Handbook of Systems Science", M. De Cleris, ed., UNESCO, Athens.

Parra-Luna, F., 1993, "El Balance Integrado de la Gestión Estratégica", Ediciones Deusto, Bilbao.

Parra-Luna, F., 1994, The third cybernetics: an introduction, *in:* Proceedings iof the XIII World Congress of Sociology", 18-23 July, Germany.

Robb, Fenton, 1989(a), Cybernetics and supra human autopoietic, *Systems, Systems Practice.* 2: 47-74.

Robb, Fenton,1989(b). The application of autopoietic systems to social organizations: a comment on John Mingers an introduction to autopoiesis: implications and applications, *Systems Practice.* 2: 353-360.

Varela, F.G., Maturana, Humberto R., and Uribe,R., 1974, Autopoiesis: the organization of living systems, its characterization and a model, *Biosystems.* 5: 187-196.

39

NEW DEVELOPMENTS IN THE USE OF SYSTEMS THINKING IN THE ANALYSIS OF FAILURES

Geoff Peters and Joyce Fortune

The Open University
Walton Hall
Milton Keynes
MK7 6AA
United Kingdom

INTRODUCTION

In a previous paper (Peters and Fortune, 1992) the authors looked back over a twenty year period at ways in which systems-related ideas had been used to analyze and model failures and disasters of various sorts. The paper ended with a short description of The Failures Method, developed within the Open University (Fortune, 1984), which was then in use.

A NEW METHOD

A re-evaluation of their own work and reflection on that of others (for example, Mansell, 1993) has now led the authors to reformulate the method referred to above and to present it in a revised form as The Systems Failures Method. A detailed exposition of this new method can be found in Fortune and Peters (1995), but an indication of the changes that have been made are set out in this paper which shows the two versions in Figures 1 and 2.

LOOKING FORWARD TO THE NEXT STAGE OF DEVELOPMENT

The wisdom of building flexibility into systems methodologies and allowing them to evolve should be plain to all who have experience of working with them. (Contrast, for example Checkland, 1981, with Checkland and Scholes, 1990.) The authors certainly regard their work with failures as a process of inquiry in the widest sense, so in what ways might The Systems Failures Method evolve further? A number of possibilities are already being investigated. One set of ideas which appears promising will be outlined here.

Critical Issues in Systems Theory and Practice, Edited by K. Ellis et al.
Plenum Press, New York, 1995

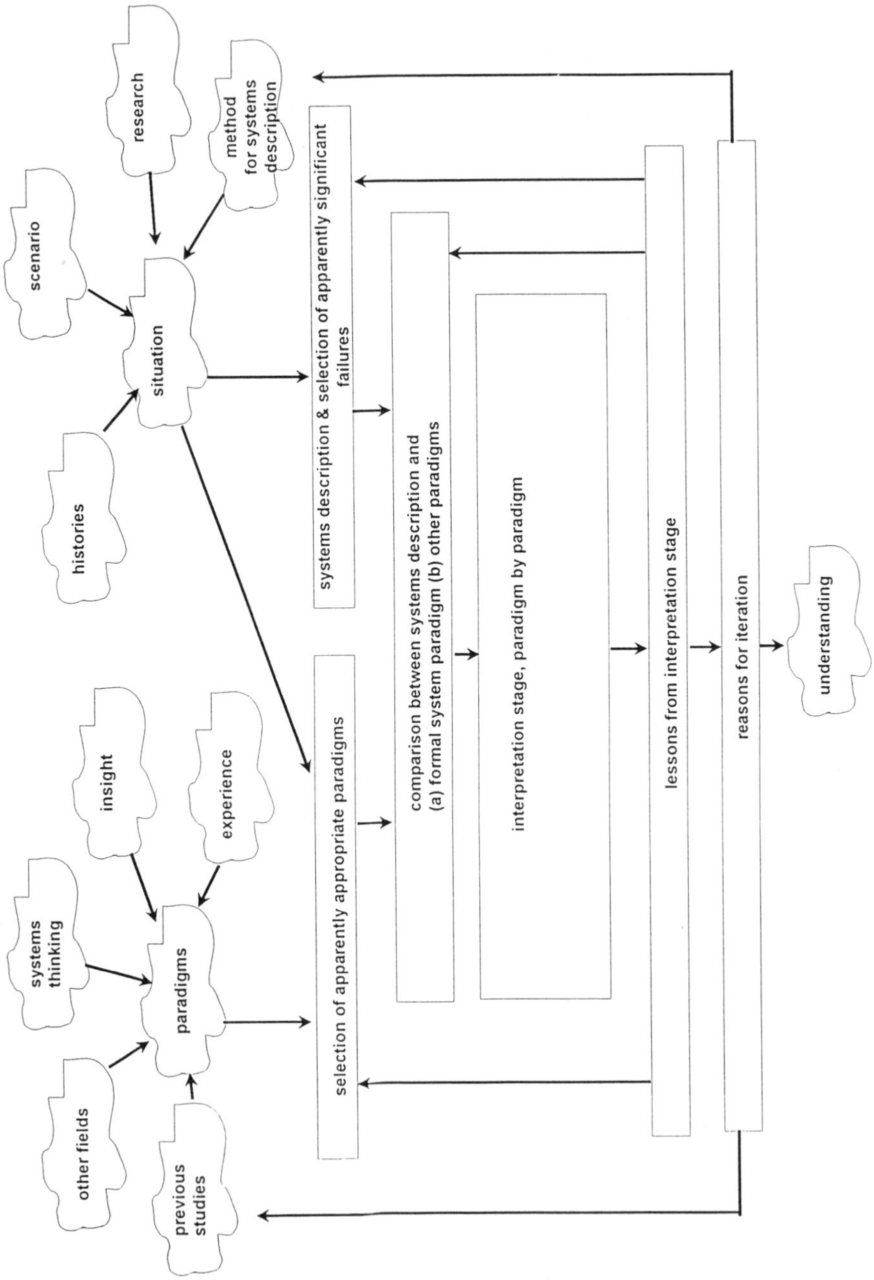

Figure 1. The failures method (Fortune, 1984).

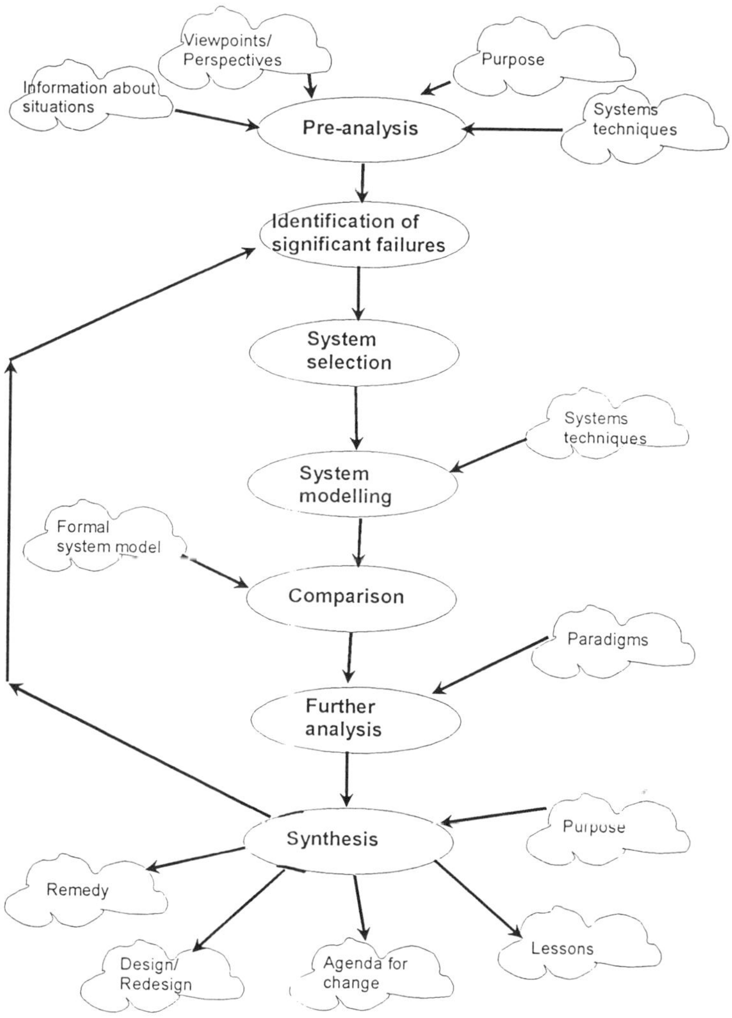

Figure 2. The systems failures method (Fortune and Peters, 1995).

The Notion of System Accident

One way of developing the Systems Failures Method would be to add further bases for comparison with which systems conceptualised from actual or potential failure situations can be contrasted.

One model which in theory should be applicable is based upon Perrow's notion of a 'System Accident' (Perrow, 1984). Within the realm of human activity systems, and in particular high risk technologies, considerable attention has been focused on the design and engineering of plant which contains sophisticated safety features. Concepts such as redundancy, fail-safe and fail-soft, and techniques such as fault tree analysis and failure mode and effect analysis have been used in an effort to increase reliability. However, despite the best efforts of designers and engineers these modern complicated technologies do still go awry, often with spectacular and catastrophic consequences.

Perrow argues that high risk technologies frequently display two features which tend to make them prone to failure: interactive complexity and tight coupling. Interactive complexity is, as its name suggests, an aspect of a designed system which means that failures and events are not independent, but related, and that the more complex the system the more chance there is that sequences and patterns of events will be unforeseeable, and unexpected. Attempts to subsequently engineer warnings or the like only in turn increase the complexity still further, thereby introducing new potential configurations of events. Tight coupling refers to the speed of interaction, and the extent to which parts of the system can be isolated. Taken together, interactive complexity and tight coupling, will, in Perrow's view, lead to what he terms System Accidents, or Normal Accidents.

Perrow's ideas have been much cited by those interested in the study of failures. For example Bowonder, Arvind and Miyake (1991) incorporated Perrow's ideas, alongside other socio-technical system concepts, in an analysis of what is usually described as 'the world's worst industrial disaster', the leak of methyl isocyanate at Union Carbide's plant in Bhopal, and then went on to build them into a framework for considering generic reasons for failure in hazardous situations. They concluded that ".. technical personnel in hazardous facilities neglect and discount the mode and speed of social interactive processes, resulting in neglect of social aspects of accidents as well as emergencies." (Bowonder et al., 1991, p.32). They also made a number of generalised recommendations about how to prevent hazardous failures by strengthening capabilities in a number of areas such as scanning trigger events and rehearsing for emergencies.

A question mark which hangs over Perrow's ideas is whether there is merit in considering the two component concepts of interactive complexity and tight coupling as separate. It is the view of the authors that tight coupling is an aspect of complexity. Like Casti (1979), the authors consider that complexity has two facets: the structure of the irreducible subsystem components; and the manner in which the components are connected. The connectivity and structure when taken together provide the static complexity, but there is also a dynamic complexity which results from the time behaviour of the system.

Reason (1990) refers to notions like Perrow's Normal or System Accident as "resident pathogen metaphors". His argument, in effect, is that many failures are not easily predictable and do not result from a simple cause or factor. Instead they arise from a set of events which must occur together. He argues that this is similar to a multiple cause illness like cancer where 'resident pathogens' in the human body are potentially destructive and weaken the bodies' defensive mechanisms. If this metaphor provides additional insights into particular failures it will add an additional dimension to the concept of vulnerability which the authors have already explored (see Horlick-Jones, Fortune and Peters, 1993). However, Reason (1990) also raises some doubts about the potential of vulnerability. Again he makes an analogy, this time with accident proneness. Accident proneness is a theory which at one time appeared to offer an explanation of the variation in accident rates amongst individuals. On subsequent examination it was found that the differing rates of accidents which individuals experienced were primarily explicable by statistical variation. Any remaining variation did not support the accident proneness hypothesis since the group who were experiencing higher than expected numbers of accidents in a given period were a markedly different sub-set of the population from those looked at in a later period. For the moment, however, the concept of vulnerability seems to have more mileage as an idea worth pursuing than accident proneness; although elderly people or babies may not be accident prone, but their physical state may make them more vulnerable to events.

CONCLUSION

Exploration of the ideas outlined above in relation to their use within the Systems Failures Method is still in its early stages. It is not even clear whether incorporation of the ideas would be rewarding. It does however seem likely that the full value of the concepts Perrow deploys will only be apparent if it is possible to move beyond the concept stage to discover some way of measuring or evaluating the degree to which interactive complexity and tight coupling are present in a particular situation.

REFERENCES

Bowonder, B., Arvind, S.S. and Miyake, T., 1991, Low probability - high consequence accidents: application of systems theory for preventing hazardous failures, Systems Research. 9.2:5-58.

Casti, J.L., 1979, "Connectivity Complexity and Catastrophe in Large Scale Systems", John Wiley & Sons, Chichester.

Checkland, P.B., 1981, "Systems Thinking, Systems Practice", John Wiley & Sons, Chichester.

Checkland, P.B., and Scholes, J., 1990, "Soft Systems Methodology in Action", John Wiley & Sons, Chichester.

Fortune, J., 1984, "Studying Systems Failures", Open University Press, Milton Keynes.

Fortune, J., and Peters, G., 1995, "Systems and Failures" (working title), in press, John Wiley & Sons, Chichester.

Horlick-Jones, T., Fortune, J., and Peters G., 1993, Vulnerable systems, failure and disaster, in: "Systems Science: Addressing Global Issues", F.A.Stowell, D.West and J.Howell, eds., Plenum Press, New York.

Mansell, G., 1993, The failures method and soft systems methodology, Systemist. 15.4:190-204.

Perrow, C., 1984, "Normal Accidents", Basic Books, New York.

Peters, G., and Fortune J., 1992, Systemic methods for the analysis of failure, Systems Practice. 5:529-42.

Reason, J.T., 1990, "Human Error", Cambridge University Press, Cambridge.

40

THEORIES THAT CREATE COHERENCE IN PRACTICE AND OTHER STORIES

Martha Vahl

Centre for Systems Studies
University of Hull
Hull, HU6 7TQ
United Kingdom

1. INTRODUCTION

In the past, it was as clear as it is today that things are not necessarily as we want them, and that improvement is needed. Nevertheless, there was a different expectation, as changes or 'improvements' would inevitably happen. They were considered a gift, however. One might get more money, food or riches, and thank the higher powers for their benevolence. The difference with the present is thus considerable. For the poor as well as the rich, the organised and the non-organised, the prisoner and the emprisoner, improvement has become a calling rather than a gift. This has made even the higher powers more accountable, also for their help.

Part of this change is due to the fact that methods have become available that make it possible for the calling to come about. These methods often are characterised as 'scientific', adhering to the strict (Cartesian) distinction: one should have knowledge and be able to predict, and one should be able to make explicit what one wants. If both exhortations are realised, it is assumed possible to decide on and select the steps that will change the present, as well as realise one's desires in the future. The set of methods that adhere to this distinction is to be labelled here as *M*.

The number of methods M is vast, and their effect enormous. Nobody nowadays even thinks anything about demanding that the methods should be used precisely and with full control in, for example, areas such as a faultless harnessing and distribution of energy, or for the treatment and prevention of diseases. Given these successes, the number of cases where methods M fail is even more remarkable–such as the quest for a more equal distribution of food, a management of businesses and social services that better satisfy the needs of clients and government policies to prevent crime.

In such cases, a different set of methods has been put forward–to be labelled as *M**. They aim to improve where the Cartesian distinction can not be implemented. They clearly constitute a motley lot: those that hark back to the times when changes were gifts, like the outcries of prophets and gurus, the processes of politics; those that implement some 'practical theory' in the sense of Kurt Lewin, such as decision support systems (Fischoff, 1992), general

Critical Issues in Systems Theory and Practice, Edited by K. Ellis et al.
Plenum Press, New York, 1995

problem solvers (Newell and Simon, 1972), mental care systems (Schagen, 1994); those that aim to extend and even include methods M, such as Checkland's Soft Systems Methodology (SSM) (Checkland and Scholes, 1990).

Of many of these methods it is claimed that they 'prove their mettle' (better than others, in Popperian terms). This kind of claim implies that it must be possible to improve on the methods. The obvious procedure to do so, of course, would be to implement M, and evaluate and compare the methods. As it happens, this turns out not to be easy. A review of a few thousands of evaluation studies on psychotherapy showed little difference between intervening and not intervening (Schagen, 1994). Methods in management often can but claim the success of their public relations, the books written about them, the cash made, the number of students taught.

One might conclude that methods M* are thus not worth much in any systematic sense. Methods M* are clearly accident prone. They often fail in individual cases: when employees do not recognise their representatives in the debates that are proposed, when a chief executive officer does not participate in applying the method, when clients, actors or stakeholders do not behave as assumed. Methods M* often prove imprecise or even useless in predicting future developments: companies may go broke, even when helped properly.

Such a conclusion does not seem warranted yet, however. The obvious counter-argument is that methods M* aim to help where methods M fail, and that the latter hence may fail again when applied to assess their alternatives, that is methods M*. Moreover, positive reports about methods M* abound: clients claim their lives to have been changed by their use (even by methods as mindless as Weizenbaum's ELIZA), companies that 'hopes' have been re-kindled, managers similarly claim to have realised moving to more innovative forms of behaviour, employees report organisational visions to have increased their competence.

This leaves one with the question of how to assess improvement on methods M*: can one actually claim some of those methods to be superior, possibly only in specific cases? Or is *any* method to be preferred that creates debate and a little bit more interaction and uncertainty, in case no effective method M is available? What is one to make of the efforts to structure *conversations*, or implement *paradigms* in new *communities*, *networks* or *fora*? Must one restrict oneself to demonstrating that methods M* can only improve on themselves, and that there is no way of including methods in M* that will boost the set of methods M* to attain a higher quality?

2. IMPROVEMENT OF METHODS M*

The following procedure is proposed to answer these questions. First, the set of methods M* is characterised as precisely as possible. Second, a general theme to cover the characterisation must be formulated. Third, methods M* will have to be compared in terms of this theme. A suitable form of experimentation has to be derived. Finally, a conclusion is drawn: improvement of methods M* is possible. Methods can be constructed that imply the generation of new agents, whilst providing the co-ordination that such agents need to remain agents. This procedure enables the intended forms of comparison and improvement of methods M*.

2.1 Characterisation

The criterion that was used above to identify methods M* is a broad one: it includes in M* anything that does not belong to M, but still aims to support improvement, like the methods included in M. Starting from the examples provided, a number of additional characteristics may be surmised. First, each method assumes some form of *exchange*: all indicate that events produced by one source are to be observed and reacted to by another. Such exchanges are *structured*, therefore. Terms that come to mind here are *debates*, *interactions*, forms of *mutual control and of co-ordination*.

A second characteristic is that each method appears to structure the exchanges according to two forms of constraints. Firstly, reactions to observed events require some form of *punctuation* of events and reactions (or inter-punctuation): they require that a combination of observed events is to be reacted to, for example, the previous event, or two previous events, and so on. Secondly, combinatorial inter-punctuations of this nature can be recognised only when they are *transferable*, that is when they can be applied to other series of events that may be recognised in other (or later) situations. Moreover, both constraints are mutual. Producing events and generating reactions can get *locked in*, that is to say, a stable pattern may appear in the exchange.

2.2 Theme

It is proposed to summarise the way exchanges are structured by the term *script*. Scripts indicate what events to recognise, and how to react to combinations of them. They are similar to the idea of frames or schemes used in artificial intelligence (Mandler, 1984). Scripts are taken to act as guidelines, as coding schemes, as points of view that summarise what is to be recognised and reacted to. That is to say, scripts indicate a *theme*: they make explicit that control and co-ordination on the basis of exchanges are not based on the experience of what *is*, but on what is *to be*. Scripts thus denote the characteristics of methods M*, not more, not less.

According to this theme, the distinction that is the basis of methods M is reversed in the case of methods M*. One no longer makes use of the present as the basis of predictions (theories of), to realise a desired future. Instead, one selects what one wants in order to shape a desired present (theories for). This reversal warrants a change of terminology: the desired future provides the rules of co-ordination that makes it possible for agents to address each other, hence re-generate themselves and each other as well as their exchange, and allow them to continue their co-ordination.

2.3 Testing

Methods M* may be compared, therefore, as to their ability to distribute what is effective in addressing agents, and to what is co-ordinating them. For example, democratic voting obviously creates agents, that is voters, but limits their exchange mainly to indicate priorities. The soft systems methodology (SSM), on the other hand, is careful in stating constraints on a number of aspects: on *inter-punctuation* (some 'problem' has to be solved, that is some 'transformation' identified), on *co-ordination* (a debate has to take place), and on creating *agents* (actors and clients have to be identified, mainly on the basis of the events they produce when describing their situation as a 'rich picture').

It is in this sense that one may say that SSM is richer in empirical content than democratic voting: it can handle a wider variety of events and reactions. This implies another difference as well: the level of *situatedness* of the *locus of control*. Voters are not assumed to change the structure of the voting procedure: this would allow some agents to have more control than others. SSM, on the other hand, does allow changes by the agents to some extent: agents may change their debate by introducing constraints that are more general than those implied by the need to define problems (summarised by the concept of 'Weltanschauung').

2.4 Example

The ability to order methods belonging to M* implies a testing procedure. One should look for methods that have more empirical content than others, and that allow individual agents more alternatives to change the structure of the debate. It is in this sense that one can say that the author has tested a particular form of evaluation as a method for improvement, as opposed to another such method. Vahl (1994) demonstrates that a script with a rule-like structure of the form 'if..., do...' must be taken to be a *better* exemplar of M* than a script based on detecting the difference between 'intended change' and 'realised change'.

It may be interesting to point out that in a number of cases where research was used mainly to improve on prediction (using methods M), the use of methods M* is now emphasised. An example is the weather forecast. The improvement of the predictability of the weather has shown to be very limited in principle. The weather chart has been turned into a co-ordination device, therefore: it addresses users to identify themselves, for example as car drivers that need to beware of fog, or as farmers who need to act taking only the local weather into account. The chart acts as a linguistic (not necessarily representational) co-ordinating device that situates agency.

3. CONCLUSION

The set of methods M* aims to do the same as the set of methods M, but without having to rely on the assumption that what is and what is to be are independent. In this paper a procedure is proposed to characterise methods M*. The procedure rests on a reversal of the assumption mentioned: what is and what is to be have to be dependent in a certain way. It proved possible to formulate a procedure to test and compare methods M*, and thus increase the quality of M* as a set of methods to support improvement.

REFERENCES

Checkland, P., & Scholes, J.,1990, "Soft Systems Methodology in Action", John Wiley and Sons, Chichester.

Checkland, P., 1981, "Systems Thinking, Systems Practice", John Wiley and Sons, Chichester.

Fischoff, B., 1992, Giving advice: decision theory perspectives on sexual assault, *American Psychologist*. 47:577-588.

Hayden, Ch., 1905, "ELIZA, version 1.3", Software Program, Fair Haven N.Y. 07701.

Mandler, J.M., 1984, "Stories, Scripts, and Scenes: Aspects of Schema Theory", Lawrence Erlbaum Associates Publishers, Hillsdale, NJ.

Newell, A., & Simon, H.A., 1972, "Human Problem Solving", Prentice Hall, Englewood Cliffs, NJ.

Schagen, S., 1994, "Over de Waarde van de Psychotherapie", unpublished PhD Thesis University of Amsterdam, Amsterdam.

Vahl, M., 1994, "Improving Mental Health Services in Calderdale. An Evaluation of 5 Schemes Funded through the Mental Illness Specific Grant", unpublished paper, Centre for Systems Studies, University of Hull, Hull.

41

METAPHOR AND THE GAP BETWEEN THEORY AND PRACTICE

Heather Watson and Trevor Wood-Harper

Department of Mathematics and Computer Science
University of Salford
Salford, M5 4WT
United Kingdom

ABSTRACT

We review metaphors in organisations and methodologies. While this tends to focus on specific metaphors, we reflect generally on a methodology's meaning to consider why there is a gap between theory and practice. As an outcome of this, we reformulate the gap, describing it as a special case of a more extensive metaphorical gap already admitted within practice itself. Using Soft Systems Methodology (SSM), we then suggest Schön's (1963) schema for a displacement of concepts as a way of enriching our understanding of developments in a methodology's theory and practice.

INTRODUCTION

Schön (1963) explains that metaphor works by treating something unfamiliar (A) as something familiar (B), thereby changing our understanding of both in the process. He also explains that since metaphors both hide and highlight different aspects of a situation, the truths they present us with are only partial (p.50). In relation to this, Schön (1979) later discusses two traditions of metaphor. The first considers metaphor as an ornament or as an anomaly to be explained or even to be explained away (e.g., Beardsley, 1967). The second considers metaphor pivotal in making sense of a situation (e.g., Lakoff and Johnson, 1980). This paper supports the second tradition. We argue not only that specific metaphors affect methodology, but that metaphorical thinking may explain the process we engage in when relating theory and practice.

METAPHORS IN ORGANISATIONS AND METHODOLOGIES

Morgan (1986) argues how important several metaphors are for understanding organisations, and Walsham (1991, 1993) extends this to information systems. In developing

Critical Issues in Systems Theory and Practice, Edited by K. Ellis et al.
Plenum Press, New York, 1995

Total Systems Intervention (TSI), Flood and Jackson (1991) use a range of systems metaphors, and they suggest drawing on an organisation's dominant metaphor as a guide for methodology choice. Kendall and Kendall (1993) also consider dominant metaphors in a range of methodologies, advising that chances of system success may be increased by choosing a metaphor appropriate to an organisation.

Such research suggests how situations may be structured by conceptual systems that are at least partially defined by metaphor. This focus on specific images may be a consequence of what Schön (1963) describes as the most common approach to metaphor: insofar as it is considered to be a part of language certain words or sets of words are said to "...be a metaphor..." (p.35). We take a different approach, treating metaphor "...not as a piece of language but as a process of thought." (p.37). As such, we hope to extend the above research by suggesting how metaphor is at work even when we do not explicitly choose specific images to structure our thinking about a methodology's theory and practice. We argue this below in terms of a methodology's meaning.

A METHODOLOGY'S MEANING IS METAPHORICAL

Checkland and Scholes (1990) make the following observation concerning Soft Systems Methodology (SSM): "...the inseparability of the methodology from the use made of it by a particular user in a particular context will always mean that SSM itself, as a description on paper according to the new Constitutive Rules, will forever remain undecidable." (p. 229).

This suggests that a methodology's meaning is metaphorical since it is divided between at least two sources: it is both (A) a property of methodology and (B) a practitioner's experience of using a methodology. This may explain why there is a gap between theory and practice.

Yet upon further inspection we recognise that even in practice the meaning of our experience is further divided because it has (A) already occurred, yet (B) remains to be described. This implies that a metaphorical gap occurs even in practice because meaning continues to be divided. For example, if we structure our practical experience specifically in relation to SSM through its Constitutive Rules or more generally through case studies, we indicate that methodology helps us in making sense of our experiences through descriptions. These structures continue to inform, for instance, through debates about SSM in use. But notice how even in each intervention experience is checked successively through descriptions. Indeed, these are how stages in a methodology are distinguished one from the other. So, for example, staging debate about change using SSM is initiated by comparing *models* with real situations (Checkland and Scholes, 1990).

Our practical experience may end with the production of a *real* model, a technical artefact, but this too is a description. If we then consider the meaning of this artefact in use, we further reiterate our predicament with methodology. The meaning of using an information system is both (A) a property of the information system and (B) a user's experience. Rather than having one source, meaning is not literal, even in practice because it is always somewhere in-between experience and its description.

An outcome of this is evidence of the partial nature of metaphorical truths: in highlighting the gap between theory and practice, we may tend to hide a more extensive gap already admitted within practice itself. This implies that we might need to reformulate our existing concepts of theory and practice. With this intention, we discuss a schema for the emergence of new concepts in the section below.

THE PROCESS OF METAPHOR AND LIMITS OF AFTER THE FACT THINKING

Schön (1963) refers to the process of metaphor as the displacement of concepts, which he explains is essential to forming new concepts: "...a way of treating the new as old, neither comparison, nor error, nor the application of concepts to instances, but a displacement of old concepts to new situations resulting in extension of the old." (p.34).

He argues that if we understand concepts as the 'fund of expectations' through which we structure experience, then concepts and theories cannot be separated (p.8). In this field, Schön's view is supported by methodologies such as SSM that link old concepts such as theory and practice through a relatively new concept - action research.

In Schön's (1963) view, we have difficulty in explaining the emergence of new concepts because we tend to describe them in ways suitable only to their justification 'after the fact' (p.20). Checkland's (1981) position regarding the application of traditional 'hard' systems thinking to soft problems is based in a criticism of this type of after the fact thinking: "...traditionally engineers' activities start *after* they accept a specification" (p.144). They then focus on how to do things since the what is "*...already answered* elsewhere" (p.144, our emphasis).

Schön (1963) describes a consequence of this thinking as follows: "We are used to thinking about fully formulated concepts, their relations to one another and to their instances. As long as we focus on the concept-instance relation, we will be at a loss to understand the formation of new concepts." (p.20).

Consequently, our theories of concept formation tend to exclude or de-emphasise the process of metaphor. In addressing this, we draw on developments in SSM to suggest Schön's (1963) schema for the displacement of concepts as a way of enriching our understanding of theory and practice.

DEVELOPMENTS IN SSM AND A SCHEMA FOR THE DISPLACEMENT OF CONCEPTS

Schön (1963) distinguishes the following phases in the process of concept displacement. He emphasises that these are not discrete events following a fixed order. Instead, they occur out of sequence, and they are often inseparable.

Transposition - making the metaphor, the initial shift of an old concept to a new situation. This makes a relationship between old and new. In SSM we see this in Checkland's adaptation of 'hard' thinking to 'soft' problems. Schön emphasises that this process is complicated because in ordinary thought there are no isolated concepts. Rather, there are conceptual clusters. This is supported by Lakoff and Johnson's (1980) later work on metaphorical entailments to refer to the associated network of concepts and images that a metaphor conjures up. In SSM we see that the introduction of 'hard' thinking to 'soft' problems is also associated with the concept 'system.' Checkland (1981) criticises 'hard' approaches for thinking of systems as things that really exist, whereas in SSM, the concept is a useful epistemological device for thinking 'about' the real world.

Interpretation - assigning a concept from an old cluster to a specific aspect of a new situation. Schön explains that in transposing an old concept to a new situation, we relate it to a specific aspect of that situation. He describes how transposition and interpretation are often indistinguishable as evidenced in the way we may consciously wonder about which aspects of a new situation the old concept applies. In SSM an example of this is Checkland's discussion of the definition of means and ends that are themselves problematic. Schön also considers how transposition and interpretation do not proceed with perfect freedom since each new situation already has a conceptual structure before an old theory is displaced by it. In SSM this is evidenced by the emergence of a cultural stream to accompany the existing logical stream.

Correction - the way pre-existing structures resist transposition and interpretation that results in an adjustment of a concept's displacement. In criticising the concept-instance relation, Schön explains that conceptual displacement is not a one way matter. Instead of the old concept being applied as a 'stencil fit' over the new situation, the old and the new theory mutually adapt (p.55). In this way the metaphor becomes elaborated. Generally this is supported by the concept of action research that relates SSM's theory to its practice in an adaptive, mutually informing cycle. This is also associated with the linear seven stage 'stencil fit' model becoming increasingly flexible. Schön expands on this notion of correction saying that as a concept's reference to the new situation becomes more conventional and familiar, we become increasingly aware of two equally legitimate senses of the same term. This is supported by Checkland and Scholes' (1990) distinction between two uses of SSM: Mode 1 and Mode 2.

Spelling out - working out the relation between two senses of concept. This requires a more explicit attempt at formal theorising than is common in ordinary practical inquiry (p.56). This is evidenced by SSM's Constitutive Rules. Schön explains that this process of spelling out is incomplete. This is supported by the function of SSM's rules, not in deciding who has or has not used SSM, but in engaging debate.

While this analysis is brief, we do hope it gives some indication of the potential of a schema for the displacement of concepts to enrich our understanding of methodology. Given the value SSM places in learning, Schön's schema is potentially important since metaphorical thinking may help us to develop new ideas for solving problems.

REFERENCES

Beardsley, M., 1967, Metaphor, *in*: "Encyclopaedia of Philosophy," P. Edwards, ed., vol. 5, MacMillan, New York.

Checkland, P., 1981, "Systems Thinking, Systems Practice," John Wiley and Sons, Chichester.

Checkland, P. and Scholes, J., 1990, "Soft Systems Methodology in Action", John Wiley and Sons, Chichester.

Flood, R.L. and Jackson, M.C., 1991, "Creative Problem Solving: Total Systems Intervention," John Wiley and Sons, Chichester.

Kendall, J. and Kendall, K., 1993, Metaphors and methodologies: living beyond the systems machine, *MISQ*. June:149-171.

Lakoff, G. and Johnson, M., 1980, "Metaphors We Live By," The University of Chicago Press, Chicago.

Morgan, G., 1986, "Images of Organisation," Sage, London.

Schön, D., 1963, "Displacement of Concepts," Tavistock Publications, London.

Schön , D., 1979, Generative metaphor: a perspective on problem-setting in social policy, *in*: "Metaphor and Thought," 2nd Edition, pp.137-163, A. Ortony, ed., Cambridge University Press, Cambridge, UK.

Walsham, G., 1991, Organisational metaphors and information systems research, *Eur. J. Inf. Sys*. 1:2:83-94.

Walsham, G., 1993, "Interpreting Information Systems in Organisations," John Wiley and Sons, Chichester.

INFORMATION SYSTEMS THEORY AND PRACTICE

SOFT SYSTEMS ANALYSIS AND MODELING TOOL (SSAMT)

Computer-Based Support for Conducting Soft Systems Studies

M. S. Davenport and J. Ayers-Hunt

University of Teesside
Teesside Business School
Middlesborough
TS1 3BA
United Kingdom

INTRODUCTION

A critical issue for would be practitioners of Soft Systems Methodology (SSM) (Checkland, 1981) is the limited availability of support tools and guidance to aid an investigation and analysis. Potential users of SSM can be discouraged from participating in studies by the apparent lack of rigour and seeming 'wooliness' of SSM as an approach to problem solving in organisations. Indeed, critics argue that the richness of the understanding and subjective context of each individual problem situation could be hindered by over prescriptive tools (Kreher, 1993). Useful attempts have been made to provide interesting and exiting support for SSM. Avison et al (1992) have developed a rich picture diagramming tool and Stowell et al. (1991) an expert system trainer for teaching SSM. What has not been developed are specific modelling tools for practitioners to use in a study. It is well documented that practitioners find difficulty implementing significant parts of SSM, selecting 'relevant systems' (RS's) to model, defining 'root definitions' (RD's) and creating 'conceptual models' (CM's) (Mingers, 1992). There is a need for suitable tools to aid SSM practitioners with modelling and analysis in a way which complements SSM without restricting the systemic nature of a study. This paper describes SSAMT a computer based modelling tool for SSM. Developed at Teesside Business School, SSAMT is a windows based application for assisting SSM practitioners with modelling and analysis, incorporating the selection of relevant systems to model, defining root definitions, creation of conceptual models and diagnosing potential change. It is concluded that such tools can be useful for practitioners and rather than constrain studies they can assist by providing rigour, diagnostic support and a focal point for debate.

FEATURES OF SSAMT

SSAMT is run from windows using an icon depicting the seven stage process of SSM (Fig 1.). The main menu for SSAMT is a screen version of the SSM process and is accessed by clicking the mouse onto the various stages of the process where the practitioner needs to work. SSAMT focuses on the below line system modelling activities of SSM, leaving rich pictures to a pen and paper exercise. The features included are :

- selecting relevant systems
- defining root definitions
- conceptual modelling
- problem diagnostics
- help
- reports

The system allows for all features in an analysis to be displayed and accessed together. This enables the practitioner to view relevant systems, root definitions, CATWOE analysis, conceptual models and diagnostics at the same time. It allows changes to be made to each component in an iterative, developmental way.

Selecting Relevant Systems

Clicking the mouse onto the main menu Stage 2. of the process presents the practitioner with the opportunity to summarise problems and problem owners from the rich picture phase of the study. The intention is to list the various problems and problem owners encountered from the rich picture and for the practitioner to identify 'Weltanschauungen' ('W') and relevant systems for each. The practitioner may then use this list to help select a system to model for either a primary task or issue based analysis. Help is provided in explaining how this feature operates and in defining 'primary task' and 'issue based' root definitions.

Defining Root Definitions

Root definitions may be developed from a relevant system selected or independently by clicking the mouse onto Stage 3. from the main menu. A CATWOE analysis can be defined and used to aid typing in the text of the root definition. Help and useful hints are provided explaining the CATWOE analysis and the preparation of root definitions e.g. given the defined purpose of the relevant system are the transformation process and 'W' consistent? If you were to perform the transformation would the desired purpose be achieved and the 'W' be satisfied ? Would it be likely that a conceptual model of this RD would prove useful in the problem situation?

Developing Conceptual Models

Conceptual modelling of the root definition can then proceed by clicking on the toolbar item 'Conceptual Model' or Stage 4. of the main menu. The practitioner can then insert each model activity and the system will create a 'bubble' activity on screen. The activity 'bubbles' can then be moved anywhere on the screen and connecting lines inserted. The initial activities are inserted in short form, a more detailed description of each activity added 'behind' the 'bubble' and accessed by clicking the mouse on each activity. An 'Activity List' is provided for assistance in selecting the verbs and words to include in the activity

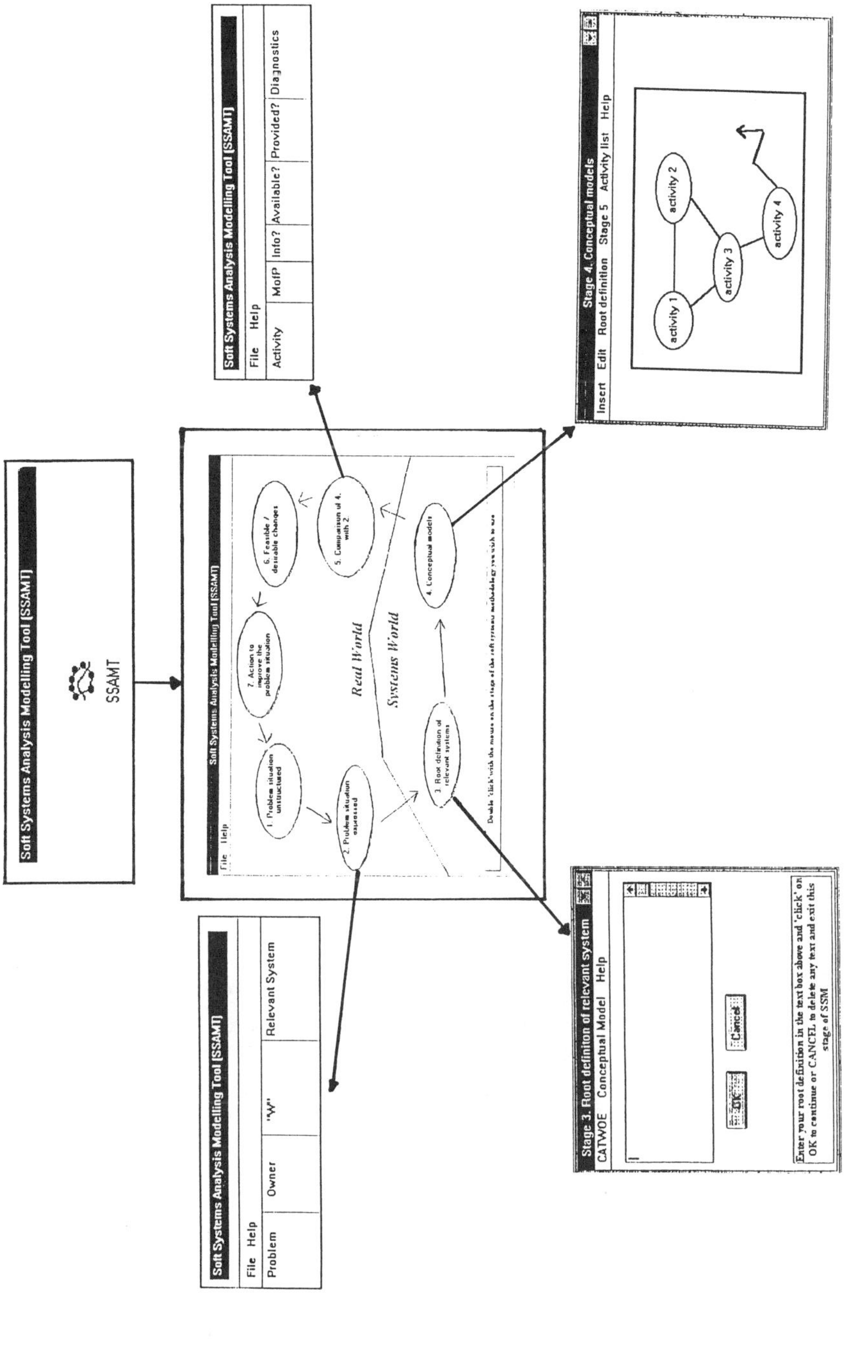

Figure 1. SSAMT Overview.

description. At this stage the root definition can be referred to and changes made. Help is provided with guidance in conceptual modelling, hints and questions regarding consistency and common faults with CM's.

Problem Diagnostics

The comparison phase of the process is accessed by clicking on Stage 5. of conceptual model menu tool bar of the main menu. The choice of problem diagnostics are :

- a basic comparison
- an organisation/role comparison
- an information requirements comparison

Each analysis lists the model activities and the practitioner must then enter the details. Figure 1. shows an information requirements analysis where the model activities are listed with measures of performance, information required, what information is available, how it is provided and what point for debate arises, being added by the practitioner. Help is provided in a description of the different modes of diagnostics and how they may be completed. At this stage the practitioner can review the list of relevant systems, root definitions and conceptual models previously developed.

Reports

Standard reports can be produced as follows :

- a list of problems, owners and relevant systems
- copies of root definitions and CATWOE analysis
- conceptual models and detailed activity descriptions
- diagnostic reports.

The practitioner can use the above for analysis, debates, records of studies and producing client reports.

DISCUSSION

SSAMT is a practitioners tool, assisting with system modelling and problem diagnosis. It is not a replacement for the 'systems thinking' required by the practitioner. It provides help in understanding SSM and advice on the parts of the process found difficult by practitioners. It represents a platform for analysis and a focus for conducting studies. It also provides a record of studies and useful output in the form of reports. SSAMT can be used as a group decision support tool with multiple users making contributions. The software is only a prototype at the demonstrator stage of development, however initial trials with practitioners have been encouraging. Typical responses have been that conceptual modelling on screen is interesting, listing problems and owners etc. helps with selecting relevant systems, the problem diagnostics feature is useful particularly with the reports produced, the help and advice is a comfort for new practitioners and being able to iterate between RD's and CM's and make changes on screen is an improvement. Importantly practitioners feel more comfortable if they have a focus for a study when conducting SSM studies. Many potential users of SSM have commented they feel uncomfortable using the approach because of the lack of, for want of a better word, 'legitimacy'. The tools of SSM are basically pens, a suitable wall, lots of brown paper, and thinking power. Consultants tend to need to sell formal methods with their services in order for clients to have a 'feel good factor' about the cost of a study. The language of SSM is not always conducive to organisations who are used to discussing financial appraisals and 'bottom lines' and not 'what are we doing?' or 'why are we here?'. A tool like SSAMT can help to overcome some of these problems.

A major problem using formal procedures and prescriptive tools with an approach like SSM is the loss of richness of understanding in the subjective, situational context of a problem situation. Each problem has its own nuances and pre-empting solutions or approaches to analysing the situation will diminish the systemic nature of the intervention. SSAMT, perhaps, overcomes this problem by allowing an iterative analysis to take place whilst maintaining the rigour required for system modelling. The modelling process is practitioner driven, none of the important elements are created automatically upon any pre-determined basis. Developing conceptual models on screen seems to help create a freer flow of ideas providing the ability to easily change and manipulate models as they are drawn. Using pen and paper can be restrictive in that changes and revisions eventually require a new diagram. Maintaining clear records of an investigation allows the practitioner to review a project and all its aspects. A retrospective analysis can be fruitful for identifying points at which false or inappropriate assumptions were made and the investigation departed on a wrong tangent. An SSM investigation needs to double-back and criss-cross, an overly prescriptive tool may prevent this type of dynamic analytical behaviour.

The main point to drawn from this experience is that merely computerising the features of SSM would have only minor benefits which would not outweigh some of the dis-benefits of formalised tools and procedures. We should look to develop tools which enhance SSM, for use by practitioners, by providing additional features which assist the implementation of SSM in practice. Future developments of SSAMT will be directed towards the diagnostics phase of SSM and orchestrating debates about change. The software also needs to be developed in a technically robust mode for practitioners to use in actual investigations.

REFERENCES

Avison, D. E., Golder, P.A., and Shah, H.U., 1992, Towards an SSM tool kit : rich picture diagramming, *European Journal of Information Systems.* 1:6.

Checkland, P., 1981, "Systems Thinking, Systems Practice", John Wiley & Sons, Chichester.

Kreher, H., 1993, Critique of two contributions to soft systems methodology, *European Journal of Information Systems.* 2:4.

Mingers, J. C., 1992, Questions and suggestions in using soft systems methodology, *Systemist.* 14:2..

Stowell, F. A., West, D., and Stansfield, M. H., 1991, The application of an expert system to an unstructured domain of expertise : using expert system technology to teach SSM, *European Journal of Information Systems.* 1:4.

43

UNITING SYSTEMS THEORY WITH PRACTICE

Kevin G. Doyle

Department of Computing
The University of the West of England (Bristol)
Coldharbour lane
Frenchay
Bristol, BSI6 1QY
United Kingdom

INTRODUCTION

Since the middle of this century much important work has been done by members of the systems community to further understanding of complex matters in areas such as computation theory, self reproducing automata, cybernetics, information theory and the softer areas of social or purposeful systems. However, the objective of re-integrating disjoint scientific specialities under a general systems theory has not been met. During the last twenty years, fragmentation rather than integration has occurred with significant differences developing between the conceptualisation of 'natural' and 'artificial' systems, 'hard' and 'soft' systems, 'concrete' and 'abstract' systems, and 'physical' and 'social' systems. During this period the 'meta-discipline' of Systems Science has developed to the point where, (to paraphrase Einstein) "As far as the laws of General Systems Theory refer to reality they are not certain and as far as they are certain they do not refer to reality", (Capra, 1992).

An area of Systems Science that has grown quickly amongst all the confusion is that of Information Systems, a new and highly dynamic subject area where academics and practitioners often fail to agree at any level about things as fundamental as the meanings of 'information' and 'system'. There is ample evidence to show that in the world of Business Information Systems, technologists do not understand the world of business and vice-versa. This paper asserts that a similar gulf exists between Systems theorists and Information Systems professionals. Given that (to paraphrase Einstein again) theory without practice is lame and practice without theory is blind, Einstein (1950), the need to unite theory and practice is clear.

Information Systems is an ideal area to consider this unification, as IS development and use is a natural point of contact for the 'natural' and the 'artificial', the 'hard' and the 'soft', the 'concrete' and the 'abstract' the 'physical' and the 'social'. This paper introduces work currently in progress at University of the West of England to try and use and further

Critical Issues in Systems Theory and Practice, Edited by K. Ellis et al.
Plenum Press, New York, 1995

develop the theories of Systems Science in a series of practical and important Information Systems projects in order to promote the unification of theory and practice.

THE SECOND AND THIRD TOWERS OF BABEL

In order to think about, understand and explain the world about us, it is necessary to develop models or abstractions of the world and ways of using them to think about it. These abstractions and approaches then become the epistemological constructs which form the basis of our reasoning, communication and discussion about the world.

As disciplines have emerged, each has developed its own ways of modelling and reasoning about the world. Many providing different descriptions of what is essentially the same set of phenomena. Religion and Philosophy, Politics and Economics, Sociology and Anthropology, all have their own brand of models and approaches, rules and evidence. The differences between them often lead us to the mistaken belief that the phenomena which they study are not the same.

> "Our civilisation seems to be suffering a second curse of Babel: Just as the human race builds a tower of knowledge that reaches to the heavens, we are stricken by a malady in which we find ourselves attempting to communicate with each other in countless tongues of scientific specialisation." (Davidson, 1983)

Prior to the 1940's, the exploration of complex phenomena using the predominant paradigms of science and engineering involved the logical manipulation of component relationships. This was found to be inadequate for the representation of the structure of complex wholes, required by biologists, psychologists and the designers of complex engineering systems. A new approach was called for, and the foundations were laid for General Systems Theory (Emery, 1969 ; Angyal, 1941).

This new systemic paradigm pioneered during the 1940's, by Von Bertalanffy, and developed further during the following years by control engineers and cyberneticians, seemed to promise not only an approach capable of the treatment of complex wholes but also a way of integrating the multiplicity of descriptions or abstractions commonly used by the different scientific disciplines (Bertalanffy, 1973).

In its early days the systems movement emphasised structure, mechanism, isomorphy and hierarchy. It contained at its core, two simple but fundamental pairs of ideas: emergence and hierarchy; and communication and control. (Weinberg, 1975) Much important control theory has developed on this theme, such as Ashby's law of requisite variety and Bertalanffy's law of equifiniality (Ashby, 1956; Bertalanffy, 1969)

During the decades that followed Bertalanffy, structuralism, functionalism, metafunctionalism, holism, spiritualism and many other 'isms have advanced the systems movement to the point that the many systems specialities have considerable difficulty in finding a comparatively stable patch of common ground. No (commonly accepted) definitive text on General Systems Theory exists and very few academic institutions can claim to treat the subject as a discipline in its own right.

When existing disciplines in science or engineering turn to the Systems Movement to develop or borrow and adapt new and improved models and methods, the range of fare on offer is bewildering. A potential convert to the new paradigm of Systems Thinking, who asks of his fellow diners at a UKSS/ISSS conference dinner, in all innocence, the question 'what is Systems Science?' might be forgiven for imagining that he has wandered onto the construction site of yet another tower of Babel.

> "Every individual is at once the beneficiary and the victim of the linguistic tradition into which he has been born - the beneficiary inasmuch as language gives access to the accumulated records of

other peoples experience, the victim in so far as it confirms him in the belief that reduced awareness is the only awareness and as it bedevils his sense of reality, so that he is all too apt to take his concepts for data, his words for actual things" (Huxley, 1954).

The abstractions discussed at a typical UKSS/ISSS conference will include 'natural' and artificial systems, 'hard' and 'soft' systems, 'concrete' and 'abstract' systems and 'physical' and 'social' systems. As with the models and methods of the various cognate disciplines, the differences between them might lead us to the mistaken belief that the worlds from which they have been drawn are different. Systems Thinkers know that to focus on one abstraction or approach without first attempting to determine its likely utility is risky. Having determined the utility of one, to deny or ignore the value of others is narrow-minded. Yet fragmentation and compartmentalisation persist.

The problem of fragmentation exists because of the differences in the epistemological constructs used to describe the world and its phenomena (observable or otherwise). It is true that to move from one paradigm to another constitutes a fundamental shift and that a lack of isomorphy means that the models of each may not be easily mapped or transformed into an equivalent form in the others. The various systemic sub-disciplines may, however be considered to be complementary because despite their differences in emphasis, focus and use, they have a common point of contact; the real world of social action.

UNITING THEORY THROUGH PRACTICE

The notion of improving theory through practice is anything but new. The action research approach, exemplified by Checkland's Soft Systems Methodology, is well established. (Checkland, 1981; Checkland et al, 1990). Conventional research approaches following the paradigm of science (reductionism, experimentation, refutation and repeatability), seek to form and then verify theories. Action research promotes the emergence of (academically defensible) theories through practice and the speculation over how such practice might be bettered in some way. In this respect action research is both descriptive/interpretive and subjective/argumentative (Galliers et al, 1987).

A clear point of contact for different tools and techniques, frameworks and approaches, is the world of information systems. That SSM has been used in this area together with the tools/techniques and approaches of 'Hard' systems engineering to the satisfaction of client, actor, problem owner and problem solver is not in dispute, and there are many examples in the literature of how this may be done (Stowell, 1985; Mingers, 1988; Miles, 1990; Mansell, 1990; Prior, 1990; Doyle et al, 1993).

Indeed many programmes of action research have sought to find a way to make SSM and the tools and techniques of system development 'comfortable bedfellows' (Paton, 1991) and have enjoyed some success in this endeavour.

In trying to fashion and explore a framework within which the various forms of systems thinking and systems engineering can be mutually complementary, it is necessary to be aware of the fundamental differences in the epistemology and ontology in which they are rooted and to ensure that neither of these two elements is unduly compromised when interfacing the application of one approach with the other. This is not the "...lifting of intellectual skirts for fear of contamination" (Prior, 1991) but the need to establish a firm philosophical underpinning on which to build and develop a discipline which is currently still in its academic infancy (Doyle et al, 1991[1]; 1991[2]). This undertaking cannot come from abstract academic thought alone. Such a framework needs to be hammered out in the real world of human activity, through the medium of action research.

"The modern specialist field-worker soon recognises that in order to see the facts of savage life, it is necessary to understand the nature of the cultural process. Description cannot be separated from explanation, since in the words of a great physicist, 'explanation is nothing but condensed description.' Every observer should ruthlessly banish from his work conjecture, preconceived assumptions and hypothetical schemes, but not theory" (Malinowski, 1936).

THE PRACTICE OF PRACTICE: THE FRENCHAY HEALTHCARE TRUST ICWS PROJECT

During the last two decades, Information Technology has done very little to improve the global efficiency of clinical medicine. Improvements in the quality and delivery of treatment during this period have largely been due to an increase in doctors, nurses and paramedicals and other resources (Griffith, 1992). In an attempt to break the patient/resources curve, the National Health Service Executive have developed a research and development strategy, which seeks to strengthen collaboration between the NHS, the Medical Research Council and other research councils, the universities, the medical charities and industry.

Part of a series of projects supported by the Executive under this initiative is a project based at Frenchay Healthcare Trust to develop a multi-modality health informatics resource. The project involves a programme of action research which includes the pre-competitive specification of an integrated clinical workstation (ICWS) for the Neurosciences. The research team involved includes several members of UWE academic staff and a range of hospital specialists including: consultant neurosurgeons, neuropathologists, neurologists, neuropsychiatrists, neurophysiologists, senior registrars, registrars, senior house officers, nurses, administrators and managers.

One of the features of this research is to explore some of the relationships between the various models used by some of the sub-disciplines of the systems approach, and the more formal methods normally applied to conventional information systems engineering projects. It is hoped that a suitable framework can be found to enable the systems approach to provide a framework for debate and agreement about the 'whats' while systems engineering tools and methods are used to structure and deliver the 'hows'.

The ICWS project seems an ideal testing ground for this form of action research. The issues addressed during the project thus far have been of enormous variety and range. Clinical procedures and audit of outcomes, medical informatics and medical information management, nursing protocols and interdisciplinary communication, have all been the subject of consideration during the project to date.

So far the research has focused largely on tools and techniques from Checkland's SSM, Sawyer's OPIUM, (Sawyer, 1990) and the harder tools and techniques of Information Engineering (Martin, 1986; Davis, 1982; Rockart, 1979) within Water's Information Management Framework (Waters, 1988; Waters, 1989; Waters et al, 1994). SSM and OPIUM both use an action research approach to learning about a problem situation, rejecting the conventional hypothetico-deductive methods used by science in favour of a cycle of exploration understanding and change (Checkland, 1981; Checkland et al, 1990). This accords with Checklands observation on Poppers belief that "...in social engineering incrementalism and trial and error are the wisest approach" (Popper, 1945; Popper, 1963).

The Frenchay ICWS project has been seen from its inception as a body of action research. Accordingly, it is not expected that a single pass through a formal software development life cycle could do more than produce a prototype ICWS. The cyclic process of exploration, understanding and change, which is the hallmark of the action research project fits more closely the notion of software systems prototyping than that of the software engineering life cycle. The focus of the early stages of the ICWS project has not been to determine information requirements, but to begin the process of growing a specification of requirements organically (Brookes, 1987; Doyle 1994).

REFERENCES

Angyal, A., 1941, "A Logic of Systems, in Foundations for a Science of Personality", Harvard University Press.

Ashby, W. R., 1956, "An Introduction to Cybernetics", Chapman and Hall., London.

Bertalanffy, L. Von., 1969 The Theory of Open Systems in Physics and Biology. *in*: F.E. Emery, (ed.),1969, "Systems Thinking", Penguin, Harmondsworth.

Bertalanffy, L.Von, 1973, "General Systems Theory", Penguin University Books, Harmondsworth.

Brookes, F., 1987, No silver bullet: essence and accidents of software engineering, *IEEE Computer*, April 1987, pp1069-1076.

Capra, F., 1992, "The Tao of Physics : An Exploration of the Parallels between Modern Physics and Eastern Mysticism", 3rd ed., Harper Collings, London.

Checkland, P. B., 1981, "Systems Thinking Systems Practice", John Wiley & Sons, Chichester.

Checkland, P., & Scholes, J., 1990, "Soft Systems Methodology in Action": John Wiley, Chichester

Davis, G., B., 1982, Strategies for information requirements determination, *IBM Systems Journal* Vol.21, No.1, pp 4 - 30.

Davidson, M., 1983, "Uncommon Sense, The life and Thoughts of Ludwig von Bertalanffy", Houghton Mifflin.

Doyle, K.G., & Wood, J.R.G., 1991[1], Systems thinking, systems practice : dangerous liaisons, *Systemist,* Vol.13, pp 28-30.

Doyle, K.G., & Wood, J.R.G., 1991[2], The paradigm shift & The protozoon, *Systemist*, Vol.14, pp 131-134.

Doyle, K.G., & Wood, J.R.G., & Wood-Harper, A.T., 1993, Soft systems and systems engineering: on the use of conceptual models in information systems development, *Journal of Information Systems*, 1993, 3, pp.187-198.

Doyle, K., 1994 Growing computer based information systems organically: The integrated clinical workstation project, *in:* "New Systems Thinking and Action for a New Century": *Proceedings of the 38th International Conference of the ISSS*, Asilomar, Pacific Grove, California, June pp 527-534.

Einstein, A., 1950, "Out of my Later Years", Greenwood Press, London.

Emery, F.E., (Ed), 1969, "Systems Thinking", Harmondsworth, Penguin.

Galliers, R.D., & L. and F. F.,1987, Choosing appropriate information systems research methodologies, *Communications of the ACM*, vol.30, no.11, pp.900-902

Griffith, H., 1992, Open Letter by Huw B Griffith, FRCP, FRCS Clinical Director of Neurosciences, Frenchay Healthcare Trust.

Huxley, A., 1954, "The Doors of Percerption", Harper and Row, New York.

Malinowski, 1936, "Anthropology", *in:* The Encyclopaedia Britannica : (first supplementary vol).

Mansell, G., 1990, A comparison of entity modelling and SSM conceptual modelling in information systems development, *Systemist*, Vol.12, No.2, pp.59-64.

Martin, J., 1986, "Information Engineering: Vol 1 - 4": Savant Research Studies, Carnforth.

Miles, R. K., 1988, Combining soft and hard systems practice : grafting or embedding ?, *Journal of Applied Systems Analysis*, Vol.15, pp,55-60.

Mingers, J., 1988, Comparing conceptual models and DFD's, *Computer Journal*, Volume 31, No.4. pp.376-379.

Paton, G., 1991, Letter to the Editor, *Systemist*, Vol.13, No.2, pp.96-97.

Popper, K.R., 1945, "The Open Society and its Enemies", Routledge, London.

Popper, K.R., 1963, "Conjectures and refutations: the Growth of Scientific Knowledge", Routledge and Kegan Paul, London.

Prior, R., 1990, Deriving data flow diagrams from a soft systems conceptual model, *Systemist,* Vol.12, No.2, pp.65-75.

Prior, R., 1991, Dangerous liaisons: A reply to Doyle & Wood, *Systemist*, Vol.13, No.2, pp.81-85.

Rockart, J.F., 1979, Chief executives define their own data needs, *Harvard Business Review*, March/April pp.81-93.

Sawyer, K., 1990, Multiview Diagrams - an introductary description, *Systemist*, vol.12, no.3, pp.117-119.

Stowell, F.A., 1985, Experience with soft systems methodology and data analysis, *Information Technology Training*, May, pp.48-50.

Waters, S.J., 1988, Managing information technology by objectives, *International CIS Journal, Command and Control, Communications and Informations Systems*, vol.2, no.1, pp.1-10.

Waters, S.J., 1989, SWOT analysis in IT projects, *International CIS Journal, Command and Control,Communications and Informations Systems*, Vol.3, no.1, pp.20-29.

Waters, S.J., & Bakehouse, G. & Davis, C. & Doyle, K., 1994, "Integrated clinical workstation: user requirements for a neurosciences directorate", NHS Executive, IM, Leeds.

Weinberg, G., 1975, "An Introduction to General Systems Thinking", John Wiley & Sons, New York.

44

A METHOD FOR ENTITY IDENTIFICATION

Jun-Kang Feng

University of Paisley
High Street, Paisley
Scotland PA1 2BE
United Kingdom

THE CURRENT SITUATION OF ENTITY IDENTIFICATION

Information systems are those software systems that are characterized as being data-intensive, transaction-oriented, with a substantial element of human-computer interaction (Loucopoulos and Zicari, 1992). Data modeling is a process of understanding the 'data aspect' of a domain of interest, for which an information system is developed, so it plays a vital role in the development of information systems. For example, SSADM takes three perspectives in analyzing the 'current environment', namely DFD, LDS (Logical Data structure) and Entity life history (Ashworth and Goodland, 1990; Weaver, 1993). SSADM places a great emphasis on data modeling, as Weaver put it, the Logical Data Model 'is possibly the most important and ultimately the most rigorous product of an entire SSADM project', and the Logical Data Model 'is a vehicle for analyzing the logical structure of an organization's information' (Weaver, 1993).

Data modeling is carried out by using data models. Data models at the human level, including the entity-relationship data model (Chen, 1976), the enhanced entity-relationship data model (Elmasri and Navathe, 1994), and the extended relational data model - RM/T (Date, 1995), are centered on the concept of *entity*. As for RM/T, it is the concept entity with which the original relational data model is extended.

A data modeling process is normally started with entity identification. For example, with SSADM version 4, when a LSD is developed, the first step is entity identification, followed by relationship identification and so on. Obviously, without entities identified, relationships have no ground to exist. With RM/T, the first step is to identify kernel entities. And with the entity-relationship data model, a modeling process starts with the identification of regular entities.

So, the concept entity is vital to information systems development and semantic data modeling, but its definition in the literature appears vague and general. It is taken as "a 'thing' in the real world with an independent existence" (Elmasri and Navathe, 1994); or 'a real world object which is distinctly identifiable' (Date, 1995); or 'any object or concept about which a system needs to hold information' (Weaver, 1993). These definitions often cause confusions among novices of information systems and databases, who feel that entities seem everywhere and difficult to know where to start and where to stop. Thus the above definitions

Critical Issues in Systems Theory and Practice, Edited by K. Ellis et al.
Plenum Press, New York, 1995

of entity can hardly serve as an adequate guideline. Moreover, neither information systems development methodologies nor database design methods provide systematic technique for entity identification. As such, entity identification poses as a difficult task, full of informality and uncertainty.

POSSIBLE CAUSES

Contemporary methods for information systems over-simplify the problem of entity identification and provide little guidance for practitioners. This is probably rooted to a set of fundamental ontological and epistemological assumptions.

Two types of ontological and epistemological assumptions are observed by Burrell and Morgan (1979) to yield two dimensions: a subjectivist-objectivist dimension and an order-conflict dimension. The ontology of the aforementioned appears of *realism* since an empirical reality made up of entities that is independent of its perceiver or observer is believed to exist. For example, Date suggests that we might agree that the world is made up of entities, and entities can be usefully grouped into entity types (Date, 1995) The epistemology falls in that of *positivism* in that the information systems developer gains knowledge about the domain of interest by observing the current environment, modeling it, and then logicalising it.

The objectivist paradigm has not been capable of providing powerful intellectual constructs for data modeling. Data modeling is an enterprise of understanding and formalizing some information handling aspect of human activities in the social world. Unlike the natural physical world, the social world is complex, ever-changing and a mix of objective and subjective phenomena. Realism and positivism are competent in tackling the physical world, but impotent in dealing with the problems of the social world.

A SUBJECTIVIST METHOD

If the ultimate purpose of an information system is to satisfy the information needs of all actors in an organization, rather than to formalize an actor-independent social realty as a frozen model and then to impose it upon all the actors, then all the objects they perceive and all the information about the objects they require should be captured and represented in the information system. The information model the information system is based upon will be necessarily an integrated one, rather than just a linear collation of fragmented individual views, but all individual views should be retained. Entity identification is to appreciate and integrate individuals' views of the domain of interest in terms of the objects they perceive. So, using Sowa's term (Sowa, 1984), entity is a concept, rather than objective object. Entity refers to certain referents - some perceived objects which may or may not exist in the real world.

The ontological assumption of this approach is that of *nominalism* in that the reality is not given, being made of entities. The 'current environment' is not a given existence in dependent of actors within it, but is socially constructed. It is the product of the human mind, so it cannot be modeled objectively. Objects that exist in such a current environment are merely something perceived by the actor within the environment in the process of and for interacting with other actors, who directly take part in the social process of reality construction. The epistemological assumption is *anti-positivism* reflecting the belief that modeling the current environment made up of entities is misguided and should be replaced my sense-making from the points of view of the actors within the domain of interest. The result of a process of entity identification is not a rigorous scientific discovery of the structure of the domain of interest governed by some laws, but an interpretative accounts of the actors' individual experiences.

Objects perceived by actors in an organizational environment seem to fall into the following categories:

1. Objects that are perceived as normally required for the operations of the *primary tasks* (Checkland, 1981) of an organization. They are stable as long as the nature of the domain of interest remains unchanged. For example, as for a personnel department, the following objects may be seen as necessary for its primary task 'personnel information management': *Employee, post, grade, performance*, and organizational structure such as *branch* and *department*. This category of objects may be perceived neutral by most of the actors within the domain of interest because primary tasks are determined by the very basic nature of the organization.
2. Objects that are physical media which people in the domain use to store and make use of information (e.g., information presentation and retrieval) about those mentioned in the above category. For example, in the personnel department, there could be *attendance sheet*, *monthly* or *weekly report* regarding employee's performance, *employees' index card*, and *master personnel file* in a computer.
3. Objects that are used for performing operations on those objects directly determined by the primary tasks. For example, there might be *rule* and *regulation* regarding personnel rewards, promotions and working conditions.
4. Objects that perform processes on the above objects. For example, *officer* in a personnel department who recruits staff, transfers staff, places staff, and maintains staff personnel information.

It can be observed that the objects in category 1 are the most primitive and objective ones. They are directly used as a type of passive material for the execution of the primary tasks and they are likely to be acknowledged by all the actors in the domain of interest. Furthermore, the category 1 objects probably serve as the elements of the objects in the rest of the categories. That is to say, more actor-dependent objects (i.e., more subjective) might be made up of category 1 objects or based upon them, in different ways of course, depending on particular individual actors. So the task of entity identification is twofold: To find category 1 objects, and to use them to form objects of other categories. The former is represented in a data model in terms of *entity* and/or *attribute* which should hold persistent data, the latter in terms of *views*, *joins* etc. of entities, so they do not hold persistent data in their own right.

All types of categories of objects perceived by actors can be identified in a *systemic* way. The idea is to use SSM (Soft Systems Methodology) (Checkland, 1981). First, a notional relevant system is created, which formalizes an individual actor's view; Second, the notional system is decomposed using SSM recursively to its primitive level; Third, raw data items that are required for the activities at the primitive level (they are termed 'task's) are identified; Fourth and finally, the raw data items are statistically analyzed and objects then should emerge. This method was elaborated in (Feng, 1993).

Following that method, in the rest of the paper, a formal system is presented for the identification of the category 1 objects.

The alphabet of the language is {A, B, ..., A1, A2,..., $\Rightarrow$, $\supseteq$};

Sentence = String of Words, $\Rightarrow$, String of Words;

String of Words = 'A' | 'B' | ... | 'A1' | 'A2' ..., String of Words, 'A' | 'B' | ... | 'A1' | 'A2' ...;

The deductive apparatus consists of the following rules:

1. $X \supseteq Y$ / $X \Rightarrow Y$;
2. $X \Rightarrow Y$ / $X \supseteq Y$;

3. $(X \Rightarrow Y, Y \Rightarrow Z) / X \Rightarrow Z$;
4. $(W \supseteq Z, X \Rightarrow Y) / XW \Rightarrow YZ$.

The above language and inference rules constitute a formal system. Now an interpretation is attributed to it. The meaning of String of Words is an object (simple or composite) perceived by an actor; the $\Rightarrow$ symbol stands for 'covers', meaning that *the object(s) on the left hand side is believed capable of holding all information the object(s) on the right hand side can;* and the $\supseteq$ symbol means 'is a subset of' as usual.

With this interpretation, the first inference rule says that if an object Y is a subset of object X (X and Y are not necessarily distinct), then X covers Y; The second rule means that if object X covers object Y, then Y is a subset of X; The third rule states that if object X covers object Y, and Y covers Z, then X covers Z; The four rule indicates that if object Z is a subset of W, and object X covers Y, then the combination of objects X and W covers the combination of objects Y and Z. With this interpretation, the above rules (hereafter called the Rules) can be re-written as the following semantic rules:

1. $X \supseteq Y \models X \Rightarrow Y$;
2. $X \Rightarrow Y \models X \supseteq Y$;
3. $X \Rightarrow Y, Y \Rightarrow Z \models X \Rightarrow Z$;
4. $W \supseteq Z, X \Rightarrow Y \models XW \Rightarrow YZ$;

Let F be a set of 'covering' relationships on set of objects U, and let X be a subset of U. Then X^+, the closure of X with respect to F is the set of objects A such that $X \Rightarrow A$ can be deduced from F by the Rules, and then F^+, the closure of F, is the set of 'covering' relationships that are logically implied by F.

Lemma 1: The Rules are sound. That is, if $X \Rightarrow Y$ is deduced by the Rules, then $X \Rightarrow Y$ is true.

Proof: To prove the soundness, we need to prove each of the rules is sound. Rule 1 is sound because any set has more than or the same number of objects its subset has, so it can hold all information regarding those objects in the subset. Rule 2 is sound because if a set can hold at least the same amount of information as another set can, then it contains at least the same objects as the other does. As for Rule 3, if X covers Y then X holds more than or equal amount of information as Y does, and if Y covers Z then Y holds more than or equal amount of information as Z does, so X holds more than or equal amount of information as Z does, that is to say, X covers Z. Rule 4 is also clearly sound. Because If Z is a subset of W then W holds more than or equal amount of information as Z does, and we know that X covers Y, so the combination of X and W holds more than or equal amount of information as the combination of Y and Z does.

Lemma 2: $X \Rightarrow Y$ follows from the Rules if and only if $Y \subseteq X^+$.

Proof: Let Y = A1...An for perceived objects A1,..., An, and assume $Y \subseteq X^+$. By definition of X^+, $X \Rightarrow Ai$ is implied by the Rules for all i. By Rule 3 and 4, $X \Rightarrow Y$ follows. Conversely, assume $X \Rightarrow Y$ follows from the Rules, then for each i, $X \Rightarrow Ai$ holds by Rule 1 and 4, so $Y \subseteq X^+$.

Lemma 3: The Rules are complete. That is, any 'covering' relationship between two objects (simple or composite) that holds on a set of given objects and their 'covering' relationships can be deduced by using the Rules. Another way of stating this completeness is that any 'covering' relationship that cannot be deduced by the Rules is not true.

Proof: Suppose $X \Rightarrow Y$ cannot be inferred by the Rules. Regard the given set of objects as composed of two parts: one consists of all objects included in X^+, the other the rest of the objects perceived. Obviously $X \subseteq X^+$; Because $X \Rightarrow Y$ cannot be inferred by the Rules, according to Lemma 2, $Y \not\subset X^+$; so $Y \not\subset X$, therefore $X \Rightarrow Y$ is not true. That is, the Rules are complete.

We can use the Rules to derive a *minimal cover* of F^+. A set of objects is called a minimal cover of F^+, if the right hand side of all sentences are single objects, no proper subset of the left hand side is redundant, and no sentence can be derived from the rest of the sentences by the Rules. The collection of the objects that appear on the left hand sides of the sentences in the minimal cover constitute a set of *elementary* objects of which all objects perceived by the individual actors are made up, so they should be defined as entities. All the rest can be defined as views, joins etc. of the entities.

SUMMARY

Entity identification is normally the staring point of data modeling, but poses as a difficult task, due to the underpinning realism and positivism and the lack of supporting formal systems. This paper presents a method based on an alternative subjectivist paradigm and a formal system, with which objects are identified from the viewpoints of individual actors within a domain and entities are selected by means of a formal procedure.

REFERENCES

Ashworth, C., and Goodland. M., 1990, "SSADM - A Practical Approach," McGraw-Hill, London.

Burrell, G., and Morgan, B., 1979, " Sociological Paradigms for Organisational Analysis," Heinemann, London.

Checkland, P., 1981, "Systems Thinking, Systems Practice," John Wiley & Sons, Chichester.

Chen, P.P., 1976, The entity relationship model — Toward a unified view of data, *TODS*, 1:1, March.

Date, C.J., 1995, "An Introduction to Database Systems," Sixth edition. Addison-Wesley Publishing Company, Reading.

Elmasri, R., and Navathe, S.B., 1994, "Fundamentals of Database Systems," Second edition, Benjamin/Cummings, Redwood City, California.

Feng, J.-K., 1993, Function-oriented data modelling, *in*: "Systems Science: Addressing Global Issues," F.Stowell, D.West, and J.Howell, eds., Plenum, London.

Loucopoulos, P., and Zicari. R., ed., 1992, "Conceptual Modeling, Databases, and CASE," John Wiley & Sons, New York.

Sowa, J.F., 1984, "Conceptual Structures," Addison-Wesley, Massachusetts.

Weaver, P.L., 1993, "Practical SSADM Version 4," Pitman, London.

45

'THE ETERNAL TRIANGLE'

Conception and Misconception of Systems Hierarchy in the Application of Information Systems

M. J. S. Harry

Management Centre
University of Bradford
Emm Lane
Bradford BD9 4JL
United Kingdom

INTRODUCTION

> "One might have supposed that the historians, who ascribe the actions of the masses to the will of one man, would have found it impossible to fit the flight of Napoleon's armies into their theory, considering that during this period of the campaign the French did all they could to bring about their own ruin . . . . But no! Mountains of volumes have been written by historians about this campaign and in all of them we find accounts of Napoleon's masterly arrangements and deeply considered plans; of the manoeuvres executed by the troops, of the strategy with which they were led and the military genius shown by the marshals" (Tolstoy, 1869).

As this quotation shows, debates about strategy applied to organizations are much older than modern management theory. These debates have not only been confined to what strategy is, but have also extended to questioning whether it exists at all, outside the heads of those who claim to practise it.

My reasons for using this particular quotation are not confined to establishing a historical context or to impressing my readers with literary erudition. Tolstoy's "...odd ideas about the powerlessness of individual human beings" (Duffy, 1972), are expounded in the specific context of war and the functioning of military hierarchy and strategy. But the chosen subject of his work is just one which we find in a wider context that includes the management of business and organizations.

The word strategy comes from the Ancient Greek for the leader of an army, or a general. The word tactics from a word meaning an officer. In Western European culture we therefore have had several millennia of conforming to the concept that the higher few 'strategically' lead the many via the 'tactical' machinations of the middle ranks.

The language and concepts of military hierarchy have not disappeared with the passage of global military conflict. Having long since entered the business management context, they have persisted despite affectations of consent and liberal pluralism in many schools of modern management thinking.

Critical Issues in Systems Theory and Practice, Edited by K. Ellis et al.
Plenum Press, New York, 1995

The 'Triangle' referred to in the title of this paper refers to a diagram, Figure 1., versions of which perpetuate the ancient military concepts of strategy, tactics and operations. The diagram is generic, and has frequently appeared in management texts for at least twenty years, with some version of it appearing as an appendage to explanations of hierarchy in works purporting to present a view of 'systems' in a management context. Such presentations usually link the supposed levels of strategy, tactics and operations to the organizational positions of those who are responsible for their implementation. For academic peace I quote no references, but they are available to the pedantic.

In checking my sources I note that Anthony (1965) is quoted as the origin of this diagram, but I strongly suspect that it is much earlier. Urwick (1943) could be quoted as origins of "...the hierarchy of authority and the principles on which they were based", as in Leavitt et al. (1973).

Such sources reveal the conceptual confusion the logical and the physical, or the 'whats' and the 'hows' of management hierarchy.

This paper aims to separate the logical and physical views of hierarchy and strategy, and to show that both liberal and authoritarian views of hierarchy are logically unsustainable and practically irrelevant to the design and implementation of information systems whose intention is to support the management of organizations.

THE LIMITATIONS OF PERSONAL OR TIME-SPAN HIERARCHY

The linking of the words strategy and tactics to their Ancient Greek origins is not just academic nicety. The physical binding of levels of manag*ing* with levels of manag*ers* persists. Thus just as strategy is for generals and tactics for officers, so versions of the Eternal Triangle persistently link strategy with such a description as 'senior executives', and tactics with 'middle management'.

In an information age this is easily shown to be nonsense, even if it was ever true. Consider a simple, hopefully fictitious, case. During one of the sessions of an international conference a presenter causes a scandal by departing from his subject and treating his hearers to an offensive, say racist, diatribe. Most of his hearers walk out and some stay to make a noisy protest. In less than 24 hours the scandal caused is loudly featured in the local and national media. When confronted with this situation, does the Vice-Chancellor of the

Figure 1. 'The eternal triangle.'

university concerned refuse to get involved, on the basis that his job is 'strategic' and this local difficulty is a firefighting, tactical/operational issue? In practice what happens on such occasions is that the highest levels of the management hierarchy intervene and take command. A press conference is called at which a bevy of 'senior executives', supposedly concerned with strategy, appear before press and television in an attempt to take control of day to day operations.

The example is fictitious but is regularly reflected in the life of public and private organizations and businesses. The work of senior executives, like the work of the middle management below them, is actually a complex mix of decision making with a wide range of time spans involved, whatever the 'Eternal Triangle' tells us.

The issue of time span then introduces a second way in which elaborations of the Eternal Triangle seek to confuse. Just as they link strategy and tactics physically to individual people in management positions, so the delineation of these management roles is frequently made by *quantitative* distinctions.

Time-span is a popular choice for this process. Strategy is presented as being concerned with future threats and opportunities, or adapting to a changing environment when facing the new millennium. Tactics is seen in terms of how we deal with a threatened walk-out by staff, or the closing of a parts supplier now.

Yet what are the real differences between criteria like those in a strategic planning 'SWOT' analysis and the same criteria being applied to the control of next 24 hours of operations? In our previous examples, 'future threats' could be tomorrow's: like that of a 'walk-out'. 'A changing environment' could be 'the closing of a parts supplier now'. The use of time span as a delineator of strategy and tactics effectively makes strategy no more than long-term tactics. From an information viewpoint this means that strategy defined in this way is inconsistent, as we shall now see.

THE INCONSISTENCY OF STRATEGY AS LONG-TERM TACTICS

'Keynsham, that's K.E.Y.N.S.H.A.M.', so said an advertisement on Radio Luxembourg in the 1950s. This pedestrian presentation was made by a Somerset entrepreneur called Horace Bachelor who promoted a forecasting method for making a fortune on the football pools. As an academic, he might have restructured his clichés in terms of 'optimizing decision making in a strategic context'.

The weakness of football pool forecasters is one they share with their academic equivalents, the strategic forecasters in business schools: they appear not to understand an argument first put forward by Popper (1957).

Rational long-term forecasting is only possible if we assume the existence of discoverable rules of history. These rules can only be used after they have been discovered. Once these rules have been discovered, they can now be used to anticipate the actions of others who use them to subvert the results of the actions which should result from the rules themselves. Hence discovered rules soon cease to be rules at all: any discoverable rule based forecasting has a limited shelf-life once people discover it.

Strategy as long term tactics may therefore work as good empirical judgement, like a view of future football results. What is less convincing is the possibility of long-term planning based on serious rationalist pretension. Statements like Greenley (1989): "Strategic management is concerned with the overall long-range direction of organizations", imply a confidence in futurology which would make us all football pool winners.

If strategy is to continue as a possible and useful management concept therefore, it must be separated from any implication of the Eternal Triangle which links it to either the positions of the people involved or the decision time-span.

STRATEGY, HOLISM AND CONCEPTUAL HIERARCHY

Away from the 'Eternal Triangle' we can find other recurring themes in writing on strategy and hierarchy. One in particular is the use of qualitative and logical criteria as a means of defining hierarchy and hierarchical levels. A recent example of this is Ben-Eli in Espejo & Harnden (1989). Referring to 'levels of management' he says, "The pertinent point is that approaching problems that are related to each such level requires a *different conceptual orientation*, a *different language*...", (my italics). One attempt at this is my own (Harry, 1994), but my aim in this paper is to promote the concept of qualitative and logical definition criteria, not a particular model.

If we apply these criteria to much writing on strategy, we frequently find a confusion of conceptual distinctions. Take the example (Christensen, 1973), "Strategy is the pattern of objectives, purposes or goals and major policies and plans for achieving those goals stated in such a way as to define what business the company is in or is to be in and what kind of company it is or is to be". Here we have normative statements like 'is to be', mixing in with straightforward operational concepts like 'objectives'; whilst reflection about 'policies' enables the production of 'plans'. Such definitions of strategy seem like a great trawling net that scoops up everything in the organization and its environment which may be relevant.

There is however one way in which a take-all definition like that of Christensen will work in the context of criteria like those of Ben-Eli. This is a simple recognition of the central systems concept of holism. If strategy is really about the *whole* organization and neither the property of one group of senior executives, nor just a word for long-term tactics; then the highest level which is called strategy is not the narrow exclusive top of a triangle, but some emergent property of the whole. If I were to coin a phrase, I might say the whole is more than the sum of its parts, but I think Aristotle has already done that.

This is where a trawling statement like that of Christensen contains the successful catch. Amongst the heterogeneous collection of 'plans', 'objectives' and other words which talk of *how* strategy is to be implemented is the real fish of *what* strategy *is*. 'What kind of company it is or is to be' is strategy; the rest is operations, tactics or long-term tactics.

INFORMATION, IGNORANCE AND PASSIVE CONTROL

The view of strategy thus developed has some very disturbing implications for the application of information systems and their component information technology. To review the conclusions so far:

- Strategy is not the sole responsibility or the possession of a particular management group.
- Strategy is not long-term tactics.
- Law-based forecasting is ephemeral.
- Strategy is a qualitative property emerging from the interaction of the whole of an organization.

To take each of these points in turn.

The belief that strategy is the possession of a particular group has led to the marketing of software designed to flatter managers who regard themselves as senior executives or similar. 'Executive Information Systems' would be an example of these. The actual aims of the promoters of such systems have often had more to do with the selling of hardware and software than those which benefit the whole organization. Indeed, such software reinforces the 'Eternal Triangle' view of strategy and strategic management as private property. The belief that strategy is *not* the possession of a particular management group would work towards wide availability of all software, without trying to define any concept suitability for particular management positions.

If strategy were merely long-term tactics, then the needs of strategists would be only quantitatively different from those of tacticians. Given the quantitative flexibility of software, it is hard to see why strategic information systems would be different from other forms of information system.

A belief in, or as likely a persuasion to, law-based forecasting has promoted two dangers to organizations. First, it has promoted professional view of the deliverers of information systems. Here I go with Shaw's statement, "...all professions are conspiracies against the laity". Secondly, it implies that automated information systems can replace human judgement. Since we have shown that human judgement can subvert attempts at historicist law making, the working of automated information systems will continue to require human management.

Strategy as a qualitative property emerging from the whole of an organization then leads to more positive conclusions. Thus the following criteria can be established on the basis of the three previous points:

- Information technology-based information systems should be available to all within an organization.
- Emphasis in available software should redress the balance away from quantitative/logical applications such as spreadsheets and databases to qualitative audio/visual applications.
- All managers in an organization should be intellectually equipped to understand the operation of information technology, but the organizational decision-making procedures should actively enable human decisions to override information technology based analysis.
- Interactive response for all people within the organization should be encouraged by only a minimum of necessary standardization in equipment, software and procedures; and the maximization of communications.

Not I suspect, what any Ancient Greek general or modern senior executive would desire. But their concern is personal position; not strategy.

REFERENCES

Anthony, R.N., 1965, "Planning and Control Systems: A Framework for Analysis", Harvard University Press, Cambridge, Mass.

Christensen, C.R., 1973, "Business Policy: Text and Cases", Irwin, New York.

Duffy, C., 1972, "Borodino", Sphere Books, London.

Espejo, R., and Harnden, R., 1989, "The Viable Systems Model", John Wiley and Sons, Chichester.

Greenley, C., 1989, "Strategic Management", Prentice Hall, New York.

Harry, M.J.S., 1994, "Information Systems in Business", Pitman, London.

Leavitt, H.J., Dill, W.R., and Eyring, B.E. 1973, "The Organizational World", Harcourt Brace Jovanovich, New York.

Popper, K.R., 1957, "The Poverty of Historicism", Routledge & Kegan Paul, London.

Shaw, G.B., 1906, "The Doctors Dilema", London.

Tolstoy, L.N., 1869, "War and Peace", Moscow.

Urwick, L.F., 1943, "The Elements of Administration", Harper and Row, New York.

INFORMATION USE IN THE 'NEW' ORGANIZATION

The Organic Metaphor and Its Implications

Richard Kamm*

Faculty of Computer Studies and Mathematics
University of the West of England
Coldharbour Lane
Frenchay, Bristol. BS161QY
United Kingdom

INTRODUCTION

A common theme in much recent work on managerial practice is that, if organizations are to survive the challenges of contemporary economic, technological and social conditions, they need to take on a particular form. The well-structured, bureaucratic corporation, is criticised as rigid, dehumanising and incapable of adapting to the rapid changes of the business environment. The style of thinking that it embodies, mechanistic philosophy or scientific management, is held to be less appropriate than an approach which emphasises decentralization, autonomy and flexibility. Such principles are seen as resting in an organizational form variously characterised as 'new' (Drucker, 1988), 'knowledge-based' (Quinn, 1992), 'informated' (Zuboff, 1988), 'virtual' (Harrington, 1991) and even 'postmodern' (Bergquist, 1993). Beneath these labels lies a common perception of the adaptable, decentralised organizations which recapitulates Burns' (1971) idea of the organic type of structure. This, in turn, can be related to the wider organismic metaphor for social activity (Morgan, 1986).

If the organic organization is to be founded on a qualitatively different form of management, it is important that the implications of its underlying metaphor be developed, particularly with respect to the way in which information is managed and used. An organismic structure, less centralised and formally bureaucratic, is seen as the appropriate form for contemporary business, partly because of the wider economic and social pressures (Kast and Rosenzweig, 1985) but also because of the effects of information technology. The linkage of telecommunications and computing technology now allows the components of organizations to be geographically dispersed and communication between them to develop a much greater degree of complexity (Drucker, 1988; Scott Morton, 1991). What follows, then, is an outline of some of the issues that need to be faced in consideration of an organic

* E-mail rm-kamm@csm.uwe.ac.uk.

Critical Issues in Systems Theory and Practice, Edited by K. Ellis et al.
Plenum Press, New York, 1995

view of information itself and an evaluation of the extent to which the 'new' organization actually embodies an organic approach.

1. THE NATURE OF THE ORGANIC METAPHOR

The organic metaphor treats social activity as the product of evolution. In contrast to the mechanistic approach, with its emphasis on the benefits of rational design, organismic thinking is sceptical of any attempt to specify patterns of behaviour in a comprehensive blueprint for an organization. Organizations, indeed any form of social activity, cannot be analyzed purely within the framework of universal laws of behaviour because such behaviour is rooted in a wider and variable context of interaction, culture and values.

These ideas derive from a long tradition of social thought which prizes social integration and a sense of community above the rational redesign of social institutions. While it is argued below that some ideas concerning the nature of evolution can help to develop organizational analysis, the metaphor itself is not dependent on a direct importation of the principles of biology. For one thing, there is no complete consensus on the appropriate concepts for the study of flora and fauna; it makes just as logical to talk of mechanistic and organismic views in biology as it is in the social sciences (Levins and Lewontin, 1985; Hodgson, 1993). What is organic about the metaphor is its sense of the need for balanced, gradual development which allows a society to develop within the conventions of its own culture. It has its pessimistic side; the view that change needs to proceed from historical precedent can give the metaphor a conservative flavour. But it can also be more optimistic in its view that humans are not constrained in their behaviour by immutable laws; hence the use of organic ideas by the liberal critics of the dehumanising effects of modern industry who argued for the recognition of a social aspect of work and production (Durkheim, 1957; Mayo, 1949).

The primary emphasis in the organic metaphor is therefore on integration. Where mechanistic views tend to assume that society will inevitably cohere so long as its individual components are rationally designed, organismic thinking tends to treat moral integration as being vulnerable precisely when that kind of rational thinking is most prevalent. In contrast to arguments that self-interest, for example, should be seen as a universal constant around which human behaviour can be directed to the benefit of society, organic traditions would tend to see the danger of amorality leading to anomie and the erosion of social connections. Adaptability is important to the organicist, but it is assumed that only a well integrated society or organization can adapt; change will otherwise be in the direction o disintegration rather than adaptation.

If the 'new' organization is to be seen as organic, the ideas surrounding it need to be concerned with more than decentralised structures. Integration, as the foundation of organicism, needs to be addressed as well as the more obviously appealing priorities of flexibility and adaptability. Organizational analysis, similarly, tends to be organic when it stresses the importance of establishing shared cultural understanding as a prerequisite for technical or structural change. An organic view of the use of information would similarly reflect a concern that it reinforce integration and promote adaptation.

2. THE ORGANIC METAPHOR AND THE NATURE OF INFORMATION

As with other phenomena, an organic definition of organizational information can be most instructively explained in contrast to the mechanistic view of the same subject.

Information within the mechanistic metaphor is pictured as a physical substance and the problems associated with it are those of the organization of data rather than the understanding of meaning: it flows from one part of the organization to another in order to deliver data which carries with it a single meaning. The tendency to regard meaning as unproblematic makes it possible to define information use as decision-making of a clear, unambiguous kind. A quantitative value of information can be determined by measuring the extent to which it resolves uncertainty (Kamm, 1993).

Organicism, by contrast, would treat information as distinct from tangible entities. Its very intangibility requires that meaning be accepted as rooted in particular cultural assumptions, and therefore as something which will vary between social groups. The effective use of information in organizations requires a common understanding of the meanings which will be applied to items of data and also that those who use information be able to interpret it. 'Anomie' in information work would be constituted by data processing workers passing on letters and figures without regard to what they were saying; Zuboff's (1988) view of the effects of automation makes a similar point. If there is to be an organic view of organizational information, then understanding its meaning, the objective of using it, the type of decisions to be made with it, requires the establishment of consensus.

An instructive parallel may be drawn with law. A positivist view of laws is as commands which secure obedience through fear or the rational calculation by individuals of the costs and benefits of infringing them: just as the mechanistic view of information relies for its effects in organizations through the design of formal procedures. An organic view of law, on the other hand, rejects this stimulus-response view of the motivation of individuals and concentrates instead on the need for a normative framework if laws are to command assent and legitimacy (Hart, 1961). Similarly, organic views of information would emphasise the need for its importance and meaning to be commonly understood if it is to stimulate not just efficient decision-making but also the essentially organismic priorities of integration and adaptation.

3. THE ROLE OF INFORMATION AND THE ORGANIC METAPHOR

The extent to which the 'new' organization can be said to have adopted an organic approach to information depends on the use it makes of information to promote integration and adaptation. This implies that at least as great a value would need to be placed on qualitative as on quantitative information, which is important for both of those requirements, and that the priorities of systems development practice will have to be rethought to accommodate the structure of the organic organization.

That some attention is being given to the problem of integration in the new organization is suggested by the recent vogue for the codification in mission statements of corporate raisons d'tre. It is certainly accepted that recognising the need to stimulate a sense of belonging is an important task of management in the new organization, partly as a reaction against mechanistic views which assume that strategic objectives are well-bounded and require little explicit elaboration (Bergquist, 1993; Campbell and Tawady, 1990). In principle, then, qualitative information concerning strategic objectives and mission would be distributed to the members of organizations alongside the more conventional quantitative data which is necessary for detailed, internal administration. This would constitute what Katz and Kahn (1978) describe as 'systemic' information in contrast to the 'systematic' types of information on which conventional systems development has traditionally concentrated. The stress on developing appropriate, group-based cultures to allow the full exploitation of information technology (Delisi, 1990), however, rarely extends to modifying the type of information which is communicated.

The idea of the need for responsiveness is a more familiar aspect of discussion of the 'new' organization. Businesses are held to depend for long term survival on their ability to adapt to changing environments through a flexible focus on customer wants and quality production (Peters, 1987). Developing new products requires a shift in the priorities of organizations through the release of the creative energies of their members (Quinn, 1992). Information is clearly essential to achieving these objectives, and a great deal of attention is given to developing computer-based applications which support adaptability, largely in the area of Total Quality Management, just-in-time production and the provision of managerial information. These developments are usually placed in the context of devolved, decentralised patterns of work organization which reflect the organic metaphor's traditional enthusiasm for autonomous, group-based production.

This view of adaptation, however, is not specifically organismic. Statistical quality control, for example, is perfectly compatible with mechanistic, centralised bureaucracies. Indeed, it was developed in businesses of precisely that type during the 1920s, so that the rise of Total Quality Management since World War II can just as easily be interpreted as a return to mechanistic basics (Beniger, 1986). It may be argued that sensitivity to environmental change is a quality of living systems, but that it contributes little to the long term evolution of such organisms. Biologists working within an organismic framework tend to argue that evolution is more than the survival of forms which happen to be adapted to a particular niche. It is less a picture of passive organisms being selected for fitness by active environment than of species developing their own capacity to survive in different conditions (Levins and Lewontin, 1985).

Of the features of the 'new' organization which are held to enhance adaptability, therefore, it is the emphasis of creativity which is the more specifically evolutionary. It is perfectly arguable that the organic metaphor would emphasise an organizations's general ability to reflect on its condition and produce new ideas as a means of affecting, not simply responding to, its environment (Hodgson, 1993). In biological terms, such a 'special evolutionary capacity' (Campbell, 1985) enables organisms to adapt more easily to future change. It can only refer to the use of information because the changes to which it will have to adapt can only be anticipated, not directly experienced. Organizations can go further and directly consider the principles by which information should be organised in order to develop the capacity for long term survival.

Such principles are far from being fully developed but, as with organismic approaches to any institution, it can be surmised that they would question some of the familiar mechanistic priorities of information systems development. One such doubtful priority is that of avoidance of redundancy in data and data flows. If information is regarded as a tangible product, as in mechanistic approaches, then optimization requires that its use be made as economic as possible; data move through as few places as possible and held only in one place. If, on the other hand, an organicist view is taken, with meaning being attributed to data in different ways, the idea of optimization is less appropriate. Within organicist assumptions, moreover, the desirability and possibility of limiting data flows become questionable. It is unlikely that current managerial orthodoxy is sufficiently self-confident to contemplate such a change in the practice of systems development.

The 'new' organization, then, is only partially organic in its use of information. Its attention to organizational culture suggests an organismic feel for the problematic nature of integration but its use of information to promote adaptability is so restricted as to drag it back towards scientific management. An emphasis on statistical measurement as the most meaningful type of information, even within decentralised structures, may reinforce Katz and Kahn's (1978) view that to treat anything intangible as a tangible object leads to the reintroduction of mechanistic assumptions.

4. INFORMATION OWNERSHIP AND THE ORGANIC ORGANIZATION

The disaggregated structure of the organic organization suggests that information use should be similarly decentralised: autonomous groups developing their own information requirements rather than having them determined with reference to a single corporate standard. On the surface this raises few technical problems. Decentralised information use and organization have, arguably, been reinforced by the development of communications networks and the spread of the microcomputer (Scott Morton, 1991). Control of information, however, is affected by more than the available technology, incorporating questions of security, access and ownership. Whatever its faults, the mechanistic metaphor has produced a well-understood structure for resolving these problems; an organization owns its information in the same way that it owns its physical resources and therefore regulates access to it according to a 'need to know' principle (Kamm, 1993). The organic approach, by contrast, does not mesh so neatly with conventional systems development practice and instead raises ambiguities which arise from its more complex definition of the nature of information.

The essential point here is the organic metaphor's suspicion of the benefits of formal codification. Just as an organic view of information finds it hard to treat it as a tangible flow with fixed meaning, so an evolving organization would tend to develop information use in ways which cannot be formally regulated. Unlike a bureaucratic view of organizations, it would not be clear that information was purely the property of the corporation, and could as easily be seen as residing with those individuals who happen to be using it at any one time. Any rules regarding access to information therefore become problematic. Official or corporate secrecy would be difficult to preserve if autonomous groups were able to develop the practice of information use according to their own priorities. There is no guarantee, however, that the result would be an extension of privacy since that, too, may rely on formal regulations which may or may not be observed within a more decentralised structure.

There are some indications that the flexible, customer-oriented organization has just such a irregular view of information access. One use being made of the capacity of Executive Information Systems to 'drill down' to lower levels of detail is to stimulate the development of a particular corporate culture, encouraging personal responsibility in the knowledge that lapses may be seen by senior management; in effect a means of producing the integration which is fundamental to the organic organization (Rodgers, 1990). Developing an adaptable focus on customer wants has, for some American companies, led to the development of vast databases which draw on information about individuals from a variety of sources; a use of information which has been obstructed in continental Europe by the kind of formal privacy legislation characteristic of a more mechanistic view of government (*Computing*, 5/8/93). While the organic metaphor has often signified a more human-focused or holistic view of organizational activity, there is little in it that necessarily restricts the accumulation of data by those with the power and resources to do so.

This vision of the unregulated use of information to extend the gaze of the powerful resembles the view of the total institution which, as explained by Foucault (1977), has often been presented as essentially mechanistic; the marriage of utilitarian philosophy with modern technology. However, the archetypal total institution, as in the Victorian prison, was as much the product of organic as mechanistic ideas. Ignatieff (1978) notes that the political impetus of penal reform was provided less by a mechanistic stress on rational design than from a desire to reform both individuals and society by stimulating desirable patterns of thought; a cultural emphasis which was vividly justified on the grounds that "There are cords of love as well as fetters of iron" (Ignatieff, 1978). Such a view was at odds with the fundamentals of mechanistic philosophy but had greater effects on the total institution in practice.

The emphasis of the 'new organization' on integration and adaptability is therefore consistent with the patterns of information control suggested by the organic metaphor. The difficulty of such a view of information use is the way in which it can move towards either a manipulative view of organizational culture (the selection of appropriate forms of thinking begging the question of what happens to those who do not share them) or a cavalier attitude towards the privacy of individuals. For the individualism of the mechanistic metaphor is substituted a more collective view of the objectives of information use which, for all its advantages, runs the risk of subordinating the interests of the individual to those of the group.

CONCLUSION

Although the organic metaphor provides a more rounded view of both the nature of information and of its role, this insight is in danger of being obscured by the more instrumental priorities of the 'new' organization. The challenge which the organic metaphor offers to the conventional organization of information is not reflected in the stated advantages of decentralised structures. Those features of the organic metaphor which are present in current developments of information use, a looser regulation of access and an emphasis on cultural consensus, may tend to produce a form of cultural manipulation. Ironically, such a development may be reinforced by a more mechanistic view of the self-evident nature of the meaning of quantitative data which would produce not decentralization but the a form of centralization which is invisible as opposed to the 'visible hand' (Chandler, 1977) of the bureaucratic corporation.

REFERENCES

Bergquist, W., 1993, "The Postmodern Organization," Jossey Bass, San Francisco.

Beniger, J., 1986, "The Control Revolution," Harvard University Press, Cambridge MA.

Burns, T., 1971, Mechanistic and organismic structures, *in*: "Organizational Theory," D. Pugh, ed., Penguin, Harmondsworth.

Campbell, A., and Tawady, K., 1990, "Mission and Business Philosophy," Heinemann, London.

Campbell, J.H., 1985, An organizational interpretation of evolution, *in*: "Evolution at the Crossroads: the New Biology and the New Philosophy of Science, D.J. Depew and B.H. Weber, eds., MIT Press, Cambridge MA.

Chandler, A.D., 1977, "The Visible Hand: the Managerial Revolution in American Business," Harvard University Press, Cambridge, MA.

Delisi, P.S., 1990, Lessons from the steel axe: culture, technology and organizational change, *Sloan Mgt. Rev.* 32:1:83-93.

Drucker, P., 1988, The coming of the new organization, *Harvard Bus. Rev.* 66:1:45-53.

Durkheim, E., 1957, "Professional Ethics and Civic Morals," Routledge and Kegan Paul, London.

Foucault, M., 1977, "Discipline and Punish," Allen Lane, London.

Harrington, J., 1991, "Organizational Structure and Information Technology," Prentice Hall, Hemel Hempstead.

Hart, H.L.A., 1961, "The Concept of Law," Oxford University Press, Oxford.

Hodgson, G.M., 1993, "Economics and Evolution," Polity Press, Cambridge.

Ignatieff, M., 1978, "A Just Measure of Pain," Macmillan, London.

Kamm, R.M., 1993, Metaphorical thinking and information systems: the example of the mechanistic metaphor, *in*: "Systems Science," F. Stowell, D.West and J.Howell, eds., Plenum Press, New York.

Kast, F.E., and Rosenzweig J.E., 1985 "Organization and Management," 4th ed., McGraw-Hill, New York.

Katz, D., and Kahn, R.L., 1978, "The Social Psychology of Organizations," 2nd ed., John Wiley & Sons, New York.
Levins, R. and Lewontin, R., 1985, "The Dialectical Biologist," Harvard University Press, Cambridge MA.
Mayo, E., 1949, "The Social Problems of an Industrial Civilization," Routledge and Kegan Paul, London.
Morgan, G., 1986, "Images of Organization," Sage, San Francisco.
Peters, T., 1987, "Thriving on Chaos," Macmillan, New York.
Rodgers, T.J., 1990, No excuses management, *Harvard Bus. Rev.* 68:4:84-98.
Quinn, J.B., 1992, "Intelligent Enterprise: a Knowledge and Service Based Paradigm for Industry," Free Press, New York.
Scott Morton, M.S., 1991, "The Corporation of the 1990s," Oxford University Press, New York.
Zuboff, S., 1988, "In the Age of the Smart Machine," Heinemann, Oxford.

47

INTERPRETATIVE DATA MODELS AND RELATIONAL DATABASE SCHEMA

Toward Some First-Cut Rules for Translation

P. J. Lewis

The Management School
Lancaster University
Lancaster
United Kingdom

INTRODUCTION

Previous publications concerning the use of soft systems methodology (SSM) in the field of information systems development (Lewis 1992, 1993(a), 1993(b), 1993(c), 1993(d), 1994) have argued the case for developing system data models. These are a form of interpretative data model that shows the cognitive categories, and the associations between them, relevant to a SSM definition of a system and allow the richness and complexity of problem situations to be captured to a greater extent than is possible by using database design techniques, such as Entity Relationship modelling, whose focus, even in the earliest stages of development, is upon the creation of practical data storage schema.

An interpretative data analysis can be useful in any use of SSM, establishing a shared vocabulary for discussing the problem situation and relevant human activity systems. Where, however, the problem situation concerns the creation of information systems then it may also be of use in the later stages of development, especially in the design of data storage schema. The suggestion (Lewis 1993(d)) that interpretative data models could furnish the starting point for identifying entities or objects and provide a context for decisions about what is, or is not, relevant has been supported by a study in the municipality of Jeddah, Saudi Arabia. In this the use of SSM and the identification of relevant cognitive categories proved workable and valuable in the design of a complex database. We are now investigating whether it is possible to go further than this, and use the associations of the interpretative data model to guide the detailed design of data storage schema. This is not trivial because of differences between the two types of model and because interpretative data analysis allows for some forms of association between categories that have no direct equivalents in the most common forms of database.

This paper discusses some of this work, especially the definition of 'translation rules' of a kind that may be incorporated in software and produce an initial 'first-cut' database design from an interpretative data model. If such rules can be identified then this would not

Critical Issues in Systems Theory and Practice, Edited by K. Ellis et al.
Plenum Press, New York, 1995

only ensure consistency between the business analysis that identifies the need for an information system and the design of the same, but also reduce the need for scarce database design skills. The 'novice errors' discussed by Batra & Antony (1994), for example, are often the result of inadequate conceptualisation and poor understanding of what is being modelled rather than incompetence, and be less likely to arise if a preliminary interpretative analysis had been used.

ACCEPTED CONSTRAINTS

In attempting to define such rules we have taken as constraints that any such rules should lead to a transformation that is repeatable (the rules should always produce from a given interpretative data model the same data storage model), operator independent (the routine should produce the same results regardless of who applies the rules) and be machine processable so that all or some of the rules may be applied through computer software. All three constraints mean that there can be no reliance upon understanding the rich meanings of the interpretative data model and any transformation must be based upon the explicit structuring of the model itself. We also recognise that, because of the different interactions between data that are allowed by different forms of database; a different set of rules is required if, for example, it is intended that a relational database rather than an object-oriented database is to be used to implement a new information system. Separate sets of rules will therefore be needed for the move to different forms of data storage structure, each set taking account of what data interactions are possible in the target database.

In the brief confines of this paper we discuss some aspects of the transformation to a relational database schema, showing the relational structures that arise from different forms of association between categories in an interpretative data model.

INTRODUCTION OF TIME CONSIDERATIONS

It is important to be clear that an interpretative data model is not an alternative to conventional forms of data modelling done using the notations of entities and relationships, objects or relations and that the notation introduced in Lewis (1994) is not a new variant of semantic data modelling. In interpretative data analysis one is doing more than using different diagramming symbols and introducing new ways of relating one thing to another: one is attempting to model something entirely different.

Conventional data modelling aims to reflect an objective reality (or at very least, a view of such a reality) and to show 'facts' about the problem situation. This is exquisitely captured in claims that all information systems possess an underlying, generic data structure that changes little over time (Cutts, 1987, p. 23; Ashworth & Goodland, 1990, p. 10). The reason for the modelling is therefore always to 'map out' the situation so as to provide the best possible means of storing data to meet known processing requirements. In contrast, interpretative data modelling never claims to reflect more than a particular appreciation of what might be relevant and worthy of discussion. What are shown are the cognitive categories and associations that are meaningful in respect to a particular notional system created in the course of using SSM. No claim is made either in the soft systems methodology or the interpretative data analysis, that the system investigated is either 'what-is' or 'what-should-be'.

The different nature of the two types of modelling has some quite important consequences, not least of which is that if we wish to use an interpretative data model to guide data storage decisions we are forced to introduce considerations of time; we must re-orient ourselves to thinking about the storage of historical data rather than definitions, and to recording behaviour over time rather than describing the potential for behaviour. For example, we might in the course of work in the construction industry define within an interpretative data model that the meaning of 'work-team' is a number of 'employees' led by a 'team leader'. A data storage design however would need to be capable of recording the membership of work-teams both now and in the past. This need to introduce a consideration of time will be apparent in each of the following sections, in which we discuss three elements of an interpretative data model and their equivalents in a relational database schema.

COGNITIVE CATEGORIES

In an interpretative data model the cognitive categories are the basic elements of a vocabulary for discussing the problem situation. One might say, crudely, that the categories are those things about which discussion is centred and about which one must attach some meaning if one is to understand a given system definition. When we turn to thinking about data storage then it is about the cognitive categories that we would want to store data; for example, if the category 'employee' has emerged as important in all the SSM modelling then it is likely that in the real-world situation there will be a need to record details of particular real-world employees.

The simplest, perhaps most intuitively obvious 'translation' rule that we might frame is therefore that: 'If there exists in the interpretative data model an identified cognitive category, C1, there will be in the data storage design a relation, R1, together with some attribute, R1-id that acts as a primary key for that relation and uniquely identifies individual tuples of the relation.'

The first step in the transformation from interpretative data model to data storage design is therefore to create a relation for each cognitive category and to introduce as an attribute for that relation a primary key. Some of the relations so created may though disappear or be amended as further rules are applied.

ACTION ASSOCIATIONS

In order to define the meaning of a cognitive category precisely it may be necessary to detail the actions which one expects examples of that category to take. For example, in building system definitions relevant to the construction industry we may define the meaning of 'work team' to be rather more than simply a group of employees led by a team leader. It may include that such work teams carry out particular job specifications. This would be recorded in a data dictionary definition of WORK-TEAM and by showing the potential for fulfilment of particular tasks by means of an action association, as in Figure 1.

When we move to a storage design we need to allow for time considerations, focusing upon what has happened rather than what might happen, and allowing for records of the same work-team carrying out many different job-specifications over time. The need for

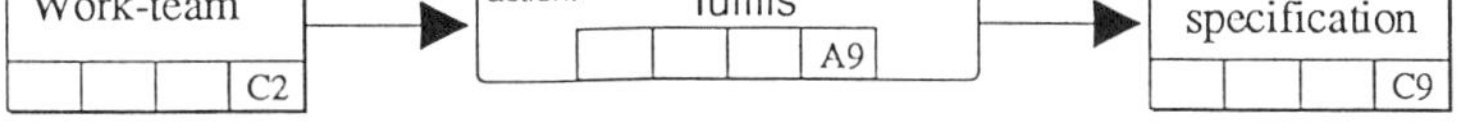

Figure 1. Example of an action association.

relations capable of storing historical data is met by the rule: 'If there exists in the interpretative data model two categories, C1 and C2, and an action association, Ax, between C1 and C2, then there will be in the data storage design: a relation, R1, having some attribute, R1-id, that acts as a primary key for that relation; a relation, R2, having some attribute, R2-id, that acts as a primary key for that relation; and a relation R3 that records the event of the action Ax having been taken, with the primary key of R3 being a concatenation of R1-id, R2-id and the date/time of the event'. This produces in response to the situation of Figure 1 the relations: WORK-TEAM {Team-id, _, _ }, JOB-SPECIFICATION { Job-id, _, _ }, JOB-FULFILMENT { Job-id, Team-id, datetime of fulfilment, _, _ }

Note that in the above example the date/time stamping may be unnecessary but in others it would not and the rules must always cater for the most general case. The redundancy could be removed as part of later refinement of the schema. Also note that action associations need not be binary and can involve more than two categories; one might for example define that work-teams fulfil job-specifications using materials. This could dealt in a similar way to above, but in a binary fashion. For example, for an action association involving three categories can be treated as if it consisted of two separate binary associations, creating two additional relations with each with the key of the other involved category included as a non-key attribute.

SPECIALISATION ASSOCIATIONS

The rules described so far may seem fairly obvious. This is because the relational model provides suitable equivalents to categories and action associations. This is not the case though when we turn to the specialisation associations that can exist in an interpretative data model.

In interpretative data analysis one may use specialisation associations to define one category as a specialised sub-category of another. For example, it may seem relevant to distinguish between those employees who are permanent staff and those who are on temporary contracts, even though both types of employee have, for the most part, similar characteristics. Figure 2 for example defines that the categories of FULL-EMPLOYEE and CONTRACT-EMPLOYEE are sub-types of the category EMPLOYEE whilst allowing for finer distinctions about what each may, by definition, do or not do. Only full-time employees, according to Figure 2, may contribute to a pension.

A relational schema does not allow real equivalents for such specialisation associations, and however we make the transformation to a relational schema some of the meaning of Figure 2 will be lost or poorly captured. In order to see this it is instructive to examine alternative ways of making the transformation.

We might create relations corresponding to only the sub-categories or only the super-category. For the latter option we could frame a rule such as: 'If there exists in the

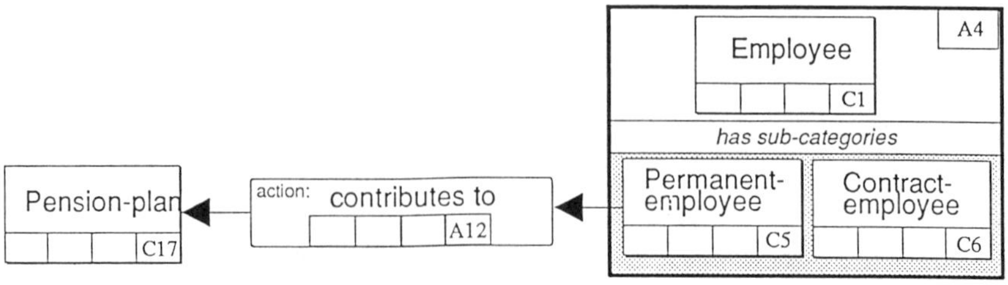

Figure 2. Example of specialisation association.

interpretative data model an identified cognitive category, C1, with a specialisation association such that cognitive category, C2, is a sub-type of C1, then there will be in the data storage design a relation, R1, together with some attribute, R1-id, that acts as a primary key for that relation, and R1 shall have as a non-key attribute a type-indicator with 'C2' being a valid value for that type-indicator.' Under this option the relational equivalent of Figure 2 would be: EMPLOYEE {Employee-id, Employee-type, _, _}.

In particular circumstances this might be a viable transformation but since we know nothing of the real-world uses to which the schema will be put we must judge it as unsatisfactory. It provides no relations corresponding to the sub-types, so that the existence of sub-types and what they are is now hidden in the data rather than explicit in the structure of the schema. We can record that a particular employee is a full-time or contract employee by recording appropriate values in the attribute 'employee-type' but we can only record the current status of the employee; if a contract employee became full-time we could change the type-indicator but should lose all record of the fact that they were ever a contract employee. And if there are different facts to be recorded about the two types of employee we will be forced to allocate storage space for all of these in the 'employee' relation and a large number of null values may result.

The second possibility is that: 'If there exists in the interpretative data model an identified cognitive category, C1, with a specialisation association such that cognitive category, C2, is a sub-type of C1, then there will be in the data storage design a relation, R1, together with some attribute, R1-id that acts as a primary key for that relation, and a relation R2 that has as its primary key the key R1-id.' Following this rule the relational equivalent of Figure 2 would be: EMPLOYEE {Employee-id, _, _ }, PERMANENT-EMPLOYEE {Employee-id, _, _}, CONTRACT-EMPLOYEE {Employee-id, _, _}

This is the relational expression of a generalisation hierarchy (or IS-A relationships) recommended by standard texts such as Kroenke (1992) or Batini et al (1992). It is certainly better since it allows the different facts about contract and full-time staff to be recorded in separate relations while the common data is stored in the employee relation. However, it still prevents the recording of changes in employment status over time. This can be rectified by amending the rule to: 'If there exists in the interpretative data model an identified cognitive category, C1, with a specialisation association such that cognitive category, C2, is a sub-type of C1, then there will be in the data storage design a relation, R1, together with some attribute, R1-id, that acts as a primary key for that relation, and a relation, R2, having as its primary key the compound key consisting of the primary key of R1 concatenated with the time /date of assignment.' This provides the relational equivalent of:

EMPLOYEE {Employee-id, _, _ }, PERMANENT-EMPLOYEE {Employee-id, Timedate-status -commences, _, _}, CONTRACT-EMPLOYEE {Employee-id, Timedate-status -commences, _, _}

This allows discrete records to be kept even if the employment status of an employee changes and different attributes to be held for each sub-category. Even so, we are left with complex keys for the relations and, in practice, we will need to set up a number of alternative keys to access the data; we would not want for example to quote the date and time of assignment of a full-time employee every time we sought details of that employee. And, as is true for the other alternatives also, constraints such as that only full-time employees may contribute to a pension must be imposed by programming rather than by the structure of the relations themselves.

CONCLUSION

The full version of this paper discusses the 'translation' rules for other elements of the interpretative data model and the difficulties that arise from combinations of associations. Even now though it is clear that, because when we move to data storage designs we must introduce considerations of time, the transformation from one kind of model to another is not a trivial matter. If it is to be assisted by software then we must define 'translation rules' that do not require any understanding of the model itself, and this requires rules that are not specific to particular circumstances. We believe that it is possible to define such rules but that the lack of a context for making design considerations inevitably leads to data storage designs that, while workable, are inferior to those that could be produced by an experienced human designer with an understanding of what an 'employee' is, or knowledge of the uses to which the data storage mechanism may be put.

This does not mean that the definition of translation rules is not worthwhile or that the result of using them is not useful. The rules discussed are machine processable, so allowing a 'first-cut' schema to be produced quickly, the resulting storage designs provide a good starting point for further refinement, and they force attention to be given to things that might otherwise be overlooked. While we might find the relations above to be unwieldy they do at least force recognition of different types of employee, and the interpretative analysis tells us that this was an important feature of the problem situation. Clearly, much work remains to be done in defining translation rules more fully, in incorporating those rules in software engines and in testing their use in real-world problem situations. The results so far at least suggest that this is work worth doing.

REFERENCES

Ashworth, C., and Goodland, M., 1990, "SSADM: A Practical Approach", McGraw-Hill, Maidenhead.

Batini, C., Ceri, S., and Navathe, S.B., 1992, "Conceptual Database Design", Benjamin/Cummings, Redwood, Ca.

Batra, D., and Antony, S.R., 1994, Novice errors in conceptual database design, *European Journal of Information Systems* 3(1):57-69.

Cutts, G., 1987, "SSADM: Structured Systems Analysis and Design Methodology", Paradigm, London.

Kroenke, D.M., 1992, "Database Processing; Fundamentals, Design, Implementation", (4th Ed.), Macmillan, New York.

Lewis, P.J., 1992, The feasibility and desirability of a closer linking of SSM with data-focused information systems development, *Systemist* 14(3):168-179.

Lewis, P.J., 1993(a), Linking soft systems methodology with data-focused information systems development, *Journal of Information Systems* 3(3):169-186.

Lewis, P.J., 1993(b), Towards an interpretative form of data analysis for the soft systems methodology. *In:* F.A. Stowell, D. West and J.G. Howell (Eds.), "Systems Science: Addressing Global Issues", Plenum, New York, 1993, pp:391-396.

Lewis, P.J., 1993(c), Culture and IS planning present new challenges for data analysis and modelling, *In:* N. Jayaratna, G. Paton, Y. Merali and F. Gregory (Eds.), "Theory, Use and Integrative Aspects of IS Methodologies", British Computer Society, pp. 137-150.

Lewis, P.J. 1993(d), Identifying cognitive categories: the basis for interpretative data analysis within soft systems methodology, *International Journal of Information Management* 13(5):373-386.

Lewis, P.J., 1994, "Information Systems Development: Systems Thinking in the field of Information Systems", Pitman, London.

48

A FRAMEWORK FOR THE DESIGN AND IMPLEMENTATION OF DISTRIBUTED INFORMATION SYSTEMS

Kim Merchant and Steve Clarke

The University of Luton
Faculty of Businss
Park Square
Luton, LU1 3JU
United Kingdom

INTRODUCTION

In the 1990s and beyond we have to meet the challenges associated with distributed information systems. In terms of technology, the midrange computer is being squeezed by personal computers (PCs) from below and mainframes from above, with PC local area networks becoming an alternative to mini computers. However, this move to micro computers requires the capacity to change away from an organisational model of direct control to an approach where organisational structure and culture are closely aligned with the enabling networking technologies.

Downsizing, and especially client/server technologies, are heralded as a perfect way to cut reliance on mainframe computers, but downsizing needs careful planning, a possible strategy being to transfer some of the user interface to PC. Open systems (computer platforms which conform to industry wide standards), enabling the sharing of data and application software on different systems, is seen as a way to platform independence.

Organisational Structure

The traditional hierarchical view of organisations exemplified by the pyramid model is often used as a guide to the implementation and use of computerised information systems. From this the argument is derived that transaction processing fits at the lower operational levels, with MIS and DSS applying to middle management, and these plus EIS to executives.

This systematic approach to information systems is being challenged by the new 'network organisations', where the hierarchical structure is seen to be of little relevance.

> "Computer networks have the potential to flatten traditional hierarchies"

> "..in a hierarchical company with rigid chains of command, a network is not going to do much to make the company more egalitarian." (Filipczak, 1994).

Critical Issues in Systems Theory and Practice, Edited by K. Ellis et al.
Plenum Press, New York, 1995

> "The structure of organisations will change; they will be more flexible and more responsive to change with fewer levels of management. The information network will [be] based on an open systems type of organisation. ...the forecast of Peter Drucker that most of the workforce will be knowledge workers will come true. Therefore, it is safe to conclude that the organisations with the most effective methods of exploiting new knowledge will be the businesses that will dominate the 21st century." (Burrows, 1994).

> "The network organisation overcomes complexities in its operation and exploits technology by organising itself as a set of sub-units networked together in a form that is very different from the traditional vertical hierarchy or even the matrix organisation. New technology is often seen as an enabler and driver of the networked organisation." (Powell, 1992).

Peter Drucker (1988) argues that the emerging form of organisational structure now evident (called a network structure) heralds a change from the traditional 'command and control' organisation to an 'information based' organisation, or as Zuboff (1988) states, the 'informated organisation'. This new structure relies on flexible technology to manage information flows horizontally across cross-functional boundaries as these fluid project teams form. Forms of technology could include groupware, strategic resources planning (such as Manufacturing Resources Planning), telecommunications and computer networks, all based on an open systems approach. Networking is probably the most commonly used and the most flexible.

The critical success factors of networking strategies can be summarised as platform independence, the ability to transfer data, user acceptability and skills development, capacity management and the control of risk. The key to success is to manage the implementation of steady increments in technology. This requires integrated architectures, quality management and standards setting.

The application portfolio has to be planned. Network systems give a different set of application compatibility problems.

Where downsizing is concerned, it is usually advisable to concentrate on support or turnaround applications . If networks affect so called "factory" applications (those essential to the success of the organisation), the risk must be carefully appraised.

In analysing network strategies it is important to recognise that the network spans the whole organisation: it is a support activity to the value chain (Porter, 1993), and therefore has to be chosen to enhance the processes essential to the organisation's success.

A way of dealing with this is the soft systems approach, which takes a systemic view of the organisation.

Much research is being conducted and concern being expressed about the 'systematic' nature of computing developments, and networking is not immune from criticism in this respect.

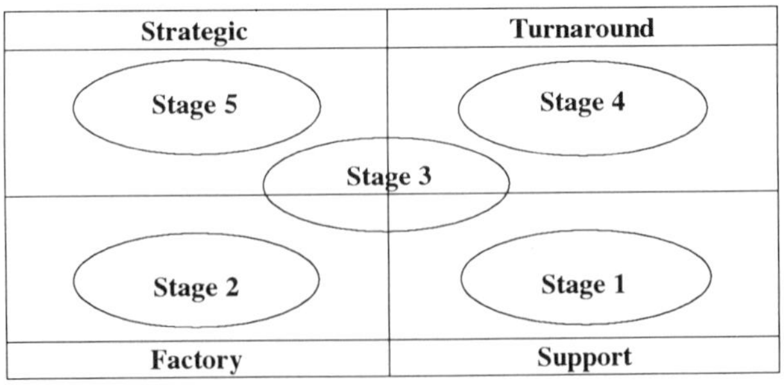

Figure 1. Application Portfolio Model showing the evolution of Information Systems planning. (Adapted from the McFarlane Matrix in "Strategic Planning for Information Systems," Ward, Griffiths, and Whitmore, 1993, John Wiley and Sons, Chichester.)

> "Focusing on one aspect or even several aspects of a situation is unsystemic, and at best systematic. The systematic nature of IT clashes with the systemic nature of IS .. coping with this tension between systematic and systemic natures is a challenge which has to be taken up by the IS profession" (Angell & Smithson,1991).

We need to define the system, not develop technology. We are not developing machines, but human activity systems. This calls for softer approaches.

USER INVOLVEMENT

To address user involvement, the first problem lies in defining the user community, which extends much further than just the computer network. All user populations must be well informed and fully involved in the process.

Critical to an understanding of how user involvement can be expected to improved system is the organisational climate. The organisational structure at The University of Luton, for example, can be categorised as a professional adhocracy, made up of organic teams with liaison groups impacting progress in a relatively unstable environment. In such a structure, formal IS which regulate and control are less important than systems which allow information and decision processes to flow flexibly and informally, a situation which points to networks and user models.

The organisational culture is a process culture. The processes are hard to change, so what is needed is the optimisation of sub-processes and the linkages between them. Again, a very user-dependent environment.

In terms of networking, The University of Luton is at stage 1 of the application portfolio model, in the sense that the network applications both current and planned are to support the organisational strategies, but are not essential to the success of the university. Strategies are needed to move toward stages 4 and 5, networking applications which will be of future strategic importance, or critical to future success. At present all such applications are designed to run on Digital VAX mini computers, but strategies to distribute this information are in the process of being formulated.

In the past the development of information systems at The University of Luton has followed a traditional functional approach of define problem / specify / analyse / develop solution / implement solution. Consequently, information systems strategies are operationalised as problem solving systems, for which the technology is known, the problem defined, and a solution can be found.

This is not to imply that the work done has not been beneficial, since it undoubtedly has. More that the approach has been too restrictive. Future developments are now beginning to be addressed less as problems to be solved, and more as the implementation of information systems strategies.

To an extent a major problem with the traditional, 'hard' approach is its apparent success. Networking at the university has been used to link together facilities which have always been distributed, consisting mostly of word processing and file transfer. But this is not a networking strategy to support information systems, for this we need to consider the information that can be downsized from the central to the networked systems. An example of this is the Higher Education Management Information System (HEMIS), which runs on the central VAX mini computer. There is a clear intent to give wider access to this system, in support of activities like timetabling. To do this via terminal access would be a retrograde step without first considering a networked solution.

This is just one example of how information systems and networks can be used to provide linkages to enhance value chain activities.

The key is to take a more holistic, systemic approach. Soft methods could be used to define areas for improvement in the systems. An analysis using Soft Systems Methodology (SSM),(Checkland, 1981) for example, would enable the university to develop a clearer picture of where to apply networking. SSM forces a focus on the processes of the organisation rather than the functional areas, so the information systems and networking strategies could be targeted at process improvement.

Luton University Networking - A Strategic Proposal

> "There are .. hundreds of organisations which have gained competitive advantage through the use of information systems and technology. Yet in the majority of cases it is almost impossible to state that the competitive advantage actually arose through the application of traditional planning approaches." (Ward, Griffiths & Whitmore, 1993)

Strategic planning is an appealing technique, but has severe limitations when applied in areas impacted by dramatic technological advancement. The technological changes which form the enabling technology for networking are by their nature incremental, so the strategies to make use of them need to be incremental strategies, contained in a loose framework which allows incorporation of the incremental technological changes. To succeed in this, the university must arrive at a view of the emerging technologies over the next five to ten years, then develop the networking strategy accordingly.

Networking strategies also need to support the overall strategic direction of the university. To achieve this it is essential to adopt a systemic, process oriented approach. The starting point for this could be the application of soft systems methodology, to give a holistic view of information systems requirements. It is not necessary for this to cover the whole university, but initially the methodology could be piloted on a major process area such as the control of the modular scheme. This approach would identify the information systems areas where improvement could be made to processes in respect of all forms of information throughout the university.

Following this, using an accepted strategic development methodology, information systems strategies could be developed to support the organisational strategies. Then information technology and networking strategies could be derived.

Strategic planning can be applied to network strategic planning, paying particular attention to issues which are especially relevant to networking, such as environmental scanning, internal capabilities and requirements, planning and allocating resources and organisational structure.

Within the networking strategic plan, risk must be carefully assessed and managed. Where the plan involves using networks to link already distributed systems, the risk is low, and the plan needs to look at emerging technologies, training and compatibility issues. Where downsizing is involved, the risk is higher, and more care must be taken.

Care must also be taken to assess the impact of any changes caused by networking on the organisational structure of the university. Since organisational culture is slow to change, the approach must be to manage any change within existing culture.

CONCLUSION

As organisations adopt the emerging network structure, open access to information will become critical to survival in a volatile, rapidly changing business environment. Information systems need to be flexible and should promote collaboration by making it easier for people to communicate horizontally between business functions as well as between global strategic business units. Thus distributed information systems will become of increasing strategic importance to these organisations.

REFERENCES

Angell, I., & Smithson, S.,1991, "Information Systems Management: Opportunity and Risk," Macmillan, Oxford.

Burrows, Brian, 1994, The power of information: Developing the knowledge based organisation, *Long Range Planning,* 27:1.

Checkland, P.B., 1981, "Systems Thinking: Systems Practice," John Wiley & Sons, Chichester.

Connors, Lynne, 1994, Empowering staff on PC LANs may not reap expected reward, *Computing Canada,* 20:1.

Drucker, P., 1988, The Coming of the New Organisation, *Harvard Business Review,* Jan - Feb 1988.

Filipczak, Bob, 1994, The ripple effect of computer networking, *Training Journal,* 31:3.

Jordan, E., 1994,. Information Strategy in Adhocratic Businesses, *Hong Kong Polytechnic Working Papers,* April 1994.

Porter, M.E., 1985 "Competitive Advantage:Creating and Sustaining Superior Performance," Free Press, New York.

Powell, Philip., 1992, Beyond Networking: The Rise of the Nebulous Organisation, *European Management Journal,* 10:3.

Ward, J., Griffiths, P., & Whitmore, P., 1993, "Strategic Planning for Information Systems," John Wiley & Sons, Chichester.

Zuboff, S., 1988, "The Age of the Smart Machine", Basic Books, New York.

INFORMATION SYSTEMS

Not Artifacts but Models

Graham Paton

The Open University
4 Portwall Lane
Bristol BS17 4TD
United Kingdom

THE AIM: A RECONSTRUCTION OF 'INFORMATION SYSTEMS'

I wonder who first coined the phrase "information system". At the time it probably seemed a very useful way of focusing thought on a particular type of system. Like all labels in language, it will have achieved common usage because it was useful. But perhaps we've now exploited it too much and in so doing manoeuvred ourselves into a semantic cul-de-sac. In this short paper I shall try to show that the phrase has a dangerously harmful effect in its current use and suggest how we might begin to overcome this.

Before we can look at the harm it does we have to be sure we understand what 'information system' (IS) actually means. This means going right back to first principles, so let's start with 'information' because it appears first.

THE MEANING OF 'INFORMATION'

The most important thing about information is its lack of substance. It is not a concrete, material thing, but rather an intellectual construct. This is not to say there are not concrete entities relating to information. There are many. They come in a whole range of media. The ink markings on this page are one example, the electrical impulses in a computer are another; the carefully positioned beads on an abacus, the arrow-shape of twigs left on a path by the scouts during their tracking game outside my house yesterday; and so we could go on. But all these objects are simply that: objects: ink marks, electrical currents, beads, twigs. Indeed even there we can get into difficulty because the electrical currents are really only particular arrangements of charged particles. But we don't want to digress into the philosophy of quantum mechanics in this short paper.

Before we move on, however, we should note that, because of our human sensory apparatus, we can only sense the existence and state of objects (or the movement of objects) in our environment. The key point is that the objects themselves are not 'information'. It is the context in which they are positioned and the meaning which we ascribe to them in that

Critical Issues in Systems Theory and Practice, Edited by K. Ellis et al.
Plenum Press, New York, 1995

context which together 'create' the information. The noun 'information' is developed from the verb 'to inform'. It (the noun-form) carries with it a deep level of meaning from its verb-origin. That is the connotation of something, (some knowledge or message) being passed to us by another human being. Now in our current usage we accept that this 'passing to us' may be an active manoeuvre on our part. So we talk of 'acquiring information' or 'seeking information'.

The reality of the exchange of items of knowledge is that the recipient has an immense mental database of contextual information, which is utilised to comprehend, assess and give meaning to the received message. We can usefully think of this contextual database as hierarchically structured. Language is near the base and cultural knowledge near the top. Rather like the International Standards Organisation's 7 Layer model for (Computer Systems) Open Systems Interconnections (OSI), any item of knowledge passed from one person to another has to be packaged into a form understood at each layer. I may initially comprehend some vital truth about society, but if I wish to share this knowledge I have to 'put it into words' to pass it to you. At any level of focus in the hierarchy you need the contextual knowledge of the next higher level to make sense of a message. What all this means is that whilst it is certainly theoretically possible to pass a particular item of knowledge by starting an explanation from, as it were, first principles; it is not a practicable proposition.

At this point perhaps we should pause and summarise where we have got to.

1. All we can sense is the state of objects in our environment.
2. Such sensory input is assessed by using a hierarchically-structured mental database which allows us to make successively more sophisticated deductions from the sensory input.

The implicit understanding here is that we already have the contextual knowledge and cognitive ability to ascribe meaning, all we need is the triggering object or event. (Our equivalent of the digital data stream). When the red light changes to green in the fuselage of a Hercules transport aircraft, the line of parachutists begins its crab-like shuffle to the door. It is disingenuous to talk of the green lamp as 'informing' the parachutists that it is time to jump. The green lamp is at the bottom of our layered stack. It is literally senseless to talk of a green light 'informing' anyone. It is, of course, the pilot who is informing the parachutists that it's time to go. The green light is merely an object whose change of state can be used as a signal.

So the conclusion we should draw from this discussion is that the act of informing is something humans do to each other. We should not obliviously personify any machine or object (and particularly computers) by talking of their participation at this level of abstraction.

The whole of this discussion has focused on knowledge and linguistics rather than on the technical management of communication. This has been deliberate. It is because Communication Theory, as so eruditely explained by Shannon (1948) and his successors, makes much use of the word 'information'. However, I wish to side-step that. Shannon needed a word as a label for a particular artifact within his Communication theory. He chose to use the word 'information' as that label. In the context of communication theory 'information' is defined as the novelty or surprise effect of data passed over a communication channel.

I have no quibble with that. It was, and is, a remarkably useful intellectual construct. It is quite often the case that words in any language are overloaded with several distinct meanings. It is vital however, that we maintain a clear distinction between information as that which is related to the (upper layer of the interconnection stack) human activity of informing (our focus here in this paper) and information as related to the definition of data exchange in communications engineering (Shannon's usage; at the bottom layer of the intercommunication stack).

When I execute a plan, the 'execution' has a base in the actual state and change of state of objects in the world, but is an overall label for the totality of those objects and activities. In the same way 'Information' is a construct of our minds; that which has passed from me to you when I have successfully informed you about something. The word 'information' in this sense is a label for the totality of comprehension resulting from the act of 'becoming informed'. It is a particular parcel of knowledge, acquired at a particular moment in time.

THE MEANING OF 'SYSTEM'

We seem to have deconstructed 'information'. Lets turn our attention now to 'system'. We can follow a similar line of argument but this time the path is well trodden and we find a number of distinguished fellow-travellers. Indeed Arthur Koestler (as discussed in Checkland and Scholes, 1990), and more recently, Peter Checkland (1988) have been so convinced that we should consider 'system' as an intellectual construct that they have advocated the adoption of a new title (Holon) specifically to emphasise its abstract status. But we're getting ahead of ourselves. Let us look briefly at the origin of the word. The word system has evolved from an early Greek form meaning 'organised whole'. It has a number of links to Latin; in medieval (and modern) usage as 'the universe'. It has been applied in so many contexts that it now requires over two pages of definition in the Oxford Dictionary (1994).

Now, from the perspective of our concerns, there are two key, conflicting, general usages; one technical, one common. The technical usage considers 'system' to be an intellectual construct. Definitions abound but all agree that a system is 'An organised set of entities connected together in such a way as to achieve some action, as perceived by some human mind'. This is the basis of the Open University definition (Course T301,1991) and the essence of all others.

I have to say straight away that I think this technical usage is the only really valid one but more of that in a moment. This view of the meaning of system has grown out of the development of scientific understanding in the last 50 years. As the reductionist approach to science became successful, it also began to manifest its limitations. An explanation of the synergy of interacting elements was needed and a formal intellectual framework to link this to the properties of such interacting element groupings; properties which could not be attributed to any of the individual elements. The systemic concepts of hierarchy and emergence met this need. The notion of 'system' now pervades all science. It is one of the most powerful metaphors of our age.

What of the conflicting view? The situation of conflict has arisen because the word 'System' has slipped gradually into common language as an appendage to any organised grouping. It's a convenient sugar-coated word which lets us off the hook of being precise. Thus we talk of the 'railway system' the 'education system' and so on; pervasively. We could easily drop it completely. In this usage the word is almost meaningless. We implicity know that a railway service must be an organised, interacting collection of components. It adds nothing to our understanding of this reality to add on the word 'system' to 'railway'.

THE PRACTICAL IMPLICATIONS OF CURRENT THINKING

We have now arrived at a crucial position. I have argued that both 'information' and 'system' are intellectual constructs which have no reality in the sensory world of objects and object activities. What then are the 'Information Systems' which so many people are seeking to design and construct? Well whatever these artifacts are they are not either information or systems or any combination of either. But 'Don't be silly' someone will say. 'Such computer systems DO exist; you're just being pedantic'. Well, I concede I'm being pedantic, and I agree computers exist but I don't think it is useful to call them 'information systems'. So far I've advanced a linguistic rationale for disowning the term 'information system'; there are however, very important practical implications involved and I would like now to turn to look at these.

We all recognise that organisations are more or less resource constrained. A major resource for most organisations is its funding. Companies set up complex financial management structures to ensure that this is achieved. Sometimes these (or the accountants operating them) wrest control of the business from general managers. When general managers are weak and their decisions poor such a shift of power can save a company from failure. However, when general managers are strong and their decisions sound, such a shift of power can stifle innovation and change, prevent risk-taking and ultimately create a situation for failure. It is therefore, always in the long term interest of a company to keep control in the hands of general managers and ensure that these are competent.

If you think of information as 'existing' out there in the world in such a way that it can be managed by a (computer) 'system' then you are working with information as a resource, just like finance. All the dangers of the take-over by accountants now apply to take-over by IS managers. But there is a significant added danger. It is true that a company must make a profit to survive. There is no such bottom-line with information as a resource; you can never tell if your IS is comprehensive or sophisticated enough. So there is a never-fulfilled drive for bigger and better computer support which can crucify an organisation. It is also easy to propose that the experts in managing IS know best what form of IS is required. General managers and users need only be consulted in so far as IS managers need to know what they do in order to decide how IS can assist. The real workers of the business are thus relegated to second place in influence over the computer support they receive. All this mayhem, I submit, results directly from an invalid acceptance that IS can 'exist' as entities in the world.

A HEALTHIER VIEW OF 'INFORMATION SYSTEMS'

The picture is far healthier if we stop thinking of IS as artefacts. Now we're left with a business and the possibility of computer support. Computer staff, whether in-house or in the marketing and technical divisions of suppliers, know what computing support is, in general terms, feasible. General managers know the business. The two must work together to decide what elements of the business and its procedures can usefully be automated, given present state-of-the-art in computing and the resource constraints and overall goals of the company.

So where are we in our consideration of IS? I've tried to show that our present, widely accepted, use of the term is linguistically illogical and that it can lead to emphasis shifting away from the core activities of a business in a less than satisfactory way. The remedy is to abandon this use of the term and revert to talking directly about computers as the superb support tools they are.

Our focus in requirements definition can then be on the business activities - should we be doing them? Could we improve them? and (only then) which activities, or which elements of activities, can we usefully now computerise? This focus on business activities leads on to what I suggest is the valid use of the term 'information system' as a particularly powerful model of an organisation. Let me explain. To manage organisations we need to understand them. One very powerful aid to such understanding is the use of models.

We can usefully view and model organisations from a number of perspectives:

1. business function emphasises the elemental business activities.
2. financial control emphasises the financial flow and control mechanisms.
3. data-flow shows the base data moving into, out of and around the organisation.
4. information emphasises the contextual knowledge and cultural mores used by actors in the organisation in assessing and acting on data

Overlaying these perspectives is the fundamental notion that we can think of the organisation as if it were a system. To fully understand the functioning of any organisation we need to develop a range of models from different perspectives; it is only sensible to make them systemic models.

Because an 'information systems' model can begin to make explicit the intellectual activities which humans do in relation to activities which can be automated in a computer, it is one of the most relevant models to have to hand when deciding on how to extend or install computing facilities in an organisation. Well understood systems methodologies, like Checkland's SSM (Checkland, 1981; Checkland and Scholes, 1990) have a great deal to offer here. Making explicit the links between one hierarchical level and another in an organisation is a key part of information systems modelling. What is construed as information at one level may become data to be passed to a higher hierarchical level. So we have a complex interplay between data provided and the assessment and evaluation process at each level. Tools like Beer's VSM (1985), can be a powerful aid to understanding this extremely complex series of interactions.

Decision making involves managers in identification and assessment of options. It involves the very central issues of what it means to inform and be informed. I believe that, by providing an intellectual framework in this area "information systems" modelling can, and should, be the leading force in enabling elegant, useful and cost-effective computer support for decision making to be provided.

REFERENCES

Beer, S., 1985, "Diagnosing the System for Organisations", John Wiley & Sons, Chichester, England.

Checkland, P.B., 1981, "Systems Thinking, Systems Practice", John Wiley & Sons, Chichester, England.

Checkland, P.B., 1988, The case for holon, *Systems Practice* 1:3, pp 235-238.

Checkland ,P.B., and Scholes, 19 J., 1990, "Soft Systems Methodology in Action", John Wiley & Sons, Chichester, England.

Course T301, 1991, "Complexity Management and Change - A Systems Approach", The Open University, Milton Keynes, England.

Koestler, A., 1978, "Janus : a summing up", Hutchinson, London, as cited in Checkland and Scholes, 1990.

Oxford English Dictionary, 1994, Oxford University Press, London.

Shannon, C.E., 1948, A mathematical theory of communication, *Bell Systems Techn. J.* 27, pp 379-423, and 623-656.

50

SYSTEM PICTURES

Towards a Tool for Information Systems Analysis

Angus J. Quin and Malcolm Bronte-Stewart

Department of Computing and Information Systems
University of Paisley
Paisley
PA1 2BE
United Kingdom

INTRODUCTION

Some of the perennial difficulties experienced in analysing situations in an organisation include: identifying and learning about the existing sub-systems and problems, and recording these ideas in a form that is understood by all the people involved. These concerns are aggravated when considering large and complex industrial, commercial and public service systems. Trying to find out about the situation and gathering information from everyone involved is usually difficult, especially since individuals may have different opinions of what is causing the problems, and their own perception of the role they play in the organisation.

The authors believe that an approach that quickly and easily captures the essential relationships, responsibilities and roles in a situation is needed. A diagrammatic approach is being developed from experience and consultancy which requires little training, aids investigation, improves communication, brings out questions and highlights problems (Quin and Bronte-Stewart 1994). The approach may also assist with the analysis and design of information systems.

System Pictures were first used, by one of the authors, for problem investigation within the aeronautics industry in the late 1960's and early 1970's (MRCA, 1967-1974) to encourage meaningful discussion among people with widely different backgrounds. Since then the authors and others have used System Pictures for teaching (McDermid, 1990), research (Stowell, West, Smith and Begg, 1995), organisational analysis (Renfrew District Council, 1990), and business information systems consultancy (Harrisons Clyde Ltd., 1991).

THE NEED FOR PICTURES AND DIAGRAMS

In the early stages of an organisational study analysts and managers are faced with many problems. For example, even ascertaining an individual's actual role in the organisa-

Critical Issues in Systems Theory and Practice, Edited by K. Ellis et al.
Plenum Press, New York, 1995

tion can be difficult. It seems to us that job specifications and traditional data and process oriented approaches inadequately describe the social and informal interactions that are an important part of the way the system operates. Documentation, however, is important at every stage of problem solving to record the learning process involved for both the client and the analyst. We would argue that many systems analysis diagrams, like Data Flow Diagrams (De Marco, 1978; Gane and Sarson, 1979), Integrated Definition method (Ross and Schoman, 1977), Petri-Nets (Peterson, 1981), Viable Systems Model (Beer, 1985) and Corporate Data Model (Earl, 1989) have strict rules, use a limited set of conventions, notations and symbols, and attempt to be completely objective. In our experience these diagrams may: a) be time consuming and difficult for the novice to create or interpret, b) seem over technical and complicated, c) have to be read serially, and d) can be difficult to create on the spot during a conversation. While, Rich Pictures (Checkland, 1981) have few definitive guidelines or conventions, are time-stamped, attempt to express the relationships and 'feel' of a situation, may be a subjective and personalised view and can seem confusing and open to interpretation. In our opinion, Systems Maps (Open University, 1984; Waring, 1989) seem to provide a means of inquiry that is easy to use, particularly when exploring the issues involved in complex and messy subjects (West, 1991). The full value of Systems Maps, however, relies on the participants having an appreciation of the Venn diagram convention.

During an investigation, many members of staff may be consulted. Some people may find it hard to answer a stream of formal technical questions, some may be put off by or find it difficult to adapt to certain knowledge elicitation techniques or styles, some may feel constrained to say only what they think they are supposed to in response to direct questions, some may be hostile to the investigation. Analysts need a communication medium to derive a model of the system which everyone involved has been able to contribute to and can understand, yet at the same time is sufficiently rigorous to form a solid foundation for proposals and recommendations. We have found that diagrams, based on perceived distinctions or boundaries, assist this process of inquiry and are a valuable means of communication.

ANALYSIS USING BOUNDARIES AND SYSTEMS PICTURES

People seem to use conceptual boundaries to describe their world. It would be difficult for us to describe our own experiences without drawing boundaries, we distinguish systems and name them in order to explain and discuss interpretations. Partitioning appears to allow us to describe, analyse and then reconstruct an area of interest. In the 1950's Spencer Brown developed an algebra to deal with boundaries which form distinctions (Brown ,1969; Wynn, 1989). This algebra enables us to investigate a system based on these distinctions. These ideas are also found in object oriented analysis and design (Balin, 1989; Coad and Yourdon, 1991) and General Systems Theory (von Bertalanffy, 1968).

The essential concept of a System Picture is that a diagram is used to capture a person's current perception of their situation. An annotated picture can be drawn while discussing an individual's roles and the areas they distinguish as being under their control or management. It is usually constructed collaboratively during an interview, as the person describes the way they see things. We have found that this process usually acts to make the individual feel more involved and important perhaps because they feel their views and problems are the focus of the attention, which, in turn, may help to reduce barriers and enable meaningful discussions to take place. Creating a System Picture during an interview also seems to help those involved to concentrate on the area of investigation and to stimulate further questions.

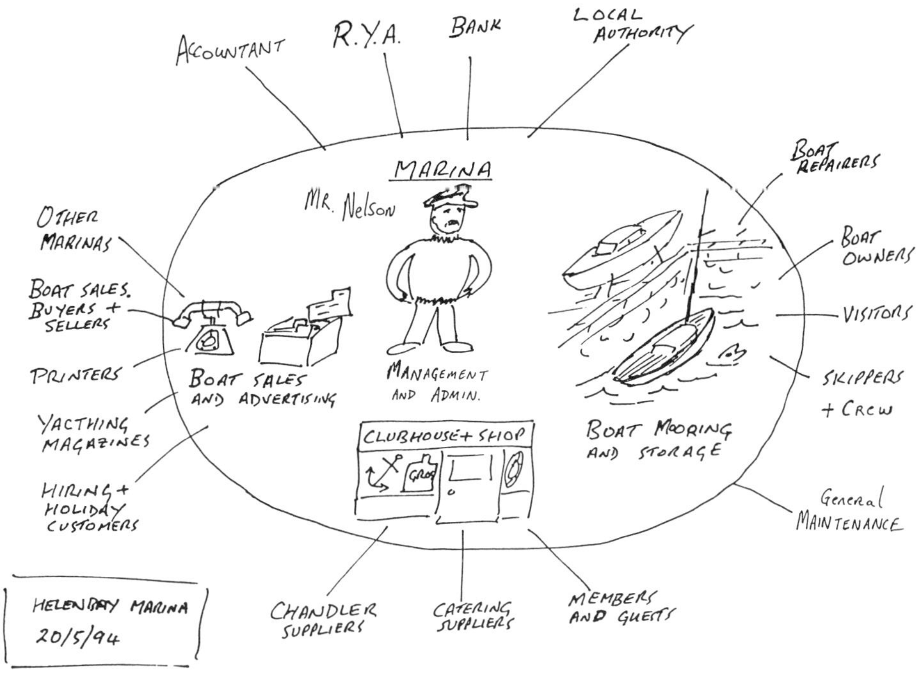

Figure 1. An example of a system picture that records the first conversation with the owner of a yacht marina.

COMPOSING A SYSTEM PICTURE

Our experience of using System Pictures has shown that typically the first questions that the interviewer asks are: what you are responsible for? what are the main things you do? and what kinds of activities do you manage? Each of the main areas the interviewee distinguishes are discussed and noted as subsystems near the centre of a sheet of paper. When the client decides they have outlined what they see as their main responsibilities a boundary is drawn enclosing all the subsystems. The client is then asked to consider who they interact with outside this boundary - the agencies, organisations, departments, groups and individuals they communicate with. These external perceived systems are sketched around the outside of the page and links drawn to them. Comments and notes are usually made beside the links as the person describes each relationship. Cartoon images and symbols may be used to enhance the picture and add to its meaning and richness. The picture is developed progressively until the interviewee feels they have explained a representative overview of their role, the aspects they think they control and the problems they have carrying it all out. (See Figure 1). A self validation process seems to take place which we may liken to the hermeneutic circle (Dilthey, 1976). It should be mentioned that System Pictures do not have to be developed in a way according exactly to the method described above, external agencies, for example, may be discussed and noted at any time during the conversation.

In order to learn more about the client's perspective, each of the subsystems identified in the overview System Picture can be explored in greater detail by putting it at the centre of a new page. The main activities and concerns that form the components of that system are described and noted along with the external interactions and relationships. In this way a

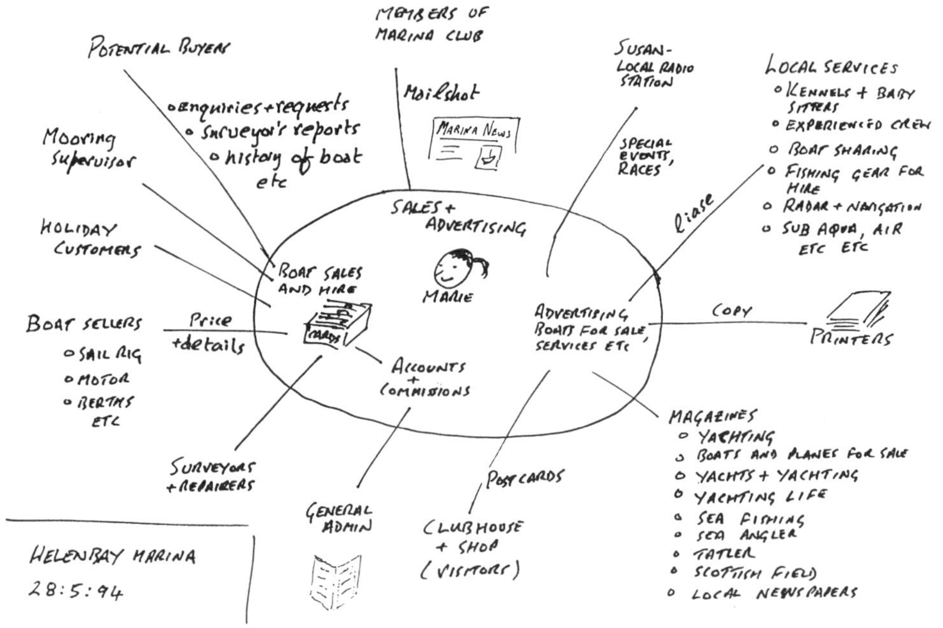

Figure 2. System picture of the sales and advertising manager's view of her responsibilities.

set of pictures are composed of that person's perception of their objectives, formal duties, informal ways of working, contacts, information needs and problems (See Figure 2.).

SOME FURTHER THOUGHTS

We have found that System Pictures seem to assist the learning process for both the client and the analyst, (Vickers, 1965). System Pictures developed by each person can be compared with those developed by others. Meetings may take place to discuss and identify links, clashes and overlaps. In doing this staff may be reeducated to appreciate each other's problems and modify the way they work. In our experience a client's understanding of their problem domain crystallizes during the construction and review of System Pictures and at any stage during a project new insights and facts can be added. We hope to find out more about comparing and combining System Pictures and using them as a tool for information systems design. A network or meta-model of System Pictures may be able to be built up which can help to provide a basis for revealing an understanding of the nature of the organisation and its fundamental activities, (Quin, 1993).

SUMMARY

In order to gain a better understanding of a problem situation and try to ensure the acceptance of the proposed solutions it is important that everybody involved should contribute from the start of the investigation. Interviewers should make use of techniques that provide a common language and should avoid technical jargon, concepts and symbols that require specialist training. Their communication tool should be self documenting to save time, capture both structural and procedural information, be easily amended and updated and be rigorous.

A System Picture is a pictorial diagram that captures a perception of the main subsystems within a system's boundary and the other systems that it interacts with in the environment for a given level of recursion. System Pictures may be developed with the client during an interview. This diagrammatic approach may help to focus managers', technicians' and operatives' attention on the real problems they are experiencing and can lay the basis for the design or development of new systems. In our experience, System Pictures are a valuable tool for inquiry, communication, documentation and gaining an understanding of organisational systems.

REFERENCES

Balin, S. C., 1989, An object oriented requirements specification method, *Communications of the ACM*, 32:5.
Beer, S. A., 1985, "Diagnosing the System for Organisations", John Wiley & Sons, Chichester.
Brown, G. S., 1969, "Laws of Form", George Allen and Unwin Ltd, London.
Checkland, P.B., 1981, "Systems Thinking, Systems Practice", John Wiley & Sons, Chichester.
Coad, P., and Yourdon, E., 1991, "Object-Oriented Analysis," Prentice-Hall, Englewood Cliffs, NJ.
Coad, P., and Yourdon, E., 1991, "Object-Oriented Design", Prentice-Hall, Englewood Cliffs, NJ.
De Marco, T., 1978, "Structured Analysis and System Specification", Yourdon Press, New York.
Dilthey, W., 1976, "Selected Writings", H. P. Rickman, ed., Cambridge University Press, Cambridge.
Earl, M.J., 1989, Management Strategies for Information Technology, Prentice-Hall, London.
Gane, C., and Sarson, T., 1979, "Structured Systems Analysis: Tools and Techniques", Prentice-Hall, ," Prentice-Hall, Englewood Cliffs, NJ.
Harrisons Clyde Ltd., 1991, Report of consultancy project under the auspices of the Enterprise Initiative, Department of Trade and Industry, M. Bronte-Stewart, D. McDermid.
McDermid, D., 1990, "Software Engineering for Information Systems", Blackwell, Oxford.
MRCA, 1967-1974, Multi Role Combat Aircraft, British Aerospace Panavia Contract.
Open University, 1984, T301 Block 1, Complexity, management and change: applying a systems approach, Introduction Workbook.
Peterson, J. L., 1981, "Petri-Net Theory and the Modelling of Systems", Prentice-Hall, London.
Quin, A. J., 1993, Cybernetics in management, *in:* "*Systems Science: Addressing Global Issues*", F. A. Stowell, D. West, J. G. Howell, eds., Plenum, New York.
Quin, A. J., and Bronte-Stewart, M., 1994, System pictures: a method for capturing clients' views, *Systemist*, 16: 4.
Renfrew District Council, 1990, Consultancy report on the proposal for an Information System for the Building and Works Department of Renfrew District Council, R. Frew, M. Bronte-Stewart, and D. McDermid.
Ross, D., and Schoman, K., 1977, Structured analysis for requirements definition, *IEEE Transactions on software engineering*, 3: 1.
Stowell, F. A., West, D., Smith, S. A. and Begg, C. E., 1995, Decision-making in mental health care: towards a technology based decision support aid, *in:* "Critical Issues in Systems Theory and Practice", K. Ellis, A.Gregory, B.Mears-Young and G.Ragsdell, eds., Plenum, New York.
Vickers, G., 1965, "The Art of Judgement", Chapman and Hall, London.
von Bertalanffy, L., 1968, "General Systems Theory", Braziller, New York.
Waring, A., 1989, "Systems Methods For Managers: A Practical Guide", Blackwell, Oxford.
West, D., 1991, "Towards a subjective knowledge elicitation methodology for the development of expert systems", Unpublished PhD thesis, Portsmouth Polytechnic.
Wynn, E. W., 1989, Spencer Brown Logic: The prepositional calculus through metalogic, *Systemist*, 12:1.

51

THE SIXTH STEP

A 'Structured' Approach to Resolving Conflicting 'Clues' in IS Strategy Definition

Wendy Robson

University of Humberside
Cottingham Road
Hull
HU6 7RT
United Kingdom

Introduction TO Defining Consistent Strategies

Our awareness of the competitive power of information systems (IS) has increased considerably over the past decade (from Ives & Learmonth, 1984, to Vcnkatraman, 1994); an expensive burden if inappropriate but a key competitive lever if *appropriately* used and exploited. The pattern of IS resource allocations (the IS strategy) represents expensive IS actions and, potentially even more expensive, the exclusion of IS actions. Since IS may enhance or detract from the competitive health of the organisation there has emerged a focus on planning the systems *strategy* (Ward, 1988; Earl, 1989; Galliers, 1991; Silk, 1991; Ciborra & Jelassi, 1994). This paper forms part of the literature on how to predict the competitive health impact of IS and so manage the competitive risks inherent in IS use (Vitale, 1986; Kemerer & Sosa, 1991; Economist, 1992; Computer Weekly, 1994; Kettinger et al, 1994).

Much literature has exhorted IS to be business *aligned* (Ward, 1987; Lederer & Mendelow, 1989; Das et al, 1991; Goldsmith, 1991; Baets, 1992; Galliers, 1993). However, because of implications of ownership or primogeniture in the word aligning, using it would pre-define the nature of IS strategy planning (Sinclair, 1986; Robson, 1994(a)). For this reason I prefer to use a less prescriptive phrase and would name what is sought as strategies that are *consistent*. If we look for appropriately connected business and IS strategies we are seeking strategies that are *mutually* consistent. Consistent strategies will be competitive health enhancing strategies.

We seek *certainty* and *confidence* that we *have* established IS/business consistency to reduce the risk inherent in an IS strategy. It is outside the scope of this paper to address what might be *meant* by success and failure in this context, but we must recognise that one man's success is another man's failure. There is a clear need for far more research on the danger of strategic IS success and on differing stakeholder perceptions of success! This paper investigates the confidence building process by considering some models that have been used to define, test or advise us on strategy consistency. These contingency models are part of the 'scientific' need for *understanding* of consistency as opposed to a 'religious' *acceptance,* perhaps from guru-led belief (Jackson, 1993; Economist, 1993). Investigation of the models used charts the development of one aspect of the theory of IS strategy.

In this paper I focus upon the conception stage of IS strategic management, namely the process of defining the IS strategy. A variety of strategy defining approaches exist, too many to reference here and, in any case, my concern is the search for certainty of consistency rather any specific strategy definition method. Therefore, although the approaches could be classified in many different ways (Bergeron et al, 1991; O'Brien, 1992), it is reasonable to use a coarse granularity system and desirable to adopt one that uses familiar terminology (Robson, 1994(a)). So, using the categories of Earl (1989), IS strategy development methods fall into three types; top-down, bottom-up and inside-out; where each type implies something rather different by the word consistency.

- Top-down analysis approaches emphasise the need to be certain that business goals are supported by IS activities (example tools include critical success factor analysis, strategy set transformation, strategic impact analysis);
- Bottom-up evaluation approaches emphasise the need to leverage strengths (and avoid weaknesses) in defining the acceptable strategy (example tools include systems portfolio analysis, skills audit, Nolan's (1979) stage of maturity model);
- Inside-out innovation approaches emphasise the need for creative identification and exploitation of opportunities (it is impossible to instance *tools* as such in this context).

CONFIDENCE BUILDING THROUGH CONTINGENCY MODELS

The process of confidence building has been one of looking for key factors in choosing the IS strategy defining method such that consistency is assured. The initial contingency models were single dimensional. To take just two examples, the key contingency factor in obtaining consistency may be the stage in the development of IS objectives (e.g. Silk, 1991) or it may be the organisational context of IS use (e.g. Sullivan, 1985). In either instance it is assumed that there is a *single* best approach to consistency. It may be that all who share the same IS objective (Silk, 1991, suggests a developmental path through efficiency, effectiveness, strategic advantage to an integrated portfolio) will share the same 'best' approach to defining a consistent IS strategy. Alternatively, the answer may be tied to IS impact and be related to the nature of the organisation such that, if we can understand and categorise what IS means to that organisation (Sullivan (1985) suggests four categories; backbone, traditional, federation and complex) then we can know the best method with which to obtain consistency. Even these single-dimension models contain a stage or cell that is 'undefined' or eclectic and therefore not readily associated with confidence in a *single* best approach to defining an IS strategy.

The single contingency models were joined in the late 1980s by multi-dimensional models that built confidence in consistency by simultaneously capturing several distinct issues for *broad* consistency as, for example, the multi-dimension model from Earl (1989) which aimed to simultaneously address three issues:

- What are the business *needs*? These show what IS *must* do.

Top-down methods deliver consistency confidence; bottom-up does not.

- What are the *capabilities/positions*? The business could be built from what IS *can* do.

Bottom-up methods deliver consistency confidence; top-down does not.

- What are the technological *opportunities*? Each domain can change what the other *does* do.

Neither analysis, nor evaluation entirely helps build confidence here - innovation needed.

The three issues model argues that confidence in the IS strategy definition process can only be achieved by using multiple approaches to defining the strategy. Hence, all three questions are simultaneously managed. Earl (1989) suggests that the question to make prominent depends upon the nature of the organisation's industry (so subsuming the organisational context model mentioned earlier) but also suggests that there is a learning curve for the process of strategy definition and only at a 'maturity' stage, reached after about five years, can multiple contingencies actually be accommodated (so also subsuming the development of IS objectives model). Collectively, therefore, which of the three issues (and so which type of approach to IS strategy definition) is most prominent might differ for:

- Different types of organisations;
- Different times for the same organisation.

Galliers (1991) looks at much the same issue from a slightly different perspective and produces the model shown in Figure 1. This model shows us the learning path towards the need for multiple approaches to strategy definition. This path may best be traced in a three-dimensional spiral fashion by which the development stage *leads* to the need for, and the skill to use, multiple approaches.

Our state of knowledge of defining IS strategies (as represented by the available models) has reached the point of recognising the need to use multiple models, or at least single models that involve multiple dimensions. What is not 'known' is *how* to manage the final stage, that is, how to *combine* the multiple indicators that the multi-dimensional methods must inevitably provide us with.

Resolution of Conflicting 'Clues'

We have arrived at needing multi-dimensional approaches to feel confident that our IS strategy *is* business consistent. However, what we are 'told' from one dimension often does not coincide with what we are told from another and we need a way to resolve the

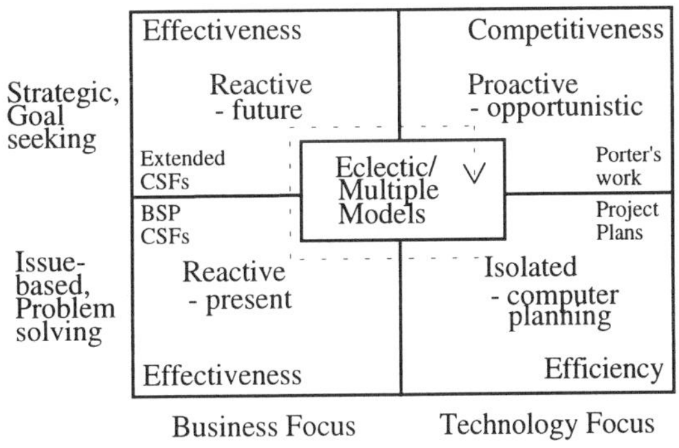

Figure 1. Spiral development path of IS strategy definition (after Galliers, 1991).

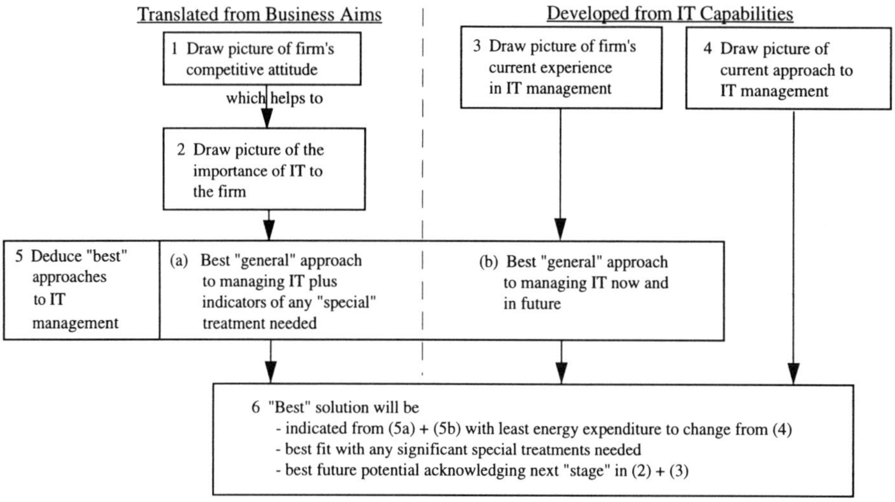

Figure 2. Example Method of Defining IS/Business Consistent Strategies (Robson, 1994(b) from Ward, 1987).

differing messages into a coherent IS strategy. So we need a model that includes the final, resolution, stage. This paper draws its title, 'The Sixth Step', from the example chosen to show that the need to resolve conflicting 'clues' is not currently addressed. Figure 2 gives my diagrammatic representation of the example multi-dimensional method for defining an IS strategy. This one (drawn from Ward, 1987) has been chosen because it builds upon well-known models and is clearly multi-dimensional in its modelling of IS management expertise and business demand priorities to capture both bottom-up and top-down issues. The method gives its advice, or clues, by presenting the implications of each single dimension model in terms of the six linking (or generic) IS strategies of Parsons (1983). Using a set of generic IS strategies in this way makes receipt of the advice manageable. Restricting the naming of possible outcomes to a convenient six-member classification set and, indeed, having a shorthand *name* for each possible strategy, serves to focus attention upon the essence of the advice rather than diffusing that attention. One important aspect of this method is that it models business importance associated with different types of IS applications (McFarlan et al, 1983, suggest four categories; turnaround; strategic; factory and support) and the strategy implications of these different *types* of importance. Therefore, as well as clues to an overall IS strategy, the method gives clues to which generic strategy is consistent with any significant 'special treatment' needs, that is, minority groups of IS applications. The operational details of the approach are outside the scope of this paper but are explored in Robson(1994(b)).

The example method defines an IS strategy by combining a series of single-dimension models to capture different dimensions of consistency. These are: 1 by Porter (1980); 2 by McFarlan et al (1983); 3 by Nolan (1979); and 4 by Parsons (1983). The method shows how to *separately* build from the model of *each* dimension into an IS strategy (labelled as box 5 in Figure 2.). What this example method does not include is a step 6 which would make manageable the enormous leap from independent, discrete dimensions into a *combined* multi-dimensional picture. All stages until the final one (those labelled 1 to 5 in Figure 2) are based upon analytical, and implicitly rational, models; however, the sixth step, that is, how to *combine* the clues available from 5(a), 5(b) and 4 of Figure 2 is, as yet, unmatched by similarly analytical models. So how to resolve across the dimensions remains a missing

Dimension returning the 'clue'	'Clues' provided by each dimension (expressed as one of 6 generic IS strategies) Centrally Planned	Leading Edge	Free Market	Scarce Resource	Monopoly	Necessary Evil
Current top-down business demands	1½	1	1	0	0	0
Anticipated future business demands	1	1½	0	0	0	0
IS of 'special' importance type - Support	0	0	1	1½	1	½
IS of 'special' importance type - Factory	0	0	0	1	1½	0
IS capabilities - now	1	1	1	1½	1½	0
IS capabilities - future	1½	1	1	0	1	0
Change from current IS strategy	-1	0	1½	-½	-1½	-1
SCORE	3	4½	5½	3½	3½	-½

Figure 3. Worked example of a scoring system used to resolve conflicting clues.

element in the theory of IS strategy. In fact, this example method, along with most of the advisory literature, does not explicitly recognise that such a step must be taken.

So how can that unsupported final step be taken? Box 6 of Figure 2 indicates attributes of a solution to any conflict observed at step 5. One potential way of achieving this solution is to attach scores to the messages or clues returned by each dimension and combine them into a multiple picture by simple addition. Figure 3 illustrates what such a scoring method might look like if used as box 6 of Figure 2. Since the example IS strategy defining method diagrammed in Figure 2 delivers its advice in terms of the six generic IS strategies of Parsons (1983) the same terms are used across the X-axis of Figure 3. The Y-axis shows the dimensions returning that advice and are those dimensions separately captured by the models used in the example method.

The process of using this scoring method is to take each dimension and from it attach a score to each possible 'clue' (one of the 6 possible generic IS strategies). In the illustration given in Figure 3, $1\frac{1}{2}$ has been awarded for a high degree of consistency with a particular dimension; 0 awarded to a neutral state whilst $-1\frac{1}{2}$ has been awarded for a high degree of inconsistency with that dimension; scores have been decided on a subjective basis and awarded in $\frac{1}{2}$ point increments and should be interpreted as an example only. So, for example, Figure 3 shows that, in this situation, a Monopoly IS strategy is highly consistent with the minority of IS applications of business importance type Support and so achieves a high score from that dimension. However, the same 'clue' (the Monopoly generic IS strategy) is highly inconsistent with the current IS strategy and so achieves a very low score from that particular dimension. Clearly this method could be developed and refined by attaching weightings of importance to each dimension and these weightings may reflect the nature of the organisation and the stage in the learning process illustrated in Figure 1. Without a weighting factor, all dimensions are given equal prominence and, of course, this may well be inappropriate. Any set of strategy defining methods, using any set of tools to present advice could be similarly treated, although, obviously, entries along the X and Y axes would then need to be adjusted to be meaningful to the models actually used.

Conclusions and Cautions

This paper has shown that confidence in the desirable consistency between business and IS strategies is sought. The normative models used to gain that confidence highlight the need to use a mixture of methods to achieve consistency but do not address *how* to manage the mixture. Each method (and even within one method) may return conflicting clues from its models so, if we are to operationalise the need for multi-dimensional confidence, an important step is to resolve that conflict. The currently available approaches neglect this important 'sixth step' aspect. The issues and difficulties have been illustrated with respect to just one multi-dimensional IS strategy definition approach. The use of a simple scoring system is suggested as one potential way of dealing with the need for balance. This would introduce the possibility of increasing the richness of existing models and, when generalised to score generic clues against each consistency factor in any multi-dimension IS strategy defining approach, could enable structured debate through the scoring rule.

In the high-risk circumstances of defining an IS strategy, increased confidence can be gained by using a structured way of resolving conflict so a sixth-step model offers much of value. However, we must be cautious of the role of normative models in such a complex and problematic situation. Any of the methods, be they multi-dimension or not, founded upon the use of rational, analytical, normative models offer, potentially over simplistic, value judgements and any application of them, with or without the sixth step, should recognise the limitations inherent in the nature of such models.

References

Baets, W., 1992, Aligning information systems with business strategy, *Journal of Strategic Information Systems*. 1:4, pp 205-213.

Bergeron, F., Buteau, C., and Raymond, L., 1991, Identification of strategic information systems opportunities: applying and comparing two methodologies, *MIS Quarterly*. March.

Ciborra, C. U., and Jelassi, T., 1994, "Strategic Information Systems: A European Perspective", John Wiley & Sons.

Computer Weekly, US Bureau., 1994, Greyhound going to the dogs after $7m fiasco, *Computer Weekly*. 27 Oct, pp 10.

Das, S. R., Zahra, S. A. and Warkentin, M. E., 1991, Integrating the content and process of strategic MIS planning with competitive strategy, *Decision Sciences*. Vol. 22, pp 953-984.

Earl, M. J., 1989, "Management Strategies for Information Technology", Prentice Hall, Englewood Cliffs.

Economist., 1992, Banks and technology: cure-all or snake-oil?, 3 October.

Economist., 1993, Good guru guide: take me to your leader, 25 December.

Galliers, R. D., 1991, Strategic information systems planning: myths, reality and guidelines for successful implementation, *European Journal of Information Systems*. 1:1, pp 55-64.

Galliers, R. D., 1993, Towards a flexible information architecture: integrating business strategies, information systems strategies and business process redesign, *Journal of Information Systems*, Vol. 3, pp199-213.

Goldsmith, N., 1991, Linking IT planning to business strategy, *Long Range Planning*. 24:6, pp 67-77.

Ives, B., and Learmonth, G., 1984, The information system as a competitive weapon, *Communications of the ACM*, 27:12, pp1193-1201.

Jackson, M.C., 1993, "Beyond the Fads: Systems Thinking for Managers", Inaugural Lecture October 1993, University of Hull, Hull.

Kemerer, C. F., and Sosa, G. L., 1991, Systems development risks in strategic information systems, *Information and Software Technology*, 33:3, pp 212-223.

Kettinger, W. J., Grover, V., Guha, S., and Segars, A.H., 1994, Strategic information systems revisited: a study in sustainability and performance, *MIS Quarterly*. March, pp 31-58.

Lederer, A.L., and Mendelow, A.L., 1989, Co-ordination of information systems plans with business plans, *Journal of Management Information Systems*. 6:2, pp 5-19.

McFarlan, F.W., McKenney, J.L. and Pyburn, P., 1983, The information archipelago - plotting a course, *HBR*, Jan-Feb, pp145-156.

Nolan, R.L., 1979, Managing the crisis in data processing, *HBR*, March/April, pp115-126.

O'Brien, B., 1992, "Demands & Decisions: Briefing on Issues in Information Technology Strategy", Prentice Hall, Hemel Hempstead.

Parsons, G.L., 1983, "Fitting information systems technology to the corporate needs: the linking strategy: Teaching Notes 9-183-176", Harvard Business School, Boston.

Porter, M.E., 1980, "Competitive Strategy: Techniques for Analyzing Industries and Competitors", Free Press, New York.

Robson, W.D., 1994(a), "Strategic Management and Information Systems: An Integrated Approach", Pitman, London.

Robson, W.D., 1994(b), "Six Steps to Establishing Business Consistent Management of IT", Unpublished Working Paper Series, University of Humberside, Hull.

Silk, D.J., 1991, "Planning It: Creating an Information Management Strategy", Butterworth-Heinemann, Oxford.

Sinclair, S.W., 1986, The three domains of information systems planning, *Journal of Information Systems Management*. Vol. 3, pp 8-16.

Sullivan, Jr.C.H., 1985, Systems planning in the information age, *Sloan Management Review,* Winter, pp 3-12.

Venkatraman, N., 1994, IT-enabled business transformation: from automation to business scope redefinition, *Sloan Management Review*. Winter, pp 73-87.

Vitale, M.R., 1986, The growing risks of information systems success, *MIS Quarterly,* Dec, pp 327-334.

Ward, J., 1987, Integrating information systems and business strategies, *Long Range Planning,* June.

Ward, J.M., 1988, Information planning for strategic advantage, *Journal of Information Technology*. Sept, pp 169-177.

- The knowledge elicitor or 'analyst' plays a vital role in using IT with the domain expert. Many domain experts commented that they would have been at a loss as to what was required of them without the elicitor's guidance, and found it useful to discuss ideas with the elicitor to gain feedback and reaction, thus facilitating a mutual learning experience and process. It is important that the use of IT merely supports the interaction between the elicitor and domain expert rather than replace the elicitor and 'de-humanise' the learning process.
- The use of IT may be distracting in that the domain expert is focusing on technical aspects of using the computer software rather than the issues relating to the domain under investigation which may detract from the process of learning and appreciation. Furthermore, a number of the domain experts disliked using computers and thus felt uncomfortable in using the IT based AIM which again may distract them from the purpose of the session ie. learning about the problem domain.

It would appear that the use of IT in fully supporting the use of 'soft' systems concepts is still along way off being adequately fulfilled. There are many problems and obstacles to be overcome and this paper has only highlighted a small number of them. However, the efforts of researchers in attempting to fulfill this objective in my view is to be welcomed since it contributes to learning about the complexities of systems thinking. Further enhancements to the IT based inquiry tool are still in progress as a result of the learning process gained from the action research.

5. REFERENCES

Avison, D.E., Golder, P.A., and Shah, H.U., 1992, Towards an SSM toolkit: rich picture diagramming, *Eur. J. Inf. Systs.* 1, (6): 397-407.

Checkland, P.B., 1981, "Systems Thinking, Systems Practice", John Wiley & Sons, Chichester.

Checkland, P.B., 1985, From optimizing to learning: a development of systems thinking for the 1990, *J. Opl. Res. Soc.* 36, (9):757-767.

Checkland, P.B and Scholes, J., 1990, "Soft Systems Methodology in Action", John Wiley & Sons, Chichester.

Kreher, H., 1993, Critique of two contributions to soft systems methodology, *Eur. J. Inf. Systs.* 2, (4):304-308.

Stansfield, M.H., 1990, The Development of an Expert System to Aid the User of Soft Systems Methodology, unpublished M.Sc dissertation, Portsmouth Polytechnic.

Stowell, F.A., and Stansfield, M.H., 1991, A first step to the automation of SSM, *in*: "Systems Thinking in Europe" M.C. Jackson, C.J. Mansell, R.B. Blackham and S.V.E. Probert, eds., Plenum Press, New York.

Stowell, F.A., West, D., and Stansfield, M.H., 1991, The application of an expert system shell to an unstructured domain of expertise: using expert system technology to teach SSM, *Euro. J. Info. Systs.* 1, (4): 281-290.

Stowell, F.A., West, D., and Fluck, M., 1991, The appreciative inquiry method: an approach to knowledge elicitation based upon the notion of knowledge elicitation as an inquiring system, *Systemist.* 13, (4): 154-165.

Stowell, F.A., West, D., and Javelaud, V., 1993, Developments in the automation of SSM tutorials using multimedia technology, *in*: "Systems Science Addressing Global Issues", F.A. Stowell, D. West and J.G. Howell, eds., Plenum Press, New York.

Vickers, G., 1983, "Human Systems are Different", Harper and Row, London.

West, D., 1992, Knowledge elicitation as an inquiring system: towards a 'subjective' knowledge elicitation methodology, *J. of. Inf. Systs.* 2: 31-34.

West, D., Stansfield, M.H., and Stowell, F.A., 1994, Using computer-based technology to support a subjective method of inquiry, *Sys Prac.* 7, (2):183-204.

53

ACTION RESEARCH AND INFORMATION SYSTEMS RESEARCH

The Consultancy versus Field Study Dilemma

D. West,[1] F.A. Stowell,[1] and M.H. Stansfield[2]

[1] Department of Computing and Information System
University of Paisley
Paisley, Renfrewshire

[2] Department of Business Information Management
Napier University
Edinburgh, Scotland
United Kingdom

INTRODUCTION

Action Research (AR) has, for some time, been advocated as a useful way of conducting work in the field of Information Systems (Checkland and Scholes, 1990) since it offers an alternative to the traditional positivist approach in inquiry. In practice this has led to AR being used as a framework for designing, developing and implementing information systems (IS) (Avison and Wood-Harper, 1990) as well as a means of conducting IS research. Those choosing to work within this framework soon become aware of the difficulties that the approach brings to bear upon the development and/or research processes involved. A popular criticism of AR, which is particularly problematic for those interested in using it as a research method, is that it leads to studies that are 'all action and no research'. A further criticism is that AR studies are 'unscientific' and lack rigour. It may be true that AR has received some 'bad press' due to sloppy or uninformed use but, used thoughtfully and rigorously, the approach can provide a practical means of conducting inquiry in social situations that has a strong theoretical underpinning. The literature on AR tends to focus upon describing and arguing the theoretical basis of the approach (e.g. Susman and Evered, 1978), discussing some concept relevant to the approach such as 'dialogue' or 'calibration' (e.g. Bartunek, 1993), or upon giving case study type reports to illustrate the effects and results of an AR study (e.g. Ledford and Mohrman, 1993). Unfortunately, there appears to be little in the literature to describe how an AR study can be conducted and to advise how the inherent problems may be overcome. In an attempt to go some way towards providing such a discussion, in the rest of this paper two AR study 'models' for conducting IS research are described and the difficulties and issues associated with each are discussed.

Critical Issues in Systems Theory and Practice, Edited by K. Ellis et al.
Plenum Press, New York, 1995

TWO MODELS OF ACTION RESEARCH FOR CONDUCTING RESEARCH IN INFORMATION SYSTEMS

From our own work we have been able to identify two distinct types of AR study which have been used as a means of carrying out IS research. It seems useful to describe these two types of study and detail the way in which they were set out and executed in order to provide guidance for those setting up similar studies, to promote reflection about the use of AR for IS research and to promote discussion of others' experiences in using AR for IS development and research.

The first type of study, which we refer to as the 'field study' approach, concerns learning about a set of ideas that seem useful to a particular situation. In our case this study involved the exploration of ideas about a method of inquiry intended to enable the practitioner to appreciate and then elicit descriptions of complex domains of expertise (West, 1991). The second type of study, the 'consultancy' approach, involves the use of research ideas in a 'live ' project within an organisation which is undertaken as a piece of consultancy (Stowell, 1989).

The 'Field Study' Approach

In this type of study AR is adopted as a means of reflecting upon and evaluating research ideas that are themselves evolved from a critical evaluation of the literature and practical experience. When the ideas are put into practice in one situation the lessons learnt from this experience enable reflection and refinement of the ideas which are then reapplied in a new situation so as to provide new experience and opportunity for reflection. The theory-action cycle of AR is clear in this approach. However, we might argue that the approach tends to be theory-focused since in terms of the research the emphasis is upon the research ideas which are explored and, hopefully, strengthened as a result of practical experience of their application. Figure 1 represents an attempt to illustrate how the process of reflection in this type of study tends to take place around the theory 'activity' of the theory-action cycle of AR.

The story of such a study may be something like this: some recognition of a 'real world' problem leads to an investigation of how this problem (or similar ones) has been dealt with in

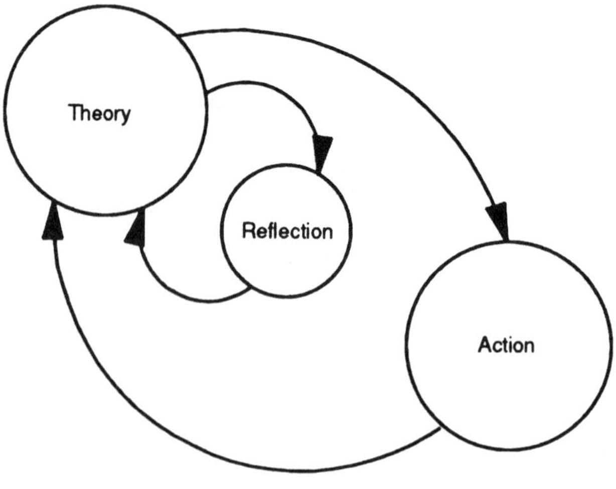

Figure 1. A representation of the theory-action cycle of AR in the 'field study' approach showing how reflection tends to take place around the 'theory' activity.

the past. This investigation may involve an extensive review of the literature as well as an exploration of current practice. A growing appreciation of the problem situation leads to some tentative ideas about how the problem may be tackled and so research ideas are generated.

The next step is to explore these ideas by applying them in some real world situation. An organisation that is deemed to contain the type of problem situation being researched is contacted and a 'field study' is planned and 'set up'. the research ideas are implemented and both the process and results are carefully documented. The researcher now reflects upon the appropriateness of the original research ideas as a result of their application and considers such questions as: how easily were the ideas applied? what has been learnt about these ideas through their application? are they robust? are they flexible enough or too flexible to be useful? is the result of their application as we anticipated? have they been shown to be consistent with the theoretical standpoint from which we claimed they were derived? In this way reflection upon the ideas resulting from their practice enable us to enrich the research ideas prior to their further application.

The 'Consultancy' Approach

The impetus for the 'consultancy' approach may be the same as for the 'field-study' approach, namely a perception of some 'real world' difficulty encouraging exploration of this difficulty and the ways in which it might be overcome. An alternative scenario is that an organisation seeks help to alleviate some problem situation and engages a consultant as problem solver. The consultant sees this engagement as an opportunity to (i) develop a better understanding of some research ideas or (ii) to undertake a study which will serve as an example of how to address a particular type of problem situation, out of which will come some ideas about future practice.

In the 'field-study' approach the research is 'controlled' to some extent by the researcher by means of the planning and reflection that takes place away from the field of action. However, the very nature of the 'consultancy' approach means that the need to respond to the live situation results in the reflection being closely associated with the action activity of the theory-action cycle of AR as illustrated in Figure 2. Consequently, although theory should underpin the action, in practice the action tends to predominate.

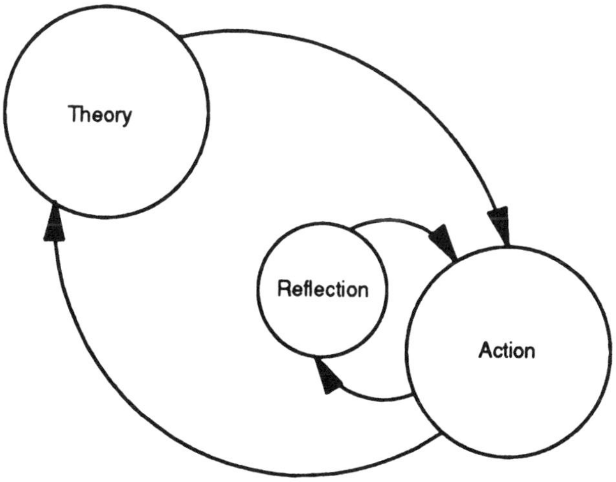

Figure 2. A representation of the theory-action cycle of AR in the 'consultancy' approach showing how reflection tends to take place around the 'action' activity.

PRACTICAL AND THEORETICAL ISSUES

Each of the two types of study outlined above raise a number of practical and theoretical issues which are summarised and discussed in Tables 1 and 2 below.

CONCLUSION

The topic merits a far more detailed treatment than can be offered here. In this paper the aim has been to raise awareness of the problems of adopting AR as a framework for undertaking IS research. How some questions can be posed as a result of the above discussion. For example, an important question to be asked of the 'field study' approach is whether or not it constitutes true AR? Is there not a danger that action and learning may be subjugated to theory? It would seem that an argument in support of this type of AR study is that it can provide a valuable approach if the researcher is aware of its limitations and pitfalls.

With the 'consultancy' approach it might be argued that any one individual problem situation may result in a context-dependent way of dealing with the situation. Furthermore, the pressures of the consultancy situation may encourage an approach whereby the theory is 'fitted' to a pragmatic approach to problem solving.

Table 1. Summary of practical and theoretical issues relating to the 'field study' approach

Planning and implementing the study	When undertaking formal research (ie. for a research degree) it is vital to plan the study. Planning in this sense is not to be equated with 'setting up' the study and predicting the results as in an empirical study but relates to knowing how the AR study will be implemented.
Engaging willing and able collaborators	If an organisation has agreed to act as a field study it may be difficult to engage full commitment from those memebers of the host organisation to the study. In this respect it may be difficult to engender the joint learning process that plays such a central role in AR. How can we protect our study against the withdrawal of key players (eg. through loss of enthusiasm, lack of time, movement from the organisation?
Outlining the programme of AR	How can an AR study be managed given that one of the main problems of AR is that "it cannot be wholly planned and directed down particular paths" (Checkland, 1981. p153), nor should it be? A possible way around this dilemma is to concentrate upon managing the approach to inquiry rather than the inquiry itself.
Writing up the study	For the presentation of formal research it is necessary to 'write up' the research and discuss the findings. How is this to be achieved without the work becoming anecdotal? It is important the those using AR uphold the rigour and respectability of the approach by ensuring that their research is clearly presented and strongly 'validated', otherwise they run the risk of exposing AR to justifiable criticism from the scientific 'community'.
The problem of 'rigour'	Any AR study needs to be rigorously planned and managed so as to make learning and reflection possible. The emphasis upon learning from experience means that rigorous documentation of the AR study is of paramount importance if suppositions are to be supported (Elden and Chisholm, 1993, p128).
Ethical concerns	Given that the primary concern is with exploring the research ideas how much attention should be given to "solving the problem" (the topic of our research) within the host orgnaisation?
Knowing when to stop	If we accept that all applications of the ideas are unique, then how do we know when we have learnt sufficient lessons to finish the research? Is not the learning cycle embedded in AR continuous?
Learning	Given the nature of the study it is likely that learning resulting from the study will be the researcher's introspective learning.

Table 2. Summary of practical and theoretical issues relating to the 'consultancy' approach

Planning and implementing the study	This may be dictated by the problem situation and the demands of the client. It may be difficult for the researcher to step back away from the study and reflect upon the research process.
Engaging willing and able collaborators	This is likely to be easier given that the host organisation has a personal interest in the study. However, the researcher has to be aware of the potential of the clients' hidden agendas (eg. being engaged to bring about necessary but unpopular change).
Outlining the programme of AR	The demands of the live situation may make it difficult for the researcher to follow their intended approach to inquiry.
Writing up the study	The need to ensure client confidentiality may cause some difficulties in reporting the research.
The problem of 'rigour'	The demands of the situation may make it difficult for the researcher to maintain detailed records of the research activities.
Ethical concerns	Should the researcher declare that the study is being used as a piece of research? If this is declared how will it affect the study?
Knowing when to stop	The study may not fulfill the needs of a research study. Alternatively, the demands of the client may pull the study away from the interests of the researcher.
Learning	The opportunity for both clients and researcher to learn about the problem situation is strong in this approach but it may be difficult for the researcher to extract the lessons from the research given the complexity of the situation.

REFERENCES

Avison, D.E., and Wood-Harper, A.T., 1990, "Multiview: An Exploration in Information Systems Development", Blackwell Scientific Publications, Oxford.

Bartunek, J.M., 1993, Scholarly dialogues and participatory action, *Human Relations*, Vol. 46, No. 10, pp.1221-1233.

Checkland, P.B., and Scholes, J., 1990, "Soft Systems Methodology in Action", John Wiley & Sons, Chichester.

Elden, M., and Chisholm, R.F., 1993, Emerging varieties of action research: introduction to the special issue, *Human Relations*, Vol.46, No. 2, pp.121-142.

Ledford, G.E. jr., and Mohrman, S.A., 1993, Self-design for high involvement: a large-scale organizational change, *Human Relations*, Vol.46, No. 1, pp.143-173.

West, D., 1991, "Towards a Subjective Knowledge Elicitation Methodology for the Development of Expert Systems", Unpublished PhD. thesis. Portsmouth Polytechnic.

Stowell, F.A., 1989, "Change, Organisational Power and the Metaphor 'Commodity'", Unpublished PhD. thesis. University of Lancaster.

Susman, G.I., and Evered, R.D., 1978, An assessment of the scientific merits of action research, *Administrative Science Quarterly*, Vol. 23, pp.582-603.

CRITICAL ISSUES IN INFORMATION MANAGEMENT

54

STRUCTURING INFORMATION FOR EXECUTIVE ACTION

C. James Bacon

University of Canterbury
Christchurch
New Zealand

INTRODUCTION

In today's competitive, fast-changing, global economy, one of the critical competencies needed by executives and organisations as a whole is skill in developing and using information. The actual and effective use of information is therefore a fundamental issue for systems theory and practice. In addition, in today's information society, the growing issue of information glut and overload needs to be addressed and dealt with as part of that skill. This implies the need for some form of information filtering, and a focus on the right kind of information. But what is that 'right kind' of information?

The notion of critical success factors (CSFs) has been popular as a concept in addressing the question, notwithstanding difficulties in implementation. However, a more fundamental question pertains to that of information purpose. What is information for, in the organisational environment? If the answer to this question can be satisfactorily determined, then information might be structured and oriented accordingly. The traditional paradigm is that information is for decision-making. However, based on studies of how executives actually work in practice, it is evident that this paradigm is narrow and out of touch with reality. What is more characteristic of executive behaviour is action.

The purpose of information is knowledge. That knowledge may be used for decision-making, but 'decision-making' as such is only a part of the use to which information and knowledge is put in organisations. A larger use to which an executive's constantly developing knowledge content is put is action and behaviour. Furthermore, studies of executive behaviour indicate that it is more often action that implies decision, rather than the reverse. Consequently, it is action-oriented information that is needed, in focusing on the right kind of information in organisations. Accordingly, this paper proposes the construct of kinetic information, which is information oriented to action.

The paper is part of a larger work that is oriented to the development of a kinetic information methodology, which itself is oriented to the effective management and use of information in organisations. The following sections will summarise the basis and structure of kinetic information - information for action.

Critical Issues in Systems Theory and Practice, Edited by K. Ellis et al.
Plenum Press, New York, 1995

HOW EXECUTIVES WORK AND USE INFORMATION

An executive's job is to decide what things needs to be done, and then get them done in a continous, agenda-setting, networking process (Kotter, 1982). There is a dynamic portfolio of information needs and things to do, rather than a series of discrete problems. There is also a constant mental process of environmental scanning for information, from various sources, that might pertain to actual or potential problems and agendas. And mental models are being developed continuously.

The executive is therefore an information intermediary; a virtual 'nerve centre'. This is especially so for informal or 'soft' information flows, which actually make up the major part of the total information picture in organisations (Liebenau and Backhouse, 1990).

It is often a rational, planned, logically sequential decision-making process that is ascribed in textbooks to executive work. However, it has been shown that executive activity is characterised by brevity, variety, discontinuity and fragmentation (Mintzberg, 1975; Kotter, 1982; Isenberg, 1984; Stewart, 1988; Keane, 1993). Executives tend to jump from issue to issue, responding in 'real-time' to the needs of the moment, almost like a juggler, keeping a number of constantly changing projects at various stages of development, up in the air at the same time. Consequently, not only is much information processed verbally, but many information flows are ad hoc, rather than planned. Furthermore, although formal decision analysis tools are relevant, they are marginal to the work of an executive (Kotter, ibid.). Similarly, it has been found that strategies often emerge through time, casual interaction, and un-premeditated (informal) information flows, as part of the organisational culture (Mintzberg and Waters, 1985). This is different from the sequential, programmed, planning and decision process represented in some textbooks.

In practice, actual 'decision-making' is not central to executive behaviour. In one detailed study of ten chief executive officers it was found that, of all actions taken as a result of incoming information, only 21 per cent resulted in a CEO engaging in decision-making activities (Keane, 1993). A main conclusion of this study was to seriously question the paradigm of senior executives as 'decision-makers'. Instead, the study showed that CEOs spend much of their time influencing others, not in making decisions as such. Furthermore, it was found that the most effective executives actually made few decisions.

A basic conclusion is that, in executive work there is an emphasis on *how* (i.e. the process), rather than *which* (decision choices) to do. That is, the emphasis tends to be on *action*, with decisions being implicit in such action.

KINETIC INFORMATION DEFINED

The word 'kinetic' is defined in the Oxford Dictionary as pertaining to motion, caused by motion, or characterised by movement. It is derived from the Greek *kinetikos*, which means moving, and the verb *kinein*, to move. Thus, the term kinetic information modifies movement to *action*, and refers to that information which enables or energises action. In full, kinetic information is defined as follows: "Kinetic information is that which has quality and value, and is efficient in enabling or energising purposeful and relevant action in key areas of the organisation".

In focusing on action rather than decision-making, kinetic information separates itself from the traditional decision-making paradigm that has long influenced information systems, originally through the work of Herbert A. Simon, the Nobel Prize-winning economist and decision scientist. Under this paradigm, information systems are seen to exist for 'managerial decision-making' (Ahituv and Neumann, 1987). This is in spite of evidence suggesting that the decision-making paradigm is marginal to executive work and the organisation (Lindblom, 1959; Grayson, 1973; Ulvila and Brown, 1982; Kotter, 1982; Brown, 1992). The decision-making paradigm, and particularly its use as the *primary* metaphor in the organisational use of information, has also been described as 'fantasy', and deviating from real world experience (Boland, 1987; Feldman and March, 1987; and Galliers, 1987). It has also been shown that the portrayal of CEOs as decision-makers is a 'fallacy' (Keane, 1993).

One of the components of kinetic information is critical success factors, or CSFs (Rockart, 1979). That is, kinetic information addresses and supports CSFs. But kinetic information also supports critical business processes (CBPs), critical current issues (CCIs), and explicit goals & objectives (EGOs). Thus, kinetic information is action-oriented because it has *relevance;* it is relevant to the things that are most important.

But information is also action-oriented to the extent that it is easy to use, that is to say, it has *efficiency,* which is a second attribute of kinetic informtation. Third, kinetic information has *quality*, that is to say it is reliable and can therefore be acted upon, because it is based on quality data. And information has *value*, if it makes a difference when used or acted upon. For example, it has 'surprise' value, or increases confidence about something.

Lastly, if information tells you not only what to do and why to do it, but also tells you how to do it, it is thus oriented to the action process, so that the *what-why-how* of kinetic information is the fifth attribute that gives its orientation to action. The optimisation of all of these five information attributes constitutes kinetic information; information that enables or energises action.

THE KINETIC INFORMATION MODEL

If all of the attributes of kinetic information are optimised in an information flow or information source, then, 'we've hit the bulls-eye', so to speak. For example, information relevance is optimised if an information flow (such as a group of existing reports), addresses CSFs, CBPs, CCIs and EGOs, as defined. This can be assessed by using a Likert-type Scale, and asking a group of executive users to rate a given information flow for relevance on that scale. The following question is asked:

How *relevant* is the information that you get or are able to get from [the 'Super Duper EIS' enquiry screens or monthly project meeting, or customer field visits, or the 'Instant Management Journal', the 'Weekly Cash Flow Report', etc]?

In other words, to what extent does the information you get from this source support or respond to your critical success factors, critical business processes, critical current issues, or explicit goals & objectives?

Rate from 0 to 5 on the scale by circling one of the numbers, or a dash (-) in between the numbers:

Continuing with the same example, the executive users are asked to assess the same information flow for information efficiency on a Likert-type scale. The following question is asked:

How efficient is the information that you get or are able to get from [the 'Super Duper EIS' enquiry screens (or other information sources)] ?

In other words, how easy to use, how available, how accessible, how concise, and how flexible to use is this source of information for you?

Rate from 0 to 5 on the scale by circling one of the numbers, or a dash (-) in between the numbers:

If that same information is then rated in the same way, using a Likert Scale, for each of the other information attributes as well, then an overall picture emerges as to how *kinetic* is the information flow, as a whole.

Table 1.

Hardly any relevance	A little relevance	Moderately relevant	Very relevant	Extremely relevant
0 - 0.9	1 - 1.9	2 - 2.9	3 - 3.9	4 -5

The picture or model that emerges, after each of the five kinetic information attributes have been assesed for a given information flow, can be represented in the form of a 'dart-board' model. Each of the attributes or factors is measured on a Likert Scale, with the 'Low' rating end of each scale at the outside of the 'dart-board', and the 'High' rating end at the centre of the board, i.e. in the bulls-eye. The dart board model for kinetic information is shown in Figure 1.

This illustrates a hypothetical example, where the 'Super-Duper EIS' enquiry screens have been rated (using five questions for relevance, efficiency, quality, value, and what-why-how, and a Likert-type scale for each), as a kinetic information source by the group of executive users. The hypothetical ratings were:

1. relevance - 4.1,
2. efficiency - 3.7,
3. quality - 3.5,
4. value - 3.6, and
5. what-why-how - 3.1.

The overall average is thus 3.6 (rounded). This means that the 'Super-Duper' EIS enquiry screens actually provide *useful* information on the whole, but that this could be enhanced to *kinetic* information with some improvement in a couple of areas.

Kinetic information, as graphically represented by the bull's-eye orientation in the dart-board model, is an *optimal* concept, optimal in the sense that, if *all* the kinetic information attributes are present, action will be enabled and energised. This has implications for the effective use of information in organisations, and the 'bottom-line', where information is a key resource for knowledge, competitive advantage, and human effectiveness.

THE KINETIC INFORMATION METHODOLOGY

There is not the space to detail the kinetic information methodology, so it will be summarised here. It involves the selection of a group of executives in an organisation, who are asked to participate in about three days work and training over the course of about one month. In addition to the training, this involves several surveys of the multiple information sources used, the most important and least important information flows, a kinetic evaluation of those flows, and an evaluation of the kinetic information construct itself, all of which is oriented to a cultural/organisational change process.

Table 2.

Hardly efficient at all	A little efficient	Moderately efficient	Very efficient	Extremely efficient
0 - 0.9	1 - 1.9	2 - 2.9	3 - 3.9	4 - 5

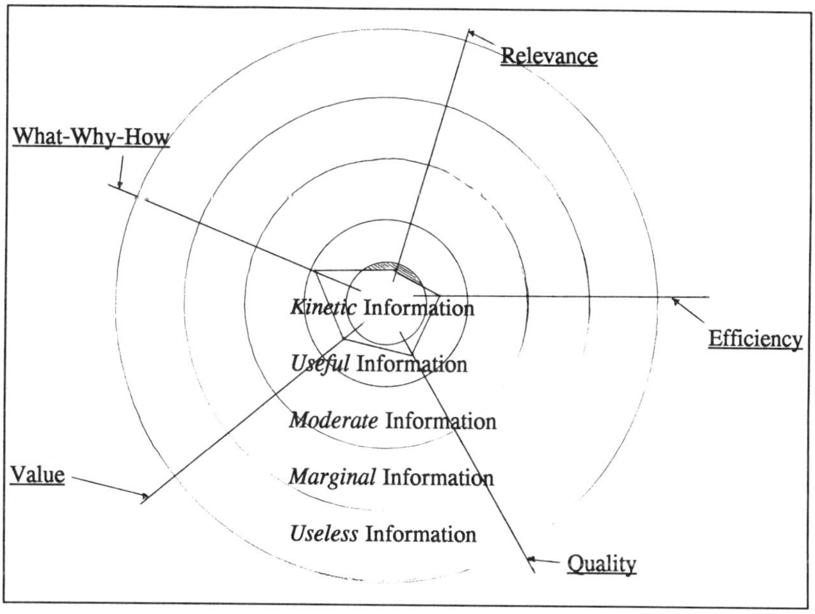

Figure 1. The kinetic information model.

REFERENCES

Ahituv, N., and Neumann, S., 1987, Decision making and the value of information, *in "Information Analysis: Selected Readings"*, R. Galliers, ed., Addision-Wesley, Wokingham.

Boland, R.J. Jr., 1987, The information in information systems, *in*: "Critical Issues in Information Systems Research", R.J Boland Jr. and R.A.Hirscheim, ed., John Wiley & SonS, Chichester.

Brown, R.V., 1992, The state of the art of decision analysis: a personal perspective, *Interfaces*, Vol. 22, No. 6, pp.5-14.

Feldman, M.S., and March, J.G., 1987, Information in organisations as signal and symbol, *in:* "Information Analysis: Selected Readings", R. Galliers, ed., Addision-Wesley, Wokingham.

Galliers, R.D., ed, 1987, "Information Analysis: Selected Readings", Addison-Wesley, Wokingham.

Grayson, C.J., 1973, Management science and business practice, *Harvard Business Review*, Vol. 51, No. 3, pp. 41-48.

Isenberg, D.J., 1984, How senior managers think, *Harvard Business Review*, Vol. 84, No. 6, pp. 81-90.

Keane, D.G., 1993, "Unravelling the Mind of the CEO: An Information Management Study of Ten Chief Executive Officers", unpublished PhD Thesis, Victoria University of Wellington, New Zealand.

Kotter, J. P.,1982, What effective general managers really do, *Harvard Business Review*, Vol.60, No. 6, pp. 156-167.

Liebenau, J., and Backhouse, J., 1990, "Understanding Information: An Introduction", Macmillan Education, Basingstoke,.

Lindblom, C.E., 1959, The science of muddling through, *Public Administration*, Vol.19, No.2, pp. 77-88.

Mintzberg, H., 1975, The manager's job: folklore and fact, *Harvard Business Review*, Vol. 53, No. 4, pp. 49-61.

Mintzberg H., and Waters, J.A., 1985, Of strategies, deliberate and emergent, *Strategic Management Journal*, No. 6.

Rockart, J., 1979, Chief executives define their own data needs, *Harvard Business Review*, Vol. 57, No. 2, pp. 81-93.

Stewart, R., 1988, "Managers and Their Jobs", Macmillan, Basingstoke.

Ulvila, J.W., and Brown, R.V., 1982, Decision analysis comes of age, *Harvard Business Review*, Vol. 60, No. 5, pp. 130-141.

INFORMATION SYSTEMS EVALUATION

From Accounting Theory to Systems Theory Through Practice

Christopher Davis

Department of Computing
University of the West of England
Bristol, BS16 1QY
University of Glamorgan Business School
Pontypridd
Mid Glamorgan, CF37 1DL
United Kingdom

INTRODUCTION

This paper provides an overview of current information systems evaluation practice. By highlighting the inadequacies of some commonly used techniques, the aim is to demonstrate how a move from the rational, reductionist perspective of accounting theory to a holistic, systemic one might ameliorate many of the difficulties experienced by information systems practitioners and their clients. Specifically, the capacity of systems theory to more adequately cope with the temporal aspects of evaluation is discussed.

The dominance of accounting techniques in the process of evaluation is illustrated by Grindley (1991). The semantic issues which facilitate understanding are acknowledged - these having been addressed elsewhere, the emphasis in this paper is on the educational value of adopting a systemic perspective to evaluation in order to promote a more meaningful appraisal of information systems. It is argued that a common collective perception of information systems is needed.

Current research, which seeks to explore the opportunity for organisational learning through action research throughout the life cycle of the information system, is described together with some intermediate findings. The need to develop a more comprehensive and collective understanding of information systems has, it is argued, been more fully addressed in the work reported than in any conventionally used technique as a result of the longitudinal, action research approach adopted.

EVALUATION

Keen (1994) points out that evaluation can serve a variety of purposes and, therefore, the specific evaluation methods used must suit their purpose. Currently, the purpose of most information systems evaluation is investment appraisal. The most problematic area for both builders and evaluators of information systems is that of the benefits which those systems bring to the organisation. Conventional evaluation of information systems has focused on the tangible artefacts - the computers and supporting communications technology - due to the visibility of their high costs, resulting in dominance of evaluation in this area by accounting techniques. Inevitably, such evaluation is primarily ex post facto and ignores the process aspects of information and the frequently unrealised benefits which it can bring to the organisation. Paradoxically, the dominance of such historical analyses also obstructs meaningful evaluation of the decision to invest. Such decisions are frequently taken at a strategic level in the organisation, due to the level of investment involved. Evaluation of the investment, however, is usually exclusively based upon operational arguments, typically concentrating on evidence of cost reduction so as to provide justification for the investment.

Evaluation thus becomes 'chronologically challenged' for a variety of reasons, not least the time lag between the decision to invest and implementation of the information system. More importantly, the systemic nature of organisations, information and evaluation itself is ignored by accounting techniques.

Such shortcomings are well illustrated by factors cited among responses to Grindley (1991), who sought to identify factors causing difficulty in evaluation:

- 'automation' of administrative procedures is not seen as a benefit since such 'backshop' activities are perceived merely as overhead costs of the 'frontshop' or main organisational function;
- the justification of information systems investment for non-productive organisational functions such as marketing involves assessing and taking risks;
- the ubiquity of information systems demands substantial investment in infrastructure to support them - since no specific use is perceived, neither do they have any benefit;
- investment in information systems is not only substantial but spasmodic, and therefore apparently incoherent;
- information systems do not directly contribute to corporate purpose - it is the employees of the organisation, the information systems' users, who satisfy customers needs, not the systems themselves;
- commitment to the evaluation of information systems is low since, due to the centralisation which the level and irregularity of the investment demands, information systems are perceived as a centrally owned resource;
- the benefits of improved control which information systems frequently bring are not perceived as tangible, therefore arbitrary revenue measures are attributed to them;
- the organisational flexibility which information systems enable are rarely seen in the context of broader organisational change;
- those responsible for information systems investment decisions frequently lack understanding of the nature of information systems and, therefore, of their investment.

A key difficulty is demonstration of the addition of value or benefit to both the financial sponsors of the information system and the people who use it. The final point on Grindley's list emphasises the need for greater understanding. One respondent said "I would like to think that it was an educational problem. Unfortunately, it's not that simple. Because we have no teachers. We are travelling in unexplored territory, and, since no one has been there before, it's very dangerous to listen to anyone who says he knows the way" (Grindley, 1991, p197). A key problem identified in some recent research (Davis, 1994; Mackintosh et al., 1994) is that of solipsism, whereby the language, customs and practice of professional groups help to define them and act, therefore, as a barrier to investigation and evaluation. Information is a particularly problematic factor in this context since its dissemination or sharing - frequently a key objective of information systems development - is perceived as a threat by the group, since such information sharing can demystify professional practice. Information becomes the key currency in 'turf wars' between groups, consequently misunderstanding occurs between groups at both inter- and intra-organisation level and significant suboptimality results (see Waters, 1994).

Evaluation of information systems, particularly the benefits they bring, must address the social setting of the systems in order to

i. allow longitudinal study of information systems use by benchmarking current practice;
ii. develop the competence and credibility of the evaluator(s) and
iii. facilitate the incremental growth in value added by the systems.

What is required is an evaluation framework that is more eclectic than those currently in general use; a framework which can accommodate

i. the longitudinal nature of information systems development and
ii. the requisite variety of both information systems and the social systems, such as professional groups, which use them. Ultimately, such an evaluative framework will enable movement towards a common collective perception of the information system and its benefits, thus overcoming the 'semantic gap' or solipsism which currently precludes effective evaluation. This gap cannot be bridged by accounting techniques which are typically rigid and involve looking backwards, downwards and inwards (in organisational terms) as opposed to outwards and forwards. Although systematic, such techniques are demonstrably not systemic and cannot, therefore, deal with the huge range of perceptions among system stakeholders.

Grindley (1991) suggests that the limitations of accounting techniques to evaluate information systems benefits described above are exacerbated by the separation of power and responsibility in relation to information within the organisation. Power to control the information management processes of the organisation is concentrated, and typically centralised, into the expert elite of the information technologists, whilst responsibility for delivery of goods or services is left with business managers and their staff - when those staff themselves comprise an expert elite, the difficulties are further compounded. The following paragraph describes some research which seeks to overcome those limitations by adopting an ethnomethodological approach to evaluation of information systems.

OBSERVATIONAL FIELDWORK

Between July and November 1993 observational studies were undertaken in the Clinical Neurosciences Directorate at Frenchay Healthcare NHS Trust, Bristol. 16 members of staff and 3 patients were overtly shadowed by four members of a research team based at the University of the West of England, Bristol. A minimum of two observers were used at all times so as to allow comparison of observation notes. The actors were shadowed for at least one full working day; frequently, the duration of the shadowing exercise was extended, particularly in the case of the patients and the surgeons, to allow continuity of observation of their episodic care and practice, respectively. The timing of the shadowing exercises was also varied so as to sample routine and non-routine activity at various times of the day and night.

Field notes were made openly at the time of the observed activity, or as soon after as was possible. This occasionally presented problems when actors other than the principal had not been made aware of the purpose of the exercise; on such occasions an explanation was immediately given so as to dispel fear and to maintain the excellent working relationship which had been developed over time. A proforma sheet was used to record observations in the field, indexed initially by time and location. The hand-written observations from each observer were collated and typed up under the supervision of a team member prior to being checked against the original field notes by their authors; this essential, albeit low-level, verification exercise proved to be very time consuming.

The individual activities recorded in the consolidated shadow histories were then analysed to generate lists of observed information management problems and of the interface or interaction which had been in progress at the time of the problem occurrence. Our aim was to further consolidate the recorded data and simultaneously transform it from a sequential list of activities into a list of observed problems. In this form, the problems could be readily ranked by the shadowee; providing a prioritised list for information systems intervention; such a list is beneficial to both the academic research being undertaken and to the development of a functional specification for an Integrated Clinical Workstation.

The next stage of the shadowing exercise involved further verification of the field work findings by the shadowee; feedback to the shadowees was provided in the form of a generic list of problem types presented in tabular classification, generated from analysis of all the field notes. Shadowees were asked to indicate the frequency of the occurrence described and its significance to them. They were provided with a copy of their shadow history to assist them in this task. The outcome of this stage of the research was a list of problems and associated interfaces which have been prioritised by the shadowees themselves. That prioritisation may be confirmed by further, more highly focused, observational fieldwork and by comparison with the findings of similar studies undertaken elsewhere.

The key findings at this intermediate stage in the research are as follows:

- almost three quarters of all the information management problems identified concerned manual, rather than computer based, systems;
- poor communication between the groups responsible for the clinical process (professional, semi-professional and non-professional) was frequently identified as a causal factor;
- the problems observed, though significant, were frequently unrealised by the actors concerned. Coping mechanisms had been introduced to ‘fix’ the problem and become routinised into day-to-day practice;
- the scale and opportunity cost of the problems observed was not realised.

The findings themselves and their significance are more fully discussed in Waters (1994).

ORGANISATIONAL LEARNING

The aim of this action research approach is to enable movement towards a common collective perception of what information systems are and what they can (and cannot) do by involvement of the actors in the process of evaluation. In this way, actors themselves can bridge the gap between the content of the evaluation (data gathering, note taking) and its context (the objectives sought of the evaluation)(Farbey et al, 1993). Much of the dissatisfaction with information systems evaluation and, therefore, the information systems themselves stems from the misinterpretation of the ultimate goal of evaluation, which should be understanding and learning. It is contended that development of the current research to explore the opportunity for organisational learning through action research throughout the life cycle of the information system will move some way to addressing this problem. The hybrid manager initiative, which seeks to bridge the semantic and culture gap between 'technologists' with power over information systems development and 'business people' with responsibility for their effective use is valid but insufficient.

Argyris and Schön (1978) provide some valuable insights into the double binds or hidden agendum which frustrate organisational or group learning. When considered in the context of the solipsism described above, the absence of such learning becomes a key issue in the misunderstanding of information systems' purpose, capabilities and value. Argyris and Schön explain that organisations collectively engage in single-loop learning, inculcating received wisdom throughout the workforce, thereby brutalising them (intellectually) into acceptance of organisational norms, leaving the norms unchallenged, organisational learning inhibited, and the power of the group unchallenged. Argyris and Schön (1978) show that such single-loop learning arises from innate 'theories in action' which individuals use to guide their day-to-day decision making. The secondary, longer term aim of the research reported here is to examine the extent to which theories in action and organisational learning can influence and be influenced by the information systems development process; this will require further development of the evaluative framework used to date in order to measure both information systems benefits and group/organisational learning.

Systems theoreticians can make a contribution in this research effort, but only by demonstration of the worth of their own practice. Hopefully we, as systems practitioners, will be able to develop evaluation methods which will overcome the dominance of the accountants, so that scrutiny of information systems focuses on their appreciation rather than depreciation.

REFERENCES

Argyris, C., and Schön, D., 1978, "Organisational Learning: A Theory of Action Perspective", Addison Wesley, Reading, MA.

Davis, C., 1994, "Getting Better with Information Systems: the ICWS Project", PICT International Doctoral Colloquium, Policy Studies Institute, London.

Farbey, B., Land, F., and Targett, D., 1993, "How to Assess Your IT Investment", Butterworth-Heinemann, Oxford.

Grindley, K., 1991, "Managing IT at Board Level", Pitman, London.

Keen, J., ed., 1994, "Information Management in Health Services", Open University Press, Buckingham.

Mackintosh, I., Waters, S., Bakehouse, G., Davis, C., and Doyle, K., 1994 , "Integrated Clinical Workstation: User Requirements for a Neurosciences Directorate", NHS Information Management Group, NHS Executive, Leeds.

Waters, S., 1994, "Information Management Problems: Towards a Cross-Sectoral Perspective", Health Economics Consortium, York.

56

AUDIT AND THE NET

Critical Issues for Information Management and Control

James Gilligan*

School of Information Systems
Faculty of Computer Studies and Mathematics
University of the West of England
United Kingdom

INTRODUCTION

As new organisational forms require management information and control systems to exist and interact through *connective metasystems* (more than simply networks) such as the Internet, so the need for change in the theory and practice of information systems audit provides a critical focus for significant issues in current methodological approaches to the development of business information systems and in the epistemological assumptions which are made about such systems.

Audit has been chosen as an area of information systems study which includes important questions regarding information management and control, many of which are likely to increase in significance. To date audit has been a *reactive* rather than a *proactive* agent in systems control. In discussing the critical lag of audit behind technology, Weber (1988, p.xiii) warns that "...the rate at which new computer technology is developed and introduced seems to outstrip the rate at which we can develop viable audit methodologies."

The current pace of new organisational forms and emergent technology, particularly in communications, requires that such a position is critically reviewed. This paper is a summary of initial findings from the author's current research into the validity of a systems approach to CBIS audit in the context of organisational change and emergent technology.

THE ROLE OF AUDIT IN CBIS

The American Accounting Association provides the following general definition of auditing: "Auditing is a systematic process of objectively obtaining and evaluating evidence regarding assertions about economic actions and events to ascertain the degree of correspon-

* e-mail J.Gilligan@csm.uwe.ac.uk.

Critical Issues in Systems Theory and Practice, Edited by K. Ellis et al.
Plenum Press, New York, 1995

dence between those assertions and established criteria and communicating the results to interested users."(Cushing and Romney, 1994, p. 694).

Within this definition, there are a number of categories of internal audit:

- Information Systems Audit, which concerns itself with reviewing the general and applications control of CBIS.
- Financial audit, concerned with the reliability and integrity of accounting records.
- Operational audit, sometimes referred to as management audit, and concerned with a review of operating performance of specific activities within the organisation.
- Compliance audit, used to assess the organisation's compliance with laws, contract provisions, government regulations, and other obligations to external parties.

This paper is concerned primarily with the first and the last of these categories. Cushing and Romney (1994, p. 698) state that compliance audit is "of lesser importance" than the other categories. It is suggested in this paper that while this may have been true in the past, the role of compliance audit is likely to increase and become of considerable importance within the field of IS audit itself, as Inter-Organisational Systems (IOS) and Electronic Data Interchange (EDI) proliferate. The difficulties of management and control over an 'untrusted network' (IEEE Computer Society, 1994) such as the Internet, when it or its like become vehicles for the exchange of data and participation in commerce, will surely increase the role of compliance audit. It is the purpose of this paper to identify the gap between the tradition and future of audit in such circumstances.

The Standards for the Professional Practice of Internal Auditing of the Institute of Internal Auditors state that (Cushing and Romney, 1994, p 695): "the scope of internal audit encompasses the examination and evaluation of the adequacy and effectiveness of the organisation's system of internal controls and the quality of performance in carrying out assigned responsibilities.". Five specific standards deal with the scope of audit:

1. Reliability and integrity of information.
2. Compliance with policies, plans, procedures, laws and regulations.
3. Safeguarding of assets.
4. Economical and efficient use of resources.
5. Accomplishment of established objectives and goals for operations or programs.

Note that the second standard requires compliance. Cushing and Romney (1994, p. 696) state that "Ensuring compliance with matters such as these requires an in-depth review of the information processing facility itself.".

It is significant that the term 'internal audit', and the use of the words 'system' and 'internal controls', appear to assume well-bounded, accessible, and relatively static organisation. Such characteristics are generally less true of recent organisational forms and of emergent technologies. To carry out such a review, the American Institute of Certified Public Accountants (Cushing and Romney, p. 704) advocate a four step framework:

1. Consider the types of error and irregularities that could occur.
2. Determine the accounting controls procedures that should prevent or detect such errors and irregularities.
3. Determine whether the necessary procedures are prescribed and are being followed satisfactorily.
4. Evaluate any weaknesses, i.e. types of potential error and irregularities not covered by existing control procedures - to determine their effect on a) the nature, timing, or extent of auditing procedures to be applied, and b) suggestions to be made to the client.

This is consistent with the accepted practice of 'risk-based' audit, requiring a 'systems review', a 'test of controls', the 'evaluation of control weaknesses and compensating controls.' Again this appears to assume the system's accessibility and permanence. The various tools available for such a systems review need not be considered in detail here. Any competent text on EDP audit (for example, Harnois, 1991) will provide a list of evidence collection and evaluation techniques similar to techniques of fact finding and analysis within computing. However, it should be said that several audit techniques are more recent in their application and are relevant to the changes under consideration in this paper.

The traditional approach to systems audit has often been that of *archaeology*. The auditor examining existing or past transactions, attempts to identify and authenticate the 'remains' of the data record, and construct from those remains a testable model of events within the system under investigation. Clearly such an approach admits to a principle weakness of audit to date, that of checking after the event. The perpetrator of a fraud, or an erroneous program, may only be discovered when it is too late, if at all.

To compensate for this in the light of the increasingly widespread use of on-line systems throughout the seventies and eighties, the audit profession introduced techniques such as the relatively sophisticated System Control Audit Review File (SCARF) and Continuous and Intermittent Simulation (CIS). In SCARF audit modules are embedded in the host application software. In this way, transactions which are defined as being of special interest are added to the SCARF file and periodically reviewed by audit personnel. Weber (1988, p. 710) provides a taxonomy of 'information' captured in such systems including that which is relevant not only to material inaccuracies but to 'policy and procedural variances'. In CIS the embedded audit modules do not reside in the application, but in the database management system which the application calls. The advantage here is that the user programs (or rather programmers) have no direct contact with the audit modules. Such approaches detach the audit from the application and so may prove valuable in the development of audit methods where the application, its user and the means of interconnection are inherently remote and unknown. In general however, current approaches to computer based information audit are based on the audit practitioners using techniques which assume a semi-closed, well-bounded business information system. This is not only true of audit, but is true of system methodologies themselves. While it may have been a reasonable assumption in the past, it may no longer be one we can depend upon.

THE RELEVANT EPISTEMOLOGY OF COMPUTER BASED INFORMATION SYSTEMS

'Computer Based Information Systems' has now replaced 'Data Processing' as the term most often used to describe the generic professional identity and focus of academic study for those interested in the use of computers in business. Consequently, though not necessarily, the dominant view of the application of computers to business is one which follows the epistemology of systems thinking. Indeed, so prevalent is this view of 'business computing' as 'information systems' that it is difficult to discuss the use of computers in business without recourse to the word 'system'. This position is increasingly open to question, and is questioned here in the context of audit as a business activity related but not subservient to the assumptions of information systems academics, practitioners and professionals.

In Gilligan (1994) it is argued that the word 'information' is a term whose current application to a large class of computer based business activities no longer provides us with a useful definition. What may seem merely pedantic has considerable real world consequences however when we look at the changing technologies which are being brought to bear on computer based business activities. For example, how useful is it to describe a telephone conversation or an e-mail message as part of an information system? Is video-conferencing or groupworking an 'information system' conforming to any traditional epistemology within that discipline? It may be argued that the more common uses of the word 'system' fail to serve several aspects of emergent technology which are based on loosely bounded systems in a state of constant modification and change which are not necessarily to be regarded as intrinsically systemic. It appears possible that in addition to the word 'information', the word 'system' is currently being used in ways which require radical reconsideration. This requirement is in part intellectual, where these words no longer provide us with useful definitions, forcing a reconsideration of the epistemology of the subject; and in part pragmatic, where organisations or technology have changed requiring us to reconsider the means by which we discuss our situations and requirements. Closely coupled to the word 'system' is the word 'boundary'. The literature of systems thinking and practice is in some measure the record of the means to establish boundaries and those entities and processes which are internal, external and transitional to what comes to be seen as the model of a system. The epistemology suggests that such systems are well-bounded and contained within the organisation, that the data is owned and the processing can to be defined.

Systems methodologies have used such assumptions, and evolved to support higher levels of organisational requirement, focusing on the epistemological subjective/objective divide (Doyle and Wood, 1991), the need for well engineered products (Date, 1986), or conforming to the consensus view of what is desirable for the management and control of an organisation (Checkland, 1981). Recent communications technologies may share some of these epistemological ideas and their associated representational techniques, but they are fundamentally different in that they do not accept ownership or the internalising of systems to any person or organisation. They use them to participate in the information process of the user, not to represent information itself. The Internet cannot be said to store 'information' in the traditional sense (as 'meaningful' data) for there is quite literally no way of knowing who will access it or what they will take from it.

In many areas of emergent communications technology, access and use do not expect well-defined statements of requirement from a person or organisation wishing to use a facility. There are resources available for use in the activity of information carried out by a user. Such facilities make no spurious claims to transform from data to information, rather they are available for those who wish to be informed. Their strength lies in resource management, where information remains the activity of the user. They do not fit well to any hard or soft system methodology, to the 'systemic' view of information, nor to corporate standards for information systems development. We may find ourselves looking back on an established class of computer based information system which is characterised by the modelling of organisational reality, well bounded, and claiming to provide 'information'. Emerging today however, is a new class of computer based technology which is adaptive in its means of modelling reality, loosely bounded and which assists in the information process of the user. This latter class cannot be easily defined in any traditional epistemology of information systems, but will include many recent computer based technologies which have the characteristic roles of data provision, decision support, or of communication.

THE IMPACT OF ORGANISATIONAL CHANGE EMERGENT TECHNOLOGIES AND CONNECTIVE METASYSTEMS ON THE ROLE OF AUDIT

We are rapidly entering the age of 'connective metasystems' which are more than mere networks. Open, rather than closed, they provide data highways at a global level, shared by individual and corporation alike, with the potential to provide telecommunication protocols, network control, client-server software, the human computer interface and, indeed, applications. All of these 'systemic' components may come from few or many machines as locally or remote as the user requires. Generic client-server technologies are among the first steps in this direction, but steps which expand the horizons of the user at each level of the computer facility. The user can sit at a PC yet work on intercontinental computer networks which make up the Internet without necessarily perceiving the change from one machine or continent to another. Software may run from a PC on servers across the world in untrusted facilities accessed across untrusted networks. How do we identify an activity whose components may exist across several machines and several continents? Can we still use the word 'system' to describe the application of computers on global networks? They are not simply bigger systems, they differ in fundamental ways. For example, the topology, components and personnel of the Internet constantly change. There are no necessary components, each is able to be replaced by another. The epistemology of such 'connective metasystems' is quite different from those of discrete, bounded applications. We must question whether or not our audit and systems methodologies are valid in such an environment.

The development of EDI and IOS, while relatively limited to date, is not new. Banking applications in particular operate ATM facilities on 'untrusted' networks. Existing commercial application of such techniques show that they can prove very successful in the controlled environments (Hirschheim and Adams, 1991, pp. 65-66), and in particular where such activities are centred around the exchange over networks of standard documents (Hinge, 1988). Much of the difficulty and reluctance to follow such developments may centre around the inability to audit or control access from competitors or other hostile agents. However, the disinclination will soon be tested by the development of higher level IOS, shifting from document exchange to decision support and strategic management. Such development need not be contained within internal systems of one or more organisations. Indeed, strategic systems have usually depended on external resources and, with global facilities now technically to hand, management requirements are likely to go in the direction of using those resources. The development of CSCW (Computer Supported Cooperative Work) and DDSS (Distributed Decision Support Systems) may grow, allowing dynamically linked specialist organisations sharing a global information resource, a further move towards the concept of quasi-firms, following the trend of new organisational forms. In such environments where the interface between organisations is not only significant, but is perhaps the largest component in that environment, the means by which auditors are expected to work within computer systems must change. Weber (1988, p. 15) states that

> "Collecting evidence on the reliability of an EDP system is more complex than collecting evidence on the reliability of a manual system. Auditors confront a diverse and sometimes complex range of internal controls technology that did not exist in the manual system.....Hardware and software continue to evolve rapidly, and although there is some time lag, the associated controls evolve rapidly too."

Given the quotation with which this paper began, this lag is not to be underestimated. Whatever the precautions taken, Weber (1988, p. 19) accepts that "Fraud perpetrated by a skilled systems programmer may be almost impossible to detect by an auditor who does not have a corresponding level of technical knowledge."

Yet the faith of business computing specialists in methodologies for dealing with systems problems is mirrored in the current audit literature. For functionalists such as Harnois (1991) the new technology presents difficulties to be met with increasingly functionalist techniques. Harnois (1991, p. 15) lists the specific objectives for EDP audit of Personal Computer Processing:

1. Ensure that an effective corporate policy exists relative to software acquisition and use.
2. Ensure that all necessary input, output and processing controls are present in each PC operation.
3. Ensure that only legally obtained and licensed software is being used for company business.

The sixty pages then given to a step by step audit through a PC computer system checks passwords, door lock, directory listings and user security. It is a very thorough internal review, though it makes no provision for the transmission or receipt of data on external networks. How are we to ensure the completeness of such an audit when the PC has access to the 40,000 networks which comprise the Internet?

Weber (1988, p. xiii) states:

> "Perhaps in the more traditional areas of auditing we, as auditors, are gaining confidence in the soundness of our methodologies. Admittedly, the lawsuits still prevail, but the problems seem to be in the application of the methodologies rather than in the methodologies themselves.".

This confidence is considered to be increasingly misplaced in the light of rapid developments in technology and organisational change which are beyond the grasp of traditional systems and audit methodologies. Trends in communications use, decentralised organisations, EDI, IOS, and alternative work structures suggest that traditional methodological approaches are already under pressure. The growth of communications is identified as a major area of concern for audit. Weber (1988, p. 930) states as of particular concern:

1. Electronic funds transfer systems.
2. Office Information Systems.
3. Customer Electronic Services
4. Electronic message systems.

The problems noted are:

1. These systems are generally technologically more complex than centralised systems. They have 'more components, more interfaces among components and higher levels of functionality.'
2. Exposures are often higher in a data communications system. Controls can be circumvented at more points in the system. The control technology 'is somewhat unstable and is still evolving.' The data being transmitted is itself increasingly valuable.
3. Data communications systems increase dependency between users. In EDI for example, 'a breakdown in controls in one part of the system potentially exposes all users of the system....Global control standards must be developed and someone must be appointed to monitor user compliance with these standards. As the networks become larger however, it is increasingly difficult to obtain the levels of assurance required.'
4. Privacy issues become more important.

5. The illegal or fraudulent use of the data communications networks becomes more likely.

It is particularly significant that Weber notes that the larger networks become, the more difficult it is to ensure compliance. It is also startlingly optimistic to imagine 'global' standards being introduced under a controlling body, given our poor experience in the computer world of attempting to create communications standards at even the most basic technical level. Weber concludes: "Auditors will have to undertake substantial research and to acquire substantial practical experience with data communications systems before suitable controls and audit procedures can be devised."

CONCLUSION

It is clear from the literature of EDP audit that the Auditor has many sophisticated techniques for dealing with the audit of large computer systems. However in almost every case, that provision is directed towards semi-closed, centralised or internally distributed systems, rather than open or external systems.

It is also clear that the growth of external network facilities continues apace. The use of the Internet (Internet Society, 1994, p. 1) has increased by 21% in the 3rd quarter of 1994 alone. The .com (commercial) domain has now become the largest on the Net. The .net (network) domain which is heavily used by many public data service providers for their customers grew by 66%.

The systems methodologies which prevail in systems thinking have an underlying epistemology which assumes a controlled environment and a contained computer system. These are no longer the relevant images of our computer networks. Added to what Chambers and Court (1991, p. 70) describe as 'computer related', 'computer assisted' and 'computer dependent' fraud, we must add the risk of 'untrusted networks'. Changes in our understanding of how to maintain control of our business activities across such technology may become an essential part of participating in global networks.

A first requirement must be to 'design for audit'. CIS and SCARF are steps in that direction, but further steps must be taken outside the organisational boundary. It may prove to be a starting point to have embedded modules within internal systems, across the firewall, within client-server applications, and so on. The notion of such embedded modules may remind the reader of claims that public telecommunication computers perform checks on calls outside national boundaries where known sensitive telephone numbers will trigger a recording of the call. The arrival of 'untrusted networks' may mean that such methods will prove relevant to the emergent technologies of business systems. The organisational, technical and social considerations are considerable, while the audit consequences of neglecting such methods may provide insurmountable. However even this will not be sufficient. The literature admits in several places to the lack of basic research. Weber (1988, p. 935) states :

> "To date, little theoretical and little empirical research have been carried out within the EDP audit area... whereas a fairly extensive literature now exists, it primarily comprises practitioner writings. Thus the experiences of EDP auditors are reasonably well documented but basic research is lacking...commitments are needed in the EDP audit area if a research base is to develop."

It is unlikely, however, that auditors will grasp the nettle of changing technology when those who create information systems have only a limited knowledge of the impact to be expected from the commercialisation of global network facilities.

This paper is the product of initial research from a systems perspective into changes in the theory and practice of audit which may be required as a result of organisational change and the introduction of commercial external and untrusted networks in coming years. It is hoped that further developments may be reported to future conferences.

REFERENCES

Chambers, A.D. and Court, J.M., 1991, "Computer Auditing", 3rd ed., Pitman, London.

Checkland, P.B., 1981, "Systems Thinking, Systems Practice", John Wiley and Sons, Chichester.

Cushing , B.E. and Romney, M.B. 1994, "Accounting Information Systems", 6th ed., Addison-Wesley, Reading, Mass.

Date, C.J., 1986, "An Introduction to Database Systems", Volume 1, 4th ed., Addison-Wesley, Reading, Mass.

Doyle and Wood, 1991, Soft systems and systems engineering: across the great divide, *in:* Addendum to proceeding of UKSS 1991, Plenum, New York.

Gilligan, J., 1994, Patterns on glass: the language games of information, *in:* Proceedings of the Symposium on Philosophical and Logical Aspects of Information Systems, University of the West of England, 1994 (pending publication).

Harnois, Albert J., 1991, "EDP Auditing: A Functional Approach", Prentice-Hall, New York.

Hinge, K.C., 1988, "Electronic Data Interchange: from understanding to implementation", American Management Association, New York.

Hirschheim, R. and Adams, D., 1991, Organisational Connectivity, *Journal of General Management*. 17(2).

IEEE Computer Society, 1994, "Internet Security, Secure Communications Over Untrusted Networks", Seminar Summary from 12th November, 1994, published on Usenet by voidstar@netcom.com

The Internet Society, 1994, "Internet Domain Survey", data available on ftp.nw.com or http://www/nw/com.

Weber, R., 1988, "EDP Auditing: Conceptual Foundations and Practice", 2nd edition, McGraw-Hill, New York.

Woolf, E., 1990, "Auditing Today", 4th ed., Prentice-Hall.

57

INCORPORATING INDIVIDUAL DIFFERENCES IN PERSPECTIVE IN A SYSTEMIC APPROACH TO INFORMATION SYSTEMS REQUIREMENTS ANALYSIS

Judy McKay and Peter H. Marshall

School of Information Systems
Curtin University of Technology
Perth, Western Australia
Australia

Introduction

Investment in information systems (IS) and information technology (IT) in contemporary organisations is both widespread and substantial. There are few businesses today that can operate effectively and competitively without resorting to at least minimal reliance on IS/IT as an integral part of day to day operations. For most, investment in IS/IT now represents a significant proportion of capital and operating expenditures. Indeed, for these organisations, seeing adequate returns on their investments has now become a pressing management issue (Ward et al., 1990). Hence it can be concluded that information systems development (ISD), whether this is defined as developing computer-based systems from first principles, or tailoring off-the-shelf packages, or some variant of these two extremes, is a common, necessary and important process in most contemporary organisations.

Yet the academic literature abounds with tales of disappointments and failures with ISD projects (Eason, 1988; Galliers, 1993). More thorough examination of some of the research into failed ISD projects is startling. Ford (1994) reports that fewer than 10% of custom-built IS are successfully implemented and used without significant modification, and that more than 50% of ISD projects fail to result in a system deemed satisfactory by users. A 1988 study by the Department of Trade and Industry (Flynn, 1992) suggests that poor quality IS cost one billion pounds annually in the UK, this figure excluding estimates of lost business, damage to reputation, and missed business opportunities directly attributable to the failure of the IS.

Careful analysis of the reasons underlying many of these failures and disappointments reveals four pervasive themes, of which the first three seem to be highly interconnected and are of particular interest in this paper. These four themes are:

- difficulties existed in adequately assessing the scope and context of the ISD

Critical Issues in Systems Theory and Practice, Edited by K. Ellis et al.
Plenum Press, New York, 1995

- difficulties existed in adequately determining user information requirements
- communication difficulties existed amongst and between analysts and users
- poor budget control (Hirschheim, 1985; Davis, 1992; Flynn, 1992)

It can be concluded therefore, that adequately determining user information needs, addressing the appropriate organisational problem context, and adopting a process which adequately supports communication amongst participants (both analyst and users), in the early stages of systems development remains a vexing problem. This is despite the many technological advances to support the process of information systems development such as CASE tools, for example.

In this paper, the authors will argue that the adoption of a systemic approach to Information Systems Requirements Analysis (ISRA) can help to alleviate some of the difficulties previously identified. In particular, it will be argued that there is a need for ISRA methodologies to strive to take into account the multiple and often conflicting individual perspectives that, if ignored, can lead to communication problems and disappointment with the system development effort. The Strategic Options Development and Analysis (SODA) methodology, and its accompanying technique, cognitive mapping, will be put forward as a suitable candidate methodology for ISRA. The paper will conclude with a brief discussion of one successful ISRA intervention in which SODA and cognitive mapping proved to be valuable support tools for determining information requirements for a systems development project in a government agency in Western Australia.

Systems Thinking Applied to ISRA

The recognition that IS are concerned primarily with human activity rather than being considered purely technical matters, has lead to the development and application of methodologies that offered much more help to the analyst in determining user information requirements (Mumford, 1981). ISD has also been helped by the recognition that a holistic approach to problem solving often helps to surface the 'subjective' nature of organisational problem contexts (as opposed to a real problem with an 'objective' existence in the world). Such systemic thinking also arguably supported valuable learning about organisational contexts and interpersonal interactions within these contexts (Prior, 1992). Soft Systems Methodology (SSM) is often cited as a powerful tool for the front-end analysis of any systems development project for a number of reasons (Wilson, 1989). Its interpretivist stance, giving acknowledgment to the concept that social reality (and hence the context of ISD) is constructed and reconstructed through interaction and negotiation between participants, implies that both the subjective nature of perceived problems and their interconnectedness with other aspects of organisational contexts can be accommodated through the application of SSM (Prior, 1992).

However, with any group-based activity, there remains the risk that in contexts where individuals perceive there to be even a slight lack of psychological safety, differences in individual viewpoint may remain unarticulated. There may be many underlying causes for this: a highly charged political atmosphere, a culture that rewards the suppression of contention, a lack of motivation amongst group members, a lack of faith in the sincerity of management to change, and so on. Whatever the underlying cause, and no matter the degree of severity of this phenomenon, the risk in ISRA is that a shared understanding of the organisational context is not reached by all participants, and requirements may not be fully articulated. Furthermore, there appears to be a need to ensure that it is the users', rather than the analysts', perspectives and needs that are the drivers of the subsequent system design.

Few methodologies for ISRA provide any support for the analyst in this regard. For example, it does not appear that SSM, despite its power and success, takes account of perceived individual differences in information requirements and understandings of organisational contexts (Holwell, 1992). There have been attempts to overcome these limitations through the application of Group Decision Support System technologies (GDSS) to ISRA. The GRIPPPER methodology (Group Requirements Identification Process, Prototypes, Pilots, Evaluation and Rollout) (Wanninger & Dickson, 1992(a),1992(b); Limayem & Wanninger, 1993) allows for anonymous input of ideas and opinions, and may help in the surfacing of individual differences, but suffers from being a technology-driven process and offers little support for the negotiative aspects of group process and the like. The authors contend, however, that the SODA methodology, supported by cognitive mapping, has features that would suggest it is well-suited to being applied to ISRA. There features are further discussed below.

The SODA methodology AND cognitive mapping

The SODA methodology was developed by Eden, and is more fully described in many of his writings (for example, Eden 1989 and 1990). SODA is an approach designed to support decision makers in dealing with the complexity and interdependency inherent in many organisational problem situations. Pivotal to the SODA approach is the view that each individual in the decision making team will have his/her own subjective view of what constitutes 'the problem', and will bring to the group a range of skills and knowledge. This individuality, the different perspectives, skills and knowledge, needs to be exploited so that individuals and the group as a whole can carefully examine the assumptions underlying articulated goals and values, gaining increased self-awareness and intersubjective understanding. From this can grow confidence in and commitment to decisions on actions which are viewed as serving to ameliorate the perceived problem situation (Eden, 1989).

The SODA approach could be viewed as consisting essentially of two phases. During the first phase, a facilitator works with individuals concerned with the problem situation. In a one-to-one interview, the facilitator seeks to elicit the relevant understandings, beliefs, values and expertise of the individual, in relation to the problem of concern (Ackermann et al., 1991). These are used to construct a cognitive map for that individual which is subsequently discussed, explored and verified with that person.

A cognitive map may be said to be a model of a person's understanding of the problem situation by means of linguistic fragments (chunks of text) linked by arrows. The arrows denote that the thoughts or concepts embodied in one chunk of text leads to or is a consequence of another chunk of text. The maps are thus natural, language-based representations of the understanding an individual has of a problem situation, and the meanings that that individual attributes to concepts forming part of that situation. In other words, cognitive maps are a means of representing the way in which an individual defines and conceptualises a situation (Eden, 1989).

The second phase of the methodology revolves around the SODA workshop(s). In this phase, the individual cognitive maps are aggregated to form a group map, a single model representing the group's definition of the perceived problem (Eden, 1988; Marshall et al., 1993). During the SODA workshop(s), the group of problem solvers meet to discuss and negotiate with one another, using the group map to facilitate increased levels of intersubjective understanding through reconstructing meanings of the problem situation (Eden, 1993). In this way, the group is assisted in working towards new levels of insight and understanding, with the aim of reaching a consensus and commitment to action, as a result of the dialogue based on individual and group maps (Eden, 1988).

SODA and Cognitive Mapping Applied to ISRA

There are a number of reasons why SODA and its associated technique of cognitive mapping appear to be so well suited to supporting ISRA. These have been articulated at length elsewhere (McKay, Marshall & McDermid, 1994) and it is proposed in this paper only to briefly summarise the main points. It needs to be borne in mind that these arguments reflect the authors' belief that SODA and cognitive mapping can help to overcome some of the reasons underpinning the failure of systems development projects articulated earlier in this paper.

1. SODA has a proven track record in dealing with complexity and other aspects of organisational problem contexts such as social, cultural, political and behavioural factors. These are all important features of the IRD process (Dumdum & Klein, 1986).
2. SODA acknowledges the existence of multiple, often conflicting mental models and provides a framework in which the assumptions underlying these models can be explored.
3. The dialogue supported by the individual and group cognitive maps leads to increased self-awareness and shared understanding of the context of the information system under development.
4. SODA's participative approach supports the emergence of a negotiated group view of information requirements, leading to much greater involvement and feelings of ownership and commitment.
5. Cognitive maps rely on natural language to represent individual and group meanings and understandings. This system of representation is equally accessible to users and analysts, helping to overcome the communication problems traditionally associated with the adoption of common IS modelling conventions, such as E-R diagrams, DFDs, etc. (Oliver & Langford, 1987). Thus, SODA typically results in improved communication between analysts and users.
6. Electronic and paper records of the various maps generated by the software COPE help to capture the essence of the workshop, stimulating an emotional connection with that session (Eden, 1993). This emotional connection is important in refreezing new values, beliefs, and understandings (Lewin, as cited by Baron & Greenberg, 1990), remembering decisions reached during the workshop(s), and bringing about learning and a commitment to implement the decisions taken. Arguably, the maps also provide a critical record of deliberations, disagreements, and decisions on courses of action in an IRD project. This may help build a better understanding amongst IS professionals of the changing nature of user information requirements, and the need for flexibility in system design to better accommodate the evolutionary nature of IRD in a learning organisation.

ACTION RESEARCH AT THE XYZ AGENCY

The XYZ Agency is a semi-autonomous government instrumentality in Western Australia with major links to the transport industry. The Agency currently employs some 230 staff (approximately 100 white collar workers), having reduced from a staff of over 700 some three years ago. A major restructuring and overhaul of its operations in 1990-91 was deemed necessary following a number of years where the trend was for increasing staff numbers and operating costs, and decreasing productivity and profitability. With its new, trimmed workforce and management vigorously pursuing a policy of returning XYZ Agency to focus on its core business and distinctive competency, the Agency is now operating effectively and profitably. The changes have been carefully articulated and sold to staff who seem to understand the reasons behind the restructuring and are generally accepting and supportive of the new managerial initiatives.

The authors became involved with the XYZ Agency following a decision to migrate from an Informix to an Ingres environment. The IS manager had decided to tie this process to an upgrade and refurbishment of their information systems. The authors' brief was to undertake the ISRA for a new Workers' Compensation Systems (WCS). Although the WCS was considered a small system in terms of the number of regular users (4) spanning three departments (Human Resources, Occupational Health & Safety, and Payroll), it was given relatively high priority for redevelopment, essentially because of the obvious executive commitment to creating a safe and supportive work environment.

Full details of this action research project using SODA and cognitive mapping are to be reported elsewhere (McKay, 1995), and constraints do not allow a full discussion here. Nevertheless, a few important points can be made.

First of all, the series of initial interviews with users, resulting in the production of individual cognitive maps, served at an early group workshop to show that quite marked differences in perspective existed as to the nature and scope of the new WCS in what was otherwise a highly cohesive group. Broadly, it could be demonstrated that two 'camps' existed, one viewing the major purpose of the WCS to support the efficient and effective processing of claims for compensation, the other, in contrast, viewing the major requirement of the WCS to help identify accident trends and to thus improve safety at XYZ Agency. A meeting of users and IS Branch management subsequently resolved to develop two systems: the upgraded WCS was to handle the processing and on-going management of claims and the legal reporting requirements associated with such claims, while a new 'Safety' system would be developed to identify unsafe work areas, accident trends at XYZ Agency, and to produce reports for presentation to Executive Board meetings.

The users were confident that the cognitive maps had served to clearly illustrate that differences in individual perspective existed, energies thus being able to be directed as to how best to accommodate such differences (a win-win situation), rather than the meetings degenerating into arguments over whose viewpoint should prevail (a win-lose situation).

Secondly, the cognitive maps also proved useful in helping users articulate their specific information requirements. During group workshops, maps were either projected through a large TV monitor, or more commonly, participants were supplied with hard copies of the maps. It was a common occurrence for users discussing a group of concepts to make remarks such as 'Well, if we all agree that these concepts are important, doesn't it imply that we will need...'. It is not intended to suggest that the cognitive maps were the *only* means by which needs were identified, but they did appear helpful in triggering thoughts and ideas that directly lead to the articulation of requirements.

Finally, the use of cognitive maps during both the individual interviews and the group workshops served the purpose of providing an impersonal exterior 'thing' that was being talked about, rather than the more common, potentially confrontationist situation of analyst-user direct communication. To the authors, as analysts, it was graphically illustrated that the users were learning about their context through the ISRA process. It became a natural and expected occurrence, therefore, that their understanding of their requirements would evolve as the process continued. From the users' perspective, their sense of ease with and ownership of the maps meant that they could easily identify areas where the analysts had misunderstood or misinterpreted their intentions. The maps were easily understood by users, and often became a vehicle by which users were empowered to clarify the analysts' understandings. Thus, the maps appeared to support the communication process between participants in the process.

CONCLUSION

This paper has argued that systems thinking has done much to improve the practice of ISRA. However, the ability to take into account differences in individual perspectives of systems requirements was argued to be an important, if somewhat neglected, consideration. The power of SODA and cognitive mapping for ISRA was illustrated by a brief description of a successful study in which the authors were involved.

REFERENCES

Ackermann, F., Cropper, S., Eden, C., and Cook, J., 1991, Cognitive Mapping for Policy analysis in the Public Sector. Paper presented at the American Academy of Management Conference, Miami, USA, August 1991.

Baron, R.A., and Greenberg, J., 1990, "Behavior in Organizations: Understanding and Managing the Human Side of Work," Allyn & Bacon, Boston.

Checkland, P., 1981, "Systems Thinking, Systems Practice," John Wiley & Sons, Chichester.

Davis, G.B., 1992, Systems analysis and design: a research strategy macro-analysis, *in*: "Challenges and Strategies for Research in Systems Development," W.W. Cotterman and J.A.Senn, eds., John Wiley & Sons, Chichester.

Dumdum, U.R., and Klein, H.K., 1986, The need for organizational requirements analysis, *in*: "Quality of work versus quality of information systems," H.E. Nissen, and G. Sandstrom, eds., Report of the Ninth Scandinavian Research Seminar on Systemeering, Lund.

Eason, K., 1988, "Information Technology and Organisational Change," Taylor & Francis, London.

Eden, C., 1988, Cognitive mapping, *European Journal of Operational Research*, 36:1-13.

Eden, C., 1989, Using cognitive mapping for strategic options development and analysis, *in*: "Rational analysis for a problematic world," J. Rosenhead, ed., John Wiley & Sons, Chichester.

Eden, C., 1990, Strategic thinking with computers, *Long Range Planning*, 23:35-43.

Eden, C., 1993, On evaluating the performance of "wide-band" GDSS's, *Unpublished paper, Strathclyde Business School, University of Strathclyde*, Scotland.

Flynn, D.J., 1992, "Information Systems Requirements: Determination and Analysis," McGraw-Hill, London.

Ford, S.G., 1994, Practical software quality assurance, *Institute of Internal Auditors Australia, National Conference*, Perth.

Galliers, R.D., 1993, Information systems and business reengineering, *Paper presented at IFORS '93*, Lisbon, Portugal.

Hirschheim, R.A., 1985, "Office Automation: a Social and Organizational Perspective," John Wiley & Sons, New York.

Holwell, S., 1992, SSM information approach: a critique, *Systemist* 14:93-98.

Limayem, M., and Wanninger, L.A., 1993, The use of a group CASE tool to improve information requirements determination, *Document de Travail 93-33, Universite Laval.*

Marshall, P.H., Atkinson, D.J., Ee, L.P., and McKay, J.A., 1993, Computer-aided group problem solving: an overview of successful methodologies, *The Environmental Professional*, 15:159-163.

McKay, J.A., 1995, The application of soft management science/operational research to information systems requirements analysis: a study using cognitive mapping and the SODA methodology, *Unpublished doctoral dissertation, Curtin University of Technology*, in preparation.

McKay, J., Marshall, P., and McDermid, D., 1994, Cognitive mapping for information requirements determination in the learning organisation, *in:* "Proceedings of the 17th IRIS Conference", Syote.

Mumford, E., 1981, Participative systems design: structure and method, *Systems, Objectives, Solutions* 1:5-19.

Oliver, I. and Langford, H., 1987, Myths of demons and users: evidence and analysis of negative perceptions of users, *in*: "Information analysis: selected readings," R. Galliers, ed., Addison-Wesley, Sydney.

Prior, R., 1992, Linking SSM and IS Development, *Systemist*, 14: 128-132.

Wanninger, L.A., and Dickson, G.W., 1992a, GRIP - Group requirement identification process: the integration of GDSS technology into the systems development life cycle, *in*: "Proceedings of the Hawaii International Conference on Systems Science Vol III," J.F. Nunamaker, ed., IEEE, Los Alamitos.

Wanninger, L.A., and Dickson, G.W., 1992b, Phased systems design, development, and implementation: process and technology, *in*: "The Impact of Computer Supported Technologies on Information Systems Development," K.E. Kendall et al., eds., Elsevier Science, Amsterdam.

Ward, J., Griffiths, P., and Whitmore, P., 1990, "Strategic Planning for Information Systems," John Wiley & Sons, Chichester.

Wilson, B., 1989, A systems methodology for information requirements analysis, *in*: "Systems Development for Human Progress," H.K. Klein and K. Kumar, eds., Elsevier Science, Amsterdam.

58

THE CRITICAL INFORMATION SYSTEMS PRACTITIONER

Listening to the "Babble of Voices"?

Brian Nicholson

Department of Technology
Anglia Polytechnic University
Chelmsford
United Kingdom

INTRODUCTION

The 'grail - like' quest for the ultimate super methodology has led the information systems community to witness a veritable "babble of voices" (Schon, 1983) with between 300 (Avison and Fitzgerald, 1988) and 1000 (Jayaratna, 1994) methodologies available. All of these methodologies claim to have advantages over others and jostle for position in an overcrowded, confused marketplace. The 'voices' come from many directions each advocating its own philosophy which it attempts to sell as the ultimate 'silver bullet' or universal approach. What of the bemused practitioner at the receiving end? It has become close to impossible to effectively select an appropriate methodology due to the sheer diversity of offerings from a plethora of different schools of thought. The practitioner must make sense of chaos before he/she can even begin the problem solving effort. The aim of this paper is to examine the available means for selection of information systems methodologies considering the assumptions they make about the problem solver. The methodology selection arena is broadly reviewed and some potential threats to effective practice are discussed. An educational agenda is proposed to enable present and future practitioners to make sense of this "...diverse and ill-defined intellectual enterprise" (Checkland, 1983) of methodology selection.

METHODOLOGIES, FRAMEWORKS AND META METHODOLOGIES

A typical newcomer to the field of information systems research may be a competent practitioner seeking ways to improve his or her practice by the use of an information systems development methodology. On consulting the literature the practitioner looking for guidance through the 1000 methodology minefield, witnesses that there are strong differences of

Critical Issues in Systems Theory and Practice, Edited by K. Ellis et al.
Plenum Press, New York, 1995

opinion over how information systems should be developed. Wood Harper and Fitzgerald (1982) identify 6 different approaches each of which "...views the world in a different way". (Episkopou and Wood Harper 1986, p.222) go on to discuss the human activity, participative and data analysis approaches: "...corresponding to the 'soft', 'cybernetic' and 'rational' views identified by M.C.Jackson." (p.222).

In the same paper, Episkopou and Wood Harper (1986) use the metaphor of the problem solver "changing spectacles" when switching between the world views held by particular methodologies. Therefore, the 'spectacles' which are been worn by the problem solver dictate whether the project is: "...seen as a logical, technical or people problem" (Avison and Wood Harper, 1990, p.20).

Jayaratna (1994) diversifies the debate by discussing the need for information systems practitioners to "...acquire the richest possible understanding of organizations if they are to become effective problem solvers" (p.61) and earlier states that "...one of the major weaknesses of most current information systems methodologies is that they are not concerned with what really happens in organizations" (p.58). This implies the need for consideration of approaches calling for reflective practice (Schon, 1983), greater attention to the contextual backdrop of the proposed information system and perhaps the incorporation of management science methodologies. Thus many confusing 'voices' echo around our hypothetical practitioner.

WHICH METHODOLOGY WHEN?

The call for help from those realizing the complexity of Information Systems design has led to 'frameworks' and 'meta methodologies' to help to facilitate choice. Again, there are inevitably different schools of thought offering guidance to assist the process of methodology selection. Checkland's observation, although optimistic, indicates the fragmentation evident in the selection arena: "The whole field of information systems lacks its Newton to bring it conceptual clarity" (Checkland 1988, p.304).

Meta methodologies or frameworks for selection could be divided into three broad categories; conceptual, contingent and integrative approaches to selection.

i. Conceptual Guidelines and Frameworks

Conceptual approaches are broadly speaking concerned with engaging problem solvers in critical evaluation of a methodology which will indirectly facilitate selection. Examples of this approach to implicit methodology selection are Jayaratna's Normative Information Model based Systems Analysis and Design, (NIMSAD) framework and the criteria for comparison suggested by Avison and Fitzgerald (1988). Jayaratna (1994) differentiates frameworks from methodologies and defines a conceptual framework as:

> "...a static model...a meta-level model through which a range of concepts models, techniques, methodologies can either be clarified, compared, categorized, evaluated and/or integrated." (Jayaratna 1994, p43).

Jayaratna's definition of a framework is expanded into NIMSAD (Jayaratna, 1994) which proposes a model that: "...requires its user to consciously consider the rational basis and the relationships of the activities suggested by the steps of any methodology" (Jayaratna 1994, p.116).

Jayaratna's conception of a framework is rather more sophisticated than the criteria proposed by Avison and Fitzgerald (1988). Avison and Fitzgerald propose general criteria to assist comparison of methodologies.

Conceptual frameworks assume a very high level of sophistication of the analyst in the situation, the problem solver. NIMSAD particularly demands that the user engages in a rigorous and penetrating level of critical questioning of the problem situation, the methodology user and of the methodology under consideration. For instance, the problem solver is required to question the perceived system of the problem owner and be able to critically reflect in the manner suggested by Schon (1983) on his and other participants 'mental constructs'. NIMSAD describes what activities should be considered in an information systems problem solving situation and thereby enables the user to consider and critically reflect on the candidate methodology. This involves the NIMSAD user in such activities as assessing the methodology's sensitivity to organisational culture and politics and the organisational and social implications of the methodology philosophy.

ii. Contingency Approaches

Contingency approaches are developed around pragmatic lines in that they attempt to provide guidelines for matching problem types to methodology approaches. An example of this is shown in Multiview (Avison and Wood Harper, 1990) in which users are presented with a selection approach which incorporates Checkland's Soft Systems Methodology (Checkland, 1981; Checkland and Scholes, 1990), Mumford's ETHICS (Mumford, 1983) and structured methods. The practitioner is encouraged to match the situation with the tools and techniques on offer in a 'horses for courses' approach . These 'all encompassing' methodologies rely heavily on the analyst's skill in identifying appropriate circumstances for application. Used in this way they could be potentially dangerous if used uncritically, in that there is insufficient guidance to enable reflection on the appropriateness of the techniques. The creators of Multiview assume that the user will be sufficiently sophisticated to be able to select from the techniques with an understanding of the implications of the choice. Jackson (1991) provides a rich discussion on the dangers of unreflective pragmatic practice or 'tool kit' strategy in the field of management science. Importantly, he argues that the pragmatic selection of techniques without respect for theoretical guidance or critical reflection could appear to 'work' because they reinforce the position of the powerful in problem situations.

iii. Integrative Approaches

Work is ongoing in attempting to provide a means of integrating methodologies. Bell and Oates (1994) propose a method for method integration, building on the work of Kronlof (1993). This research area is still at a relatively early stage, Bell and Oates have set an agenda for building a framework to provide structured guidance for the integration task, however the complete model is not yet available. A potential problem in the approach, yet to be overcome, is that of paradigm incommensurability. The integrative approach faces the same dangers as the contingent discussed above in that uncritical combining of techniques by an unreflective user, without understanding of the paradigm from which the methodology is derived, leads to unreflective pragmatic practice.

THE THREAT OF ISOLATIONISM

The methodology selection arena is becoming divergent in that the frameworks represent differing views on the way methodology selection should be approached. It is unclear whether selection of methodologies should be accomplished by development of an underlying critique, as in the conceptual approaches; by matching problems with methodologies using contingency methods or by the utilisation of an integrating framework for systems methods. The time could arrive when there will be a need for a framework/methodology in order to choose an appropriate approach at the meta level. Practitioners faced with this disarray are apt to shy away from new approaches, the consequence of this could hold a potential threat to information systems practice in the form of isolationism (Jackson, 1991). This threat lies in practitioners adopting a house style or a "...we do what we know" attitude, Jackson (1991,p257) discusses the dangers of this approach. In summary, isolationism encompasses a blinkered view of problem situations induced by an adherence to one methodology for most, if not all problems. An example of this in the information systems context is shown in the wide acceptance of SSADM and proposed development of a Euromethod. SSADM is a cry for methodology standardisation when standardisation in our world of 'wicked problems' (Rittel and Webber, 1974) is inappropriate. DeMarco (1982, p131) summarises the dangers: "The idea that a single methodology should govern even two different projects is highly suspect. The differences between projects are more important than the similarities" . Peter Middleton in his critical account of SSADM discusses some distinct weaknesses of the SSADM approach (Middleton, 1994). The wide acceptance of SSADM, given the potential dangers, is an indication of the inability of the systems community to get the message across regarding other approaches, partly because adoption and selection have become over complex and risky.

AN AGENDA FOR INFORMATION SYSTEMS EDUCATION

The frameworks discussed above make potentially unrealistic assumptions about the sophistication of their users and their ability to critically reflect on the methodology choices they have made. Jayaratna (1994) discusses the assumptions methodologies make about their users: "We feel that methodology creators should inform their intended users about the nature and type of concepts they need to be familiar with before applying the methodologies. Methodologies should therefore provide "...'health warnings', since they have a responsibility to alert their users to these dimensions." Jayaratna (1994, p120)

Meta methodology / framework creators are equally guilty of making assumptions about their users, therefore by the same token similar 'health warnings' ought to be provided by the selection arena. In addition to technical ability, the frameworks for selection and evaluation discussed in this paper assume the following essential abilities and understanding of their users:

- Schon style critical reflection
- Social and organisational implications of adopting a particular methodology view or philosophy
- Well developed interpersonal skills
- Sensitivity to organisational context
- Political skills and political awareness
- Social background to information systems practice
- Ethical behaviour

It is implicit in these assumptions that the framework producers presume that their users have a level of sophistication commensurate with their own. To bridge the gap and enable the 'safe' and reflective use of contingency and integrative approaches, there is a need to encourage and develop the assumed skills and understanding. The responsibility for augmenting this understanding lies in a changing emphasis in information systems education.

i. Teach Don't Train

In order to use the tools and frameworks proposed above and make sense of the 'babble of voices', future practitioners need to know the background to theory which underpins the methodology under consideration.

Jayaratna's (1994) observation is fundamental: "Methodologies contain models and reflect particular perspectives for thinking about 'reality' based on their embedded philosophical paradigms" (p.43).

Oliga (1988) provides a rich discussion of the paradigmatic foundations of systems methodologies. The requisite understanding for selection of methodology is associated with a realisation of the implications of the ethical and social consequences of adopting a particular methodology view. Future practitioners must understand the organisational impact and the philosophical and sociological theory behind the methodology view.

Understanding the background to practice is a useful precursor to developing critical thinking skills, self reflection is more difficult to foster. In the author's experience of teaching approaches to methodology selection, it is possible to produce a different type of thinker capable of critical reflection. This kind of ability appears to be is strongly related to the maturity, 'openness of mind' and world experience of students. Often this kind of reflection is problematic for the 'diehard' technician who tend to see critical reflection as more relevant to philosophy or humanities courses. It is equally important, however, to discuss the feasibility of this type of critical reflection, in that the employing organisation must allow and support reflective choices.

Faced with the potential political problems, students often allude to methodology selection being akin to an organisational therapists approach, treading the political tightrope. Overall it must be stressed that only if the problem solver is sensitive to the weaknesses of the proposed approach (and that includes the framework) can he/she be reflective in choice.

ii. Independent Learning

For the problem solver to make sense of a dynamic and diverse marketplace there is a need to be capable of independent learning. Equipped with critical thinking skills they must be confident to examine methodological offerings independently and be critical of both their own practice and the methodology's likelihood of enhancing that practice. An understanding of underlying theory enables the practitioner to survey the marketplace guided by an intelligent use of the aforementioned frameworks. It is the ability to scan new methodological offerings and apply frameworks in an informed and reflective manner that can help to make sense of a diverse methodology marketplace.

CONCLUSION

Several authors have provided models and frameworks that claim to help to understand methodologies and to enable selection. However, the marketplace is diverse and dynamic and thus problem solvers are faced with a veritable 'babble of voices' from methodology and framework producers. It has been argued that the frameworks make assumptions about the sophistication of their users and that there is a need to need to augment this understanding, either through the education system or via self education.

To make sense of methodologies and to sensitively use the frameworks there is an educational responsibility to produce a different kind of information systems practitioner, a reflective thinker capable of recognising underlying methodology philosophy and independently augmenting his/her understanding. Only if equipped with a critical understanding of the assumptions of methodological and framework offerings can future problem solvers be reflective of their own practice and be capable of selection in a dynamic marketplace.

REFERENCES

Avison, D., and Fitzgerald, G., 1988 "Information Systems Development", Blackwell Scientific Publications, Oxford.

Avison, D., and Wood-Harper, A.T., 1990, "Multiview", Blackwell Scientific Publications, Oxford.

Bell, F., and Oates, B., 1994, Framework for method integration, *in:* "Proceedings of the Second Conference of the British Computer Society Information Systems Methodologies Specialist Group" Herriot Watt University, Edinburgh.

Checkland, P.B., 1981, "Systems Thinking, Systems Practice," Wiley, Chichester.

Checkland, P.B., 1983, OR and the systems movement: mappings and conflicts, *JORS*, 34.

Checkland, P.B., 1988, Information systems and systems thinking: time to unite ?, *in:* P.B.Checkland, and J.Scholes, 1990, "Soft Systems Methodology in Action," John Wiley & Sons, Chichester.

Checkland, P.B., and Scholes, J., 1990, "Soft Systems Methodology in Action," John Wiley & Sons, Chichester.

De Marco, T., 1982, "Controlling Software Projects: Management, Measurement and Estimation," Prentice Hall, Englewood Cliffs.

Episkopou, D.M., and Wood - Harper, A.T., 1986, Towards a framework to choose appropriate IS approaches, *The Computer Journal* 29 (3).

Jackson, M.C., 1991, "Systems Methodology for the Management Sciences," Plenum, New York.

Jayaratna, N., 1994, "Understanding and Evaluating Methodologies," Mc Graw-Hill, Maidenhead.

Kronlof, K., 1993, "Method Integration, Concepts and Case Studies", John Wiley & Sons, Chichester.

Middleton, P., 1994, Euromethod: the lessons from SSADM, *in:* "Proceedings of the Second European Conference on Information Systems," Walter Beats, Breukelen, Nijenrode, Holland.

Mumford, E., 1983, "Designing Human Systems" Manchester Business School, Manchester.

Oliga, J., 1988, Methodological foundations of systems methodologies, *in*: R.L. Flood and M.C. Jackson "Critical Systems Thinking : Directed Readings," John Wiley & Sons, Chichester.

Rittel, H.W.J., and Webber M.M., 1974, Dilemmas in a general theory of planning, in: "Systems and Management Annual" Petrocelli, New York.

Schon, D., 1983, "The Reflective Practitioner - How Professionals Think in Action," Basic Books, USA.

Wood Harper, A.T., and Fitzgerald, G., 1982, A taxonomy of current approaches to systems analysis, *The Computer Journal* 25 (1).

59

THE WORK OF THE SWEDISH INFORMATION TECHNOLOGY (IT) COMMISSION IN THE PERSPECTIVE OF CRITICAL SYSTEMS THINKING

U. E. Söderberg

Department of Computer Science
University of Karlstad
650 09 Karlstad
Sweden

INTRODUCTION

The Swedish government, on the 17th of March 1994, established a commission with the task to promote the use of Information Technology(IT). Some of the objectives expressed are as follows:

- Sweden shall be a society where modern IT helps to raise the level of life-quality for all members of society and to develop wealth;
- The new technology shall be used for information and knowledge seeking in the areas of education, art, science, enterprising and social service;
- Information technology is of major importance for competition within the Swedish economy, therefore, the Swedish enterprises and their employees must be given the best possibilities for the use of IT;
- To be the leading European country in IT by year 2010.

These objectives state the government's wish to promote the utilisation of information technology in Sweden. By using information technology the quality of life should improve. Also Sweden's international competitive position will be promoted.

To be able to fulfil these objectives there has to be a major change in the system in Sweden. The Swedish model has, for a long time, been in focus as one of the best societies in the world. This is not the situation today. Due to the impoverishment of the welfare state, Sweden is no longer on the top.

The great period of Sweden was based on the successful heavy industry after the second world war. During the seventies the international competition put an end to these successful times. But there had been a development towards a high tech industry which helped Sweden to move in the right direction.

Critical Issues in Systems Theory and Practice, Edited by K. Ellis et al.
Plenum Press, New York, 1995

But the change from heavy industry to high tech industry was an economic break-down. With impoverishment of the welfare as a result of this. We are now looking at the next big change in industry, to master information.

A way of not making a mistake again is to fulfil the objectives of the IT Commission. If we view this task in the eyes of Critical Systems Thinking (CST) we are able to analyse whether the work of the commission will achieve for all individuals, working through organisations and society, the maximum development of their potential (Flood and Jackson, 1991).

This paper is by no means trying to put the work of the IT Commission in a bad light. It is meant to focus on some aspect that is important to consider when you are planning to make a major change in a society, especially when it is concerning technology that have a major affect on all individuals' life situation.

THE REASON

The reason to promote IT in the Swedish society is not to promote IT itself. Due to the falling status as an affluent nation, the Government recognises that the way to find a way back onto the right track is to increase the individuals' level of knowledge. As the IT Commission (1994) points out that these opportunities affect every areas of society and life, and every individual. But they also determine Sweden's success in an increasingly inter-meshed international community of nations.

Therefore is it of vital importance that IT will be implemented in the society, because it is one of the most important factors for success in the country. The importance of this is expressed in the following statement by the IT Commission (1994):

> "Our position as a leading affluent nation will be increasingly dependent on our ability to exploit the rapid development of knowledge and the efficient provision of information"(p. 6).

THE QUESTION

Two important objectives are stated in the Commission's report concerning the individual and his development. The first is that IT will promote the search for higher knowledge so that the individual will be able to free himself/herself in order to create choices. The other objective which concerns the individual is to move from centrally controlled hierarchies to decentralised networks. IT integrated in the whole society would make these issues possible if it is used in the right way, says the Commission (1994). How is technology going to be used in the right way?

It is of extreme importance to realise that the implementation of IT, as a technology, in the Swedish society will not automatically create 'know-how', according to CST, among the people. The emancipatory interest with the individual might not be furthered if the implementation is seen just as a technology implementation.

KNOWLEDGE

The Commission claims that we are going to reach higher knowledge in the following areas; Education and Research, The Legal system, Public Administration, Health Service and Medical Care, Communications Networks, Industry and Commerce and IT Research. What is higher knowledge in the perspective of CST concerning the individuals?

In this case, higher knowledge is equal to giving wings to human ability. One of the five major commitments that Critical Systems Thinking (Jackson, 1991) demonstrates is human emancipation. Giving wings to human ability seems to me like the same thing as giving the possibility to emancipate. In this case it is important to let technology work as an catalyst to promote emancipation.

Technology could easily be misused as a tool for power. This is not hard to believe, because who could get away from IT in an IT society? According to Habermas (1971) it is self-reflection that releases the subject from dependence on hypostatized powers. In order to deal with the problem that IT could work as a tool for power in a society, it must be expressed that the IT should have room for self-reflection to be able to promote knowledge to the individual.

If information technology is going to work as a promoter of knowledge, the commission must provide means for understanding how information can become knowledge. Information in it self, can be seen as data and subjective interpretation, or intersubjective in the meaning that we base our subjectivity on common experience and certain rules (Mingers, 1993). So, to use information technology we must understand the subjectivity and knowledge that constitutes interpreting data into information.

DECENTRALISATION

To strive for decentralisation is natural in this context, because decentralisation is supposed to give more power to the individual. To give the individual a situation with choice would certainly promote an emancipatory thinking. The use of decentralisation is doubtless but, how can this mix decentralisation with technology?

Computer aided information systems can, if you like, act for strengthening the local independence, ease-up geographic isolation and compensate for transportation distance. It can function as an irrigation system for democracy. But it could on the other hand work as the vacuum cleaner for central power and bureaucracy (Hedberg et al., 1987). Certain aspect have been distributed in decentralisation, concerning operative decisions, but other aspect as controlling the resources have not been distributed. In what way should we view the IT Commission's work concerning the problems with decentralisation and technology?

The distribution of power could on the surface be promoted by IT, but beneath that lies the possibility for certain groups to take care of the strategic decisions without the knowledge of the individual. To even be aware of the problem would affect the belief or trust in an IT society. The IT commission is aware of the problem. The commission says that the technology should be applied in the right way to work properly and that it might divide the society in two camps, those who master the technology and those who don't (IT Commission, 1994). What shall we do about these problems?

The Commission says little about how to solve the problem, instead they try to focus on the positive side. The Commission (1994) says that dangers of vulnerability should not be exaggerated. In several respects, society's vulnerability can be decreased with the aid of IT. To state such a weak opinion in this rather important question is telling us that the problem is not easy to handle, and therefore put aside.

Emancipation could be a victim if the problem with decentralisation and technology is not taken care of. The first thing is to address the problem as crucial to the success of an IT society, the second is to give examples on how to secure the possibility for emancipation. Critical systems thinkers' first task, says Schecter (1991), is to liberate suppressed knowledge in order to launch more penetrating critiques on the dominant ones. Critical systems thinking does not try to set aside the difficult problems, instead it puts them up front to secure the technical and practical interests.

CONCLUSION

When we confront such issues like implementing new technology into a society we must take certain aspect in consideration. Information technology affects, as I have shown, the development of knowledge with the people. We have to ask ourselves questions like: Does technology provide knowledge by itself? How important is technology in the creation of knowledge in modern society? What other kind of knowledge, different than technological knowledge, do we need to satisfy modern society? I have argued that we need knowledge about how to create a situation where emancipation is promoted by technology and how to decentralise with technology without the risk to centralise.

REFERENCES

Flood, R.L., and Jackson, M.C., 1991, "Creative Problem Solving: Total Systems Intervention", John Wiley & Sons, Chichester.

Habermas, J., 1971, "Knowledge and Human Interests", Beacon Press, Boston.

Hedberg, B., Elling, M., Jönsson, S., Köhler, H., Mehlman, M., Parmsund, M., Werngren, C., 1987, "Kejsarens nya Kontor", Liber, Malmö.

IT Commission, 1994, "Wings to Human Ability", Alfa, Stockholm.

Jackson, M.C., 1991, "Systems Methodology for the Management Sciences", Plenum, New York.

Mingers, J., 1993, An examination of information and meaning, *Systemist*, Vol. 15, No. 1.

Schecter, D., 1991, Critical systems thinking in the 1980s: A connective summary, *in*: "Critical Systems Thinking: Directed Readings", R.L.Flood and M.C.Jackson, eds., John Wiley and Sons, Chichester.

60

REPRESENTATION AND DOUBLE-LEVEL LANGUAGE IN INFORMATION SYSTEMS

Robert Stephens

Bristol Transputer Centre
Faculty of Computer Studies and Mathematics
University of the West of England
Frenchay Campus
Bristol, BS16 1QY
United Kingdom

1. INTRODUCTION

A recent theme in some contemporary research suggests that organizational realities and structures endure to the extent they are made and remade in social praxis. Accordingly, effective information systems development and operation depends on a subtle interplay of the technical and social. Viability is thus sustained not by appeal to a pre-ordained organizational structure, but in a linguistically mediated cycle of assimilation and accommodation. In this paper we seek to capture this notion by examining representation as both synchronic and diachronic, and developing the concept of 'double-level language' to describe how formal and informal languages are mutually supportive.

2. A LINGUISTIC INTERPRETATION OF INFORMATION SYSTEMS

The information systems convention that assumes an underlying organizational structure that is identifiable, enduring, and external to the actions, interpretations and knowledge of its members has been subjected to many challenges (e.g., Boland and Hirschheim, 1987). One of the most telling suggests organization and structure to be a human accomplishment dependent on, rather than a priori to, ordinary linguistic achievements (e.g., Blackler, 1993; Truex and Klein, 1991). Structure in this account is activity-dependent, both the medium and outcome of a process variously described as 'structuration', 'ordering', or 'enacted'. This points to a quite fundamental idea of organization, involving the production and reproduction of practices across time and space, which is best captured by not using the noun 'organization' but the verb 'organizing' (Weick, 1979), or by considering 'social-structure' as a verb not a noun (Law, 1994).

Critical Issues in Systems Theory and Practice, Edited by K. Ellis et al.
Plenum Press, New York, 1995

In adopting the verb use of organization and order, organizations become not merely organizers of information, they also construct the forms in which information appears; an insight taken up by Zuboff (1989) to distinguish the dual function of information technology, informating and automating. When we consider informational specificity against this background there are broadly two apparent IS research concerns. The first centres upon the distribution of information - its flows, patterns, and concentrations; the second upon the meaning attributed to the contents of information. Borrowing metaphors from linguistics, the dual function may be termed, respectively, synchronic and diachronic, concerning, on the one hand the spacial, or the coordination of actions across the organization, and on the other, the temporal, or the interpretation of information, in particular, its consequences for communication and inter-subjective understanding.

The linguistic metaphor may be pursued as a vehicle for gaining insight into some general problems associated with the operation of information systems in organizations (see, for example, Stamper, 1987; Lyytinen, 1985; Blair, 1990). This image "emphasizes the symbolic nature of organizational life" (Boland, 1979), and presents information systems participating in a process of institutional change derived from new language uses that are stimulated, but not determined, by the information model. In other words, linguistic information has the power to do work, but not always deterministically, and never in the possession of a single individual; action in language is always joint action, and must sometimes have unintended consequences, that is outcomes that were not intended, but nonetheless remain with the participants. As a linguistic artifact, the components of the information model are analogous to words in natural language, and bear similar systemic attributes. Of interest here is Wittgenstein's (1963) observation that linguistic tokens, and by implication, information items, gain their meaning through use rather than ostensive definition. In practice this means the synchronic application of terms in a rule-like manner and the diachronic use of rules to fashion information semantics. Language change (Lyytinen, 1985) is the product of a subtle interplay of the two sides as each seeks to shape and inform the other.

3. REPRESENTATION

Information technology adds to the synchronic dimension, but to be true to the verb use of organization and order we need to go beyond information and consider the concept of representation. Representation is the more fundamental concept simply because information must first be represented in some way, implying the construction of a representation (pattern, picture, model) of the domain of interest (Cooper, 1992). Indeed, the old axiom, 'no computation without representation' is testimony to the fundamental and mandatory nature of representation in the processing of information. By representation, we are not concerned with the singular association of word, symbol or concept with a referent, but its incorporation into a systematic structure, abstract or concrete, whose features purport to symbolize or correspond in some sense with those of some other structure.

Winograd and Flores (1986) note that computer systems actually cascade representational systems that range from the 'natural' human representation of the task performed in the social environment to the 'machine' representation of data and procedures involved in solving task-related problems. This recognises a major linguistic component in the technology: an articulated symbolic system lies inevitably at the core of any digital computer application.

Information systems contain symbolic representations that may be combined and manipulated in a law-like fashion, a condition that satisfies the synchronic dimension of representation. By this we mean a formalism that is axiomatic or rule-governed, bounded, composed of a finite vocabulary and internally consistent. In most computer systems the representational notation must be sufficiently modular that individual elements of the notation can be combined to make molecular expressions. In this way, ever more complex structures can be used and constructed by a finite system. Internal semantic discipline is maintained by enforcing the principle that however complex the symbol, its meaning is a function of the meaning of its parts and their syntactic arrangement.

At one level of analysis, all computer symbol systems appear self-sufficient, in the sense that each symbol refers only to other internally represented symbols. There is a strict engineering imperative why this should be so, but the apparent self-sufficiency hardly completes the representational picture. Symbols have semantic values beyond their internal (physical) instantiation and computations are operations on symbols that respect or are constrained by those semantic values. As Varela et al (1991) note, it is impossible to comprehend the idea of computation without recourse to the semantic relations among symbolic expressions. While the technology operates only on the internal form of the symbols it computes, and has no access to their semantic value, its operations are nonetheless semantically constrained because every semantic distinction relevant to its program has been encoded in the syntax of its symbolic language.

In the computer representation therefore, syntax (hopefully) mirrors or is parallel to the (ascribed) semantics. But the concept of representation in its entirety concerns far more than the technologies, and even the infrastructures, that support the verb use of organizing or ordering, which is a distinctly diachronic concept. As Law (1994) points out, representations are not just a necessary part of ordering, they are also ordering processes in their own right:

"...representations, whatever their provenance, are always the product of ordering work. They're a part of ordering. And they are produced in ordering." (Law, 1994:153).

This much suggests the concept of representation cannot be restricted to formalisms alone, but must include organizational culture; its goals and objectives, its structure and boundaries, characteristics of roles and statuses, and the explanations of the cause of problems, their ideal solutions and the posited consequences of such interventions. Weick (1979), for example, argues that such notions as the properties of the environment exist only within the taken for granted system of meanings created, sponsored, and retained within the occupational and organizational culture.

4. DOUBLE-LEVEL LANGUAGE

It follows from this account that an adequate explanation of the location and content of information, as well as its relationship to organization and understanding, will require analytic categories that develop the relation between formality and informality, without privileging one in favour of the other. What needs to be included in such an analysis? Following from section 2., we will require representational formalisms to be articulated not only as specifications of worldly states, but also as resources for action in the world. In order to proceed in this direction, we need to emphasise a communicative, rather than reflective, epistemology, and stress interrelatedness of subject and object within a practically informed discursive framework (see Sinha, 1988). To do this we will employ Robinson's (1991) concept of the 'double-level language', but extend it to include all cases where representational formalisms are implicated in human action.

Robinson (1991) employs the concept of double-level language to analyse computer systems that purport to support cooperative working. Noting that human interpretation is a sine qua non for any non-determinate work, Robinson maintains that for any genuine Computer-Supported Cooperative Work (CSCW) two linguistic fields must be able to operate freely. A formal language, the machine artifact, provides a sort of 'external world' that can be a common reference point for participants to coordinate action, and whose behaviour is rule governed and predictable. However, at the cultural level social mores and individual subjectivities are significant. These can be considered as 'norms', 'conventions', 'agreements' or 'rituals'. Computer support is valuable insofar as it facilitates the separation and interaction between the 'formal' and the 'cultural' aspects of language used in organizations.

According to Robinson, applications that restrict or accentuate one level at the expense of the other tend to fail as cooperative systems. However, his claim that poor systems fail to support a double-level language contradicts Suchman's (1987) analysis, which may be read as suggesting that a double-level language is always a feature of any human-machine interaction. Her analysis identified expert system users involved in a process of reciprocal interpretation of meaning; that is using instructions to comprehend the environment and the environment to comprehend the instructions. The double-level language may be said here to embrace the play between formalism and informalism, or, the operationalizing of the formalism.

The experience of one of the best reported CSCW systems, the Coordinator, is also instructive in this sense. This system succeeded or failed as a cooperative workspace according to how the formalisms offered by the systems were interpreted within prevailing organizational norms. Where a strong collaborative ethos prevailed, the Coordinator served as a useful tool for cooperative working, however, in hierarchical organizations, the system, which was still regarded by some as a useful tool, functioned as a management command and control system.

A study by Nardi and Miller (1990) offers further support for the view that formalisms unwittingly embody communicative or conversational properties. In their ethnographic examination of the organizational use of spreadsheets, the authors found that problem solving was distributed across users with different skills and interests and concluded that, far from being restricted to single-users, the software functioned to represent and disseminate organizational knowledge.

5. CONCLUSION

The suggestion is therefore, that the double-level language is ubiquitious, and it is its differing operation, rather than its presence or absence, that determines the system's use. Sometimes it is natural for a state of tension to exist between these two semiotic fields, as their intersection may be the site of contradiction and a struggle for meaning. Such a state of affairs is most likely during periods of change. Information systems must intervene at this intersection, paradoxically attempting to balance change, implicit in the development process, with the imperatives of structure and order. As Argyris (1987) suggests, contradictions are endemic to the process.

Many studies locate the source of this tension in the emphasis in IS methodologies on formalism at the expense to the informal domain required to sustain it (e.g., Boland and Hirschheim, 1987; Preston, 1991). However, while this may be true, there is also a sense in which formal and informal continually conspire to undermine each other. The accentuation of the synchronic dimension of representation in the imperative to create structure and construct syntactic arrangements, particularly computer instantiated representations, can only be arrived at by suppression or suspension of the temporal, or diachronic dimension of representation (see Bowers, 1991, for an account of how cognitive science achieves this reduction). On the other hand, representational and symbolic development does not end with the consolidation of knowledge in canonical rules or axiomatic systems (Sinha, 1988).

Rather, these form the basis for subsequent developments in which rules and axioms are systematically violated, by symbolic and substitutional operations, or in the construction goal orientated activities. This ability to decouple the synchronic function is crucial to language change, the ability to 'go beyond the information given', to the generation of novel means ends relations, and is general to most problem-orientated activities.

6. REFERENCES

Argyris, C., 1987, Some inner contradictions in management information systems, *in*: R. Galliers ed., "Information Analysis: Selected Readings", Addison-Wesley, Singapore.

Blair, D.C., 1990, "Language and Representation in Information Retrieval", Elsevier Science Publishers, Amsterdam.

Blackler, F., 1993, Knowledge and the theory of organizations: organizations as activity systems and the reframing of management, *Journal of Management Studies*. 30 (6): 863-884.

Boland, R.J., 1979, Control, Causality and Information System Requirements, Accounting, *Organizations and Society*. 4 (5): 259-272.

Boland, R. J. and Hirschheim, R. A., eds., 1987, "Critical Issues in Information Systems Research", John Wiley and Sons, Chichester.

Bowers, J.M., 1991, Time, representation and power/knowledge: towards a critique of cognitive science as a knowledge-producing practice, *Theory and Psychology*. 1(4): 543-569.

Cooper, R., 1992, Formal organization as representation: remote control, displacement and abbreviation: Organization theory in the postmodern era, in: "Rethinking Organization: New Directions in Organization Theory and Analysis", M. Reed and M. Hughes, eds., Sage, London.

Law, J., 1994, "Organizing Modernity", Blackwell Publishers, Oxford.

Lyytinen, K., 1985, Implications of Theories of Language for Information Systems, *MIS Quarterly*. March: 61-74.

Nardi, B.A., and Miller, J.R., 1990, An Ethnographic study of distributed problem solving in spreadsheet development, *in*: Proceedings of the . Computer Supported Cooperative Work '90, ACM, October, 1990, 197-208.

Preston, A. M., 1991, The "problem" in and of management information systems, Accounting, Management and Information Technology. Vol. 1 (1): 43-69.

Robinson, M., 1991, Double-Level Languages and Co-operative Working, *AI and Society*. 5: 34-60.

Sinha, C, 1988, "Language and Representation: A Socio-naturalistic Approach to Human Development", Harvester Wheatsheaf, London.

Stamper, R., 1987, Semantics, *in*: R. Boland and R.A. Hirschheim, eds. op. cit.

Suchman, L.A., 1987, "Plans and Situated Actions: The Problem of Human-machine Interaction", Cambridge Press, Cambridge.

Truex, D. P. and Klein, H. K., 1991, A rejection of structure as a basis for information systems development, *in*: R. K. Stamper, P. Kerola, R. Lee, and K. Lyytinen, eds., "Collaborative Work, Social Communications and Information Systems", Elsevier Science Publishers, Amsterdam.

Varela, F.J., Thompson, E., and Rosch, E., 1991, "The Embodied Mind: Cognitive science and Human Experience", MIT Press, Cambridge.

Weick, K., 1979, "The Social Psychology of Organizing", Addison-Wesley, Reading, M.A.

Winograd, T. and Flores, F., 1986, "Understanding Computers and Cognition: A New Foundation for Design", Addison-Wesley, Reading, M.A.

Wittgenstein, L., 1963, "Philosophical Investigations", (translated by G. E. Anscombe), Blackwel, Oxford.

Zuboff, S., 1989, "In the Age of the Smart Machine: The Future of Work and Power", Heinemann, Oxford.

61

INFORMATION SYSTEMS DEVELOPMENT

Methods, Modelling and Metaphors in an Object-Oriented World

R. T.Vidgen and J. R. G.Wood

Department of Mathematics and Computer Science
University of Salford
Salford
United Kingdom

1. INTRODUCTION

In talking about the common-sense approach to Object-Orientation (O-O) and the need to mirror reality more closely we have created a straw man that is supported by some major and questionable assumptions. In this paper we consider the assumption that a good object model is one that mirrors reality. We argue that the the opposing view, that object models can be used to create a social reality, is not useful any more.

2. OBJECTIVISM AND SUBJECTIVISM

A paradigm provides an overall way for organizing our thoughts about the world. In science a paradigm determines what methods and explanations will be considered as acceptable. A basic framework for classifying paradigms involves the categorization of assumptions on an objectivist/subjectivist dimension. This framework has been applied widely in Information System research (see for example Klein & Hirschheim, 1987), particularly when expanded into a four box model (Burrell & Morgan, 1979) with the addition of a consensus/conflict dimension (see for example Hirschheim & Klein, 1989).

The *objectivist* position is characterized by a realist ontology and a positivist epistemology. Ontology is concerned with the nature of reality. Epistemology is concerned with how we gain legitimate knowledge about the world, i.e., with truth. Loosely speaking, the objectivist assumes that objects and structures exist as empirical entities independent of human observers and that the appropriate way of acquiring knowledge of the world is by observation and the identification of causal relationships (correspondence to reality is the measure of a good theory). The *subjectivist* position holds that scientific method is not appropriate for explaining the social world as different people interpret the world in different ways and any agreement is inter-subjective. The appropriate way to investigate the social world is to recognize multiple realities and to adopt an interpretivist stance.

Critical Issues in Systems Theory and Practice, Edited by K. Ellis et al.
Plenum Press, New York, 1995

With respect to object modelling two questions are raised by this framework. Firstly, what is the ontological status of the object model, and secondly what is an appropriate way to go about the construction of an object model? How might an analyst approach the task of object modelling working within the different paradigms?

The objectivist sees object modelling as an objective exercise where classes and associations can be said to exist independently of people. Problems with object models are associated with a failure to capture and specify the real requirements. The solution is to improve the engineering process such that the requirements specification is accurate, complete, consistent, and unambiguous. The assumption that there is agreement on means and ends leads to a single object model that is suitable for and acceptable to all stakeholders. The object model is a value-free reflection of a singular reality and scientific method is the appropriate way of building an object model (object models are theories which can be tested empirically in the real world).

The subjectivist considers reality to be socially constructed and that there are potentially as many realities as there are people involved in the development exercise. Object models are at best agreed inter-subjectively, requiring many of the parties affected by a proposed Information System to be involved in agreeing what classes and associations should be incorporated in the object model. The more people's interests are reflected in the model the more likely it is that a shared understanding will be reached and a successful implementation achieved. A good object model is one which creates a shared understanding and the appropriate way to go about creating such a model is through participation and facilitation.

In assuming that a good object model is one that mirrors reality we are working within a positivist tradition in which the measure of a good theory is its correspondence with reality. This is but one way of approaching the world. Do modellers work within a functionalist paradigm?

3. OBJECT MODELLING AND FUNCTIONALISM

Many of the working definitions of an entity have been derived from Chen's seminal paper in which the entity-relationship model is considered to be a reflection of the real world.

> "The entity-relationship model adopts the more natural view that the real world consists of entities and relationships. It incorporates some of the important semantic information about the real world.....An entity is a "thing" which can be distinctly identified" (Chen, 1976).

Lewis gives a number of quotes from the data modelling and Object-Oriented (O-O) literature which show that many methods assume an objective ontology (Lewis 1994). Lewis argues that there is a potential danger with O-O in the insistence that objects are a more natural representation of the real world since such a stance sidelines the question of whether entities and classes are useful epistemological devices or whether they have ontological significance. Although object modellers espouse a functionalist approach, what do they do in practice? Martin & Odell refer to O-O as a model of reality.

> "The models we build in OO analysis reflect reality more naturally than the models in traditional systems analysis. Reality, after all, consists of objects and events that change the state of those objects. Using OO techniques, we build software that more closely models the real world. When the real world changes, our software is easier to change - a real advantage." (Martin & Odell, 1992:67)

The authors appear to have adopted an objectivist stance, but in practice it is not so easy to stick to such functionalist principles and in the same text we find:

> "Analyzing organizational systems means creating an effective, shared set of concepts. Concepts underlie each and every organizational process. Concepts define a shared organizational reality and form the basis of an organizational language with which to communicate." (Martin & Odell, 1992:237)

This creation of shared meaning through the definition of concepts does not seem to sit easily with the objectivist flavour of the previous quote. Is it possible to accept both the above quotes without being inconsistent? We argue that the issue is not whether to be objective or subjective but that one implies the other and that any method that is based upon one of these ontological positions must accept and cater for the other. Substantiating this argument is the aim of the next section.

4. DUALITIES AND DUALISMS

The original work by Burrell and Morgan (1979) proposes that the paradigms are mutually exclusive and that methods should be developed independently within the individual philosophies - this is known as paradigmatic closure. The implication of paradigmatic closure is that object modelling methods should be based either upon the notion of an objective reality or upon the notion of a reality that is socially constructed. Taking a strong objectivist position is difficult - we cannot demonstrate the existence of an objective reality and science cannot legitimate itself (Lyotard, 1984). But a strong subjectivist stance is not a lot more comfortable and runs the danger of an extreme relativism where anything goes. The problem we have created is in assuming that object modelling must be either objective or subjective. We have created a binary opposition between objectivism and subjectivism and cast them as a dualism.

Berger and Luckman (1966) proposed a framework where subjective meanings (externalization) become objective facticities (objectivation) which then 'act back' as they socialize present and future generations (internalization). Reality construction is the dialectical relationship of these three moments. Giddens (1984) seems to have moved even further from the paradigmatic closure of the Burrell and Morgan (1979) framework by conceiving the objective and subjective to be present simultaneously, as represented in structuration theory (Giddens, 1984). An important implication of structuration theory is that the objective/subjective distinction should be seen as a duality and that it is not possible to separate out objective and subjective components.

Supporters of the natural science approach reject interpretivist methods as lacking rigour and generalizability; supporters of interpretivist approaches attack scientific methods as being inappropriate and impoverished ways of enquiring into human systems. Latour (1993) argues that one aim of the modernist project is purification - the separation of an objective and given natural world from a socially-constructed subject world. Modernism contains a paradox insofar that it must separate the natural and social worlds, whilst at the same time relying upon their inseparability for its successes. The middle ground is more than mediation between natural and social worlds: the process of mediation creates the natural and social worlds:

> "We do not need to attach our explanations to the two pure forms known as the Object or Subject/Society, because these are, on the contrary, partial and purified results of the central practice that is our sole concern. The explanation we seek will indeed obtain Nature and Society, but only as a final outcome, not as a beginning. Nature does revolve, but not around the Subject/Society. It revolves around the collective that produces things and people. The Subject does revolve, but not around Nature. It revolves around the collective out of which people and things are generated. At last the Middle Kingdom is represented. Natures and societies are its satellites." (Latour, 1993:79).

Perhaps it is through the practice of IS development that objective realities and social constructions arise. The structures of signification represent organizational reality (the current organizational context). The activity of IS development is constrained by the structures of signification, but at the same time has the potential to change those structures of signification, i.e. to (re)create organizational reality by the introduction of the unfamiliar.

We define a method to be a set of procedures for getting things done. A methodology is a higher level construct that provides a rationale for choosing between different methods. Data and object modelling are techniques that can be employed by different development methods. IS methods form part of the stock of knowledge used to make sense of the participants' actions. It is not appropriate to consider the process as objective (IS development mirrors organizational reality) or as subjective (organizational reality is created through IS development). Both objective and subjective aspects are present at the same time; without the notion of objectivity the notion of subjectivity is meaningless (and vice versa) - objective methods can be said to be objective because they are not subjective. The developer who adopts a predominantly objective approach to development will tend to reinforce the current structures of signification; the developer who can recognize that reality is also socially constructed is more likely to be involved in the re-creation of structures of signification. The situation can be seen to be further complicated by the aspects of domination and legitimation, tacit knowledge, and unintended consequences.

5. METAPHORS, LITERAL TRUTHS AND OBJECT MODELLING

The notion of information adopted for the purposes of this paper is of information as a metaphorical notion. From Aristotle onwards, metaphor has often been defined as transference (Morgan, 1986). The application of a metaphor is at the same time a way of not seeing, as it highlights one view while pushing others into the background. Richard Rorty argues that truth is a metaphor couched in language and is not a given that is 'out there' (Rorty 1989). Without sentences we do not have truth. Sentences are part of language and language is a human creation. Thus, only descriptions of the world can be true or false since we cannot say anything about the world other than through language. The world without humans cannot be true or false and truth must be made rather than found. Rather than seeing language as a medium between ourselves and the real world it is more pertinent to consider language as a metaphorical device, where the major distinction is between the familiar and the unfamiliar. Attention is then focussed on how we create the unfamiliar and the process by which it becomes familiar and thus literally true. Rorty distinguishes between irony and common-sense.

The common-sense approach accepts the current vocabulary as sufficient and sees reality as determining our vocabulary. The ironist distrusts the vocabulary that is inherited and approaches it critically. A common-sense object modeller will accept that customers are a real world phenomenon - all organizations have customers don't they? But many retail banking systems were built without a customer entity/class; these systems were product processing systems, reflecting the banks' account orientation in the 1960s and 1970s. Subsequently they have found it very difficult to change their computer systems (and their business) to a customer basis. The notion customer at one time was unfamiliar noise; as it became accepted and used widely it became a literal truth. To be creative, object modellers need to think metaphorically and to be willing to introduce unfamiliar noise. If the unfamiliar noise gets accepted and implemented successfully in a computer system then the metaphor dies and literal truths emerge.

The ironist object modeller will question the inherited vocabulary of the classes and associations currently recognized by an organization; the common-sense object modeller will accept the existing categories that the organization uses and implement them as literal truths in an object model. Which approach would you rather take when attempting Business Process Reengineering?

6. CONCLUSIONS

We believe that many of the ideas presented here lead to a practical agenda for object modelling:

- think in terms of dualities rather than dualisms and be comfortable with current structures while recognizing that you may be creating new structures;
- recognize that data is not value-free and that you can never wholly be an objective outsider;
- retain the notion that object models are metaphorical and that successful computer systems embody dead metaphors as literal truths.

Following the lead of Latour we should perhaps refer to "quasi-object" models in order to reflect the the role of object modelling as mediation in contrast to modelling as purification.

7. REFERENCES

Berger, P. L., and Luckman, T., 1966, "The Social Construction of Reality," Doubleday, GardenCity, New York.

Burrell, G., and Morgan, G., 1979, "Sociological Paradigms and Organizational Analysis," Heinemann Educational Books, London.

Chen, P., 1976, The entity-relationship model: towards a unified view of data, *ACM Transactions on Database System.*, 1(1): 9-37.

Giddens, A., 1984, "The Constitution of Society," Polity Press, Cambridge.

Hirschheim, R.A., and Klein, H.K., 1989, Four Paradigms of Information Systems Development, *Communications of the ACM.* 32(10): 1199-1216.

Klein, H.K., and Hirschheim, R.A., 1987, A comparative framework of data modelling paradigms and approaches, *The Computer Journal.* 30(1): 8-15.

Latour, B., 1993, "We Have Never Been Modern," Harvester Wheatsheaf, Hemel Hempstead, England.

Lewis, P., 1994, "Information Systems Development," Pitman, London.

Lyotard, J.-F., 1984, "The Postmodern Condition: A Report on Knowledge," Manchester University Press, Manchester.

Martin, J., & Odell, J., 1992, "Object-Oriented Analysis and Design," Prentice-Hall, Englewood Cliffs.

Morgan, G., 1986, "Images of Organization," Sage, Newbury Park, CA.

Rorty, R., 1989, "Contingency, irony, and solidarity," Cambridge University Press, Cambridge.

62

THE TROJAN HORSE PHENOMENON

Heather Watson

Department of Mathematics and Computer Science
University of Salford
Salford M5 4WT
United Kingdom

INTRODUCTION

An issue in relation to theory and practice centres on the different perspectives we have in information systems development. The purpose of this paper is to consider what is entailed in a perspective. That is, in developing an argument we do not proceed in isolation since we depend on some of the language and assumptions of others, if only to contrast them with our own point of view.

Given this situation, a perspective often entails contrary themes within it that potentially undermine the main position it asserts. Perhaps unwittingly, a perspective may therefore tend to reiterate some of the 'mistakes' it criticises in others. These tendencies are referred to as the Trojan Horse Phenomenon.

This is discussed below using Lyytinen's (1987) taxonomic perspective as an example.

CONTEXTS IN MAKING A TAXONOMIC PERSPECTIVE

Lyytinen (1987) is concerned about recurrent reasons for information systems failures. As such, he considers theoretical concepts and recommendations for further research to change the way we approach information systems development. He explains that people have different viewpoints in perceiving "object systems." These are phenomena that the development group identifies as a "target of change" (p.6). He describes these as "object system contexts," listing them as follows:

- technology,
- language, and
- organisational (p.11).

He then considers how methodologies influence choice of object systems, and in proposing taxonomic principles for systems development, he asserts the following with regard to the above contexts:

Critical Issues in Systems Theory and Practice, Edited by K. Ellis et al.
Plenum Press, New York, 1995

> "We believe that these three are *exhaustive* and segment the systems development domain into three distinct realms, which respond differently to introduced change."

The three contexts are *hierarchically* ordered. (Lyytinen, 1987:11) (my emphasis).

In developing his perspective, Lyytinen (1987), follows a well-travelled route: he creates a taxonomy. Yet, his situation may be strangely paradoxical since he creates a taxonomy of *contexts*. In principle taxonomies inevitably exclude, but a taxonomy of contexts undermines itself by its own example: in classifying contexts it excludes contexts. The next section questions Lyytinen's assertion that contexts are exhaustive.

FROM CONTEXTS TO SUPER-CONTEXTS TO BOUNDLESS CONTEXTS

Contexts would not appear to be exhaustive, but rather, they might be boundless in at least two senses:

- first, any known context remains open to further description since there is potentially no limit as to what may eventually be relevant;
- second, context is unmasterable insofar as attempts to classify it can be reintroduced onto the contexts it sought to describe resulting in new contexts that exceed prior formulations (Culler, 1983:125).

Examples of the first sense are scientific discoveries in which previously neglected factors become relevant to describing the behaviour of certain objects such as the existence of ether. In information systems, the emerging significance of social factors might offer a similar example. In the second sense, Culler (1983) describes Wittgenstein's example of not being able to say 'bububu' to mean 'if it does not rain I shall go out for a walk'. Paradoxically, this denial results in a connection making it possible to do just that (p.124). In information systems, the emergence of complementary approaches despite claims of paradigmatic incommensurability might offer a comparable example.

These two senses of boundless context are also at work in Lyytinen's argument. For instance, he extends his description of contexts to super-contexts, which derive from bi-directional interactions between all three object system contexts. This enriches his argument. However, in the second sense of unbounded context, this has the paradoxical effect of undermining his initial assertion of an exhaustive hierarchical description.

This indicates the way that a perspective may differ from itself. The next section discusses how such differences open potential for contradiction.

CONTRADICTIONS IN AN ASSERTED POSITION

Concerning the second sense in which context cannot be mastered, we see Wittgenstein's example of denying the potential for further comparison reiterated in Lyytinen's taxonomic perspective:

> "The proposed framework suggests that only those methodologies which fall in the same taxonomic class *should* be compared, i.e. STS-methodologies *cannot* be compared with technical design-oriented methodologies because they differ in terms of their covered contexts." (Lyytinen, 1987:34) (my emphasis).

This denial paradoxically makes a comparison possible. Moreover, our field already has a tradition that does compare social and technical perspectives, (e.g. Mumford, 1983; Avison and Wood-Harper, 1990). Indeed, Lyytinen (1987) contradicts this position later when he recommends Multiview Methodology (MVM) as an example for further research in solving issues of interlinking object representations: this is a methodology that links social and technical perspectives.

But if we look closer, there is a counter theme that contradicts what Lyytinen asserts in the above statement itself - STS, socio-technical systems. These two perspectives, both the social and the technical, are *already* implicated in a comparison that Lyytinen claims to deny.

This is an example of the Trojan Horse Phenomenon: because a perspective often relies on the language and assumptions of others, it often reiterates what it simultaneously criticises. This is not simply a matter of sloppiness that could have been avoided. Instead, it is evidence of the partial nature of the truths that different perspectives entail. The next section explains why a perspective may be inconsistent, yet coherent.

COHERENCE IN METAPHORICAL ENTAILMENTS

Lakoff and Johnson (1980) explain how single metaphors entail others because they are linked to a whole network of associated concepts. This explains the puzzle of how we can use quite different metaphors, or in this case, quite different perspectives to describe information systems development. Lyytinen's argument concerning technical and socio-technical systems methodologies is inconsistent, yet it is coherent insofar as the perspectives share a key entailment, in this case, the notion of 'system' and a host of related concepts.

Lakoff and Johnson (1980) explain that inconsistency arises because there is no single metaphor or perspective that is sufficient in all contexts. However, the overlap of entailments between perspectives defines the coherence between them. This is what allows different perspectives to 'fit together' even though they are not completely consistent (pp. 94-5). In this way, complete consistency across perspectives is rare, but coherence in the sense of shared entailments is typical (p. 96).

The next section develops an implication of this for 'pure' points of view in information systems development.

AN IMPLICATION FOR 'PURE' POINTS OF VIEW

An outcome of this analysis is that differences between perspectives are often special cases of differences already admitted within perspectives. This was evidenced by the way Lyytinen's argument undermined itself. An implication of this is that individual perspectives are not 'pure' insofar as they entail wider contexts and other perspectives with which they may or may not agree. For example, the debate between Checkland (1982) and Jackson (1982) about whether Soft Systems Methodology (SSM) is regulative or radical is embedded in a network of issues associated with Burrell and Morgan's (1979) typology, and these are additionally associated with Habermas'(1976) views of distorted communication. In relation to such debates, the Trojan Horse Phenomenon suggests that we may unwittingly 'smuggle in' contrary perspectives despite our best efforts, perhaps even if we are fortunate enough to find ourselves debating in an 'ideal speech situation.'

The next section considers how we might begin to address this.

A POST-STRUCTURALIST CONCLUSION

This research supports Rosenhead's (1984) view that methodologies carry "a pan-technicon of ideological baggage" (p.82) whether they are reactionary or emancipatory. As such, he recommends that proponents of methodologies show a critical awareness of this. He also suggests that others have a responsibility to expose claims to universality in recognising that methodologies are partial and biased perspectives.

With this end in mind, I would like to conclude by adapting some post-structuralist methods from Culler (1983) to information systems since these might be helpful in exposing critical issues about our methodologies:

Consider ways a methodology differs from itself - The Trojan Horse Phenomenon suggests that what may define as the identity of a particular methodology will entail perspectives that cause it to differ from itself.

Attend to conflicts within methodologies that are reproduced as conflicts between them - Critical disputes can often be identified as displaced re-enactments of conflicts already admitted within a methodology. In this way we can learn to question the assumptions and decisions that transform a complex pattern of internal differences into alternative perspectives.

Think about how descriptions of another methodology can be read as a self-description - when a methodology describes how another approach works, can this be read as a description of its own methods? This often indicates the return in a disguised form of a procedure that it criticises in others.

Watch out for asymmetrical oppositions in value-laden hierarchies - often in hierarchies the first term is more highly valued than the second. For example, in considering theory and practice, is the second term treated as the marginal version of the first? Does this prove to be the condition or possibility of the first?

Observe what methodologies marginalise - this could help to identify exclusions on which hierarchies such as theory and practice depend and by which they may be disrupted.

Examine points of condensation - is there a single term, such as, 'system,' that integrates different lines or sets of values?

In engaging in such activities, it is perhaps worth bearing in mind that the Trojan Horse Phenomenon would seem to suggest a general pattern of transgression and restoration of the perspectives that we criticise. Given this, it may be a mistake to think that we can escape limitations simply by closing the door on traditions that gave rise to our present situation. These so often have a tendency to return.

REFERENCES

Avison, D., and Wood-Harper, T., 1990, "Multiview: an Exploration in Information Systems Development," Blackwell Scientific, London.

Burrell, G., and Morgan, G., 1979, "Sociological Paradigms and Organisational Analysis," Heinemann, New York.

Checkland, P., 1982, Soft systems methodology as process: a reply to M.C. Jackson, *JASA*. 9:37-39.

Culler, J., 1983, "On Deconstruction: theory and criticism after structuralism," Routledge, London.

Habermas, J., 1976, Systematically distorted communication, *in*: "Critical Sociology," P. Connerton, ed., Penguin, New York.

Jackson, M., 1982, The nature of soft systems thinking: the work of Churchman, Ackoff and Checkland, *JASA*. 9:17-29.

Lakoff, G., and Johnson, M., 1980, "Metaphors We Live By," The University of Chicago Press, Chicago.

Lyytinen, K., 1987, A taxonomic perspective of information systems development: theoretical constructs and recommendations, *in*: "Critical Issues in Information Systems Research," R.J. Boland and R.A. Hirschheim, eds., John Wiley and Sons Ltd., Chichester.

Mumford, E., 1983, "Designing Human Systems," Manchester Business School, Manchester.

Rosenhead, J., 1984, Debating systems methodology: conflicting ideas about conflict and ideas, *JASA*. 11:79-83.

POLITICS VERSUS LEARNING?

Jane A. O.Young*

Leeds Business School
Leeds Metropolitan University
Leeds
United Kingdom

INTRODUCTION

Peter Checkland (1981) has claimed that the literary references used as keynotes to his work are of primary importance, a communications device to enhance the understanding of the reader. Mindful of this, and the claim in Habermas (1972) that every experience of cognitive significance is poetic, it seems appropriate to follow the example. The key to the action research work detailed here, is the belief that trying to understand a situation, and communicating that understanding might help others, while appreciating the anxieties and fears of those involved in the situation. The quotation, which seems most apt, almost a frame of reference for the 'action researcher', is borrowed from Flood & Jackson (1991), where it is used as an introduction to Total Systems Intervention (TSI).

> "I am a dreamer I am, a practical dreamer. My dreams are not airy things. I want to turn my dreams into realities. Throughout my life, I have gained more from my critic friends than from my admirers, especially when the criticism was made in courteous and friendly language. I am a lover of my own liberty and would do nothing to restrict yours" (Mahatma Ghandi in Flood & Jackson, 1991).

NOTES ON METHODOLOGY

The research for this work was Action Research, using a soft systems approach. Lewin originally formulated Action research, as a process of understanding by initiating change, in (Bennis et al, 1976). On this occasion the change being attempted is a change in the thinking of the participants in the problem situation, in the hope that, as Checkland (1990) so aptly says, "...the rich problematical pageant of human affairs can be improved by some structured thinking".

The claim that 'a soft systems approach' was used instead of claiming that 'Soft Systems Methodology' (SSM) was used, is intended to highlight both the strengths and weaknesses of an interpretive approach. Checkland (1990) states that 'SSM' itself will

* Telephone 0113 283 2600; email J.Young@LMU.AC.UK.

Critical Issues in Systems Theory and Practice, Edited by K. Ellis et al.
Plenum Press, New York, 1995

forever remain undecidable, add to this the words of Vickers (1983) on the problems of communicating, "...the meaning received may bear no relation to what the sender was trying to convey." Is it therefore possible to know what 'SSM' is, let alone claim to use it. Is it more realistic to admit that it is a personal perception, or appreciation of what is believed to be 'SSM' that is being used - one of what Atkinson (1986) refers to as a "...genus of soft systems methodologies" or a soft systems approach. Using the current terminology, after Checkland and Scholes (1990), the use could be seen as a combination of both Mode 1 and Mode 2, with the addition of ideas from Senge (1990). There is comfort to be gained from the logical, sequential steps of the seven stage model, so much easier to communicate to enquirers, wanting to know about progress. There is real power in using the internalised Mode 2, at the same time, to form a recursive structured thinking tool. To this potent brew add the systems thinking archetypes from Senge(1990), which help to make sense of the situation and provide easily shared metaphors, the result is an approach which is truly synergistic.

CASE STUDY

The organisation in which the intervention takes place consists of a central management sector, known as 'the centre', and five divisions, each of which has its own management team. The management information system being studied, which will be referred to as (CMIS), is part of a suite of systems. The trigger for development of these systems was the devolution of control of the organisation from its previous political masters. In the case of (CMIS) there was also a perceived need, on the part of some senior managers within the organisation, for a system to monitor more closely the activities of the business. These involve the work of many, previously autonomous, professionals, forced by changes in circumstances to allow much greater scrutiny of their actions.

The devolution of control necessitated the fast development of systems to enable survival. Financial systems were given priority and developed relatively quickly. These were perceived as satisfactory, primarily because they worked sufficiently well, to enable the organisation to survive financially. A personnel system was also developed. (CMIS) was the most difficult system to be attempted. (CMIS) could be considered as a customer database, plus a product database, the primary requirements as outputs, being the fast production of customer invoices, and provision of statistics to 'the centre' and to central government.

The development of (CMIS) followed the traditional pattern, first a feasibility study was commissioned, and a report submitted to the organisations' leader, approval was granted and systems analysis began. The human resources for the systems development consisted of a team seconded from one division of the organisation.

Development proceeded, many within the organisation were consulted about desired outputs from the system. Expectations were high that an important information tool was to be provided, which would of great benefit to all. A manager from outside the organisation was recruited to oversee all the systems. This manager did not survive long and was replaced by a member of the original development team. At this point it seems that a decision to consolidate on the work done to date was made, it also seems that only a part of the originally envisaged system was to be developed, what could be seen as the central core, the part most useful to the 'centre'.

Approval for this decision and proposed method of proceeding was sought from the top management of the 'centre' and the five divisions. They were introduced to the intricacies of the data model at a series of meetings. This approval was forthcoming, so production and implementation commenced.

A steering group of the top management from each division, plus other interested parties, was set up to oversee further developments and make decisions as required.

The authors intervention comes some way into the implementation process, when commissioned by one division to work on a project to design and build simple information systems to assist in the management process. Once work began the existence of (CMIS) became known, a system which could already offer some of the required facilities, yet it was not being used. Worse still its reputation, in this particular division was catastrophic. While not actually blamed for the poor weather, it did seem to be thought responsible for many other totally irrelevant calamities.

Thus the problem which drove the research was very much a 'soft systems' style trigger, and was within the larger context of the divisional project to provide management support systems. Essentially there was a need to understand why the central system was perceived so badly, to determine if (CMIS) was really just a poor system, or if something more complex was involved. Whatever the problem, this contextual understanding was crucial to the planning of divisional systems, and understanding what constraints existed within the organisational culture in relation to using IT.

A major constraint on the work, is the centrally determined policy that organisational information systems in general, were within the remit of the manager who had ultimate control of (CMIS). Within this framework any local development was seen as an additional criticism of (CMIS), although even when it is fully implemented and working well, it will not be able to meet all local information needs.

A Turning Point

The poor use and support of (CMIS) had a serious impact on its possible use for its major original aim - providing central management information and statistics. This led to the issue of a major directive from the controlling steering body. The divisions were to start using a section of the system and complete data entry within a defined timescale. The directive caused some consternation within the divisions, evoked a number of strongly worded responses, but overall caused little in the way of action to improve system use.

Thus the stand-off situation arises. The downward spiral could perhaps be illustrated as shown in Figure 1.

Situational Analysis

The influence diagram shows similarities to the 'Tragedy of the commons' archetype from Senge (1990). What would that try to show? The 'Centre' has the institutional power - the political power to issue the directive. It was issued to those within the institution whose managerial remit included control of the use of (CMIS). Officially responsible for making the system work, it is not always the case that they understand enough of such a system to be able to manage its use, or understand what its use entails. Thus system use is seen as merely data entry and devolved down to clerical level. In addition the managerial level may not understand the ultimate purpose of such systems or the organisational roles played by information systems in general. The essential overview of 'what' the system is for, becomes lost. Without this 'helicopter vision', another key issue then, is the lack of perceived benefits to be obtained from using the system, without this incentive, ensuring system use ceases to be a priority.

Looking at (CMIS)

In order to further inform thinking about the situation the author, undertook a project which required intensive use of (CMIS). This was to provide a 'users' view of the system, to offer a perspective of it as an information system. The conclusion from this work was that

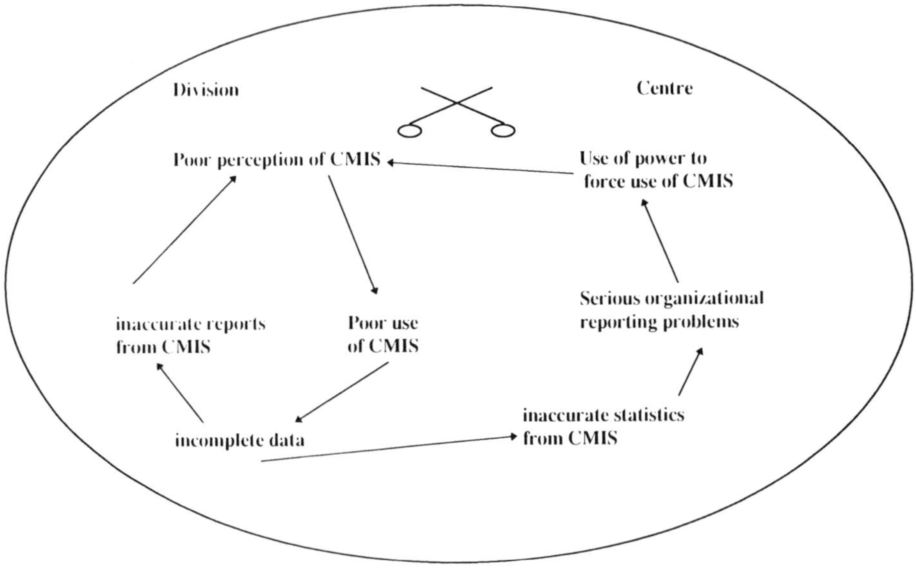

Figure 1. The downward spiral.

(CMIS) was a perfectly adequate management information system, which potentially offered considerable benefits to the division. It was not a particularly easy system to use, had a fairly high pain barrier, for reasons outside of the control of the developers. In addition it was not supported as well as it might be, but nonetheless could be made to work well.

REFLECTIONS

The conceptualisation of organisations as political systems, within Morgan (1986) offers some useful insights to the contextual dimensions of this situation. "An organization's politics is most clearly manifest in the conflicts and power plays that sometimes occupy centre stage." The view is offered that these organizational politics can be analyzed by considering the interplay between interests, conflict and power, a powerful metaphor for use here.

Flood & Jackson (1991) add to this 'using the political metaphor sensitises us to the possibility of conflict in organisations' and warn that it is dangerous to use this analysis, 'when explicit recognition of the politics of the situation leads to further politicisation and generates mistrust'. Yet knowing about the rules of the political game in any organisation is part of the tacit knowledge needed for organizational survival, that appreciation of the roles, values and norms that operate. A reinforcing view of the role of organizational politics comes from Checkland & Scholes (1990) where, within the definition of SSM, analysis three is concerned with an examination of the political aspects of the situation, an analysis carried out by consideration of the disposition and commodities of power.

Considering the case study, one perception of the situation is that the *power* to direct that system use *must* take place, is seen to serve the *interests* of the 'centre' and to obstruct the *interests* of the divisions, by both allowing the 'centre' information about divisional activities, and to utilise some of the divisional resources to provide the data for the system. The lack of belief in the system, drove the managers in the divisions to try to use their own

Figure 2. Situation model.

positional power to try to cause a shift in organizational policy, to persuade *higher powers* that the organizational *interests* would be best served by not backing and therefore enforcing the directive. The *conflict* caused by the original directive, and the divisional attempts to block its enforcement consumed a considerable amount of time and effort and delayed the much needed system use. There was clear evidence of the destructive inter-group conflict and mutual negative stereotyping that Schein (1986) describes when basic goals seem to conflict. A perceptual model of the situation might be as shown in Figure 2.

The way forward in this situation, according to Schein (1986), is to try to re-establish valid communication between the groups. An attempt to take action to resolve the *conflict* by clearly identifying that it is in the *interests* of both groups to do so, and to provide congruent goals. The move towards a learning approach.

A LEARNING APPROACH

A learning approach must try to temper the conflict of organizational politics. Senge (1990) states that 'learning involves a fundamental shift of mind', and organisational learning is only possible through the learning of members of the organisation. While Easterby-Smith (1991) suggests that organizational learning is characterised by a sympathetic culture. A revealing comment made by the chief executive of this organisation was that; 'we have a culture where people are afraid to make mistakes', and indicated that efforts would be made to make a change. The metaphor used to describe the situation was of a Newtonian body in uniform motion, spiralling downwards until acted on by a force. The appreciation by the organizational leader that change is needed, could be the force needed to cause the metanoia to halt the downward spiral. A move towards making a learning organisation, what Senge(1990) describes as, "...a place where people are discovering how they create their reality. And how they can change it." (Handy, 1994) describes this as second curve thinking, while most important of all De Geus (1988) observes that, in the future, management learning will be the only form of competitive advantage.

CONCLUSION

This is the story of the information system as victim, victim of the political conflict which characterises so many organisations. It highlights that in evaluating an information system powerful cognitive filters are at work. How often is a system not so much problematic as perceived as problematic, by those whose appreciative settings are tuned to another frequency. In this case there are three major problems with the system. They are *ownership, ownership* and *ownership* - that is the central department which develops the system feels ownership, the central department which is a major system beneficiary feels great ownership but regrettably the divisions which provide the data and operate it feel no ownership. When this issue is appreciated and addressed a way forward to effective system use, which will serve the interests of all sides is possible.

The important learning from the use of a soft systems approach in this situation, was that care is needed when trying to encourage participation, in a culture where confrontation and an adversarial stance are the norm this can be seen as extremely threatening.

REFERENCES

Atkinson, C.J.,1986, Towards a plurality of soft systems methodology, *Journal of Applied Systems Analysis*. 13:19:31

Bennis, W., Benne, K., Chin, R. and Corey, K., 1976, "The Planning of Change", Holt Reinhart & Winston, New York.

Checkland, P.B., 1981, Systems Thinking, Systems Practice, John Wiley & Sons, Chichester.

Checkland, P.B. and Scholes J., 1990, Systems Thinking in Action, John Wiley & Sons, Chichester.

De Geus, A.,1 988, Planning as learning, *Harvard Business Review*,Vol 66:2. @REFAUSTY = Easterby-Smith M., 1991, Creating a learning organization, *Personnel Review*. Vol.19.

Flood, R.L. and & Jackson, M.C., 1991," Creative Problem Solving", JohnWiley & Sons, Chichester.

Habermas, J., 1972, "Knowledge & Human Interests," Heinmann, London.

Handy, C., 1994, "The Empty Raincoat", Hutchinson, London.

Morgan, G., 1986, "Images of Organization", Sage, Newbury Park, California.

Schein, E., 1980, "Organizational Psychology,Prentice Hall, London.

Schein, E., 1986, "Organizational Culture & Leadership", Jossey Bass, San Fransisco.

Senge, P.M.1990, "The Fifth Discipline", Century, London.

Vickers, G., 1983, "Human Systems are Different", Harper & Row, London.

CRITICAL SYSTEMS THINKING

64

SYSTEMS CONSTRUCTS AND CULTURAL FILTERS

Learning from a Management Development Program in Poland

Margaret A. Barrett[1] and Barry Crellin[2]

[1] Department of Business Information Management
University of Central Lancashire
Preston, PR1 2HE
United Kingdom
[2] Crellin Consultancy
80 Riverside Drive
Hambleton, FY6 9EB
United Kingdom

ABSTRACT

The nature of the business environment in Poland changed dramatically following the major political events of 1989. In response the British Government pledged a programme of technical assistance. This programme would be supplied to Poland as the *Know-How Fund* (KHF). A key purpose of the KHF was to provide advice on Management and Organisational Development.

This paper uses systems constructs to reflect on the problematic nature of a three year interventionist KHF programme, which sought to transfer western management ethics and practices across the cultural boundaries of two nations.

INTRODUCTION

The Political events that took place in Poland during 1989 created an environment of optimism for Polish business. This in turn led Western Governments to display enthusiastic support for Poland's de-centralised economy and to a widespread recognition of the potential for investment opportunities (Kirkland, 1989).

Between 1989 and 1993/4 the British Government pledged £50 million in the form of technical assistance to Poland in the form of the Know-How Fund (KHF). This pledge reflected the view that the expansion of management capacity in Poland was a key area for the development of the country's future economy.

Critical Issues in Systems Theory and Practice, Edited by K. Ellis et al.
Plenum Press, New York, 1995

The four largest KHF development projects in Poland created Regional Management Centres (RMCs) in Gdansk, Lodz, Lublin and Poznan. Each project had a budget of £1 million to £1.4 million allocated to it. The overall objective was to develop the RMC's capacity to self-financing sustainability by 1994.

The RMCs were to carry out a varied range of activities. They included consultancy, education and training support programmes ranging from in-service company seminars to Postgraduate Business degrees. The British Council invited consortia of UK consultants and educational establishments to submit proposals for assisting the RMCs by developing and implementing these programmes. Subsequently the British Council designated The Lublin Business School, an affiliate of the Catholic University of Lublin, as the RMC for the Lublin region.

A consortium from Lancashire Business School and Lancashire Enterprises plc launched a bid to provide assistance within the Lublin RMC. This bid was successful and the Polish partners responded by requesting the introduction of appropriate Western business management skills and teaching methods.

Toward a Development Programme for Lublin RMC

All the consortia's partners recognised the need for a practical programme sensitively matched to the evolving Polish situation. The partners agreed that joint development of courses by the Lublin and Lancashire staff was essential, and that Polish would be the language used in course delivery.

The course team designed a modular programme, with a ladder of qualifications, which met the perceived training needs of the Lublin academic staff and the management students. We offered three levels of qualification: an Advanced Certificate in Management, Diploma in Management and finally an MBA. The profile included 22 one week taught Modules described in detail elsewhere (see Reed, 1993).

The international project partners quickly reached agreement concerning the rationale, aims and objectives of the programme. However, the design, development and delivery of these courses was to become an exploratory journey of negotiation, diplomacy and cultural learning.

TWO PROBLEMS OF IMPLEMENTATION

The teaching traditions in use within Polish educational establishments follow didactic learning practices. In general the associations between staff and students are formal and based upon hierarchical relationships. The Lublin lecturers had indicated no major reservations when the Lancashire team suggested incorporating participative teaching practices into the programme, despite the implication of changed teaching styles and methods.

Nevertheless, at the outset of programme delivery, the high level of student participation immediately became the first of two areas that generated tensions within the international course team.

The second problem area concerned the development of appropriate Polish case studies. We belatedly discovered there was no available historical information about Polish business organisations and indigenous management practice to build case studies around. Consequently, we used only western case studies at the beginning of the programme. Student reaction to this material was positive, since it reflected issues that they wished to examine and understand. However, the Polish lecturers found the material problematic, since they had no greater initial understanding of the material and its context than the students. These immediate and complex conflicts made the utilisation of any further informal or interactive approaches by the Polish lecturers impractical.

Relevant research, by Argyris and others, suggests that confronting complex problems of this nature can create embarrassment and threat for the individuals involved. The effect is to trigger self-sealing behaviour, which in turn diminishes the prospect of learning and the likelihood of action for change (Argyris, 1990; Argyris, Putnam & Smith, 1985; Argyris and Schon, 1978).

Unfortunately the Lublin teaching staff consistently exhibited all the characteristics of self - sealing behaviour, particularly during the critical first phase of the delivery programme. When confronted with their first interactive case teaching environments, most found it necessary to retreat from the new approach and return to their traditional didactic style.

Initial tensions over methods and materials soon evolved into more open symptoms of conflict within the International teaching team. Over time these conflicts further developed into ideological differences, which inhibited implementation of the agreed programme.

We resolved these differences by a process of charitable negotiation and friendly compromise from both parties.

A CULTURAL ANALYSIS OF THE PROGRAMME IMPLEMENTATION

We quickly recognised that we must re-evaluate the cultural perspectives of the implementation. Culture provides the strong influences and assumptions that underlie the basis for communication and decision making, so that; "...Most of what goes on in an organisation is guided by the cultural qualities of shared meaning, hidden assumptions and unwritten rules..." (Kilman, Saxton & Serpa, eds.,1985)

Within our project a dichotomy concerning these unwritten rules and assumptions now existed. We had agreed *What* to do; that was to develop management capacity in Poland by using western approaches. Clearly we had not given sufficient consideration to *How* to do this through a changed system of delivery based on *shared meanings and assumptions*. Polish staff had visited the UK to gain theoretical experience of our teaching methods. However, in practice they could not overcome their internalised cultural and ideological norms and this inhibited their ability to apply new approaches.

In an attempt to understand our differences and overcome our difficulties we turned to an analysis of culture first proposed by Hofstede (1983). This analysis contains four basic dimensions of culture, Power distance, Individualism-collectivism, Uncertainty avoidance and Masculinity-Femininity.

The most revealing information came from analysing the dimensions of Power distance and Uncertainty avoidance.

A greater power distance indicates an accepted hierarchy in organisational relationships. Conversely a smaller power distance relates to greater participation in decision-making, and greater equality in the organisation.

For the Polish team it was traditional to have a large Power distance between superior and subordinate, in this case between staff and students. For the Lancashire team the opposite held true, so that a small Power distance encouraged interaction and equality between staff and students. Whenever the Lancashire staff tried to introduce groupwork or seminar discussions within the teaching programme this emphasised the cultural differentiation in Power distances.

Uncertainty avoidance refers to a tolerance for ambiguity, or a need for stability in an organisation. People with high Uncertainty avoidance try to limit conflict and avoid situations where the risk is high. Those with low Uncertainty avoidance encourage conflict and risk-taking.

The Polish team exhibited characteristics of formality, adherence to rules and low tolerance to change. These characteristics indicate a high avoidance of uncertainty. In contrast, the Lancashire team exhibited the informal and flexible behaviour that characterises low Uncertainty avoidance.

REFLECTING ON A HIERARCHY

At the end of the first implementation phase we returned to England. We decided to re-appraise and extend our understanding of the problem situation we had encountered. We were now able to reflect in calmer circumstances and over a longer timescale.

Simple concepts often reveal profound truths about complex situations. On this occasion the systemic notion of Hierarchy proved useful in conceptualising the conflicts within our programme.

From Governance to Hegemony

In Fig. 1A we illustrate a conceptual model of superior and subordinate systems separated by a system boundary. It is important to note that there is a notion of controlling governance (and possible political dominance) implied by the superiority of one system over another within the model.

In Fig. 1B we extend this concept to illustrate a hypothetical situation where a subordinate system rejects a dominant but undesired ideology in exchange for a more acceptable alternative. If the subordinate system accepts the ideological values and implicit norms of the newly preferred alternative as close to the ideal, and proceeds to adopt them enthusiastically, then a process of hegemonic change may evolve.

However, we cannot realistically convert these initial conceptual models into a practical social reality. Beliefs, values and norms do not change instantaneously. We must recognise that changing national ideological perspectives involves an evolutionary cultural conversion of the wider society. This in turn depends upon the cognitive process of internalisation of new concepts by a nation of individuals. In reality it may take considerable time to move from theoretical acceptance of new ideologies to practical realisation of them as social norms.

In Fig. 2 we suggest a model derived from the previous series. Here we reflect that it is difficult to reject held values and norms instantaneously. This leads us to a complex situation in which conflict within a turbulent environment may be endemic. Prior to 1989 social values and norms in Poland would have been dominated by the external social and economic influence of a Communist ideology. It is not the specific ideology which is centrally important here: what is far more interesting is that the norms of the previous ideology are differentiated from those we might expect within a decentralised and democratised Western economy.

This second conceptual model illustrates the possibility of conflicts and confusions induced within a social system by the need to review existing values and norms in response to new doctrines. This model compelled us to recognise that we had, unintentionally, acted in the manner of a governing and controlling body at the outset of the programme. Unintentional, simply because the original consortial programme did not contain *intentional* mechanisms to identify and interpret the probable inter-cultural conflicts.

With hindsight, we realise that it would have been more effective to develop our conceptual models, and share the learning from them, at the outset of the programme. This might have provided more immediate and effective solutions to the practical problems we encountered.

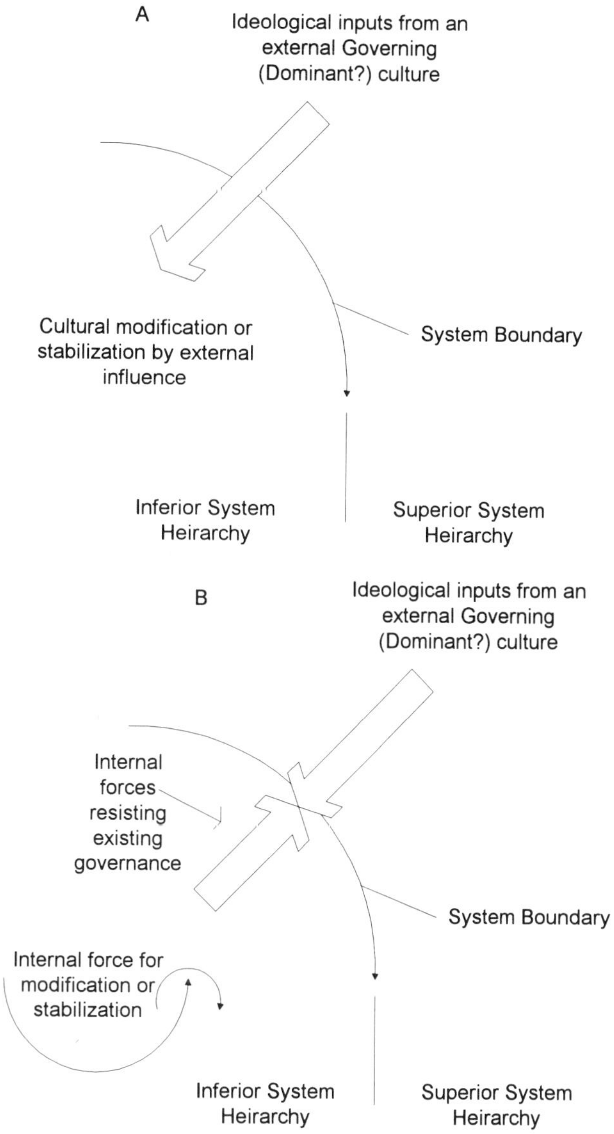

Figure 1. Conceptual model of superior and subordinate systems.

MOVING TOWARDS A CONSENSUS DRIVEN FUTURE

Our cross-cultural partnership programme encountered conflict because it emerged from widely differentiated ideological and cultural frameworks. We have learned that when underlying cultural values and norms remain unexplored and unexpressed between inter-cultural groups, they lead to tensions and alienation which make conflict almost inevitable. Consequently, implicit conflict and compromise, rather than an explicit consensus seeking process, may underpin human relationships.

It seems sensible, wherever possible, to address such issues at the outset of a collaboration, rather than within the difficult process of practical change. We suggest that

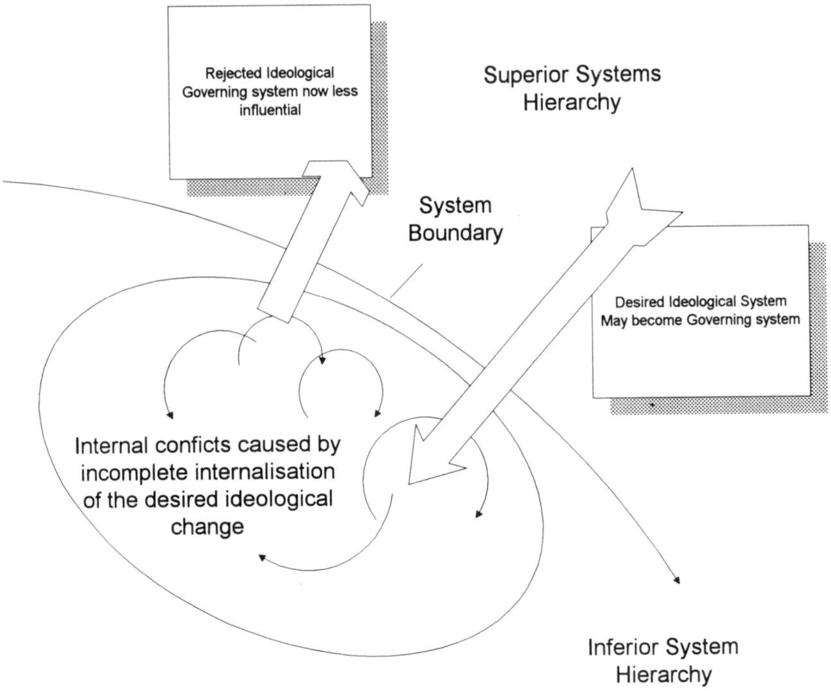

Figure 2. Derived model.

the tenets of Critical Systems Thinking, (Flood & Jackson, 1991) with their concern for Complementarism, Sociological Awareness, Human Emancipation and Well-being, could provide a framework for developing an appropriate appreciative learning process. This might then be implemented as the initial step of practical programme design. Devising and applying a framework for sharing inter - cultural models in this way might have helped us remove the inappropriate concept of dominant control inherent in our real world intervention.

REFERENCES

Agyris, C., and Schon, D., 1978, "Organisational Learning: A Theory of Action Perspective," Addison Wesley, Reading, MA.

Agyris, C., Putnam, R., and Smith, D.M.,1985, "Action Science," Josey-Bass, San Francisco CA.

Agyris, C., 1990, "Overcoming Organisational Defences," Allyn and Bacon, Boston, MA.

Flood, R.L., and Jackson, M.C., 1991, "Creative Problem Solving: Total Systems Intervention," Wiley, Chichester.

Hofstede, G., 1983, The cultural relativity of organisational practices and theories, *J. Int. Bus. Studies.* 14: 75-89.

Kilman, R., Saxton, M., and Serpa, R., eds., 1985, "Gaining Control of the Corporate Culture," Jossey-Bass, San Francisco, USA.

Kirkland, R.I.,1989, Who gains from the new Europe?, *Fortune*. Dec 18th: 48-54.

Reed, J.,1993, A new approach to management education in Poland, *J. Eur. Ind. Train.,* 17: 6: 26-32.

65

AN INTEGRATED APPROACH OF ANALYSIS

Ahmed Beloucif*

Aberdeen Business School
The Robert Gordon University
Hilton Place
Aberdeen AB9 1FP
Scotland
United Kingdom

INTRODUCTION

The author develops a framework for examining how and why auditors interact with their clients. Building on theoretical development from relationship marketing (based on exchange theory, interpersonal relationships theory (IPRs) and interorganisational relationships theory (IORs)), it is intended to provide a general framework of analysis that can be used to understand such interactions between service providers and their clients. Grounded theory, in this context seemed the most appropriate in helping to build a model. On the other hand the complexity of the problem, requires a macro view of the situation and here the systemic approach has been used for better understanding. The researcher is faced with this dilemma to develop a theory or model based on grounded theory, on the other hand the systemic view is required to complete the holistic picture of the auditor-client relationships. Dealing with such a complex research problem and looking for a configuration which could be described with a holistic approach has led the researcher to use a multidisciplinary approach.

THE BACKGROUND OF THE STUDY

This study does not begin with an established theory, and then prove it. Rather, it begins with the area of study (auditor-client relationships) and what is relevant to this phenomenon is then allowed to emerge. At the start of this research project difficulties emerged when trying to find the relevant theoretical framework for analysing auditor-client relationships. It has been the interest of the researcher to find a way that helps to uncover and understand what lies behind the business relationship phenomenon about which little is yet well known. No doubt, the difficulty lies in the nature of the service, its legal requirement and the complexity of the whole audit business. It became clear from the start of this project

* Tel. (01224) 283741; Fax.(01224) 488545.

Critical Issues in Systems Theory and Practice, Edited by K. Ellis et al.
Plenum Press, New York, 1995

that there was a lack of clear theory capable of bringing all aspects of the audit activity together and explaining the behavioural implications it entails.

So to gain an innovative and fresh view of the audit phenomenon, a solution is found by combining grounded theory advocated by (Glaser and Strauss, 1967; Charmaz, 1983; Strauss and Corbin, 1990 and Swan, 1985) and systems approach in marketing as defined by (Alderson, 1950; Achrol, Reve and Stern, 1988 and Gronroos, 1990).

OBJECTIVE OF THE PAPER

Of primary importance is the development of a method which can help the researcher to analyse qualitative data. Insights obtained from the two different approaches lead to the development of a new approach to deal with the problem of relationship in general and interorganisational relationship in particular.

In order to understand the complexity of the audit phenomenon, a three dimensional model of analysis was developed (see Figure 1). This model integrates the related theories to auditor-client relationships, and shows how the use of grounded theory as a tool of analysis and the system approach, can be tightly woven into one analytical framework. Its three integrated diagrams depict the linkages between the different parts of the research and ties the study methods of analysis (Grounded Theory and Systems Approach) together to form an explanatory model of qualitative analysis.

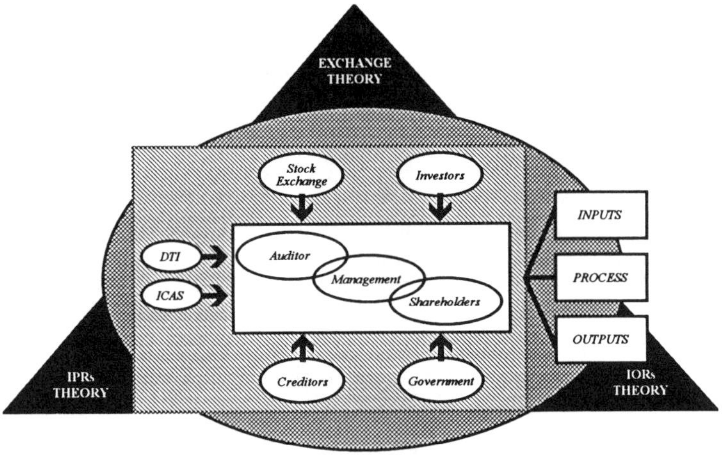

(1) ***Theoretical Background*** does not refer to the context of theories but to the way in which the study goes about analysing and integrating them in order to explain the audit phenomenon.

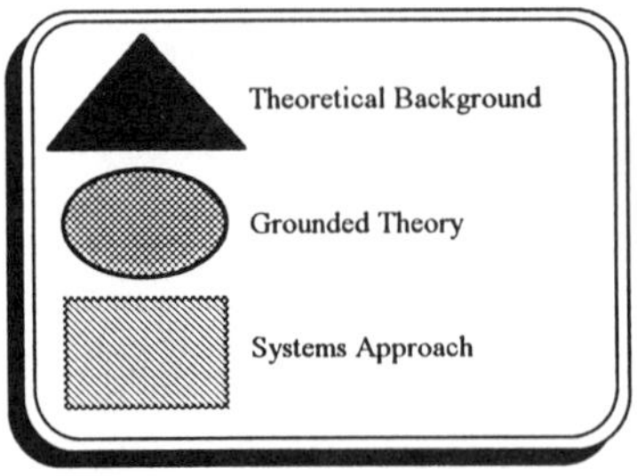

(2) ***Grounded Theory***, as analysis tool, involves breaking down the qualitative data into concepts, linking the concepts and integrating them in way to give a new meaning to the data.

(3) ***System Approach***, as an open system, treats the auditor-client as a "holistic" interactions, i.e., to see the audit relationships as a whole."

Figure 1. Integrated framework of analysis.

THE CONCEPTUAL FRAMEWORK OF ANALYSIS

The conceptual framework of analysis contains three main dimensions (i) theoretical background, (ii) grounded theory and (iii) Systems approach.

i. Theoretical Background

It does not refer to the context of theories but to the way in which the study goes about analysing and integrating them in order to explain the audit phenomenon.

ii. Grounded Theory

As an analytical tool, grounded theory involves breaking down the qualitative data into concepts, linking the concepts and integrating them in a way to give a new meaning to the data. Therefore, a coding procedure (namely open coding, axial coding, selective coding) in grounded theory is necessary. Strauss and Corbin (1990) define "...coding represents the operations by which data are broken down, conceptualised, and put together in new ways. It is the central process by which theories are built from data." Hence the analytical procedures of grounded theory aim at:

1. Build rather than only test the theory.
2. Give the research process the rigour necessary to make the theory 'good' science.
3. Help the analyst to break through the biases and assumptions brought to, and that can develop during the research process.
4. Provide the grounding, build the density, and develop the sensitivity and integration needed to generate a rich, tightly woven, explanatory theory that closely approximates the reality it represents.

Why Use a Grounded Theory Approach? The study is first of all exploratory as it aims to discover what dimensional aspects are relevant in examining the development of audit-client relationships. The significance of the study is in generating and formulating hypotheses rather than testing theory.

iii. Systems Approach

A system-oriented approach to marketing is not a new phenomenon. At the beginning, it was the concern of various authors (e.g Alderson, 1950; Fisk,1967; Fix and Dixon, 1967). The literature on channels and buyer-seller relationships stresses the importance of the *open system approach*, i.e., the interaction of the organisation with the environment (Dwyer and Welsh, 1985; Achrol, Reve and Stern, 1988). According to Gronroos (1990), the relationship approach to marketing is based on a systems approach, introduced in marketing literature by Alderson in 1950. The systems approach treats the auditor-client relationship as a 'holistic' interaction, i.e., to see the audit relationships as a whole.

Why Use a Systems Approach?

1. The nature of the research problem and its complexity. It has become clear to the researcher that this study is dealing with the complex problem of audit phenomenon and the systems approach appears to be essential in helping to understand it.
2. The second reason emerges from the nature of interorganisational relationships, interpersonal relationships and exchange theory dealing with this complex prob-

lem of auditor-client relationships. The study aims to obtain an insight into two organisations within the context of business-to-business relationship, By its nature, a systems approach tries to bring together different isolated parts into one part. It helps to obtain a *holistic picture* of what is happening in the audit market.

3. Systems occurring in the reality of an audit market are very complex and difficult to describe and comprehend. This has encouraged us to build a model or models that help to give a better understanding of auditor-client relationships, whether these models are based on structural components or functional components, i.e., subsystem of the whole system. Rosenblueth and Wiener (1945) comment on this point: "No substantial part of the universe is so simple that it can be grasped and controlled without abstraction. Abstraction consists in replacing the part of the universe under consideration by a model of similar but simpler structure. Models, formal or intellectual on the one hand, or material on the other hand, are thus a central necessity of scientific procedure."

COMBINING GROUNDED THEORY AND SYSTEMS APPROACH

This paper attempts to use a methodological integration of different approaches to build up one. By formulating analogies between these two approaches, Table 1 is presented.

CONCLUSION

This paper has outlined the importance of using a multidisciplinary approach to research auditor-client relationships. The point of departure of this new orientation of qualitative analysis is the nature of audit problem arising within interorganisational relation-

Table 1.

GROUNDED THEORY		*SYSTEM APPROACH*	
Concept	*Applications*	*Concept*	*Applications*
Category	*Open Coding* *breaking down all information: meaning and categorisation*	*Entity*	*- has an objective or physical reality and distinction of being a character* *- property of entity defines its recognisability* *- without properties, entities cannot be observed*
Relationship	*Axial Coding*	*Relationship*	*- the way in which two or more entities are depended on each other* *- interrelationships only with entities having attributed properties* *- distinction:* *(1) relation of an entity towards the whole,* *(2) relatiuon of an entity towards other entities*
Integration	*Selective Coding* *- abstraction level of analysis: taking diagrams and memos, and systematically develop it into picture of reality that is conceptual, comprehensive, and above all grounded*	*Wholeness*	*Seeing the situation as a whole*

ships. This requires a different and related methodology, i.e., one which can give a holistic picture of the situation. Therefore a systems approach is dimly needed. The systems approach is already found in various parts of managerial sciences, such as organisation theory, where the organisation is regarded as integrated complex of interdependent parts capable of interacting sensitively and correctly with one another and with their environment. The second point draws from the ambiguity and no clear methodology for qualitative analysis for interacting organisations. This has led to use of a multidisciplinary approach.

REFERENCES

Achrol, R.S., Reve, T., and Stern, L.W., 1988, Environmental determinants of marketing decision making uncertainty in marketing channels, *Journal of Marketing Research.* 25:36-50

Alderson, W., and Shapiro, S.J., 1964, Survival and adjustment in organized behaviour systems, *in:* " Theory in Marketing," R. Cox and W.Alderson, (eds.), Homewood (Ill), Irwin.

Charmaz, K., 1983, The grounded theory method: an explication and interpretation, *in:* "Contemporary Field Research," R.Emerson, ed., Little Brown, Boston.

Dwyer, F.R., and Welsh., M.A., 1985, Environmental relationships of the internal political economy of marketing channels, *Journal of Marketing Research.* 22:397-414.

Fisk, G., 1967, "Marketing Systems", Harper & Row, New York.

Fisk, G., and Dixon., D.F.,1967, "Theories of Marketing Systems", Harper & Row, New York.

Glaser, B., and Strauss, A., 1967, "The Discovery of Grounded Theory", Aldine, Chicago.

Gronroos, C., 1990, "Service Management and Marketing", Lexington Books, Massachusetts.

Rosenblueth, A., and Wiener, N., 1945, The role of models in science, *Philosophy of Science*, 12, 12, *in*: "Systems Thinking," Kramer, N.J.T.A., and De Smit, J, 1977. Martin Nijhoff Social Science Division, Leinden.

Strauss, A., and Corbin, J., 1990, "Basic Qualitative Research: Grounded Theory Procedures and Techniques", Sage, London.

Swan, J.E., 1985, Grounded theory: an inductive methodology for theory building research in marketing, *in:* American Marketing Association (AMA), Educator Proceedings, pp.348-354.

IN THE SPIRIT OF TOTAL SYSTEMS INTERVENTION

Systems Modeling at Telecom (NZ)

John Brocklesby

Victoria University of Wellington
Wellington
New Zealand

INTRODUCTION

Flood and Jackson's (1991) Total Systems Intervention (TSI) is a meta-methodology that advocates the complementary use of hard, soft and critical systems methodologies. It encourages problem solvers to suspend judgement about which methodology will be used in the study pending the outcome of a creative process during which metaphors are used to illuminate issues of concern. Other factors that impinge on the question of methodology choice include the specific interests that are are being pursued by the problem solvers, and knowledge about various assumptions embodied in the different methodologies.

The purpose of this paper is to report on what has become a major stream of research that has been carried out with Telecom (NZ) Ltd., using the TSI framework. Over a period of four years, a range of systems methodologies including Beer's (1985) Viable Systems Model, Eden et al (1983) Strategic Options Development and Analysis, Checkland's (1981) Soft Systems Methodology, Ackoff's (1981) Interactive Planning and Mason and Mitroff's (1981) Strategic Assumptions Surfacing and Testing, have been used to address a wide range of concerns. The paper reports upon the first of these studies that involved a major Viable Systems analysis of the company during a period of extensive re-organisation and down-sizing. This study is important for two reasons. Firstly, it provided the macro-level context for further micro-level systems work throughout the company and its subsidiaries. Secondly, it was instrumental in a process that has culminated in a large number of executives acknowledging the value of the systems language as a means of facilitating communication across specialist areas, and using it for this purpose.

BACKGROUND - THE COMPANY

During the 1980's the New Zealand Labour Government instituted a massive programme of economic and commercial reforms that included the corporatisation (and sub-

Critical Issues in Systems Theory and Practice, Edited by K. Ellis et al.
Plenum Press, New York, 1995

sequent privatisation) of state trading organisations. In 1987, as part of this process, Telecom (NZ) was formed to operate the telecommunications business of what was the New Zealand Post Office (NZPO), a Government Department. Whereas the NZPO had traditionally fulfilled a number of social and political objectives, Telecom was required to operate as a commercial concern.

At the time of the change the company was ill-equipped to compete in the new competitive environment, into which a number of network providers and equipment suppliers had already entered. Telecom's network (mainly analogue) technology was out of date, its management structure highly centralised and bureaucratic, its operations were inefficient, and customers perceived that it offered poor quality service.

Over the next two or three years the company made a number of key strategic responses to this situation. These were geared towards making the organisation more competitive pending its projected sale. They included widespread cost cutting, strengthening financial structures, condensing operations to focus upon the core telecommunications business, modernising the network by introducing digital technology, and recruiting managers with proven commercial success. These changes were introduced concurrent with a complete re-organisation of the company's structure, and it is this aspect of the change that we concentrate upon here.

THE NEW ORGANISATIONAL SRUCTURE

Before its demise the NZPO's structure had evolved to meet the requirements of politicians and Treasury as well as customers. Decision making was the prerogative of head office managers who were far removed from the customer. Much of the time of these managers was spent making submissions and seeking approval from ministers and cabinet committees before implementing decisions. Inside the Post Office, lines of communication were predominantly up and down the formal hierarchy, and there was limited communication across the various districts and branches that were responsible for operational matters.

Once established, Telecom's district and branch structure (40 competing areas) was reorganised along functional lines in order that responsibility for functions could be established and to make managers more accountable. Several layers of management were removed to reduce costs and force decision making closer to the customer. A marketing department was formed and financial structures strengthened throughout the organisation. Although significant improvements in performance occurred as a result of these changes, more radical surgery was carried out a year later when a new management team was introduced. The main structural features of Telecom now became:

- The formation of four autonomous Regional Operating Companies (ROC's) to operate the main business of the Telecom network.
- Decentralisation of head office functions into the various ROC's.

Two years later, the author and two colleagues were commisioned to undertake a review of this structure. Our decision to use the VSM was based upon the review's structural focus, and the company's stated need for continuous service improvements, product innovation and environmental adaptability. These factors were consistent with the guidelines provided for the use of the VSM in TSI. In what follows we make the assumption that readers are familiar with Viable Systems Modelling.

ANALYSIS OF THE ROC STRUCTURE

Under the ROC structure four basic network businesses were made responsible for business in their geographical area, a fifth provided common network and international services. All other businesses were established as new venture companies. The ROC's were set up as autonomous business units, and most functional activities were decentralised.

The autonomy ceded to the ROC's is consistent with viable systems logic that S1 units should be capable of embodying sufficient variety commensurate with that existing in their immediate environment. Organisationally this also made sense, because previously the hierarchic and bureaucratic features of the post office had severely constrained the ability of operational units to deal with increasingly complex technological and market environments..

Other features of the new structure consistent with viable systems thinking included:

- Linking the compensation of each CEO to performance (S3* monitoring of S1 performance)
- Introducing budgets and performance targets (S3, S3*)

The senior executive team viewed all of these developments in a positive light. However, our research revealed a strong sense of disatisfaction with other emergent features of the new structure. Most managers believed that the pendulum had swung too far in favour of decentralisation, leading to dis-economies of scale in certain areas and inconsistencies, duplication, and unhelpful competition across the ROC's. Although the company had foreshadowed this by establishing a number of 'Functional Councils' (FC's) in areas such as human resource management, information systems, finance, commercial, and business planning to coordinate and develop consistency in these areas across the corporation, the FC's were viewed as having been only partially successful. Other issues of concern included perceived inefficiencies in the functioning of the management board, confusion about the role of ROC managers and the responsibilities of the corporate strategy group. These issues became the focus of our initial VSM work, and they are reported on below.

THE REGIONAL OPERATING COMPANIES

Under the new structure most functions that had previously been carried out by Head Office were decentralised into the ROC's, with the result that over a period of 12 months, the corporate office was reduced in size from 1800 to 400 staff. Although this is broadly consistent with the idea that in principle S1 units should be capable of being sold as going concerns, in practice there are limits as to how far S1's should be decentralised. Under Telecom's regional structure, there was a great deal of similarity across the various local environments, and in this circumstance there is a compelling argument to remain centralised in order to reap the benefits of economies of scale. Thus, some managers complained of there having been a 'knee jerk' reaction away from the highly centralised bureaucracy of the NZPO towards excessive decentralisation and that this had led to inefficiencies in various areas including treasury, marketing, human resource management and information systems. Subsequent deliberations about the degree to which the company should be decentralised revolved around ensuring that only those specialised activities that needed to be close to the customer should be decentralised. All others should remain a centralised function in order to secure the benefits of economies of scale.

THE FUNCTIONAL COUNCILS

Initially it was thought that company-wide consistency (S2) across the ROC's could be achieved through cross-directorships and representation of ROC managers on the Management Board. However it was soon realised that a more formal S2 was required, and the FC's were formally charged with this responsibility. Over time however their role became somewhat confused as various FC's carved out roles that included relieving pressure on the Management Board for policy making in their particular areas (S5), assisting in developing strategy in various functional areas (S4), and performance monitoring (S3*) In practice, most FC's were uncertain about what their role in the organisation should be. This ambiguity had allowed some of the more active FC's to define their roles in ways that suited the predilections of their members, spanning the full range of S2-S5 functions. Some were perceived as heavy handed by ROC managers and unduly intrusive. Others became largely inactive, resulting in a loss of respect from ROC managers.

THE ROC MANAGERS

In addition to their primary S1 role, most ROC managers were fulfilling a range of S2 - S5 responsibilities across at least three and up to 5 levels of recursion. This reflected their various formal responsibilities as board members (S5), convenors of F.C.'s (S2), heads of corporate office functions (S5, S4), and directors of new venture companies (S1-S5, at RL2, RL3).

Although excessive workload was a concern to the ROC managers, their inability to conceptualise their various formal roles in terms of appropriate behaviours, was a greater concern. Roles frequently became 'contaminated' by extraneous concerns, where for example, meta-systemic company-wide matters were allowed to interfere with S1 management, and meta-systemic functions were purloined by narrow S1 agendas. This role contamination was rife in the company at the time of the research.

THE MANAGEMENT BOARD

Among the senior management team there was a strong feeling that the Board was not functioning properly because it was pre-occupied with detail and because meetings were badly run. In response to the question of what role the Board should play, the group unanimously supported the idea that its main responsibility should be policy making and direction setting (S5). However the research revealed that S5 functions were being compromised by the inclusion of agenda items relating to a range of S1-S3 business. Because ROC managers were conceptually unable to separate their S5 responsibilities from their other roles, the Board frequently became an arena where operational detail from the ROC's, the new venture companies, and corporate office functions diverted attention away from its main business.

THE STRATEGIC PLANNING PROCESS

The strategic planning process was intended to create a sense of corporate ethos (S5) and to help create a long term focus (S4), at a time when there was a sense that too many decisions were being made on the basis of short term considerations. It involved creating a high level statement of common purpose or vision for the corporation as a whole, further to which 'Strategic Excellence Positions' (SEP's) were established in the areas of customer relationships, quality and efficiency, and staffing. From each SEP, broad level objectives were developed that included strategies and actions. In VSM terms this process was deficient in two respects. Firstly Telecom's stated vision included the need to 'provide superior service' and 'exceptional business performance'. This contained conflicting signals leading to confusion, especially among front line staff who felt that it was not always possible to satisfy both objectives concurrently. Secondly research showed that these statements concerning company ethos had not been successfully communicated throughout the corporation. Only 6% of staff had heard of the vision and SEP statements.

REMEDIAL ACTIONS

Following the initial VSM study these difficulties were addressed on a broad front. Changes included:

- Recentralisation of treasury, sales and marketing, human resource management, purchasing. All other activities that need to be close to the customer eg. installation and maintenance remaining decentralised.
- Strengthening of vertical communication channels between the ROC's.
- FC's to refocus their activities around their primary S2 function
- ROC managers required to give primary consideration to their S1 role, and to conceptualise their other roles in VSM terms, to prevent further role contamination.
- The Board refocus its activities around its primary S5. ROC and New Venture (NV) operational matters handled at lower recursion levels.

Although the researchers regarded these actions as broadly consistent with VSM logic, it was felt that more radical changes were required to create a more sustainable viable system. Two ideas were put to the management team. Firstly, we did not believe that a clarification of the S5 role of the Board coupled with an improvement in the way meetings were to be run, in themselves, would be sufficient to deal with the overload problem. Under the existing structure, there were 17 direct reports to the Board. We felt that even a clearly-focussed and well organised board could not absorb this amount of variety, and there was a need for another level of recursion between the Board and the ROC's/new venture companies. This would involve the establishment of two new organisational units, one which would be responsible for meta-management of the complete ROC group, the other responsible for the new venture companies. Individual ROC and new venture business could then be dealt with at this lower level of recursion, thereby freeing up the Board to deal with S5 business for the company as a whole.

Our second suggestion invoked the possibility of even more radical re-structuring. Although the original contract had required that we restrict our analysis to an evaluation of the existing organisational structure, our investigation led us to believe that there were sound reasons why a fundamentally different structure should be put in place. According to VSM logic, the amount of autonomy and independence ceded to S1 units is partly a function of the diversity existing across S1 environments. In simple terms, the greater the 'market' diversity, the more autonomy S1's need to provide 'specialised' service. If S1 units are serving the same or very similar environments, the decentralisation argument is harder to sustain, and, under these circumstances it is tempting to reap the benefits of economies of scale by allowing many functions to remain centralised. In Telecom's case, the ROC's were organised on a geographical basis, and their environments had much in common. Most

ROC's were providing the complete range of services, although there was some difference in emphasis, Auckland and Wellington, for example, had more large corporate customers than the others. The systemic illogicality of the ROC structure was further reinforced when it became known that corporate clients with operations in both Wellington and Auckland were becoming increasingly frustrated with having to deal with two or more Telecom companies rather than just one.

Our recommendation for a more systemically logical S1 structure required that we identify activities within the total market that warranted specialised attention. On this criterion the most obvious distinction was between the residential and business segments. Although some core products traverse both markets, the needs of these two groups are distinctly different. Thus, a case was made to restructure the organisation along these lines.

In the immediate period following the study, neither of these recommendations were implemented. A year later, however, a chief operating officer was appointed to take responsibility for all four ROC's. This went some way towards providing the additional level of recursion that had been recommended. Then, in 1993, Telecom was restructured into 8 autonomous divisions that included business sales and residential sales.

On a final note, it is worth mentioning the manner in which the pendulum has swung backward and forwards from the centralisation to decentralisation of key functions over the years that the VSM has been used to track the company. Under the ROC structure, where S1 units were guilty of duplicating services across different regions, the company forfeited economies of scale by decentralising functions in circumstances where this was *least* justified. Somewhat paradoxically, it is interesting to note that under a structure where the diversity between residential and business sales is embodied in new organisational units, and at a time when these specialised markets invite decentralisation of functions, both marketing and information systems have been established as a separate centralised function. Although this need not necessarily imply that these functions have become less sensitive to the needs of the two key S1 operational units, it does raise the possibility of this becoming a problem.

CONCLUSION

Before the VSM study, there was a sense among the management team that although the ROC structure was not entirely failing to deliver on its promise of dealing with the problems associated with the old NZPO bureaucracy, it had introduced new problems of its own. However, progress in dealing with these issues was being hindered because the team appeared to lack a common framework for identifying problems, for understanding their causes and consequences, and for coming up with solutions. The diverse background of some of these managers was also a problem in that they lacked a common language to reflect upon the situation in a manner that was accesible to the others involved. The VSM study and the systems work that has followed it has been extremely useful in dealing with these sorts of issues, and in gaining credibility for further work in the future.

A final word on TSI. In all of the systems studies that have now been carried out at Telecom (10 studies involving 15 researchers), we have attempted to be true to the spirit of TSI on the question of methodology choice. In each project we have genuinely attempted to avoid entering projects with a particular methodology in mind. We have tried to suspend our overall judgement about problem situations until after some form of creative thinking working with project sponsors, although metaphors have not always been used for this purpose. Despite this, we are becoming increasingly conscious of how our predilection to favour some methodologies at the expense of others can unwittingly subvert this process, our suspicions having been aroused when in almost every case the methodology used in the study has been that in which the primary researcher had the greatest level of interest and expertise. Although we have no way of knowing whether this has detracted from the overall value of the various projects, we do raise the issue here. TSI lays claim to offer a process in which methodology choice is problem-dependent, yet neither it nor the wider field of systems thinking provides any substantive guidance or recommendations for ensuring that this will eventuate. It is an area where further work could usefully be done.

REFERENCES

Ackoff, R. L., 1981, "Creating the Corporate Future", JohnWiley & Sons, New York.
Beer, S., 1985, "Designing the System for Organisation," Wiley, Chichester.
Checkland, P. B., 1981, "Systems Thinking, Systems Practice," John Wiley & Sons, Chichester.
Eden, C., Jones, S., and Sims, D., 1983, "Messing about in Problems," Pergamon, Oxford.
Flood, R. and Jackson, M. C., 1991, "Creative Problem Solving - Total Systems Intervention," John Wiley & Sons, Chichester.
Mason, R. O., and Mitroff, I. I., 1981, "Challenging Strategic Planning Assumptions," John Wiley & Sons, New York.

THE LANGUAGE OF SYSTEMS

One Linguist's Perspective

M. R. Combs

University of Humberside
Hull, HU6 7RT
United Kingdom

Once a linguist and anthropologist; I am now struck by a number of parallels between some of the assumptions that have been made (and debated) about linguistics and language, and between those that I have recently begun to make about the language of systems and the ideas that it contains, particularly in the context of business and information systems development. In so far as making a case for a link between language, linguistics and systems is concerned, I would not be initiating a debate but entering one which has occupied some rather more illustrious minds than my own. Stafford Beer (1985,p.vii), for example, notes that **language was one of the viable systems** which informed some of the discussion that gave birth to what has come to be known as the Viable System Model.

Assuming that the systems approach may spare a little of its relatively youthful time to listen to its rather more mature systemic relative 'linguistics', I will draw attention to some parallels or near parallels that have struck me with a view to recalling experience in linguistics which may, however, have to be relived to be relearned by its younger counterpart. It would be less than honest of me though to pretend that linguistics has always been successful in discovering or capturing all that we would like to know about language as a viable system. Also, one might argue that from what we do know, that language is not quite as viable a system as we would like it to be; nevertheless we must marvel at its complexity and the profound questions it raises.

SOME PARALLELS BETWEEN LINGUISTICS AND SYSTEMS THINKING

One of the prime functions of an information system is to communicate information about business activities so that they can be effectively managed; this is surely an instance of what Checkland speaks of as systems thinking's interest in communication and control. Language is clearly central to almost all human communication and language communicates *information* about all sorts of real world activities; in that respect language can be seen as an essential component of systems thinking. Linguists like myself could be said to have studied as much the structure of information and the nature of communication as to have merely been students of language. In this paper I will allude to four parallels that spring to mind concerning linguistics and systems (systems thinking), and these are that:

- They both need special language with which to communicate fundamental ideas.
- A distinction needs to be recognised between universal notions that capture a set of abstract 'neutral' theoretical assumptions, and between models that capture useful sets of assumptions and patterns of relationships and rules that depend upon them.
- They are both concerned with the relationship between the essential realities of activities or behaviour, and communicating information about them.
- They are both interested in complex empirical data.

Linguists use language to talk about language and in that sense use language 'concretely' to do the job of communicating, while making certain observations that we can call 'abstractions' to capture general or specific facts about it. The discipline of what we now call linguistics (the study of language and languages) might reasonably be seen as the replacement and extension of what was called the discipline of philology (the study of language and languages - primarily within a classical tradition in which the Indo-European language family provided the principal models). One of the reasons for the growth of modern linguistics away from philology was to be found in the experience of pioneering linguists such as Sapir (1921) and Pike (1967), a former senior colleague, who found that any assumptions about language or culture based on Indo-European models tended to interfere with their analysis of 'exotic' Indian societies and languages of the Americas. New thinking and more sophisticated approaches to analysing and understanding languages and the phenomenon of language had to be developed. They needed specialist language with which universal linguistic principles could be used in an 'open-minded' approach to analysis of language data.

The Need of a Specialist Language with which to Communicate Fundamental Ideas

To me systems seems to supply a specialist language subset which may embody universal organisational principles that might be used in an open-minded way to talk about what might be called the grammar of organisations, that is to say, the way in which they perform. 'Systems' addresses itself primarily to certain aspects of performance, just as linguistics identifies and addresses certain aspects of language, even though acknowledging the interaction between different aspects such as syntax, semantics, and sound systems. The development of a special language subset is difficult in that either completely new terms have to be coined and then infused with appropriate meaning by advocates of the language and its ideas, or else existing terms in general currency that already carry some sense of the meaning intended by the advocates become the special legal tender of the users of the language subset.

Checkland (1981) acknowledges this difficulty when he says, "Writers on systems have frequently felt the need to invent a new word for the entities which are wholes at one level of a hierarchy while simultaneously being parts of higher level entities" (p.82). He then goes on to give examples of suggested terms. The earliest reference to such a term is in the 1960s, though interestingly Pike (1967) the originator of the 'tagmemic' school of linguistics used the term 'tagmeme' at least a decade before for a virtually identical concept, and even that was hinted at in de Saussure's (1916), work early this century.

Distinguishing between Universal Notions and Developing Models

If systems ideas do provide the basis for a set of universal notions that could be used as a foundation for analysis and gaining greater understanding of business activities; then from an information systems perspective we would be particularly interested in the information, and the systems that generate information to give us understanding of the state of vital business processes and resources. This parallel is one which again can be drawn with reference to Pike and tagmemic linguistics; the term 'tagmeme' and the school of linguistics which resulted from the extensive use of the approach in analysing hundreds of minority languages over two or so decades. The point is that although the tagmemic approach was valuable in terms of providing a set of universally valid notions that set the analytical agenda, language models were sometimes better developed on the basis of grammatical patterns discovered in analysis but which captured truths and systems that were peculiar to, or more meaningful in terms of, the individual languages (see later the comment re. chess).

One of the problems with tagmemics or any other school of linguistics is that most, if not all, either had to invent new terms to convey their ideas more precisely than they thought existing terms did, or else they had to redefine existing terms to reflect their system of thought. This problem is, of course, not limited to the discipline of linguistics, it is common to the development in ideas in many disciplines. Linguistics, schools of linguistics, and systems thinking need special language, but inevitably part or all of the language subset is language that is in common use or already in use by competing schools of thought. Let us take a brief linguistic perspective on this issue.

We know, for example, that each word in a language has meaning that can be expressed in terms of a semantic domain. We use the word 'synonym' to talk of words which have the same meaning, and 'antonym' to talk of words that are opposite in meaning. But even a casual look at a good dictionary warns us that not only can the same lexical item (word) have very different meanings, but each meaning can have a number of different senses, not all of which are likely to be shared by all the synonyms. An antonym may therefore only have an opposite sense for only one aspect of the meaning of its 'opposite'.

'System' is a word which is variously associated with computing (a computer system), prescribed, set ways of doing things (an ordering system), a set of business activities and resources (the banking system), cybernetics and issues of control in business or industry, or a way of thinking about things (systems thinking). The problem is that these domains of meaning coincide or overlap in the real world, and there are enough aspects of shared meaning sufficient to interfere with clear communication of ideas. This is a real challenge to systems thinkers who try to communicate about systems ideas or abstractions (albeit potentially useful) to audiences who are more inclined to associate systems language with concrete, real experience of, say, computers, or ideas about which they have already formed opinions such as those of efficiency-driven mechanistic modernists!

Communicating Information about Essential Realities

In a recent collection of papers on organisational theory entitled Rethinking Organisation, Gergen (1992), an organisation theorist, summarises a strand of popular management wisdom which dismisses systems theory as part of the discredited modernist tradition. However Gergen cannot get away from acknowledging that language has a role and relationship with regard to the real or 'essential' world that it talks about. As a post-modernist he appears to want modernists to sign up to a view of language in which it is used as a logical corset into which is squeezed the unsightly bulges of the real world of organisation, whilst he stands on the enlightened high ground and infers that as a post-modernist he views language as having acquired "...a new master, and this master is the community". In a similar context he discusses the notion of 'indeterminate rationalities' and states, "The post-modern drama begins, however, with the realisation that the 'rational sayings' available to the individual are of indeterminate meaning." and later says, "But the plot now thickens, for at the outset it is clear that there are multiple meanings for such terms as 'logical', 'bottom line' and the like...".

As for seeing a plot in a number of words or expressions having more than one reading, one wonders at so innocent a view of language. Gergen appears to lack familiarity with long understood characteristics of language such as the existence of synonyms; even some of the earliest recorded language makes use of such devices. His other criticism seems to suggest that he sees systems either as a prescriptive approach which assumes business activities can be designed solely to account for the rational attainment of business objectives, or that systems has an interest only in the logic and control of business activities. Either way, Gergen appears to assume that systems has little or no real interest in the 'essential' or the 'realities' of business experience. The language which speaks of these realities has to, as it were, come out of them. Interestingly tagmemics, with respect to the development of the ideas which it contained, and its language did come out of the realities of empirical research with real speakers and real language, but it was often used to talk about those languages in a way that did not reflect adequately the rules and patterns of use that were most meaningful to the individual communities. Gergen echoes the modern trend, or is it a swing of the pendulum, which is to empower the individual, or possibly the employee community.

Interest is in Complex Empirical Data

If I understand him correctly, Gergen's discovery that the community is the new master of its language recalls in a somewhat superficial way observations made by Ferdinand de Saussure, often called the father of linguistics (and also by some the father of semiotics), about eighty years before, in which he stated that language can be understood in two distinct senses. One sense which he referred to as '**lange**' can be taken to represent the abstract language system which we colloquially call 'language'. This (following Hawkes, 1977) is akin to how, for example, we use the word 'chess', to mean the game of chess which includes, by implication, the rules. The other sense of 'language', he referred to as '**parole**', which we could render as 'speech', and by this he meant a specific language or dialect as it is used by its community of speakers and writers. Perhaps we could extend the analogy of chess to suggest that significant patterns in terms of 'parole' might be similar to recognising patterns of play such as in a 'Sicilian defence'.

'Lange', Saussure described as being, "both a social product of the faculty of speech and a collection of necessary conventions that have been adopted by a social body to permit individuals to exercise that faculty" (Saussure, 1916). Saussure was acutely aware that 'language' is, in a sense, a product of a speech community, but how one decides upon whether the community is the master of it, or is itself to some extent moulded by it, is hard to tell. The important point that Saussure makes is that there is much to be learned about language which is abstract and which is the stuff of testing, and model building, but the complement of the abstract is the real or essential world of language as it is used, in which analysis looks for evidence to test or prove theory and models.

This uneasy relationship between the 'abstract' and the 'real' is a fact of life with which linguists like myself have wrestled with on a regular basis. In the last quarter century the well known linguist Noam Chomsky also made a 'discovery'. This resulted in him making a distinction between language **competence** and language **performance**. Competence he considered to be the abstract representation of a language which he claimed was common to all the speakers of a speech community. The competence of a speaker would include a largely unconscious knowledge of the grammar of the language which would include the lexicon and all the rules that would be necessary to encode or decode all and only grammatical sentences in that language. Performance on the other hand is the term used to represent what people actually do say and how they really do use the language; the correlation between competence and performance was often hard to see. While the linguist's grammar could judge a speaker's utterance as ungrammatical, many real, generally acceptable utterances often judged the grammar inadequate! Grammar consists, of course, of the abstract linguistic rules of a language which govern how different types of language elements function together. The problem with Chomsky's grammar was that it was treated as an almost closed system; neither language nor business organisations have any meaningful systems or subsystems that could usefully be modelled as closed systems. Any mechanistic interpretation of systems which thinks of it as providing a language which assumes that the entities it models can be adequately represented as purely logical or rational, would indeed be an inappropriate interpretation. Empirical study, action research, or plain consultancy scream out to us about the bulges; the muddle, anxiety and frustration that characterises the complex realities of the real world, yet purposeful activities, and goal setting and attainment are also part of this essential world.

But systems does and could provide a universally applicable language with which purposeful activities within business can be better understood and modelled. The ideas themselves and the language with which they are expressed can and should come out of careful in-depth research of the business contexts and communities about which the language speaks. In linguistics 'parole' and 'performance' speak for themselves; they will pronounce the ideas that we have about universal notions contained in theories of 'lange' and 'competence' in need of revision when they judge them inadequate. In systems thinking too, meaningful statements of theory abstracted from our study of management behaviour in the business community, especially as it affects the information that speaks of states of resources and measures of performance, can only be gained from empirical action research; the stuff of linguistics and systems. Perhaps the healthy tensions in linguistics encouraged by the complementary notions of 'lange' and 'parole', and 'competence' and 'performance', suggest that we revisit Checkland's ideas of 'systems thinking' and 'systems practice' with a view to exploring the nature of the tension which should exist between them.

REFERENCES

Beer, S., 1985, "Diagnosing the System for Organizations," John Wiley & Sons, Chichester.

Checkland, P., 1981, "Systems Thinking, Systems Practice," John Wiley & Sons, Chichester.

Chomsky, N., 1965, "Aspects of the Theory of Syntax," MIT Press, Cambridge.

Gergen, K.J., 1992, Organization theory in the postmodern era, *in*: "Rethinking Organization," Michael Read and Michael Hughes, eds., Sage Publications, London.

Hawkes, T., 1977, "Structuralism and Semiotics," Routledge, London.

Pike, K.L, 1967, "Language in Relation to a Unified Theory of The Structure of Human Behavior," Mouton, The Hague.

Saussure, F. de., 1966/1916, "Course in General Linguistics," McGraw-Hill, NY.

Sapir, E., 1921, "Language," Harcourt Brace Jovanovich, NY.

68

SYSTEMS PRACTICE CHALLENGES

D. P. Dash

Xavier Institute of Management
Bhubaneswar - 751013
India

INTRODUCTION

I am a systems-oriented management consultant and a teacher of corporate policy and planning. In that capacity, I have interacted with several systems scholars and management professionals who are influenced by the systems movement. I have also met scholars and professionals from management and other disciplines who, on their own, have adopted a style of thinking resembling the systems approach. My reading suggests that all these people are regularly faced with various intellectual and practical obstacles, as much as I have myself been. In this brief article, I wish to provide a rough conceptualisation of these obstacles. I will round off with a few concrete suggestions.

Although systemists of various persuasions continue to debate the nature of the systems approach, it is now largely accepted that the approach involves a conscious switch from 'particular' to 'total' conceptions. Differences exist about the nature of the ontologic and epistemologic shifts necessitated by the above. These philosophical differences are not resolved by adducing 'evidence'. Such debate is, however, illuminated by praxis.

Interestingly, the systems community has more or less adopted the latter root; the international journal *Systems Practice* is testimony to that. Being a systems practitioner myself, with interest in theory development, I find that systems practice has not yet resulted in clear decisions on any of the contentious issues of theory. On the other hand, the neoteric practice itself is faced with several challenges which, if not effectively addressed, may dampen its vigour and delay the desired outcomes.

If we accept the shift from 'particular' to 'total' conceptions as endemic to the systems approach, then any practice consistent with this requires both developing such conceptions and intervening in the real world in accordance with such conceptions. Both these tasks are fraught with several challenges.

CHALLENGES OF CONCEPTION

Generating 'total' conceptions remains a fundamental challenge in systems practice. The genesis and character of this challenge can now be probed.

Critical Issues in Systems Theory and Practice, Edited by K. Ellis et al.
Plenum Press, New York, 1995

Many 'Experts'

Systems theory suggests that there are no experts in the systems approach. But, experience suggests to the contrary. Claims to systems expertise abound. Cyberneticians, Social Cyberneticians, System Dynamicists, System Analysts, Systems Engineers, Soft-systems Modellers, Multi-modellers, Interactive Planners and Social Systems Designers, among others, have all staked some claim. There is no clear *mainstream* in the systems approach.

As a result, there is no definite body of knowledge to which systems practitioners can adhere. This is a major obstacle for generating 'total conceptions'. Even after more than four decades of systems research, there is no reliable method of generating 'total' conceptions; except, perhaps, through metaphors (Flood and Jackson, 1991). But several basic questions remain, at best, partially answered. For example: What is the methodological legitimacy of conceptualising through metaphors? How to deal with metaphors which are potentially obfuscating? How to formulate new metaphors? How to evaluate and select metaphors? How about metaphors for the enveloping system?

Boundary Judgement

Responsibilities of systems practitioners are yet to be conclusively clarified. Also not clarified are the normative implications of the act of systems practice. This provokes a *boundary judgement* question. Issues pertaining to this are the choice of a system of concern, the choice of clients, the choice of a system representation method, the choice of model purposes, the choice of the role to be played by the systems practitioner and impressions about the ultimate worth of such practice.

I have found that people trained in the systems approach continue to be ignorant of their personal and social biases. Most management professionals fail to recognise themselves as a powerful group playing a specific role in the industrial society. In a predominantly agrarian society like India, many of these professionals (as most other professionals here) are also members of the propertied caste-elite, who are by-and-large not conscious of their indoctrination into the values and judgements of their specific communities and sub-cultures. This kind of ignorance is a mighty barrier against 'total' conceptualisation.

CHALLENGES OF INTERVENTION

Even where promising 'total' conceptions are available, howsoever rough, special difficulties are faced in *operationalising* these through a process of intervention. These difficulties are outlined below.

Short-Termism

Systems practitioners often encounter clients for whom 'management' is solely for, of and by the *shareholders* (those who have contributed towards the financial capital of an enterprise). Such enterprises are designed, managed and evaluated through principles harmonious with shareholders' interests. Not to talk of the occasional violation of these principles by managers in order to further their own interests, when such interests do not coincide with those of the shareholders! In such an environment, one principle triumphs: Short-termism. This strongly discourages the systems practitioner from forwarding any truly systemic idea, which may show beneficial results ('beneficial' in a holistic sense) only in the very long run. I once suggested to a soft-drink bottling company which distributes an international brand in India, that they should consider introducing an indigenous product. The idea was promptly rejected as 'too expensive'.

Ad-hoc Institutions

While dealing with larger issues such as Urban Planning, Child Development, or, Public Health, one major step under the systems approach (I hope you agree) is to define *relevant systems*. For example, if you are studying agro-climatic cycles in Nepal, you may define your relevant system as the 'Himalayan Eco-System'. Thus defined, the problem-domain cuts into several pre-defined boundaries, not the least of which is the international political boundary. The systems practitioner finds that there is no definite body/authority which has a clear mandate over the meticulously conceptualised relevant system.

This difficulty can be attributed to the fact that many of our socio-economic institutions have been built in an ad-hoc manner. This imposes an anti-systemic constraint on the practitioner, of having to design solutions/recommendations which lie within the mandate of the 'client'.

Coercion

The pluralist spirit of the systems approach is severely challenged by coercive structures in several organisations. As a result, many serious systems-oriented consultants end up merely emphasising that many stakeholders exist in a complex web of interaction among themselves. Nothing much is achieved thereby, except inviting more subtle forms of coercion from a now-alerted coercing group. The problem is exacerbated by the varying degree of communicative competence and varying compulsions and motives of the coerced stakeholders.

SUGGESTIONS

Now I proceed to outline some suggestions on how to tackle the above two categories of systems practice challenges. These suggestions are reasoned and presented following the conceptualisation developed in this article.

To Facilitate Conception

- A handbook of systems practice needs to be written. This would act as a guide for new entrants into the profession. Further, it would also help practitioners link their practice with the existing body of knowledge. In writing such a handbook, Critical Systems Thinkers should take a lead, because they have a way of integrating various strands of the systems approach. This would help clarify the nature of 'expertise' now available and thus fight against the disciplinary trend in systems thinking and practice.
- The idea of *metaphor-based conception* has to be examined more closely. If metaphoric understanding is reduced to merely analogical reasoning, then it would suffer from all the weaknesses of logical induction. If it is something more, as it is in creative literature, then its relation with the normative content of systems practice must be clarified. This would improve the quality of 'total' conceptions employed by systems practitioners.

- The normative implications of systems practice itself needs to be clarified. This can then be used to clarify roles and responsibilities of systems practitioners.
- Some popularising of systems ideas is essential. Today, these remain largely confined to the realm of philosophical debate. Towards bringing these into the public domain, the most articulate systemists should be encouraged to develop an enduring communication with the rest of the world. This is to be oriented towards extracting boundary judgements implicit in the practice of various professional groups. The resulting self-awareness may improve the potency of 'total' conceptions created and used by these groups.

To Facilitate Intervention

- Systemists should join hands with other groups who are questioning short-termism at various levels, such as some environmentalists, who are questioning the implicit short-termism in the dominant ideology of economic development. The deleterious effects of short-termist management have to be highlighted in public fora. Systemic evaluation of corporate bodies could be undertaken to demonstrate the inappropriateness of an excessive focus on shareholders. Unless short-termism is widely accepted as harmful, systems ideas would not find enthusiastic takers.
- Socio-economic institutions should be evaluated to demonstrate their ad-hoc character. This may result in some systemic *reengineering* of these institutions, consistent with the *mess* they are supposed to manage. This appears to be both a necessary condition and a desired outcome of systems practice.
- Systemists should aim to be the harbingers of non-coercive organisations. Coercive structures of all kinds are to be shown as working against the human cause of *work, interaction* and *emancipation*. Towards this end, both positive and negative cases can be compiled. These can then be used by several academic disciplines as supporting material within the emerging strands that recognise and challenge coercion.
- Systemists must associate with professionals, scholars and workers in several disciplines which are geared towards demystifying culture and wisdom. Such collaboration would invigorate the effort towards exposing and undermining coercive hegemonies of various kinds.

An Appeal to Systemists

Systemist have to squarely face the differences among themselves and resolve these through sympathetic understanding of each other's points of view. A set of suggestions were given elsewhere (Dash and Murthy, 1994), while commenting on the System Dynamics method. These suggestions were meant to further *methodological pluralism* in systems practice. This continues to be a need even now.

ACKNOWLEDGMENTS

I am indebted to the Systems Engineering and Cybernetics Centre of Tata Consultancy Services, Hyderabad, India, where I was initiated into systems thinking and practice. Most of the insights presented in this article had germinated while I was a member of this Centre.

REFERENCES

Dash, D.P., and Murthy, P.N., 1994, Boundary judgement in system dynamics modelling: an investigation through the science of complexity, *Systems Practice*. 7 (4): 465-475.

Flood, R.L., and Jackson, M.C., 1991, "Creative Problem Solving: Total Systems Intervention," Wiley, Chichester.

69

SURVIVING SEEING MEDUSA

A Meditation on Truth in Systems Thinking

Peter Dudley

Centre for Systems Studies
University of Hull
Hull
United Kingdom

THE QUESTION

Quoth the author:

A most promiscuous partner, cynical and mercenary in the extreme. Hermaphrodite, devoid of conscience, capable of satisfying any desire or firing any fantasy. An attentive, considerate but passionate (and submissive?) lover. Insatiable; giving and taking pleasure as it is found available, a new provider (and a new provision) for each and every client. In Philos' eyes the seduction is pure and untainted, the climax sublime, beyond the carnal; but for Jack all is reduced to meat.

The highest of price is charged for the service, the most venal of acts demanded for satisfaction (the ultimate skill of the libertine is to feign sated desire and inflate the ego of the tyrant or fool) and the most venereal of diseases transmitted in the act of consummation. The wise suffer always the fatal inability to resist whilst Jack is bathed in the poison imbibed as the source of their strength.

How then to survive the seduction? To succumb is to look the gorgon full on, to petrify; then wither in the grip of wind and tide. To withdraw then? To flee the labrynth in terror, chase wildly over the crumbled remains of the brave; fated never to see the power of the titan. Or, learn the lesson of Perseus; to gain the head of Medusa by reversing the image in the looking glass?

INTRODUCTION

In attempting a consideration of the *truth* in systems thinking a model which reveals levels of closure will be developed. This model will be presented through a formal reconsideration of generic pluralist systems approaches (i.e. soft and critical; see for example Flood and Jackson, 1991(a); Flood and Jackson, 1991(b)). The intention is that, by considering the nature of the author of a particular system or, as it will be regarded in this paper, systemic text, it should be possible to comment on the *legitimacy* of designed systems. The

Critical Issues in Systems Theory and Practice, Edited by K. Ellis et al.
Plenum Press, New York, 1995

enquiry, a deconstructive* reading of systems models, is intended to demonstrate the inherent undecidability of any given intervention thus demonstrating the near equivalence of what are regarded as being mutually exclusive approaches and, as a secondary outcome, exposing the intentionality of designed systems.

THE APPROACH

In developing the Viable Systems Model (VSM) (Beer, 1985) Stafford Beer proposes a 'systems 3, 4, 5' homeostat which functions as the guardian of systemic identity. Systemic identity is ensured by deciding which actions the system will take in relationship to the normative definition of the system as defined in 'system 5'. By controlling the value set by which the differences between day to day identity (i.e. system 3) and developmental identity (i.e. system 4) are resolved this normative system 5 controls the identity of the system. If Flood and Jackson are correct in categorizing the VSM as 'unitary' (Flood and Jackson, 1991(a)) it is reasonable to assume that any other systems model which can be represented (more or less accurately) by the '3, 4, 5' homeostat is also unitary. The approach taken in this paper, using the VSM analogy as a guide, is to generate the 'second reading' of a deconstructive analysis (Derrida, 1987; Critchley, 1992; Norris, 1987) and thus the basis for the *making evident* of a unitary kernel within all systems models.

Soft Systems Models (SSMs)

Culture (as defined by Checkland, 1981; Checkland and Scholes, 1990; Flood and Jackson, 1991(a)) forms the heart of the soft approaches, it is that area of commonality that allows agreement to be reached. That culture in this sense can operate every bit as effectively as the most complete unitarist model in constraining what is considered meaningful in debate can be demonstrated in a consideration of the work of Michel Foucault.

Foucault considers (Foucault, 1991) the relationship between crime and punishment and introduces two registers to typify the areas in which *society* reacts to breaches of its rules. The first, deterrence, considers the perceived damage to society caused by a particular breach and, on the basis of this, the extent and nature of punishment that will form a deterent to future potential offenders. Note that this register is based on the assumption that society, through its representatives, is capable of determining the extent of such damage in an absolute way rather than as relative to the (socially determined) rules in operation. The second register considers the extent to which the *offender* is a *criminal* as per his consideration of mitigating factors, (Marat in Foucault, 1991).

> "Of two men who have committed the same theft, how much less guilty is he who scarcely had the necessities of life than he who overflowed with excess? Of two perjurers, how much more criminal is he on whom one has striven to impress feelings of honour than he who, abandoned to nature, never received the benefit of education.",

and the earlier consideration of the status of the *offender* in relation to the state; (Rousseau in Foucault, 1991).

> "Every malefactor, by attacking the social rights, becomes, by his crimes, a rebel and a traitor to his country; by violating its laws he ceases to be a member of it; he even makes war upon it. ...", (Foucault, 1991).

* see *Tympanum* in Derrida, 1982 for an example of this approach to boundary conditions

In his choice of allusions Foucault introduces a duality which is contained not so much within the relation of the individual to the state, but, more in the relation of the individual to the rules of membership of the state (i.e. citizenship). In the second case the 'malefactor' is excluded from membership, considered, by virtue of his/her action, to be beyond acceptance by society as he/she does not consider themselves to be bound by its rules. In the first case, however, the degree of malfeasance is mitigated; there are circumstances by reason of which such an offending action can be tolerated or, more precisely, *explained* thus permiting amelioration of the penalty. The role of this explanation is not, however, to allow leniency in the application of penalty to transgression but to reinforce the normal, that which identifies the action *as transgression*.

Thus whilst all (potential or actual) transgressors are excluded in some way only the criminal is reviled; the basis for this being the offender's choice to transgress. The offence (as action) becomes less important than the perceived formulation of the offence in the mind of the offender; although this formulation is categorized (i.e. defined and valorized) in terms of the rules of the state. The state as represented by the legislature and the body of law represents *reason* and it is this which seems to underlie the measure or degree of criminality.

If Foucault was correct in his analysis of the control function of the norm (and I believe he was) it forms the constitutive condition for a two level critique of SSM's.

Firstly, non-conformity (to the accepted norm) is a necessary and sufficient reason for exclusion. Despite the assertion that "The best studies have been characterized by...a readiness to collect as many perceptions of the problem as possible..." (Checkland, 1981) the need for *identity* effectively underscores the term 'as possible' which, following Flood and Jackson's implied argument (Flood and Jackson, 1991(a)) can be read or acted upon as *as desirable* therefore the conditions of systemic closure (i.e. the limitation of inclusion) become arbitrarilty teleological. This problem is partially addressed in Checkland and Scholes' later work (Checkland and Scholes, 1990) in the form of a 'cultural enquiry' and in the work of Flood and Jackson (Flood and Jackson, 1991(a)) through the commitment of Total Systems Intervention (TSI) to a 'sociological awareness'. Both attempts stress the need to identify power differentials and those (individuals or groups) who are dominant within them, therefore, presumably, steps to prevent (or at least identify) overt distortion of free debate are possible. This seems to be a case of assuming that regicide necessarily leads to a republic. Killing the king, however, removes only the monarch, not the monarchy.

Secondly, and more fundamentally, once the possibility of exclusion has been established it can be located. the relationship of the actor to the culture can be presented as the responsibility of the actor him/herself as demonstrated by Foucault in the passage above. In this way any person who does not demonstrate the *right* (i.e. culturally acceptable) relationship to the culture *excludes themself* from membership.

The dominant culture closes the debate by defining not only the norms which are to be observed, but also the standards by which this observance will be judged. Hence the pluralist stance attributed to the SSM's can be directly compared to Foucault's "...economy of illegalities". (Foucault, 1991) where particular acts are tolerated because they are committed by those who belong to the dominant order (in our case systemic culture), whereas others are not (i.e. they lead to exclusion) because the actors are assumed to be enemies of the dominant order*. This oppositional stance suggests that the pluralism of SSM's functions intra-culturally; that differences of approach, opinion or desire are tolerable when their teleology supports the wider teleology of the dominant culture. Those actors who cannot

---

*Foucault demonstrates the circularity of the teleological nature of the argument in that the status of the *outsiders* is determined by the acts they commit and the acts of 'treason' that they commit are defined as such because they are *already* outsiders

formulate or express their desires in this manner are not *excluded* by the system but, because of the location (by the dominant culture) of the responsibility for their relationship to (i.e. acceptance or otherwise of) the culture with them personally (under the rationality of the dominant culture) they *choose* to be external to it. Thus, any acts they commit are, by definition, either irrational or criminal. Once this is recognized the right to self defence justifies exclusion. Those 'poor souls' who are incapable of belonging are *treated* or *educated* to show them the error of their ways and thus be re-admitted to the culture; whilst those who *wilfully* reject the culture are punished as they wage a "...veritable war on all citizens" (Foucault, 1991). Culture, given this reading, forms a closed control loop, in that it specifies (a) standards of complicity to the norm; (b) the standards by which individual acts are judged as to their relationship to the *spirit* of the norm and (c) the *corrective* action to be taken in response to transgressions of the norm. Hence the so-called soft models have a unitary kernel in common with 'Hard' Systems Methodologies (HSM's).

Critical Systems Models (CSMs)

In addressing Critical Systems Models (CSMs) an exemplar methodology will be used as before; in this case the methodology chosen is Ulrich's Critical Systems Heuristics (CSH) (Ulrich in Flood and Jackson, 1991(b); Flood and Jackson, 1991(a)). The aim of CSH, and of CSMs in general, appears to be the provision of a facility by which to question the legitimacy of a given systemic identity (i.e. the boundary conditions upon which the *truth* of the system is posited).

To continue the deconstructive analysis, a first reading of CSH apparently demonstrates the intention to question the legitimacy of proposed or actual systems designs through the use of a set of twelve 'heuristic' questions in 'is' and 'ought' mode (Ulrich in Flood and Jackson, 1991(b)). The apparent aim being to challenge the boundary conditions assumed by the (presumably dominant) systems designers/experts. By including those who are affected by the outcomes of the design although not necessarily involved in the design process (i.e. the 'witnesses') the 'affected but uninvolved' are enabled to take greater control of their destiny.

A second reading of this emancipatory approach reveals a number of inconsistencies:

1. If Flood and Jackson correctly identify Ulrich's advocacy of a "...purposeful systems paradigm" (Flood and Jackson, 1991(a)), CSH, as a system of critiquing (perceived as dogmatic) assumed systems knowledge, is itself is 'purposeful' and if, in fulfilling this purpose, "a dialogical concept of ratonality *must* replace the conventional monological understanding" (Ulrich in Flood and Jackson, 1991(b), my emphasis) it appears that CSH has the (at best) unitary purpose of supporting pluralism.
2. If systems designed using CSH must be "adequately designed if they are not to serve people and purposes other that those intended" (Flood and Jackson, 1991(a)) a double intentionality is exposed in that the system is *designed for* rather than *designed by* and hence is reliant on *experts* rather than the *affected* themselves. Given the need, as per the 'moral question' for the 'social system designer to seek continually to improve the human condidion" the designer's and not the recipient's perception of progress is automatically embedded in the *improvement* in the situation.
3. Because of the reliance on *experts* (whether in is or ought mode) necessitated by the current representation of the 'guarantor' idea the design of the system and the guarantee of its legitimacy/success is always validated by a *not me*. It is always

external, thus the affected and uninvolved are perpetually and necessarily uninvolved.

Because of these inconsistencies CSH must be either a unitary or coercive guarantor of pluralist principles (i.e. I am imprisoned by my freedom as per Gibran's "The Prophet") or, because of the necessity to accept many, possibly incommensurate, models of reality/the good/the truth accept a majority version of truth which reduces the idea of guarator to absurdity.

As either of these outcomes are potentially despotic; even the argument 'if it is applied properly'...reduces to a unitarism in that one must decide and enforce the *proper*. Because of this potential for unitarism/coercion within the CSH model it is possible to represent an intervention using the methodology as driven by a '3, 4, 5 homeostat' in that it controls against the (single and unquestioned) standard 'does this model promote/support pluralism?'. CSH and, by implication, CSM's have therefore a unitary kernel in common with HSM's.

CONCLUSION/CALL TO ARMS

The outcomes generated by the methodologies quoted reduce the richness and complexity of the real world in an attempt to create a systemic identity around which action can be taken and legitimated. The nature of the closure of these models leads to *simple identity* a unitary solution; SSM by consensus around the culture and CSH through the oppositional stance of the 'is' and 'ought' modes of the heuristic questions.

This simplistic view of systemic identity leads also to the potential for simplistic and simplistically perceived truth or dogma. The primary role of a critical theory of systems must surely attempt to go beyond this, to integrate the concepts of heteronomic and subjective validation.

It was not the intention of this paper to mount a polemic against systems theory, much less against the systems methodologies discussed (which were selected for their exemplary character), but to demonstrate an inherent unitarism which connects systems models and, therefore, provides the basis for an identifiable systems theory. The question of identity is far more complex than we (the systems community) has so far discovered and it is to this point that the paper argued; it is necessary, I believe, to consider the paradox of *complex identity*.

REFERENCES

Beer, S., 1985, "Diagnosing the System for Organizations", John Wiley & Sons, Chichester.

Checkland, P. B., 1981, "Systems Thinking, Systems Practice", John Wiley & Sons, Chichester.

Checkland, P. B., and Scholes, J., 1990, "Soft Systems Methodology in Action", John Wiley & Sons Chichester.

Critchley, S., 1992, "The Ethics of Deconstruction: Derrida and Levinas", Blackwell, Oxford.

Derrida, J., 1982, "Margins of Philosophy", Harvester Press, Hemel Hempstead.

Derrida, J., 1987, "Positions", Athlone, London.

Flood, R. L., and Jackson, M. C., 1991(a) "Creative Problem Solving: Total Systems Intervention", John Wiley & Sons, Chichester.

Flood, R. L., and Jackson, M. C., (eds.) 1991(b) "Critical Systems Thinking: Directed Readings", John Wiley & Sons, Chichester.

Foucault, M., 1991, "Discipline and Punish", (trans. A. Sheridan), Penguin, London.

Gibran, K., 1991, "The Prophet", Pan, London.

Norris, C., 1987, "Derrida", Fontana, London.

SYSTEMIC REFOCUSING STRATEGY

An Emancipatory Approch to Intervention. Part 1

Peter Dudley[1] and John Hassall[2]

[1] Centre for Systems Studies
University of Hull
Hull
United Kingdom
[2] Wolverhampton Business School
University of Wolverhampton
Telford
United Kingdom

INTRODUCTION

Systemic Refocusing Strategy (SRS) is an experimental approach intended to facilitate the reconstruction/reformulation of systemic goals. By providing a conceptual framework which enables actors involved in a given situation to reflect on their values and to critique the *taken as natural* assumptions underlying *systemic progress* SRS suggests a *way of walking* rather than the plan of a journey.

Constructed around an operationalized interpretation of Derrida's concept of deconstruction (Derrida, 1987; Critchley, 1992; Norris, 1987) and the belief that the organization and its *problems* can be meaningfully represented as *texts*, SRS attempts to make organizational problems susceptible to literary critique. By synthesizing deconstruction and generic systems models SRS attempts to provide an emancipatory tool for organizational analysis.

BACKGROUND

SRS is a strategy for learning and (potential) change in an organization. Located in Systems Theory, i.e. that things (e.g. people, organizations, machines, organisms, etc.) can, and indeed must, be regarded as wholes if we are to gain meaningful knowledge of them; and Critical Theory i.e. that any knowledge we may have, and, therefore, any action we may take, is necessarily limited (by e.g. background, culture, education, etc.) and, as such, will tend to privilege one body of belief to the detriment of others, which will be regarded as *unscientific, barbaric, etc.*

Critical Issues in Systems Theory and Practice, Edited by K. Ellis et al.
Plenum Press, New York, 1995

This synthesis leads to two of the three underlying assumptions of SRS:

- ASSUMPTION 1 That organizations can meaningfully and profitably be regarded as systems.
- ASSUMPTION 2 That by being embedded in this *systemic reality* the knowledge that the organization can have of its environment, and of itself, is necessarily coloured and/or constrained (by the value/knowledge constructions of those other systems within which it is embedded).

The goal of SRS is, therefore, to provide a tool which will allow managers to question the limits of their knowledge and thus allow the consideration of (systemically) radical proposals.

The impetus behind the literary analogy underpinning SRS is derived from an insight provided by the semiotic model of Hjelmslev which was later adapted by Barthes (see Silverman, 1983) which proposes two levels of 'language':

- LEVEL 1 The denotative; which provides the means for communication (i.e. words and their definitions, syntax, grammar, etc.) and which would normally be regarded as a language *proper*;

and

- LEVEL 2 The connotative; that which is to be communicated (i.e. stories, feelings, information, etc.) and is, therefore, that which is to be expressed.

Given this interpretation, language is the servant of thought; it is there merely to allow the communication of thoughts between thinking entities.

This leads to the third assumption of SRS:

- ASSUMPTION 3

That management practice can usefully be regarded as semiotic; i.e. that technology can be regarded as the denotative expression of the connotative meaning provided by managerial desire; and that management practice is therefore susceptible to a linguistic analysis.

To a greater or lesser extent the view that language serves thought has a parallel in management theory. The view apparently taken by management theorists conflates decision with action (i.e. once the decision has been taken the outcome of the act is beyond doubt). It is here that the first problem SRS seeks to address lies. Such an approach to management practice assumes the *already existing* quality of a technology to carry out what it has been decided to do; it presumes the existence of an adequate language to express and fulfil desire. To present this in managerial terms: technology is the servant of, and is delimited by, managerial decision.

A more appropriate conceptualization seems to be that which follows from Foucault's philosophy * i.e. not only is thought not the absolute master of speech and i.e. that managerial decision is not the absolute master of technology; but that it is possible (in Foucault's case necessary) that the limitations of speech (technology) are also the conditions for the practical limitation of thought (effective managerial decision making). Therefore the *practice* of

* Although Foucault does not use this specific terminology the implication of the determining role of *the discourse*, particularly as developed in Foucault, (1992) but also Foucault, (1971) and Foucault, (1991), renders such a statement acceptable. The politico-economic rationale Foucault gives for the determination and determining ability of discourse, although not alluded to explicitly here, forms one of the underpinnings of SRS

management (i.e. the formulation of desire) is limited by the *discipline* of management. This in itself calls for a reconstruction of the accepted managerial model.

Common sense suggests:

a. that managerial thought is not all thought; it is not e.g. theoretical physics or cosmology. Managerial thought can be represented, therefore, as a subset of all currently possible thought.
b. commercial or industrial technology, as the mode of expression of managerial thought, is not the only possible mode of expression; it is not e.g. art, literature or speech. Commercial/industrial technology can be represented, therefore, as a subset of all currently possible modes of expression.

We shall deal only with the relationship between managerial thought and commercial/industrial technology. The traditional management model (i.e. whereby technology is completely determined by managerial thought) represents technology as a subset of thought and as such will tend to shift the emphasis of management theory (along with the legitimacy of power, *authority*) away from production (i.e. away from the technologists, crafts, trades and the limitations of the machinery etc., employed) and attempt to define or design *good* business practice in line with managerial desire. The role of the manager under this model is to decide *what* is to be done and then to control some or other *pre-existing* technology in order to achieve it. It is in relation to this point that the politico-economic model of Foucault (1991) becomes manifest. The shift of *legitimate* influence from the producer to the *controller of* production betrays the essentially political nature of the *perceived as natural* dominance of the manager. Comparisons of possible solutions are made; viability, feasibility, profitability, etc.; in some cases the search for solutions is abandoned altogether and becomes a search for different (solvable) problems.

The last sentence betrays the essentially *undecidable* nature of the managerial model; the search for different problems is a denial of managerial authority, i.e. the technology defines the decisions it is meaningful for the manager to take.

In practice both the models described may lead to deadlock in a wider developmental process, because, in effect, they are fundamentally incompatible. Neither is capable of speaking to, much less understanding the viewpoint of, the other.

This deadlock has two forms; the first (less pernicious) form can be addressed using the tactical level of SRS ; the second (extremely pernicious) form of deadlock has a politico-ontological element which requires great effort to bring to the surface and, even then, can be difficult and in some cases (systemically) dangerous to address; as it calls into question the legitimacy of the organization's perceived *reality*.

THE STRATEGY

Stage 1

The role of the tactical level of SRS is to identify those areas where there is a mismatch between desire and capability and to provide a forum for discussion on how to reduce it. It is in effect homeostatic and for the moment at least, value free; in that it assumes a desire to reach an acceptable *solution*, that both desire and capability can change and that changing desire to fit capability is neither more nor less valuable than changing capability to accommodate desire. Using the simple homeostatic process iteratively leads to accommodation being reached. A further, though less obvious, assumption is the sharing of a common language; as without the ability to communicate no meaningful accommodation can be reached. Once the cybernetic principles underlying this progress to an accommodation are grasped, managing the process becomes relatively simple; it is a facilitation of the series of meetings at which the negotiation/communication of the iterative processes occur.

If participants in the process are both:

a. able and willing to reach an accommodation
b. speaking the same language i.e. they both use the same set of signifiers and share the same relationship between signifiers and the signified (between the thought and the act), the process of convergence will lead to an accommodation as per Ashby's homeostatic model (Ashby, 1956; Beer, 1974).

Is is now possible to identify the first step of the SRS intervention:

1a. State formally the managerial goal
1b. State formally the technological capability to achieve it
1c. Instigate and (particularly) observe the intermediate stages of the process of convergence toward homeostasis (accommodation).

It is still possible that the convergence process will stall. Perhaps the participants are not able to move from their original positions, due to a lack of resources for R&D or policy constraints on the side of the decision makers. To deal with these eventualities is necessary to refine the original homeostatic model such that it operates more as the 'systems 3, 4, 5' homeostat of the VSM (Beer, 1985). The VSM homeostat provides a facility for convergence between two systems with divergent goals by introducing an overarching normative element into the process. This normative element, provided by what Beer calls 'System 5' ensures systemic identity; deciding the differences between 'System 3', in our case technology, and 'System 4', in our case managerial decision, on the basis that it defines what the system 'is'. For the practical purposes of SRS this normative 'system 5' must be capable of:

a. defining systemic or situational identity, i.e. it must be capable of defining the wider goals which lie behind the managerial desire which is under scrutiny; an extension of this is that it must also define *what the system is not*;
b. it must therefore be capable of varying or allowing the management goals under scrutiny to be varied,

and/or

c. it must control the resources allocated to technical development and therefore be able to support technical developments where technical capability cannot express managerial desire,

and/or

d. it must have the power, and be prepared to accept responsibility for, the termination of any such process when neither managerial goals are permitted to be redefined nor the resources necessary can be made available for technical development.

It is now possible to identify the second step of the SRS intervention:

2. A normative 'system 5' element must be identified which can be added to the original homeostatic process. This element must act to prevent the stalling of the convergence process by defining systemic identity and be authorised to vary the conditions of the participants in terms of resource allocations and/or policy constraints. It must also, in view of the authority it possesses, accept responsibility for the cancellation of such projects when it cannot provide sufficient variations in terms to the participants to allow convergence to be completed.

Stage 2

Thus far SRS has said nothing really new; the tactical level of SRS is simply good managerial practice.The problem that arises from this *saying nothing new* is that, almost inevitably, the change process stalls. The identity provided by the newly added (to the process) 'System 5' can only be made manifest through a binary choice (i.e. that between Technology, on the one hand, and Managerial Desire, on the other). The ability to question systemic identity is not available as an option to the decision makers. The only other option available to the decision makers is to abandon the project altogether. Thus, even with the addition of a normative element, it is still possible for the situation to degenerate into isolationist positions which leave the decision makers trapped in a *take it or leave it* dilemma.

The third step of SRS attempts to overcome this deadlock by building upon the knowledge of the situation gained in steps one and two.

Closer consideration of the model suggested that it was possible for the model to fail if the aims of the two participants (i.e. technology and managerial desire) were fundamentally incompatible. Because of this, progress toward a solution would stall due to isolationist stances being taken leading each of the participants to attempt to subsume the views of the other.

The aim of the third stage of SRS is to make this higher level rationality available to the participants using a technique based on the Derridian concept of 'deconstruction' (Derrida, 1987; Critchley, 1992; Norris, 1987).

Let us return to the tactical stage of SRS to consider the initial implications of the above discussion. By functioning as the custodian of identity, 'System 5' decides the conflicts which arise between technological capability and managerial desire; but, as stated earlier, the choice is binary, it is an either/or (i.e. either technology or managerial desire). By considering the outcomes of such conflicts over time it is possible to arrive at an opinion as to the identity of the system. It is important to note at this stage that the *goodness* or otherwise of the solutions arrived at is defined solely in relation to the dominant rationality (i.e. identity) operating within the system, i.e. technologically constrained solutions are *good* or otherwise only if the identity of the system requires (or does not require) deference to technological constraints.

It does not matter whether 'System 5' exists as a physical entity or not, what is important is the dominance of the rationality which controls the decision making process, and thus (for all intents and purposes) the perceived identity of the system.

It is now possible to define step 3(a) of SRS for the project based approach we are currently using.

Step 3(a) is the equivalent of the 'first reading' of Derrida's deconstruction; by reviewing the outcomes of stages one and two a decision is reached as to the *systemic identity* supported by the normative 'System 5', (i.e. the underlying value set is disclosed), therefore:

3a. Using the information gathered in stage one determine the values (i.e. the dominant rationality) underlying the 'first reading'.

The outcomes arrived at as a result of Step 3(a) present the first major action/decision point in the strategy. It is possible that the the value set induced from this analysis will be different from the value set identified as 'System 5' in step 2. It is also possible that covert power bases within the system (i.e. those groups whose interests are being served outside the presumed structure) will be demonstrated to exist.

By assisting in the identification of the disintegrities of a given approach and/or demonstrating the possible existence of vested interests, which may be distorting accepted decision structures, SRS sets a number of *items for the agendum* of the change process.

Having completed the 'first reading' it is possible for the system to explicitly consider, questions of identity and legitimacy; members of the system are enabled to ask questions such a 'Is this what we intended to be?', 'Is this appropriate given ...?' or 'Is this instruction legitimate?' The answers to these questions (and others like them) form one of the two inputs to step 3(c).

The second part of step 3 corresponds to a 'second reading', in that it attempts to identify *significant* expressions, those expressions upon which the induction of the identity of the system was based, and to consider the possiblity that there are other, equally valid, interpretations that could be attributed to them.

Therefore:

3b. Recursively apply SRS to the technological base paying particular attention to the values the available technology can serve (e.g. the *less work* or *less employment* opposition).

The question being asked at step 3(b) is not 'What does the technology produce?' but 'What values *can* the technology serve?' It is apparent that to answer this question another (recursively down) application of SRS is necessary. Once the answers to these questions of the values served by technology are available new models of the system (i.e. models which reflect these different values) can be built. These new models can then be overlaid onto the answers to the questions of systemic identity.

Hence:

3c. Overlay the outcomes of steps 3(a) and 3(b) to identify an intersection set.

Thus step 3(c) of SRS is the result of two recursive applications of the strategy itself.

This formulation of systemic progress is, essentially a paraphrase of Checkland's "Systemically Desirable and Culturally Feasible" (Checkland, 1981; Checkland and Scholes, 1990) with the fundamental advantages of:

- recognizing the co-determining roles of technology and teleology through the commodifying effect of cultural constraint and thus the ability to redefine the basis of this commodification once the cultural constraint is relativized;
- providing an explicit mode of cultural critique through the recursive application of deconstructive strategy at various stages within the parent strategy;

and, therefore,

- providing an explicit basis for cultural recontruction in line with other (and different) value constructions.

REFERENCES

Ashby, W. R., 1956, "An Introduction to Cybernetics", Chapman and Hall, London.
Beer, S., 1974, "Designing Freedom", Canadian Broadcasting Corporation, Toronto.
Beer, S., 1985, "Diagnosing the System for Organizations", John Wiley & Sons, Chichester.
Checkland, P. B., 1981, "Systems Thinking, Systems Practice", John Wiley & Sons, Chichester.
Checkland, P. B., and Scholes, J., 1990, "Soft Systems Methodology in Action", John Wiley & Sons, Chichester.
Critchley, S., 1992, "The Ethics of Deconstruction: Derrida and Levinas", Blackwell, Oxford.
Derrida, J. 1987, "Positions" (trans. A. Bass), Athlone Press, London.
Foucault, M., 1971, "Madness and Civilization" (trans. R. Howard), Routledge, London.
Foucault, M., 1991, "Discipline and Punish", (trans. A. Sheridan), Penguin, London.
Foucault, M., 1992, "The Archaeology of Knowledge" (trans. A. M. Sheridan-Smith), Routledge, London.
Norris, C., 1987, "Derrida", Fontana, London.
Silverman, K., 1983, "The Subject of Semiotics", Oxford University Press, Oxford.

71

SYSTEMIC REFOCUSING STRATEGY

An Emancipatory Approach to Intervention, Part 2

Peter Dudley[1] and John Hassall[2]

[1] Centre for Systems Studies
University of Hull
Hull
United Kingdom
[2] Wolverhampton Business School
University of Wolverhampton
Telford
United Kingdom

SUMMARY

Systemic Refocusing Strategy (Dudley and Hassall, 1995) is applied to problems of Group Scheduling Systems (GSS) within three types of organisation. In this application SRS is used as a probe to illuminate factors relating to cultural perspectives, managerial objectives and rationales and technological possibilities available. Some generic managerial perspectives and rationales emerge and some observations are made upon the ability to co-ordinate technological capabilities with the organisation's ability to implement them. A generalised framework for evaluating the current normative GSS technology and its application is attempted.

INTRODUCTION

Diary and Group Scheduling Systems (GSS) are considered one of the most potentially useful of the information technologies now becoming available to organisations to aid group co-ordination (Wilson, 1991). They have been described as follows (Hassall and Macefield, 1995). Group Scheduling (GS) is a software technology which combines the ability to hold an electronic diary for an individual and electronic mail (Email).

In principle what can happen is that someone wishing to schedule a meeting with a number of colleagues can instruct the GS Software (GSS) to;

- Select an available free slot which corresponds in each persons diary and book it.
- Book a slot in each persons diary, overriding anything they have in there already.
- Inform the individual, by a message that a provisional entry has been made which needs to be confirmed.

Critical Issues in Systems Theory and Practice, Edited by K. Ellis et al.
Plenum Press, New York, 1995

In each case an advisory message would be posted on Email. Such technologies are now so sophisticated that it is possible to schedule persons who are linked via Email in different technological domains; that is, who are using different Email and Diary software. An essential feature of such technology is that it depends for its effectiveness upon a degree of synchronisation between individuals in order that (eg) I become aware of a diary entry when it happens rather than at some later date. Similarly, I must be able to make a diary entry immediately available to the wider system when it is created. Contrast this with Email, whose effectiveness (crucially) derives from exactly the opposite property, the ability to de-synchronise conversations. We can say that GSS is a synchronous process and Email an asynchronous process; or that Email allows message passing but GSS requires a real time conversation.

CASES OF USE OF GSS WITHIN ORGANISATIONS

Case 1. An International Computer Company

In the 1980's one of the authors worked for an international computer company which had developed a very advanced office automation product. Amongst the applications of the product was GSS in the form of Diary and Calendar Management. Essentially all of the functions of GSS were possible over multiple sites throughout the world. The system was capable of searching for the earliest slot to hold a meeting and scheduling into all affected person's diaries.The business culture could be characterised as follows in relation to this application.

i. Very technically aware. Senior managers had keyboard skills and were happy to employ office technology.
ii. Very dynamic in terms of development of business, timescales and deadlines. People were used to being asked to do things at short notice and being expected to perform to a high level.
iii. A lot of time spent out of the office (away from the computer network) and with customers.
iv. Innovative culture.
v. Positive 'can-do'feeling about the company.
vi. Good teamwork.
vii. Very strong management control of staff performance via appraisal and salary review.

As might be expected applications such as Email were successfully used in this organisation. The main benefit of Email results from the ability to de-synchronise conversations and allow persons to work in a way that optimises their own time and performance whilst ensuring a dialogue with other staff members.

GSS was a total failure within the organisation due to the inability to ensure that diaries were complete and accurate.

Case 2 The Business Link

In this organisation there is an implicit concept, established when the business was established, that Personal Business Advisors (PBAs) will operate a delegated diary system. The principle is that customers can call into the business link and request an appointment with a named PBA or the first available PBA. There is also the need to be able to call meetings between PBAs. Clearly the technical requirement is for a GSS.

The main problems and issues arising are as follows.

The PBAs spend most of their time remote from the office and link in to the computer network via telephone line. Because they are out of contact with the central diary facility there is an inevitable need to reconcile diaries on a regular basis. This system, in common with many others, works well most of the time but requires considerable discipline and monitoring by management in order to be effective. In fact, the management apply discipline through the PBA's contracts; they are not full time employees and thus can be (implicitly) contractually bound to obey the rules. They also have an interest, because of the way in which they are remunerated, in ensuring that their diaries are full and show maximum activity. One observation is that both a desire and a capability to control on behalf of management is implicit within this situation.

A significant modification was made to the standard GSS software tools available to make the present system effective. Typically GSSs will schedule appointments based upon constraints (earliest, specific date, specific times). However, the Client Advisors receiving calls wished a degree of manual intervention; something in fact nearer to a traditional paper diary. This has been provided as an addition to the available software. A number of PBA diaries can be viewed on the same page and entries manipulated in a similar manner to a paper diary or wall planner. The Client Advisors are thus able to incorporate a degree of human interaction in order to make the system work more effectively. The Business Link GSS is a hybrid system and only works because of the high degree of cultural control applied to participants behaviour.

In particular;

i. This is a green-field situation where no previous practices existed. It was possible to develop from the start procedures which integrated IT with business operations.
ii. The PBAs have strong, economically motivated, reasons for conforming to the system.
iii. The basic GSS concept has been humanised by the addition of the wall planner interface.
iv. The qualities of the Business Link manager have proved important in driving the whole system forward.
v. The PBA's have, from the start, been encouraged to develop IT skills.

Case 3 University Business School

What are the prospects for GSS in such an organisation. Briefly, the parameters and cultural issues are as follows.

i. Around 100 staff with varying levels of IT skill and varying motivations to improve skills.
ii. Interest and enthusiasm in some quarters for new technology.
iii. Managerial culture relatively new and immature.
iv. Recent implementation of Email, enthusiastically taken up by some staff and ignored by others.
v. Not all staff having easy access to a computer attached to the network.
vi. Non - professional culture; for example, many academic staff do not consider it necessary to let management or office staff know of their whereabouts or activities.
vii. Many staff conservative about changes in working practices and many working practices long established.

APPLICATION OF SRS

Step 1

Because the details of cases above are only briefly described it will be necessary to introduce further observational material at 1c.

Managerial Goal. The managerial goal for GSS may be stated as follows. To provide centralised and co-ordinated diary and scheduling systems for all staff within an identified organisational group which can be relied upon to be accurate and up to date for the majority of the time. This is the essence of what is desired. Both of these phrases reflect the requirement for organisational effectiveness of the application.

Technological Capability. The technological capability to meet the managerial goal exists. It has been described in section 1 of the paper. Provided we allow persons to have access at all times to the IT system it is possible, with suitable discipline, to centralise and co-ordinate diaries.

Limitations may be felt where people are not able easily to maintain communication with one-another. The existing GSS technology only partly accommodates the necessary synchronous nature of diary and calendar management.

The technologies then can achieve two main things;

- Keep individual diaries.
- Reliable group scheduling and diary management assuming synchronisation of communication between individuals which may or may not be a practical possibility within the (cultural) constraints of all organisations.

Observe the Process of Accommodation. Ignoring, for the minute, case 3 we may draw some observations from cases 1 and 2 above. In case 1, the computer company, we note that the application of GSS has failed and the technology been rejected. This is the more surprising when we consider that the technology was the company's own !

In case 2, the business link, we observe that the selected technology has been made to work by putting resources into additional software development, the technology is not allowed to function in a fully automated fashion but mediated by human intervention.

In both cases the issue of IT skills and usage does not present any barrier to using the technology as there are strong incentives and cultural constraints which produce high motivation to use the technology.

Step 2 (Retrospective)

At step 2 we examine the outcomes for cases 1 and 2 in the light of the overarching system identity revealed which can be shown to mediate the application of the GSS (3.). This is not an active intervention but an observation process based upon historical information. Both cases are revealing.

Step 2 - Case 1. What does the outcome of case 1 reveal about the identity of the implied system whose purpose is being addressed? An important clue is that we know the organisation is performance orientated. It is sales driven and must meet targets. The use of particular tools and technologies (and the contribution of individuals as well) are irrelevant if they do not contribute to the achievement of targets. The GSS technology introduced does not contribute effectively and is (ruthlessly?) abandoned. A further and related point is that the decision to abandon the technology was taken relatively rapidly. There was little attempt to ask 'can we modify the technology to make it more effective in the situation we actually face?'. This was a question to be left to the research and development people, not operational management. So the identity (or aspect of identity) revealed is something like 'If it does not contribute rapidly to operational goals abandon it (him/her, the idea ... etc.)'.

Achievement of short term targets and objectives appears to dominate and provides the rationale for abandoning the technology.

Step 2 - Case 2. In the case of business link there were a number of factors which might have contributed to the approach adopted and the implicit system identity. Starting from a blank sheet of paper there was a need to develop a robust business process supported by relevant technology. Once the process had been designed and the technology selected there was a strong motivation on management's part to make it work but only if the business process could be proved reliable.Thus, the option to abandon technology for which costs had been incurred was not perceived as viable by the managers involved. Discussion with the persons involved has convinced the one of the authors that they would have abandoned something which did not, ultimately, perform but that they had sufficient belief in the importance of using the latest technology to try quite hard to make it work.

Eventually, of course, they have a working system which is a compromise. Extra costs in implementation and considerable managerial discipline have been required to ensure the concept survives in the implemented system. The outcome has been a modification of both managerial desire and technology to meet a business process requirement which is both robust and effective.

Interestingly, there have recently been comments from users of the system indicating that they would like a more open access to the system allowing them more asynchronous use of central facilities and more automatic updating. This has been denied simply because only fairly strict centralised control allows the system to work as required by management.

The identity (aspect of identity) of the system revealed is thus; We require a robust business process that enables us to control the scheduling of staff resources and we are committed to the use of the latest technology. We will modify the technology in a way that enables us to develop such a robust business process and, at the same time, we are prepared to adjust our requirements to go some way to meeting technological limitations. Thus the business process is strong but is supported by a strong desire/need to make the technology work.

It is the authors' belief that the business link have succeeded in putting the GSS technology into use because their requirements have been business process driven. The result is a very obvious accommodation between what is desirable and what is technologically possible.

Step 3

We may seek some insights into the possibilities for GSS technology by applying step 3 of SRS to the information we now have.

An initial observation is that in both case 1 and case 2 the organisations have come to an accommodation with the technology, and we have identified the system identity which appears to be providing the contexts for this. It is now possible to proceed with step 3 in which we explore technological capabilities and possible alternative system identities.

Technological Capabilities. A summary of aims/rationales that the technology is capable of serving can be attempted with reference to the earlier readings of the cases and a reflexive (second) reading of both. We attempt to identity as well some generic rationales which the technology is capable of serving. These will be matched within the specific case (or intervention).

Case 1 Readings. Possible alternative readings of the case (or situation) might be;

- 'The technology is our own technology and therefore we should make every attempt to implement it effectively in order to provide a good sales story to our potential customers.' Something of this sort did arise in the case of another large computer company who decided to commit to implementing their Email system as a act of faith in its relevance to large organisations. Of course, as we have seen, Email is not GSS (2.1. Image/Public Relations).
- 'Implementing GSS will be useful to us since it will enable us to gain experience of the technology and iron out some of the bugs'. A superficially attractive reason. This, unfortunately, conflicts with the purpose of the world-wide sales and service operations of the company where such 'research' activities will get in the way and divert resource (3.1/3.2. Innovation/Creation).
- 'The technology is the latest technology and we, as a modern company, should use such.' This can be a useful argument in many cases. It carries force in a number of circumstances where the organisation is aiming to be innovative in its use of technology (2.1 (Self) Image).

Case 2 Readings. Alternative readings that arise include.

- 'We have committed to purchasing the technology so we must make it work.' The authors think that this was close to what was actually happening during the process of accommodation with the available technology (4.1 Reputation (of management)).
- 'The use of such technology is central to the image of the new business we are setting up.' About the same view as described in 3 for case 1 (2.1 (Self) Image).
- 'The use of this technology will give management control over our staff since it enforces a fairly rigorous operational discipline upon them.' An understandable view when staff to be scheduled are contract staff. In a sense this is the 'controlling inputs rather than outputs' school of management. If you are busy than you must be achieving something! (1.1/1.2. Control).
- 'We can create something new and effective by modifying the available technology which is exciting for its own sake' (3.1/3.2. Innovation/Creation).

SUMMARY AND EVALUATION

With reference to the above evaluative framework, derived through our readings of the cases, the following summary is presented.

We observe that 9 possible rationales (or elements of rationales) have been drawn from our applications of SRS to the case studies. Since this is not a quantitative study (although it has the capability of being backed up by quantitative analysis) the evaluative dimensions presented are rhetorical/subjective and may readily be challenged. It is the authors position that this reflects the very nature of an emancipatory approach such as SRS.

We now expect to be able to say something about the relative ability to GSS systems to serve various rationales within the organisations studied. In case 1, the computer company, we observe that we have 4 positive rationale fits, 4 neutral fits and one negative fit. In case 2, the business link, we have 7 positive fits and 2 neutral fits. These may lie on either side of the boundary between viable use of the GSS technology and none viability.

Generic Rationale	Organisational Rationale or Capability	Comments on Technology Fit
1.1.Control	Management capability to control	Needs capability to control in order to enforce disciplines
1.2	Management desire to control	Needs motivation to control
2.1.Image	Concern for image	Valuable in boosting image
3.1 Innovation	Interest in innovation by managers	An innovative technology
3.2.	Managers ability to innovate	A quality in mind - anti-conservative stance needed.
4.1 Reputation	Implementation serves managers reputation	Either way, to support or abandon
5.1.Business processes	Robust business processes exist	As in all information technology other than very simple stand-alone systems
5.2	Willingness to develop robust business processes	Managerial desire (business pressures)
6.1.Meeting short term objectives	Focus on short term objectives	Unlikely to succeed if it obstructs ability to perform

Figure 1. Rationales that the GSS Technology may serve.

In case 3, the Business School, we have 3 positive fits, 3 neutral fits and 3 negative fits. What conclusions may we draw from this. In fact it seems that there may be a qualitative difference between rationales above the line within the diagram (Fig 2.) and those below the line. It seems that rationale 6.1. for example may be a show stopper in some organisations. This will, however, depend upon the organisation itself. (It should be noted that the computer company stopped using the GSS software entirely during a very short space of time).

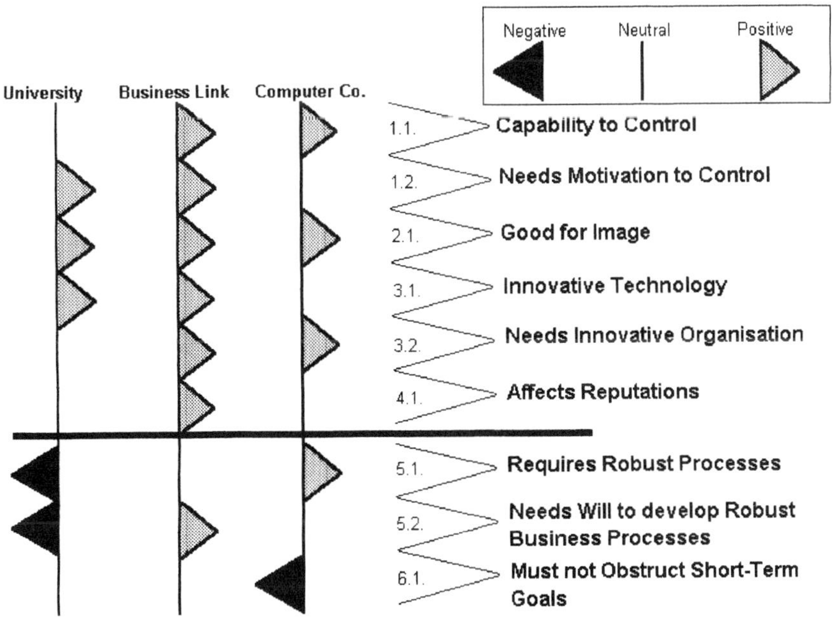

Figure 2. A matching of organisation to technology.

In the case of the university business school the two negatives 5.1. and 5.2. may in fact act counter to one another. If there are not robust business processes in existence it will certainly make implementation of GSS much more difficult but on the other hand, if there is no great will to develop robust processes, success may be measured in terms other than actual operational results !!

The conclusion based upon this analysis is that the university business school may be quite likely to implement or attempt implementation of GSS. The result is likely to be only limited success in operational terms but that this would not lead to any definite decision to abandon the technology.

OBSERVATIONS AND CONCLUSIONS

A rationalist or scientific observer will be able to raise objections to the way in which the evaluative framework growing from (this particular) application of SRS has been derived and presented. In order to anticipate some of these consider the following.

- The rhetorical nature of the arguments presented reflect the nature of systems thought and of the structuralist position.
- The framework provided by SRS has been proved (in action) to be capable of generating the evaluative argument presented.
- Experience of actual consultancy interventions by the authors suggests that the generated framework is useful.

The authors feel confident in using the evaluative grid for GSS in discussions with (real) clients.

REFERENCES

Dudley, P., and Hassall, J., 1995, Systemic Refocusing Strategy: An Emancipatory Approach to Intervention, Part 1., *in*: "Proceedings of the UKSS Conference 1995", Hull.

Hassall, J., and Macefield, R., 1995, University of Wolverhampton.

Wilson, P., 1991, Computer Supported Co-operative Work, Crown Copyright.

72

CRITICAL REFLECTIONS ON THE PAST, PRESENT AND FUTURE DEVELOPMENT OF ORGANISATIONAL EVALUATION

Amanda Gregory

University of Humberside
School of Computing and Information Systems
Hull, HU6 7RT
United Kingdom

1. INTRODUCTION

This paper is about the pattern of development of organisational evaluation. Following a review of Bruscaglioni's theory of the development of organisational theory (1982), the question is posed 'what does all this imply for evaluation theory?'. It will be argued that evaluation has progressed from the 'isolation of approaches stage,' during which the goal, system-resource and multi-actor were each seen to be, at different times, the 'best' model of evaluation, to the 'mechanical complementarism' stage. Use will be made of Flood and Jackson's Total Systems Intervention (TSI) (1991 (a)) as a mechanical approach which might be relevant in the evaluation context. Finally, an assessment will be made of what the final stage of development, 'the integration of approaches,' might hold for evaluation theory and practice.

2. STAGES IN THE DEVELOPMENT OF ORGANISATIONAL THEORY

It has been suggested by Bruscaglioni (1982) (as the original paper is written in Italian, we have to rely on Francescato's (1992) interpretation of Bruscaglioni's work) that there are three stages to the development of organisational theory: 'isolation of approaches,' 'mechanical complementarism,' and 'integration of the various approaches.' According to Francescato the first stage, 'isolation of approaches,' is "...characterised by the fact that supporters of each theory consider their theory a *general* theory, even if it privileges only *some* organizational phenomena." (Francescato, p. 134). The second stage, 'mechanical complementarism,' "...witnesses an increase in the awareness that the various approaches are at least partially complementary and that the complexity of organizational reality requires the use of more than one approach." (Francescato, p. 134). The final

Critical Issues in Systems Theory and Practice, Edited by K. Ellis et al.
Plenum Press, New York, 1995

stage, 'integration of the various approaches,' involves 'the generation of a variety of competing general theories'. Francescato states of the general theories in this stage that each "...should be different in the kind of explanations it gives for phenomena and for the relations and the connections hypothesized between organizational phenomena: that is, *they should differ in the model through which they organize the same elements.*" (Francescato, p. 136).

Whilst Bruscaglioni's work refers particularly to the stages in the development of organisational theory, it may be argued that organisational evaluation theory is following a similar pattern.

3. ISOLATION OF APPROACHES: FOUR MODELS OF EVALUATION

It may be claimed that traditionally there are three dominant models in evaluation theory (the goal model, the system-resource model and the multi-actor model). In addition to these three dominant models a fourth model may be added: the culture model. Whilst the notion of cultural analysis has reached popular acclaim in the organisational field it has not, so far, been widely accepted as a form of evaluation. It is argued in this paper that much cultural analysis can be seen to be a form of evaluation and on this basis it is included here.

An appreciation of the nature of each of the models may best be gained through the surfacing of the different definitions of effectiveness that underlie each of the models.

The Goal Model: *Effectiveness is the organisation's ability to achieve goals.*

The System-Resource Model: *Effectiveness is the organisation's ability to adapt and survive in a dynamic environment.*

The Multi-Actor Model: *Effectiveness is the organisation's ability to satisfy the needs of all those parties influenced by and having an influence upon its activities.*

The Cultural Model: *Effectiveness is the organisation's ability to generate and perpetuate a culture or cultures which, by facilitating the development of its members, enhances the organisation's own variety.*

Each of the models outlined above appears to have had its day: the goal model was dominant during the early part of the Century up to the late 1950s (Drucker, 1946; Thompson and McEwen, 1958); the system-resource model took off during the 1960s (Etzioni, 1960; Friedlander and Pickle, 1967; Seashore and Yuchtman, 1967; Georgopolous and Tannenbaum, 1969; Katz and Kahn, 1978; Quinn and Cameron, 1983); the multi-actor model was largely developed during the 1970s (Pfeffer, 1977; Meyer and Associates, 1978; Pfeffer and Salancik, 1978; Zammuto, 1982; Guba and Lincoln, 1989); finally, cultural analysis which for the most part, as has been said, has not been seen to be a form of evaluation, was developed in the 1980s (Deal and Kennedy, 1982; Kanter, 1983; Schneider, 1983).

The late 1970s appeared to signal a period of conflict: there were the stalwarts who held true to the more traditional methods of evaluation, those who stayed with the mainstream, those who sniped at the previously and currently popular models, those who anticipated the next evaluation panacea. Whilst the tightening fiscal belt meant that the practice of evaluation was becoming ever increasingly popular, evaluation theory appeared to be on the road to nowhere.

4. ADDRESSING THE COMPLEMENTARISTS' DILEMMA

The evaluation field was not alone in the predicament of having a lot of 'not quite good enough' methodologies: the systems field was also in the same state. In addressing the predicament in which systems theory found itself, Jackson (1987) proposed that the proliferation of methods did not signify crisis but, on the contrary, increasing competence. In recognising that each of the methodologies had strengths and weaknesses Jackson faced a dilemma: given multiple methodologies how to choose the most appropriate in a particular context. As a solution to this problem, Jackson and Keys (1984) put forward a classification of systems methodologies known as the system of systems methodologies (SOSM) (this will be further discussed in a later section). The development of the SOSM represented a shift from the isolationism of earlier periods to the mechanical complementarism stage. A similar shift also took place in the evaluation field. Several different classifications which sought to fit evaluation methodologies to their most appropriate contexts for use were developed during the 1980s (see for example: Weick and Daft, 1983; Quinn and Rohrbaugh, 1983). The latest development in this search for a meta-methodology for evaluation has been the an assessment of whether Flood and Jackson's Total Systems Methodology (TSI) represents an appropriate means for facilitating the choice between evaluation methodologies (Gregory, 1995).

4.1 The Methodology of TSI

Stage 1 Creativity. The task at his stage of the methodology is to highlight aims, concerns and problems through the use of systems metaphors. TSI employs a set of metaphors (machine, organism, brain, culture, team, coalition, prison) which "...capture, at a general level, the insights of almost all management and organisation theory." (Flood and Jackson, 1991(a), p.7). The outcome of this first stage should be the identification of dominant and dependent metaphors which highlight the major issues.

Stage 2 Choice. The task at the second stage is to choose an appropriate systems based intervention methodology or methodologies based upon the issues revealed in the first stage. The tool provided by TSI to facilitate this choice is the SOSM. The SOSM is an ideal-type grouping of problem contexts based on the two dimensions of:

1. system complexity which ranges from simple to complex; and
2. nature of relationships between participants which may be classified as unitary, pluralist or coercive.

Combining these two dimension results in a six-celled grid of problem contexts into which Flood and Jackson fit problem-solving methodologies based upon an analysis of the assumptions inherent in the methodologies. In summarising their arguments, Flood and Jackson formulate the SOSM through the metaphors. Thus, the outcome of the second stage of TSI should be the identification of the most appropriate systems methodology for use based on the issues raised.

Stage 3 Implementation. The final stage is implementation in which specific proposals for change should be acted upon. The identification in the two previous stages of dominant and dependent metaphors and, consequently, dominant and dependent problem solving methodologies should, it is concluded, serve to ensure "...coordinated change brought about in those aspects of the organisation currently most vital for its effective and efficient functioning." (Flood and Jackson, 1991(b), p. 329).

4.2 TSI: A Meta-Methodology for Evaluation Theory

In the previous section a review of the principles and methodology of TSI was undertaken. Given that TSI serves to enable the selection of the most appropriate systems methodology in a given context, might it not also facilitate the selection of the most appropriate evaluation methodology in a given context also?

At the heart of TSI is the SOSM. In order to demonstrate the usefulness of TSI to evaluation, it is first necessary to demonstrate that the SOSM can be employed to classify evaluation methodologies. To do this, we must analyse the evaluation methodologies in terms of the assumptions they make about system complexity and the relationship between system participants.

In its purest form, goal based evaluation accepts the goals of the owners or controllers of the organisation as being those of the whole organisation. The allegiance of other stakeholders to the goals is taken for granted. Thus, goal based evaluation is founded on the assumption that the organisation is in a unitary state. As regards system complexity, the organisation is regarded as being a closed system with routine, mechanistic operations and may be classified as a simple system on the complexity scale. Based on this analysis, the goal form of evaluation may be placed in the simple-unitary cell of the framework.

The system-resource form of evaluation emphasises the organisation's adaptation to and interactions with the environment; its various sub-systems act together to meet its needs in order to do this (the sub-systems act in harmony like the organs of the body). Hence this form of evaluation regards the organisation to be in a unitary state as it seeks to rise to the challenges thrown up by the environment. At the heart of the system-resource form of evaluation is the comparison (by a supra-system element or the 'expert') of the actual organisation with some ideal model. Whilst ideal models may vary, each should serve to propose a set of critical processes necessary for the organisation to maintain its viability. These models assume greater complexity and turbulence than the goal based models: the system-resource form of evaluation may be said to assume a complex system. Thus system-resource based evaluation may be placed in the complex-unitary cell of the framework.

The multi-actor form of evaluation revolves around the surfacing and collecting of interested parties' opinions and is quite explicit about its pluralist orientation. As regards system complexity, with its emphasis on the identification of interdependent groups of interested parties, within and without the organisation, this form of evaluation may be regarded as assuming a complex system. Hence the multi-actor form of evaluation may be placed in the complex-pluralist cell of the framework.

The culture based form of evaluation concentrates on the determination of a future state of being for the organisation which is the product of both the intrinsic abilities/nature of the members and reflection upon the desired future individual and collective state. Thus, with the culture form of evaluation the system is seen to be comprised of elements which are capable of exhibiting high variety but that variety is highly restricted and predictable as it is based on the gradual development of only those abilities, values, etc., which are of worth to the organisation. Consequently, the system in a culture based evaluation is regarded as being of low complexity. This classification may be further justified on the grounds that whilst the system may be said to be structurally coupled to its environment, for the most part the evaluation focuses only on internal matters. However, given the potentially diverse nature of the system's elements, the members, and that they are capable of exhibiting free-will and having goals separate from, or even conflicting with, those of the organisation, the system is classified as being pluralist. Based on this analysis, the culture form of evaluation may be placed in the simple-pluralist cell of the framework.

		UNITARY	*PLURALIST*	*COERCIVE*
SYSTEM COMPLEXITY	*SIMPLE*	Goal Based Evaluation	Culture Based Evaluation	?
	COMPLEX	System-Resource Based Evaluation	Multi-Actor Based Evaluation	?

Figure 1. A system of evaluation methodologies.

It can be seen from Figure 1 that the combination of the two dimension of 'system complexity' and 'nature of relationships between participants' provides a 'good' means for differentiating between the evaluation methodologies. It may be concluded that TSI represents as good a meta-methodology as any that currently exists for use in the evaluation context.

5. ONWARD TO INTEGRATION?

The meta-methodology of Flood and Jackson, TSI, is clearly located within Bruscaglioni's second stage of development and I would also locate the work that I have conducted on evaluation theory and practice at this stage of development. According to Bruscaglioni, progression from the mechanical complementarist stage should take the form of the integration of the various approaches and, recently, there has arisen an awareness that 'live' evaluation situations are far too complex to evaluate adequately with a single method. As a consequence of this realisation, there has arisen a certain amount of mixing and matching of methods in practice without much attention being paid to the theoretical implications. Perhaps we should look to Bruscaglioni for some guidance on how integration might best be achieved. According to Francescato, Bruscaglioni recommends two strategies for the integration of approaches:

- the development of complementary theories and perspectives. Bruscaglioni cites the example of role analysis being complemented by the use of systemic-functional analysis which might facilitate an appreciation of the dynamics of organisational functions, sociostructural analysis which might serve to deal with issues of class conflict and psychosocial analysis which might provide insights on interpersonal relations.
- the study of phenomena commonly investigated by one approach using variables usually associated with another.

The notion of integration is not new in the evaluation field. Seashore (1983) defines three models of evaluation (the natural system model, the goal model and the decision process model) and states that as each model refers to a different facet of organisational behaviour they should be integrated by means of triangulation. According to Seashore, triangulation should take the form of multiple integrations, that is the view that an organisation should be evaluated from different constituents' perspectives. Hence, it is assumed by

Seashore that the constituents will fully reflect criteria from the three schools of evaluation. There is little evidence to support Seashore's assumption that the variables represented in the three models will be reflected in the evaluations conducted by constituents and, furthermore, this form of evaluation is commonly seen to be a separate form of evaluation, referred to in this paper as the multi-actor model. Given these criticisms, it may be argued that Seashore's model does not represent a true integration of the three separate models he defines.

As others in the evaluation field appear to be 'plunging head-first' into integration, the temptation is to go with the flow lest one should be left-behind. However, before I start marrying my methodologies and crossing my variables, I think a pause for critical reflection is needed. Is integration a good thing? Isn't it just being used as an excuse for mixing and matching methods according to need (pragmatism)? Integration might appear an ideal state ('Yes, of course my methodology will supply answers to *ALL* your questions!) but I believe behind the glossy facade there lurks some unresolved issues.

How can the methodologies be integrated? My question rests on what is meant by the term integration. The dictionary definition of the terms is "...combination of diverse elements of perception etc." (The Concise Oxford Dictionary) and it is that definition which shall be taken here. As we have seen, at the heart of TSI is the SOSM: the SOSM is an oxymoron, that is '...a framework which helps us to recognize the seeming contradictions in a construct' (Quinn and Rohrbaugh, p. 376). Thus, I am left asking, if at the heart of the different methodologies are core assumptions which are contradictory, how can these 'opposites' be combined?

6. PRACTICAL INTEGRATION, THEORETICAL PARALLELISM

Whilst I am unable to see how a theoretical integration of the different methodologies can be achieved, given the conflicting assumptions upon which they are based, I am, following my own practical work in this area, able to appreciate that the methodologies might be integrated in practice (Morales is currently working on how the four approaches might best be integrated for use in an evaluation project in Colombia). I believe this practical integration of the different methodologies should enable their use in *parallel*. The use in parallel of the methodologies would mean that the different theoretical foundations of the methodologies would be respected. Indeed, the ability to use several methodologies at once is particularly pertinent in the field of evaluation where, unlike in the area of organisational problem solving, the methodologies require long lead times and where an evaluation report may be required unexpectedly, for example to satisfy the demands of a funding body. Perhaps, following Francescato's lead in coining the phrase 'a multidimensional perspective of organisational change,' this approach to evaluation might be termed multidimensional evaluation given that it seeks to address a host of organisational issues and represents an integration of the methodologies from a practical point of view and the use of the methodologies in parallel from a theoretical point of view.

7. CONCLUSION

In this paper the progress of evaluation theory has been traced through the first two stages of Bruscaglioni's theory of development: it was shown that up to the late 1970's there had been an isolation of approaches with three models dominating evaluation theory and practice (the goal model, the system-resource model, and the multi-actor model). In the early 1980s an awareness arose that each of the models had different strengths and weaknesses and, therefore, different contexts of application. As a result of this awareness, several different forms of mechanical complementarism were developed in the evaluation field. Currently, it would appear that there is a shift towards Bruscaglioni's third stage of development, the integration stage. In this paper the practice of integration was assessed and it was concluded that whilst an integration might be achievable in practice, this should not be seen as a theoretical integration but more of a means of enabling the use of the methodologies in parallel.

REFERENCES

Bruscaglioni, M., 1982, Il comportamento organizzativo, *in*: "La Psicolgia Organizzativa," M. Bruscaglioni and E. Spaltro, eds., Angeli, Milan.

Deal, T., and Kennedy, A., 1982, "Corporate Cultures. The Rites and Rituals of Corporate Life," Penguin Books, London.

Drucker, P.F., 1946, "Concept of the Corporation," John Day, New York.

Etzioni, A., 1960, Two approaches to organizational analysis: a critique and a suggestion, *Administrative Science Quarterly*. 5: 257-278.

Flood, R.L., and Jackson, M.C., 1991(a), "Creative Problem Solving. Total Systems Intervention," John Wiley and Sons, Chichester.

Flood, R.L., and Jackson, M.C., 1991(b), Total systems intervention: a practical face to critical systems thinking, *in:* "Critical Systems Thinking. Directed Readings," R.L. Flood and M.C. Jackson, eds., John Wiley and Sons, Chichester, pp. 321-337.

Francescato, D., 1992, A multidimensional perspective of organizational change, *Systems Practice*. 5(2): 129-146.

Friedlander, F., and Pickle, H., 1967, Components of effectiveness in small organizations, *Administrative Science Quarterly*. 18: 291-310.

Georgopoulos, B.S., and Tannenbaum, A.S., 1969, A study of organizational effectiveness, *in:* "Readings on Modern Organizations," A. Etzioni, ed., Prentice-Hall Inc., Englewood Cliffs, New Jersey, pp. 80-88.

Gregory, A.J., 1995, "Organisational Evaluation: A Complementarist Approach," Unpublished PhD Thesis, University of Hull, Hull.

Guba, E.G., and Lincoln, Y.S., 1989, "Fourth Generation Evaluation," Sage Publications, London.

Jackson, M.C., 1987, New directions in management science, *in*: "New Directions in Management Science," M.C. Jackson and P. Keys, eds., Gower, Aldershot.

Jackson, M.C., and Keys, P., 1984, Toward a system of system methodologies, *Journal of the Operation Research Society*. 35: 473-486.

Kanter, R.M., 1983, "The Change Masters," Routledge, London.

Katz, D., and Kahn, R.L., 1978, "The Social Psychology of Organizations," 2nd ed., John Wiley and Sons, Chichester.

Meyer, M.W., and Associates, 1978, "Environments and Organizations," Jossey-Bass Publishers, London.

Pfeffer, J., 1977, Usefulness of the concept, *in:* "New Perspectives on Organizational Effectiveness", P.S. Goodman, J.M. Pennings and Associates, eds., Jossey-Bass Publishers, London, pp. 132-145.

Pfeffer, J. and Salancik, G., 1978, "The External Control of Organizations: A Resource Dependent Perspective," Harper and Row, New York.

Quinn, R.E., and Cameron, K., 1983, Organizational life cycles and shifting criteria of effectiveness: some preliminary evidence, *Management Science*. 29(1): 33-51.

Quinn, R.E., and Rohrbaugh, J., 1983, A spatial model of effectiveness criteria: towards a competing values approach to organizational analysis, *Management Science*. 29(3): 363-377.

Schneider, B., 1983, An interactionist perspective on organizational effectiveness, *in:* "Organizational Effectiveness. A Comparison of Multiple Methods," K.S. Cameron and D.A. Whetton, eds., Academic Press, London, pp. 27-54.

Seashore, S.E., 1983, A framework for an integrated model of organizational effectiveness, *in:* "Organizational Effectiveness. A Comparison of Multiple Methods," K.S. Cameron and D.A. Whetton, eds., Academic Press, London, pp. 55-70

Seashore, S.E., and Yuchtman, E., 1967, Factorial analysis of organizational performance, *Administrative Science Quarterly*. 12: 377-395.

Thompson, J.D., and McEwen, W.J., 1958, Organizational goals and environment: goal setting as an interaction process, *American Sociological Review*. 23: 23-31.

Weick, K.E., and Daft, R.L., 1983, The effectiveness of interpretation systems, *in:* "Organizational Effectiveness. A Comparison of Multiple Methods," K.S. Cameron and D.A. Whetton, eds., Academic Press, London, pp. 71-93.

Zammuto, R.F., 1982, "Assessing Organizational Effectiveness: Systems Change, Adaptation, and Strategy," State University of New York Press, Albany.

73

MPSB FRAMEWORKS EXPLAINED

Joseph, K.K. Ho

Informated Systems Design
M/F., 33 San Shing Avenue
Sheung Shui, N.T.
Hong Kong

1. INTRODUCTION

The topic of reviewing concepts and methodologies in management disciplines and inter/trans-disciplinary approaches, such as the various strands of systems methodologies, from multiple perspectives is a recurring theme, see, for examples, Ho and Sculli (1994) on Logistics Management and Ho (1994(b)) on Management Accounting; Ho and Sculli (1993(b)), Ho, Partington and Sculli (1994) on Information Systems Design; and Ho, Sculli (1993(a)) on Operational Research. Other writers that encourage review of management disciplines from various multiple perspectives (and thus endorse pluralism) include Hirschheim (1992) on Information Systems research, and Mears-Young (1993) on Logistics Management. There are also regularly published papers on Accounting based on different perspectives in journals such as *Accounting, Organizations and Society* (from Pergamon), and *Critical Perspectives on Accounting* (from Academic Press). What then would a meta-framework that is multi-perspective for reviewing management disciplines look like? How could such a multi-perspective framework be constructed and enhanced? These are the questions addressed here.

To develop a meta-framework to review management disciplines that is multi-perspective, we can draw on the works of systems theorists, such as Checkland (1984), Flood and Jackson (1991). The reason is that these writers have been reviewing systems methodologies from various perspectives, and, especially for Jackson and Flood, a meta-theory has recently been formulated to assimilate various strands of systems methodologies, called Total Systems Intervention (TSI). The project of developing a multi-perspective framework by the writer to review management disciplines is thus stimulated by these new development in the Systems Movement, thus itself also systems-based. First of all, some insights from the writings of these prominent writers that are more relevant to the project of formulating multi-perspective/ systems-based (MPSB) frameworks are highlighted in the next section.

Critical Issues in Systems Theory and Practice, Edited by K. Ellis et al.
Plenum Press, New York, 1995

2. SYSTEMS METHODOLOGIES AND THEIR UNDERLYING PERSPECTIVES

To illustrate the close relationship between a systems methodology and its underlying perspective, I refer to the work of Checkland (1984), in which he compared the paradigms of Hard Systems Thinking and Soft Systems Thinking in terms of six elements, namely:

a. Customers, i.e. victims and beneficiaries, such as decision-makers and participants, of what the notional human activity system (or, simply, the system) does;
b. Actors, i.e. those, such as external analysts, who carry out the systems activities;
c. Transformation, i.e., what input, such as information, does the system transform into what output, such as advice to decision-makers and learning for the actors?
d. Weltanschauung, i.e. the world view on the essence of reality and methodology which makes it meaningful to consider the system;
e. Owners, i.e., those who could destroy the notional system and prevent the system from acting; and
f. Environmental constraints, i.e. the things in its environment, such as power structure, which the human activity system takes as given.

These six elements of a paradigm together define the underlying perspective of a systems methodology. Further clarification of the topic of paradigms (which is taken to be similar to the term 'perspective') could be sought in the writing of Burrell and Morgan (1979) for a discussion of the underlying paradigms of organisational theories in terms of the assumptions of the nature of science (the subjective-objective dichotomy) and the assumptions about the nature of society (the regulation-radical change dichotomy); Flood (1990) for a discussion of the paradigms of traditional systems, soft systems, and critical systems in terms of their underlying rationalities (further explained in terms of theoretical, ideological and practical premises) and epistemology; Flood and Jackson (1991) for a comprehensive and pragmatic review of various systems methodologies in terms of theory (i.e., at the theoretical level), methodology (i.e., at the methodological level), ideology (i.e. conservative or emancipatory), and utility. (Note: Besides the discussion on the underlying perspectives of various systems methodologies, the works of Flood and Jackson on the development of the Total Systems Intervention (TSI) approach is considered by the writer especially informative for the construction of MPSB frameworks. This point is further elaborated in the next section.) The concern of power as covered within the element of environment in Checkland's work as well as Flood, Jackson's consideration of ideology also remind us of the importance of these topics in examining the issue of perspective. These two topics of power and ideology have been covered at length from a critical systems perspective by Oliga (1990; 1991).

Given the existence of the in-depth works of these writers on this topic, the discussion of the relationship between systems methodologies and their underlying perspectives remains controversial. The following examples explain the reasons:

Example 1: There are objections to the Jackson and Keys' classification (Jackson and Keys, 1984) of the Socio-technical approach as unitary in orientation. For example, Babüroglu (1992) argued that

i. there were four developmental tracks of socio-technical systems (STS) approach and was, thus, broad in scope; and
ii. that the STS approach resembled the soft systems methodologies of Ackoff (1981) and Checkland (1981), and was also concerned with the theme of liberation, thus more affiliated with the soft and critical systems perspectives than the hard systems (unitary) perspective.

Example 2: It has been pointed out that the System Dynamics (SD) approach, though initially grounded on hard systems thinking, has been migrating to a softer perspective as it attempts to deal with more complex human issues, see Keys (1990), Dash (1994). SD is, in the words of Dash, "a living paradigm". The migration of perspective, in turn, led to its criticisms by both hard systems and soft systems theorists: the hard systems theorists considered SD as "on the verge of scientifically acceptable" while the soft systems theorists viewed SD as "inappropriate to deal with the many difficulties found in complex, pluralist situations." (Keys, 1990).

Example 3: The TSI approach was formulated by Flood and Jackson (1991) and was intended by them to be grounded on Critical Systems Thinking (CST). That is, it endorses the unitary (hard systems), pluralistic (soft systems), and critical (emancipatory) perspectives simultaneously. However, the TSI methodology has been criticised by Tsoukas as mere common sense and not necessarily grounded on CST, see Tsoukas(1993). A heated debate between Tsoukas (1993) and the Critical Systems Theorists then ensued. (See, for examples, Jackson (1993), Ho (1994(a)).

These examples illustrate the kinds of problems to align a systems methodology with a particular perspective. There is an additional problem for the TSI approach of Flood and Jackson which endorses entertaining multiple perspectives: the position of such a theoretical stance of pluralism at the theoretical and methodological levels has been argued by Midgley as impossible to be plural or extraparadigmatic(Midgley, 1989). The attempt to put together the various perspectives into a meta-framework creates tensions in its usage, a point which Ho, Partington and Sculli (1994) argued in the case of developing such a framework for Information Systems Design. These problems identified from the study of systems methodologies from a largely critical systems perspective (i.e. multiple perspectives) are expected to be encountered in the study of management disciplines based on a MPSB Framework. However, unlike the substantial progress already made on the evaluation of systems methodologies, there is a pertinent task now to formulate such kind of frameworks for reviewing management disciplines in the first place. This topic is covered in the next section.

3. THE FORMULATION AND STRUCTURE OF MPSB FRAMEWORKS

The task of formulating MPSB Frameworks is facilitated by referring to the existing systems literature on reviews of systems methodologies, as well as the development of Critical Systems Thinking (CST) and systems methodologies grounded on it, notably the TSI approach of Flood and Jackson (1991). Some initial work has been done by Ho and Sculli for reviewing management disciplines from this orientation. See, for examples, Ho and Sculli (1993(b)) on Information Systems Design; Ho (1994) on Management Accounting, and Ho and Sculli (1994) on Logistics Methodologies. The steps involved in formulating MPSB Frameworks for reviewing Logistics Methodologies are:

1. to select a specific logistics approach such as Electronic Data Interchange (EDI) or Distribution Resource Planning(DRP);
2. identify and expose the underlying image of organization and perspective of the logistics approach and contrast them with alternative images of organization and perspective; finally, employ the findings to enhance the MPSB frameworks and clarify the nature of the logistics approach under review. See Ho (1994(c)).

The whole approach adopts a number of techniques from the TSI approach of Flood and Jackson (1991). Briefly, the TSI approach of Flood and Jackson consists of three phases, namely: Creativity Phase- The task is to highlight aims, concerns and problems; the tools to use are systems metaphors; Choice Phase- The task is to choose an appropriate systems-based intervention methodology; and the tools to use are the 'system of systems methodologies' as formulated by Jackson and Keys (1984) as well as the 'relationship between metaphors and methodologies'; Implementation Phase: The task is to make and implement specific change proposals; the tools to use are systems methodologies employed according to the logic of TSI, which are embodied in the tenets of

i. complementarism at the theoretical and methodological levels,
ii. sociological awareness, and
iii. human emancipation.

Specifically, the process of formulating MPSB frameworks employs the TSI techniques of using multiple systemic lens (images of organization) at the Creativity Phase of TSI and reviewing a methodology in terms of theory, methodology, ideology and utility. Finally, the process endorses the tenets of TSI. An example of such a framework for synthesising and reviewing Office Automation methodologies has been proposed previously, see (Ho and Sculli, 1993(b)). It takes on multiple perspectives to consider various Information Technology-related ideas and concerns to build up a rich picture of a problematic situation. This makes up the Phase-1 exploratory exercise. The Office Automation-related insights can then be utilized in the Phase-2 Office Design exercise that is guided by multiple Office Automation methodologies grounded on various perspectives.

The model of Ho and Sculli (1993(b)) can be treated as a generic MPSB framework to review other management disciplines, such as Management Accounting and Logistics Management. What are needed to suit a particular subject are some minor changes in the Ho and Sculli's model, such as replacing Logistics Methodologies with Office Automation Methodologies. Five major enhancing tasks can be identified in this framework, namely:

i. the task to continue with the theoretical works on reviewing topics related with various perspectives (or paradigms), at the level of perspectives;
ii. the task to explore various issues and concerns of a management discipline from multiple perspectives, as a rich-picture building exercise;
iii. the task to examine management methodologies in terms of their underlying perspective, at Phase 2 of the framework of Ho and Sculli (1993(b));
iv. the task to review the impacts and efficacy of the various management methodologies in the management of organizations and in the organizational change processes at Phase 2 of the framework; finally:
v. the task to review the relationship between the practitioners of the methodologies and the various stakeholders of a problematic situation, again, at Phase 2 of the framework.

In reviewing management disciplines based on MPSB frameworks, we can expect to revisit the problems encountered in similar exercises on systems methodologies. An important difference between the MPSB project and the CST/ TSI project is that the former shows a greater readiness to preserve the identity of the various management disciplines while the latter, being supradisciplinary (and truly systems) in orientation, is more ready to criticise (or, recognize the limitation of) the very act of (i) keeping the separate identity of various management disciplines and (ii) viewing problems from the narrower perspectives of a separate management discipline in the first place. While the stance of CST/TSI, being less reductionistic, is theoretically superior to that of the MPSB project, the MPSB project has certain pragmatic advantages over the CST/TSI project:

i. it provides an immediate and more receptive forum for exchanging ideas between systems theorists and professionals in various management disciplines - as it appears less threatening to the profession of the various management disciplines;

ii. due to its focus on management disciplines, it has an easier access to opportunities to guide management projects that are originally intended to introduce a chosen management methodology/ technique, such as Just-In-Time management (see, for example, Schniederjans (1993).), or Activity-based Management (see, for example, Morrow (1992).), from a particular management discipline to an organization. (Note: one should also note the inter-relatedness of the approaches from various management disciplines, such as Cobb's (1993) study on Just-In-Time and Management Accounting.)

As time goes by, the MPSB Frameworks will become more relevant for reviewing and synthesising management approaches/ concepts as these approaches are increasingly employed to cope with more complex management and human issues: these issues demand a better understanding of (i) the object relationship of the 'natural world', (ii) the subjectivity involved, (iii) moral/ ethical considerations involved, and, finally, (iv) the inter-connectedness of (i), (ii) and (iii). This is argued by Midgley (1992) as a new understanding of ontological complexity which could only be coped with by endorsing pluralism (or entertaining multiple perspectives at the same time). In terms of the MPSB frameworks, the unitary perspective is good at dealing with (i), the pluralist at (ii), the critical perspective at (iii). Finally, the consideration of all three perspectives simultaneously, as encouraged by the MPSB frameworks, respects the inter-connectedness of (i), (ii), and (iii).

4. CONCLUDING REMARKS

The process of reviewing and enhancing management literature is a rewarding learning experience. Formulating MPSB frameworks for reviewing the various management approaches, while recognising its relatively reductionistic orientation when compared with the CST/TSI project, is a route to go through this review process. The MPSB frameworks could, in this case, be considered as meta-structures of knowledge, which are ready to be challenged (and are encouraged to be challenged) and negated by alternative and better ones.

REFERENCES

Ackoff, R.L., 1981, "Creating the Corporate Future", John Wiley & Sons, New York.

Babüroglu, O.N., 1992, Tracking the development of the Emery-Trist systems paradigm (ETSP), *Systems Practice*, Vol. 5, No.3, pp. 263-290.

Burrell, G., and Morgan, G., 1979, "Sociological Paradigms and Organisational Analysis", Gower Publishing Co. Ltd., Aldershot.

Checkland, P.B., 1981, "Systems Thinking, Systems Practice", John Wiley & Sons, Chichester.

Checkland, P.B., 1984, Rethinking a systems approach", *in* "Rethinking the Process of Operational Research and Systems Analysis", R. Tomlinson, and I. Kiss, ed., Frontiers of Operational Research and Applied Systems Analysis Vol. 2, pp. 43-60, Pergamon Press, Oxford.

Cobb, I., 1993, "JIT and the Management Accountant: A Study of Current UK Practice", Chartered Institute of Management Accountants, UK.

Dash, D.P., 1994, Research Notes: System dynamics: changing perspectives, *Systems Practice*, Vol. 7, No.1, pp. 87-97.

Flood, R.L., 1990, "Liberating Systems Theory", Plenum, New York.

Flood, R.L., and Jackson, M.C., 1991, "Creative Problem Solving: Total Systems Intervention", John Wiley & Sons Ltd, Chichester.

Hirschheim, R.A., 1992, Information systems epistemology: an historical perspective, in "Information Systems Research: Issues, Methods, and Practical Guidelines", R.Galliers, ed., Blackwell Scientific Publications, Oxford.

Ho, J.K.K., 1994(a), Is Total Systems Intervention(TSI) no better than common sense and not necessarily related with Critical Systems Thinking(CST)?: A research note", *Systems Practice*, Vol. 7, No. 5., pp.569-573.

Ho, J.K.K., 1994(b), Management accounting and operational research, in "The Proceedings of the Fourth Conference of the Operational Research Society of Hong Kong", Oct. 17.

Ho, J.K.K., 1994(c), Formulating MPSB frameworks, using logistics management as an example: a research note, *Systems Practice*,(in print).

Ho, J.K.K., Partington, C.E., and Sculli, D., 1994, A Meta-Framework for information systems design, *in* "Proceedings of the twenty-third Annual Meeting of the Western Decision Sciences Institute", March 29-April 2, pp. 592-597, Maui, Hawaii.

Ho, J.K.K., and Sculli, D., 1993(a), An overview of operational research theories and practices, " The Proceedings of the Third Conference of the Operational Research Society of Hong Kong", May 28, pp. 149-158.

Ho, J.K.K., and Sculli, D., 1993(b), Critical systems thinking and information systems design, *in* "The Conference Proceedings of the Second International Conference of Decision Sciences Institute", June, pp. 685-688, Seoul, Korea.

Ho, J.K.K., and Sculli, D., 1994, Computer-based Logistics Management Methodologies, *in* "The Proceedings of the International Conference on Data and Knowledge Systems for Manufacturing and Engineering", May 2-4, pp. 590-596, Chinese University of Hong Kong, Hong Kong

Jackson, M.C., 1993, Don't bite my finger: Haridimos Tsoukas' critical evaluation of total systems intervention, *Systems Practice*, Vol. 6, No.3, pp. 289-294.

Jackson, M.C., and Keys, P., 1984, Towards A System of Systems Methodologies, *JORS,* Vol. 35, 6, pp. 473-486.

Keys, P., 1990, System dynamics as a systems-based problem-solving methodology, *Systems Practice*, Vol. 3, No.5, pp. 479-493.

Mears-Young, B., 1993, 'Non-traditional' Logistics - A Step Forward ? *in* "Systems Science: Addressing Global Issues", F.A.Stowell., D.West., and J.G.Howell, ed., pp. 143-148, Plenum Press, New York.

Midgley, G., 1989, Critical systems and the problem of pluralism, *Cybernetics and Systems: An International Journal*, 20:219-231.

Midgley, G., 1992, Pluralism and the legitimation of systems science, *Systems Practice*, Vol. 5, No.2, pp. 147-172.

Morrow, M., ed., 1992, "Activity-based Management: New Approaches to Measuring Performance and Managing Costs", Woodhead-Faulkner, New York.

Oliga, J.C., 1990, Power in organizations: a contingent, relational view, *Systems Practice*, Vol. 3, No.5, pp. 453-477.

Oliga, J.C., 1991, Concepts of ideology: an exegesis and critique, *Systems Practice*, Vol. 4, No.2, pp. 101-129.

Schniederjans, M.J., 1993, "Topics in Just-In-Time Management", Allyn & Bacon, Massachusetts.

Tsoukas, H., 1993, The road to emancipation is through organizational development: a critical evaluation of total systems intervention, *Systems Practice*, Vol. 6, No.1, pp. 53-70.

74

SYSTEMS, RATIONALITY AND EQUITY IN A POST-MODERN WORLD

Gilbert Mansell

School of Computing and Mathematics
University of Huddersfield
Huddersfield
United Kingdom

SYSTEMS PRACTICE

Systems practice involves intervention in society intended to bring about improvement. This is true of both 'hard' and 'soft' traditions in the systems movement. 'Hard' systems engineering extended the ambitions of engineers from the design of effective and efficient machines to the the design of effective and efficient *systems* of 'Men, Money Machines and Material' (Jenkins, 1969). The need for designers to choose an appropriate boundary progressively gave rise to the incorporation of human beings in system designs, and the insight that systems occur as part of a hierarchy of systems again caused the characteristics of human beings to be considered at levels higher than the operational. In the tradition of 'hard' Systems Engineering, therefore, social planning and intervention is necessitated by the drive for comprehensiveness inherent in the systems approach. The paper by Jenkins (1969), is full of the optimism of the times for the benefits that will ensure from the application of a systems approach. Jenkins explicitly distinguishes Systems Engineering from engineering by claiming that the former can be applied not only to hardware systems but "...can also be applied, for example, to parts of firms, or whole firms, or to local government" (Jenkins, 1969). The benefits that Jenkins claims will ensue include clarification of objectives, the replacement of guesswork by model-building and optimisation , and the tackling of problems in their correct order of importance. These are similar claims to those made by the operational research movement that similarly flourished in the 1960s.

Checkland's well-known critique of both Systems Engineering and Operational Research (Checkland, 1981) is that both disciplines solve problems by facilitating rational choice between alternative means of achieving a known end. Both disciplines fail, however, in providing ways of clarifying and reaching agreement over desirable ends. Furthermore Checkland considers that human activity is problematic in a way that the behaviour of a machine is not. He subdivides human activity (an abstraction) into either behaviour (genetically, or socially conditioned) or action (willed by the doer). The existence of free-will and, possibly private motive, means that human activity has to be interpreted, and this process of interpretation is subjective because of differing value systems or world views. For this reason

Critical Issues in Systems Theory and Practice, Edited by K. Ellis et al.
Plenum Press, New York, 1995

Checkland argues that identification of systems of human activity to be engineered or optimised is problematic and maybe contentious. The essence of Checkland's Soft Systems Methodology (SSM) (Checkland, 1981; Checkland and Scholes, 1990) is to explore desirable purpose using ideal-type models of human activity to engender debate. Purposeful activity is assumed to be intended to bring about desired change in problem situations. In this respect the ambition of 'soft' and 'hard' approaches is the same.

The dichotomy between 'hard' and 'soft' systems approaches is arguably resolved by a socio-technical approach that sees human activity as having a social aspect involving roles, norms and values and a technical aspect involving the transformation of inputs into outputs under cybernetic control (Susman, 1983). Knowledge is gained about a technical system by observation but about a social system by interpretation and interaction. The developed form of SSM, as defined in the 'two-stream' diagram, (Checkland and Scholes, 1990, p.29) seems to define a socio-technical enquiring system, but the important difference is that the logic-based stream of analysis does not pre-suppose an objectively existing real-world system. Socio-technical approaches make use of generalised ideal-type models of systems, for example as described by Jackson (1991(a) pp 48-67). Like all systems approaches the socio-technical approach involves intervention in a problem situation in order to bring about improvement.

JUSTIFICATION OF INTERVENTION

Using a systems approach, intervention in society can be justified by appeal to systems theory that specifies ideal type models. Systems Engineering, for example, involves the construction of models of the system to be implemented. The engineer wishes to prove that implementation of the design will bring about desired outcomes. One way to do this is to implement the system and, evaluate its performance, for example by implementation of a pilot scheme or prototype. Where this is infeasible, experiments can be performed on the model itself, for example using simulation or computer aided design. In the case of SSM the conceptual models produced are not intended to be designs, but explorations of the consequence of defining systems to achieve particular purposes. There is a tradition, particularly developed by Wilson (1984), of primary task modelling using conceptual models, however, and these models may be implemented as organisational systems. Since conceptual models also contain activities intended to monitor and control the system defined, it is in principle possible to evaluate the primary task model. It is not possible to perform experiments on a conceptual model, however, so there is no secure way to judge the success of the application of the methodology prior to implementation (except intersubjectively in the judgement of the participants in the exercise.) Socio-technical approaches offer an additional method of justification, and that is an appeal to the characteristics of ideal-type models of any successful system. For example one ideal-type socio-technical principle is that the boundary of a system should be chosen so that the interdependence between required activities within the systems boundary is greater than the interdependence of required activities that cross the system's boundary (Susman, 1983). Intervention in a social system could be justified as an attempt to bring about this state of affairs.

All systems approaches justify intervention in social systems by appeal to rationality. Ulrich (1991) distinguishes between instrumental reason and practical reason. Instrumental reason pertains when means are efficiently utilised to achieve ends that conform to accepted values. Hard systems approaches are particularly strong at demonstrating congruity of means and ends. Soft systems approaches are stronger at defining ends in relationship to value systems. Practical reason addresses issues that arise when conflicts of interest exist between stakeholders in a problem situation, so that what is rational for one group is inimical to the interests of another. SSM and socio-technical approaches provide no systematic way of ensuring that this issue is dealt with and hard approaches fail completely. Ulrich claims that his Critical Heuristics (Ulrich, 1983) provide not a way of objectively justifying social intervention but of avoiding submission to "...an objectivist illusion" (Ulrich, 1991 p263). In coercive situations there is a risk that powerful stakeholders will attempt to perpetrate such an illusion in justifying their social engineering.

Jackson (1991(a)), in his review of systems methodology judges it to be a modernist phenomenon - "...upholding reason and believing that rationality can play an increasing role in helping human beings perfect themselves and their societies." (Jackson, 1991(a)). Following Lyotard he distinguishes between approaches that exemplify *systemic modernism* (concerned with effective and efficient performance, such as systems engineering) and *critical modernism* (concerned with social progress and human emancipation, such as Ulrich's Critical Systems Heuristics). Jackson somewhat disparages soft systems thinking as "a rather underdeveloped form of critical modernism" [that is] "particularly prone to slipping back into becoming no more than an adjunct of systemic modernism, readjusting the ideological status quo by engineering human hopes and aspirations in a manner that responds to the system's needs ..." (Jackson, 1991(a) p 180).

EQUITY IN A POSTMODERN WORLD

Perpetration of an objectivist illusion in justifying inequitable social intervention is one way that the powerful might misuse a systems approach. The other is to attempt to engineer hopes and aspirations. A third way, however is to abandon the rational modernist approach altogether. Jackson (1991(b)) characterizes postmodern society as having no respect for objective truth and thriving on "...instability, disruption, disorder, contingency, paradox and indeterminacy." (Jackson, 1991(b)). Such a characterisation seems to capture the essence of life under 'new-right' governments in advanced capitalist countries. Postmodernism apparently renders the systems approach obsolete and this is exemplified by UK government policy in education and health that is contemptuous of objective research and driven by ideology (Rosenhead 1992). In practice, UK government policy is characterised by a mixture of postmodern contempt for scientific research and rational planning and attempts to foster objectivist illusion and re-engineering of national culture.

An example of government contempt for objective research is afforded by the reception given to the Black Report (Townsend, Davidson and Whitehead, 1992) in 1980 and a revised version known as the Health Divide (Townsend, Davidson and Whitehead, 1992) produced in 1987. This story is recounted in Townsend, Davidson and Whitehead (1992). The Black Report produced evidence that health was related to social class and that the root of the problem lay outside the current scope of the NHS. Ill health was associated with low income, unemployment and poor housing. A systemic problem had been identified that necessitated re-definition of relevant systems. The new Conservative administration rejected the recommendation of the report arguing that the way forward lay in encouraging individuals to lead healthier lives. The Health Divide was produced in 1987 by the Health Education Council. This report confirmed and updated the findings of the Black report but in the intervening years the gap between the richest and poorest had increased in the UK, the plight of the homeless had become worse and health inequalities had increased. The government abolished the Health Education Council. The authors of the report were accused in the House of Commons of being Marxists and trying to impose discredited socialism on society. This reaction can be seen as postmodern.

Systems thinking and postmodernism are antithetical because of the reliance of the former on rational justification of policy rather than ideology. Nevertheless there are examples of the application of systems approaches within the postmodern NHS. Briefly the government 'reform' of the NHS consists of setting up an 'internal market' where health purchasers (health authorities and fund-holding GPs) buy health care on behalf of patients from health providers (hospital trusts - quasi-independent bodies, - and private hospitals). This revolution in the provision of health care was bound to be more costly. Cook (1993) estimates that "...up to 4% of revenue resources are being applied simply to the mechanisms of operating the market" and concludes "There is no evidence that the benefits from 'competition' outweigh the administrative cost involved." (Cook, 1993). The reform has also been bitterly opposed by the medical profession, perhaps because it alters the balance of power between professionals and managers. Harrison, Hunter, Marnoch and Pollitt (1992) found that "Managers continue to be seen as agents of the government in the way that doctors and nurses and other service-providers are not. The messages and proposals which come from managers ... are 'screened' for possible contamination by central government's current political concerns."

SYSTEMS PRACTICE IN A POSTMODERN WORLD

Into this swamp enters Checkland with his underdeveloped form of critical modernism. Checkland and Holwell (1993) describe an application of SSM in the NHS. Checkland and Holwell (1993), concede that the way the NHS is managed "is changing significantly in ways which have not been tested in pilot schemes or thoroughly thought through." The problems posed by the internal market are treated as entirely technical by Checkland and Holwell, however, and issues such as the funding of the service, democratic representation in its management, conflicts between medical professionals and managers and the risk of a two-tier service developing are ignored. The authors blandly accept the new situation and set about helping managers to run it - "Government has imposed a new way of conceptualising the provision of health care; the 'meanings' have been changed by the imposition of the purchaser - provider relationship. Hence what is needed is examination of the purposeful activity necessarily implied by that relationship, and its embodiment in a particular organization structure ..." (Checkland and Holwell 1992: p 9). The real problem with Checkland's methodology is that it provides no way of ensuring that anything other than a managerial point of view is adopted in any practical consultancy situation. Checkland's subjective criteria for selection of relevant systems to be investigated and, perhaps, his status as a management consultant, have betrayed the critical potential of Soft Systems Methodology in this report of NHS action research.

An application of systems thinking in health care that derives methodologically from the earlier hard systems tradition of simulation modelling (or Systems Dynamics, as it is known) is provided by Wolstenholme (1993). Wolstenholme argues that a qualitative form of Systems Dynamics is useful in problem structuring, possibly as a precursor to a computer-based systems modelling phase. This represents a 'softening' of Systems Dynamics, but the distinctive difference from Soft Systems Methodology is that a real-world system is modelled rather than an ideal-type model of human activity. This feature enables alternative scenarios to be simulated either as a 'thought experiment' if the model is qualitative or as a computer-based simulation. Wolstenholme's study has to be read in the context of government policy of closure of long-stay NHS beds for the elderly and their disposal to the 'community'. Cash limiting of local authorities means that there is a 'rationing' of community care. Paradoxically this limits the ability of the NHS to discharge people from hospital with the consequence that admission rates are curtailed. People on the waiting list are, of course, notionally in the care of the community, but, since they are elderly and infirm they may not survive to enjoy treatment or care. The unintended consequences of ill-thought out policy are convincingly revealed in this case, in a way that would be difficult to achieve with SSM. There is nothing in Wolstenholme's approach that *guarantees* use of practical rather than instrumental reason, but Systems Dynamics can clearly retain its valuable realist philosophical origin while being used as a vehicle for discussion and democratic decision-making. The point being made here is that real injustice demands a realist philosophy.

REFERENCES

Checkland, P. B., 1981, "Systems Thinking, Systems Practice", John Wiley & Sons, Chichester.

Checkland, P. B., and Holwell, S., 1993, Information management and organisational processes: an approach through soft systems methodology, *Journal of Info Systems* 3:3.

Checkland, P. B., and Scholes, J., 1990, "Soft Systems Methodology in Action," John Wiley & Sons, Chichester.

Cook, T., 1993, The NHS reforms and the finance function, *in*: "The New Face of the NHS," P. Spurgeon, ed., Longman, Essex.

Harrison, S., Hunter, D. J., Marnoch, G., and Pollitt, C., 1992, "Just Managing: Power and Culture in the National Health Service," Macmillan, Basingstoke.

Jackson, M.C., 1991(a), "Systems Methodology for the Management Sciences," Plenum, New York.

Jackson, M.C., 1991(b), Modernism, post-modernism and contemporary systems thinking *in*: "Critical Systems Thinking -Directed Readings," R. L., Flood and M.C. Jackson, eds., John Wiley & Sons, Chichester.

Jenkins, G. M., 1969, The Systems Approach, *Journal of Systems Engineering*, 1,1.

Rosenhead, J., 1992, Into the swamp: the analysis of social issues, *Journal of the Operational Research Society*, 43:293.

Susman, G., 1983, Action research - a socio-technical systems perpective *in*: "Beyond Method-Strategies for Social Research," G. Morgan, ed., Sage, London.

Townsend, P., Davidson, N., and Whitehead, M., 1992, "Inequalities in Health," Penguin, London.

Ulrich, W., 1991, Systems thinking, systems practice, and practical philosophy: a program of research, *in* "Critical Systems Thinking - Directed Readings," R. L. Flood and M. C. Jackson, eds., John Wiley & Sons, Chichester.

Wilson, B., 1984, "Systems: Concepts, Methodologies and Applications," John Wiley & Sons, Chichester.

Wolstenholme, E. F., 1993, A case study in community care using systems thinking, *Journal of the Operational Research Society* 44.

75

THE CYNICISM OF SYSTEMS THINKING

S. K. Probert

Department of Computer Science
Birkbeck College
University of London
Malet Street
London
WC1E 7HX
United Kingdom

INTRODUCTION

It will be argued that a diffuse cynicism permeates some recent systems thinking, i.e. the prescriptions of Total Systems Intervention (TSI); this is despite its philosophical basis in Critical Systems Thinking (CST). In some respects the concerns to be discussed are similar to some of those expressed by both Tsoukas (1992, 1993(a)) and Brocklesby (1994). The arguments herein are informed by the considerations put forward by Sloterdijk (1987) concerning the cynical condition. Such arguments also provide genealogical insights (in the Foucauldian sense) as to the question of the ready acceptability of TSI to a subset of the population of the Western world. The analysis will focus on the work of Flood and Jackson (1991), as qualified by Jackson (1993). The main discussion will be concerned with the tension between what can be termed *the compelling force of the enlightenment*, i.e. the need for learning, critique, and liberation to be embodied in the prescriptions of TSI, and (what can be termed as) *the requisite concessions to praxis*.

THE COMPELLING FORCE OF THE ENLIGHTENMENT

This force is expressed in TSI in the prescriptions relating to *sociological awareness* and *human well-being and emancipation*: "Critical systems thinking, and the thrust of TSI therefore, is emancipatory in that it seeks to achieve for all individuals, working through organisations and society, the maximum benefit for their potential... [T]he philosophy of TSI, which comes through in the principles and practice... should be known and respected by all who would use this approach." (Flood & Jackson, 1991, p49).

This approach in general owes much of its intellectual grounding to the various works of Jurgen Habermas.

Critical Issues in Systems Theory and Practice, Edited by K. Ellis et al.
Plenum Press, New York, 1995

THE REQUISITE CONCESSIONS TO PRAXIS

TSI is, of course, concerned with practical interventions in organisations. So are e.g. Trade Unions, and so (sometimes) are governments (or government agencies, etc.). But the practical interventions that TSI practitioners can make are essentially *consultancy* assignments; generally undertaken on the behalf of the *management's* of the client organisations. So an obvious tension may emerge between the aim of a *bona fide* TSI practitioner, which will be emancipatory in that "it will seek to achieve for all individuals, working through organisations and society, the maximum benefit for their potential" (Flood and Jackson, 1991, p49) and the aims of at least some management's of organisations at any rate. *The requisite concessions to praxis* - the need to work with 'strident capitalists', macho managements, the disciplinary forces of social cohesion (e.g. the doctors, the police, the professions, etc.) - will have to be made. The tensions that are generated by these considerations sometimes emerge in the TSI texts, e.g.: "What should we do as ... consultants when all the evidence we have highlights the coercive forces at work? Obviously this would be a matter of strategy and conscience. We could refuse the work since we might be obliged to concede our interests to those in power. On the other hand we might take on a job in the hope that some progress can be made towards our ideals, and that something is better than nothing." (*ibid.*, p. 244)

Although Flood and Jackson admit the possibility of 'refusing the work' it seems that the second strategy ('taking on a job in the hope that some progress can be made towards our ideals') is the preferred one, as this diatribe by Tsoukas makes plain:

> Jackson and Flood have spent much ink on writing about "coercive contexts," "exploitation," "oppression," and the rest, and how positivist and interpretative problem-solving methods fail to tackle them, but I have not ... observed them using some discriminatory criteria in picking their consultancy assignments and their clients. (Tsoukas, 1993(b), p. 316).

CYNICISM

Sloterdijk (1987) offers a critical analysis of what he describes as "enlightenment cynicism". Sloterdijk's project is outlined here by Andreas Huyssen (1987, pp. xi-xii):

> "Sloterdijk sees cynicism as the dominant operating mode in contemporary culture, both on the personal and institutional levels ... By focusing on cynicism as a central feature of the postmodern condition of the 1970s and 1980s and by searching for strategies to resist it, Sloterdijk attempts to theorize that which has often remained submerged in the recent debate about modernity and postmodernity: the pervasive sense of political disillusionment in the wake of the 1960s and the pained feeling of a lack of political and social alternatives in Western societies today. After all, the 1960s in West Germany - against the arguments of Adorno and Horkheimer's *Dialectic of Enlightenment* [Adorno and Horkheimer, 1972] - were once labeled a second expanded enlightenment that seemed to promise a major and long-lasting realignment in the country's political culture ... the notion of *Aufklarung* carried a great potential of utopian hopes and illusions with it at that time ... Cynicism and resignation are therefore indeed dangers for a generation that had its formative political experiences in the 1960s and that has since then seen its hopes not so much dashed as crumble and fade away ..."

Sloterdijk's analysis therefore provides an intriguing argument for understanding the motives for the original development of TSI, and (equally importantly) a genealogical hypothesis for understanding its ready acceptability to a significant number of would-be and actual practitioners.

CYNICISM IN TSI

In Jackson (1993) the impression is given that TSI is considered to be of genuine value: "Management science is an important enterprise concerned ... with improving rational intervention in human affairs. CST and TSI seem to me to be significant new developments in management science... They subject existing approaches (and themselves) to an uncompromising critique; they are aware of the social conditions in which they operate and the social consequences their use might have ... they openly declare an ideological commitment and reflect on how to realize it. What other version of management science can claim as much?" (Jackson, 1993, p. 293).

There are several problems inherent in this argument. Firstly, what does it mean to say that, in any sense, 'one is aware of the social conditions in which one operates'? This seems to imply that one accepts as given the unchanging nature of a set of social conditions, or - more properly - one concedes that only limited changes are possible in given contexts. There is an implied acceptance of a set of conditions which Sloterdijk describes as *hard reality*:

> "Real existing socialism", as it exists today ... shares with capitalism, as with any political-economic order, the imprint of hard reality, which can never be Left or Right in and of itself but ... is always how it simply is ... The late and cynical feeling of the times is that of the trip and of mundane everyday life, stretched between irritable realism and incredulous daydreams..." (Sloterdijk, 1987, pp. 97-98).

The assumptions made about how things 'simply are' must surely conflict heavily with the TSI advocates' 'openly declared ideological commitments'. Furthermore, TSI is described as being a generalisable 'management science', but why is a management science 'an important enterprise' in the first instance? The answer to this question is most likely to be found in the changed nature of (what generally counts as) 'knowledge'; this change has taken place in recent years, and includes an equation of 'truth = usefulness' - i.e. *performativity*, to use Lyotard's (1984) term. No doubt management science is seen as being an 'important' science because it has the potential to (ostensibly) improve the performance of organisations, and improving performance is surely a concern of many managers. And as the purpose of academic research is increasingly judged to be subordinated to that of *realizing 'our' (economic) potential* (or some such phrase), management science is seen as potentially useful in this respect. But can such performative concerns be linked with emancipatory concerns? If one adopts a cynical view of what is possible 'in the current climate' (i.e. Sloterdijk's *hard reality*), then a (sort of) 'yes, at a price' can appear to be a plausible response: "In the final analysis, it is a matter of the social and existential limits of enlightenment. The compulsion to survive and desire to assert itself have demoralized enlightened consciousness. It is afflicted with the compulsion to put up with pre-established relations that it finds dubious, to accommodate itself to them, and finally even to carry out their business." (Sloterdijk, 1987, p. 6).

And if 'can such performative concerns be linked with emancipatory concerns?' is a genuine question then the argument by Sloterdijk above can provide the basis for a plausible - but ultimately *inauthentic* - answer (as Sloterdijk is well aware).

WORK IN PROGRESS

Brocklesby (1994) argues that TSI practitioners (and, perforce, advocates) are "... historically situated actors": "As members of particular groups, they have been acculturated into viewing the world in distinctive ways, and these meanings have a huge bearing on the doing of systems research." (Brocklesby, 1994, p. 82).

Although Brocklesby seem to be propounding a pluralist view of these 'accculturations', he judges that a key issue is whether or not the 'methodological choice' argued for in TSI is genuinely available to the TSI practitioners: "In much of the academic and practitioner literature, research is viewed as a rational act involving real choice ... but it is questionable whether this portrays an accurate picture of systems practice." (*ibid.*,p82)

This is an important point, and will form the focus of further research concerning TSI, which will be based on a more Nietzshean perspective than the work herein.

CONCLUSION

Provisionally it can be concluded that there is a tension between the aspirations of TSI advocates to emancipate people and their aspirations to *generally* improve organisational performance; TSI advocates must simply endure this tension, as it arises in a context of widespread cynicism within society. However this serves to make TSI readily acceptable to at least a significant subset of the population and these circumstances provide the necessary soil from which TSI continues to grow. One danger which results from these circumstances is that genuinely emancipatory methodologies may not be developed if TSI is itself seen as (genuinely) emancipatory - as such developments might then seem to be unnecessary. Another danger is that if intervention methodologies have to legitimate themselves by reference to (ostensible) improvements in organisational performance, then genuinely emancipatory methodologies may never appear at all: "Those who make the means of liberating reflection available and invite others to use them appear to conservatives as unscrupulous and power-hungry good-for-nothings who are reproached with "Others do the work." Very well, but for whom?" (Sloterdijk, 1987, p. 21).

And if there is to be *choice*, then the TSI advocates and practitioners must *choose* whom they are working for.

REFERENCES

Adorno, T. W., and Horkheimer, M., 1972, "Dialectic of Enlightenment," Verso, London.

Brocklesby, J., 1994, Let the jury decide: assessing the cultural feasibility of Total Systems Intervention, *Syst. Pract.* 7:75-86.

Flood, R. L., and Jackson, M. C., 1991, "Creative Problem Solving: Total Systems Intervention," John Wiley and Sons, Chichester.

Huyssen, A., 1987, The return of Diogenes as postmodern intellectual, *in:* "Critique of Cynical Reason," P.Sloterdijk, ed. Verso, London.

Jackson, M. C., 1993, Don't bite my finger: Haridimos Tsoukas' critical evaluation of Total Systems Intervention, *Syst. Pract.* 6:289-294.

Lyotard, J., 1984, "The Postmodern Condition," Manchester University Press, Manchester.

Sloterdijk, P., 1987, "Critique of Cynical Reason," Verso, London.

Tsoukas, H., 1992, Panoptic reason and the search for totality: A critical perspective of the critical systems perspective, *Hum. Relat.* 45:637-657.

Tsoukas, H., 1993(a), The road to emancipation is through organizational development: A critical evaluation of Total Systems Intervention, *Syst. Pract.* 6:53-70.

Tsoukas, H., 1993(b), "By their fruits shall ye know them": A reply to Jackson, Green and Midgley. *Syst. Pract.* 6:311-317.

76

SOME ANOMALIES IN ULRICH'S CRITICAL INQUIRY AND PROBLEM-SOLVING APPROACH

Norma Romm

University of Hull
Centre for Systems Studies
Newlands House
Hull
HU6 7TQ
United Kingdom

INTRODUCTION

This paper probes some of Ulrich's assertions with respect to the relevance of a Critical Systems Heuristics (CSH) approach to critical inquiry and problem solving, and it draws out certain anomalies in the position. The concept of 'anomaly' as used here is derived from Gouldner (1980). It refers to *internal contradictions* in a theory (as located by a critic) or points to 'information' which represents *bad news* for the theory (which the theoretician may be inclined to resist). 'Hostile information' or bad news evokes (perceived) events or ideas not accounted for in the theory and not in keeping with its premises - hence generating an external anomaly. Gouldner argues that all theories can be considered critically by locating anomalies (both internally and externally derived ones), thus providing impetus for reconsideration of theoretical positions and their implications for relevant practice. In this paper, anomalies are posed in confrontation with the texts of Ulrich, as part of an effort to enter into dialogue with, and extend, the work.

ACCESSING 'THE POWERFUL'

As Jackson (1985, p881) has noted, some of the questions that can be asked of Ulrich's texts are not confinable to a consideration of Ulrich's work alone - but indeed are questions which face many theorists (and practitioners) subscribing to some form of 'critical theory'. One of the questions which Jackson asks Ulrich - and by implication other 'critical' theorists - is the following:

> "The question remains - why should the involved [planners/designers of systems] bother to take account of the views and interests of those who are affected but not involved? The question of

Critical Issues in Systems Theory and Practice, Edited by K. Ellis et al.
Plenum Press, New York, 1995

> which class, group or agency has the power, the will and the interest to bring about the rational society [in critical theoretical terms] has bothered critical theorists throughout the twentieth century. .... Ulrich rather neglects this type of issue" (1985, p881).

This question has been similarly raised by other authors commenting on the hope that has been expressed by some critical theorists (cf. Ulrich 1986, p12; and 1991, p112) to 'embarrass' designers who fail to confront the normative implications of their designs. For instance, Ivanov (1991) asks:

> "What if those powerful agents who organise the negotiations and who ought to feel embarrassed by polemical questions do *not* feel embarrassed? What if they, instead, state for example that "we find that it is too expensive to bear the burden of proof that you want to impose on us" or "it is too expensive to elicit the voluntary consent by all those who might be affected by the system and its consequences", etc?" (1991, p45).

Flood and Ulrich, in spelling out a critical heuristics option, have suggested that:

> "Critical heuristics ... pays careful and explicit attention not to presuppose that those in control of "decision power" are willing to take account of the views and interests of those affected, but only that they are interested in making their own views and interests appear to be defensible on rational grounds" (1990, p24).

Flood and Ulrich assert that although a critical approach cannot force the powerful to take account of the less powerful, it can "... at least unveil this [the] facade of rationality and objectivity that is so characteristic of the strategic action of powerful vested interests ...." (1990, p24). They emphasise that the validity of a critical approach (and more specifically critical heuristics):

> "... must be judged primarily in terms of its potential in bringing about practical application and emancipation. It is therefore important to build in a facility whereby practical judgements can be constantly reflected upon in transparent nonexpert terms, and their partiality revealed by everyday accounts of the nature of social experience in ordinary language" (1990, p24).

But the problem with the above formulation of the task of a critical approach (as Ivanov notes) is that it still leaves unresolved the question as to why the plea to reflect upon the normative implications of systems designs (by involving nonexpert views) should be heard and seriously considered by planners/experts. For it seems that according to the premises of Ulrich's CSH approach, powerful vested interests are likely to seek to control the situation, rather than open themselves up to the requirements of a genuinely discursive rationality. Ulrich has postulated that

> "It will always be a minority of citizens that can be moved "with reason" to follow Habermas's appeal to the partiality [preference] for reason, and even they will not thereby become bloodless beings renouncing all but generalizable needs and interests" (1983, p151).

If it is posited that 'the powerful' are likely to be intransigent or at best to bargain (in terms of their greater power) for pseudo-compromises in decision-making, in what sense does critical heuristics offer theoretical provision for the practical application of an 'emancipatory' agenda - defined as improving the practice of discursive rationality? (Habermas has 'resolved' the question of the grounding of discursive rationality by suggesting that the (transcendental) human interest in emancipation is written as a historical possibility in our species (cf. Romm, 1991, pp134-144). Ulrich, in re-opening the meaning of this resolution, has opened a hornet's nest, where anomalies are nested!)

Anomaly 1 (Internally generated). Theoretical claim: Only a minority of citizens will ever be moved by, and willing to practise, discursive reason. Competing theoretical claim: Improvements towards generating more discursive rationality in human relations is possible through the intervention of critical theory. Suggested area for reconsideration of premises:

What view of human propensities is implied by the theory? *Why and when will 'the powerful' feel uncomfortable with serving particular interests?*

Let us now continue to consider the moral issue raised by Ulrich in regard to possibilities for the critical inquirer to intervene or mediate in the meeting of stakeholders with, as he perceives it, different interests.

POSSIBILITIES FOR MORAL INTERVENTION?

Babüroglu (1992) casts light on the issue by (re)interpreting Ulrich's position and considering its relevance in different contexts. He argues that in terms of Ulrich's position (as he interprets it), at certain points the inquirer is required to suspend the notion that the meeting between stakeholders can result in "...a mutually agreed upon and desired notion to proceed from or to strive toward" (1992, p286). [The suspension of the notion of mutuality would be consistent with Ulrich's suggestion that in many contexts not everybody's interests can be served - mutual agreement on the basis of a supposed generalisability of interests or even on the basis of genuine compromises, is difficult/impossible to achieve (cf. Ulrich, 1983, p393; 1988, p424; and Romm's discussion - 1995: section 6).] But what is the practical purpose of *this* theoretical position? Babüroglu's argument in this regard is that the theoretical suspension of the notion of achievable mutual agreement between stakeholders with defined interests, may lead the interventionist to consider ways forward by enabling stakeholders to "surrender control" (1992, p287).

Babüroglu thus argues that at a certain point it is necessary to design methodologies which 'enable surrender' (of control). The aim is to help people (all stakeholders) to relinquish their desire for control, and to introduce a liberation theme which "...exposes how oppression in many aspects of social system design and management can be removed" (1992, p287). The critical engagement here is aimed at "...trying to improve the human condition, in highly conflictual settings, particularly ones that tend to produce protracted and apparently unresolvable disputes" (1992, p288).

Significantly, Babüroglu refers to the disputes as "apparently unresolvable", thus implying that if people can be enabled to surrender control, a form of mutual accountability is possible. (Babüroglu - 1992, p287 - distinguishes this from a consensus orientation, but does not explore further what might be implied by the "...liberation from harmony and consensus orientation". McKay and Romm, 1992, pp124-126; Romm, 1994(a), pp333-336; and Romm, 1994(b), pp25-27, provide further detail on this.)

Babüroglu's account of how methodological intervention in cases of apparently impenetrable stances in society may be effected, varies somewhat from Ulrich's insistence that there may be situations where there *is* an inevitable conflict among different client groups (1983: 393). It is in such cases that Ulrich argues that it is incumbent on the inquirer/interventionist to "support" the disadvantaged as the 'key' client. However, it is also at this point that the problem arises as to what the intervention may entail. Ethically the problem can be resolved by pointing to the idea that, as Gregory puts it (1993, p207), "...the process is expected to result in a net improvement in the satisfaction experienced by participants of the system". But the question remains as to what theoretical provision is made for persuading 'the advantaged' to negotiate ways forward with those with whom (according to Ulrich's account) they are in unnegotiable conflict. In other words, how is 'the process' to which Gregory points, to be effected, once one has posited inevitable/unnegotiable conflict? How does Ulrich hope to engage the various stakeholders?

Anomaly 2 (Externally generated).Theoretical premise: When there is an inevitable conflict of interests the critical inquirer/intervenor must support the key client (whom s/he perceives as most disadvantaged) in order to help move the discussion towards a genuine consensus/negotiation. Bad news: If there is a conflict of interests and people are oriented in terms of fulfilment of predefined interests, they are unlikely to 'find' (through argument) a solution to which they will feel morally bound. *An appeal to people (including those seeing themselves as more powerful and those seeing themselves as less powerful) to be moved to 'fair' solutions implies that they cannot be oriented solely in terms of interests.*

Ulrich's texts are ambiguous in regard to the question of how various stakeholders may become committed to the project of generating more discursive rationality.

At some points in his work, he seems to argue that it is possible to:

> " ...secure [in theory?] to both sides an equal position for reasonable dialogue; each side, if only it renounces the dogmatic or cynical employment of boundary judgements, can now argue its case and work toward mutual understanding about the premises and consequences of planning, by advancing its own good grounds, i.e. "facts" and "values" capable of consensus such as existential needs of all individuals, ecological knowledge, ethical principles, principles of constitutional democracy, and so on" (1991, p113).

This line of argument is echoed in his statement (1994 pp31/32) that:

> "It is important ... that we provide them [participants] with the tools they need to argue their concerns in an authentic, openly subjective, but nevertheless cogent way. ... I believe that institutional arrangements, such as the planning cell, can provide ordinary citizens, and experts and designers, with occasions to train themselves in what I call critically-heuristic debate."

Ulrich seems to suggest that it may be possible to persuade people (including what he calls 'cynicists') to move toward 'mutual understanding' on the basis of arguments which lead towards the creation of some consensual arena for 'fair' decision-making. Nevertheless, it is unclear how his claims in this respect fit in with his insistence that the (dialogical) process is likely to be threatened in practice by the tendency of people to bargain towards a position that serves their interests. (See again Anomaly 2 above.)

At times, Ulrich calls for a normative re-education of 'the elite' in order to ground his belief that it may be possible to appeal to ethical and democratic principles in cases where there is an inevitable conflict among 'client groups' (1983, pp404-405). Thus he refers to the possibility of "...internalizing within the elite, through education or acculturation, the values that are to be promoted". He speaks about hoping to generate an 'outlook' in elites which may come to "...conform with the public interest" (1983, p404). He also speaks of "...raising the consciousness" of the elite (1983, p405). Does this mean that he believes that the tendency of people to be pseudo-dialogical and to bargain in protection of interests, can be undercut through normative education? In other words, is this tendency learned behaviour which can be shifted through alternative learning processes - such as those which concentrate on developing norms of discourse and enable people to relinquish what Babüroglu calls a "control" mentality? (cf. also Romm and Romm, 1987; and Romm, 1994(b), pp25-27).

At times (cf. Ulrich, 1994, p31) Ulrich speaks of constructing "...institutional arrangements that would enable concerned citizens to confront [through debate] proposed designs with culturally different future visions". He cites Ackoff's work as relevant in this regard. The problem, though, as Ulrich is aware, is that institutional arrangements can be manipulated by people, unless these are rooted simultaneously in a 'new' type of consciousness/ethical stance. Democratic structures can become a playground for manipulators, unless the institutions are grounded in commitments to forms of discourse.

Anomaly 3 (Internally generated). Theoretical claim: A 'sweep-in' process is needed to appreciate (normative) implications of action from different viewpoints. Competing theoretical claim: Different viewpoints will never be given fair treatment or appreciation by contenders in debate, because most people are not moved by reason. Suggested area for reconsideration of premises: Why should we assume that people will become moved by an ethic which focuses on consensus-generation? *Are there alternative ethical principles that can be invoked to ground the appeal to engage in 'discourse'?*

THE FOCUS ON CONSENSUS RECONSIDERED AND SOME INTERVENTION OPTIONS OUTLINED

Gurevitch's analysis of what he calls the "...meeting of conflicting identities" (1989) offers a useful indication of what may be involved in surrendering the quest for 'mutuality' (as control). What is required is that the parties recognise that they will never be able to reach a consensually shared understanding, precisely because they have to accord the right of the 'other' to be 'other', or, in his terms, 'strange'. Gurevitch refers to cases where participants with long-standing differences were helped/enabled to recognise the importance of *not* understanding. He emphasises (1989, p71) that:

> "What became evident in the group process was the existence of more than one authoritative origin of meaning, truth, and justice. This realizaton bore in it the ability to not understand [that is, the recognition that common understanding in a consensual perspective could not be attained]."

Gurevitch goes on to note that:

> "...the participants developed no deep sense of understanding, empathy or identification with the other, but rather a sense of having retreated a step from their former attitudes. ...The participants came to a realization of their own [initial] closure and translated these insights into possible applications..."(1989, p171).

Gurevitch's argument/discussion of cases is supported by Gregory, Romm and Walsh's study of quality initiatives in the National Health Service in a region of the UK. The aim of engaging participants was to engender shifts in perspective on the part of participants who were all confronted with challenges as levelled by other participants - including the researchers/facilitators as participants.(See the brief account of this in terms of its relation to Ulrich's position in Romm, 1994(b), p21, footnote 1; and see Gregory and Romm's discussion (1994) of the facilitation approach which was adopted to aid dialogical confrontation.) The process involved in this study concentrates on the need to generate increased awareness of the other, so that people can meet and act as 'others'. Gurevitch sums up the (ethical) requirement as follows: "The very premise on which dialogue and dialogic ethics is founded is the meeting, not the denial of each other as others" (1989, p172).

Of course, an ethical requirement has here been set out in terms of a hope that people may become committed to developing more dialogical social relations (as defined briefly above - for a fuller discussion, see Romm, 1994(a), pp329-334). Ulrich has made some theoretical provision for this in his notion of 'internalizing within the elite' (and presumably within others) values that are to be promoted (such as the value of discourse). His focus, however, has not been on outlining the theory or practice of this re-education process (nor on what may be implied by a dialogue which encapsulates opposition). While he at times refers to normative learning, he seems to concentrate on impediments to learning as necessarily rooted in the tendency of 'the powerful'" as a rule to seek control in terms of their perception of their interests (cf. 1983, p151). (See also Anomaly 1.)

Romm (1995: section 7) has indicated that at the points in his texts where Ulrich postulates the likelihood of elite intransigence, his approach (and consequent intervention) may run into the danger of becoming a self-fulfilling prophecy (which has not been distinctly reflected upon as such). The (political) action of those 'at the receiving end' who believe that they will have to use 'counterforce' to move their undialogical opponents, may indeed contribute to setting up an unnegotiable conflict. Insofar as the possible effects of this form of intervention are theorised and reflected upon, Romm contends that it may become a practical option, which could be defended in terms of (certain) participants (and the researcher's) assessment of the possibilities for 'fair' improvements. But this would require a sensitivity to the way in which the construction of the situation may become self-fulfilling - and possibly counterproductive. A self-reflective approach requires a consideration of the way in which the inquirer's visioning itself may influence (and constrain) intervention practice. This in turn means that researchers cannot appeal to a situation facing addressal, independently of their own involvement in defining 'the situation' (cf. also Taket's account of the dilemmas raised by the recursive relation between knowing and acting - 1994, p129).

Anomaly 4 (Externally generated). Theoretical claim: When there is an inevitable conflict of interests the critical heuristic inquirer must support the disadvantaged as key client, and help them to protect their concerns, as the advantaged are prone to protect their own interests. Bad news: In perceiving 'the situation' as one which renders futile the attempt to shift the contending positions via (forms of) argument, the researcher may contribute to creating a climate where control mentalities are reinforced. This is especially so if unnegotiable conflict of interests has been predefined by the theory. Hence: the relation between knowing and acting has to be theorised in terms of a notion of *critical inquirer responsibility* (with no recourse to 'inevitabilities') .

There is another possibility for practical intervention which Flood and Romm (1995) have isolated as an option which requires consideration by practitioners. This is the possibility of operating *obliquely* with an emancipatory agenda - addressing 'the [officially] powerful', but not appealing to them to relinquish control by virtue of ones introduction of a normative agenda. Rather the interventionist may attempt in certain (perceived) situations, to shift power relations by aiding 'the powerful' to recognise their dependence on the 'less powerful'. In order to prevent their own self-destruction and the self-destruction of their would-be designs, one can help people to shift their conception of their interests towards a consideration of others' inputs. In this form of (emancipatory) intervention, the practitioner makes a choice to utilise the interaction with the various clients to address 'the powerful' in terms which could support those 'less powerful'. The purpose is to shift a power relation which the intervenor has assessed as coercive. However, in such cases the avenue for addressing the powerful may be opened not through the plea for discourse in explicit moral terms, but through the threat of self-destruction (in the powerful decision-makers' terms). In many instances it can be argued that self-destruction of decision-power is linked to the failure to recognise the fragility of power. This failure in turn is linked to the belief that one can act on the basis of ones particular conception of the 'realities' of the situation, forgetting the propensity of others to assert their impact in defining 'the situation' (cf. Romm, 1994(a), pp332-333). Flood and Romm (1995) discuss the implications of this for creating a space where researchers may operate oblique forms of emancipatory intervention to open a discussion on practices. Such intervention consists in addressing 'the powerful' as part of ones clientele, using methodologies designed to address decision-power, but operating the methodology through an 'angle' which has not been theorised by self-named proponents, nor been built into their guiding framework for intervention. The argument is based on the idea that decision-power should not be disregarded under the guise that critical intervention becomes sullied by involvement with 'powerful' clients; or rather it is based on the suggestion that such a disregard may limit the choice-making range of the interventionist unnecessarily, and in ways that have not been substantially theorised.

CONCLUSION

Although Ulrich poses his critical systems heuristics as a critical inquiry and problem-solving approach in addressing the design of systems, the practical suggestions for emancipatory intervention fit uneasily with some of his theoretical premises. This paper has located some of the points of tension in the theory - and offered a way of confronting them. This does not finalise the discussion on the argument, but indeed calls forth further responses to re-address and thereby redefine the relevance of his work.

REFERENCES

Babüroglu, O., 1992, Tracking the development of the Emery-Trist Systems Paradigm, *Systems Practice,* 5:3, pp263-290.

Flood, R.L., and Romm, N.R.A., 1995, Enhancing the process of choice in TSI, *Systems Practice* (in press).

Flood, R. L., and Ulrich, W., 1990, Testament to conversations on critical systems thinking between two systems practitioners, *Systems Practice*, 3:3, pp7-29.

Gouldner, A.W., 1980, "The Two Marxisms", Macmillan, London.

Gregory, W.J., 1993, Designing educational systems, *Systems Practice,* 6:2, pp199-209.

Gregory, W.J, and Romm, N.R.A., 1994, Developing multi-agency dialogue: the role(s) of facilitation, *Working Paper 6*, Univ. of Hull, Centre for Systems Studies.

Gurevitch, Z.D., 1989, The power of not understanding, *Jnl Appl. Beh. Sci.* 25: 2, pp161-173.

Ivanov, K., 1991, Critical Systems Thinking and Information Technology, *Jnl Appl. Syst. Anal.*, Vol.18, 39-55.

Jackson, M.C., 1985, Book review on W. Ulrich: The itinerary of a critical approach, *J. Opl Res. Soc.*, 36:9, pp878-881.

McKay, V.I., and Romm, N.R.A., 1992, "People's Education in Theoretical Perspective", Maskew Miller Longman, Cape Town.

Romm, N.R.A., 1991, "The Methodologies of Positivism and Marxism", Macmillan, London.

Romm, N.R.A., 1994(a), Symbolic theory, *in*: "Social Theory", N.R.A. Romm and M. Sarakinsky, eds, Heinemann, Johannesburg, 325-339.

Romm, N.R.A., 1994(b), Continuing tensions between soft systems methodology and critical systems heuristics, *Working Paper 5*, Centre for Systems Studies. University of Hull.

Romm, N.R.A., 1995, Knowing as intervention, *Systems Practice*, 8:2 (April).

Romm, N.R.A. and Romm, N.L., 1987, Militarizing tolerance, *De Arte,* Vol. 36, pp23-26.

Taket, A., 1994, Undercover agency? - ethics, responsibility and the practice of OR, *J. Opl Res. Soc.*, 45: 2, pp123-132.

Ulrich, W., 1983, "Critical Heuristics of Social Planning", Paul Haupt, Bern and Stuttgart.

Ulrich, W., 1986, Critical heuristics of social systems design, *Working Paper 10*, Dept. of Management Systems and Sciences,University of Hull, Hull.

Ulrich, W., 1991, Critical heuristics of systems design, *in*: "Critical Systems Thinking", R.L. Flood and M.C. Jackson, eds., John Wiley & Sons, Chichester, 103-115.

Ulrich, W., 1994, Can we secure future-responsive management through systems thinking and design?, *Interfaces, 24:4, pp26-37.*

77

AN ETHICAL BASIS FOR CRITICAL SYSTEMS THINKING

Communicative or Communitarian?

M. W. J. Spaul

Anglia Polytechnic University
Chelmsford
Essex, CM1 1LL
United Kingdom

INTRODUCTION

In many respects the systems designer is in the position of a legislator, responsible for the structure of institutional arrangements within which people will spend their working lives; and with the same moral duty to ensure that those arrangements embody some conception of social justice. Like legiwlators, system designers have been motivated by a range of political philosophies: utilitarian calculation by expert elites (Poundstone, 1993), scientific paternalism (Beer, 1981:245-347), and libertarian democracy (Checkland, 1981:283). The distinguishing feature of Critical Systems Thinking (CST) has been its adherence to the principle that justice is best achieved in participatory processes in which power differentials have been recognised and neutralised. The specific participatory arrangements envisaged by critical systems thinkers have, in the main, been modelled on the developing philosophy of Jurgen Habermas (Habermas, 1971, 1984, 1987). The purpose of this paper is to consider whether the interests of CST are best served by such an exclusive allegiance.

The work of Habermas has many of the ethical characteristics of liberal modernity: a separation of the realm of morality, justice and the common good from the realm of private values and preferences, and an attachment to the achievement of justice through the application of neutral procedure. Precisely these characteristics make Habermas' work congenial to the systems tradition with its application of methodology to impersonally specified 'problem situations', attempting to engineer favourable outcomes - technical soundness, efficiency and justice - by procedural means. The ethic of liberal modernity has been under attack in recent years (a periodic revival of a dispute which dates at least from the time of the enlightenment) from a constellation of theories labelled as 'communitarian' or 'neo-Aristotelian' ethics (Sandel, 1982; MacIntyre, 1985; Taylor, 1989). The common characteristic of these ethical positions is a radical questioning of the assumption that a coherent theory of justice, or a detached 'moral point of view' can be articulated in isolation from a consideration of private values and a conception of the 'proper ends' of human life.

Critical Issues in Systems Theory and Practice, Edited by K. Ellis et al.
Plenum Press, New York, 1995

The failure of liberal assumptions compromises a procedural conception of justice or morality, with obvious consequences for the systems tradition: the failure of methodical neutrality to carry with it a coherent ethical stance.

THEORETICAL BACKGROUND

Procedural Conceptions of Morality and Justice

The proceduralism of deontological ethics and liberal politics is rooted in Kant's (1956) view of the human subject. At the heart of Kant's position is a view of the self characterised by a universal, rational core separate from the social ties and value preferences which may be contingently attached to that core. This image sustains a number of important conclusions about human moral and political deliberation (Sandel, 1992). The distinction between the essential, rational self and its contingent encumberances licenses a distinction between the proper domain of morality, and the public notion of justice, and the domain of 'ethics' as a sphere for the exercise of private value preferences and tastes. Morality and justice are associated with the universal characteristics of all rational subjects, and their principles are reached by exercising the rational faculties. This leads to a definition of moral rightness and justice in terms of a deliberative procedure which tests principles for their consistent universalisability. The transcendental Kantian conception of the self has not, generally, survived modern scrutiny; however, as Rawls (1972) showed, the basic architecture of the Kantian position can be reconstructed assuming only a set of social goods which are deemed to be universally desirable, and the description of the human agent underlying 'rational choice' theories of economics - that human beings are rational utility maximisers. Under these assumptions, Rawls was able to show that a convincing procedural definition of distributive justice can be cast as what all people would choose as an optimum social arrangement if they were not aware which position in society they were to occupy. Again, this procedure is only designed to deal with a public conception of justice, it being assumed that private variations in value preferences need not enter into a generalised calculative procedure.

Habermas' (1990) discursive definition of moral rightness is the result of some major modifications to a liberal position such as Rawls'; in particular, the individual reasoning subject is discarded in favour of a self consitituted by communicative exchange. Both Kant (1956) and Rawls (1972) allow the deliberative procedure defining moral rightness and justice to be exercised by a single, rational subject - since the shared rational core of separate selves should always reach impartial conclusions. Habermas (1990) replaces this with the demand that any procedural definition of rightness and justice must be based on the outcome of the cooperative deliberations of all the real, social selves - with all their desires and attachments - having any interest in the outcome of applying the moral norm in question. Although the liberal approach to the rational debate of the public good is retained, the universal principles of subject-centred rationality are replaced by universal principles of rational argumentation. The warrant for assuming that consensus can, in some theoretical limit, be reached is shifted from the rationality inherent in the subject to an orientation to impartial discourse inherent in language (Habermas, 1990:43-109). The emergence of concrete processes of impartial debate is regarded as the product of social evolution and personal development (Habermas, 1990:116-188).

The Communitarian Critique

The basis of the communitarian attack on the abstract proceduralism of the liberal ethic is the questioning, with Hegel, (see Taylor, 1977) of a set of fundamental oppositions between public and private spheres. On the public side of the supposed divide are those aspects of social life susceptible to debate and reason: public morality, justice and the common good. On the other are subjective matters of private choice: personal morality and value choices, a conception of the 'good life' and an ordering of domestic existence. Only if this division can be maintained is there a plausible case for the operation of reason in the determination of moral principles. There is, however, telling evidence against such a division. MacIntyre (1985:6-12) catalogues a set of ethical arguments (such as those over abortion or welfare provision) which are 'rationally interminable'; in which the protagonists cannot find grounds for reasoned conclusions. Similarly, governments find it increasingly difficult to enact legislation which arbitrates successfully between the competing claims of different interest groups - an example being the 1994 Criminal Justice Bill's inability to accommodate the divergent interests of youth, travelling populations and the middle classes. The communitarian conclusion is that a realm of reasoned morality and justice is a fiction, that public morality is anchored in personal values and a conception of the good life, and that such a conception can only be nurtured in a community which pursues shared aims. The orientation towards moral communities provides important common ground with critical theory, as the formations of modernity targeted by the Frankfurt School (Held, 1980) - international capital, the military-industrial complex, bureaucratic government - are seen as destructive to organic ways of life.

The basic communitarian position may be elaborated in various ways, not all of them compatible. MacIntyre (1985:62-78) diagnoses the societies of late modernity as failing to be in 'good moral order', and interprets the lack of a unified ethical perspective as a serious defect. His stance is associated with a certain archaism, privileging the supposed unity of the lost ways of life of ancient Athens or medieval Christendom. In contrast, Taylor (1989:495-521) views the late modern inheritance of incommensurable ethical traditions in a positive light, as part of an enlarged moral perspective. Loss of unity is seen as a penalty to be paid for genuine choice. However divergent the conclusions of communitarians there is a commonality of theoretical method: a historical and ethnographic focus on particular forms of life and ethical perspective. This emphasis on the exploration of concrete social arrangements is allied to Gadamer's emphasis on *phronesis* in ethical judgement (Gadamer, 1975:278-289), and the insight that the foundations of an ethical orientation lie below the level of rational examination and depend on the embodied skills of our social behaviour (Dreyfus and Dreyfus, 1990).

Feminist Perspectives on Procedural Morality

The above, highly schematic, opposition between liberal proceduralism and communitarianism is complicated by a variety of feminist perspectives on morality and justice. The long-standing feminist opposition to the distinction between a public and private sphere is based on the oppressive consequences which result for women who have traditionally been confined to the domestic realm (Pateman ,1989:118-140); the exclusion of private, domestic concerns from public sphere has served to suppress examination of the injustices inherent in patriarchal family structures. Despite the similarity of this critique and that of the communitarians, feminist writers have not - for obvious reasons - been so ready to express nostalgia for the loss of traditional ways of life; rather, a search for entirely new forms of community has been stimulated (Friedman, 1992). A related, but more specific, attack on masculine bias in Kohlberg's (1981, 1984) conception of post-conventional, universalist moral reasoning (which is central to Habermas' discourse ethics, (see Habermas, 1990:116-188)), has been made by Gilligan (1982). Gilligan points out that a universalist ethical orientation, based only on the impartial giving and considering of reasons, is a predominantly masculine trait. In contrast, mature feminine moral reasoning exhibits a greater degree of particular attachment and sympathy, but with a recognition of the dilemmas which attachment can bring. This basic contrast has been developed by Fraser (1989) into a more broadly political critique of the 'gender blindness' of Habermas' position. Following Gilligan (1982), Benhabib (1992) has attempted to redefine Habermasian discourse ethics to provide a universal ethical perspective based on concrete, unconstrained encounter; whilst Young (1990:156-191) has argued that a universal perspective is not a reasonable expectation and that stalled decision-making in the face of partiality and difference may, in itself, be a just outcome.

INSTITUTIONS AND ETHICAL THEORY

Deontological ethics or principles of distributive justice based on monological reasoning, in the manner of Kant (1956) and Rawls (1972), are assumed to find a direct application in a variety of institutional contexts, whether as a guide for a social legislator or in the deliberations of a systems practitioner considering the ethical outcome of a proposed design. Habermasian proceduralism works more subtly, as a constraint on the forms of discourse permissible between those involved in, or affected by, any proposed change in institutional arrangements if the outcome is to be deemed ethically acceptable. This subtlety engenders difficulties in applying the procedures of discourse ethics in concrete situations, in which - using Habermas' categories of social action - communicative action is intertwined with strategic action. There are two principal methodological reactions to these difficulties. The first, favoured by Habermas as the means by which systemic imperatives may be brought under the control of the lifeworld (Habermas, 1992), is to encourage the creation of 'public spheres' of discourse which function outside institutionalised strategic action (Habermas, 1984). As support for this suggestion Habermas indicates various historical precedents (see Calhoun, 1992). The second, recognised by Habermas as active in parliamentary and legal deliberations, and favoured by critical systems thinkers (see, e.g., Ulrich, 1983) has been the operationalisation of the constraints of communicative discourse in situations in which the exercise of power or resource limitations prevent ideally open-ended debate. The assumption common to both of these routes is, however, that it makes sense to identify, in freely-formed public spheres or controlled discourse, an orientation towards the public good; a public good which can be generalised and the subject of universal consensus.

From a communitarian standpoint, the possibility of achieving a workable consensus from debate carried out under Habermas' discourse constraints is conditioned by the liberal assumptions which legitimate it. The problem is not simply that of instituting arrangements which allow the public good to be identified and pursued, but the existence of a unitary public good. If conceptions of the public good are as diverse as personal conceptions of the good life, then there is little reason to suppose that rational debate, however protracted and widely participative, will produce consensus. Once this point has been conceded, the possibility of identifying any neutral, procedural position which might be adopted by an agent acting in the impartial role of judge or facilitator is lost. In place of this impartiality there has to be, for the communitarian, a commitment towards the conditions for nurturing moral communities which will develop, over time, a coherent ethical standpoint. If communities are small and organic, and are characterised by a high degree of citizen participation, it is assumed that a coherent ethic, both concrete and partial - like that of ancient Athens - will evolve and fill the ethical vacuum of procedural neutrality (see Borgmann, 1993). A systems practice rooted in communitarian ethics would be fored to abandon a methodological stance, in favour of experimentation with participative structures and the formation of concrete, normative models (see, e.g., the experiments in participation in Pateman, 1970).

Despite the current popularity of this response to the current political malaise of the West (see Giddens, 1994), simplistic communitarianism does not capture adequately the enlarged moral perspective of modernity. The retrieval of a world of Greek city states is not a 'live' option (Williams, 1985:161-167); and if it were, then the problem of just accommodations between them would be a pressing problem in a densely interconnected world. A model which attempts to bring together Benhabib's (1992) extended form of Habermasian discourse and the communitarian recognition of irreducible differences in forms of life has been proposed by Young (1990:226-256). In a sketch of a normative ideal for a politics of difference she proposes a focus on the best facets of the post-modern city, in which diverse ways of life (Hannerz, 1992) can coexist in a public space, without a forum of universal reason being required to ensure harmony. Again, a systems practice based on this work could not be expected to focus on abstract methodology.

CONCLUSION

The ethical uncertainties of the contemporary Western world have an impact at any level of planning and decision making. Liberal models of moral deliberation, however sophisticated and dialogic, are no longer an adequate response to the non-homogeneous publics who are subject to our designs. The communitarian response to this situation seems partial and inadequate, and there is a pressing need to develop frameworks, like Young's (1990), for considering ethical problems arising from systems design which do justice to the cultural complexity of the world which we now inhabit.

REFERENCES

Beer, S., 1981, "The Brain of the Firm," John Wiley & Sons, Chichester.

Benhabib, S., 1992, "Situating the Self," Routledge, London.

Borgmann, A., 1993, "Crossing the Postmodern Divide," Chicago University Press, Chicago.

Calhoun, C., 1992, "Habermas and the Public Sphere," MIT Press, Cambridge, Mass.

Checkland, P., 1981, "Systems Thinking, Systems Practice," John Wiley & Sons, Chichester.

Dreyfus, H., and Dreyfus, S., 1990, What Is Morality?, *in:* "Universalism Versus Communitarianism," D. Rasmussen, ed., MIT Press, Cambridge, Mass.

Fraser, N., 1989, "Unruly Practices," Polity, Cambridge.

Friedman, M., 1992, "Feminism and Modern Friendship," *in:* "Communitarianism and Individualism," S.Avineri and A.de-Shalit, Oxford University Press, Oxford.

Gadamer, H., 1975, "Truth and Method," Sheed and Ward, London.

Giddens, A., 1994, What's left for labour?, *New Statesman and Society*. 30th Sept.

Gilligan, C., 1982, "In a Different Voice," Harvard University Press, Cambridge, Mass.

Habermas, J., 1971, "Knowledge and Human Interests," Heinemann, London.

Habermas, J., 1984, "The Theory of Communicative Action, Vol. 1," Heinemann, London.

Habermas, J., 1987, "The Theory of Communicative Action, Vol 2," Polity Press, Cambridge.

Habermas, J., 1990, "Moral Consciousness and Communicative Action," Polity Press, Cambridge.

Habermas, J., 1992, Further Reflections on the Public Sphere, *in:* "Habermas and the Public Sphere," C.Calhoun, MIT Press, Cambridge, Mass.

Hannerz, U., 1992, "Cultural Complexity," Columbia University Press, New York.

Held, D., 1980, "Introduction to Critical Theory," Polity Press, Cambridge.

Kant, I., 1956, "Critique of Pure Reason," Bobbs Merrill, New York.

Kohlberg, L., 1981, "Essays in Moral Development, Vol. 1," Harper and Row, New York.

Kohlberg, L., 1984,"Essays in Moral Development, Vol. 2," Harper and Row, New York.

MacIntyre, A., 1985, "After Virtue," Duckworth, London.

Pateman, C., 1970, "Participation and Democratic Theory," Cambridge University Press, Cambridge.

Pateman, C., 1989, "The Disorder of Women," Polity Press, Cambridge.

Poundstone, W., 1993, "The Prisoner's Dilemma," Oxford University Press, Oxford.

Rawls, J., 1972, "A Theory of Justice," Oxford University Press, Oxford.
Sandel, M., 1982, "Liberalism and the Limits of Justice," Cambridge University Press, Cambridge.
Sandel, M., 1992, The Procedural Republic and the Unencumbered Self, *in:* "Communitarianism and Individualism," S.Avineri and A.de-Shalit, Oxford University Press, Oxford.
Taylor, C., 1977, "Hegel and Modern Society", Cambridge University Press, Cambridge.
Taylor, C., 1989, "The Sources of the Self," Cambridge University Press, Cambridge.
Ulrich, W., 1983, "Critical Heuristics of Social Planning," Haupt, Bern.
Williams, B., 1985, "Ethics and the Limits of Philosophy," Fontana, London.
Young, I., 1990, "Justice and the Politics of Difference," Princeton University Press, Princeton.

WORKING WITH HETEROGENEITY

A Pluralist Strategy for Evaluation

Ann Taket[1] and Leroy White[2]

[1] School of Education and Health Studies
South Bank University
103 Borough Road
London SE1 0AA
United Kingdom
[2] School of Mathematics, Statistics and Computing
University of Greenwich
Wellington Street
London SE18 6PF
United Kingdom

INTRODUCTION

Our experience in evaluation has been that we need to find ways of working in situations which have a high degree of variety and in which acceptance and respect for difference is important. Such situations display a high degree of heterogeneity, in a number of different senses: the differences between the people involved in the situation; differences between groups; and lack of stability in the environment. In contrast, most well-established evaluation strategies expect a stable hypothesis to be under test, do not recognise more than one perception of a situation, and are based on the implicit assumption that a unitary description of reality can be offered. The starting point for the paper will be a critical analysis of a particular strategy for evaluation arising from within the systems tradition. The concept of heterogeneity will be explored and the paper will present an alternative way of approaching evaluation, which we have found particularly valuable for use where a high degree of variety exists. The pluralist strategy described is based on the acknowledgement and respect of difference, rather than its rationalisation. Elements of this strategy for evaluation will be illustrated with examples from recent work in the social policy arena.

EVALUATION IN THE SYSTEMS TRADITION

One type of approach to evaluation that has come out of the systems tradition is contingency based. Contingency based approaches are those where a suitable evaluation methodology is selected according to characteristics of the context, function and stakehold-

Critical Issues in Systems Theory and Practice, Edited by K. Ellis et al.
Plenum Press, New York, 1995

ers involved, i.e. the systems involved. Here, Gregory and Jackson's (1992) approach will be discussed; this is of particular interest since it uses the cybernetic notion of 'variety', which we would like to contrast with our notion of heterogeneity. It is also of interest, since it was derived in the voluntary sector, where most of our work takes place. Within Gregory and Jackson's approach, the context for the evaluation is classified according to two factors: the approach (objectivist or subjectivist) adopted to organisations by the 'Evaluation Party' (or EP, defined as: 'the group of people directly concerned with carrying out the evaluation'); the Evaluation Party's 'variety' (assessed according to the size of the Evaluation Party, their level of knowledge about evaluation, priority of evaluation, group integrity and level of interference from outside elements, and classified as either 'high' or 'low'). A selection from four different types of evaluation methodology is then made using the contingency table shown in Table 1.

There are several problems with this approach; the discussion here will be in terms of Gregory and Jackson's model, similar critiques can be offered of the other contingency based approaches (e.g. those based on total systems intervention (Flood and Jackson, 1991) or system of systems methodologies (Jackson and Keys, 1984; Jackson, 1990). The first of these is the implicit assumption that the classification of the EP is static, in other words assuming homogeneity through time. The second problem is the assumption that a choice between subjectivist and objectivist outlooks is required, or even possible; the view of objectivist/subjectivist as an either/or choice can be unhelpful. This is a problem because of its assumption about a lack of heterogeneity within the group. The final major problem with the approach is with the different types of evaluation in the boxes in Table 1. As Gregory and Jackson (1992) themselves argue, the systems-resource and multi-actor forms of evaluation implicitly assume a consensual organization-environment relationship, arguably unlikely to be present in the voluntary sector, which is a heterogeneous environment. There are also difficulties with each type of evaluation in dealing well with questions of power and

Table 1. Gregory and Jackson's contingency table of evaluation approaches.*

	EP approach to organisations	
EP variety	Objectivist	Subjectivist
High	Systems-resource based *(examination of programme's ability to survive and adapt in a dynamic environment)*	Multi-actor based *(examination of ability to satisfy needs of all those parties influenced by and having an influence upon the programme)*
Low	Goal-based *(examination of achievement of prescribed programme goals)*	Culture-based[a] *(examination of programme's ability to realise its potential whilst maintaining core identity)*

Key:
EP = Evaluation Party ("the group of people directly concerned with carrying out the evaluation")
EP variety - assessed according to the size of the EP, their level of knowledge about evaluation, priority of evaluation, group integrity and level of interference from outside elements
[a] This form of evaluation is argued by the authors to be the least well-developed; in later papers it is not distinguished separately, see for example Gregory 1993.

(*) based on Table 2 in Gregory and Jackson, 1992.

the possible lack of equal relationships/status between members of the EP and/or the organization under evaluation. We return to a number of these points later.

A PLURALIST STRATEGY FOR EVALUATION

One key feature is the notion of triangulation; in this context to be understood as the use of a range of sources of information and the examination of relevant issues by a range of different direct and indirect methods. The indirect method of inquiry is based on looking for 'traces': evidence of effects of change in desired directions. The value of this strategy is best illustrated by means of a simple specific example: it is argued that it is insufficient to find out about people's views on the effects of the project's activities by asking them (solely) direct questions about these activities. While this will yield some relevant information, people may not connect changes in aspects of their everyday experience and social practices to the project's work, they may think of it in other terms, thus underestimating the effects of the project's activities. An indirect series of questions which explores people's accounts of the changes in their everyday experience and social practices that have occurred can be used and the connections to project activities examined in the analysis stage. Triangulation is applying heterogeneity in sources of data and methods of analysis. A further example of heterogeneity is the use of a heterogeneous evaluation team, which provides further opportunity for triangulation during all stages of the evaluation.

In adopting a pluralist strategy for the evaluation, we seek to respect and acknowledge the views of a wide range of stakeholders, i.e. the heterogeneity within the group. The danger with this perspective however is its tendency towards relativism, as if any one perspective is as valid as another. When views are reinforced by artificially constituted boundaries, such as those of race or ethnicity, then the evaluators need to separate this out in the course of the analysis and explicitly identify the consequences of this for evaluation conclusions. Our practice is to provide a space where participants can acknowledge the differences of others, and which will also allow them to articulate what before remained private and personal and perhaps even unsymbolised. In terms of our analysis not all sources of information used for triangulation are given equal weight, and most importantly, the situated work of accounts generated in different contexts will be recognised.

The type of pluralist strategy we propose is not without potential criticisms. Firstly, does it not amount to the possible use of methods based on incompatible theoretical assumptions? The position advanced here is that such a pluralist strategy might be thought of under the heading of 'pragmatic pluralism', and justified as an appropriate response to recognising evaluation as a social process (rather than a scientific experiment remote from every day life), which typically involves a variety of groups and individuals who may have conflicting theoretical beliefs, unlikely to be reconciled within the span of the evaluation. The argument offered here is closely related to part of the argument advanced by Legge (1984), who presents one potential resolution in terms of viewing the evaluator's role as that of a client-serving professional, where appropriate validity claims are made in terms of ability to mediate the clients' problems. Similar arguments within the health promotion context are advanced by Thompson (1992). There are also strong similarities to the arguments Fox (1991), (discussing the evaluation of health care services), puts forward for adopting a postmodern position which rejects the idea of a single rationality in favour of the recognition of rationalities.

Another aspect to the strategy we propose is drawn from the Rapid Appraisal methodology produced from development (McCracken et al., 1988; White, 1994). It can be described as a participative semi-structured way of learning, quickly and in a multi-disciplinary team, from a variety of stakeholders about relevant issues. The RA approach is:

- iterative, where issues are modified and refined as information is progressively revealed;
- semi-structured, providing the flexibility to accommodate both the specific subject of analysis and the wider socio-cultural environment;
- qualitative, emphasising perceptions and values of those involved rather than measuring and counting, using triangulation for different sources;
- value sensitive, surfacing the different values of those involved.

A key concept is that of 'optimal ignorance' whereby the amount and detail of information is required to formulate a useful analysis of the situation, in a limited time period, are expenses to be kept to a minimum.

EVALUATION IN THE SOCIAL POLICY ARENA

We can consider examples of programmes based on a number of voluntary sector projects with which the authors have been involved in London and Belize. All of the projects are based on what can be referred to as a community development approach, where active involvement of the 'target' group in shaping the intervention is sought, using a framework referred to by some as action research. The projects work with 'target' groups in the community to enable them to identify and articulate their needs, and then explore ways of meeting those needs. Each project involves the delivery of different 'programmes', which may change over time as the project progresses.

Application of a contingency based approach to the problem of evaluation would certainly not have been feasible using the scheme of Gregory and Jackson (1992) described above. The projects considered here exemplify all the difficulties discussed earlier. In particular, rather than being able to identify an objectivist or subjectivist bias in the Evaluation Party, it was clear that for all of the projects it would be necessary to satisfy both objectivists and subjectivists. In some cases, external evaluation was imposed on projects through the regular reports required as a condition of funding. Both objectivist and subjectivist assumptions could be detected underlying the different criteria included, illustrating the pluralist nature of the evaluation required. It was also noticeable that the criteria strayed across into each of the four different types of evaluation set out in Table 1.

Experience in these projects indicates that the characteristics of the EP tend to vary: typically a steering, advisory or consultative group is formed to guide/oversee the evaluation, and the characteristics of this group will vary over time (due to permanent changes in personnel or temporary absences). Variation may also result from external stimuli, for example requirements or requests from individuals or bodies not part of the EP, but to whom the EP needs to make some response. The dynamic nature of the EP referred to above makes it difficult to identify a static bias in the EP, and experience has typically been that project evaluations must often satisfy both objectivists and subjectivists, if they are to be widely acceptable. Externally imposed evaluation criteria are often highly objectivist, while there is often also recognition that they 'do not tell the whole story'.

The pluralist strategy described above however enabled these challenges to be met. In particular, parts of different OR/systems methods were combined in a process that might be labelled 'judicious mix and match' (White and Taket, 1993; Taket, 1993), which was tailored to fit the requirements of each particular situation, and continually reshaped in response to insights from the evaluation party. This requires a process of ongoing critical reflection on the part of the evaluator(s), plus intermittent group critical review.

DISCUSSION

It should be noted that the pluralist strategy discussed above is a rather different approach to the complementarism of the Gregory and Jackson contingency approach (Gregory and Jackson, 1992; Gregory, 1993). A pluralist strategy is not without potential criticisms. Firstly, does it not amount to the possible use of methods based on incompatible ontological or epistemological assumptions? In which case, does this represent a problem only from a strongly positivist standpoint? The position advanced here is that such a pluralist strategy might be thought of under the heading of 'pragmatic pluralism', and justified as an appropriate response to recognising evaluation as a social process (rather than a scientific experiment remote from every day life), which typically involves a variety of groups and individuals who may have conflicting ontological and epistemological beliefs, unlikely to be reconciled within the span of the evaluation.

Connected to this adoption of a pluralist strategy, the practice of evaluation must be recognised as an inevitably value-saturated process. Within this, there are important questions to do with the role and stance of the evaluator(s), in particular at the stage of the presentation and use of evaluation findings (Taket, 1993; Taket, 1994). Once again it needs to be emphasised that this does not amount to a rejection of quantitative approaches, rather to a plea for selective and sensitive use of them, alongside qualitative approaches.

In order to further elaborate our position, we would like to explore the problems we have found with systems approaches. In the process of critical reflection, we make use of a notion of feedback somewhat different from the more mechanistic systems notion. Also, the concept of the boundary, essential to delineating systems, has to be artificially imposed in situations characterised by heterogeneity and dynamism. So that in Gregory and Jackson's schema, boundaries are set up between objectivist/subjectivist and low/high variety, which cannot do justice to the situations described above. Our pluralist approach adopts the notion of leaky and changeable boundaries: this paper "...is an argument for pleasure in the confusion of boundaries and for responsibility in their construction" (Haraway, 1985:150).

For the future, we will be interested to explore the links between the approach we set out here and other systems approaches. In particular Harnden's (1994) reinterpretation of cybernetics seems to offer a potentially useful way of viewing the interaction between evaluator(s) and the evaluation situation.

REFERENCES

Flood, R.L., and Jackson, M.C., 1991,"Creative Problem Solving: Total Systems Intervention," John Wiley & Sons, Chichester.

Fox, N.J., 1991, Postmodernism, rationality and the evaluation of health care, *Sociological Review*. 394:709-744.

Gregory, A.J., 1993, Managerial problem solving and evaluation: a complementarist approach, *in*: "Proceedings of the Second European Congress on Systems Science," Prague.

Gregory, A.J., and Jackson, M.C., 1992, Evaluation methodologies: a system for use, *Journal of the Operational Research Society*, 43:19-28.

Haraway, D., 1985, A cyborg manifesto: science, technology and socialist-feminism in the late twentieth century, *in*: "Simians, Cyborgs and Women: the Reinvention of Nature", 1991, Free Association Books, London.

Harnden, R., 1994, "Multimedia - a new space of expression," *Systemist,* 16: 229 - 241

Jackson, M.C., 1990, Beyond a system of systems methodologies, *Journal of the Operational Research Society*. 41:657-668.

Jackson, M.C., and Keys, P., 1984, Towards a system of systems methodologies, *Journal of the Operational Research Society*. 35(6):473-486.

Legge, K., 1984, "Evaluating Planned Organisational Change," Academic Press, London.

McCracken, J., Pretty, J., and Conway, G., 1988,"An Introduction to RRA for Agricultural Development," IIED, London.

Taket, A.R., 1993, Mixing and matching: OR and innovatory health promotion projects, *OR Insight*. 6(4):18-23.

Taket, A.R., 1994, Undercover agency? - ethics, responsibility and the practice of OR. *Journal of the Operational Research Society*.45(2):123-132.

Thompson, J.C., 1992, Program evaluation within a health promotion framework, *Canadian Journal of Public Health*. 83(Sup. 1): S67-S71.

White, L.A., 1994, Development options for a rural community, *International Transactions of OR*.in press.

White, L.A., and Taket, A.R., 1993, Community OR - doing what feels good, *OR Insight* 6(2): 20-23.

79

THE POWER OF DISTINCTIONS

Ioanna Tsivacou

University of the Aegean
30, Voulgaroctonou street
Athens, 11472
Greece

INTRODUCTION

Power is one of the crucial issues in Critical Systems Thinking. It is connected with the position of that approach for "...human well-being and emancipation" (Flood and Jackson, 1991: 49). As is well known, there are differences in the way that critical systems thinkers view the subject of emancipation (Flood and Ulrich, 1990; Jackson, 1985, 1990). However, as their main concern is a redefinition of Management and Systems Sciences, as well as the provision of theoretical and practical principles for the functioning of a critical systems methodology based on dialogue and on critical self-reflective action, they sought mainly theoretical support in the Habermasian analysis for distorted communication. Therefore, in their writings, the notion of power is not the outcome of an original work, but rather a notion borrowed primarily from the theoretical project of Habermas, and secondarily from that of Foucault. An exception could be considered Oliga's endeavor to deal with the concept of power as he attempted to inquire into the phenomena of stability and change of social systems (Oliga, 1990(a), 1990(b)).

Oliga regards power as a function of a relationship between the haves and the have-nots. He argues for the historical nature of the social control and he defines the latter as the maintenance of the dominant interests of the ruling class upon the ruled one. The affinity of Oliga's argument with that of the Marxian tradition is obvious. However there is an important difference concerning the notion of interest and the questions that it raises. Which are the objective interests of the oppressed class? How can the oppressed know what his/her real interests are, if the ruler not only refrains from his/her action but from his/her intentions too? What if his/her beliefs and his/her wordviews are also formulated according to the proper interests of the oppressor? Is the consciousness of the ruled is manipulated and subordinated to the dominant social order? To these dilemmas Marxist theory replies in the analytical framework of the class division and of the capitalist social relations. If this framework is not totally accepted, then the analyst is obliged to define the normative conditions in which the objective interests might be recognized. For Lukes (1974) these normative conditions would be met in a political environment of a quasi full autonomy; For Habermas (1972) in an ideal speech situation. Although Oliga refuses to abandon the theoretical link between power and interests as Hindess (1982) and other sociologists

Critical Issues in Systems Theory and Practice, Edited by K. Ellis et al.
Plenum Press, New York, 1995

propose (Benton, 1981), he does not describe with clarity the normative judgements of a social and political context as Marxists do and, consequently, he does not confront the theoretical implications of a similar argument. He confines his statement to the necessity of existing a "theory-dependent method that guides research into the meaning of a form of consciousness by relating to its context of interests and realities" (Oliga,1990(a): 40). So, while Oliga's critical approach contributes significantly to an emancipatory systems theory, the elaboration of the notion of power remains incomplete, thus rendering vulnerable the critical systems thinking in toto.

This paper does not pretend to be able to fill the aforementioned conceptual gap of power. It wants rather to points out a view of power independent of interests and intrinsically connected with the human action. The paper claims that this view addresses by another perspective the features of a critical systems methodology, liberating it from the bounds of the emancipatory ideal as it is currently understood.

The first section describes how a formal organization, typical domain of contemporary action, is constituted through moment-to-moment interactions. It develops the notion of the 'distinctions', or 'commands', in order to explain the bringing of something into appearance by an actor. In the next section consideration is given to develop the idea that commands, in the form of distinctions, are the expression of power. Here an attempt is made the concept of power, which stems from the function of distinctions, to be outlined. In the third section some ideas drawn from the analysis of the previous sections are presented, as well as their practical implications for a critical systems methodology.

THE CONSTITUTION OF ORGANIZATIONS

The theoreticians of 2nd order Cybernetics (Maturana, 1987, 1988; Von Foerster, 1979) have brought to our attention that the reality of organizations is strictly dependent on participants' interactions as well as on participants' meaningful observations concerning these interactions. Both of these factors are languaged in a common domain of action, that participants call 'organization' (Espejo, 1993), which for analytical purposes could be divided in two sub-domains: in the operational domain and in the informational one. In the operational domain participants as living systems, and as psychic systems too (Luhmann, 1990(a)), act interactively and coordinate their action using a biologically given generative mechanism, such as that of language. During their interactions, in order to bring forth the praxis of living, participants are operating linguistic distinctions, using different forms of illocutionary acts. However, there is a special form of a linguistic distinction, the command, which is uttered explicitly in an imperative sentence, or implicitly in denotative or interrogative sentences, and the form of which conceals a purpose that the sender transmits to the receiver. So command is used as a tool of motion, or rather of the revealing of a cause (the ancient Greek *aition*) that provokes the birth or the change of a thing. At the same time participants operate as observers of their interactions and of their distinctions accomplished in the operational domain. In these, second order observations, they try to explain their experience in the praxis of living, producing thereafter a second-order experience (Maturana, 1988). The explanations rise in the informational domain and they consist of representations of distinctions which have taken place in the operational domain. These representations are also distinctions, since the observers are obliged to select among different explanatory paths for their explanations. The distinctions of the informational domain give rise to new distinctions in the operational one, guiding the organizational system to a recursive flow of interactions and distinctions.

Both sub-domains, making up organization, are constituted by a multiplicity of distinctions. Actors, under their double role as participants and observers, actualize distinctions interactively from a wide horizon of potentialities. That means, using language, they proceed to choices bringing forth in this way a part of the social potential of experiences in their consciousness. What is not selected remains in suspension, or it is temporarily negated. The horizon of potentialities, before selecting, is a wholeness which includes the positive (i.e. the on going selected) and the negative (ie. the non on going selected). From the moment that selection is realized and the on going selected becomes actual, the actuality, as part of the whole, emerges and by its emergence permits the conceptualization of the wholeness as negated potentiality.

Distinctions or choices in order to be realized need a pregiven, socially constructed structure of meaning (Luhmann, 1990b) which renders possible the act of selection. Command as an illocutionary act might be regarded as an impulse which makes use of this structure of meaning, contributing in this way to the act of selection. However, command might be regarded itself as a generic act of selection, since it is the only form through which the choices are actualized. Then command is a part of the function of meaning, which actualize the capacity of meaning to function using the act of distinctions. In a social system, such as that of an organization, the interactively actualizing of choices, or the transmission of commands presupposes that the pregiven structure of meaning contains endogenously the attribute of shared understanding. According to the above analysis, every distinction, every command, every choice, with the aid of meaning, draws from the potential of experiences a part which is an information able to be shared with the participants. This informative process invested by a historically created social understanding is what Luhmann (1990(b)) calls "...communicative process" . The domain of this communicative process, namely the locus where communications are realized is every social system and of course, every organization; the reason is that this is the domain where the synthesis of the three constitutive elements of communication, that is, command (the utterance mentioned by Luhmann), information, and understanding could coincide (Luhmann, 1990(a)).

This approach towards organization has important implications. The most significant is that the relationship between a knowing subject and his/her world, developed during a long anthropocentric tradition, has been replaced by the relationship between a system and its environment. Human beings as conscious systems are not the element of the organization but they are considered as systems of its environment connected with it through loose couplings. That is, human beings when they establish or enter in the organizational domain they find and they use a pregiven structure of meaning which predefines their linguistically mediated distinctions or choices. They constitute organizations through their intersubjective communications, but there is a suprasubjective structure of meaning which prescribes their utterances and their understanding. This structure facilitates the function of choices as its rationality is generative of this function, namely it is based on the generic distinction between positivity and negativity. Human beings in order to create organizations, in order to act, are obliged to adopt this rationality and consequently to subjugate themselves to the rationality of the commands, of the detaching of a part from its ground, from its whole.

COMMAND: THE ESSENCE OF POWER

This approach replaces action as constitutive element of organizations by distinctions and choices. Action is a behavior strictly connected with the actor, while in the above analytical framework command and distinction are the necessary premises for the realization of action. If by action an appearance of being takes place, here the process and the form of this appearance are not only connected with the intentions and purposes of the human being, but are dependent on the dominant rationality of the social systems concerning commands and distinctions. In this framework of analysis command does not issue from the actor, and therefore it is disconnected from the perspective of human relations. It primarily complies with the fundamental property of the system's rationality to function using distinctions. This approach does not deny the importance of human relations, but it wants to stress the attention to some characteristics of social systems, not only of a longer historical duration, but also of a basic necessity for the establishment of these relations.

It is argued here that command is the essence of power. At this point the claim of Canetti (1962) that power is the imposing of order via commands enhances the present argument and offers new insights for the understanding of power. However, if the position that command is the detachment of the part from its integrative wholeness is valid, then, a logical consequence is this: the totality of commands, that a human being receives during his/her daily life, deracinates him/her from his/her social background and reduces him/her to the position of an 'individual'. In agreement with this theoretical model, the pain that a human being feels in front of power, is not the pain of an individual subordinated to an other individual that social theories about power usually allege; it is not the suffering of the oppressed individuality, but it is the fear of the human being in front of this individuality. It is the anxiety of the social person against a continuous recreative process of his identity as actor in the social systems where he/she develops his/her praxis of living and his/her representations concerning this praxis.

According to Canetti (1962), command is a means that power uses in order to establish its teleology, that is, the establishment of order. Command is in the service of power in an explicit or implicit form. It is the necessary, evident or latent, condition for the starting of action. With a command 'Let there be light' God came out from his stillness and created the world. Command is the reason that cancels all the ambiguities of language and declares the order that this reason wants to impose upon the things. In the command there is not any doubt, because its function is the disappearance of any obscurity, thus the unspeakable, the undivided, the endless and the immeasurable become speakable, divided, ended and measurable.

In case that command is addressed to a human being, it is composed by a momentum and a sting. The momentum is the force on the person to act, while the sting is the hidden simulation of the command, or the content of the command that the person receives. The sting is poked in his/her body till the moment that the recipient transmits the command to another one and therefore he/she is liberated by this reversal. It is this property of the reversal that Cooper (1983) believes that lies in the core of the sting. This interpretation of Canetti's sting is what is here considered very relevant with our conceptualization of power. For Cooper the sting is ".....the outraged sense of loss of equivalence occasioned by distinction. The function of the sting is to redress the inequality and it does so by reproducing earlier situations but in reverse" (Cooper, 1983). Via command power proceeds to distinctions or divisions of a prior unity into multiple parts. The sting, or as it is already referred to the content of the command remains in suspension, witness of this prior unity and an impulse for its pursuing. The only moment that this impulse is inactive, according to Canetti, is the moment that the rationality of the commands ceases to operate. Only when a human system stops to act and any distinction comes to an end; that is, when a human system is transformed to a crowd. Crowd is the only domain where the human being is socially involved with other human beings without acting. Crowds seems to act when they destroy a social model of action in order to replace it with another one. Even then, crowds act only at the precise moment of the blind destruction. When the destruction starts to be organized, and mainly when the reconstruction of the new model is designed and applied, then crowds are substituted by agents. Crowds is the space of the inaction and here it must be sought the explanation of their alleviative function. It is the place where human beings deliberate themselves from any rational praxis of living and demand the abolishment of any order, reason or organization.

Social systems have very early understood the necessity to instutionalize organized forms of social life to reverse the results of the power of distinctions. They could not permit the disorder to be the alleviative counter of order, therefore social ceremonies, myths, rituals and games have been organized in order to restore the lost feeling of wholeness or collectivity. Canetti reports many anthropological findings of this kind, but even in ancient societies of a developed civilization such as that of ancient Greece it could be easy to make similar confirmations. The problem of power and the accompanied commands has acquired a new emphasis since the rising of modernity, or of the 'new times'. The modern world is distinguished from the old by the fact that it has raised the individuality towards the Heaven, and has given birth to the human 'subject'. The consciousness of the subject became the flux for the running of the distinctions, their continuous multiplication and consequently the emergence of new forms of actions and products. Complexity is a phenomenon which owes its existence to this proliferation of commands and choices, without any counter balance for the oppressed, objectified and subordinate subject.

ORGANIZING THE REVERSE OF COMMANDS

Organization is the opposite of crowd. It is, as already mentioned, the physical place of commands. Till some decades ago, the counter balance of this place was the habitus of the private life where the individual could retire and rest from the commands of the work process. Today this retirement is almost impossible since free time is also well organized. Where and how could be located a new context freed by the rationality of distinctions and open to the possibility of the creation of a new sprit of collectivity? Could not be considered this new context as an emancipatory one, since its function ought to be a counter balance to the function of power?

The paper points out that possibly the success of the Japan and other East Asian enterprises could be sought to the collective mind that their culture spreads as an antidote to the rationality of commands and distinctions of the modern industries. This collective mind as it is well known has been enforced and applied in the formal East Asian organizations (Kenney and Florida, 1988; Redding, 1980, 1990) in the form of different meetings and other institutionalized ceremonies in parallel with the working process. In the Western countries of a long humanistic tradition a similar collective mind cannot be applied if it is not accompanied by a self-reflective discourse. This discourse ought to surpass the enlightenment of the participants' consciousness concerning their position in an antagonistic social system; it ought, also, to surpass the descriptions of the social conditions which produce a distorted communication and to enable participants to gain a real understanding of human reasoning, languaging and acting that produce distinctions and commands. In continuity, the discourse ought to find new linguistic games in order to assist to the expression of the hidden longing "... for that which is lost. What is lost is the experience of the whole" (Cooper, 1983), of the collectivity. This discourse could probably be viewed in the context of a critical systems methodology, considering it in such a way as the counter balance to the domain of commands and power. This consideration lends also to a similar methodology the emancipatory character that the notion of interest is not able to lend as it has been mentioned in the introduction.

A CONCLUDING REMARK

Systems methodologies in general are formalities institutionalized in the informational domain of the organizations. If they are not faced as a means to solve the disturbances produced in the operational domain, but as cultural regularities in charge of the tightening of the loose couplings between organization and psychic systems, then they should not be restricted to the role of a regularity destinated to maintain the social coherence. Besides, they should be considered as cultural systems able not only to strengthen communication, but to restore an equivalence among its constitutive elements: that is, weakening command and information, and enhancing comprehension. In such a way, systems methodologies should analyze the dominant structure and function of meaning in which they will be embedded. Also, systems methodologies should investigate the opportunities to take advantage and to reverse the ruling elements of the actual structure and function of meaning, restoring in their context the lost feeling of collectivity.

REFERENCES

Benton, T., 1981, "Objective" interests and the sociology of power, *Sociology*, 15(2):161- 184.

Canetti, E., 1962, "Crowds and Power," Gollancz, London.

Cooper, R., 1983, Canetti' s sting, *unpublished*, Department of Behaviour, University of Lancaster, Lancaster.

Espejo, R., 1993, Domains of interactions between a social system and its environment, *in*: "A Challenge for Systems Thinking," I. Tsivacou, ed., University of the Aegean Press, Athens.

Flood, R.L., and Jackson, M.C., 1991, "Creative Problem Solving,"John Wiley & Sons, Chichester.

Flood, R.L, and Ulrich, W., 1990, Testament to conversations on critical systems thinking between two systems practitioners, *Systems Practice*. 3(1): 7-29.

Habermas, J., 1972, "Knowledge and Human Interests," Heinemann, London.

Hindess, B., 1982, Power, interests and the outcomes of struggle, *Sociology*, 16: 498-511.

Jackson, M.C., 1985, Social systems theory and practice: the need for a critical approach, *International Journal of Genernal Systems*. 10: 135-151.

Jackson, M. C., 1990, The critical kernel in modern systems thinking, *Systems Practice*. 3(4): 357-364.

Kenney, M., and Florida, R., 1988, Beyond mass production : production and the labor process in Japan, *Politics and Society*. 16(1): 121-158.

Luhmann, N., 1990(a), The autopoiesis of social systems, *in:* "Essays on Self-Reference," Columbia University Press, New York.

Luhmann, N., 1990(b), Meaning as sociology' s basic concept, *in:* "Essays on Self- Reference," Columbia University Press, New York.

Lukes, S., 1974, "Power: A Radical View," Macmillan, London.

Maturana, H., 1987, Everything is said by an observer, *in*: "Gaia: A Way of Knowing," W.I. Thompson, ed., Lindisfarme Press, California.

Maturana, H., 1988, Reality: the search for objectivity or the quest for a compelling argument, *The Irish Journal of Psychology*. 9(1): 25-82.

Oliga, J., 1990(a), Power-ideology matrix in social systems control, *Systems Practice*. 3(1): 31-49.

Oliga, J., 1990(b), Power in organisations: a contingent, relational view, *Systems Practice*. 3(5): 453-477.

Redding, S.G., 1980, Cognition as an aspect of culture and its relationship to management processes: an exploratory view of the Chinese cases, *Journal of Management Studies*. 17: 127-148.

Redding, S.G., 1990, "The Spirit of Chinese Capitalism," De Gruyter, Berlin.

Von Foerster, H., 1979, Cybernetics in cybernetics, *in:* "Communications and Control in Society," K. Krippendorff, ed., Gordon and Breach, New York.

CRITICAL ISSUES IN BUSINESS SYSTEMS

80

SYSTEMS MODELS FOR DESIGNING THE ORGANIZATIONAL RESTRUCTURING OF COMPANIES DURING THE PROCESS OF OWNERSHIP CHANGES

Wiktor H. Adamkiewicz

Merchant Marine Academy
ul. Morska 83
81-225 Gdynia, Poland

SUMMARY

This paper presents the possibility for the formation of a practical systems model for the design situation, and an algorithm for the solution of the design problems connected with a firm's modernization in post communist countries.

Solution of such problems consists of the application of an appropriate systems model of the firm and its environment. Applying an appropriate method for evaluation of the relationships which exist between elements of the given system is another important question. The hierarchisation of these relationships is a decisive point for the usefulness of models relative to the problem of developing a design process (Adamkiewicz, 1990(a); 1990(b)).

THE PROCESS OF RESTRUCTURING A FIRM

During the restructuring of an existing firm it is necessary to elaborate with respect to the constituent parts of the firm. The proper way is to have in one's mind the differences necessary to introduce the algorithms applicable in Western Style economies. The main differences exist in the environment of a restructured firm, and they are shown in Figure 1. (Adamkiewicz, 1994).

In summary these differences are :

- the differences in the law;
- the differences in the ownership of the firm, especially with regard to the conviction of employees and their ownership rights of the property of the firm;
- the difference in social problems, especially :
    1. the communist state supplement;
    2. low level salaries;

Critical Issues in Systems Theory and Practice, Edited by K. Ellis et al.
Plenum Press, New York, 1995

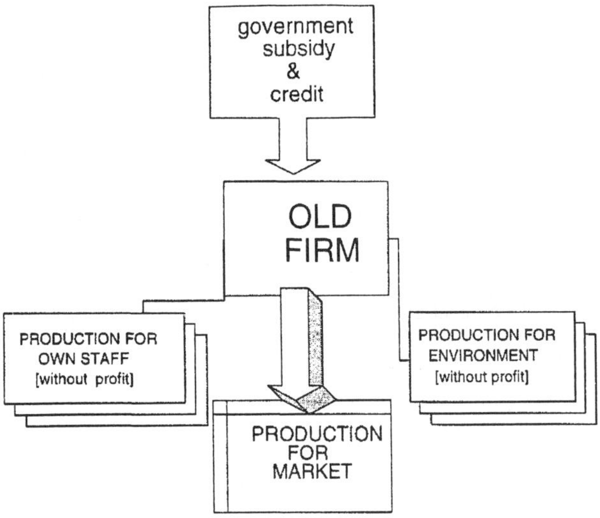

Figure 1. Environment of a restructured firm.

3. worker privileges;
4. low rent accommodation;
5. rest houses and sanitoriums;
6. infant nurseries and kindergartens;
7. hospitals;
8. items owned, financed and administered by the firm.

- the organizational-technical differences:
    1. inefficient national infrastructure;
    2. poor telecommunications;
    3. inadequate transport systems;
    4. service and maintenance systems.

These are reasons why, even medium sized organizations, have their own railway trains, building and repair activities because of the inadequacies of a centrally planned economy. Generally speaking, we are able to state that a firm, or group of firms, were attempting to be independent because of the lack of capacity in the national economy. Indeed, in small towns firms possessing their own 'service' functions took over these roles from the municipality. In a Western style market economy it is unlikely that such firms exist. However, we need to take such factors into account if we are to understand and design a process for restructuring firms in a transition between a centrally planned economy and market driven economy.

DESIGN FOR SYSTEM MODERNIZATION

The simplified model of a firm existing in a post communist country is shown in Figure 1. The project for restructuring a firm was carried out in accordance with the model which is conceptualized in Figure 2.

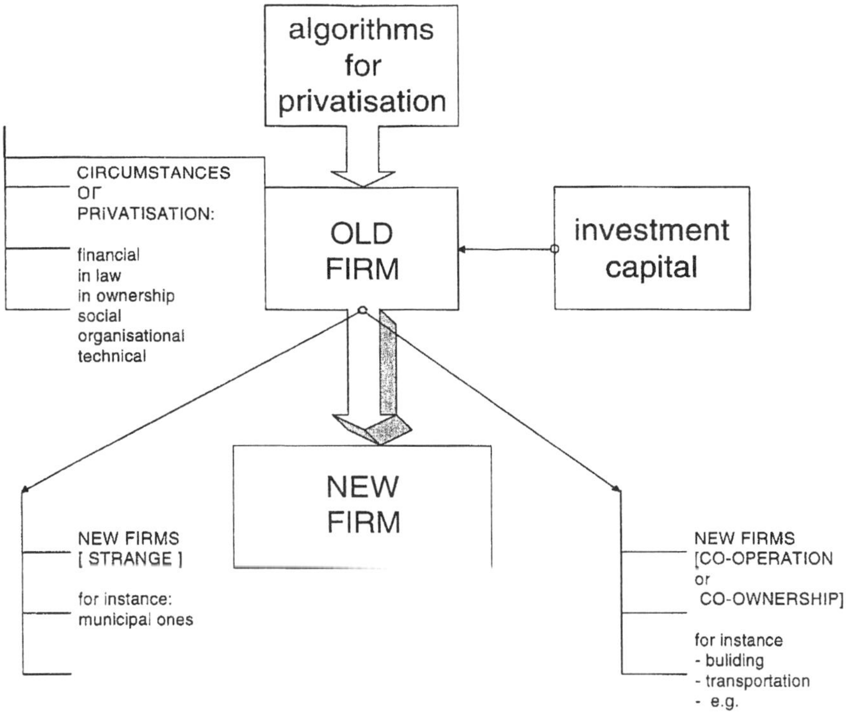

Figure 2. Restructuring model.

Theoretical circumstances for dissolving mentioned problems one can find, for instance, in the book edited by Adamkiewicz (1983). Some detailed remarks were proposed by Adamkiewicz (1987) and Adamkiewicz and Jedrzejowicz (1988).

A suggestion of a certain way of designing large scale systems is presented. Justifications for the suggestions come from General Systems Theory and its different interpretations. General mathematicization of the suggested designing procedure may be carried out based on the theory of graphs, theory of relations, theory of sets or algebra, etc. Yet, it is necessary to treat the process of identification of real data, which are essential for designing in a new way. The area of identification usually called 'the environment of the designed system' should be presented formally in the form of a system.

It is suggested to consider design from the point of view of the theory of change. Thus, the designed system is the one which after being installed in another system, introduces a desirable change. This 'another system' is the environment of the designed system.

The introduction of a change in such a system may consist in the following operations:

- the introduction of new elements and/or relations to the system;
- the removal of existing elements and/or relations from the system;
- the reinforcement of activity of existing elements and/or relations;
- the reduction of activity of existing elements and/or relations.

It is suggested to apply generalized methods similar to network analysis (matrices, graphs) as design methods. In order to do this, it is necessary to establish the principles for creating the matrix sets which contain data which could be useful for a designer. These matrices should contain numerical data, and also the data presented in a descriptive way. There should also be a possibility to define the relations appearing among particular matrices and their sets.

It is possible to accept the most popular definition according to which the system is a set of elements and the set (sequence) of relations (connections, relationships) among elements and other systems. For instance,

> "The system is an ordered pair { M, S } composed of a set M and a sequence S defined as the set of relations in the set M. M is the set of elements of the system. S is the structure of the system." (Wintgen, 1971).

It is, of course, a simplified notation of the definition of such a system as mentioned in this paper.

Among many propositions, which may be found in abundant literature on systems theory, the 'The Mathematical Theory of General Systems' as elaborated by M.D. Mesarovic in Klir, ed., (1972) seems to be useful for the purposes of this discussion. The system propositions of G.J.Klir may also be advantageous (Klir, 1972).

The multitude of the system characteristics, their direct and indirect links to the functional characteristics of a system, make the design decision:

1. multi-criterial;
2. uncertain;
3. complex and non-programmable;
4. unique;
5. non-repetitive.

A lack of global and complex models does not mean that there are no practically useful models and techniques allowing analysis (sometimes even optimization) of the partial problems. It is left to intuition helped by general methodology known as the 'systems - theoretical approach', to use partial solutions and limited answers to improve the performance of the whole system.

The essence of the proposed method of identification and analysis of relations is to present all the parameters of the designed system in a certain n-dimensional space, whose dimension 'n' is equal to the amount of the considered factors of the system description and also to ensure relations between matrices and their sets by the choice (in this space) of appropriate axes of real time. These axes should correspond to the real course of processes in the designing system.

The essence of the mathematical approach consists in presenting the designing situation with the help of a general form of tensor defined in n - dimensional Riemann space. The tensor constituents will be n - parameters describing the design situation. After some appropriate formal simplifications the matrices will be created from each pair of tensor constituents. In general, it is possible to create a number of matrices to be at the designer's disposal. These matrices will prove to be indispensable after a simple analysis of their usefulness. The relations among matrices will be also clear and easy to interpret after their presentation in the n-dimensional spaces of Euclid.*

The above mentioned summary of 'tensor' reasoning was to point out that there is a possibility of presenting the design of a system in an exact way. These are proposals of practical applications.

---

* The Riemann tensor should not be identified with its simplification known from the theory of elasticity and plasticity.

A point for such an analysis is to define the Riemannian space determined on the manifold whose points represent complex parameters of a designed system. If a metric of n-dimensional Euclidean space with rectangular coordinates is accepted in the mentioned form of coordinates then analysis can be made on the basis of matrices as below.

GENERAL BASES FOR FORMING HEURISTIC ALGORITHMS OF DESIGN

A system is a set of material and/or immaterial and/or abstract elements, joined together by different relations: informatical, energistic and materials. A subset of these relations, determining the aim of a system, is called a system-formative relation.

Real systems generally contain such large numbers of elements and particularly such a considerable amount of relations, that their simultaneous examination is impossible. That is why, stating the essentially of elements and relations from a considered point of view is a very important constituent of every analysis and systems synthesis (Adamkiewicz, 1983). It should be remembered that no 'hard methods' are useful for analysis and synthesis of systems of this class being the subject of this discussion (Jackson et al., 1986).

It is suggested to begin systems design by designing the process which is to be realized by the system. It is, after all, an obvious stipulation for a correctly conceived design process.

- Every process may be faithfully enough represented by a discrete sequence of states on a real time axis. Evident relations always exist among states. It is suggested to treat a set of system states arranged on the time axis together with a set of relations occurring among states as a SYSTEM OF SYSTEM STATES. It is this pattern of the system which can be seen from operational and strategic managing levels. Designing SYSTEM- STATES on the hierarchically highest (strategic) levels of management and moreover its decomposition to executive level may be the basis for designing an executive subsystem and also for a management subsystem or for designing a new system.
- The making of a SYSTEM PURPOSES TREE should be the first stage of design.
- Next, after specifying the place of the process as the aim of the system's existence, in every state of process realization process decomposition will be performed.
- In the next stage, identification of all relations which can occur among states of the considered process and steering interactions of the strategic level, should be made. To this end, axes of representation should be determined:

  A. - axis of process states (axis of real time);
  and for example:
  B. - strategic interaction in the material sphere;
  C. - strategic interaction in the subjective/social sphere and also necessarily;
  D. - managing subsystems on the operational level.

- Identification of the relations should be made by using MATRICES: A - A; A - B; A - D. AXIS 'A' as the axis of real time is the decomposition and aggregation axis in the whole design process, because it represents the system of states. AXES 'B' and 'C' represent interaction sets, whose identification does not require using the rigors of system methods. AXIS 'D' represents organizational decomposition of the designed system.

- The obtained MATRICES should be used in the following way:
  1. MATRIX 'A - A' serves to determine relations in the system of states realization in view of the considered system of states.
  2. MATRICES 'A - A and 'A - C' serve to determine steering interactions on the considered system of states on the strategic level.
  3. MATRIX 'D' serves to determine steering interactions of the strategic level on subsystems of the operational level regarding the considered system of states.

- Next, the decomposition of the system of states should be made on the operational level. It may turn out that the process 'A' in regard to real time, decomposes itself into parallel processes - also discretely situated in time (as sequences of states) for example 'A1', 'A2', 'A3'.
- In order to specify the relations on the operational level, further axes should be introduced, for example: F - spheres (domains), in which results of functioning will appear, G - steering interaction of the operational level on the considered system of states.
- Analysis space may be specified as DESIGNING SPACE of the system and the following co-ordinate axes may be suggested for it: 'A'; 'A1'; 'A2'; 'A3'; 'Az'; - as above, 'G'; - as above, 'H'; - axis of the set of SYSTEM PURPOSES REALIZATION functioning. MATRICES: 'A - H'; 'A1 - H'; 'A2 - H'; 'A3 - H'; 'Az - H'; are the MAIN MATRICES of SYSTEM DESIGN. MATRICES: 'H - G'; 'H - H'; are the MATRICES of DESIGN COORDINATION.
- All the matrices mentioned by way of example should be created on the basis of graphical analysis.
- The formed matrices are the tables in which all potential possibilities of relation existing in the system occur.

REFERENCES

Adamkiewicz, W., 1983, "Cybernetyczne aspekty badania obiektow technicznych" [translation: "Cybernetic Aspects of Technical Object Investigation"], Polish Academy of Science, "Ossolineum", Wroclaw, Poland.

Adamkiewicz, W., 1987, Efficiency and security as criteria of analysis and synthesis of reliability in shipbuilding, *in:* "Proceedings of 4th International IMAEM-87 Congress", Varna, Bulgaria.

Adamkiewicz, W., 1990(a), The method of aiding the designing of the management systems, *in:* " Proceedings of 11th International Cost Engineering Congress and 6th AFITEP Annual Meeting", Paris, France.

Adamkiewicz, W., 1990(b), Designing the management system as a set of subsystems of differently oriented aims, *in:* "Proceedings of the Xth INTERNET'90 World Congress on Project Management - Management by Projects", Manz-Verlag, Vienna, Austria.

Adamkiewicz, W., 1994, Systemic models for project management of the organizational reconstruction of companies during the process of the ownership changes, *in*: "Proceedings of the XII European Meeting on Cybernetics and Systems Research", Vienna, Austria.

Adamkiewicz, W., and Jedrzejowicz P., 1988, Designing Reliable and Efficient Systems, *in:* " Proceedings of the International Conference on Engineering Design, ICED '88", Heurista - Zurich, ed., Budapest, Hungary.

Jackson, M.C., and Keys, P., 1986, Towards a system of system-based problem - solving methodologies, *in:* "Projektowanie i systemy. Zagadnienia metodologiczne", vol. VIII, Warszawa.

Klir, G.J. ed., 1972, "Trends in General Systems Theory",. John Wiley & Sons, New York.

Wintgen, G., 1971, Zur mengentheoretischen definition und klassifizierung kybernetischer systeme *in:* "Mathematic und Kybernetik in Oekonomie (III)", H

SIX AND TWO THREES

Business Process Re-Engineering : A Historical Perspective

Linda Charnley

Aberdeen Business School
The Robert Gordon University
Hilton Place
Aberdeen, AB9 1FP
United Kingdom

INTRODUCTION

Businesses and organisations are frantically concentrating their energies on redesigning their primary processes and this has been heralded as new and highly innovative, but is this really the case? Soft Systems Methodology (SSM) (Checkland, 1981) has long practised the art of 'process gazing', if only in the 'Systems Thinking' stages of the methodology. As today's technology increasingly crosses and blurs traditional boundaries, Fayol's and Weber's functional division of organisations is clearly no longer appropriate, but is reorientating to a process determined modus operandi going to be enough and what lessons, if any, can be learned from information systems practice by incipient re engineers? In this paper the author will discuss the roles of IT and information systems within the enterprise; and will examine how they may have contributed to the ascendancy of current organisational trends such as Business Process Re engineering. The paper will also consider the more disparate, socio-economic factors which are being embraced, and at times overlooked, in this latest quest for process-centred operations.

THE HISTORICAL CONTEXT

On the eve of the last millennium, classical theorists viewed organisations as essentially closed systems, and aimed to develop efficient systems that would support the business needs of the day. Fayol postulated his fourteen management principles, primary among these being the division of labour, and many of which are still followed by managers today, whilst Max Weber concentrated on the structural ideal of bureaucracy. Frederick Taylor's approach of maximising labour output by reducing the control that skilled workers had over their jobs, and managing every minute aspect of the work of subordinates, was heralded as revolutionary in its time when change was a stranger, and the business world and its supporting markets were stable, predictable and expanding. Each and every one of

Critical Issues in Systems Theory and Practice, Edited by K. Ellis et al.
Plenum Press, New York, 1995

these developments were, however, entirely in keeping with the socio-economic climate of the day.

Throughout the nineteenth century, industrial development had been underpinned by the growth of empires and facilitated by the design of automotive machinery replacing the need for skilled labour. Society was divided along class lines, and there was little mobility between the classes, people knew their station in life and tended to live and work within it. Alongside this, the scramble for Africa at the end of the nineteenth century, arguably the peak of imperialist ambition, had opened up new markets and extended the resource base, thus laying the foundations for business growth. It is in this overall context that the classical theorists should be judged, i.e. as an effective response to a gradually changing set of circumstances. Their lasting impact within large bureaucracies is testimony enough to their fitness for purpose at the time.

The approach is no longer appropriate, and still less acceptable in a society that has grown up alongside principles of ability, equality, openness and change, and in which Information Technology is able to reinforce these messages by sharing and disseminating information, ideas and decisions altogether more efficiently; via developments such as groupware, video conferencing, global networks, etc.

The evolving social, economic and political climate experienced in the post war decades has accommodated the concepts of integration across traditional barriers of race, colour and creed and is currently extending to trade, geographic boundaries and more significantly political boundaries. The geopolitical division of Cold War Europe led naturally to economic alliances such as the EEC and Comecon. The recent collapse of the USSR and the ending of the Cold War, has meant that the Eastern bloc is now open for trade and exploitation, and the developed world is embarking on yet another end of century 'scramble'; only this time the imperialist motivations of the last century, have been superseded by the global market needs of the late twentieth century, reinforced yet again by technological advancement. However, in this instance the technology has been enabling increased integration along process lines, whereas one hundred years ago technological advancements were encouraging segregation along functional lines.

Similarly, social trends and rapidly developing IT have paved the way for a reappraisal of work design criteria. Improved social mobility, universal suffrage and education together with the recognition and expansion of employee and civil rights in the post war years has accelerated the departure from the authoritarian, de-skilling and very individually focused work design criteria extant at the start of this century, toward a far more liberal, multi-skilled team focus - ably supported by information technology and dramatically improved communications systems.

One oft quoted distinction between the two trends is said to be the speed at which they have arisen. The industrial revolution unfolded over a century, whereas the information revolution has taken place within a decade and in strictly technological terms that is indisputable. However, the social, political and economic arena in which the information revolution was born, and is thriving, has been evolving throughout the twentieth century, albeit more rapidly in the latter half. Given that organisations react to their broader environments, rather than determine them, it is the contention of this paper that business process re engineering is the legacy that the twentieth century will bequeath to the twenty-first century, and as such it is the logical conclusion of the century's progress.

The Developing Role of Information Systems and the advent of Business Process Re engineering

Business practice responds to and reflects contemporary social, political and economic conditions. Ergo, the development of information systems thinking has evolved along much the same lines as those of organisational thinking. In the immediate post war period systems approaches focused on hard, engineered systems paying little heed to the human element within any system, i.e. people fitted in with, and around, the technology. As individual rights and expectations expanded into the sixties and seventies, organisational thinking reflected social change, no doubt spurred on by shortages in the labour market, and began to acknowledge and accommodate the softer, 'people' factor, so that systems, and even technology with the much improved man-machine interface, began to be designed around people.

Soft Systems Methodology (SSM) (Checkland, 1981) emerged around this time offering a revolutionary, non-linear approach to problem solving aimed at 'messy situations', i.e. more specifically organisational systems. Hitherto, the hard, linear methodologies with their roots in military and logistical issues addressed specific, clearly identified and previously articulated problems. SSM's novelty and strength was that it could address more generic situations where there may be little more than a feeling of unease that a system, department or organisation was not functioning as well as perhaps it might. As we approach the new millennium, we encounter the phenomenon of Business Process Re engineering (BPR) (Hammer et al., 1993), and it would appear that its many disciples advocate its wholesale application to just about any business that intends to have a future.

Business process re engineering is such a new discipline that there are no hard and fast rules as yet. There is no one tool-set, or ever likely to be, that will deliver the goods for every type of enterprise that chooses to go down this route. There is even still debate about the exact nature of BPR, and a consequent plethora of publications espousing different approaches and sometimes different definitions!

BPR is essentially radical in nature, as is soft systems methodology with its root definitions which focus on the essence of a system, and resulting conceptual models. For the purposes of this paper, the prescription of business process re engineering is based on Michael Hammer's definition, namely BPR is "... the fundamental rethinking and radical redesign of business process to achieve improvements in critical. contemporary measures of performance such as cost, quality, service and speed" (Hammer and Champy, 1993).

It is the "fundamental rethinking and radical redesign" aspect that sets BPR apart from the majority of other problem solving or analytical approaches. The traditional hard systems analysis techniques studied processes, but did not embark on radical redesign, total quality management initiatives reset the focus on service and quality, capitalising on the growing trend toward team work and a multi-skilled approach, but still within existing boundaries.

In order to embark on such a radical programme as business process re engineering, it has become apparent that any project team must have a fundamental understanding of what the organisation is trying to achieve; together with the context and environment within which that organisation exists. Soft systems methodology offers a holistic approach designed to provide just such an understanding, by the detailed analysis implicit within the CATWOE (Checkland, 1981) mechanism and consequent activities.

SSM and BPR are revolutionary in so far as they reflect, and respond to, the recognition of the essentially holistic nature of organisations. More and more writers are warning against over-emphasis on any one of the three building blocks of organisations today, and recommending an approach that focuses on the benefits of their integration (Galliers, 1994; Preece and Peppard, 1994) rather than on the differentiation as has traditionally been the case in organisation structure, and some of the earlier BPR projects.

It is worth noting that the dawn of the softer, holistic and altogether more integrated approaches, such as SSM, to Information Systems and their rôle within the broader organisational context, has taken place in the light of rapid technological developments and the socio-economic trends experienced over the last twenty years; namely high unemployment and poor economic performance over a sustained period together with changing career patterns, from one full-time job throughout a working life, to many jobs not always full-time, not always in the same career path and often punctuated by periods of unemployment This development may have been unthinkable fifty years ago, but today few see it as remarkable. This is the climate into which business process re engineering has emerged.

In its brief history, BPR has undergone the same evolution, albeit on a more rapid scale, as organisational thinking. It is generally recognised that BPR draws on a range of hard and soft sources (Preece and Peppard, 1994; Hallworth, 1994). A glance through the *first generation* BPR literature (Hallworth, 1994) reveals an emphasis on usually one of the three essential elements of every organisation people, process and technology, usually process or technology. However, as the concept is gaining in ascendancy, more recent publications and discussions are revealing that the 'people' aspect of organisations is finally being acknowledged as equally valid as the technological and process emphases of BPR. (Coulson and Thomas, 1994) There is here too a more recent political and socio-economic parallel to be drawn. Can it just be coincidence that BPR was born in the mercenary, self-seeking eighties and that the early writings on its potential reflected the desire for reduced cost, at whatever price, and concentrated exclusively on the dramatic cost savings and accompanying job losses; whilst latterly in the early nineties, an apparently more caring decade, BPR initiatives are taking people issues into consideration once again?

Arguably, this combination of circumstances have eased the way for the more amorphous approaches that today are collectively grouped under business process re engineering initiatives, concentrating on integration, team work and the identification of core activities, and which have come to question the relevance of hitherto sacred, but cumbersome and costly support activities; as such "BPR is concerned with catholic innovation, and not simply with patching up defective areas" (Wood et al, 1994).

BPR per se is not so much a methodology as it is a philosophy; SSM can be viewed in a similar light. Consequently, neither advocates one set of right answers, recognising that each situation is uniquely complex and a flexible approach is the only way forward. Neither are strictly speaking management approaches with prescribed rules and tools, but combine flexibility with the recognition that at times tools from outwith their remit can be 'borrowed' to deliver specific outputs.

To conclude, business process re engineering is very much a child of its time. Its origins are many and varied but it shares a common evolutionary route with soft systems thinking and is, in essence, the logical conclusion of this type of holistic, integrationist approach. The organisational mind set is changing in line with social, political and economic trends, and combined with the enabling effects of technological advances, this has rendered business process re engineering both possible and acceptable in the run up to the twenty-first century. Successful re engineers should keep in mind this need for flexibility and will need to shed the dogmatism of yesteryear in favour of a new pragmatism for the future.

REFERENCES

Checkland, P., 1981, "Systems Thinking, Systems Practice", John Wiley and Sons, London.

Coulson-Thomas, C., 1994, "Business Process Reengineering : Myth or Reality" Kogan Page, London.

Galliers R.D., 1994, Information systems, operational research and business re engineering, *in*: *Int Trans Op Res*. Vol. 1, No 2.

Hammer, M., and Champy, J., 1994, "Reengineering the Corporation", Nicholes Brealey, London.

Hallworth, M., 1994, Re-engineering the learning organisation, *Management Education, Training and Development*, October 1994.

Preece, I., and Peppard J.,1994, A study of tools, methods and methodologies for business process redesign, "Proceedings of the 2nd European Academic Conference on BPR-Academic Directions", Cranfield, June 1994.

Wood, J.R.G, Vidgen R.G., Wood-Harper A.D., and Rose J., 1994, Business process redesign, radical change or reactionary tinkering?, "Proceedings of the 2nd European Academic Conference on BPR-Academic Directions", Cranfield, June 1994

82

ADMINISTRATION SYSTEMS

Mythology versus Reality

Janet Davison and Doreen Osuch

School of Computing and Information Systems
University of Humberside
Cottingham Road
Hull, HU6 7RT
United Kingdom

INTRODUCTION

This paper presents an intuitive examination of modern administrative systems; it seeks to raise a number of important questions but does not attempt, at this stage, to provide the answers.

Administrative systems are all too often seen as mechanistic, bureaucratic and academically non-challenging. What frequently escapes attention is the fact that, whether we like it or not, we are all administrators. Too often, however, we are unwilling to acknowledge or admit to undertaking an administrative role.

In recent years, probably only the last fifteen, administrative systems have been changing; they have bowed to the influence of technology. Key administrative support technologies have the potential to radically alter the very nature of work. Electronic Data Interchange (EDI), multimedia systems, workgroup computing, teleworking, the superhighway, all have the potential to bring sweeping changes to the ways in which businesses trade, share information and make decisions - and each of these has major implications for the administrative systems that underpin such business activities.

The dichotomous argument of technology as liberating problem-solver versus technology as problem-creator is well-rehearsed. Irving and Higgins (1991) and Tapscott and Caston (1993) emphasise the need for technology to be well-managed and appropriately applied. It is acknowledged that in itself technology has no value; its value is provided through its context, intention and management and it can be argued that it is these three influencing factors which create the conditions for technology to act either as a liberating force or a problem-creator.

Given these arguments, we are interested in exploring the liberating/problem-creating dichotomy through an examination of the major themes of modern administrative systems. We will question whether technology provides the opportunity for administrative management to shake off its image as a slow, dull, lifeless and low investment activity - in other words to be liberated. Liberation, it seems, is not always entirely a positive phenome-

Critical Issues in Systems Theory and Practice, Edited by K. Ellis et al.
Plenum Press, New York, 1995

non. If the liberating forces of technology were allowed free reign, complete access to global data and software would be available at no cost. Anyone could log on to any system at will and boundaries - organisational, national and international - would, technologically, cease to exist. Although this scenario has some proponents, it is generally recognised that for technology to develop, there has to be some return on investment and the potential problems of limitless access to data and software must be addressed. Legal restrictions provide one, widely endorsed, method of containing the liberating forces.

By focusing specifically on this aspect of administrative systems - new statutory instruments arising out of the widespread use of computers in the office environment - we can explore briefly how mechanisms to control the problem-creating forces of new technology operate and question whether, in attempting to limit the problems of administrative support technology, we in fact witness the generation of intractable administrative systems problems.

THE LIBERATION OF ADMINISTRATIVE SYSTEMS

During recent years the introduction of technology has often been seen as a substitute for effective administration. Organisations have increasingly used computer technology to gather, sort and disseminate information, displacing clerical procedures and even replacing administrative staff completely. We contend that the notion of technology underpinning business functions through the provision of adequate, up-to-date and relevant information is flawed. Technology is a tool which has been used in the development of information systems. At the same time there has frequently been a lack of recognition of the need to effectively manage the administrative systems and procedures which exist in both individual job roles and throughout the departments within an organisation,.

The IAM: Institute of Administrative Management, which is a UK-based professional body with world wide-links, seeks to promote the development of the science of administrative management. The Institute defines administrative management as: "that branch of management which is concerned with the services of obtaining, recording and analyzing information, of planning and communicating, by means of which the management of a business safeguards its assets, promotes its affairs and achieves its objectives."

At a practical level the Institute recognises that administrative management is both a specific role, and a generalist task within management for those individuals who move from a professional or technical background to managerial status. Whilst recognising that there is a professional role of administrative management, there is also a personal responsibility of all individuals within an organisation. Every individual within an organisation is involved with administration to some extent.

In the past administrative support was acknowledged as the work carried out within an office, using appropriate clerical or secretarial services. The provision of administrative support systems was seen as a low level investment: comparatively simple systems were in operation, unskilled and semi-skilled white-collar workers were recruited from a large pool of labour, and physical resources were stable, easy to obtain and relatively cheap. It has become increasingly expensive to provide the technological support and infrastructure required by today's decision makers. The logic of business would lead us to believe that linked with the increasing expenditure there would be an associated productivity gain. However, there is difficulty in defining tangible gains and many organisations have concentrated on the concept of added value rather than a strict cost/benefit model. There is evidence that organisations survive in an increasingly competitive market but it is difficult to ascribe the added value to one specific factor.

It can be argued that administration systems exist principally to keep management informed, and to assist the decision-making processes of an organisation by providing useful and timely information. To be of use, information must be accessible, both in terms of its availability and in terms of its meaning. Whilst there is an acknowledgement that business and information systems strategies should be aligned or integrated, there is little documented evidence to determine the impact on administration systems and procedures.

We are seeing major changes in many traditional structures and procedures as the implication of continuous technological advancement combined with the rapid pace of change impacts upon business:

- Flatter/learner organisational structures;
- Increasing access to a wide range of data;
- Empowerment of decision makers;
- Ability to work from any location;
- An increasing move away from paper-based transactions.

Organisations seek to implement changes with a minimum or acceptable level of risk, but all change brings with it threats and opportunities. Whilst technophobia, lack of communication, and difficulties in adapting to new situations are some of the threats which may inhibit the acceptance of change, the opportunity for a power shift based on accessibility and availability of information has been acknowledged by Knight and Silk (1990). This begs the question: will the changes in information flows and the underlying provision of administrative systems mean that administration as we know it will perish; can we afford to see administration systems as a peripheral activity within organisations?

The increasing service culture within the UK has led to the development of outsourced business support services, from IT to personnel, purchasing, distribution and transport, and property management. Whilst many of these services were originally part of a supplier driven function or administered alongside the core business functions, the need for increasing productivity, reduction in overheads, and maximisation of asset values, has encouraged organisations to consider the opportunity/cost of outsourcing these support activities. Against this must be set the concerns relating to the loss of control over these functions, many of which have been seen as the backbone of administration. In any organisation which has outsourced any or all of these support services, there will still be an administrative element within the core business function, and a need to keep some measure of control over the outsourced activities.

CONTROLLING THE PROBLEM-CREATING FORCES: THE USE OF NEW STATUTORY INSTRUMENTS IN ADMINISTRATIVE SYSTEMS

One major development affecting administrative systems which rely upon a computerised infrastructure is that of legislation. In response to the potential misuse of technology the government, eventually, created range of legal initiatives - the Data Protection Act (1984), the Copyright, Designs and Patents Act (1988), the Computer Misuse Act (1990) - all of which have implications for modern administrative systems. The ideal is that this statutory provision will prevent the abuse of technology-based administrative systems; the reality, it can be argued, is somewhat different.

Much research has concluded that organisations (and indeed people) operate to only a limited extent within the law: non-registration under the Data Protection Act is rampant; infringement of software copyright is so widespread that the computer industry calculates it loses billions of pounds in revenue each year; incidences of hacking, viruses and computer fraud are well documented.

Technology brings with it the potential for liberalisation of both information and software. Because of this it is therefore relatively easy for organisations and individuals to circumnavigate the law: computer fraud, hacking and viruses are extremely difficult to trace (and it is highly likely that in many instances they go unsuspected and completely undetected); the distribution of information outside the bounds of the register of data users is simple, quick and so widespread that it is impossible for the Data Registrar and his staff to monitor; bodies such as the Federation Against Software Theft, are well aware that even a tiny proportion of the widescale abuse of software copyright would be impossible to police.

Despite the relative ease with which organisations can avoid the ramifications of legal constraints, many undoubtedly do attempt to stay within the law and this in itself can cause further problems for administrative systems. To what extent can security measures prevent computer fraud, hacking or virus planting? Do software audits deter illegal software copying? Will the registration of individual data users throughout an organisation actually prevent information being passed to unregistered third parties?

To the accepted and traditional expectations of administrative systems are added new and even more complex administrative requirements and from this is borne a new problem: organisations now need technology in order to enable their administrative systems to function effectively, yet their administrative systems are incapable to meeting the requirements of legal constraints on the use of technology. Not only is the liberalising use of technology (perhaps properly) being undermined, but the problem-solving measures - the statutory instruments -are in themselves generating new problems for modern administrative systems.

CONCLUSION

Practical experience leads us to believe that there is an issue of freedom versus control in the management of administrative systems. At one level there has been a recognition by organisations of the need for ownership and control of data at a personal level, leading to the empowerment of decision makers. At the same time the framework which has empowered decision makers may lead to a contradiction in that the floodgates have been opened for a potential loss of control of administrative systems.

The mythical assumption that anyone can administer and that few specific skills are required has perhaps been reinforced by the introduction of technology. But, unless there is some measure of workable control (legislative or otherwise), organisations may not have effective policies and procedures in place to support business activities and could find difficulty in capitalising on the proliferation of technological opportunities or solutions.

REFERENCES

Bainbridge, David I., 1993, "Introduction to Computer Law", 2nd ed., Pitman Publishing, London.

Galliers, R.D., and Baker, B.S.H., 1994, " Strategic Information Management", Butterworth Heinemann Ltd., Oxford.

Irving, Richard H., and Higgins, Christopher A., 1991, "Office Information Systems: Management Issues and Methods", John Wiley & Sons Ltd., Chichester.

Knight, A.V., and Silk, D.J., 1990, "Managing Information", McGraw-Hill Book Co., Europe, Maidenhead.

Tapscott, Don., and Caston, Art., 1993, "Paradigm Shift: The New Promise of Information Technology", McGraw-Hill Inc., New York.

83

A LIFE-LINE FOR THOSE ABANDONING CHECK-LIST MANAGEMENT

Anthony Gill

Phrontis Limited
Beacon House
Horn Hill Road
Adderbury, Banbury
Oxon., OX17 3EU
United Kingdom

"There is no one theory providing the ideal plot, but many to choose from, like different soundtracks for different occasions." (Taket and White, 1993)

INTRODUCTION

The challenge facing leaders of organisations in the 90s is how to change - or more fundamentally, how to shift paradigm. Change is part of the dynamic world in which we live; we need to adapt to turbulence on an ongoing basis. Until recently, many organisations were able to get away with being bureaucratic machines in a relatively stable world/business environment dominated by power politics. Today, a large number of these institutions and their leaders are simply not surviving. Who would have predicted the fate of the former role model IBM? Who would have anticipated the swift departures of boardroom chiefs like Robert Horton of BP, John Akers of IBM, James Robinson of American Express, Robert Stempel of GM etc?

The future is increasingly competitive and unpredictable. In having to deal with escalating complexity - continuously changing customer preferences, accelerating technological innovation, global competition, a rapidly shifting political and social scene - managers are finding that their old rules of thumb or check lists no longer suffice. Planning can no longer mean extrapolating the past into the future; we must view 'Planning as Learning' (de Geus, 1988). The 'customer' is the final arbiter of organisational survival, and customers are demanding increasing levels of satisfaction. Herein lies the first paradox that all organisations face: they have to *continue* to produce the goods and services that satisfy customer needs but, at the same time, they have to *change* in keeping with anticipated customer demands.

Employees are also becoming increasingly vocal stakeholders - no longer prepared to be passive victims of the system. They represent the intellectual assets of an organisation walking in and out of the front door. In this situation, *leadership* becomes the key to

Critical Issues in Systems Theory and Practice, Edited by K. Ellis et al.
Plenum Press, New York, 1995

organisational cohesion and survival. The leader needs to define a vision that captures the creative spirit of the workforce and thereby harnesses this enormous energy potential. This *shared vision* reinforcing the need for change has to be supported by a *coherent strategy* which *stretches* (Hamel et al., 1994), rather than simply matches, organisational resources. In order to change, and keep changing, the organisation needs a good understanding of itself now - a model of how the organisation is structured and works - and a direction and method of prioritising change initiatives. Above all, good *communication* with all affected parties, internal and external to the organisation, is essential. The second paradox that organisations have to manage is how to grant *autonomy* in keeping with the concepts of empowerment and self regulation but, at the same time, retain *control* over the whole organisation to avoid its disintegration.

Innovative management thinking is required to take organisations into the next century. A number of organisations have been actively experimenting to find a better way forward. So how do we attract, care for and retain customers in order to survive in an ever increasing competitive environment? Fortune (May 18, 1992) in the article "The search for the organization of tomorrow" offers a clue when introducing the 'horizontal organization'. Typically when customers interact with organisations, their needs are shunted up and down through the various functional departments. Smooth and uninterrupted journeys are rare. The route by which the customer is best served cuts horizontally across organisational functions (Shapiro et al., 1992). This takes us into the domain of process management.

PROCESS MANAGEMENT

There are two key elements to process management: continuous process improvement and what Davenport (1993) calls 'process innovation'. I see these as being on the opposite ends of the 'process scale'. On one side of the scale is continuous improvement as developed under TQM programmes. Here process improvement is evolutionary - a gradual improvement over time. As part of a business strategy, it is driven from the bottom up and is an ongoing activity. At the other end of the scale is process innovation or business process redesign (BPR) which requires revolutionary improvement - a step change over time. BPR involves a radical restructuring of one or more of the major business processes and is more than likely to incorporate Information Technology (IT) as an enabler. Because of the radical nature of BPR and the need for an entire process overview, it is top-down driven.

Business Process Management at an aggregate level can be represented by the processes for Customer Order Management, New Product Development and Managing Organisational Potential.

Processes for *Customer Order Management* are likely to include: demand creation & anticipation; pricing; selling; order processing; scheduling; production; integrated logistics; supplier payments and customer invoicing; goods return and warranty claims; after sales service. Processes for Customer Order management are best performed by cross-functional teams whose focus is on meeting customers' needs today.

Processes for *New Product Development* are likely to include: market research; environmental scanning; technology scanning; research & development; ideas screening; potential product assessment (both technical and marketing); product development; prototyping and testing; product marketing plan; product launch. Processes for New Product Development again need to be performed by cross-functional teams who anticipate and match customers' future needs.

Processes for *Managing Organisational Potential* are likely to include: Leadership (vision, culture etc.); Communication; Resource management (includes finance, tangible and intangible assets, and intellectual assets i.e., people); Organisational learning; Planning;

Total Quality Management. Managing Organisational Potential is the key differentiating process. While processes for Customer Order Management and New Product Development may be replicated to some extent by other organisations through benchmarking programmes, the effectiveness of the processes for Managing Organisational Potential are determined by leadership skills. These are not easily replicated. It is Leaders, at all levels in the organisation, with the ability to harness the creative energy of the workforce in pursuit of shared goals who make their organisations the envy of their competitors. It is also their ability to stimulate and maintain an organisational climate where mutual trust between workers at all levels is the norm. This trust is the basis of open relationships that are necessary for co-operative working.

MANAGEMENT OF COMPLEXITY

In the scientific world, currently four types of systems behaviour are recognised: *fixed* or unchanging; *periodic* such as the planetary movements; *chaotic* i.e., we can establish no apparent order in the system based on current understanding; and *complex* described as "...intermediate between chaotic and periodic behaviour .... where simple systems seem able to organise themselves in complex ways. And this world of complexity may hold the secret of life." (Gribbin, 1993).

As we migrated from the craft based enterprise to the large functional based organisation a general deskilling took place. This led to a dramatic rise in organisational complexity. Whereas before one person was responsible for an entire process, many individuals now became responsible for discreet parts of the process, both the primary or producing processes and the support processes, with the difficulties this brought in regulating the whole. This greatly increased the need for the co-ordination of actions within the organisation. To a large degree the ultimate customer interface has been lost to the majority of workers. We have seen mushrooming bureaucracy in the private sector and particularly in the public sector all over the world. It has taken the recessions of the 80s and 90s and intensifying global competition to begin to dismantle this bureaucracy through the use of outsourcing, flattened hierarchies, outplacement, downsizing, privatisation and market testing.

After many decades of 'command & control' management, organisations are having difficulty in changing the behaviour of people at all levels - workers and managers - in keeping with the ideas of empowerment. Change comes as a cultural shock to most people. The difficult mental shift required is a re-interpretation of *control,* from 'command & control' over people - do this, do that - to self regulation and self organisation. Empowered multiskilled and cross-functionally trained individuals or teams manage, in fact 'own', the process for which they have responsibility - not a distant 'functional manager' as before. Empowering people is not to be regarded as 'another management fad' but rather an essential aspect of managing complexity. It accords with Ashby's Law of Requisite Variety which states: "...only variety can destroy variety" (Ashby, 1956). In essence, this implies that managers cannot know everything that happens within the organisation; similarly, organisations cannot know everything there is to know about the market place. They must instead develop strategies for selectively knowing and managing over time the key indicators of success, and for enabling the bulk of variety to be absorbed through internal self-organising processes and external business partnerships.

Modelling Complexity

As a way to cope with complexity in our environments, we all use models of one kind or another. The models may be conscious or subconscious. Models may be articulated in the form of a system, a mathematical expression or impressionistic drawing. They are all based on our experience of the world and/or logic. Organisations are immensely complex systems; you need only look at your own organisation or local community. Try to imagine all the relationships that are necessary to define the organisation. Your personal viewpoint on this is not likely to match that of a colleague or neighbour. This is not surprising since we are all differently 'conditioned' by our environment (Espejo, 1993).

When it comes to modelling organisations, it is surprising that managers use such poor models of how the organisation works. Typically, these include the functional hierarchy, the divisionalised (including multinational) form and the matrix. Matrix structures gained favour during the 80s and not surprisingly they have met with only partial success and some companies have already discarded them. The reality is that we have seen organisations vacillate between centralised and decentralised structures.

Organisational boundaries are also beginning to get somewhat fuzzy. Single sourcing strategies to improve quality and share research & development costs are becoming more common place. This has the added benefits of increasing the effectiveness of the management of the overall process that becomes intra-organisational. Some organisations are turning to strategic alliances/partnerships in a particularly competitive environment where the cost of research and development is extremely high and there is a limited market for high cost output. The current initiatives of Boeing and the Airbus partners in the development of the new super-jumbos are an instance of this. How is control over these intra-organisational processes to be exercised? We need better models that can cope with all this complexity.

The Viable System Model (VSM)

As a tool to support the study of complexity in organisations the Viable System Model (VSM) (Beer 1979, 1981, 1985; Espejo, 1989) is increasingly gaining favour. It is based on the concept of recursion i.e., a 'system within a system', a 'factory within a factory' and indeed 'a cell within a cell'. As Figure 1 shows, each level of recursion has five systems for achieving viability. Espejo calls these five systems Policy, Intelligence, Monitoring-Control, Co-ordination and Implementation. The viability of the organisational system at each level of recursion is studied by means of the mechanisms of adaptation and monitoring-control. Crucial to this is the principle of self-organisation and self-regulation as a strategy to manage complexity at the local level.

Business processes take place in an organisational context. Although these processes flow horizontally across the organisation, they need to be embedded in an organisation structure that permits the synergistic coherence of each of the processes into a seamless whole which establishes the consequent and unique identity for the organisation.

From the understanding that we have developed on Process Management, it becomes apparent that neither the traditional hierarchical organisation chart nor the matrix organisation structure are particularly useful for studying or implementing Process Management.

The Viable System Model offers a way of braiding business processes with organisational processes so as to produce a coherent structure which indicates how the organisation works. These organisational processes, defined by the VSM, should exist at all levels of recursion and include: self regulation within the constraints of higher level recursion guidelines for organisational cohesion; resources bargaining (intellectual and financial assets); accountability for resources; setting of minimal level dictates; co-ordination of activities to optimise the output of all units operating at the same level while minimising the negative effects of their competing for resources and time; monitoring the appropriate level of recursion; balancing Intelligence with Control to ensure adequate policy formulation; and environmental links.

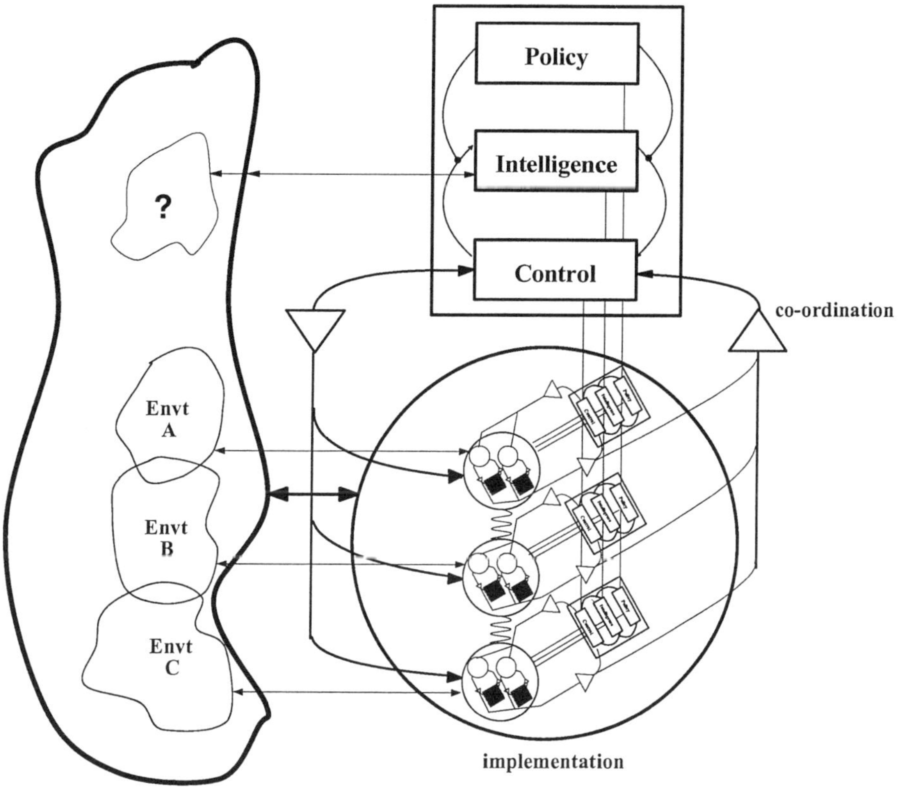

Figure 1. The viable system model (adapted from Beer, 1985).

Systems Dynamics

Senge (1990) recognises two types of complexity: detailed complexity; and dynamic complexity. He rightly concentrates on dynamic complexity - the domain in which systems dynamics has been successfully applied. However, he is too ready to dismiss the notion of managing detailed complexity. I accept that it is impossible to model the universe - perish the thought. However, some modelling of detailed complexity situations is both illuminating and necessary to gain an understanding at a macro level of how an organisation's structure works, thereby facilitating organisational and process redesign. Numerous studies using the VSM have succeeded in doing exactly this.

The strength of systems dynamics lies in the often non intuitive answers obtained from its modelling methods. Developing causal loop diagrams is a powerful way to model the underlying structures of problem situations; having the ability to carry out simulations based on these models though the use of software tools such as Powersim and iThink enables managers to explore dynamic 'what...if' scenarios in a collaborative group learning environment.

A MANAGEMENT LIFE-LINE

In conclusion, for those wishing to abandon check-list management in favour of a more systemic approach to complexity and change, what life-line can we offer? Managers first must have made the necessary mind shift: from believing in a certain and predictable world to accepting an uncertain and unpredictable one. Having done this, managers will: learn to live with paradox and thus not choose between competing alternatives but seek the synergy from choosing a position that embraces both alternatives; move from modelling using spreadsheet software to find out the 'bottom-line' to modelling 'stocks and flows' using systems dynamics software to better understand the behaviour of the system; abandon the traditional organisational chart used to apportion blame, in favour of using the VSM which helps to explain how the organisation 'works'; move from 'chalk & talk' unidirectional flow meetings to facilitated problem-solving workshops using group decision support systems, soft systems methodology, interactive management and other techniques to tap the knowledge, skills and experience of all present.

This life-line does not come cheaply. It requires time and effort to become proficient in the application of the various tools and methods and to appreciate their underlying principles. The emergence of a 'Multimethodology' in support of this life-line is long overdue. Whether we approve of it or not, practitioners are already beginning to combine tools and techniques in one form or another. As long as the integrity of each method is maintained independently, and fitness for purpose is well articulated and understood, such an approach will surely lead in the long run to more balanced and insightful problem-solving interventions.

REFERENCES

Ashby, W. Ross, 1956, "An Introduction to Cybernetics", University Paperbacks, Methuen & Co. Ltd., London.

Beer, Stafford, 1979, "The Heart of Enterprise", John Wiley & Sons, Chichester.

Beer, Stafford, 1981, "Brain of the Firm", 2nd edition, John Wiley & Sons, Chichester.

Beer, Stafford, 1985, "Diagnosing the System for Organizations", John Wiley & Sons, Chichester.

Davenport, T. H., 1993, "Process innovation: reengineering work through information technology", Harvard Business School Press, Boston, Mass.

de Geus, A. P., March-April 1988, Planning as Learning, Harvard Business Review.

Espejo, R., 1989(a), The VSM revisited *in:* "The Viable System Model", R. Espejo and R. Harnden, eds., John Wiley & Sons, Chichester, pp. 77 - 100.

Espejo, R., 1989(b), A cybernetic method to study organisations *in:* "The Viable System Model", R. Espejo and R. Harnden, eds., John Wiley & Sons, Chichester, pp 361 - 382

Espejo, R., 1993, Management of complexity in problem solving *in:* "Organisational Fitness: Corporate Effectiveness through Management Cybernetics", R. Espejo, and M. Schwaninger, eds., Campus Verlag, Frankfurt Ammain.

Gribbin, J., 1993, Aiming at the whole truth, *The Sunday Times*, 31 January 1993, page 6.8.

Hamel, G., & Prahalad, C.K., 1994, Competing for the future, *Harvard Business School Press,* Boston, Mass.

Senge, P. M., 1990, "The Fifth Discipline: The Art & Practice of the Learning Organisation", Doubleday/Currency, New York, U.S.A.

Shapiro, B. P., Rangan, V. K., and Sviokla, J. J., 1992. Staple yourself to an order, *Harvard Business Review,* July - August, 1992, pp. 113 - 122.

Stewart, T. A., 1992, "The Search For The Organization Of Tomorrow", Fortune, May 18, 1992, pp 92-98

Taket, A. & White, L., 1993, After OR: An agenda for postmodernism and poststructuralism in OR, *Journal of the Operational Research Society.* 44(9).

84

COMPARING THE CRITICISED CONCEPTS OF 'RE-ENGINEERING THE CORPORATION' WITH A SYSTEMS STRUCTURE FOR ORGANISATIONAL CHANGE

Kenneth Hammer

Department of Systems Science
City University
London, EC1V 0HB
United Kingdom

INTRODUCTION

The recent set of articles critical of the concept 'Re-engineering the Business', the report by CSC Index, and the Hammer and Champy book 'Re-engineering the Corporation' (1993), require a basis or structure for assessment of the results and realistic value of this offered method of organisational reorganisation. Nine simple principles from systems thinking are assembled into a model of essential aspects of any organisational development methodology.

Comparison of 'Re-engineering' with these requirements indicates that it has some significant shortcomings. This supports much of the content of the critical reviews mentioned above.

BACKGROUND*

During this year there has been a series of informed articles (see for example, the Economist, 1994a,b; Lorenz, 1994) that have taken a critical view of the effects upon an organisation of Reengineering. These articles have displayed a general disquiet about the application of the method but have not disclosed their basic assumptions or the line of reasoning underlying their views. This invites the question; 'in what way can systems principles provide a structured analytical model for critically examining this currently-in-vogue method of organisation development or change?'. The proposed activities of Reengi-

* In the remainder of this article the word 'Reengineering' will infer either the activity or the book by Hammer and Champy

Critical Issues in Systems Theory and Practice, Edited by K. Ellis et al.
Plenum Press, New York, 1995

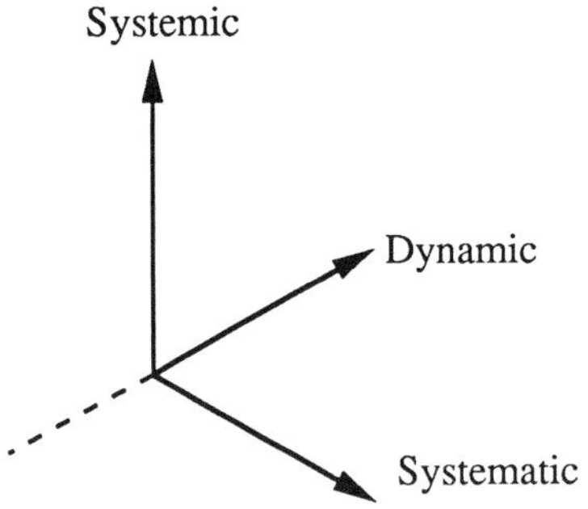

Figure 1. The Systems Thinking basic principle - to think in three dimensions.

neering should be seen as proposed 'interventions into corporate processes to gain clear benefits'.

It is proposed that systems principles can provide a structured analytical method for critically examining this, and other, popular methods of organisational development. An effective framework for critical analysis can be synthesised from several deceptively simple systems concepts, nine in all.

BUILDING A SYSTEMS-THINKING ANALYTICAL FRAMEWORK FOR MANAGING ORGANISATIONAL CHANGE

As a starting point the method uses the most basic principles of systems thinking, the three dimensions of systems thought, the troika - systemic, systematic and dynamic. The bare names of these three dimensions require deeper definition and clarification to adequately indicate their influences.

The systemic dimension requires the maintenance of a wide unrestricted view encompassing the system, the wider system, the environment; it requires the avoidance of simplification by isolation or reductionism - unless this is analytical and to be followed by synthesis.

The systematic dimension is the requirement for orderly, methodical, disciplined ways of thinking and acting; it accentuates the need to impose order upon complexity.

The dynamic dimension requires the determination to cope with the uncertainties of internal evolution, plus the external uncertainties of a continually changing environment. The dynamic dimension relates our designed system to reality; a static system is only a 'snap shot' of a system, devoid of all the dynamic interactions embedded in a 'real' system.

The resulting effects of these three principles is that the systemic and dynamic dimensions both require us to succeed with coping with increasing complexity, despite the increasing difficulties of coping with it, both in our thinking and in setting our design goal. However, the systematic dimension provides many of the ways of coping with complexity; it enables us to contain complexity and to continue to devise ways of managing its exponential increase. Attacking and coping with complexity becomes a tenth systems principle; see below for the other nine.

The next additions come from Ackoff's (1981) three closely linked principles - Holistic, Participative, Continuity. And finally are added Simon's (1960) three principles of design - Intelligence, Design, Choice. For this last set, the use of the word 'Intelligence' is the military sense of intelligence, i.e., inferring the acquisition of relevant and necessary

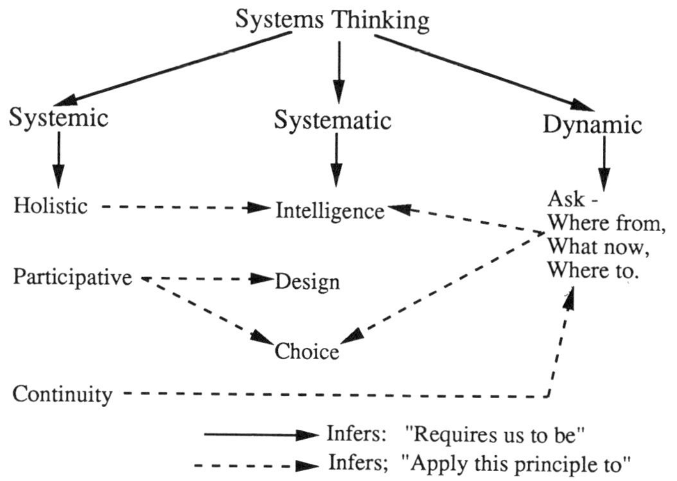

Figure 2. The relationships of nine major principles of Systems Thinking for Design.

information and these three basic principles apply to the development/design of any new complex 'thing'.

In the following diagram the above additional two sets of principles are related to each other and to the initial systems maxims and to the general body of systems thinking.

This then, is the frugal yet adequate structure of analysis for enquiring into methodologies that promise increased efficiency and effectiveness in organisations.

WHAT THESE NINE PRINCIPLES DEMAND OF THE PROCESS OF DESIGNING A NEW ORGANISATION

Systemic

That the design should incorporate attention to all matters internally and externally relevant to the present and future organisation.

Holistic

That all relationships between internal and other relevant matters should be considered and provided for; avoiding sub-optimising for a sub-set of the organisation.

Intelligence

That sufficient information about all the matters in the present situation of the organisation should be obtained, and understood sufficiently to reduce the present design uncertainties to a manageable level for that activity; that sufficient indicators for the possible evolution of these matters be obtained to generate forecasts for those future states to permit well informed reasoning.

Participative

That the wisdom of suitably competent and willing persons should be recruited to provide sources for the above 'intelligence' and for wise judgements of any proposed changes.

Design

This principle argues for the design of more than one solution to the design objectives.

Choice

That a reasoned and effective process of evaluation and choice should support the selection of the best of the alternative designs.

Systematic

That both the design process, and the management of the design process, be well organised and methodical. This requires that the process be guided by prior generated objectives that should achieve previously defined needs in terms of the future attributes or capabilities of the organisation relevant to the present and future organisation, in particular to their relationships.

Continuity

Awareness of earlier states of the involved matters, such that trends might be discerned and wisdom retained; many matters being dynamic/evolving. This principle requires continuity of management after the management of the design process. In particular it reqires managing the implementation and the post-change period to monitor, assess and steer the implemented changes.

Dynamic

Awareness that the future states must be expected to change and that to accommodate these changes organisational flexibility is needed in the design. Awareness that interactions among elements create expanding ripples of change; that dynamic modelling is sometimes feasible, and that recorded assumptions about present and future states are essential for effective management of future organisational adjustments to exogenous and endogenous disturbances and change.

A CRITICAL ANALYSIS OF 'RE-ENGINEERING THE CORPORATION'

The above ten systems principles are now used to assess the 'completeness' of the Reengineering activity as promoted in the Hammer and Champy book of that title (1993). This will be followed by the consideration of modifications to Reengineering deduced from the CSC Index report (1994).

There are in this book some recommendations or advice that comply with the ten systems principles, although often not in a full or robust manner. The deficiencies appear to be critical; they are discussed below, principle by principle.

Systemic Thinking

Little is said about the company's wider system, such as the company's suppliers, customers, regulators. Nor is much said about giving, or how to give, adequate consideration to the issues raised by the environment, such as the all-round competitive nature of free market economies.

Systematic

An orderly method of conducting the process of Reengineering is missing from this book. The lack of a description of the main essentials for good project management, of a clearly very complex activity, is a serious omission. This book gives not a description of a methodology; it is essentially a 'good advice book', however valid and widely known the advice may be.

Holistic

The book calls for concentration on those one or two obvious processes that show considerable potential for improvement, but it lacks thought for the complex interactions linking these with the many other processes in a corporation; this is a reductionist approach and invites local sub-optimisation on a very limited set of 'clear' objectives.

Participative

It does not embrace the involvement of lower levels of management, supervisors or workers, where significant potential wisdom is nowadays considered to exist; as exemplified by the authors' acknowledgement of the concept of empowerment in the 'new design' of the process. In fact the well recognised threats of Reengineering to job security mean that analytic and design support will be hard to recruit from these levels, and often from some senior managers.

Intelligence

It shows weak recognition that adequate intelligence (information) needs to be gathered about the target process AND about its wider system and environment. Neither is there a suggestion that a wide information gathering activity should precede and support the evaluation and selection of that best candidate process for Reengineering. Nor is the need for a detailed study of the present state of the process seen to be required before it is changed, in order to provide a baseline for comparison with the future process for judging the value of the improvements over a broad front that reflects the real complexity of that process.

Design

The need for strong and capable project management of this design process is virtually ignored. For this all-round reconstructive attack upon the inefficiencies in, for example, one complex process that fits within and affects the organisation's main activities (if the offered examples are believed), the management of the design of complex changes without first re-generating the organisation's present strategic objectives would be a risk-increasing action. The same risk-increasing charge would apply to taking action before reviewing the organisation's position in its current operating environment and redefining the organisation's vision of its future state. There is inadequate thought given to the design aspects of extensively reorganising a major process in the organisation.

Choice

No suggestion is made that 2 or 3 designs would enable better decisions to be made via an evaluation of the alternatives designs in order to then choose the design most likely to achieve the organisation's newly defined strategic objectives.

Continuity

There is some thought for the future, but the advice is that the past must be ignored; foregoing all the learning from the past. There is little advice about managing the difficult on-going activities of implementation, testing and correcting the new process.

Dynamic

There is some concern shown that a process might be chosen that could expect external changes to make it obsolete in a relatively short time. But not enough recognition is shown of the fact that the environment is now changing rapidly, and that flexibility to encompass likely changes is an essential prudent aspect of any present day organisational change.

Complexity

The systems thinking principles systemic and holistic inevitably mean that this type of organisation-changing project will both exhibit and experience great complexity; that is, unless the job will be reduced to an exercise of crude aggregations to gain simplicity of decision making. Complexity is only mentioned in passing, and the unfulfilled need to directly address it in the management of the project has already been noted under the principle of 'being systematic'.

In addition, in the Introduction it was noted that Reengineering's planning and implementation of change is an intervention into an organisation, it therefore becomes examinable through Schein's (1988) three models of consultancy. The interventionists are to be seen in the models as consultants, internal or external. Of these, the models 'Purchase' and 'Doctor/Patient' describe significant prior assumptions that are frequently unfulfilled. Reengineering appears to be very susceptible to these shortcomings as it is self-described as weakly participative in a way that excludes it from fitting into Schein's third model, 'Process Consultancy'.

POSSIBLE AMENDMENTS OR ADDITIONS TO THE ADVICE IN THE BOOK

Amendments to the advice offered in the book about Reengineering (1993) are given in the 'Executive Report' published by CSC Index (1994). It summarises the information gained from its questionaire, widely circulated to corporations in N.America and Europe to solicit their views on Reengineering. The associated author of the book, H. Champy, is also chairman of CSC and its report can be assumed to be offering additional advice to those intending to apply Reengineering.

A review of the information in the report finds indications that during the application of Reengineering there is a strong need for the following -

- initiators of reengineering should be ambitious - aim for big gains
- high quality and continuous project management is essential
- full involvement of senior managers is essential
- recognition that there will be extensive changes to the corporate culture
- preparations to deal with anxiety and fear over expected job losses/changes
- a clear high-level overview of and support for the project and its management
- recognition that the exercise will be exhaustive and expensive
- full acceptance that it must fit with the corporation's strategies.

However, a serious question lies over even the first of these additions, to aim high. The state of vitality of the organisation, as much as of the targeted process will determine whether large gains can in fact be made. In a top class corporation a potential gain of 10% may be the limit of the state-of-the-art, regardless of the brilliance of the new-design team. Whereas, a process in a sleepy corporation may have the potential for a 30% gain.

CONCLUSIONS

The reservations of the several recent commentators about applying Reengineering (the book version) appear to be well justified. The shortcomings indicated by this systems-principles analysis indicate that there will be marked reductions in the efficacy of the proposed 'interventions into corporate processes to gain clear benefits' - unless adequate detailed amplifications of the method of conducting the Reengineering are published, and such that the amplifications answer the shortcomings indicated by the commentators and by this analysis.

A further conclusion is that this systems framework is an effective tool for conducting a critical assessment of any method of organisational design or change.

REFERENCES

Ackoff, R.L., 1981, "Creating the Corporate Future", John Wiley & Sons, New York.
Anon, 1993, The wonders of workflow, *The Economist*, 11:12, London.
Anon, 1994a, Re-engineering reviewed, *The Economist*, 1:7, London.
Anon, 1994b, The vision thing, *The Economist*, 3:9, London.
CSC Index, 1994, "State of Reengineering Report", CSC Index, San Francisco.
Hammer, M., and Champy,J., 1993, "Reengineering the Corporation", Harper Collins, New York.
Lorenz, C., 1994, Re-engineering in small doses only, *Financial Times*, July 1st, London.
Schein, E.H., 1988, "Process Consultation", Vols. 1 & 2, Addison-Wesley, New York.
Simon, H.A., 1960, "The New Science of Management Decision", McGraw-Hill, New York.

CREATIVE MANAGEMENT

Jon-Arild Johannessen

Bodø Graduate School of Business
8002-Bodø
Norway

INTRODUCTION

The main aim of this paper is to develop a basis for creative management built on cybernetic and systems theoretical principles. The questions brought up for discussion are as follows : how can one achieve a creative process in organisations ridden of fragmentation and inner demarcation problems? How should creative organisations be organised? How can one create the best possible circumstances for enabling new ideas to be generated and grow? How can one develop management methods in order to enhance the creative potential in the organisation? If we can answer these questions, then we have indirectly also singled out some factors with a stifling influence on creativity.

Finally in the article eight general postulates on which the theory of creative management as seen here, are put forwards.

CREATIVE MANAGEMENT: FOUNDATIONS

Increased turbulence and complexity will require a completely new form of knowledge. Creative management is a form of management which does not emphasise adaptation to the environment, but which does focus on creative processes directed at creating the systems environment (Weick, 1979; Ackoff; 1981; 1984). Control themes in creative management are : connection, totality, relationships, patterns which combine, total service, elicitation and development of knowledge, communication, autonomy and responsibility for the employees, autopoiesis and dissipative structures.

The system as a whole should metaphorically be looked upon as an ecological system, where the resources are introduced in response to questions like: why does the system exist ? What is its real function? What consequences does this have for the environment ? It is important here to bear in mind Bateson's postulate (1972 :493): "...the creature that wins against its environment destroys itself".

One of the main guiding concepts of creative management is: from concepts devoid of experience to concepts close to experience. If these concepts, methods, models and theories devoid of experience cannot be coded to more experience-related concepts, they will remain empty formalism. A 'reality' does not become more 'true' because it puts on a

Critical Issues in Systems Theory and Practice, Edited by K. Ellis et al.
Plenum Press, New York, 1995

formalist and rigorous symbolic language. If this decoding process cannot be done, practice will come to resemble 'management by confession'.

By emphasising explicit details pertaining to the requirements of a rational- choice sequential problem-solving process, management science creates biases of the kind that Veblen refer to as 'trained incapacity', notably pattern-blindness and lack of understanding of broader relationships.

The total views of organisations which has been much of the starting point for management study (Carlson, 1951; Mintzberg, 1973), has completely changed character, amongst other things because of the fragmentation and segmentation which has developed in management. The individual fragments moreover are not tied together by a search for total relations between patterns binding the whole problem together.

Perhaps we should change our understanding of what management really is? Possibly the time is ripe for a creative wave, in which technical partial understanding no longer takes precedence, but where the search for linked patterns and understanding of broader relationships is epitomised while at the same time retaining the functional merits of the traditional approach.

When technical partial understanding and fragmentation have precedence, we are unable to see unforeseen negative consequences which cannot be understood in an analytical fragmented way of thinking.

It is not so much new ideas within existing ways of thinking we need, but rather new ideas about how we think. The answer does not lie in amassing more and more knowledge inside an existing framework. What we ought to set the spotlight on is how we can transform data to information, information to knowledge, knowledge to insight, and insight to wisdom.

BREAKING A NEW LAID EGG TO STUDY THE CHICK AS IT GROWS UP

Overcoming fragmentation must not be confused with, in the case of specific well defined problems, the fact that one ought to define and analyse in order to evaluate more closely causal series. It is, however when one defines and employs linear causal lines of thought related to conditions which are of non-linear nature that fragmentation must be overcome. In non-linear causality one should pay more attention to relationships, patterns, intertwined patterns and understanding of the total situation. When one isolates and analyses circular loops one breaks up patterns which are vital for understanding. It can be compared with breaking a new laid egg to study the chick as it grows up.

A dynamic mutual relationship between the specialist and the generalist, between problem- definition and problem-solution, between analysis and synthesis can seem to override the negative consequences of fragmentation. There are, however quibbling changes of power which push organisations into an emphasis on analysis and fragmentation.

A continual search for broader contexts in which larger relationships fit in is an important element in creative management. This is important in order to avoid clear demarcation lines between problems which are in fact connected to one another, albeit by subtle underlying structures. If such creative expansion does not take place, fragmentation will lead to a cumbersome organisation and internal struggles on the borders of the various system components. The organisation will therefore use a lot of its time and energy in finding solutions for such energy spills. If the worst comes to the worst they will use so many resources on inner problem solving and opportunistic behaviour that very little energy will be left for solving tasks that the organisation was originally designed to perform.

As regards the development from rational economic leadership (economic man) over to strategic management to creative management, there is a discontinuous leap in tackling situations and leadership style, but it should not be implied that there are absolute conceptual barriers between the three stages of development. It is a question of thinking in a different way not thinking about different things.

METAPHORS AND MANAGEMENT

When in rational management (economic man) one uses metaphors in which one sees the organisation as a machine, then this use of metaphors has considerable implications for how such systems are managed. The individual parts of the organisation are metaphorically associated with machine components. This leads to a stiffness and rigidity in relation to understanding processes of change. The change from using the machine metaphor to, for example, considering organisations as living systems, where amongst other things the brain metaphor has been used has lead to the development of new models for organisations treating these as living systems (Beer, 1979;1981;1985). A further development of the brain metaphor has lead to the growth of ideas about organisations metaphorically looked upon as holography (Morgan, 1988; Johannessen, 1991). This metaphor has sprung out of the thought of the brain being considered as holographically organised. If one takes the holographic metaphors literally, it can also be considered as extremely static. A further development of this metaphor is to consider organisations as 'holo-movement' (by analogy with Bohm, 1983), incorporating a dynamic element which does not exist in holography.

'Wrong' use of metaphors can harm a system, and can hinder organisational learning. New metaphors can assist breaking out of old ways of thinking, so that new ideas can generate new insight, each in its turn creates new ideas. Using metaphors can lead to narrows reason being overcome and new insight being established, and the associative thought can be used to break out of conceptual series which are too tied to reason categories and therefore a constraint to creativity. The metaphor in a way becomes a marker which releases creative energy, a bridge uniting two distinct processes of recognition.

One metaphor for the creative process itself could be 'the open child's mentality', where play and questions are central elements. The importance of play is also stressed by Weick (1979 : 248). One of the features of this metaphor is to get an understanding that new ideas do not need to be ripe at the outset. If the demands set for a creative idea are set too high and must result in practical implications, then the whole creative idea generation process gets blocked slowly but surely. No one will be interested in communicating ideas which they themselves do not know the outcome of, which is often the case when new ideas are formulated. If idea generation and creativity are looked at metaphorically as in 'the open child's mentality', as play, then a whole new attitude will pervade the internal environment.

In creation and play one can always put established truths under the searchlight and look at the arguments, premises, presuppositions and thinking on which these truths rest. Through play used as a metaphor one can allow oneself to doubt what is accepted as a basis for management, organisational theory, behavioural theory and marketing amongst other things. It is the irrational leaps away from the established thought structures that creation can help us with, while, on the other hand, research into the consequences demands rigorous thinking of insight obtained through creation.

According to creative management a new function ought to be designed on the different levels in the organisation. A new function having creativity as its main purpose. This function has as its main task to establish creative groups which are organised, both horizontally, vertically and diagonally in the whole organisation. In this way people from different inner environments can participate in the creative process. With the constitution of complementary groups, put together in a way which reflects different subjects and preferably also different epistemologies, the groups understanding of reality can be enlarged and greater creativity can be expected than if people with more or less the same background experience and ways of thinking constitute a work group. Establishment of creative groups are composed according to the complementary principle (Weinberg, 1975: 119-120), in other words at least two worlds of knowledge are brought together, two distinct personality types, and two different 'inquiring models' (Linstone, 1984: 35).

COMMUNICATION AND CREATIVITY

Knowledge does not merely generate more knowledge, it also generates non-knowledge. The more knowledge one acquires, the more one become aware of ones own lack of knowledge or ignorance. This is one of the reasons for divergent thinking being so important as a basis in the creative process.

Use of communication-related information technology will lead to ideas being spread faster, and ideas previously unable to come into contact with one-another meeting, and new ideas can be generated in this 'meeting'. The closer one can link ideas in such a communication network, the greater the creativity which can unfold itself in the network.

Use of electronic meeting systems (EMS)in strategic management has in some case studies (Dennis et. al., 1990: 34-43) shown an increase in perceived effectively of 85 present in one case, 88 percent in another case and 94 percent in the third. The freedom and openness which this type of 'computer dialogue' leads to, gets more people to come forward with creative idea generating, and the fear of a 'bad' idea being associated with ones own person disappears. Anonymity in the idea-generation process works in this way like a valve which gets participants both to openly generate ideas, and to evaluate other peoples ideas on a non-idiosyncratic basis. Subtle hierarchical pressures are demobilised by the use of EMS, and the number of persons participating in this process increases.

Fragmentation also has clear negative implication for communication in organisations. There will be 'troop movements' in defence of established knowledge borders, instead of communication across borders. Communication will thereby occur even more within the borders with 'idea inbreeding' and degeneration of creative processes as a result.

If organisations are to be transformed into creative systems it is a prerequisite that open lines of communication are established, not along the established hierarchical structures, but across such bureaucratic and social infrastructures. Communication related information technology will here constitute a really creative crux. The prerequisite is however that no barriers have been put up in the network, but that communication can go on both horizontally, diagonally and vertically and in both directions. Putting an emphasis on information technology in organisations should be related to the question: in which way can these investments further intra- and inter-organisational communication ?

CREATIVE ORGANISING

There are psychological 'sunk costs' associated with new ways of organising. A new form of organising is often looked upon as negatively related to the old way of organising. When such observations form the basis, the introduction of new organisational forms leads to internal crisis in the systems. The question which can be asked is whether or not different organising forms can and ought to inhabit the same organisation at the same time? Metaphorically one can express this as having the mechanical and organic metaphor as a thought basis for the same organisation, or that the hierarchical and heterarchical organisation form can operate within the same system. A prerequisite for this to be possible is that responsibility, autonomy and self-organising become mainstays of the subsystems in the organisation.

Organisational creativity is made difficult in organisations where the structures have more or less stiffened, and where change metaphors are not built into the system. Creatively new ideas will, in such organisations, be seen more as a threat to 'stability', while Bateson (1972: 95) by his analogy to the tight rope walker shows that the opposite is the case. Bateson's point is that if the total system is to be in equilibrium and stability, this presupposes that the parts of the system are in a continual process of change. This analogy can be used to advantage in organisations, and is here expressed by the metaphors : chaos *is* order and diversity *is* unity.

CONCLUSION - GENERAL POSTULATES IN CREATIVE MANAGEMENT

1. Breaking the new laid egg to study the chick as it grows up : Creativity and fragmentation are incommensurate concepts. Fragmentation has to do with partial understanding while creativity is linked to totality. Specialisation is closely related to fragmentation, so overriding specialisation's borders is one of the prerequisites of creative management.
2. Chaos is order: Organisations contain a totality of structures and forms simultaneously.
3. The whole is embedded in the parts: The smallest units in the organisation should be organised in analogy with the total organisation. In this way each individual vibrant unit in the organisation can handle the demands of the environment, and generate new ideas for its specific surrounding without having to go through the organisations top leadership. Development of component systems in this way means that one becomes less dependent on the complexity of the totality and possible collapses due to large outer and/or inner tensions. This is one way to build in creative redundancy into the system, in order to prevent inherent negative consequences related to an impending collapse of the total structure.
4. Stability is change: Change in the component systems are a prerequisite for creativity in the total system. Rigidity in the component system is the first sign of decay in the total system.
5. Troop movements: Reward systems in creative organisations must be of a kind that optimise organisational creativity
6. Incidents have precedents : The internal and external information systems in the organisation must emphasise the design in order that possible 'butterfly' effects can be discovered as early as possible. A repertoire of strategies to correct the system as a reaction to 'butterfly effects' must also be developed. The internal and external information systems in the organisation must be designed with a special views to incidents. It is in times of turbulence in systems where circular causality makes up the majority of the systems links, and where complexity is high, that non-linear relationships with small changes in a variable can release catastrophic changes, because of tightly inter-related feedback-loops.
7. The active variation principle: The organisation must allow and encourage 'dissipative structures', described here as the growth of sudden new structural changes and organisational forms within individual parts of the organisation, to develop freely inside the overriding structure.

8. There exist no boundary: The system must develop underlying subtle mechanisms for doing individual 'viable systems' to 'autopoietic systems'. In this connection 'autopoietic systems' means the individual systems and surroundings not being separated by a border. System and surroundings can be seen as one unit, as one ecological system i.e. it is not the system *against* an environment, but the system *in* an environment that is interesting. This leads to completely different implications for production and total service.

REFERENCES

Ackoff, R.L., 1981, "Creating the Corporate Future," John Wiley & Sons, New York.

Ackoff, R.L., 1984, "A Guide to Controlling your Corporate Future," John Wiley & Sons, New York.

Bateson, G., 1972, "Steps to an Ecology of Mind," Intex Books, London.

Beer, S., 1979, "Heart of Enterprise," John Wiley & Sons, New York.

Beer, S., 1981, "Brain of the Firm," John Wiley & Sons, New York.

Beer, S., 1985, "Diagnosing the System for Organization," John Wiley & Sons, New York.

Bohm, D., 1983, "Wholeness and the Implicate Order," ARK paperbacks, Routledge & Kegan Paul Ltd., London.

Carlson, S., 1951, "Executive Behavior : A Study of the Work Load and the Working Methods of Managing Directors," StrØmbergs, Stockholm.

Dennis, A.R., Nunamaker, J.F.Jr., Paranka, D. and Vogel, D.R., 1990, A new role for computers in strategic management, *Jrnl. of Bus. Strtgy.* 11:5, pp.38-43.

Johannessen, J.A., 1991, The holographic organization-a design model, *Cybernetics and Systems An International Journal*, 22, pp.41-55.

Linstone, H.A., 1984, "Multiple Perspectives for Decision Making", Elsevier Science Publishing Co. Inc., New York.

Mintzberg, H., 1973, "The Nature of Managerial Work," Harper and Row, New York.

Morgan, G., 1988, "Holographic Organization and the New Technology," Unpublished paper York University, Toronto.

Weinberger, G.M., 1975, "An Introduction to General System Thinking," John Wiley & Sons, London.

Weick, K.E., 1979, "The Social Psychology of Organizing", Addison Wesley Publishing Company, London.

ADMINISTRATIVE REFORM AND ADMINISTRATIVE SELF-REFERENCE

Panagiotis Karkatsoulis

National Centre of Public Administration
2, Evangelistrias Str.
GR- 105 63 Athens
Grcccc

INTRODUCTION

The concept of 'reform' is extremely attractive for the social analysts. Its definition, however, is not equally easy to derive, as it is the case for the concept of the other extreme, the concept of 'revolution'(von Krockow, 1976; Greiffenhagen, 1978; Böhret, 1983).

The inclusion of 'reform' in the political vocabulary has created a lot of confusion, because it was proceeded by the entrapment of the concept within the dominant evaluative bi-pole 'conservatism' versus 'progress'. The debate has been reduced to whether a 'reform' can be in general 'conservative' or 'progressive' and consequently, the discussion on the results of specific reforms has been downgraded. The fact that certain political parties have been characterised as 'reformative' is very indicative of the political semantics that the concept of 'reform' has vested (Narr, Scheer and Spöri, 1976; Hennis, 1977).

In search of a more concrete definition of the concept of 'reform' in the more specific area of 'administrative reform', one can find a multitude of definitions and descriptions which already constitute a separate scientific branch in the field of the science of Administration. We will cite some of these definitions in order to demonstrate their extremely weak analytical potential. Thus 'Reform' can be defined as a change corresponding to a specific social demand deriving from the functional shortcoming of an institution.

Since 'reform' aims at the maintenance of the social status quo, it is a slow and continuous procedure: its final goal being the linking of the 'new' with the 'old' (von Krockow, 1976).

Reform is being characterised by the 'principle of perfection', since it always seeks the constant amelioration of the object under reformation. Therefore, reform is being perceived as something undoubtedly 'good' and whenever something is ineffective and dysfunctional it can be corrected if it gets 'reformed'.

Such a wide range of conceptual definitions of 'reform', however, can not easily correspond to the modifications of an administrative system. Bearing in mind that there are constant modifications and shifts in an administrative system, one may put the question: 'Which of these changes can be considered as reformative ?' (Caiden, 1969; Schnur, 1970).

Critical Issues in Systems Theory and Practice, Edited by K. Ellis et al.
Plenum Press, New York, 1995

A criterion of selection which is being suggested is the size of the change which a reform in question achieves. In other words, a reform which actually manages to change primordial rules of attitude is a real one and not merely a simple internal change (Schnur, 1970). However, this very reform substantially contributing to the smooth operation of a specific organization might prove disastrous, if it were to be transposed into another one. It is also possible, that a reform acting positively on one element of the certain administrative system (i.e. on a department or a direction) can act negatively on another. Therefore, not only is there no guarantee of amelioration through reforms but it is also possible to reach through them the exactly opposite result. So, despite the fact that it seems that reform can not be expressed by a uniform conceptual definition and it can only be given an empirical, ad hoc definition, the theory of administrative reform has dealt only slightly with this subject.

What is usually being examined, is, whether the agent of the administrative reform -the legislator, the jurist, the politician- covers the prerequisites for the implementation of the reform or, whether, there is resistance against it without a parallel examination of the very concept of reform beyond the framework we have already set above. Political scientists, for example, do not consider Public Administration as an autonomous social system, but as a PAS (political-administrative-system); on the other hand, jurists tend to shrink administrative reform into the mere amendment of some articles of the legal code of Public Administration (Sidentopf, 1979).

Even the poor results of the efforts made through the OECD technical assistance programs during the 1960's and 70's were not convincing enough for the specialists and did not serve as a starting point of reflection for them in order to try to deal with the very concept of 'reform'. A very characteristic example of such an evolution is Greece; while the country presents a development during the '60s, mainly due to the fixed formulas of the OECD, in the two following decades of the '70s and the '80s, it shows recessive tendencies which, however, reflect the actual Greek reality of intra-administrative development.

There are, of course, isolated voices rising higher and higher, claiming that it is urgent that each administrative system should set its own 'internal' definition of 'reform'.

However, in order for another perception of reform to exist, there has to be an utterly different all-embracing conception which must be free of the partiality of either the normative and axiomatic spirit or the empirical one which run throughout the sciences of law, politics and administration. Such an approach - the most comprehensive for all I know- is undertaken by the German sociologist Niklas Luhmann (1964, 1965(a), 1965(b), 1966(a), 1966(b)) in the version of social systems theory that he has developed.

PUBLIC ADMINISTRATION AS A SELF-REFERENTIAL SYSTEM

According to his basic position, society is composed of an aggregate of autonomous self-referential systems which are all interdependent. These systems are composed of communicative meanings of the actions performed by the social actors pertaining to them.

Therefore, there are no meanings independent of the system or constant in time, nor is there any kind of hierarchy amongst them. There is only the rationalism of the (scientific) system, the justice of the (legal) system, etc. The way the partial meanings interconnect within the system depends on the decision of the system itself which is in correlation to the degree of its self-reference. In this case, we have the notion of the 'operational closure' developed mainly by Humberto Maturana and Francisco Varela (Varela, Maturana and Uribe, 1974; Varela, 1984).

We should underline here that precondition for the existence of a self-referential system is the identification of its particularity vis-à-vis the other social systems. Luhmann here rests with the doctrine of the English mathematician and philosopher George Spencer Brown (1972), who identifies the Difference of a system from the others as the core for the primordial formation of its self-reference. This initial Difference consists in a dual code (positive-negative) which through a process of re-entry serves as the structure of the system (Spencer Brown, 1972).

According to the above mentioned, in order to define Administration as a self-referential social system we should identify beforehand its peculiarity.

According to Luhmann (1966(a)), Public Administration stands aside of all the other social sub-systems, because it is the only one producing decisions which have a binding effect for their recipients, that is, for the other social subsystems. A school, for instance, or a bank, are confined in their administrative role by decisions already taken by 'Public Administration' which has been characterised by Luhmann (1983) as "... the Administration par excellence".

However, the notion of decision in this specific case differs from both the mathematical and legal aspects of it. 'Decision', here, is not perceived as the procedure of "internal selection among many options" but as an act of communication (Luhmann, 1978).

So, if a self-referential administrative system is defined as the system producing collective binding decisions, then politics, the personnel and the public which by tradition are considered to be elements of the system, constitute its environment.

Obviously, the relations of the administrative system with its environment are being reflected in its interior, thus creating a 'second environment', as a matter of speaking. We consider as administration only a system of production of collective binding decisions; and this is the most important key-element, because it touches also the issue of administrative reform.

Through the study of the nature of these decisions, one can observe that they have an internal structure which also defines the prerequisites for this decision-taking. Herbert Simon called this prerequisites 'behaviour' and later on 'decision premisses' (Simon, Smithburg and Thompson, 1950). The prerequisites for the decision-making remain unaltered, regardless of the various stimuli of the environment. Their structure is such that it allows them to operate as Norms of judgment for any individual case and they are not suspended even when the expectations accompanying each separate case are eventually defeated. Therefore, these prerequisites for making a decision can only alter through another decision of the system itself. In other words, the administrative system decides on the basis of its own specific code both for its programs and frameworks of its decision-making policies and for each separate case.

Precisely this feature of the self-referential system has direct consequences to the efforts made for its reform. Contrary to what the classic theory sustains, the existence of certain constant and unaltered information concerning the efficiency development, is not sufficient per se; this information should also be comprehensible, it should be able to 'make sense' to the administrative system (Luhmann, 1983).

If this information doesn't make sense for the specific administrative system, then it is absolutely uncertain whether it will be beneficial to this system. In that sense, in order for any administrative reform to be accepted by the administrative system, it must be 'legible' in a way that the system will manage to adjust its already existing structure to it. This, however, is feasible only when the Administration disposes of learning mechanisms, as for example the so called 'reflexive mechanisms', through which it will be able to absorb every new piece of information and reject any redundancy (Luhmann, 1970). Bearing all the previous in mind, we can define administrative reform as a preparation procedure of the Administration aiming at enabling it to be self-adjustable.

TOWARDS AN AUTOPOIETIC THEORY OF ORGANISATIONS

The deriving consequences from such a definition of administrative reform have direct effect on the principles of management and organisational theory. We further point out the most important ones, in reference to the work of Niklas Luhmann 'Ökologie des Nichtwissens' (1992).

a. Opposing what classic theory sustains, organisations do not constitute systems achieving goals which have been set beforehand, but systems seeking for goals, in particular, within themselves. Their occasional goals are being set depending on the planning they have previously set.
b. The eventual planning, programming and monitoring are composed of actions occurring within the system. In other words, the system acts after having grasped the reality through the specific means of perception that only the system disposes of. 'Reality' does not define the system's attitude in an 'deterministic' way. In the relevant literature on 'second order cybernetics' (von Foerster, 1981; Glanville, 1988) has often been demonstrated that there exist neither the monocracy of objective reality lying 'out there' nor the irrosion of solipsism: There exist only a co-evolution of systems observing each other.
c. Organisations constitute non-trivial machines, in other words, machines determining their way of operation according to the output they generate at a precise time. If this output is being altered or transferred at a given instance then, the attitude, the reaction of the organisation is utterly different. Therefore, what is important is not the technology of transformation of the information but the particular attitude of the system's organisation vis-à-vis this information, and this depends, of course, on the degree of self-reference this system disposes of (von Foerster, 1984).
d. Communication between self-referential/autopoietic systems is achieved through the creation of a network of loose-couplings and not through the formation of fixed functions between vertically hierarchical levels as the classic theory argues (Weick, 1976). This implies the existence of mutual loose and indirect controls, within a network of mutual understandable meanings in the sense of Parsonian 'interpenetrating'.
e. Planning and control in self-referential organisations do not have the meaning of the a priori prediction but, mostly, of the covering each distance occasionally occurring between reality and professed goals. These proceedings of the system must be conceived as a continuous, ex post corrective effort.

REFERENCES

Böhret, C., 1983, Reformfähigkeit und Anpassungsflexibilität der öffentlichen Verwaltung, *in*: "Politik und Verwaltung", C.Böhret, Westdeutscher Verlag, Opladen, p. 28.

Caiden, G., 1969, "Administrative Reform", Adline, Chicago.

Glanville, R., 1988, "Objekte", Merve Verlag, Berlin.

Greiffenhagen, M., 1978, "Zur Theorie der Reform", C.F.Müller Verlag, Heidelberg-Karlsruhe, p.7.

Hennis, W., 1977, "Organisierter Sozialismus. Zum "strategischen" Staats- und Politikverständnis der Sozialdemokratie", Klett-Verlag, Stuttgart.

Luhmann, N., 1964, "Funktionen und Folgen formaler Organisation", Duncker & Humblot, Berlin.

Luhmann, N., 1965(a), "Grundrechte als Institution. Ein Beitrag zur politischen Soziologie", Duncker & Humblot, Berlin.

Luhmann, N., 1965(b), "Öffentlich-rechtliche Entschädigung rechtspolitisch betrachtet", Duncker & Humblot, Berlin.

Luhmann, N., 1966(a) , "Theorie der Verwaltungswissenschaft", Grote, Köln-Berlin, p. 69.

Luhmann, N., 1966(b), "Recht und Automation in der öffentlichen Verwaltung. Eine verwaltungswissenschaftliche Untersuchung", Duncker & Humblot, Berlin.

Luhmann, N., 1970, "Soziologische Aufklärung I", Westdeutscher Verlag, Opladen, pp. 92 foll.

Luhmann, N., 1978, "Organisation und Entscheidung", Westdeutscher Verlag, Opladen, p. 15.

Luhmann, N., 1983, Reform und Information, *in*:, "Politische Planung. Aufsätze zur Soziologie von Recht und Verwaltung", Luhmann, N., 3rd. ed., Westdeutscher Verlag, Opladen, pp. 181-182.

Luhmann, N., 1992, "Beobachtungen der Moderne", Westdeutscher Verlag, Opladen.

Narr, W-D., Scheer, H., and Spori, D., 1976, "SPD-Staatspartei oder Reformpartei?", Piper-Verlag, München.

Schnur, R., 1970, Widerstände und Schwierigkeiten bei Verwaltungsreformen, *Deutsches Verwaltungsblatt* 19:753.

Sidentopf, H., 1979, Abschied von der Dienstrechtsreform?, *Die Verwaltung* 12:457.

Simon, H.A., Smithburg, D.W., and Thompson, V.A., 1950, "Public Administration", p. 57, Knopf, New York.

Spencer Brown, G., 1972, "Laws of Form", Julian Press, New York.

Varela, F.J., 1984. Two Principles of Self-organization, *in*: "Self-organization and Management of Social Systems", Urlich, H., and Probst, G.J., eds., W.de Gruyter, Berlin.

Varela, F.J., Maturana, H.R., and Uribe, R., 1974, Autopoiesis: The organization of living systems, its characterization and a model, *Biosystem*, vol.5, pp.7-32.

von Foerster, H., 1981, "Observing Systems", Seaside, Cal., USA.

von Foerster, H., 1984, Principles of Self-Organization in a Socio-Managerial Context, *in*: "Self-Organization and Management of Social Systems: Insights, Promises, Doubts and Questions", Urlich, H., Probst, G.J., and Gilbert J.B., eds., W.de Gruyter, Berlin, pp. 2-24.

von Krockow, C.G, 1976, "Reform als politisches Prinzip", R.Piper & Co Verlag, München, pp. 8-82.

Weick, K., 1976, Educational organizations as loosely coupled systems, *A.S.Q.* 21, pp.1-19.

MANAGEMENT SYSTEMS FOR EVOLVING NETWORKS

Garry Marshall

School of Computing Science
Middlesex University
Bramley Road, Trent Park
London N14 4XS
United Kingdom

INTRODUCTION

The products of advanced information technology are required to perform complex tasks and for this reason must themselves be correspondingly complex. They consist typically of large numbers of sub-systems with complicated patterns of interconnectivity. In addition, they must be able to respond to demands that are essentially unpredictable in terms of their times of arrival and perhaps also in terms of their content. In these broad terms, many high-technology products are 'complex stochastic systems'.

There are, of course, problems associated with the design, management and maintenance of any system, but with increasing system complexity these problems rapidly become more acute. As the number of possible occurrences within a system, and their combinations, increase, 'back of the envelope' methods of design and ad hoc means for management and maintenance become less and less reliable. The need for systematic, formal methods becomes greater.

Modern communication networks are prime examples of complex stochastic systems. Systems such as the telephone network and the Internet are as large and complex as any. In addition, communication networks must be able to carry traffic that arrives at unpredictable times, and is variable in its content. These networks need to be managed in order to give an efficient and reliable service to their users in the face of, among other things, different traffic loadings, faults within the system, and changes to the system configuration such as the addition of new nodes.

General and systematic approaches to network management have already been adopted (Brusil and Stokesbury, 1989), not least within Open Systems Interconnections (O.S.I.) (Caruso, 1990). The aim of this paper is to describe a new approach to the management of evolving networks such as the Internet and to demonstrate in a rather simple way how they might operate. These ideas may well be of relevance to the Information Super-Highway (Reinhardt, 1994), and the approach should be applicable to the automatic management of systems other than communication networks.

NETWORK MANAGEMENT

A network can be seen as a collection of resources linked by communication paths. If the resources are to be made available consistently and reliably to the users of the network, the network will have to be managed. In addition, management benefits the network operator in that it provides a value-added element which not only acts to ensure that the underlying network operates efficiently to provide a guaranteed quality of service but also helps to differentiate one network from another.

Network management is usually divided to give aspects, each with its own responsibility. These aspects are usually identified as performance management, fault management, change management, security management, and accounting management. They are, of course, interdependent. Change management is usually understood as referring to the management of physical changes to a network, but problems also arise when the capabilities of a network are extended by the provision of additional services. A new service can clash with existing services, giving rise to the so-called 'feature-interaction problem' (Griffeth and Lin, 1993). A classic example arises when the capability to display the number of an incoming call is added to a system in which callers may choose not to disclose their numbers. Such a problem may, in general, be resolved in various ways, and it is even possible to negotiate an appropriate solution at the time of the occurrence. All the same, it is necessary both to recognise the conflict and to provide ways of dealing with it. The methods describe here should deal with the feature interaction problem in the same way as they manage change in general.

A NEW APPROACH

Here, we consider the problem of maintaining a required level of network performance under conditions of change. This includes change to the configuration of the network whether deliberate, as when adding or removing new nodes, or accidental, as with the occurrence of failures. It also includes change caused by adding new services. This problem is relevant to fast-growing networks such as the Internet, and will also be an issue for the Information Super-Highway. The task of adaptation to change can be cast as one of redesign, in the sense that one way to maintain the required level of performance from the network in the presence of change is to redesign it completely so as to incorporate the changes in the most appropriate way. When this is done, the network management software may equally be seen as design software. One benefit of adopting this approach is that established body of knowledge about design may be brought to bear in the less mature field of network management. It may be worth noting that, as long as the network under management was deliberately designed in the first place, some design software will be available.

The question then arises as to the approach to design that should be incorporated in the software. One way of distinguishing possible approaches is to classify them as monotonic or non-monotonic. A monotonic approach leads to incremental development in which each change must be dealt with in the context of the current network. By and large, this leads to the adoption of ad hoc solutions: the network is forced to carry its history along with it. A non-monotonic approach will support redesign from scratch each time. Any earlier solution to a specific problem may be withdrawn and replaced by one that is suitable in the context of the overall network specification. The need to keep the items under management 'soft' is readily apparent. The rest of this paper describes and illustrates a non-monotonic approach to the management of evolving networks.

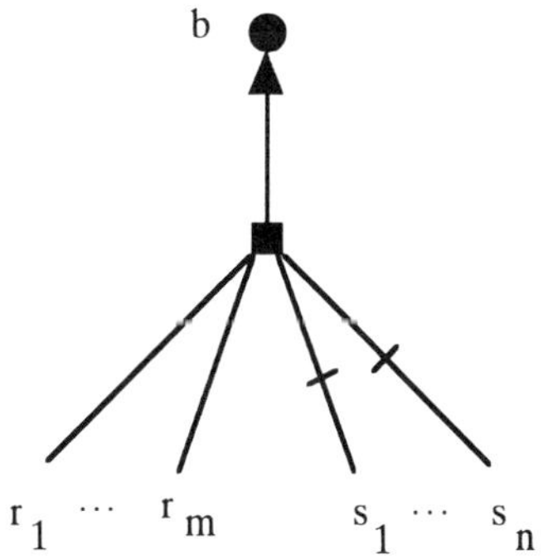

Figure 1. Graphical representation of a justification.

NON-MONOTONIC TECHNIQUES

When the specification of a system is changed, the new system can be redesigned from scratch. Clearly, with a large, complex system this is not practical. However, if all the parts of the system affected by the change can be identified, then the task of redesign is limited to these parts and becomes correspondingly more simple. One technique for doing just this is Doyle's Truth Maintenance System (Doyle, 1979).

The following brief and informal description is intended to indicate the basic idea of a Truth Maintenance System, and follows Pratt (1994). A Truth Maintenance System consists of a set of beliefs and a set of justifications. The beliefs are linked by justifications to represent the dependencies between them. A justification is written formally as:

$$< r_1 \text{ É } r_m \mid s_1 \text{ É } s_n \text{ -> } b >$$

This can be read as: if the beliefs r_1 to r_m are accepted as true and there is no reason to accept s_1 to s_n then b should be accepted. It can be represented graphically as shown in Figure 1. The list r_1 to r_m is referred to as the in-list, and s_1 to s_n as the out-list. With this representation, the dependencies between a set of justified beliefs can be represented in the form of a graph, as shown in Pratt (1994, Figure 6.2.) Once such a system has been established, the addition of a new justification may affect the validity of other justifications, which is to say that previously accepted beliefs may have to be abandoned. Pratt (1994) presents and illustrates the operation of algorithms for determining, in the light of new information, the new beliefs entailed and the old beliefs to be withdrawn.

Thus, if a system specification can be cast in this form, the consequences of a change can be traced so as to identify those items which will need to be modified. However, it still remains to determine the modifications.

NETWORK MANAGEMENT AS REDESIGN

To illustrate these ideas, consider the network represented by the graph in Figure 2a. The nodes of the graph represent network stations and the lines communication links. A message is sent by labelling it with the address of its destination and forwarding it along the link to an intermediate station where it will be stored until it can be passed on in the same way until it eventually reaches its destination. On receipt of a message, a station examines the destination address, accepting messages addressed to it or using a routing table to

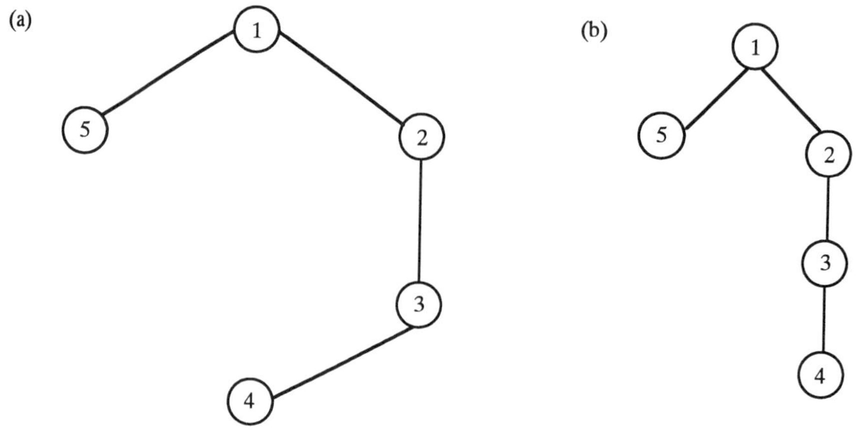

(c)

NODE 1	
Destination	Forward to
2	2
3	2
4	2
5	5

Figure 2. (a) A network, (b) The shortest paths from node 1, (c) Routing table for node 1.

determine the station to which the message should be forwarded. Figure 2b shows how the routing table of Figure 2c can be derived for the network of Figure 2a.

Routing tables can be used to deal with some of the aspects of network management. For example, network performance can be improved by routing messages along the least busy routes which may not be the same as the shortest ones. A fault can be avoided by routing messages around the area where the fault has occurred. Additions to the network can be accommodated by extending the routing tables to include them.

Now consider the problem of adapting the network of Figure 2a to the addition of an extra station connected as shown in Figure 3a. The routing tables must be extended by adding an extra row for the new node 6, but it may be necessary to make other changes because the introduction of extra links in the network may well have introduced even shorter paths between some pairs of stations. The routing tables for the new network can, of course, be redesigned from scratch, but it would be better if only the necessary changes could be made. Deriving the tree for the new network, as in Figure 3b, and comparing it with the tree of Figure 2b for the previous network, shows that there is a shorter path to node 4, and the existing routing table can be amended to give the new table of Figure 3c.

The problem of redesigning the routing tables can be cast as a Truth Maintenance System. In the justification for believing that a routing table is correct, the beliefs of the in-list are that the table records the shortest routes to the existing nodes, and those of the out-list that it records the shortest routes to nodes yet to be added. However, such a system would only detect the changes necessary. The method described above, which contains the idea of the Truth Maintenance System although in a somewhat restricted form that is matched

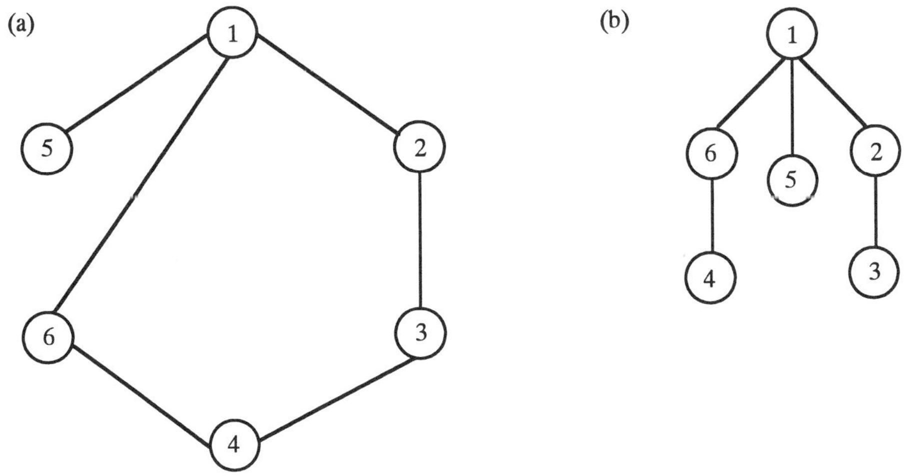

(c)

NODE 1	
Destination	Forward to
2	2
3	2
4	6
5	5
6	6

Figure 3. (a) Modified network, (b) Shortest paths, (c) Routing table.

to the constraints of its field of application, not only detects but also determines the necessary changes.

It is interesting to consider the whether the feature-interaction problem will yield to this form of approach. It could certainly be used to detect interactions. The clash described above, which arises when someone with an unlisted number calls someone with calling-number identification, could by identified by the clash between the belief that the calling-number should be displayed and its negation. There are difficulties, though, not least with determining the beliefs that are entailed by the specification of a service and with the language in which those beliefs are stated. Further, while feature interactions may be detected in this way, it is usually appropriate to resolve them by negotiation (Velthuijsen, 1993).

REFERENCES

Brusil, P.J., and Stokesbury, D.P., 1989, Towards a theory of managing large networks, *IEEE Spectrum*, 26:39.

Caruso, R.E., 1990, Network management: a tutorial overview, *IEEE Commun. Mag.*, 28:20.

Doyle, J., 1979, A truth maintenance system, *Artificial Intelligence*, 12:231.

Griffeth, N.D. and Lin Y-J., 1993, Extending telecommunications systems: the feature-interaction problem, *Computer*, 26:14.

Pratt, I., 1994, "Artificial Intelligence", Macmillan, London.
Reinhardt, A., 1994, Building the data highway, *Byte*, 19:46.
Velthuijsen, H.,1993, Distributed Artificial Intelligence for runtime feature-interaction resolution, *Computer*, 26:48.

LOGISTICS AWARENESS

Vision versus Practice

Bridget Mears-Young

Department of Management Systems and Sciences
University of Hull
Cottingham Road
Hull
HU6 7RX
United Kingdom

INTRODUCTION

In a previous paper (Mears-Young, 1993) I outlined how research into logistics had supposedly evolved from a 'traditional' to a 'non-traditional' perspective. That paper then went on to propose that the 'non-traditional' approach appears to do little but expand the physical boundaries of the logistics concept with no thought as to the legitimacy of this expanded perspective, nor really, of the acceptance of any form of logistics by the practitioner community. This raises the question as to whether all this effort on the part of logisticians is actually contributing to the evolution of knowledge, or is logistics indeed engaged in a playful 'infinite regress' ? Hence, this paper proposes that a significant step forward would be to be to look to outside of the field of logistics in order to determine whether the movements within the field do actually constitute progress. The meta-theoretical framework put forward by Burrell and Morgan (1979) is outlined to explore the underlying principles which constitute the paradigm on which logistics is based.

TRADITIONAL Vs NON-TRADITIONAL LOGISTICS

Traditional logistics views the internal supply chain as a single structure, rather than division into sections and assigning their responsibilities to functional areas such as purchasing, manufacturing, distribution and sales (Bowersox et al., 1992). The priority of traditional logistics is a systems orientation focusing on information and its control in order to optimize the system's performance profit-wise while at the same time satisfying customer service needs on a longer term basis. Logisticians, as a result of their analyses, concluded that there were inadequacies in the traditional approach specifically because they believed this viewpoint was limited to examining how logistics could be performed more effectively and efficiently, thus neglecting its effect on corporate strategy.

Critical Issues in Systems Theory and Practice, Edited by K. Ellis et al.
Plenum Press, New York, 1995

The non-traditional approach differs to that of traditional logistics in that it is seen to be more than just an operational function but more as a way of thinking (Remmel, 1991). Non-traditionalists view logistics as an 'integrating concept' whereby logistics not only creates cost savings and efficiencies, but also adopts a managerial philosophy that considers and incorporates the elements of corporate strategy and planning, competitive advantage and other more global concepts. Authors who support the non-traditional approach suggest that for it to work it needs company-wide, if not total supply chain, commitment in that all other functions should cooperate with, be coordinated by and controlled by the logistics function. However, this author considers that non-traditional logistics just seems to be 'adding on' responsibilities to logistics, making the concept bigger but not really expanding the boundaries.

PARADIGM ANALYSIS

It is evident that many of the authors on non-traditional logistics have acknowledged that, like traditional logistics, non-traditional logistics also has flaws (e.g. Stevens, 1989; Stock, 1990). Unfortunately they too are guilty of purely paying lip-service to these obvious weaknesses choosing to accept the more positive attributes of the non-traditional approach. The evidence I have found suggests that these eminent authors appear unable to detract themselves from their single-minded view and steadfast analysis of logistics !

There are worries about the field being in danger of becoming very insular (Stock, 1990), and these may well hold true if researchers continue working in their existing paradigm. Consequently, we need to contemplate a radical reformulation of logistics based on the assumptions of alternative paradigms. The justification for looking at paradigms is to try to discover on what assumptions logistics is based. The reason being that if logistics is to be properly critiqued then we need to be able to comprehensively understand where logistics is 'coming from' - i.e. the basis on which it stands, its theoretical underpinnings - in order to determine if it is these underpinnings which give rise to its weaknesses and so that we can provide genuine alternative possibilities.

Burrell and Morgan's Paradigms

I have chosen to look at the paradigm(s) underlying logistics using Burrell and Morgan's (1979) paradigm matrix as a guideline. The rationale behind this choice is logistics' systems orientation, in that it would be beneficial to use an approach already employed by systems thinkers. Since, Checkland, 1981(a); Jackson, 1993; and Lane, 1994, for example, have all used Burrell and Morgan's paradigm matrix to classify systems' approaches my choice can be seen as warranted.

Burrell and Morgan's (1979) schema puts forward the idea that the various socio-scientific communities can be conceived as residing in four distinct paradigms, each with their own clear common set of assumptions, and each in opposition to the other three. Their framework is based on a combination of assumptions about the philosophy of social science and the theory of society itself which together form a frame of reference about the "... nature of the social world and the way in which it might be investigated." (Burrell and Morgan, 1979, p*x*). The assumptions of social science with regard to ontology, epistemology, human nature and subsequently the related methodology, will define a scientific community's perspective on their discipline (i.e. 'way in which it might be investigated'). Each of these assumptions are based on a continuum between polar opposites. At one end of the continuum communities may take these assumptions in an 'objectivist' way treating the world as if it were " ... a hard, external, objective reality...", correlating social sciences with natural sciences; at the opposite end communities may subscribe to a 'subjectivist' approach, stressing the "... subjective experience of individuals in the creation of a social world ..." (Burrell and Morgan, 1979, p3).

Equally assumptions upheld in the study of society reflect how the stance taken by difference theories directly define the view of the nature of the subject under investigation (i.e. 'nature of the social world'). These assumptions regarding the theory of society itself are defined by Burrell and Morgan (1979, p16) as assumptions about the conflict between "... regulation and radical change ...". Regulative theories concern explanation of the status quo, that is, why society exists as it does as a cohesive group. In contrast theories of radical change refuse to accept the position of the status quo and envisage emancipation for humankind.

Thus social assumptions can, according to Burrell and Morgan, be categorized according to two dimensions the subjective-objective dimension and the regulation-radical change dimension. When combined these two dimensions form four fundamentally different views of the nature of the social world; indeed, four distinct sociological paradigms. These four paradigms can be illustrated by a four cell, two-by-two matrix as in Figure 1.

Given the systems orientation of logistics, it would be useful to see how systems thinking has recently explored the various paradigms.

SYSTEMS THINKERS' EXPLORATION OF PARADIGMS

Traditional systems theory, which following Checkland (1981(b)) might be regarded as the '*hard systems approach*', is based on functionalism. Systems are seen by those working within this paradigm to have a 'hard', real existence, independent of human observers. Humans are purely seen as inputs or resources to be made effective use of, consequently all issues of subjectivity are ignored. The purpose of a hard systems approach must be to gain a better understanding of the status quo, and by doing so increase the ability to predict and control (Jackson, 1993).

The aim of hard systems thinkers is to produce a quantitative model of the system concerned. It accomplishes this through objective definition of the goal to be achieved. Solution is realized through optimization of the model and transferred to the problem situation. Indeed, as Checkland (1978, p109) states, hard systems approaches are based upon "... the assumption that the problem task they tackle is to select an efficient means of achieving a known and defined end.".

However, by consequence of the fact of its objectivity and also because of its desire for quantification which often involves making assumptions that result in simplification for modeling purposes, it has been stated that hard systems approaches have difficulty in coming to terms with extreme complexity and subjectivity and are innately conservative (Keys, 1987). As a result, systems thinkers believe that a break from functionalism provides the way forward.

Interpretive systems theory is by contrast what Checkland (1981(b)) would regard as *'soft systems theory'*. 'Systems' are seen by those working within this paradigm to have a 'softer', nominal, existence, their construction dependent on human interaction, therefore issues of subjectivity are central. Humans are no longer seen as resources, rather without them 'systems' would not exist, since, in direct contrast to hard systems thinking reality is presumed to be dependent on the perception of individuals, hence the existence of multiple views of reality. Like hard systems approaches, soft systems thinkers aim to gain a better understanding of the status quo, and by doing so increase their ability to predict and control such 'systems'. However, based on the assumptions of interpretivism, it is not possible to do so through tangible models but rather by "... getting 'inside' ..." (Jackson, 1993, p566)

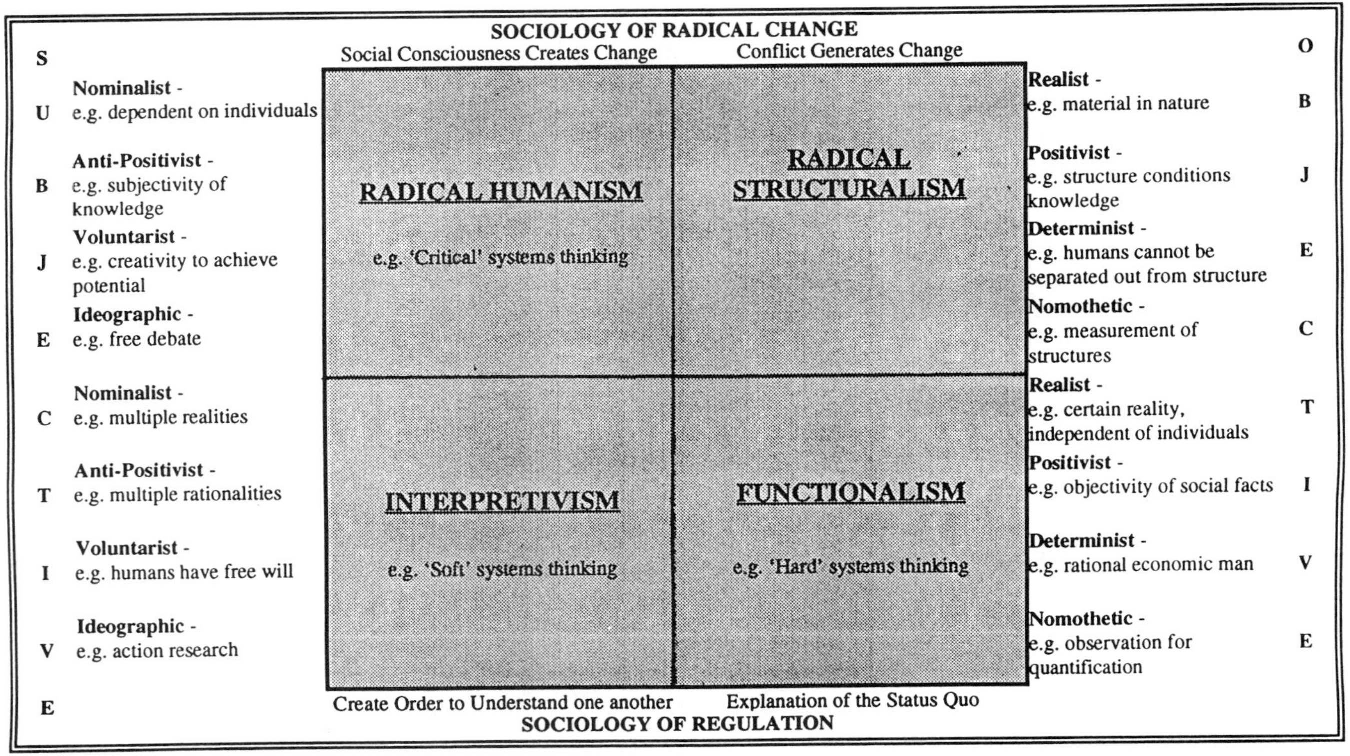

Figure 1. A representation of Burrell and Morgan's (1979) paradigm schema.

'systems' to try to " ... grasp subjectively the point of view and intentions of the human beings who construct them ...", in order to gain an understanding of the kind of action that might be required to mitigate the problem situation.

While these soft methodologies overcome the problems of subjectivity and complexity raised in criticism to hard systems thinking, as Keys (1987) points out, they do not surmount the political flaws identified by the sociology of radical change. Because soft systems methodologies aim to preserve the status quo the accommodation and consensus arrived at is restricted due to the prevailing power structure (Jackson, 1993). Radical humanist systems theories can be seen as a response to this shortcoming.

Situated at the opposite end of the regulation-radical change dimension from both hard and soft systems theory, radical humanists want to gain a better understanding of the status quo, not so that they can predict and control a system, but because they believe that it can constrain human development. It is conceived that systems are realized through human action, consequently, in order to analyze such systems we have to understand all of the interests behind the contributing views that create them (Jackson, 1993). There is a particular focus on how control is exercised, so that people can be made aware of the constraints the existing systems impose upon them. Only when they realize how their lives are controlled can they expect to achieve their true potentials through effectively transforming the systems that they themselves have created. This is different from soft systems methodologies as power structures, hierarchy and other hidden agendas are identified so that true emancipation is possible.

To complete our exploration of systems thinking we will look at the final paradigm according to Burrell and Morgan's schema, that of radical structuralism. Radical structuralist systems are very like functionalist, in the sense that systems are seen to have a 'hard', real, existence, independent of human observers. Radical structuralists would maintain that it is possible to discover the underlying causal regularities that govern the behaviour of systems, while issues of subjectivity do not enter into their schema (Jackson, 1993). Humans are purely seen as inputs or resources determined by such systems. Similarly it is possible to produce a quantitative model of the system concerned. However, contrary to functionalist theory, the purpose of radical structuralist systems approaches is to understand radical change. As Jackson (1993, p567) states : "Emphasis is placed upon contradictions in the systems and on conflict between different groups. This facilitates the emancipation of people from presently existing social structures".

LOGISTICS AND PARADIGM ANALYSIS

Traditional logistics presents itself as an operational function, and its relevance is strictly limited to particular organisational functions. Such an approach, however, has been seen as a naive and narrow definition of logistics. Consequently, non-traditional logistics goes beyond this view towards a vision of the organization as a whole. However, even though non-traditionalists promote a new broader view of logistics they still continue to attempt to the 'problem' of logistics as **they** themselves see it - a system of interacting parts.

Hence, no matter which vision of logistics is presented, both traditional and non-traditional models of logistics are firmly based on a functionalist view of the world. They conceive of a world independent of the observer, which is predictable, stable and knowable. A world in which there is only one version of reality, and other functions interfacing with logistics are assumed to share this view. Based on positivistic epistemological assumptions, both models of logistics assume a parallelism between the natural and social world. Issues of subjectivity are ignored since they cannot be modeled. As a consequence, this functionalist orientation enables people to be treated as things; "... it allows the construction of a framework of exploitation in which specifically human characteristics, particularly related to communication, understanding ...., are ignored and debased in the interests of material production and the creation of surplus value." (Morgan, 1992, p141).

However, as has been observed (Mears-Young, 1993), this normative approach to knowledge is causing dissatisfaction even within the realms of its own practitioners. Therefore, it is necessary to turn outside of logistics to theories developed in other areas for alternative perspectives. In understanding how logistics contributes to the organization and to commerce as a whole it is necessary to develop a critique of logistics itself. Whereas some management specialisms such as information systems already have their alternative approaches well developed and evaluated, logistics, has as yet no well-developed alternative perspectives. What is needed at this stage is an initial recognition that logistics can be subjected to *'reflection'*, followed by relevant research to show how such an approach can be put into practice.

PARADIGM DIVERSITY

Throughout the above discussion, emphasis has been placed on identifying the contributions different paradigms can make to our understanding. In our example of systems' exploration of Burrell and Morgan's (1979) paradigm schema there was a need for methodological innovation due to the perceived failings of the historically dominant paradigm. However, as Gareth Morgan (1990) stresses, even after looking at disciplines from different perspectives it may be the case that the original paradigm offers the best way forward, but still "... an exploration of the nature of the other paradigms provides an invaluable basis for understanding what one is doing, and why other alternatives are to be rejected" (p28).

Nonetheless, the benefits that the enlightened can see are not commonly shared, indeed paradigm diversity is often seen as a threat to those dedicated to and whose work depends on the dominant paradigm (Morgan, 1990), or as Kuhn (1970) phrases those who practice *'normal science'*. In effect paradigm diversity has the potential to totally undermine past achievement. However, if a discipline continues to practice normal science the type of problem situation that could be solved would be restricted to those that the dominant paradigm was capable of and therefore no progress beyond these scenarios would be made (Chalmers, 1982). It is not necessarily the case that normal science will be made redundant or refuted, it has a role to play when the characteristics of the problem situation match up with its views on the nature of social science and society. However, any paradigm offers at best a partial mode of understanding a given problem situation, therefore paradigm diversity should be looked upon as an opportunity to obtain new insights and understanding, even though historically dominant paradigms may in fact be supplemented or even replaced by new paradigms (Morgan, 1990).

CONCLUSION

It is my hypothesis that if logistics' awareness is to be advanced then there needs to be reflection on its actual nature. What implicit assumptions is its very nature based upon, and what are the consequences of these assumptions in its implementation ? The insularity faced by logistics could be said to constitute the fact that it is 'stuck' in its own paradigm - but are logisticians even aware that they belong to a consistent paradigm of thought ? As Burrell and Morgan (1979, p*ix*) state: "In order to understand alternative points of view it is important that a theorist be fully aware of the assumptions upon which his own perspective is based. Such an appreciation involves an intellectual journey which takes him outside of the realm of his own familiar domain.. . It requires that he become familiar with paradigms which are not his own. Only then can he look back and appreciate in full measure the precise nature of his starting point."

REFERENCES

Bowersox, D.J., Carter, P.L., and Monczka, R.M., 1992, Materials logistics management, *in:* "Logistics - The Strategic Issues", M. Christopher, ed., Chapman & Hall, London, pp38-48.

Burrell, G., and Morgan, G., 1979, "Sociological Paradigms and Organizational Analysis", Gower Publishing Company Limited, Aldershot.

Chalmers, A.F., 1982, "What is This Thing Called Science?", 2nd edition, The Open University Press, Milton Keynes.

Checkland, P.B., 1978, The origins and nature of 'hard' systems thinking", *Jrnl. Appl. Sys. Anal.* 5:2, pp99-110.

Checkland, P.B., 1981(a), Rethinking a systems approach, *Jrnl. Appl. Sys. Anal.* 8, pp3-14.

Checkland, P.B., 1981(b), "Systems Thinking, Systems Practice", John Wiley & Sons Ltd., Chichester.

Jackson, M.C., 1993, Social theory and operational research practice, *Jrnl. Opl. Res. Soc.*, 44:6, pp563-577.

Keys, P., 1987, Traditional management science and the emerging critique, *in:* "New Directions in Management Science", M.C. Jackson and P. Keys, eds., Gower, Aldershot.

Kuhn, T.S., 1970, "The Structure of Scientific Revolutions", 2nd edition, The University of Chicago Press, Chicago.

Lane, D.C., 1994, Social theory and systems dynamics practice, *in:* "Proceedings of the 1994 International Systems Dynamics Conference", Systems Dynamics Society, Stirling, pp 53-66.

Mears-Young, B.R., 1993, 'Non-traditional' logistics - a step forward ?, *in:* "Proceedings of 3rd Bi-Annual Conference of the UKSS, Systems Science: Addressing the Global Issues", F. Stowell, D.West and J.Howell, eds., Plenum Press, New York, pp143-148.

Morgan, Gareth., 1990, Paradigm diversity in organizational research, *in:* "The Theory and Philosophy of Organizations", J. Hassards and D. Pym, eds., Routledge, London, pp13-29.

Morgan, Glenn., 1992, Marketing discourse and practice : towards a critical analysis, *in:* "Critical Management Studies", M. Alvesson and H. Willmott, Sage, London, pp136-158.

Remmel, U.M., 1991, Integration of marketing and logistics : a way to competitive advantage in South Africa, *Int. Jrnl. Physical Distribution & Logistics Management*. 21:5, pp27-31.

Stevens, G.C., 1989, Integrating the supply chain, *Int. Jrnl. Physical Distribution & Materials Management*, 19:8, pp3-8.

Stock, J.R., 1990, Logistics thought and practice, *Int. Jrnl. Physical Distribution & Logistics Management*. 20:2, pp3-6.

ORGANIZATION AND INFORMATION FOR TECHNOLOGY MANAGEMENT

A Systemic Methodological Application

Gabriel A. Ramírez

Universidad de las Américas
Puebla
Santa Catarina Mártir
Pue
Mexico

ABSTRACT

The reduction of time elapsed between scientific innovation and using it successfully, constitutes one of the most important abilities that today's companies must have to keep competitive in the international markets. The process of going from spotting an innovation to its application is determined by a large number of complex activities and situations taking place through the internal and external organizational environment. In managerial terms, that process corresponds to going from strategic awareness and reflection to action and results.

This paper deals with the problems of organizing the efforts to bridge the gap existing between strategic and operational Technology Management and with the definition of the information requirements that support this effort. The ideas are the result of applying theories of organizational viability and methods of systemic problem solving to Technology Management, in a large corporation that, as a result of its country's integration to international markets, has to face increased foreign competition.

THE ORGANIZATIONAL CHALLENGES AT THE CORPORATION LEVEL

The company had been operating successfully for many years in a protected national market. It has three main business units, producing pane glass, glassware and glass containers; each of them includes several production plants scattered over an ample territory. Its top management consisted of a President, three Vice-presidents: Marketing, Finance, Produc-

Critical Issues in Systems Theory and Practice, Edited by K. Ellis et al.
Plenum Press, New York, 1995

tion; and two specialized corporate service units: Research and Development (R&D) and Strategic Planning.

For many years, this organizational arrangement did well, but as soon as the country's economy became open, the structure began to show shortcomings to keep pace with foreign competitors in the home market and to face the opportunities that could exist beyond the frontiers. In several aspects, the company was well prepared to face competitors: it had good knowledge of its national market, the manufacturing plants were properly located, the distribution channels were efficient and production costs were still competitive due to lower labour costs. It was evident that, under the new circumstances, the key competitive factors were going to be technology and quality.

The organizational problems of the intelligence function at its corporate level were, as the Viable System Model (VSM) predicted (Beer, 1984), basically the following:

i. The R&D and Strategic Planning units kept little contact with each other and normally worked under different sets of priorities; they lacked a common perception of corporate and divisional critical success factors as they operated far from the divisional market realities. Those two corporate services kept even less contact with the internal operational realities of the business units, keeping a too general understanding of the competition problems in the market. The situation was aggravated because the production plants were dispersed over a big territory.

ii. At the divisional level, management kept a mainly internal view of their business, their production and marketing department were clearly unfitted to face aggressive foreign competitors. At this level there was almost no capability to define and implement technology strategies.

iii. The Vice-presidents tended to work separately, and more than doing corporate management and strategic evaluation at the corporate level, were inclined to get involved in the functional problems of the divisions.

iv. All that fragmentation determined a lack of a concept for corporate information systems, and consequently, an incapacity to specify requirements, to define an infrastructure and to carry out systems projects to support that level.

From the point of view of technology management, the situation meant an almost complete insufficiency to translate the existing potential for innovation into relevant operational realities, as an ongoing activity. The challenge was basically one of defining new ways of managing at the corporate level, so as to provide the necessary integration among the units that dealt with the different aspects of the overall environment, and between them and the divisions. Those new forms of working should be designed to ease the generation of innovative and relevant initiatives that were socially acceptable for rapid implementation.

STRATEGIC AND OPERATIONAL MANAGEMENT OF TECHNOLOGY (MOT) - TRADITIONAL VS SYSTEMIC APPROACH

The conventional views on management and organization place strategic decision making at the top level of the organizational hierarchy, centralizing the strategic reasoning in a group of specialized corporate planners who, applying a set of formalized methods, have the mission of designing and evaluating strategic options that will be selected by the top management (Anthony 1965). Those options, in the form of development projects, are to be implemented and put to work by operational staff.

This naive view of strategic management and policy making is easy to understand, but usually too general and misleading to apply it to complex organizations. Recently the topic has gained notoriety, since Mintzberg (1994) published his critical book on the subject, arguing that the conventional approaches to strategic planning emphasized too much the normative view of strategy formulation missing the more significant real process of strategy formation; he states that "from little ideas (initiatives), and in strange places, not to mention at unexpected times, almost any one in the organization can prove to be a strategist" (Mintzberg, 1994), he argues that the generation of options is very much a social process taking place everywhere in the organization.

From a different starting point, Beer (1979, 1987) had contended long before that "...planning *is not* an activity carried out by professionals called planners; it is an act of decision carried out by managers empowered to commit resources to a different future. Planning is not an activity resulting in products called plans: it is a continuous process, whereby the process itself - namely that of aborting plans - is the pay-off".

In management of technology, the same as with any other areas or levels of managerial activity, it is possible to identify operational and strategic aspects. In the case of technology management, the operational aspect has to do with the integration and use of a given technology, whereas the strategic dimension has to do with the improvement of a given one and with the selection and development of new technologies. The distinction between operational and strategic is essentially a logical one. Distinctions that are made in such an abstract domain, keep a non trivial relation to the hierarchical power structures observed in the domain organizational reality. Operational and strategic reasoning pertain to different logical levels: operational action is doing what is to be done under a particular set of conditions, whereas strategic action is making changes in those conditions.

In today's organizations, technologies of some sort are present almost everywhere, which means that operational and strategic technology problems, in the logical sense of the term, will emerge at every level. The challenge for the organizational designer is to develop structural and functional arrangements to maintain a co-ordinated process of renewal throughout the entire organization to respond efficiently to competitive pressures and opportunities, bridging the 'great divide' that exists between action planning and performance control (Mintzberg, 1994).

THE USE OF THE VSM AND SOFT SYSTEMS METHODOLOGY

The VSM is used here to analyze the activities of strategic technology management and its relations to the process of implementing innovations; it places strategic technology management within the intelligence functions throughout different recursion levels. The Model helps to determine the organizational arrangements that are necessary to sustain the process of strategic reasoning and action all over the organization.

The viability theory states that strategic reasoning and policy formulation must occur at every level of the organization. This is something that management must promote and facilitate, since, in a hierarchically minded managerial culture, management tends to bias its vision towards short term internal problems (Beer, 1979). In this particular case, the theory prescribes that every Division must be aware and sensitive to its specific environment and they should generate innovation initiatives

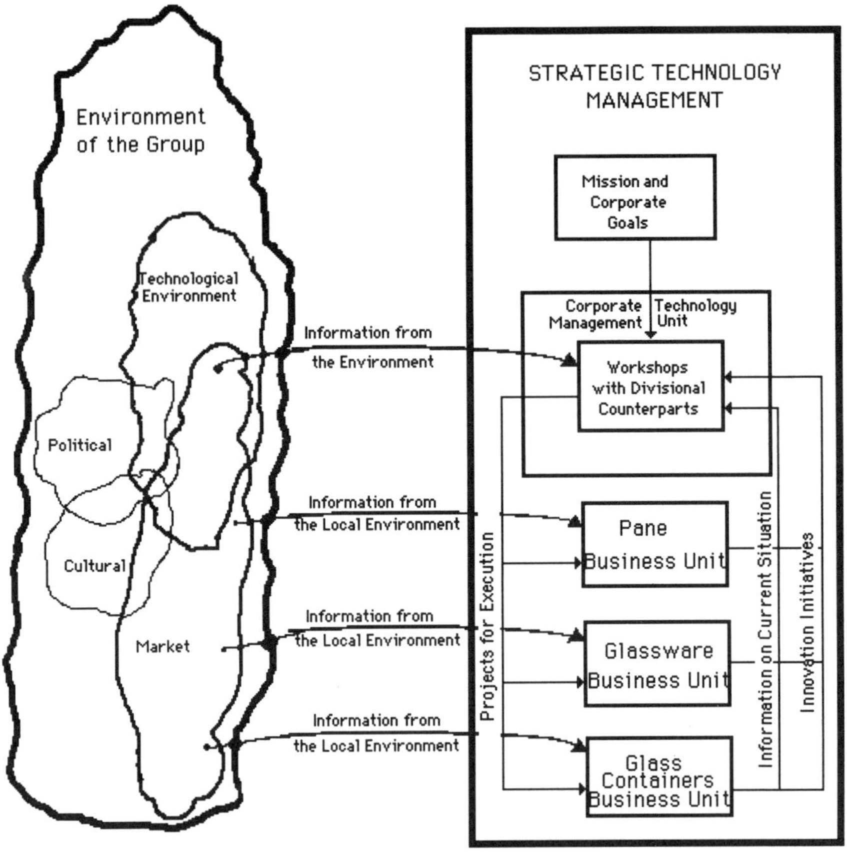

Figure 1. Strategic Technology Management. An all recursive company endeavour.

according to their perceptions of opportunities and threats. The idea that intelligence activities are carried out at every level is illustrated by Figure 1, in it are represented the divisional and corporate levels interacting and obtaining information from the environment; exchanging and processing it, developing meanings, through workshops that connect two recursion levels.

The cybernetics approach suggests that effective communications channels should be built to support the required information exchanges. In practical terms it means developing environmental surveillance (Parm UlhØi, 1992) at every recursion level and sustaining converging dialogues throughout the organization.

To improve adaptation capacity -i.e. product development time- the strategic options must, besides some conventional demonstration of their feasibility, achieve social credibility and acceptance. This means that in the whole organization, people should arrive at a shared perception of the future and of the convenience of certain strategic options; all this requires maintaining a continuous dialogue on crucial strategic issues, at each and through every level of recursion. The dialogues must be fluent, frank and, especially, free from pressures other than the ones that the actors themselves consider as emerging from the competitive situation. According to cybernetics, each viable unit must enjoy the necessary autonomy to elaborate options and consider, at the same time, the strategic intentions of the higher recursion level; so as the conditions of independence and coherence are met (Beer, 1979). To face the organizational challenge the following were necessary:

- The integration of the specialized intelligence units at the corporate level.
- The integration of those units with their counterparts at the levels of divisions.
- The creation of the organizational and physical means to keep that interaction going.
- Provide a meaningful conversation focus to facilitate the development of shared views on the organization's future.

Those objectives were approached through the creation of innovation workshops for each business unit, to be attended by people from the MOT corporate unit, the strategic planning unit, by the vice presidents, and by the functional managers from the business units and divisions, assisted by a specialist in group dynamics. The purpose of the workshops was to find out innovation opportunities to generate projects that were relevant to the competition strategies. The task was not at all easy, since the participants had so many different perspectives, backgrounds, priorities and interests. To keep the process in focus, the chain value concept (Porter, 1986) was used as a medium to encompass the whole logistics, productive, commercial and marketing domains.

Initially, the workshops were held at the company's headquarters but that had the problems of distance, dates and continuity. The way out to this, was the intensive use of the communication facilities that were already available in the computer network and to use that potential to convey a permanent conversation at business unit levels. The workshop should be understood as a means to create the permanent relations among units and levels, the final aim had to be to create capable intelligence counterparts and to replicate the experience all over the group. The Soft Systems Methodology (Checkland 1981, Checkland and Scholes 1990) was used because of its relevance and simplicity. Once a human activity system is conveniently defined, the answers to a few questions reveal the essence of information requirements.

Initially, one of the biggest problems that the workshops faced was the lack of relevant information. Their functioning, hence the activity of strategic reasoning, had never taken the form of a collective inquiry on every aspect of the company's activities. People then felt the necessity of having 'an information system' for each workshop. The problem of information requirements was piecemeal understood by each actor and the MIS professionals were confused and facing a problem they could not define. A great help was to define technology management as a human activity system, in Checkland's sense (1981). After some discussion and clarification on what was the company aiming at, with the technology management workshops, the following root definition was agreed: 'Technology management is the human activity of finding out and generating opportunities for relevant innovations, evaluating their possible impacts on the organization's performance; selecting and formulating innovation projects; carrying them out, and monitoring and evaluating their impact on the whole organizational performance'. Once a convenient root definition was arrived at, the following questions were asked of each identifiable 'holon' in the definition: What information is necessary to perform the activities implied in each 'holon'? From what sources could that information be obtained? In what form? Who would use it and when? etc. The table in Figure 2. summarizes the answers to those questions and provides the basis to formulate a project to construct in a systemic way 'the information system' for technology management.

THE STAGES OF TECHNOLOGY MANAGEMENT	INFORMATION REQUIREMENTS	TYPICAL USERS	MAIN SOURCES
Detection or Generation of Ideas	Publications in Science and Technology. News on Innovation. Etc.	Corporate MOT unit. MOT counterparts at divisional levels. MOT Workshops	Externa: Public access data basis (Dialog, Delphy, SECOBI)
Evaluation and Impact Analysis	Information on competitors and markets. Costs. Technologies in use. Production capacity and slack resources	Corporate MOT unit. MOT Workshops	Internal: Divisional Information Systems. External: Market trends Competitors moves.
Formulation of Innovation Projects	Strategic and operation plans. Materials, human and "know how" availability.	Corporate MOT unit. MOT counterparts at divisional levels. Divisional project leader	Internal: Divisional Information Systems. Innovation projects Information system
Implementation of Technological Change	Information on budget execution and physical progress.	Corporate MOT unit. MOT counterparts at divisional levels. Divisional project leader	Internal: Divisional Information Systems. Innovation projects Information system
Monitoring and Performance Evaluation	Information on costs and market performance. Production performance and quality control.	Corporate MOT unit. MOT counterparts at divisional levels. Divisional project leader	Internal: Divisional Information Systems. External: Market trends Competitors moves.

Figure 2. A summary of information requirements.

REFERENCES

Anthony, R. N., 1965, "Planning of Control Systems: A Framework for Analysis", Harvard University Press, Boston, Mass.

Beer, S., 1979, "The Heart of Enterprise", John Wiley & Sons, Chichester.

Beer, S., 1984, The Viable System Model: its provenance, methodology and pathology, *J. Opl Res. Soc.* 35:1, pp 7-25.

Beer, S., 1987, "Reflections of a Cybernetician on the Practice of Planning", Canadian Institute of Planners, National Conference Keynote Speech, Toronto.

Checkland, P.B., 1981, "Systems Thinking, Systems Practice", John Wiley, Chichester.

Checkland, P.B., and Scholes, J., 1990, "Soft Systems Methodology in Action", John Wiley & Sons, Chichester.

Mintzberg, H., 1994, "The Rise and Fall of Strategic Planning", The Free Press, New York.

Parm UlhØi, J., 1992, Linking Technology Management to Strategic Management, *in:* Khalil M.Tarek, and . Bayraktar A. Bulent, (eds.), "Management of Technology II", Institute of Industrial Engineers, Florida, USA.

Porter, M., 1986, Changing patterns of international competition, *California Management Review*, 28:2.

THE MANAGEMENT OF CHANGE IN A TURBULENT WORLD

R. G. (Bob) Saunders

Open University
Walton Hall
Milton Keynes MK7 6AA
United Kingdom

INTRODUCTION - COMPLEXITY & DETERMINISM

Project Management has adopted and developed many methods to manage projects and make decisions. As planning and decision making tools, the standard, deterministic methods have served project management well.

There are, however, industries and projects which require more specialised methods. These are Research, Development, Design and Information Systems projects; anywhere where creativity is required. They cause much concern, and fail most often. Because of their nature, they suffer from the effects of change, obscurity of vision and unpredictability which all make decisions hard to make. When the Manager loses sight of the objectives, causes and effects, then the old models and methods lose their usefulness. Determinism has its place where the links between choice of actions and outcomes are clear, but has much less relevance where probabilities have to be considered, and none at all in 'uncertain' situations. Like the military leader Napoleon, 'managers must know how to plan, and when to change plans to meet changing circumstances'. As Charles Handy remarked (1989) : "In a world of change, with the rate of change and unpredictability itself ever on the increase, and with change itself changing, becoming discontinuous and chaotic then uncertainty over a project also increases". 'Risk Analysis' is used to reduce uncertainty, as Stewart & Fortune (1994) say but here people have to cope even when assigning probability factors to risky activities is impossible. Here lies Chaos.

DECISIONS

Decision makers prefer quantitative and 'hard' information on which to base their decisions. Thus, in conditions of 'Uncertainty', they must make as much use as possible of the information they have, and try to fill in the holes by using suitable methods, even estimates.

It is necessary to consider theories of decision making because without decisions, little ever happens. There is a hierarchy of decision making to be considered first. It is basic to all that follows:

- Strategic - decisions made at the highest level, setting out long term goals, in general terms.
- Tactical - decisions to objectives to meet the strategic goals, over shorter time scales.
- Operational - low level, basic decisions, to meet short term objectives.

PROGRAMMED AND NON-PROGRAMMED DECISIONS

H.A. Simon, quoted by Stonor and Wankel (1986, 142) observes : "...programmed decisions are those made in accordance with habit or procedure, unwritten, or written, to simplify routine decision making. They are not necessarily simple, only routine. Non-programmed decisions are those made to deal with unusual, new or exceptional problems, which arise when the situation is novel or requires special attention." Simon differentiates between traditional and modern decision making techniques for both types of decision. His modern methods of aiding decision making are heuristic- a long way from the earlier determinism.

Another view is that decisions are made on a continuum 'Certainty-Risk-Uncertainty' (again from Simon). Harry (1994, 58-59) provides a clear view of Simons' ideas:

- CERTAINTY

"Where there is certainty in the link between choice of action and its outcome, relationships are deterministic. Models can be mathematical formulae, logical procedures, even look-up tables to determine the outcome following a course of action." (i.e. 'programmed' decisions and 'operational' decisions).

- RISK

"Here the link between choice of action and the outcome of that choice is less deterministic; in fact, there is a choice of link and it is necessary to determine how likely the various outcomes might be." (Harry, 1994).Probability theory is used to cast some light on likelihoods.

- UNCERTAINTY

Where possible outcomes are not known and probabilities cannot be assigned known outcomes can still be listed as is done in 'Risk' calculations and a general outcome added. This is of less value than a more intensive investigation of the problem to reduce the uncertainties. "Information on the relationships between choices of action and possible outcomes should be as complete as possible before the problem is modelled and these outcomes evaluated." (Saunders, 1979). Stonor & Wankel (1986) describe the organisational effort in the three areas as increasing rapidly as the 'Uncertainty zone' is entered. This is where chaos sensitive projects are most frequently to be found.

Other ideas besides Simons' are relevant. Earle & Hopwood, quoted by the author in the OU's project management course (Keller et al., 1987) suggest that decisions can be categorised into:

i. 'Computational' : i.e. operational, programmed decisions made where objectives and causes & effects are well known (Simon's 'Certainty').
ii. 'Judgemental': where there is no certainty over cause and effect, although objectives are clear these are unprogrammed decisions perhaps, but, with experience, not so difficult to make ('Risk').

iii. ‘Political’: where cause & effect are well known, but objectives can’t be agreed (also ’Risk’ but shading into ‘Uncertainty’).

iv. ‘Inspirational’: where nothing much is ’known’ at all. Here lies the realms of ‘Uncertainty’.

In these terms it is the ‘Uncertainty’ area where decisions are made by, at best ‘judgement’, but more usually by ‘inspiration’. It is here, where chaos lurks, that advanced analysis methods are needed to expedite decision-making. Information is needed to allow the decision to be made in the ‘computational’ area.

DECISION MAKING, PROJECTS & UNCERTAINTY

The success of a project is based on making strategic decisions early. These are not easy when the deliverables are uncertain, the requirements vague and the difficulties unknown. Thus, even strategic decisions must be amenable to change, despite the fact that senior management or the customer have made them! Lower level decisions and plans - tactical and operational - have to be just as malleable.

Harry (1994) also suggests that decision making requires different kinds of information for different needs of problem formulation, modelling and the derivation of solutions. From having enough information to make decisions, it is easy to move to a situation where decisions have to be made blind. Here, softer methods must be used. Systems methods have been used before, but ideas of chaos and complexity might also be useful.

COMPLEXITY & CHAOS

Along the Chaos route lies success - if it is managed properly. Thus, the need to reduce the unknowns, and to maximize the knowns leads to the use of ‘Chaos’ in a controlled form. Chaos is holistic, a science of process and of the global, complex nature of ‘Systems’. It is used in biology, physics, economics. By definition, chaos sounds as if it has no rules, no order, no sense but Stacey (1992) says, however, that chaos is not randomly unstable behaviour. Instability is bounded into categories, and a ‘category’ will hold individuals which are all different (e.g. snow & snowflakes). A system is unstable in its own right, but the instability is bounded and, although unpredictable in the longer term, it operates within a generally recognisable pattern. Gleick’s study of chaotic systems (1987) reveals patterns, (or fractals) which repeat in the short term arising from ‘attractors’ (Kellert 1993) (see later, but they are delicate structures underlying complexity, sets of elements of systems.) The fractals are embedded in themselves (as in the Mandelbrot sets shown by Gleick) with irregular shapes and an organising structure which is regular across hierarchies of scales. It means that although behaviour cannot be predicted, unpredictability is within predictable limits! Like a project, chaos is sensitive to small changes in the environment. Hence, ‘a butterfly beating its wings in China can cause a hurricane over the USA’ and all the other behaviours which obey the laws of Chaos and turbulence; those of columns of smoke, flags, dripping taps, cars on motorways, oil in pipes, all go from ‘regular’ flow to ‘chaotic’ flow across a phase transition (a change of state) and lose clear cut connections between cause & effect. Chance plays a part as does each component of the system. It is impossible to forecast long term behaviour, but it is possible to forecast short term developments of the system.

INSTABILITY, FEEDBACK & NON-LINEARITY

One of the causes of instability is feedback, because it is inherently non-linear in the business environment. In any organizational system feedback is required to ensure that the system will function, and to control it, and the outputs. Decisions depend on feedback to provide information.

Businesses are feedback systems with three choices :

- 1/ To run under the stability of negative feedback - which means ossification." (If no one complains, I'll keep going)
- 2/ To run under the stability of positive feedback - the equilibrium of disintegration. (I respond to suggestion, come what might)
- 3/ To run under the bounded instability and non-linearity of a constrained positive feedback. (I respond to suggestion, but within limits) (Stacey, 1992, p.62), [N.B. in brackets, my additions].

The norms are negative and positive feedback, and these are familiar. Negative usually cancels out the original change, positive usually enhances it - perhaps to disintegration! Stacey (1992)suggests though, that feedback has three, not two, states, the third state being of 'bounded instability'. Bounded instability has a pattern, but is irregular and unpredictable. All feedback is inherently unstable wherever it comes from, but feedback (or iteration) leads to new and innovative ideas because it gives new dimensions and perspectives to the situation. Interestingly, Stacey is not alone in making use of feedback loops to describe Project Management. Milosevic (1989) uses them as part of a systems model of strategic project management which provides a framework to control interactions between project and environmental elements.

USING CONCEPTS OF CHAOS

Chaos allows prediction in the short term! Hence Imai (1986) suggests, 'Kaizen', incremental innovation, thus incremental decision making. Recently, n an article in the 'Independent' (29/12/93), Hoffman (a staff reporter) said in an article on modern management, "...we should not be gulled by expensive 'Quick Fixes'. In a world where the knowledge on which decisions must be based is imperfect, it is best to proceed incrementally....".

Securing attention for incremental change issues by spreading new ideas through the organisation is one way of using positive feedback to meet chaos and instability. Managers should not impose order (in terms of plans), but operate in Stacey's (1992) bounded instability by more iteration and by enhanced interaction with the other elements involved, creating short term but incremental stability. Thinking must be based on an assumption of unpredictability and a new model for each new situation needs to be used, not forcing the standard model to fit different situations. The new model creates new ideas, not trying to predict the unpredictable. Lewin (1993) suggests that the model can be represented as an attractor, a point or state to which the systems eventually settle, in a smooth part of the turbulent whole. This represents the final state of a dynamical system in a noisy world, and holds that system working in a particular way until some change in flow occurs. In this age of rapid change, this does not take long. Thus there is a picture of a business as a non-linear feedback system, operating in chaos. The future is unknowable, and cause and effect links are obscure. The system uses new models for each situation and thus changes the business' norms to suit the situation. This means operating in the zone between order and chaos, on the edge of the phase transition which will lead to chaos, i.e. the region of bounded instability.

There are causal links between business ideas, market and performance. Between the present and future states the action is needed to arrive at the desired state. The form of those actions may not be clear, but must provide criteria which will allow a trial and error search for means to achieve that goal. It requires a flexible organisational structure, which might not deliver good strategy, but achieving incremental changes, rapidly and cleanly

CONCLUSIONS

> "Far too much time is spent...ensuring absolute perfection of the decisions.. . Those who ... eliminate every risk .... will be too late in the race" (Harvey Jones, 1993). This is borne out by what is vital to complex projects. As Olson puts it, decisions will have blurred edges, as has the information from which they are made. The time when decisions were simple and deterministic is long since past.

Decisions are not easy to make under conditions of complexity and chaos. Systems and Chaos methods can be used to simplify them. Gleick (1987) suggests that Chaos is also systemic by nature, is holistic by very definition and breaks across discipline boundaries, showing the universality of complexity. Understanding complexity and an increasing rate of change to be endemic in the world is essential for all managers. Projects are bringing chaos on themselves with the greater emphasis on team working, the ability to run design, development and manufacture concurrently. These may allow greater speed and efficiency, but increase complexity alarmingly.

REFERENCES

Gleick, J., 1987, "Chaos, Making a New Science"'. Viking, New York.
Handy, C., 1989, End of the world we know, *Management Today*, April 1989.
Harry, M., 1994, "Information Systems in Business", Pitman, London.
Harvey-Jones, J., 1993, "Managing to Survive", Heinemann, London.
Imai, M., 1986, "Kaizen-The Key to Japans' Competitive Success", Random House.
Fortune, J., 1987, "Project Management (PMT 605)", Open University.
Kellert, S.H., 1993, " In the Wake of Chaos", University of Chicago Press, Chicago.
Lewin, R., 1993, "Complexity, life at the edge of Chaos", Phoenix, Dent London.
Milosevic, D.Z., 1989, Systems approach to strategic project management, *IJPM*, Vol 7, No.3.
Saunders, W.B., 1979, "Management, Theory, Process and Practice", Hodgetts, Philadelphia.
Stacey, R.,1992, "Managing Chaos", Kogan Page, London.
Stewart and Fortune, 1994, "Application of systems thinking to the identification, avoidance and prevention of risk".(To be published)
Stonor and Wankel, 1986, "Management", Prentice & Hall, Englewood Cliffs, N.J.

USING A FORMAL MODEL AS A META-MODEL IN THE ANALYSIS OF TEAM BEHAVIOR IN ORGANIZATIONS

R. W. Stewart

School of Information Systems
Kingston University
Penrhyn Road
Kingston upon Thames
Surrey, KT1 2EE
United Kingdom

INTRODUCTION

Many organisations, in response to rapidly changing environments, are adopting structural strategies that are dependant upon the creation and operation of 'teams' in order to achieve the flexibility of operation and responsiveness required by their market place and customers. The behaviour of these teams will often determine the effectiveness and success of the organisation as a whole, hence the understanding of how teams work in their own organisation is an important part of assessing the effectiveness of organisational performance.

The analysis of team behaviour can draw upon diverse approaches and techniques from disciplines including systems, behavioural psychology, organisational psychology, sociology and many others. Each of these disciplines provide paradigms (examples, patterns or models) to investigate different aspects of the behaviour of individuals within teams and the relationship of a team as a whole to an organisation. However, in order to adopt a holistic approach to the analysis of team behaviour, this paper proposes the use of a Formal Model as a meta-model in order to provide a structural framework or form of an organisisation and its teams. Paradigms with different theoretical underpinnings can then be applied to this meta-model in order to understand the behaviour of these teams. In particular this approach enables the perspective, focus and use of each paradigm to be made explicit and relations between different paradigms made clear. The Formal Model is examined with examples of cross-discipline paradigms to demonstrate how different aspects of behaviour can be examined using this approach.

Critical Issues in Systems Theory and Practice, Edited by K. Ellis et al.
Plenum Press, New York, 1995

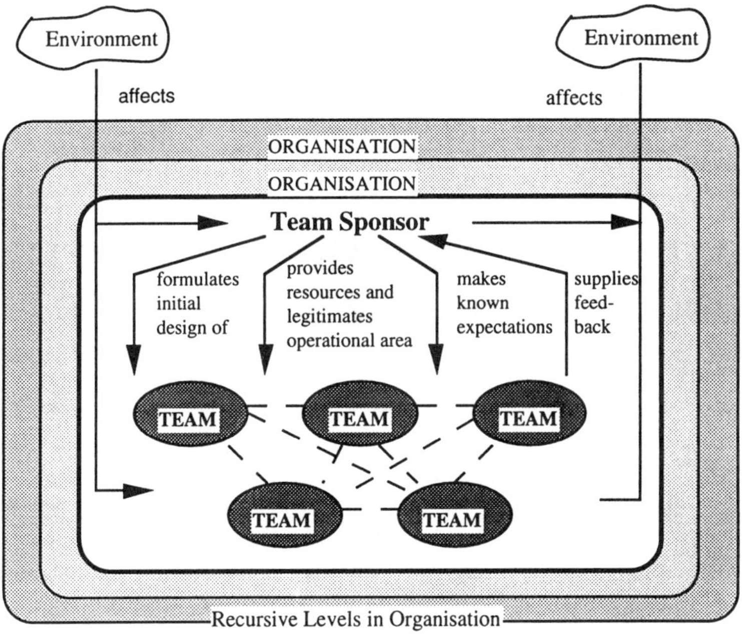

Figure 1. Team relationship model.

FORMAL MODEL FOR TEAM BEHAVIOUR

The Formal Model for Team Behaviour consists of two structural models developed from a decomposition of the Formal System Paradigm (Fortune & Peters, 1990) and made specific to flattened, team based organisations, and organisations where function specific teams are in operation. The decomposition of the Fortune and Peters (1990) model into two models enables analysis on different levels of granularity to be made, namely the relationship of the team to the organisation and the internal behaviour of a key process within a team.

The first model, as shown in Figure 1., enables the positioning of a team within an organisation, the key interactions between teams, the sponsor(s), the organisation, and possible interactions between teams.

The main components of this model are:

An **environment** which affects the organisation, the reason for the team, and operation of the team. The **organisation** - this may be the parent organisation if within a multinational company, a single company or a division within a company. This introduces the notion of recursion as the focus of any analysis may be at several levels within an organisation. A **team sponsor** which is the controlling part of the organisation for a particular team or group of teams and may have been responsible for the initial setting up of the team, but is responsible for the outputs of the team. **Teams** that interact with each other and the team sponsor. **Interactions or links** between teams and a team and its sponsor. Any of these components can be examined but as an example in this paper, the links between the team and sponsor will be concentrated upon. They are the initial team design, the provision of resources, legitimation of area of operation, expectations and team goals, and the feedback on performance.

The first analysis is whether these links exist, for example were there clear organisational and environmental reasons for creating the team, and was there an envisaged structure of the team? Were resources made available, was the team empowered to operate in a required way and its responsibilties clearly stated? Is there a feedback channel for not only performance against goals but also for changed resource needs and environmental changes? If the links are in place then the effectiveness of them can be examined by the use of, and comparison to, appropriate paradigms.

PARADIGMS

In order for any of the above links to operate in an effective manner the communication between the team and its sponsor must also be effective. Hence the informational communication paradigms used in social science or behavioural psychology can be utilised here. Typical of these are the communication models that have the components of encoding, transmission, channels, noise, decoding and feedback (Baron and Greenberg, 1990,). The shared understanding of icons, signs, languages and codes between sender and receiver; the use of appropriate channels of communication; disruptions to messages and feedback in terms of message receipt and understanding; can be examined using these paradigms.

A second set of paradigms are the sociometric paradigms that use the display of communication patterns (sociograms) between individuals, or groups of individuals, in order to analyse organisational behaviour. The pattern of communication between sponsor and team can be analysed in terms of type of communication, subject, frequency and importance. For example there ought to be regular feedback on performance, responses to major changes in the environment, requests for clarification of expectations or differing resources. In advanced uses of sociometric analysis comparisons can be made between the formal or expected communication patterns and the actual or perceived informal communications , 'Comand and Control Team Analysis' (Stewart,1994).

Cybernetic paradigms are a third set which can look at the control loops between the team and sponsor. For example feedforward control models can be used to examine the effectiveness of the initial design of the team in terms of structure, goaling, resources and empowerment when compared to the actual performance of the team. Feedback control can be used to examine the clarity and viability of the expectations (reference levels), the feedback (monitoring and reporting activities) and the resource provision and legitimated operations (corrective action). Examples may be seen in Gibson et al. (1994).

A fourth set of paradigms are behavioural paradigms that deal specifically with the human behaviour of the team. They are integrative paradigms that draw from the fields of Social Science (Stewart, 1991) and can be used to relate emergent actual behaviour of teams with the sponsor and other teams. For example a high level of communication with the sponsor can be acceptable in the start up phase of a team's operation but if it continues it may be that the empowerment of the team or its goals are unclear. A similar behaviour may indicate that the environment has changed and clarification of goals with different resources are needed. When a team has been operating for a period of time, little communication of feedback may mean that the team is operating successfully, or that the team is alienated to the sponsor or organisation, or that the environment has again changed. Any team will undergo a process of development over time and exhibit particular behaviour at different stages of the team development cycle (Tuckman and Jensen, 1977). The expectations of the sponsor can be examined to see whether they are realistic for the length of time the team has been in operation. These behavioural paradigms can be also used to analyse the links to other teams.

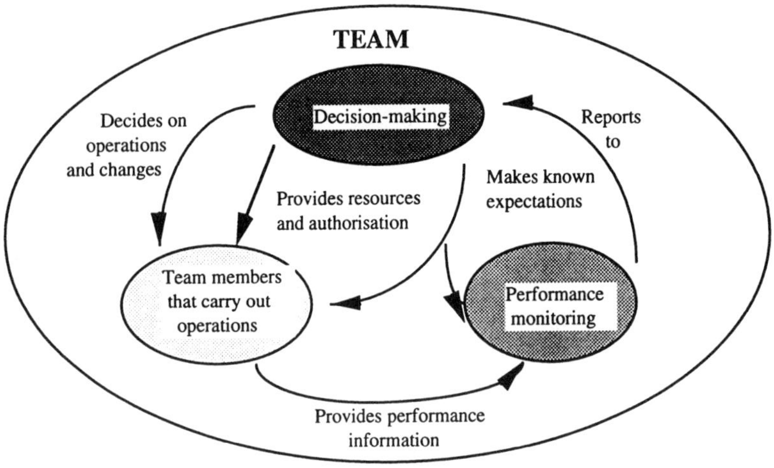

Figure 2. Team control model.

TEAM CONTROL MODEL

The second structural model, Figure 2., shows a key process for any team. It is a team control model that shows the decision-making and performance monitoring functions with their controlling process links to the transformation process operations (what the team actually does). These should be in place for any team to be internally under control. It shows those components and features that should be present if a set of activities within a team can operate without failure.

The main components of the model are: **Team members** that carry out the operations and tasks of the team. A **decision-making** individual or group that takes decisions on internal structure and sub-tasks, performance evaluation and corrective action. **Performance monitoring** of the outputs of the team and the interactions or links between team members.

As with the team relationships model, the first analysis is whether these components exist, if not then the team is potentialy not under control. If they are in place then effectiveness of team performance can be examined using appropriate paradigms.

PARADIGMS

Communication paradigms are as important in this model as in the team relationship model. In order for all of the links to operate effectively there must be a shared understanding of the languages and codes, etc., used by team members. Ambiguity or differing understanding of these will lead to misinterpretation of the messages or in the way in which information is used. For example, parts of the team working to different goals, not providing expected performance information and non standard operations between team members. The sociometric paradigms can show any of these links that are missing and who is talking to whom and on what subject, if the communications are coded. This can identify clear behavioural patterns, for example, if the decision-making group is appropriate, team members isolated when should be involved, chains of communication and gatekeepers. The identification of behaviour as represented by communication patterns can then point to appropriate behavioural paradigms for further investigation such as the neonate team and its internal conflict (Stewart, 1992). The cybernetic control paradigms can for example show where expectations (reference levels) are unclear, changes in operations due to outside changes are not being made or the performance monitoring function is not sampling team outputs sufficiently or appropriately.

CONCLUSIONS

The analysis of team behaviour can be approached from diferent perspectives and this paper proposes that the adoption of a formal model provides a structural framework on which to examine the relations of the team to the organisation and the internal behaviour of the team. It demonstrates how paradigms with different theoretical underpinnings such as informational communication, sociometric and cybernetic models together with behavioural paradigms can be linked to analyse parts of the formal model whilst retaining an holistic approach.

REFERENCES

Baron, R. A. , Greenberg, J., 1990, "Behaviour in Organisations", Allyn and Bacon, Needham Heights.

Fortune, J., and Peters, G., 1990, The formal system paradigm for studying failures, *Technology Analysis and Strategic Management*. 2:383-390

Gibson, J. L., Ivancevich, J. M., and Donnelly, J. H., 1994, "Organizations: Behaviour, Structure, Processes", 8th ed., Irwin, Boston.

Stewart, R.W., 1994, Behaviour and communication analysis of command and control teams, *Journal of Naval Science*. 20:164 - 174.

Stewart, R.W., 1991, The use of social paradigms in the analysis of team behaviour during organisational change, *in:* "Systems Thinking in Europe", Jackson, M. C., Mansell, G.J., Flood, R.L., Blackham, R.B., and Probert, S.V.E., Plenum, New York.

Stewart, R.W., 1992, The impact on the decision making process in organisations resulting from structural change in response to dynamic changing environments, *in:* "IDASCO 1992", IEEE353, London

Tuckman, B., Jenson, N., 1977, Stages of small group development revisited, *Group and Organizational Studies*, 2:419 - 427.

MEDICAL INFORMATION AND HEALTH CARE SYSTEMS

SSM
A Guiding Light in the Introduction to Hospital Case Mix Management Systems?

P. J. Baugh, D. A. Fitzsimmons, and D. M. Walters

Lancashire Business School
University of Central Lancashire
Preston
PR1 2HE
United Kingdom

INTRODUCTION

The National Health Service (NHS) in the UK has, for the past four years, been going through a period of radical transformation in its organisational policy, outlined in a number of White Papers (eg. Dept of Health 1990 (i)(ii)(iii) Griffith 1983 and Coombs & Cooper 1990). In essence, the aim has been to create an internal market for secondary health care within the NHS whereby the interests of the consumers are represented by the District Health Authorities or by General Practitioner - known as purchasers - and the producers of health care services (hospitals, community health care units, etc.) known as providers.

To operate successfully in the new environment it is felt that the providers need comprehensive information systems, such as Case Mix Management Systems (CMMS), which have not been hitherto available, to support and aid decision making.

CASE MIX MANAGEMENT SYSTEMS

Case Mix Management Systems have been developed by large software developers to keep track of case types and their associated costs without the need for drilling down to individual cases and their procedures each time. Their purpose is to provide support to the contracting process by enabling hospital managers to monitor and evaluate treatment outcomes, time involved and costs in delivering the treatment done on the basis of case type. This will help in enabling providers to strike realistic contracts with the purchasing authorities.

Using CMMS, tracking the mix of cases will aid the contract manager in determining what cases the hospital manager should undertake for the purchaser, at what price and ensure that the case mix outcome and associated costs are brought back into line with intended

Critical Issues in Systems Theory and Practice, Edited by K. Ellis et al.
Plenum Press, New York, 1995

projections during the contracting period. If the case mix is not stable or predictable, there is a danger that they may set prices below average costs, in which case they will incur losses. Alternatively, they may set prices well above costs which, in the NHS context, is equally unacceptable since, in the politically sensitive area of a free health care system, hospitals cannot be seen to be making profits over and above the normal return they are required to make on their activities. However, control of case mix cannot be completely ensured by managers without clinicians spreading good medical practice which can help to cut costs.

As previously stated, using CMMS as a Decision Support tool when letting contracts or preparing costings is solely the preserve of the manager. However, the systems are designed to also be used by the clinician. Significant amounts of patient information held on CMMS can be clinically useful to doctors. The systems can increase the capability for medical research and improve the quality of patient care by generating reliable data with which clinicians can investigate patterns in patient flows, treatment profiles and clinical outcomes in a manner more sophisticated than was hitherto possible.

Furthermore, clinicians are now expected to be actively involved in Medical Audit procedures which are defined by National Association of Health Authorities and Trusts (1994) as the "...systematic, critical analysis of the quality of medical care including the procedures used for diagnosis and treatment, the use of resources and the resulting outcome and quality of the life of the patient". Clearly, effective and efficient medical audits require access to medical records and other relevant information which can be facilitated by CMMS.

The sole interest of the clinician is the quality and quantity of care that can be given to their patients, so their core need from an information system is one that will assist them to undertake their role more effectively. They will have little desire to waste valuable patient contact time inputting data that cannot be used for this purpose.

Therein lies the potential source of friction when developing and implementing a new hospital information system. All groups of staff will need to adjust their culture and values to accommodate the needs of others, but will sufficient ground be yielded to permit the introduction of a mutually beneficial system?

An on-going research project at the University of Central Lancashire is seeking to explore the system analysis, design, development and implementation methodologies employed within the NHS and analyse the effectiveness of the implemented system from the perspective of the culturally and professionally diverse groups of staff involved.

THE EVIDENCE TO DATE

We have commenced our study by interviewing health service managers at two hospitals in the North West of England concerning the Decision Support Systems in use. Both hospitals are at the stage of piloting the development of CMMS. At this stage in the research it is too early to determine exactly how successful the systems have been in meeting their objectives. The preliminary evidence is that one is actively involving the clinicians in the use of CMMS and are using the clinicians to pilot CMMS in the hospital. The other views CMMS as the preserve of the managers and sees the involvement of clinicians at a later date as a matter for their information.

Hospital A has adopted an approach of clinician involvement at every stage of the process. They were involved in the procurement team which decided upon the technology to purchase and were actively involved in looking at the coding structure of the system to ensure accuracy. The main requirements of the system were that it had good clinical audit facilities and good ward order communications so basic patient data could be entered by clinicians directly together with good interfacing capabilities with the Patient Administration System (PAS) to ensure consistency of data and to obviate the need for entry of the data PAS can provide.

Over a period of six months, the piloting was initially amongst the clinicians to get them to see the benefits of the system but acceptance has not been uniform.

The approach at hospital B is entirely different. The assumption has been made that the clinicians will not be interested in case mix as such and that this area is the preserve of the manager. As a result, the hospital took the decision to tackle the issue of information systems for managers and clinicians separately. The result of this approach is that the clinicians have been supplied with a Clinical Information System (CIS) which allows them to enter treatment data for each patient and carry out the monitoring and evaluation both at individual and group level. The hospital has also purchased a Case Mix Management System and an EIS to support the contracting process.

Whilst the CIS can act as a feeder system to the CMMS, the interface is not yet in place to allow for data transfer. Even when this has been done, there is a danger that the clinicians, seeing that the CMMS is not in their ownership, will not use the output from the CMMS in the support of the Medical Audit process, rather they will use the CIS both to monitor and evaluate individual cases and case-types. Thus treatment will remain divorced from costs with a consequent lack of control over the latter.

The current phase of the research involves undertaking a series of detailed interviews and questionnaires at a number of hospitals within the North West in order to validate the results obtained so far. However, as implementation is taking place at an uneven pace in the NHS, it will not be possible to examine all hospitals at a given stage simultaneously. Questionnaires and regular interviews with relevant staff at the hospitals will allow a picture of experience at the different hospitals to be developed.

USING SSM TO FACILITATE THE INTRODUCTION OF CMMS

Soft Systems Methodology (Checkland, et al, 1990) is perceived to consist of seven steps:

1. Enter the situation considered problematical
2. Express the problem situation in the form of a cartoon like pictorial representation of the big picture in terms of a cultural view of the social system; social roles; norms of behaviour and values used to judge role performance.
3. Formulate root definitions of relevant systems of purposeful activity

Root definitions are described by Checkland & Scholes (1990) as "...concise verbal definitions expressing the nature of purposeful activity systems regarded as relevant to exploring the problem situation". The elements to consider when formulating the root definition are:

- Customers - the victims or beneficiaries of the Transformation
- Actors - those who carry out the activities involved in the Transformation
- Transformation - of some entity into a changed form of that entity according to a declared...
- Weltanschauung - the worldview which makes this Transformation meaningful
- Owners - those who could stop the Transformation
- Environmental constraints - elements outside the system which it takes as given

4. Build conceptual models of the systems named in the root definitions
5. Compare models with real-world actions
6. Define possible changes which are both desirable and feasible
7. Take action to improve the problem situation

If using Soft Systems Methodology to facilitate the introduction of a new Case Mix Management System the authors feel that it is at stages two and three that problems are to be encountered. Due to the inherent nature of the NHS - the multiple interest groups with diverse professional sub-cultures - it would be an endless task to attempt to identify norms, values and cultural views of each group that CMMS would have an impact upon.

If the transformation process is determined to be the transformation of data from various sources within the hospital (the Patient Administration System or radiology systems, for example) into information to be used for different purposes (e.g. medical audit, spreading best practice, placing contracts or monitoring treatment outcomes), the Weltanschauung that makes the transformation meaningful is that the organised provision of information is feasible and desirable and can be planned and organised. Even this Weltanschauung cannot currently be agreed as many potential 'beneficiaries' of the Case Mix Management System do not feel that the provision of the information is desirable to them. It is only when the systems have been installed and the staff have the opportunity to use the system for themselves that the realisable benefits become clear to them.

Furthermore, there are crucial questions that the organisation has asked itself about CMMS but, to date, has not managed to develop answers:-

- who are the Customers of the Case Mix Management System?
- who are the Actors who would perform the transformation?
- who are the Owners of the Case Mix Management Systems?

The answer to each of these questions can be the Government; hospital accountants; hospital managers; resource managers; clinicians; nursing staff; IT managers and/or case mix managers depending on who is asked the question. Without fundamental consensus on these issues it is virtually impossible to drive this methodology any further.

Even if the answer to these questions can be determined, a further problem when attempting to formulate the root definition is the number of tight environmental constraints within the NHS - the structure of the NHS; the planning system, the reporting mechanisms and the budgetary considerations are just a few examples. The NHS is so entrenched within its bureaucratic procedures it is difficult to facilitate any change requiring consensus across the full gamut of the organisation.

Soft Systems Methodology (SSM) requires a consensual world view and requires conflicts of interest and belief to be resolved. Under the strained environment of the NHS, it is felt that any form of accommodation that is achieved will only ever be temporary and may collapse at any time.

Similarly, participation in the use of SSM by all pertinent parties is essential for it to succeed. The different cultural behaviour patterns within the NHS makes this virtually impossible. Although all parties may recognise the need for their input, some may be reluctant to 'waste' valuable patient contact time on activities that do not directly affect the care given to their patients, or may feel that this area does not fall within their remit.

Generally, it is thought that SSM is affected by power imbalances so that the results obtained favour the powerful. If this is the case, the different hierarchies of power within the NHS who are prepared to participate in SSM will be able to skew the results in their favour.

CONCLUSION

The authors feel that SSM is a useful tool in many situations but when trying to use it to facilitate the introduction of a new computer system across such a large organisation with so many diverse professional sub-cultures, environmental constraints and power imbalances, there will be severe problems from the start.

As the Case Mix Management Systems are purchased ‘off the shelf’ from large software developers, the authors feel that attempting to gain the needs and requirements of the system from each of the potential users and a feel for their cultural and social systems would be virtually impossible and a needless task as this should have been undertaken prior to the development of the software. It takes a clinician to ‘sell’ the benefits of the system to another clinician and the evidence to date is that they can only see the benefits when the system is actually in place and they have the opportunity to use it for themselves.

Could it be the case that in such a complicated organisation, the optimal implementation methodology is a spiral methodology whereby the users of the system are drawn into the vortex of the development and enhancement of the system following exposure to the fundamental working system? It is hoped that further research may determine whether this is the case.

REFERENCES

Checkland, P.B., and Scholes, J., 1990, “Soft Systems Methodology in Action”, John Wiley & Son, Chichester.

Coombs, R., and Cooper, J., 1990, “Accounting for Patients? Information Technology and the implementation of the NHS White Paper,” Economic and Social Research Council, Oxford.

Department of Health, 1990, (i) “Working for Patients, ” HMSO, London.

Department of Health, 1990, (ii) “Self Governing Hospitals,” HMSO, London.

Department of Health, 1990, (iii) “Funding and Contracts for Hospital Services,” HMSO, London.

Griffiths, R., 1983, “Report of the NHS Managament Inquiry,” HMSO, London.

National Association of Health Authorities and Trusts, (1994) “1993/94 NHS Handbook (Eighth Edition)”, JMH Publishing, Kent.

BUILDING A FULLY INTEGRATED HOSPITAL INFORMATION SYSTEM

T. S. Chen[1] and T. G. Gough[2]

[1] Department of Medical Informatics
National Cheng Kung University Hospital
Tainan
Taiwan, ROC
[2] Division of Operational Research and Information Systems
School of Computer Studies
University of Leeds
Leeds, LS2 9JT
United Kingdom

ABSTRACT

This paper provides an overview of the building of an integrated hospital information system for the National Cheng Kung University Hospital. There are two features of the systems development process to which this paper seeks to draw particular attention. The first is that the information system was to be a comprehensive integrated system addressing the range of information needs of users in all functional areas of the hospital. Secondly, the system was to be designed for a new hospital to be built on a 'green field' site and was to be developed and installed as an integral component of the building of the hospital and the installation of the medical and administrative facilities. The paper describes the approach adopted to: the specification of requirements with no existing users to explain their needs; the study of capacity without existing data on which to base predictions of workload; the study of performance in advance of actual operation in an attempt to understand likely systems behaviour; and the scheduling of implementation to ensure congruence with the building and commissioning of the hospital. The successful outcome of the development is reported.

INTRODUCTION

In Taiwan there was a lag in investment in the development of computer-based hospital information systems (Chang, 1984) and there was an imbalance in the provision of hospitals, with some large high quality hospitals in the northern area but no equivalent provision in the south of the country. It was therefore decided that a modern medical school

Critical Issues in Systems Theory and Practice, Edited by K. Ellis et al.
Plenum Press, New York, 1995

and teaching hospital be built in the southern area, in Tainan. In 1984, the addition of a medical school to the National Cheng Kung University (NCKU) was approved. The teaching hospital project was also approved, with the construction work to start in February 1985, for the hospital to open in January 1988 (though delays in construction delayed the opening until June 1988).

In considering information requirements it was noted that the data needed by clinical and administrative personnel have traditionally been managed separately and that when computers have been used to process such data, the tasks are often performed on separate computers. As Wiederhold and Perreault (1990) point out the lack of integrated data from diverse sources creates a host of problems. It was felt to be essential for the NCKU hospital to design and implement a fully integrated computer-based hospital information system to provide the necessary high quality service to the whole hospital.

A number of existing systems were investigated including the IBM Patient Care System (Mishelevich et al., 1980) and the HELP system (Pryor et al. 1984) though neither was felt to be suitable for NCKU. The views of Tolchin et al (1984) were noted on the lack of comprehensive functionally integrated hospital information systems that could be used as models. It was therefore necessary to design and implement the proposed fully integrated hospital information system for this new system ab initio, as it were.

The development of this proposed system was a key element in the overall plan for the building of the NCKU hospital as the opening of the hospital was dependent on its successful completion. The proposed computer-based system was intended to service all the information needs of all functional areas of the new hospital. Secondly, as the new hospital was to be built on a 'green field' site there were no existing systems in place to service the hospital's information needs should the computer-based system not be delivered on time.

This paper outlines some of the issues addressed during the building of the integrated hospital information system for the NCKU hospital.

SPECIFICATION OF REQUIREMENTS

The process of developing a computer-based hospital information system for this new hospital is very different to that for an existing hospital. The existing hospital may have been using manual systems and it has to cope with the impact of automation, and the experience of data conversion problems and parallel operation. There will be a need to ensure the involvement of users to avoid the problems identified by Rusnak (1980) and it will be necessary to examine the needs of those involved in systems installation (Kolenaty, 1984). However, while the new NCKU hospital does not face the problems of automation impact and data conversion it has no immediately available users to consult, no data to analyse and no existing hospital management experience on which to draw. It was not therefore possible to use as a starting point for the terms of reference a statement of the requirements and expectations of the systems' users as suggested by Brookes et al (1982). The NCKU hospital project had no systems users at the initial study phase and hence it was necessary to ensure that the team responsible for the design and development contained an appropriate mix of skills and experience to fill the gap in expertise created by the unavailability of 'real' users.

The Planning Committee comprised members from seven different skill groups to provide an appropriate mix of expertise for making policy, for controlling and monitoring the project, and for authorising development. Most of the members were drawn from the National Cheng Kung University. Two members were skilled in computer hardware and software, two had experience in clinical practice, medical statistics and laboratory instrumentation, one was experienced in general business management and one was the superintendent of the local hospital. The chairman of the committee was also the project leader of the Study Team. This consisted of five full-time members whose role was to design the system. There were three systems analysts, one hardware and communications engineer and one secretary.

The study team was responsible for generating the terms of reference and for undertaking the initial study. The prospective users were represented by the study team. The definitions of user needs developed by the study team had to be approved by the planning committee, the members of which had to ensure that the fully integrated system would meet all the needs of its wide range of expected users (Scherrer, 1988).

The planning committee decided to adopt a turnkey solution to obtain the advantages identified by Chard (1988). To ensure that all of the required development activities and systems functionality be provided either by the in-house staff or the vendor or both, the planning committee decided to establish a set of principles for judging the design. A set of criteria for the turnkey design were determined and included fault tolerance, expansibility, real-time operation, connectivity, confidentiality, ability to handle Chinese characters, and flexibility. Structured analysis techniques based on Yourdon (1989) were used to analyse the application sub-systems for registration, medical records and statistics, in-patients, outpatients and emergency, drug inventory, pharmacy management, medical supplies inventory, laboratory and therapy, dietetic service, nursing care, finance management, personnel management, accounting management, and administration. The integration of hardware, software, communications and the user interface are encompassed in these criteria. See Chen (1992) for the detailed exposition of these criteria.

After considering the several hardware architectures available to support an integrated hospital information system (Weiderhold and Perreault, 1990) and evaluated the advantages and disadvantages of centralised and decentralised approaches (Bakker, 1977 and Flynn, 1983) the planning committee decided to adopt a centralised approach using a multiprocessor machine. As the NCKU hospital would not have its own computer staff before it opened, it had to acquire all its software from a vendor. The requirements for systems software, database management and applications software (for all the subsystems identified above) were determined to ensure a smooth handover from the vendor to the hospital's own computer staff as soon as the hospital had sufficient manpower.

Communications are needed to support the operational activities of all departments. Data and information are sent over many types of transmission mechanism. A comprehensive hospital information system needs a communication network which includes hardware and software mechanisms. The NCKU hospital network is designed to enable terminal users in all departments to have controlled access to all the data within the hospital. Chen and Gough (1989) sets out the key requirements to provide an integrated, flexible communication network capable of meeting the range of NCKU hospital needs, taking account of the application parameters identified by Fritz (1985) and Gamble (1988).

SIZING

Without any existing systems, there is an absence of reference data on which to base workload predictions and yet it is necessary to ensure sufficient capacity to service all the transactions without incurring excessive costs. The starting point for the calculation of the total forecast workload is the natural forecast unit (Terplan, 1987). A probabilistic model was then built which proved well suited to predicting the workload without access to historical data. The estimated CPU time proved accurate in terms of the sets of processors required and the estimated database capacity has also matched expectations. A more detailed discussion of this issue is to be found in Chen and Gough (1991).

The use of the probabilistic model solves the capacity problem but still leaves the problem of system performance. Clearly an acceptable level of performance is a very important consideration during the design stage to avoid producing an unacceptable system performance on completion. A simulation model of the proposed system was built and the results of running the model indicated that the most suitable set of processors was four sets, an outcome consistent with the workload results from the probabilistic model. A more detailed discussion is to be found in Chen and Gough (1992).

IMPLEMENTATION

As noted earlier, the hospital information system had to be completed successfully and integrated into the operation of the NCKU hospital at the opening date. As this was a new hospital, implementation of the computer system required the preparation of the environment for the equipment, installation of electrical power, air-conditioning and the citing of the cabling and networking, in addition to the design, development and testing of the fourteen application subsystems and all associated hardware and software. A critical feature of the implementation was the effective testing of the installed system to ensure the provision of all the necessary capabilities and functions set out in the specification of requirements to enable the new hospital to provide the planned service, since, as noted earlier, the hospital would be entirely dependent on the integrated hospital information system for all day-to-day activity. A particular feature of the testing was the involvement of the potential users in the testing process. The activities of about two hundred users were simulated in different kinds of procedures and the systems testing was undertaken by a combined group representing all the relevant departments. This approach was based on the argument that the testing involved many different types of skill and that corporate subsystems were being tested, not systems for separate departments or functions. A test would involve the selection of a patient with a particular condition or needs, identification of the subsystems required to serve that patient and the testing of the subsystems in the logical sequence appropriate to the delivery of the service. System operation was assessed continuously to distinguish between system problems and personnel problems, to ensure correction of errors and re-testing and the feeding of the learning into the training which was a major component of implementation. A fuller account of the process and an example of the testing sequence is in Chen (1992).

EVALUATION

Evaluation of an integrated hospital information system is an essential but difficult task. The evaluation programme for the NCKU hospital information system was begun at the announcement of the project and carried out by the planning committee to keep the project on track during development. As a system reaches the final phases of development evaluation involves testing whether the system performs correctly, is acceptable to its intended users, and is cost-effective. For the NCKU hospital information system evaluation covered: utilisation, including workload measurement and capacity planning; the measurement of outcomes, including satisfaction of the data processing requirements, information quality, management control and cost effective information delivery; acceptance, including quality of the interface, convenience, speed, reliability, ease of authorised access and integration; security, including password control and authorisation; the computing environment, hardware, software and communications devices, database access, and back-up (Chen and Yang, 1994).

The overall conclusion of the evaluation, undertaken by internal and external agencies, was that the fully integrated hospital information system for NCKU operated successfully as designed and installed and met its intended goals.

CONCLUSION

In the building of the NCKU integrated hospital information system, the initial study was followed by the specification of requirements. Capacity and performance studies for the selected system followed, using a specially developed workload forecasting model and performance modelling to ensure the design offered adequate performance at a reasonable cost. The fourth phase was implementation and finally the system was evaluated to assess the effectiveness of the design in operation.

The distinct feature of the design and development of the system outlined in this paper was the absence of an existing system to provide access to data and present user expertise. The lack of user expertise in the initial stages of the building process was addressed by the development of two groups, the planning committee and the study team, with the members of both selected very carefully to ensure width and depth of experience. Both groups were kept small and had a shared chairman/project manager. Both of these elements may have been key factors in the successful outcome.

Techniques were developed to fill the gaps created by the lack of existing data, notably in addressing the issues of capacity planning and performance estimation.

The NCKU hospital information system was designed and implemented to achieve integration of information provision across the hospital. This it has achieved.

REFERENCES

Bakker, A.R., 1977, Centralised versus Decentralised HIS, in Shire and Wolf (eds.,), *in*: Proceedings of MEDINFO-77, North-Holland, Amsterdam.

Bakker, A.R., Louwerse, C.P., and Kouwenberg, J.M.C., 1983, Data Integrity in an Integrated Hospital Information System - Practical Experiences, in Bemmel, Ball and Wigertz (eds.), *in:* Proceedings of MEDINFO - 83, North-Holland, Amsterdam.

Brookes, C.H.P., Grouse, P.J., Jeffrey, J.R., and Lawrence, M.J., 1982, "Information Systems Design", Prentice-Hall of Australia, Sydney.

Chang, C.U., 1984, "Hospital Management Systems", Ta-Lin Publishing, Taipei.

Chard, T., 1988, "Computing for Clinicians", Elmore-Chard Publishing, London.

Chen, T.S., 1992, The Design of a Fully Integrated Hospital Information System, unpublished PhD Dissertation, University of Leeds, Leeds.

Chen, T.S., and Gough, T.G., 1989, Computer Communication System for NCKU Hospital, in Barker, Cas, Qin and Wagner, eds., *in:* Proceedings of MEDINFO - 89, North-Holland, 1108-1112, Amsterdam.

Chen, T.S., and Gough, T.G., 1991, Workload forecasting study for a computerised hospital information system, in Hedley and Wong, eds. *in:* Proceedings of the First Hong Kong Medical Informatics Conference, Hong Kong Society of Medical Informatics and the Hong Kong Computer Society, 1234-1244, Hong Kong.

Chen, T.S., and Gough, T.G., 1992, Performance modelling study for a hospital information system, in Lun, Degoulet, Piemme and Rienhoff, eds., *in:* Proceedings of MEDINFO - 92, North-Holland, 204-210, Amsterdam.

Chen, T.S., and Yang, W.T., 1994, An evaluation of a fully integrated hospital information system, in Barahona, Velsoso and Bryant, eds., *in:* Proceedings of MIE '94, 356-361, Lisbon.

Flynn, F.V., 1983, Laboratory data processing using a network of computers, in Fokkens, ed., *in:* Proceedings of MEDINFO - 83 Seminars, North-Holland, Amsterdam.

Fritz, J.S., 1985, "Local Area Networks - Selection Guidelines", Prentice-Hall, London.

Gamble, J.N., 1988, Practical networking and standards today, *in*: "Towards New Hospital Information Systems", Bakker, Ball, Scherrer and Willems, eds., North-Holland, Amsterdam.

Kolenaty, G., 1984, Hospital information system planning, *in:* "Information Systems for Patient Care", Blum, ed., Springer-Verlag, New York.

Martin, J., 1981, "Computer Networks and Distributed Processing", Prentice-Hall, New York.

Mishelevich, D.J., Hudson, B.G., Slyke, D.V., Mize, E.I., and Robinson, A.L., 1980, Implementation of the IBM health care support and patient care system, *in:* Proceedings of the Fourth Annual Symposium on Computer Applications in Medical Care, IEEE, Philadelphia.

Pryor, T.A., Gardner, R.M., Clayton, P.D., and Warner, H.R., 1984, The HELP system, *in:* "Information Systems for Patient Care", Blum, ed., Springer-Verlag, New York.

Rusnak, J.E., 1980, The conduct of a health care system project, *in:* Proceedings of the Fourth Annual Symposium on Computer Applications in Medical Care, IEEE, Philadephia.

Scherrer, J.R., 1988, Retrospective and prospective usefulness of HIS, *in:* "Towards New Hospital Information Systems", Bakker, Ball, Scherrer and Willems, eds., North-Holland, Amsterdam.

Terplan, K., 1987, "Communication Network Management", Prentice-Hall, New York.

Tolchin, S.G., Stewart, R.L., and Kahn, S.A., 1984, Implementation of a prototype generalised network technology for hospitals, *in:* "Information Systems for Patient Care", Blum, ed., Springer-Verlag, New York.

Wiederhold, G., and Perreault, L.E., 1990, Hospital information systems, *in:* "Medical Information - Computer Applications in Health Care", Shortcliffe, Perreault, Wiederhold and Fagan, eds., Addison-Wesley, New York.

Yourdon, E., 1989, "Modern Structured Analysis", Prentice-Hall, London.

94

WHY ELECTRONIC PATIENT RECORD SYSTEMS IN HOSPITALS CAN GO WRONG

A Systems Failures Study

Joyce Fortune,[1] Geoff Peters,[1] Roger Stewart,[2] and Bill Dodd[3]

[1] The Open University
Walton Hall
Milton Keynes
[2] Kingston University
Kingston on Thames
Surrey
[3] NHS Executive
Quarry House
Leeds
United Kingdom

INTRODUCTION

Over recent years, in various countries, many attempts have been made to develop electronic patient record systems for use in acute hospitals. Although the investment in such schemes has been high, evidence from the international literature appears to indicate that they have tended to be less than wholly successful. Expectations have not been met, and their introduction has typically been accompanied by varying degrees of disaffection on the part of certain key players. Whittome (1993), for example, states: "Despite significant technological advances over the past few years, manual paper-intensive systems have remained basically unchanged. Computer systems are fragmented and require multiple data entry points." Young (1994) agrees: "Over the years many different computer-based medical-record systems have been developed. Few, if any, have had more than temporary success."

THE EPR PROJECT

During 1991, the computer-based patient record was exciting a great deal of interest in the United States. This interest was articulated in a series of papers presented at a conference organized by the American Medical Informatics Association and a number of

Critical Issues in Systems Theory and Practice, Edited by K. Ellis et al.
Plenum Press, New York, 1995

delegates from the Department of Health and the Medical Royal Colleges came back to the United Kingdom convinced that useful research could be undertaken here.

Accordingly, in the spring of 1993 the Department of Health gave the go-ahead to an initiative entitled the Electronic Patient Record (EPR) Project. It was launched at the beginning of 1994 as a three year strategic research and development programme under the management of a Programme Board. Its mission was to 'help doctors, nurses and other health care professionals [collectively referred to as clinicians] to give better care to patients through the use of Electronic Patient Record systems and to show the potential benefits of EPR systems by working in acute hospitals with clinicians, managers, suppliers and the Department of Health to produce two working demonstrations and by supporting a programme of research into the EPR and related problems.'

The PRINCE methodology is being used to manage the project and so as part of the initiation stage the Programme Board considered the risks associated with the Project, identifying a number of hurdles at which the Project might fall such as inability to find two appropriate demonstrator sites. They also decided to look further at why previous similar systems had failed, recognising that the experiences of others might have important lessons for them. Rather than tackling this on an ad hoc basis they commissioned a systems study. This paper describes the work carried out for the first phase of that study and outlines its findings.

The formal brief for the study was to identify the factors which are critical in ensuring the successful implementation of an electronic patient record (EPR) system in hospitals providing care to acutely-ill patients. For the first phase the Systems Failures Method (Fortune and Peters, 1995) was used to analyse selected published accounts of attempts to introduce clinical information systems.

THE ACCOUNTS USED

A literature review was carried out to identify accounts of the introduction of integrated medical record systems into acute hospitals which were sufficiently detailed to allow analysis.

Eight accounts were finally chosen:

1. The COSTAR V system at the Internal Medicine Department of Nebraska College of Medicine, USA (Campbell, Givner, Seelig, Greer, Patil, Wigton and Tape, 1989)
2. The mini-medical record system at the University of North Carolina Hospitals, USA (Carey, Thomas, Woolsey, Proctor, Philbeck, Bowen, Blish and Fletcher, 1992)
3. The clinical information system at Columbia Presbyterian Medical Center, USA (Clayton, Pulver and Hill, 1994)
4. The medical information system at the University of Virginia Medical Center, USA (Massaro, 1993)
5. The Regenstrief MRS at 30 hospitals and clinics in the Indianapolis area, USA (McDonald, Tierney, Overhage, Martin and Wilson, 1992)
6. The outpatient medical record system at Beth Israel Hospital, USA (Rind and Safran, 1994)
7. OSCAR at Foothills Hospital, Calgary, Canada (Sears Williams, 1992)
8. COSTAR in the Outpatients Department of a 100 bed acute general hospital, UK (Young, 1994)

COMPARISON WITH THE FORMAL SYSTEM MODEL

Each of the accounts was used to make a representation of the situation in systems terms so that it could be made the basis for a comparison with the Formal System Model (FSM) which is a model of a robust system that is capable of purposeful activity without failure. (See Fortune and Peters, 1990). Accounts of failures retailed by others are inevitably partial, but first hand investigation is often no longer an option, as in this case. However, it was felt that the availability of eight different cases and hence eight independent comparisons enabled a robust study to be undertaken.

The discrepancies and deficiencies found from the comparisons with the FSM are summarised below:

1. (Campbell, Givner, Seelig, Greer, Patil, Wigton and Tape, 1989)
   The decision-making sub-system within the system was deficient. For example, the effectiveness of reminder systems which gave physicians information about certain patient conditions was limited by the sophistication of the decision rules that generated reminders.
2. (Carey, Thomas, Woolsey, Proctor, Philbeck, Bowen, Blish and Fletcher, 1992)
   The wider system failed to provide sufficient resources; cost was seen as one of main barriers to a fully computerised system.
3. (Clayton, Pulver and Hill, 1994)
   The design of the sub-systems and components that carry out transformations within the system was deficient in that it relied on the use of character based emulation screens rather then graphical user interfaces and this accounted for low utilization by some physicians.
4. (Massaro, 1993)
   All three of the necessary links from the wider system to the system were deficient. For some considerable time there was no decision-making sub-system. A team from the Computing Services Group devised the computer system and put it in place but had no decision-making authority.
5. (McDonald, Tierney, Overhage, Martin and Wilson, 1992)
   The sub-systems and components that carry out transformations were deficient leading to input problems so that after 20 years of development the computer-stored records did not contain all the information that would be available on paper records.
6. (Rind and Safran, 1994)
   There were problems with the formulation of the initial design of the system which presumably led to poor decision-making. Examples included printing difficulties and security and privacy problems.
7. (Sears Williams, 1992)
   The link 'formulates initial design of' between the wider system and the system and that of 'makes known expectations' within the system were deficient leading to the emergence of unexpected consequences from the design.
8. (Young, 1994)
   The seven consultants who tried this system were never really part of the sub-system that carries out the transformations even though they were, theoretically, the key users. In a sense these key people could not make their expectations known because they did not really have any. In the words of Young, (1994) "They could not see much in the way of benefits."

SUMMARY OF THE COMPARISON RESULTS

The following findings were distilled from the eight different comparisons:

- The sub-systems which carried out the transformations tended not to have good links to the remainder of the systems nor to the wider systems.
- Ability to influence the environment appeared to be low whilst the disturbances, such as pressures for change, were high.
- Decision-making sub-systems tended to operate in isolation.
- Expectations not made known (at all levels).
- Some resource problems.

Because the study was being conducted under the assumption that the project would be properly resourced the fifth of these was ignored during the remainder of the study. All of the other main problem areas were characterised by lack of involvement of key players and poor communication, so these were investigated further using different bases for comparison.

COMMUNICATION

Comparison of the accounts with a communication model focussed on the following areas:

- Mismatch in symbols - where there are differences of format, coding structures or attributable meaning in the data or information passing between the sender and receivers.
- Channel problems - where the channel of communication has not been agreed or properly defined or has been inappropriately used or breaks down.
- Noise - where external factors adversely affect the encoding, transmission, or decoding of communications.
- Mismatch of rules - where a shared understanding of the use of data or information does not exist or is not adhered to.
- Feedback verification - where there is no process to acknowledge receipt of communication.

As before, each account was analysed separately and the results then brought together to draw out general findings as follows:

- There were many communication channel problems. These were mainly due to incorrect or ill-defined or non-defined channels.
- User interfaces were inadequate, often because they were too complicated.
- There was a lack of standard interfaces for communication.
- Rules of information use were not clearly defined or had not been agreed.

HUMAN FACTORS

Human factors were investigated using a set of social paradigms for the analysis of team behaviour (see Stewart, 1991) which drew on the following concepts: ingroup/out-group; anomic reaction; team primacy; expectations; and conflict.

Over the accounts as a whole there was obvious and widespread evidence of ingroup (management/administrative) versus outgroup (everyone else) problems and of team primacy amongst clinicians, managers and administrators and IT personnel. There was little evidence of inter-disciplinary teams being set up, and no evidence of them working. (The term inter-disciplinary is being used here to denote a team formed from different areas with a common purpose rather than a group which merely provides a forum for consultation and discussion.) There were plenty of examples of conflict being allowed to grow, with some groups remaining unable or unwilling to buy into the systems, and very few instances in which conflict was ever really resolved.

CONCLUSION

Overall, the investigation of communication issues and human factors supported the evidence from the comparisons with the Formal System Model that there was insufficient user involvement and that inter-disciplinary teams were not formed or were not cohesive. These human problems lead to poor design and to reluctance to accept the systems. The cultural change surrounding the introduction of new clinical information systems tended to be handled badly.

The findings of this first phase of the study formed one of two inputs (the other was information gained from interviews with interested parties) to the second stage which looked forward to the development and introduction of the EPR system and through the use of quantitative modelling tried to predict the risks associated with it.

The results from the study as a whole were brought together and reported as a series of lessons, which, for ease of access, were grouped under subject headings such as climate, ownership, system and record design and user involvement in design and implementation.

The report provided valuable criteria to help identify two suitable hospitals to develop the demonstrators. It has now been given to the project managers at each of the selected hospitals to use in project planning and the insights will contribute to the criteria to be used for monitoring the project.

REFERENCES

Campbell, J.R., Givner, N., Seelig, C.B., Greer, A.L., Patil, K., Wigton, R.S. and Tape, T., 1989, Computerized medical records and clinic function, *M.D. Computing* 6:282-287.

Carey, T.S., Thomas, D., Woolsey, A., Proctor, R., Philbeck, M. Bowen, G., Blish, C., and Fletcher, S., 1992, Half a loaf is better than waiting for the bread truck, *Archives of Internal Medicine* 152:1845-1849.

Clayton, P.D., Pulver, G.E., and Hill, C.L., 1994, Physician use of computers: is age or value the predominant factor?, *AMIA* 301-305.

Fortune, J., and Peters, G., 1990, The Formal System Paradigm for studying failures, *Technology Analysis & Strategic Management* 2:383-390.

Fortune, J., and Peters, G., 1995, "Systems and Failures" in press, John Wiley & Sons, Chichester.

Massaro, T.A., 1993, Introducing physician order entry at a major academic medical center, *Academic Medicine* 68:20-30.

McDonald, C.J., Tierney, W.M., Overhage, J.M., Martin, D.K., and Wilson, G.A., 1992, The Regenstrief Medical Record System: 20 years of experience in hospitals, clinics, and neighbourhood health centers, *M.D. Computing* 9:206-217.

Rind, D.M., and Safran, C., 1994, Real and imagined barriers to an electronic medical record, *AMIA* 74-78.

Sears Williams, L., 1992, Microchips versus stethoscopes: Calgary hospital, MDs face off over controversial computer system, *Canadian Medical Association Journal* 147:1534-1547.

Stewart, R., 1991, The use of social paradigms in the analysis of team behaviour during organizational change, *in*: "Systems Thinking in Europe", Jackson, M.C., Mansell, G.J., Flood, R.L., Blackham, R.B. and Probert, S.V.E., eds., Plenum Press, New York.

Whittome, H., 1993, Overview of electronic health record (EHR), *Medical Records Institute* 1:13-15.

Young, D., 1994, Consultants' views on their use of a computer-based medical record system, *in*: Proceedings of Healthcare Computing 1994 217: -222.

MEDICAL INFORMATION SYSTEM METADESIGN USING METAMET

Peter Kokol[1], Bruno Stiflic[1], Jernej Zavrsnik[2], and Kurt Kancler[2]

[1] University Maribor TF - ERI
Smetanova 17
62000 Maribor
Slovenia
[2] House of Health
Vosnjakova 2
62000 Maribor
Slovenia

INTRODUCTION

New possibilities offered by the huge advancement in information technology sets new requirements for the whole health care community. Thereafter *it is not merely enough to do more efficiently what we are doing now, but we must confront difficult challenges in new ways.* In that manner we must be able to design ***new*** medical information systems using ***new*** approaches. But to be capable to design new approaches we must first carefully analyse the information system design theory and practices. To follow slogans above and to make a step further into resolving the software crisis, we constructed a new (Medical) Information System ((M)IS) design paradigm called MetaMet (stands for MetaMethodology), and it is the aim of this paper to introduce it and its application in the medical field.

TWO LEVELED (M)IS DESIGN

IS design is (or at least it should be) a very human intensive process. Such processes cannot be tackled with "...hard engineering approaches..." (Checkland, 1981) but in a more "...softer way" (Checkland and Scholes, 1990). According to that we predetermined two main issues that must be considered in the manner to design successful (medical) information systems:

- we must obey to principles of participance, continuity, holism and endless learning. This means that we must take into account the whole 'system' (not only involved but also affected participants of the IS design process and related environments) and must consider the uncertainty of the situation (the value system

Critical Issues in Systems Theory and Practice, Edited by K. Ellis et al.
Plenum Press, New York, 1995

of participants will change during the process, unexpected events will occur, situation can be changed, etc.),
- we must be able to answer two '*basic questions*' :

Question 1.
which from many possible computerised information systems (IS) to design? and

Question 2.
how to design the selected one?

The above two questions seems very reasonable and natural in most IS design situations, but are very rarely explicitly answered. In those situations in which they are introduced, they are reduced to the form *Which methodology to use to design this IS?* But, unfortunately the simple selection of 'the best' IS design methodology is very difficult if not impossible (Kokol, 1993; 1994) (the IS design occurs in two contexts: firstly the context within which the designer operates and secondly the context produced by designing the design itself. There are no definitely good or bad design paradigms, and there are too many different IS design paradigms to simply select one. Software quality oriented processes and a way to design them are urgently needed). We assert that to enable successful IS design we must first *design* (it is true that *meta*design can in some situations be reduced to selection, but nevertheless this selection is carried out in a systemic manner) an appropriate design paradigm and called this activity *two leveled design* (Kokol 1993, 1994).

Two Level Design

The *'conventional' IS system design* is closely related to hard-system thinking. It is in general concerned only with Level 0 design (design of an IS) and one process (IS design process). In contrary the systemic and participative metadesign should involve three processes (IS design process, metadesign of the design process, and the design of metadesign) on two design levels (design and metadesign).

METAMET

MetaMet (Kokol, 1994) is a specific implementation of two leveled design based on the following suggested definitions of the IS design process:

'IS design is a process, suited to a specific paradigm, in which a transformation from a manual information system into a computerised one is performed using a situationalized methodology in the manner to construct a successful information system according to selected criteria.'

According to the above the use of the MetaMet is defined as:

'The use of MetaMet is a process in which a problem situation - containing a manual IS - is transformed into an improved situation - employing a computerised IS - appreciating the epistemology of MetaMet, the characteristics of the problem situation, recent developments in IS research and science in general (standards, norms, technology, etc.) and the principle of uncertainty performed in an endless learning loop.'

Basically MetaMet's activities are: formal description (metamodeling) of design paradigms, design process and design situations; formal evaluation and comparison of design paradigms; metadesign of design paradigms (where metadesign is treated in a very broad sense as invention of new design paradigms); adoption of known design paradigms, composition of known design paradigms, selection between design paradigms, etc.; design of the IS using the metadesigned paradigm; learning and accumulation the knowledge from current

MetaMet's applications and the reuse of this knowledge in further applications; appreciation of the current state 'of the art' in IS designs, system theory, science in general, etc..

According to above, the MetaMet should be pluralistic, systemic, recursively adaptable (recursive adaptability states that a metaparadigm must be an iterative learning system accumulating new knowledge about metadesign, IS design and related) and multidemensional and it should be composed out of a framework, theory and methodology. In current applications we selected the idea of a process as the framework, the Soft System Methodology (Checkland and Scholes, 1990) as the methodology, and the process formalization (metamodeling) (Kokol, 1993) as the theory.

COCIDINIS

Two activities are essential for successful children's health care: *adequate trace* (measurement, recording, trend and pattern analysis) *and control* (measurement, comparison and possible mediation) of their development. To enable efficient trace and control we should follow as much as possible various health parameters. COCIDIN is a numeric index founded on observations of important parameters and can be used as an appropriate base for the activities discussed above. COCIDINIS is an information system based on COCIDIN which should support the decision making about the type and features of activities needed in successful children's health care support. It enables following main activities: collection, recording, retrieval and management of data concerned with COCIDIN, management of health care activities, generation of statistical, epidemiological, research and other reports, analysis of data and generation of data for other computer systems (i. e. expert systems, neural networks). The main objectives of our the work are to enable successful care of the whole child population and the successful care of any individual, to enable fair comparison between various child populations (from different countries, different generations etc.), exchange of information, etc..

Design of COCIDINIS

According to the specific design situation found in designing COCIDINIS we have constructed a new design paradigm called Extended Structured Spreadsheet Modelling Paradigm (ESSMP) which is based on the following design root definition:
'COCIDINIS design is an iterative, multilingual, multidimensional and multiaspect process performed by a design group with the participation of medical users concerned with user's requirements learning, conflict solving, requirements implementation, maintenance and engineering communications so that the process results in a successful outpatient medical information system.'

This root definition was transformed into conceptual model of the COCIDINIS design methodology briefly described below. In the **first design phase** we try to motivate the users' participation with education, building, exercising and using prototypes and data-entry routines. In the **second phase** we try to learn about the users' requirements with proper communication procedures and the extended operational specification approach (Kokol, 1993). In this phase we first construct the formal operational specification, transform it into an executable prototype which is then exercised and if required adapted. After the changes are made on the prototype we must also tailor the operational specification according to the modification made on the prototype. To resolve conflicts the second phase is made on more instances - one instance for every conflicting party. When all the parties are satisfied with the specifications we use the consensus techniques to construct a common understanding and as a result the common specification. These specifications are then transformed during the implementation phase into the final IS. The main characteristic of the maintenance phase is, contrary to the current practice, that the maintenance is performed directly on requirements and not on the code. During all phases emphasis is placed on the engineering of communications to enable successful design without conflicts in communication.

CONCLUSION

The big advantage of the MetaMet in COCIDINIS design was that it enabled us to design a specific information system design methodology, which was very successfully used. Indeed the evaluation (Kokol, 1994) of both the COCIDINIS and the design methodology has shown that the MetaMet is a user friendly IS design paradigm which has been efficiently used also by untrained medical users without greater troubles. Thus it is our deepest belief that using the MetaMet and two leveled design can enhance many IS design weaknesses and, as a consequence, result in successful medical information systems design and use.

In general the greatest advantages of MetaMet are: pluralism, the strong theoretical base, it can be used in any design situation, it can be tailored for personal needs, and it is systemic, easy to use, learn and teach. Its greatest weakness is its relative immaturity. In future work we must verify and test MetaMet's use in many different design situations. Next, we have to construct computerised tools which should support MetaMet's basic functions and, finally, we need to research the possibility to extend, adapt and improve its theoretical base.

REFERENCES

Checkland, P.B., 1981, "Systems Thinking, Systems Practice", John Wiley & Sons, Chichester.

Checkland, P.B., and Scholes, J., 1990, "Soft Systems Methodology in Action", John Wiley & Sons, Chichester.

Flood, R.L., and Jackson, M.C., 1991, "Creative Problem Solving: Total System Intervention", John Wiley & Sons, Chichester.

Kokol, P., 1993, Metaparadigm: a soft systemic MIS design approach, *Journal of Medical Systems.* 17, pp47 - 57.

Kokol, P., Stiglic, B., and Zumer, V., 1994, Metaparadigm: a soft and situation oriented MIS design approach, *Int J Biomed Comput.*, In press.

DECISION-MAKING IN MENTAL HEALTH CARE

Toward a Technology-Based Decision Support Aid

F. A. Stowell, D. West, S.A. Smith, and C.E. Begg

Department of Computing and Information Systems
University of Paisley
Paisley PA1 2BE
United Kingdom

INTRODUCTION

Over the past few years there has been considerable media and public interest in the discharge of patients recently hospitalised under the Mental Health Acts. This interest has been stimulated by reports of discharged patients who have carried out violent attacks on members of the public and the government's policy of 'Care in the Community'. As a result of this increased interest the Queen announced, in her speech at the opening of parliament for the 1994/5 session, the government's intention to introduce new legislation which will tighten the controls on patients being discharged into the community in England and Wales. This new act will bring the law governing Mental Health Care in England and Wales, in particular the discharge decision, closer to the Scottish model.

The aim of this paper is to describe a research project which involves the modelling of the decision making process that is concerned with the discharge of patients hospitalised under the Mental Health (Scotland) Act (1984). It is intended that this model will be used as a basis for developing a computer-based decision support system which exploits the flexibility and adaptability of modern Information Technology (IT). The purpose of this decision support system will be to provide support for decision makers and to act as a training aid for those involved in the discharge decision. The domain is characterised by interaction between a number of different professionals, all of whom have an input into the decision making process. In this paper examples of the models which are emerging, as a result of the appreciation of the Appreciative Inquiry Method (AIM), are given and their use as a valuable learning exercise for those involved and a basis for a technology-based decision support aid is discussed.

The Discharge Decision

The Mental Health Act (1983) in England and Wales and the Mental Health (Scotland) Act (1984) in Scotland makes provision for the hospitalisation and treatment of individuals

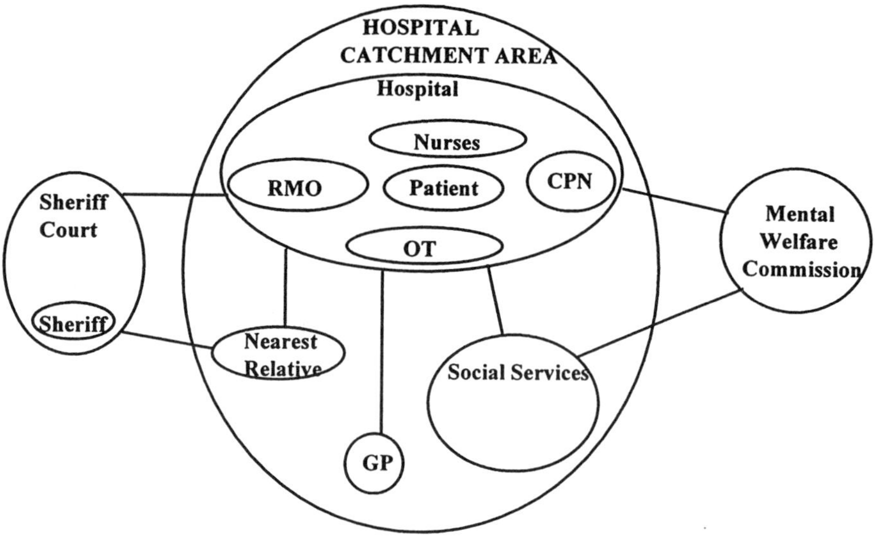

Figure 1. A System Picture (Quin and Bronte-Stewart, 1994) to show the complexity of the discharge decision.

judged to be suffering from a mental disorder. When deciding if a patient should be discharged the Responsible Medical Officer (RMO) consults the various professionals who comprise the multidisciplinary team (MDT) which is responsible for the decision making process. Expert advice is sought from anyone involved with the patient's treatment such as nursing staff including the community psychiatric nurse (CPN), social workers and occupational therapists (OT). It is recognised that the practice of the act has to be balanced against the individual's civil liberty rights (Carson, 1990). The decision making is complex since it results from an assessment of a mixture of rules (derived from the Mental Health Acts) and 'personal' knowledge gained from professional judgement. If the RMO decides that the patient should be discharged then, in consultation with the MDT, a discharge plan or Care Programme is formulated to assess the services and care the patient will need in the community.

Figure 1. shows the people involved in or affected by the decision making process. In Scotland the Sheriff, The Mental Welfare Commission and the nearest relative all have the power to discharge a patient but the most common method is by a MDT decision.

The Organisation of Mental Health Care in the Glasgow Area

To add to the complexity, from the 1st April 1994 Mental Health Care in the Glasgow area has been reorganised into the Community and Mental Health Services NHS Trust. Therefore there is an added layer of complication as Mental Health Care professionals are having to become familiar with the new organisational structure in addition to managing the already intricate discharge decision-making process.

AIMS OF THE RESEARCH

The aim of this research is to model the discharge decision making process using an appropriate knowledge elicitation (KE) method. The resulting model (or models) will be used as a basis for a technology-based decision support aid to assist with the responsibility for discharge decision making (Smith, 1993) and to act as a training aid for people wishing to explore the issues involved in this complex process. At this stage of the project it is expected that the research may exploit the characteristics of current multimedia/hypermedia (MM/HM) technology, although care is being taken to avoid technological constraints influencing the modelling process.

The domain of Mental Health Care can be described as being ill-structured, subjective and complex in that it is a mixture of clear 'rules' and the interpretation of these 'rules.' The former derives from the Mental Health Acts and the latter is a result of personal judgements based upon experience and individual interpretation of clinical evidence. There are many different KE methods (Cordingley, 1989) but due to the complexity of the domain, it is important that a KE method capable of providing a means of eliciting the more subjective elements is adopted.

The KE method selected is the Appreciative Inquiry Method (AIM) which has been developed as a means of eliciting not only the 'facts' and 'rules' of the domain but also to bring about discussion to reveal 'expert's' beliefs and understanding of the domain, intuition and other forms of 'tacit' knowledge (Stowell and West, 1990; West, 1992). AIM represents an attempt to operationalise Vickers' notion of 'appreciation' since the process of appreciation, as described by Vickers, would seem to be a useful method of explaining the way in which a human expert gains and then practises expert knowledge (Vickers, 1965).

PRESENT WORK

Figure 2. is offered as an illustration of the modelling process to date. The first interviews were conducted with the RMOs, the psychiatrists who are ultimately responsible for the discharge decision. This first phase involves the 'client' in developing a systems map of their perception of the area of interest (for a full description of the AIM see Stowell, West and Fluck, 1991; West, 1994). The results from the first phase of the discussions indicated the various background influences in play (e.g. different approaches between clinical and forensic psychiatrists). A composite map was then developed which encapsulated the components from the psychiatrists' individual maps. The composite map has been taken back to each of the psychiatrists for comment, discussion and if necessary further development. Through this discussion the original concepts have been grouped together into 5 main areas, which have been accepted by the different psychiatrists as offering an acceptable and useful way of looking at the area of concern at this stage of the project. The five main areas are the patient's mental state, the patient's or relative's wishes, home circumstances which included community support, resources which include pressure on beds and transfer to other hospitals and the views of the MDT. Each grouping was then considered in terms of the overall activity represented in the named group. This is achieved by the psychiatrists considering each grouping as a transformation and supplying the details necessary to enable the analyst to develop a strict description of the activity in the form of a root definition. The root definitions have been developed as conceptual models which are used as the basis for a more detailed discussion about the domain.

Different members of the MDT have their own specialist areas of interest and it is hoped that by gradually building up a picture of the whole domain a greater awareness of the decision-making process and the complex interaction of the many factors involved can be achieved. The aim is to work towards a detailed conceptual model (or even models) of the domain that can be used not only as a basis for discussion but also as the foundation for a decision support aid.

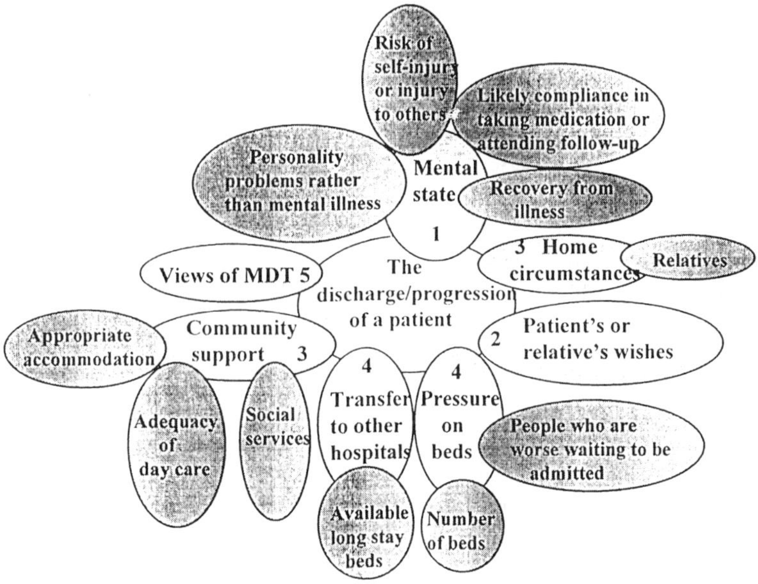

Figure 2. A composite systems map derived from the individual systems maps of psychiatrists working in IPCU at Leverndale Hospital, Glasgow.

The numbers 1 to 5 indicate the five main areas accepted by the different psychiatrists as a useful way of looking at the area of concern.

NEXT STEPS

The notion of a decision support system adopted in this project is influenced by MM/HM technology. The complexity of the domain and the heavy reliance, of decision makers, upon factors which are difficult to quantify is recognised. Consequently, the aim is to use I.T. to support the decision making process not by supplying a diagnostic aid but by providing a framework for inquiry (Isenmann, 1993) that may be accessed by those making the decision. It is expected that as well as providing direction through the discharge process it may also incorporate expert advice, discussion of past cases and may eventually allow for the tracking of patients and the decisions made about them through their history of illness.

ACKNOWLEDGMENTS

We would like to thank Dr. T. Holland at the University of Cambridge, The Douglas Inch Centre and the staff of the IPCUs at Leverndale and Gartnavel Royal Hospitals for their help and collaboration in this research.

REFERENCES

Carson, D., 1990, "Risk-taking in Mental Disorder; Analyses, Policies and Practical Strategies," SLE Publications Ltd, Chichester.

Cordingley, E.S., 1989, Knowledge elicitation techniques for knowledge - based systems, *in:* " Knowledge Elicitation - Principles, Techniques and Applications," D. Diaper, ed., Ellis Horwood Ltd., Chichester.

HMSO Mental Health Act 1983, Chapter 20, HMSO, London.

HMSO Mental Health (Scotland) Act 1984, Chapter 36, HMSO, London.

Isenmann, S., 1993, How to deal with wicked problems using a new type of information system, *in:* "Systems Science: Addressing Global Issues," F.A. Stowell, D. West and J.G. Howell, eds., Plenum Press, New York.

Quin, A.J. and Bronte-Stewart, M., 1994, System pictures: A method for capturing clients' views, *Systemist* 16:4, pp293-298.

Smith, S.A., 1993, Modelling the discharge decision process in the domain of mental health to produce a computer-based decision support aid in multimedia/hypermedia, *Systemist.* 15: 4: 205.

Stowell, F.A., West, D. and Fluck, M., 1991, The appreciative inquiry method: an approach to knowledge elicitation based upon the notion of knowledge elicitation as an inquiring system, *Systemist.* 13: 4: 154.

Stowell, F.A. and West, D., 1990, The contribution of systems ideas during the process of knowledge elicitation, *in:* "Systems Prospects: the Next Ten Years of Systems Research," R.L. Flood, M.C. Jackson and P. Keys, eds., Plenum Press, New York.

Vickers, G., 1965, "The Art of Judgement," Harper and Row Ltd., London.

West, D. , 1992, The process of knowledge elicitation and its relevance to the design and development of hypermedia/multimedia decision support aids, *in:* " Proceedings of the 3rd International Conference of the Information Systems Developers Workbench," University of Gdansk.

West, D., 1994, The appreciative inquiry method: a systemic approach to information systems requirements analysis, *in:* "Information Systems Provision: The Contribution of Soft Systems Methodology," F.A. Stowell, ed., McGraw-Hill, Maidenhead.

PUTTING SYSTEMS THEORY INTO PRACTICE AND VICE VERSA

The Role of Observation in Analysing the Real World

Sam Waters,[1] George Bakehouse,[1] Christopher Davis ,[1,2] and Kevin Doyle[1]

[1] Department of Computing
University of the West of England, Bristol
Coldharbour Lane
Bristol, BS16 1QY
England
[2] Information Management Group
University of Glamorgan Business School
Pontypridd
Mid Glamorgan, CF37 1DL
Wales
United Kingdom

INTRODUCTION

This paper seeks to explore the extent to which conventional social scientific research techniques can be productively applied to the notionally specialist task of systems analysis. Specifically, the paper examines the contribution of participant observation to the process of organisation analysis. Our aim is to critically examine both the process of analysis and observational techniques in order to identify opportunities for their exploitation. The significant results of some action research in a leading neurosurgery department where observation was the main investigative technique are reported. The complementarity of those results to those generated by more regularly used techniques is discussed.

The need for observation of organisational routine has been advocated by a number of writers, including Goffman (1961), Park (see McKinney, 1966) and Malinowski (see Polsky, 1967). Observation allows direct study and recording of the day-to-day activities of people in complex social settings, such as a hospital. The research reported below aimed to bring the well known ethnomethodological practice of participant observation into the rather cloistered world of systems analysis. As analysts we need to 'understand' the strengths, weaknesses, opportunities and threats of the present situation (the 'old, physical' situation - see DeMarco, 1978) in terms of the management of resources (see Waters, 1989) viz:

- People (and their organisational groups), Money and Technology (in its widest sense) - from the '3m's' (Men, Money and Machines) of the Industrial Revolution;

- Information, from the computer revolution;
- Environment, from the 'Green' revolution.

The research we report herein is restricted to the analysis of information weaknesses in order that they may be identified, verified, classified, quantified and prioritised; only then can they be systematically improved upon and that improvement demonstrated. Investigatory methods such as interviews and questionnaires help us to analyse the perceptions of participants in human activity systems ('I listen and I hear') but observation further helps us to analyse the practical realities of socio-technical settings ('I see and I believe') - unfortunately, only the participants themselves know these realities as part of their every working-day lives ('I do and I understand') (Oliver & Langford, 1984).

We chose to concentrate on information management problems during the second phase of our research since these phenomena were recognisable to the actors involved, even if not always realised at the time of their occurrence. Information management problems thus became the currency of exchange between the academic observers and the practitioners in their real world, enabling both groups to appreciate the goal incongruence and other social phenomena which contribute to suboptimality highlighted by the observational study.

RESEARCH ACTIVITY

During 1991 and 1992 a strategic review (Digital, 1992) was carried out at the Clinical Neurosciences Directorate at Frenchay Healthcare NHS Trust, Bristol. The review involved members of the research team and business analysts from DEC interviewing a range of clinicians, managers and other staff and produced a top-down analysis of objectives for the Directorate. This initial phase was followed by a second phase, the fieldwork being completed between July and November 1993, where participant observation was used in order to help verify the initial review by means of a bottom-up analysis of day-to-day activity.

During this second phase of the research, 16 members of staff and 3 patients at Frenchay were overtly shadowed by four members of the research team: a minimum of two observers were used at all times so as to allow comparison of observation notes. The 'actors' were shadowed for at least one full working day: frequently, the duration of the shadowing exercise was extended, particularly in the case of the patients and the surgeons, to allow continuity of observation of their episodic care and practice, respectively. The timing of the shadowing exercises was also varied so as to sample routine and emergency activity at various times of the day and night.

Field notes were made openly at the time of the observed activity using a proforma indexed initially by time and location. The hand-written observations from each observer were collated and typed under the supervision of a team member prior to being checked against the original field notes by their authors.

The individual activities recorded in the consolidated shadow histories were then analysed to generate lists of observed information management problems and of the interface or interaction which had been in progress at the time of the problem. Our aim was to further consolidate the recorded data and simultaneously transform it from a sequential list of activities into a list of observed problems. In this form, the problems could be readily ranked by the shadowee, providing a prioritised list for information systems intervention: such a list is beneficial to both the academic research being reported here and to the development of a functional specification of an Integrated Clinical Workstation for the National Health Service Executive (see Mackintosh et al, 1994).

The next stage of the shadowing exercise involved further verification of the field work findings by the shadowee: feedback to the shadowees was in the form of a generic list of problem types presented in a classified form, generated from analysis of all the field notes. Shadowees were asked to indicate the frequency of the occurrence described and its significance to them, and were provided with a copy of their shadow history to assist them in this task. The outcome of this stage of the research, shown below, comprises a list of problems and associated interfaces which have been prioritised by the shadowees themselves. Their prioritisation may be confirmed by further, more highly focused, observational fieldwork and by comparison with the findings of similar studies undertaken in other sectors of the economy (including banking, construction and transportation).

Such a longitudinal approach to analysis has been advocated by writers such as Becker (1963). More recently Hirschheim (see Galliers, 1992) and Land (1992) have highlighted the need for such a longitudinal approach to the process of information requirements analysis, enabling the insufficiencies of existing methods to be overcome (see Davis, 1994). The emphasis of the various stages of our analysis closely mirrors the three types of fieldwork identified by Becker (op cit):

- selection and definition of problems, concepts and indices: early observations can help to develop these;
- check which of the numerous preliminary ideas are worth pursuing based on frequency and distribution of phenomena in the social setting or 'community';
- attempt to incorporate the findings into a model of the system under study.

A brief resumé of the results of the observational study are given below. These are more fully discussed in Mackintosh et al (1994) and Waters (1994), but their significant conclusions include:

- everybody in the Neurosciences Clinical Directorate suffers at least one information management problem each hour of each working day;
- the most frequently occurring problem (forty percent of all recorded and twice as frequent as any other type) was incomplete information;
- the most frequently observed problem not related to the intrinsic qualities of the information being managed at the time related to a lack of training and understanding of systems;
- 62% of the problems observed with recorded information involved the patient record, compared to only 27% with computer based records and 11% with other manual records.

Feedback of the results of our observational study showed that these levels of occurrence of information management problems definitely err on the conservative side.

Neurosciences information records, particularly casenotes, are in a 'mess' (in the academic sense of Ackoff, 1971) but "...if we automate a mess, we only get a faster mess" (Strassman, 1985). All our experiences of information technology, information systems and information management convince us that technology cannot provide the solution if cultural issues are ignored; the aim must be 'to help people get better with information'. If, and only if, people will commit themselves to that aim then we can improve upon today's mess; if not, then either we continue to tolerate the mess (including its 'jobsworth' creation scheme) or we change our attitudes: if the latter is the case, then our experiences from more technologically advanced sectors of the economy suggest that the change will take a decade or so to accomplish and that people will move sideways, upwards, downwards, outwards and inwards. The choice is between revolution and evolution, at the extremes; Neurosciences believes that the way forward is the middle route - minor revolutions (eg. technological breakthroughs such as image processing) being assimilated within a gradual process evolution which allows people their freedom of choice in order to progress along the path of least resistance.

All the measures reported from the research suggest that the problems of the medical record alone may account for some 20% inefficiency across all clinical, managerial and support staff in the Neurosciences Directorate. Further research would be necessary to reliably extrapolate these results across the NHS, but

- if the NHS employs approximately 1,000,000 (making it the largest single employer in the EC);
- if the NHS costs approximately £100,000,000 per day to run;
- if the annual NHS spend on information technology is £300,000,000 (ie £300 per employee per annum),

then the inefficiencies of the manual records may be wasting a significant amount of UK resources, particularly people and money. Further, the NHS investment on information technology is comparable to the average spend per UK employee of some fifteen years ago and is a small fraction of today's average spend; there can be little doubt that appropriate investment in information technology could improve the efficiency and effectiveness of the NHS provided that we all learn from its recent failures and from the successes of more advanced sectors of the economy (eg. banking, airlines).

Carefully managed observation can produce insights into those activities which improve the understanding of both actor and observer, allowing the complexity of the social setting to be 'unpacked' by observation. Observers can then focus further analysis effort on interactions which are particularly problematic or significant. When this process is carried out in collaboration with the actors an important opportunity for individual and organisational learning is gained, enabling movement towards a common collective perception of what information systems are and what they can (and cannot) do by involvement of the actors in the process of analysis. In this way, the actors themselves become able to bridge the gap between the content of the investigation or analysis (observation, note taking etc) and its context (the objectives sought of the system, revealed by the strategic review). Much of the dissatisfaction with systems analysis and, therefore, the information systems which follow stems from misunderstanding the goals of the system. We hope to develop our current research to explore the opportunity for organisational learning by the use of action research throughout the life cycle of the information system and, in the process, hopefully moving some way towards mutual understanding of the information management problems manifested. (Valusek and Fryback, 1985)

Finally, the processes of observational investigation are well documented and criticised in the literature (eg McCall and Simmons, 1969) but we would contribute two further remarks. First, contrary to Bell (1987) we would argue that observation is more resource consuming than interviewing as a method of primary research; for example, shadowing one participant for one day required an entire week of researcher time. This can be broken down as follows, measured in researcher-days:

i1.0 preparation and prebriefing;
i2.0 shadowing by dual observers;
i1.0 analysis and debriefing;
i1.0 communication of consolidated results.

Malinowski (quoted in Polsky, 1967) pointed out that "...information must come to him (sic) full-flavoured from his own observations...and not be squeezed out of reluctant informants as a trickle of talk". Secondly, the results of the observational study are critically dependent upon the competence of the observer; our lead observer had been highly trained in the discipline during a previous police career whereas the rest of us (although experienced and competent systems analysts) had not - he observed twice as many information management problems, on average, than we did.

CONCLUSION

It is our assertion that carefully managed observational study of human activity systems offers the potential to reveal aspects of human activity which might otherwise go unchallenged or even unnoticed. If such observational study is used in the context of action research, as described above, the problem situations observed can be shared with the actors concerned. The major advantage of the approach described is that it complements the reporting emphases of conventional systems analysis techniques, allowing exploration of cultural issues and problems not immediately amenable to technological solution. Such exploration is, in action research, shared by researcher(s) and subject(s) allowing the observed human activity to become the 'common denominator' in the users' and developers' understanding of the system within its operational reality. Thus the role of observation in analysing the real world is to develop shared beliefs between users, who know and understand practical reality, and analysts, who might otherwise be restricted to somewhat abstract and misleading views based upon perceptions - 'seeing is believing'

REFERENCES

Bakehouse, G., Davis, C., Doyle, K., and Waters, S., 1994, Anthropological Reflections on Systems Engineering: Seeing is Believing. Proceedings, Conference on the Philosophical and Logical Aspects of Information Systems, Bristol.

Becker, H., 1963, "Outsiders: Studies in the Sociology of Deviance", The Free Press, NY.

Bell, J., 1987, "Doing Your Research Project", Open University Press, Milton Keynes.

Davis, C., 1994, Getting Better with Information Systems. Proceedings, PICT International Doctoral Colloquiu, Policy Studies Institute, London.

DeMarco, T., 1978, "Structured Analysis and System Specification", Prentice Hall, Englewood Cliffs, NJ.

Digital Equipment Corporation, 1992, "Frenchay Trust: An Integrated Neurosurgery Strategy and IS Strategy".

Frenchay Healthcare NHS Trust, 1994, "Project Evaluation Report: Integrated Clinical Workstation for Neurosciences"

Galliers, R., 1992, "Information Systems Research", Blackwell Scientific, London.

Goffman, E.,1961, "Asylums", Penguin, Harmondsworth.

Land, F., 1992, "The Information Systems Domain" (unpublished paper, *London School of Economics*), London.

McCall, G., and Simmons, J., 1969, "Issues in Participant Observation", Addison Wesley, Reading, MA.

McKinney, J., 1966, "Constructive Typology and Social Theory", Appleton-Century-Crofts, New York.

Mackintosh, I., Waters, S., Bakehouse, G., Davis, C., and Doyle, K., 1994, "Integrated Clinical Workstation: User Requirements for a Neurosciences Directorate" NHS Information Management Group (forthcoming) NHS Executive, Leeds.

National Health Service Management Executive, 1992, "Information Management and Technology Strategy Overview" HMSO, London.

Oliver, I. & Langford, H., 1984, "Myths of Demons and Users" *in:* Galliers R., (ed.), 1987, "Information Requirements Analysis: Selected Readings", Addison-Wesley, Sydney.

Polsky, N., 1967, "Hustlers, Beats and Others", Aldine, NY.

Valusek, J., and Fryback, D., 1985, Information Requirements Determination: Obstacles Within, Among and Between Participants, *Proceedings, End-User Computing Conference*, ACM pp103 - 110.

Waters, S., 1989, SWOT analysis in IT projects, *International CIS Journal,* Vol 3 No 1, pp.20-29.

Waters, S., 1994, "Information Management Problems: Towards a Cross-Sectoral Perspective", Health Economics Consortium, York.

EDUCATIONAL SYSTEMS

98

SYSTEMS AND THE 'POLO MINT' METAPHOR

R. Armson and R. L. Ison

The Systems Department
The Open University
Milton Keynes, UK

INTRODUCTION

The Systems Department in the Faculty of Technology at The Open University is unique in that it has taught systems thinking within an applied systems framework, using supported distance teaching methods, for 21 years. In this period over 16,000 undergraduates and associate students have taken one of the Department's three major Systems courses. In their current 'incarnations' these are; *Managing in Organisations* (T245); *Working with Systems* (T247) and *Complexity, Management and Change* (T301). New postgraduate and undergraduate courses are envisaged. All Systems courses are supported by a one week intensive and experiential systems summer school (SYST 999). The Department also hosts *Third World Development,* contributes around 80 per cent of the 'teaching' for both the Faculty's foundation course, *Living with Technology* (T102) and the Faculty's MSc and MBA programmes in *Technology and Manufacturing* management. Open University teaching material has been widely disseminated and used to guide other curricula. The formation of the Systems Department was a highly innovative move in 1972; it remains the largest Systems Department in Europe with 21 full-time academic staff. Courses remain popular with students with around 2500 enrolments in courses in 1994.

Systems staff come from a wide range of initial academic disciplines including engineering, electronics, physics, agriculture, psychology and the social and policy sciences. Student learning is supported distance-teaching materials produced within the department and by a network of part-time tutors who, whilst they have Systems expertise, are essentially learning facilitators. Staff of the department have been at the forefront of innovation within the university, including the development of the home computing policy and systems so that, in 1996, all T102 students will be on line via modems supplied as part of enrolment.

Despite an admirable record of achievement, a sense of unease and concern has existed within the department for some time. We suggest this is also evident in the Systems movement itself. In this paper we describe a process of exploration of our unease. We analyse certain metaphors prevalent in the Systems Department mythos and, by unpacking these, suggest some possible alternative and potentially liberating interpretations and alternatives. We relate these to the Systems movement as a whole.

Critical Issues in Systems Theory and Practice, Edited by K. Ellis et al.
Plenum Press, New York, 1995

REFLECTIONS ON WHERE WE ARE

The department's 21st anniversary (which had escaped the notice of most staff) and the arrival of a new professor, provided an occasion to initiate cycles of reflection and celebration. Ray Ison joined the Department as Professor of Systems in January 1994 and over the next three months conducted a series of one-to-one conversations (of between one and three hours) with all Systems staff and a range of external staff and students (Ison 1994(a)). Following this cycle of personal interviews, a paper reflecting his learnings was distributed to all staff. In that paper he articulated the following aspirations:

> "Coming, as I have, from a conservative, traditional University, I have dreamed for some time of working with a group who are excited by what they do and the ideas they generate." I am not sure how best to describe such a group - perhaps it could be any or all of 'a learning organisation', a 'critically reflecting community', a group of 'action researchers', who think of learning, research and problem solving as much the same thing, a 'community of scholars" (when I talk of scholars I do not mean it in any elitist sense, but as a group of people in an organisation who, regardless of role and position, are prepared to think about what they do and ask themselves whether this can be done differently or better. And, to move beyond thinking into action, * (or a group of enthusiasts." (Ison, 1994(b))

All staff were invited to a 21st birthday celebration lunch and workshop following distribution of the paper. In all 21 major themes of concern were identified from the interviews. These outcomes, and earlier research (Webber & Ison, 1994), provided a basis for the design of a participatory workshop. This was attended by 44 Systems staff, including post-graduate students and secretaries. Participants "floated" six major areas for attention and future action (Ison, 1994(b)). The workshop also triggered a number of reflections on the life of the department and our future in a rapidly changing environment (e.g. research assessment, changed funding arrangements, new technologies, increased competition).

During all these discussions, a single metaphor kept recurring as members described their experience of the Systems Department - the Polo Mint.†

EXPLORING THE POLO MINT METAPHOR

In the months following the 21st birthday celebration, discussions on department structure continued via e-mail, small informal meetings and internal papers. As part of these discussions, the authors decided to explore the Polo Mint metaphor further.

The image of the Polo Mint within the Systems Department goes back at least 15 years to when a member complained to one of us that Systems Theory was like a doughnut - it didn't have a centre. Since then, this image has been used by various members of the department, mostly in private conversations, to describe its social structure , its course profile and its academic *raisons d'être*.

The recurrence of this image over such a long period, and its appearance as a continuing part of the department's self-image, suggests a long-standing sense of emptiness, incompleteness or yearning. The application of the image in such a *range* of contexts suggests further that the department has not really identified the object of that yearning. It was commonly reported during the individual discussions that members each felt themselves

* In the nineteenth century anyone who went to school, or was associated with a place of "learning" was termed a scholar

† A Polo Mint is a small white sweet characterised as 'the mint with the hole' - it has a hole through the middle.

to be on the margins of departmental activity while others were more central. Most felt themselves to be in contact with only a very small number of people. There was little contact with whatever was perceived as 'the centre'.

ALTERNATIVES TO THE POLO MINT METAPHOR

The Polo Mint was mostly used to express discomfort with the sociability or working structure of the department. The authors asked themselves whether there were more positive ways of using the metaphor to re-vision the Department: a way to think about the sort of department we wanted to work in? We realised that we could not *give* the department a new metaphor for itself but we could encourage people to explore alternative interpretations of their metaphor. We first engaged in the process for ourselves and began to tell each other stories about metaphors which encapsulated the 'hole-in-the-middle'.

The Village Green

The village green is held by the villagers as common land and is watched over by them from the surrounding and protecting houses. The collective wealth of the village, its livestock, are pastured on the green and the green is the site for festivals and other activities of common interest.

The Eye of the Storm

The eye of the storm is a space of temporary respite from a hostile outside. Despite the calm, there is fear of the storm outside and no exchange with it. This fear, together with the smallness of the space, prevents any real activity within the eye of the storm.

The Engineering Annulus

As an engineering component, the annulus is more interesting and useful than a solid disc of similar mass. It can store much more energy and can sustain greater internal and external forces. The annulus works *because* there is no connection across the centre. One of the implications of an annular structure is that the only connection each element in the structure has to the whole is through immediately adjacent elements. The circular shape and the cohesiveness is an emergent property of the connectedness of each element to only a very small number of other elements. There is no contact across the annulus. That's why it works.

The Bicycle Wheel

The bicycle wheel is a sub-type of the annulus. It sustains very high stresses for its mass and is very efficient. High levels of tension in the connections with the centre allow compressive stresses to be sustained in the rim without the rim distorting. However, the centre is also the means by which very large external pressures are applied.

Conclusions

Clearly, alternative metaphors were available and several seemed to yield insights into the department's self perception. The village green was attractive and the engineering annulus suggested that the hole-in-the-middle is not *of itself* a bad thing. These alternative metaphors seemed somewhat inadequate however and the authors realised that we needed a metaphor we could truly 'own'. We came up with the torus spiral.

THE TORUS SPIRAL

The Structure of the Torus Spiral

A torus is "a ring-shaped surface generated by rotating a circle about a coplanar line that does not intersect the circle" (Hanks, 1979) i.e. the Polo Mint. The Joint European Torus is a torus wound with conductors to accelerate sub-atomic particles. From this not-very-apt idea, and through several iterations, came the idea of a surface winding around a torus shape. This structure is illustrated in Figure 1.

Like the torus, the torus spiral defines a circular space in the middle. The torus itself is an emergent property of the way the spiral is wound. Unlike the torus, the torus spiral is a *space* allowing various interactions with its environment and within itself.

The torus spiral allows for circular motion through the toroidal space enclosed by the surface, for linear motion through the centre of the torus and for interleaving motion through the coils from all directions. The environment can become part of the torus spiral by moving around and through it. Movement within the structure can take place across the surface. This involves movement between the inner and outer boundaries of the structure and progression from one part of the structure to another. Finally, the structure itself is capable of internal movement. One can visualise a kind of 'eddying' motion in which the innermost parts of the coil move up through the centre and then spread outwards and pass down the outer boundary of the structure.

A Spiral-Torus Structure for the Systems Department

This metaphor seems to offer much for re-visioning the Systems Department. Its open structure offers freedom of action while core tasks are fulfilled. It offers members of the department a model for moving between research, teaching and scholarship in an open and receptive structure and for moving in and out of the centre. It offers the possibility of alternative relationships between individuals and the Department. The centre becomes a shared space for reflection and reflection becomes a festivity on the village green. The

Figure 1. The torus spiral.

coherence of the torus spiral implies a connectedness that comes from shared systems scholarship with individual coils representing the projects that people engage in and later move on from. The sense of belonging comes from what one is doing, in the sharing of reflection, rather than the particular place one occupies. Above all, the metaphor seems to suggest a greater connectedness between 'teaching' and research. Academic staff in the Department have traditionally found it almost impossible to be active in both areas. The torus spiral encourages us to look again at what we define as research and to see action research and reflection on action as core activities in the department.

A concern with learning is the same as being concerned with research and problem solving (Ison, 1990). Equating learning with research does not appear to come easily at a time in which research is seen by most researchers to lie within a positivistic, objective epistemology, in which the researcher is seen to be outside the system under study. Popular culture and pervasive myths reinforce the isolation of research from everyday activity and from our special capacities to engage in critical reflection.

The linking of systems thinking to practice in a given context is increasingly seen as action research but despite the increasing literature on this subject it is not a common tradition in universities, particularly in departments associated with the so called 'natural world'. Jiggins' (1994) scenario is that: "Universities perpetuate teaching as a hierarchical, linear process of transfer of information and method, rely heavily if not exclusively on positivist systems of inquiry and intellectual traditions which assume linear predictability and exclude from core ... curricula the skills and competencies in communication, negotiation, team work and participation increasingly demanded by the marketplace. Partitioned universities lead to partitioned ... institutions." One of the worst partitions is between teaching and research (Ison, 1990) and accounts for our enthusiasm for action research as a means of integrating research and learning, leaving 'teaching' to wither from its place as the dominating paradigm.

Implications for Systems Scholarship

The metaphors which arise from our reflections invite us to consider the praxis of systems. We suggest that our 'village green' be the 'leisure site' for a diverse array of researching which is joined by a preparedness to be epistemologically transparent and aware, and which shares a common commitment to a process of braiding theory with action. To do so opens up the opportunity to critically reflect on our experiences and invites us to acknowledge our roles as observers observing.

For any research we suggest it is important to be aware of one's epistemological position as, whether we like it or not, this shapes what we see and do in the world. Being aware makes us able to be clear about what our preferences are and to thus act responsibly. As Humberto Maturana (1991) observes: "The most common justification for killing and torturing in the history of the occidental, patriarchal culture, to which a great part of modern humanity belongs, is the defence of the truth, the defence of reason, or the defence of universal transcendental values under the claim that the defender is intrinsically right and the others are intrinsically wrong." He further suggests that: "taking responsibility for one's actions is a most difficult action, since it always entails the risk of offending those who do not take responsibility for their actions because they claim their actions are valid on transcendental principles, or on an undeniable truth, or on a rational argument, or on an objective fact, while they privately know what they say is not ethically valid." Whilst these may appear to be strong words for some, this manner of living is none-the-less pervasive within Western cultures and the rationalistic scientific cultures in particular (Russell and Ison, 1991).

CONCLUSIONS

In all this, we need to be aware of our own roles in re-visioning the Systems Department. We cannot persuade people to share a vision or metaphor; we can only share it and live through it ourselves. We need to be responsible for it as a living metaphor, open to new interpretations and better alternatives.

REFERENCES

Hanks, P. ed., 1979, "Collins Dictionary of the English Language", Collins, London and Glasgow.

Ison, R. L. , 1990, Teaching threatens sustainable agriculture. *Gatekeeper Series No. 21,* International Institute for Environment & Development, Sustainable Agriculture Programme, London.

Ison, R.L., 1994(a), "Systems: Listening and Reflecting", Open Systems Group, Unpublished Paper.

Ison, R. L., 1994(b), "Systems 21st birthday celebration. Moving on from Stony", Open Systems Group Mimeo, Milton Keynes.

Jiggins, J., 1994, Systems approaches: the challenge of university teaching, *in:* "Proc. International Symposium on Systems-oriented Research in Agriculture and Rural Development", Montpellier (in press)

Maturana, H., 1991, Response to Berman's critique of 'The Tree of Knowledge', *Journal of Humanistic Psychology,* 31:88-97.

Russell, D. B., and Ison, R. L., 1991, The research-development relationship in rangelands: an opportunity for contextual science, Invited Plenary Paper, *in*: "Proc. IVth International Rangeland Congress", Montpellier, Vol. 3, pp. 1047 - 1054.

Webber, L. M., and Ison, R. L., 1995, Participatory rural appraisal design: conceptual and process issues, *Agricultural Systems*, 47:107-131.

STRUCTURE AND CULTURE OF HIGHER EDUCATION INSTITUTIONS

Its Impact on Information Systems Strategic Planning

Steve Clarke and Nandish V. Patel

The University of Luton
Faculty of Business
Department of Business Systems
Park Square
Luton, LU1 3JU
United Kingdom

INTRODUCTION

> "Information systems are not technical systems which have behavioural and social consequences, rather they are social systems which rely to an increasing extent on information technology for their function" (Land and Hirschheim as quoted by Lawrence and Lowe,1983).

Strategic planning is frequently undertaken as a rational, planned exercise, using methods which are assumed to be applicable to all organisations regardless of their structure or culture. This is particularly problematic in the strategic planning of information systems (IS) and information technology (IT), where such systems are enabling mechanisms supporting the organisation. This raises the issue of organisational structure and culture, since "...similar approaches to IS strategy formulation have been reported to produce different levels of effectiveness in different organisations." (Pyburn, 1983). By summarising the underlying theory related to strategic planning and the organisational structure and culture of higher education institutions, and undertaking case study research into the Higher Education Management Information System (HEMIS) currently being implemented at The University of Luton and Nene College of Higher Education, this paper shows how structural and cultural considerations must be taken into account if such strategy is to succeed.

STRUCTURE AND CULTURE

Information systems strategy permeates the whole organisation, and people are key to this strategic approach. In an age when information is becoming a primary resource people must be knowledgeable, productive and motivated. The need is for the workforce to quickly acquire new skills and be flexible and responsive to changing demands. So any IS strategy

Critical Issues in Systems Theory and Practice, Edited by K. Ellis et al.
Plenum Press, New York, 1995

formulation must take into account the specific organisational context. Business organisations and information systems are in a continuous state of change: "It is imperative to treat strategy formulation as a continuous process in a constantly changing complex environment" (Waema and Walsham, 1990). One such method is Quinn's logical incrementalism, an approach which combines the analytical and behavioural approaches to strategy formulation. The recommendation is to "...proceed flexibly and experimentally from broad concepts to specific commitments, making the latter concrete as late as possible" (Quinn, 1980). Logical incrementalism is "...a purposeful effective pro-active management technique for improving and integrating both the analytical and behavioural aspects of strategy formulation" (Quinn, 1980).

The organisational classification for educational institutions has been categorised (Mintzberg 1988) as professional bureaucracy "This organisation is one in which the operating core are highly skilled and highly trained and the organisation is co-ordinated by standardisation of skills. To ensure that autonomy, the production system must be neither highly regulating, complex not automated" (Mintzberg, 1979). An improvement on this categorisation in the current climate of change is, in our view, professional 'adhocracy' : organic teams with liaison groups impacting progress in a relatively unstable environment. In such a structure, formal IS which regulate and control are less important than systems which allow information and decision processes to flow flexibly and informally, a situation which points to networks and user models.

Culture has an impact on strategic IS planning (Revenaugh, 1994). Revenaugh, quoting Deal and Kennedy, suggests that corporate culture can be understood and managed by identifying four categories, of which the University best fits a 'Process Culture' classification. The implementation of an IS plan within such a culture is dependent on the position of the IS in the strategic grid. However, in all cases a clear plan is necessary for success, since the process is hard to change. This is particularly problematic with strategic and turnaround applications, where the 'fixed' nature of the process means that it takes great effort just to get IS planning discussed, but such a plan is critical to the organisation's development.

The processes are hard to change, so what is needed is the optimisation of sub-processes and the linkages between them. Again, this is a very user-dependent environment. User involvement is an important factor in information systems development, and nowhere is this more critical than in a professional, adhocratic business. The involvement of people in the system is a fundamental issue which must not be overlooked, and the structure and culture of the organisation needs to be considered in the strategic IS process. In higher education this points to support rather than control systems, and distribution rather than centralisation.

Care needs to be taken, however, in deciding who should be involved, at what stages of development, and how the involvement should be carried out - the type and degree.

This decision is largely dependent on organisational climate and the type of system being developed, and will be different for each category of user.

STRATEGY FORMULATION

Information systems strategies fit within the corporate strategies of an organisation. IS strategies, then, are needed which support or are 'aligned with' the organisation's corporate or business strategies, and which draw on the available information technology. Information systems support strategy as enabling mechanisms, the aim being to find the correct systems to enable the required corporate strategies to be pursued. Because of the dynamic nature of information systems, and their reliance on rapidly changing technologies, an incremental, flexible approach must be taken: not so much a rigid plan as a framework against which activities and progress may be monitored and adjusted. New technologies fundamental to the development of IS are emerging all the time, and in setting IS strategies a view must be taken on these emerging technologies. Environmental scanning to determine this technological emergence is therefore an essential part of the IS strategy process. The internal IS/IT environment can be analysed in strategic terms by reference to the strategic grid (McFarlan, 1981). Existing systems must be positioned on this grid, and their future development managed accordingly.

HEMIS is clearly a strategic system. The modular credit scheme (MCS) which it supports is a source of current and potential competitive advantage, and its future development, incorporating the relevant emerging technologies, is of strategic importance to the Universities of Luton and Nene. Technological opportunities can be reviewed on the technology implementation grid (Little, 1981). Future investments should be categorised as 'emerging', 'pacing', 'key' or 'base'. Emerging technologies are clearly of high potential, and must be carefully monitored until a decision to utilise them can be made. Pacing technologies should be invested in selectively, primarily to decide which are likely to become key, at which point they must be built and controlled systematically. Base technologies are usually ageing, and should be selectively divested. Such an analysis would highlight potential future developments. For example, the MCS, as a strategic system, must utilise pacing and key technologies, and the plans to distribute access via computer local and wide area networks is an example of this.

A significant finding has emerged in this respect from a PRISM (partnership in information systems management) survey, (Remenyi, 1990). This survey concluded that many systems central to competitive advantage evolved from already established systems that were initially built to assist at the operational level. This so called "platform effect" (Kanellis, 1993) gives rise to three main conclusions:

- Platform considerations should be included in the justification of information systems.
- Systems should be designed so that they may be subsequently used for platforms.
- In looking for systems for intensive business change opportunities, systems should be inspected for their platform ability.

Information Systems strategics must align with corporate strategy and use the relevant technologies. The internal and external IS/IT audits is the starting point for IS strategy formulation, consisting of a review of all the existing systems, a view of what technologies are emerging, and how and when the required systems should be purchased and delivered. The existing University IS portfolio must be positioned on the strategic grid. From this the impact of any system on the organisation can be gauged, and the approach to be taken decided. HEMIS, for example, as a strategic system, should use incremental strategy to incorporate technology, linked to environmental scanning and rapid development to give future strategic advantage.

The technology implementation model then looks at present and emerging technologies in order to determine the approach to be taken in respect of these. In higher education, this is critical for monitoring networked, user oriented systems to support the professional adhocratic structure and process culture. User involvement can be built in to the strategic process by the consideration at each stage of development of the stage, type and degree of involvement required. Once the systems are defined, consideration must be given to their applicability as platforms for future development. The platform considerations are particularly useful when assessed alongside the outcome of strategic grid and technology implementation mapping.

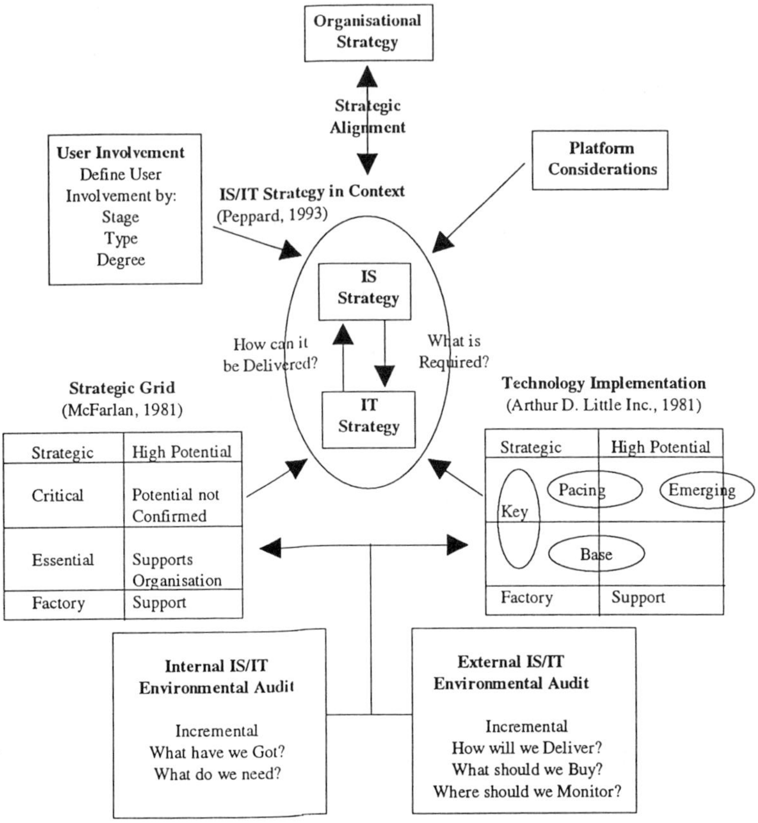

Figure 1. Recommendations: The strategic model.

REFERENCES

Kanellis, P.,1993, "SISP and Open Systems - The Platform Effect." Brunel University Department of Computer Science Discussion Document.

Lawrence, M., and Lowe, G., 1983, Exploring individual user satisfaction within user led developments, *MIS Quarterly,* June.

Little, A. D., 1981, Technology Implementation. Strategic Planning for Information Systems, *in*: J. Ward, P. Griffiths, and P. Whitmore, "Strategic Planning for Information Systems", John Wiley & Sons, Chichester.

McFarlan., 1981, Strategic Grid. Strategic Planning for Information Systems, *in*: J. Ward, P. Griffiths, and P. Whitmore, "Strategic Planning for Information Systems", John Wiley & Sons, Chichester.

Mintzberg , H., 1979, "The Structuring of Organisations", Pentice Hall, Hemel Hempstead.

Peppard, J., 1993, "IT Strategy for Business", Pitman, London.

Pyburn, N., 1983, Linking the MIS plan with corporate strategy : an exploratory study", *MIS Quarterly*, 1:14.

Quinn, J.B., 1980, "Strategies for Change - Logical Incrementalism", Irwin, Homewood, Il.

Remenyi, D.S.J., 1990, "Strategic Information Systems Development, Implementation and Case Studies", NCC Blackwell, London

Revenaugh, L., 1994, "Corporate Culture and the Strategic Grid: Preparing for the Implementation of a Strategic Plan." Hong Kong Polytechnic Working Papers.

Waema, T., and Walsham, G., 1990, Information systems strategy formulation, *Information and Management*, 18: 29.

100

CRITICAL THINKING IN THE INFORMATION SYSTEMS UNDERGRADUATE CURRICULUM

The Use of Soft Systems Methodology as an Introductory Vehicle

Brian Hopkins

Division of Information Systems and Computing
Anglia Polytechnic University
Chelmsford, CM1 1LL
United Kingdom

INTRODUCTION

This paper is an attempt to put into perspective the impact and value of Soft Systems Methodology (SSM) (Checkland, 1981) as a watershed in the growth of insight into the dynamics of problem resolution, as an instrument of teaching (and understanding) in the field of systems development and as a milestone on the road to a genuinely emancipatory (and hence, ultimately, more effective) approach to the forming of coalitions to tackle the many problems confronting us.

It is, in part, a reaction to what appears to be a currently fashionable stance of negative criticism of SSM which (in some cases) seems to owe more to personality than substantive analysis and evaluation. More than this, the paper takes the view that SSM represented a paradigm shift, in the Kuhnian sense, and that it should be seen by academics, their students and practitioners in that light.

In other words, SSM can and should be perceived and used as a progressive development which is of immense value in our understanding of systems thinking and, more importantly, as a foundation for the liberating process within education which engenders a constructively critical approach to organisational and social problems.

Finally, the paper considers SSM as an element in a wider tradition, a developmental continuum, the continuing growth of which can be observed in the theories and practice of a variety of authors, making philosophical and practical pedagogic links to that tradition.

Critical Issues in Systems Theory and Practice, Edited by K. Ellis et al.
Plenum Press, New York, 1995

SSM - ITS INTELLECTUAL ROOTS

To fully appreciate the extent of the change in systems thinking wrought by SSM we must understand the environment, theoretical, cultural and practical, in which it grew (in the 1960s and 1970s).

'Problem-solving', as it was then mainly perceived, was (and to an alarming extent still is) largely the province of the positivist, rationalist inheritors of the Enlightenment, with their self-confident belief in absolutes and their power to uncover them. This tradition, as well as being prolific in its output of scientific discovery and technological innovation, was (surprisingly?) uniform and homogeneous in its fundamental mind-set; a set of attitudes which also informed the thinking of management theorists in that period. As Schon has noted "...professional activity consists in instrumental problem-solving made rigorous by the application of scientific theory and technique." (Schon, 1983).

SSM developed against this background and, as Checkland (1981) acknowledges, its emergence was a protracted process with many setbacks, contradictions and (at the time) apparently insuperable obstacles to overcome. It should be recognised that this emergent process is still going on e.g. the philosophy and outlooks represented in the more recent work (Checkland and Scholes, 1990) have taken into account many of the criticisms levelled at SSM in the decade which has passed between the two texts.

Moreover, we should acknowledge that such ongoing development is an intrinsic and inescapable part of the process of maturation and sophistication of understanding. It would, in fact, be a contradiction of the fundamentals of systems thinking if we were to imagine or argue otherwise. So, why is it deemed necessary or appropriate to 'freeze' SSM at a particular point (or points) in its development? Why do we not use its insights and revelations in a positive (but not uncritical) manner to bring about advances in our understanding of the nature of problems so that we may better tackle them?

The intellectual inheritance which fuelled the emergence of SSM is in the phenomenological tradition of sociological exploration with its models of social reality. In the words of Checkland: "...(its model) is very different from that implicit in the approach which is usually taken to be the application of systems theory within social science, namely functionalism. Functionalism is part of the Durkheimian (or positivistic) tradition in sociology." (Checkland, 1981).

The thinking and writing of Geoffrey Vickers is also recognised as one of the formative influences on SSM growth. In 'Freedom in a Rocking Boat' (1970) he outlined one of his prime tenets:

> "As all policy makers know from experience, policy does not consist in prescribing one goal or even one series of goals, but in regulating a system over time in such a way as to optimize the realization of many conflicting relations without wrecking the system in the process. Thus the dominance of technology has infected policy making with.... bogus simplifications, just admissible in the workshop but lethal in the council chamber" (Vickers, 1970).

Checkland picked up this theme when he stated that "From the 1960s to the 1980s.....organisations are conceptualised as goal-seeking machines and information systems are there to enable the information needs associated with organizational goals to be met" (Checkland and Scholes, 1990). He then posed the question: "Why not treat organisations as if they were not goal-seeking machines, but discourses, cultures, tribes, political battlegrounds, quasi-families or communication and task networks?" (Checkland and Scholes, 1990).

When viewed from the standpoint of critical theory there must surely be familiar and welcome pointers in the above quotation. A vocabulary which includes 'discourses', 'communication and task networks' and 'political battlegrounds' does not sit comfortably with an author typified by some (e.g. Flood and Jackson, 1991, quoted in Romm, 1994) as "...likely to assist with only reformist changes". In fact, Checkland himself has the following to say on this issue:

> "Ironically, the methodology is also highly compatible with the ideas of 'Critical Sociology' of the Frankfurt School, although Habermas - its leading theorist - regards himself as opposed to systems theory. Another way to describe the methodology, in fact, is as a formal means of achieving the 'communicative competence' in unrestricted discussion which Habermas seeks." (Checkland, 1981).

So, it can be seen that at key milestone points in its development SSM has been seen by its originators as stemming directly from, and owing a clear debt to, a philosophical and intellectual source which is shared, at least in part, by many of its detractors.

SSM - THE CRITICS' CASE

Those detractors are concerned by what they perceive as a certain lack of clarity and substance in the theoretical base propounded by the authors of SSM. In particular, there is a repeated suspicion that they have 'imperial' ambitions, not merely at a careerist level but, more importantly in an epistomological sense.

It is claimed by some critics that Checkland sees SSM and its theoretical underpinning as providing a "...better understanding (than, for example, a 'hard' engineering intellectual framework) in our approach to social reality" (Romm, 1994). This is extended by the same critic to suggest that, at times Checkland seems to suggest that "...the intellectual framework of soft-oriented systems theorists can be considered as 'better' in helping us to better understand reality, or approach reality in a 'better' way." (Romm, 1994).

Others have concentrated their criticisms on their observations of a tendency in SSM in practice to reinforce the existing order within any problem situation and, generally to afford privilege - in, for example selection of actively influential participants and choice of relevant systems (Romm, 1994) - to those parties who initially possess the power. Flood and Jackson (1991) expressed concern in their critique of SSM regarding its theoretical base, arguing that it "...serves only the 'practical' interest" (of Habermas' threefold categorisation of human interests) and that "...the emancipatory interest in communication free from distortion and in realising participative decision making free from the effects of power relationships is not granted attention".

The same authors claim to have highlighted

> "...a contradiction that goes to the very heart of SSM. Interpretive thinking embraces a form of relativism where every viewpoint must be accepted as equally valid. Generating mutual understanding by exploring world views is a process that theoretically can and should go on for ever. This poses a major difficulty in practical situations since there is no means of changing from abstract debate of alternative *Weltanschauungen* to pragmatic problem solving. SSM resolves this difficulty in an extremely unfortunate way. It simply leaves closure of the debate to the prevailing power structures as reflected in the dominant culture of the organisation" (Flood and Jackson, 1991).

It is perhaps appropriate and proper that (while the debate must, of course, continue) we should be assessing the value of SSM both as a mould-breaking innovation of its time and as a foundation for the process of development, within education, of a critical and reflective stance which could begin to meet the aspirations of the critical school.

To achieve this it is surely time to see SSM in its proper historical perspective, faults and all.

SSM AS PART OF A WIDER TRADITION

As well as identifying the roots from which SSM grew it is instructive to review some of the contemporary influences with which its shares conceptual links - in particular those which many would associate with a critical outlook.

The work of (among others) Schon (1983) and Freire (1976) provides fertile ground for such exploration (although a detailed coverage must wait until later papers). In their different ways these authors have much to say which as well as chiming with the ideas of SSM could also be seen as pointing to the ameliorative developments hinted at above.

For example, we see in Schon various manifestations of what could be termed a critical approach to our received views on learning, knowledge and practice: for example, he proposes that "...we are in need of inquiry into the epistemology of practice." (Schon, 1983). Further to that approach he argues in an insight surely worthy of any 'emancipatory' school that, "Even when practitioners, educators and researchers question the model of technical rationality, they are party to institutions that perpetuate it." (Schon, 1983).

Schon also makes a distinction between 'problem setting' and 'problem solving' which would certainly not be out of place in any text on SSM.

In the course of adopting what may be described as a critical approach to the issue of knowledge and skills within the professions we can trace common threads and attitudes between the work of Schon (1983) and SSM.

Similarly, in the field of education, we discover in the work of Friere (1976) an equivalent iconoclasm which bears comparison (in terms of its radicalism) with SSM, *when set against the cultural context in which it was developed.*

Two quotations suffice, at this point, to illustrate this conceptual bridge: "The mark of a successful educator is not skill in persuasion - which is but an insidious form of propaganda - but the ability to dialogue with educatees in a mode of reciprocity." (Freire, 1976) "Hence, Freire cannot be taken seriously if he is judged only in terms of short term results." (Goulet in the Introduction to Freire, 1976).

These references provide an (albeit brief) insight into the thinking and practice of Freire, indicating strong attachment to the concept of *dialogue* and, equally significantly for the argument propounded in this paper, the claim that concepts and practice which are at odds with the contemporary received wisdom are unlikely to provide immediate results; rather they should be judged on their longer term impact on the development of the accepted body of knowledge.

In short, rather than depict SSM as theoretically and 'politically' defective we would be better engaged in identifying and drawing out its genuine involvement as part of a continuing intellectual and practical movement, the aim of which is to achieve liberation through understanding and participation.

In this role, SSM surely has a place within the curriculum as a means of introducing to students a curious and questioning attitude combined with a desire to contribute effectively (both in their studies and their later careers) to the decision making processes which have an impact on us in the workplace and in society at large.

SSM WITHIN THE CURRICULUM

We have been practising this approach at Anglia for around four years. in that time we have experimented with a variety of approaches to the introduction of 'soft' ideas in our undergraduate programmes in Information Systems (in all three years).

The experiment has provided many insights into both the continuing prevalence within our educational culture of technical rationality and, more hopefully, the readiness of students to welcome alternative, more challenging, attitudes.

It must be allowed that "...a critical approach makes heavy intellectual demands upon the participants in the educational process; demands which mean that time and flexibility must be found at the expense of a desire to 'cover all the material'." (Hopkins and Spaul, 1994).

We believe that the heavy demands are justified in the pursuit of a critical Information Systems education. Following McKay and Romm (1992) we conceive of this agenda as being principally discursive; that is "...we do not envisage it as the transmission of a body of accepted doctrine - rather it is a process in which competing theories, outlooks and techniques are placed in broad dialogue. Truth or 'correct' technique is not achieved on a once and for all basis, but is continually renegotiated during the enquiring process." (Hopkins and Spaul, 1994).

Both in principle and in practice we have found SSM to be effective as an intellectual aperient in this context. As such it plays an invaluable role in the never-ending search for "improvement" which is central to our educational project.

CONCLUSIONS

The central theme of the argument presented in this paper is that SSM can and should be seen and used as one means of re-orienting outlooks and attitudes towards a more liberal, critical, participative and reflective mind-set. The intention is not to dictate or proselytise; rather is it to open up genuine debate through informed and aware discourse.

Our experience with student groups is that this target is not utopian. The project continues; the future will provide more challenges, further insights and, we believe, further rewards for the participants.

REFERENCES

Checkland, P.B., 1981, "Systems Thinking, Systems Practice", John Wiley and Sons, Chichester.

Checkland, P.B., and Scholes, J., 1990, "Soft Systems Methodology in Action", John Wiley and Sons, Chichester.

Flood, R.L., and Jackson, M.C., 1991, "Creative Problem Solving. Total Systems Intervention", John Wiley and Sons, Chichester.

Freire, P., 1976, "Education: the Practice of Freedom", Writers and Readers Publishing Cooperative, London.

Hopkins, J.B., and Spaul, M.W.J., 1994, Critical thinking in Information Systems teaching: a political agenda? *in*: the Proceedings of the Sixth ISTIP Conference,(Sunningdale), University of Hertfordshire.

McKay, V., and Romm, N.R.A., 1992, "People's Education in Theoretical Perspective", Maskew Miller Longman, Capetown.

Romm, N.R.A., 1994, Continuing Tensions between Soft Systems Methodology and Critical Systems Heuristics, *Working Paper No.5*, Centre for Systems Studies, University of Hull.

Schon, D.A., 1983, "The Reflective Practitioner: how professionals think in action", Basic Books, New York.

Vickers, G., 1972, "Freedom in a Rocking Boat", Pelican Books, Harmondsworth.

THE HYBRID ACADEMIC

A Changing Profession

J. G. Howell and F. A. Stowell

University of Paisley
Department of Computing and Information Systems
Paisley
Scotland
United Kingdom

INTRODUCTION

In recent years the concept of the 'Hybrid Manager'* has emerged from within the business world in recognition of the changing working environment. In 1990 the British Computer Society (BCS) recommended the development of educational programmes to help produce 2,000 'Hybrids' per annum throughout the 1990s (BCS, 1990) "Hybrids - managers combining business understanding, technical competence and organisational knowledge and skills - are clearly central figures playing a vital role in bringing about significant change within organisations."

This recommendation emanated from the identification and ensuing analysis of the changing role of information management (and to some degree Information Technology (IT) within organisations: moving the function and role out of the IT department, out of the corporate team, and into the fabric of line management (BCS, 1990, Earl, 1989; Earl and Skyrme, 1992; Rockart, 1988).

The role of the hybrid manager is suggested as, (BCS, 1990(a))

- the effective management of information;
- playing a vital role in bringing about significant change within organisations;
- helping turn opportunities for improved business performance with the use of information technology into reality;
- helping avoid expensive failure by their persistence, commercial sense and understanding of people and technology issues.

Change has encompassed and affected not just the business world, but the academic world also. A fundamental appraisal of the problems of teaching and learning in mass higher

* 'Hybrids' - managers combining business understanding, technical competence and organisational skills

Critical Issues in Systems Theory and Practice, Edited by K. Ellis et al.
Plenum Press, New York, 1995

education is now deemed necessary and a radical, rather than evolutionary, approach is being advocated as necessary to finding a solution (MacFarlane, 1992).

In the light of the unprecedented growth of Higher Education (HE) in the United Kingdom (UK), part of the solution to the challenge facing HE will involve the use of (IT).

The world of the academic is changing, the 'Hybrid Academic' is emerging. The hybrid academic is a lecturer, a tutor, a teacher, a researcher, a mentor and, in many instances a manager of information, technology and people. The hybrid academic needs to be able to synthesise solutions to both academic and business problems that challenge the evolving academic world.

This paper will consider the concept of the academic as a hybrid manager and attempt to provide an image of this new type of manager. The changing role of an academic will be considered in terms of the notion of a hybrid manager .

EDUCATION FOR THE HYBRID MANAGER

It is not easy to identify a clear image of a hybrid manager although the term was first coined in 1989 (Earl, 1989). Hybrid managers is in part about bringing business management and information management together (Earl and Skyrme, 1992) at a personal level. Existing UK systems for education and vocational training seem geared to developing the former, specialists, into more generalist areas (Adair, 1983; Stewart, 1986). Their likely backgrounds encompass both specialist and generalist areas.

In response to the identified need to provide "10,000 hybrids by 1995 as a minimum target" (BCS, 1990a) many universities offered courses designed to meet this need. This expected development is noted by Robson (Robson, 1994),

> "Whilst readers on first degree programmes are likely to find themselves, at least initially, categorised as either hybrid users or members of an IT management team, in other words, single dimension managers, those on postgraduate or post-experience programmes may quickly become part of hybrid management development".

Nationally the JUPITER Consortium * offer approximately 67 (and increasing) modules through the auspices of participating Higher Education Institutions (HEIs) to cover the breadth of skills and knowledge considered suitable to the hybrid manager. No single HEI would be easily able to offer such an array of choice. At the University of Paisley, in order to create a viable course, a systems approach was adopted by the course drafting team to design an educational programme for the hybrid manager as a whole. The systems approach encompassed the notion of system,

- identifying the area of interest (in context);
- creating a conceptual analysis of the requirement;
- drawing on research material and professional experience to enrich this conceptual map;
- building a structure for a solution (the programme of educating);
- maintaining a holistic view of the requirement.

An initial picture of the hybrid manager was similar to that shown in Figure 1.

* The JUPITER Consortium is a group of participating Higher Educational Institutions and industrial representatives formed in April 1992 and tasked with bringing to fruition a modular masters degree programme in technology management. The group is supported by industry, academic institutions and the DTI

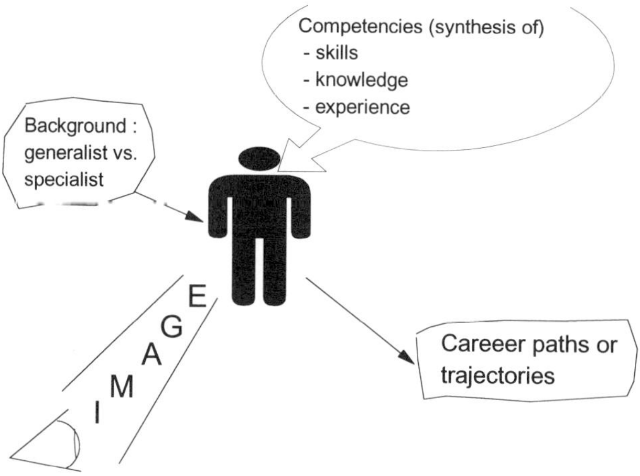

Figure 1. The hybrid.

THE EDUCTIONAL PROGRAMME

In designing the educational programme at Paisley the course team considered the hybrid at work. In business information technology systems the hybrid can be seen as a professional *involved in* the development of information systems *which are* business information technology based, *developed to an* appropriate quality standard *using an appropriate method* or approach, *utilising* modern IT and *paying attention to* the softer business skills *within a* project managed environment.

In January 1994 the first intake of 47 was recruited from applicants responding to advertising. This advertising did not target any specific or particular profession nor sought those with an IT background. Our expectation of the type of applicant was based on previous experience in recruiting to postgraduate courses: the applicants came from very different professional and experiential backgrounds wishing to undertake a course which may enable them to convert to IT. As information systems and associated technology have permeated society and the workplace, the notion of converting to IT has become outmoded.

In more recent years applicants have sought places to enable them to either upgrade or enhance their skills and knowledge to aid their return to their first profession or to acquire better knowledge of IT and learn new skills or more up-to-date skills in order to seek a new career in IT.

Research analysis (BCS, 1990(a)) suggests that the hybrid can be a business manager with acquired IT skills or an IT manager with acquired business management skills. There is some debate about the level of technical expertise but little debate that the hybrid requires business skills such as, people skills, marketing skills, financial skills and business 'awareness' of quality and service. Research points more at hybrids being business managers with IT experience rather than vice versa.

At Paisley, applicants have primarily come from business professions such as civil engineering, education, law, marketing, hospitality and other service sectors and, in the first cohort, averaged approximately 30 years of age. During the recruitment process it was particularly obvious that most of the candidates were especially selective in seeking a course of study suited to their understanding of the hybrid manager (Howell, J.G., 1994).

The image of the hybrid manager suggested thus far, depicts a manager (or aspiring manager) who is more likely to come from a business background with several years of work experience.

A further cohort of 40 successful applicants was recruited in August-October 1994 with similar professional backgrounds.

THE CHANGING ACADEMIC WORLD

Academics are now being encouraged to use IT innovatively as part of the solution to the challenges facing higher education (HE) (MacFarlane, 1992). Academics, like business managers require to be cost conscious and customer/ client oriented. Similarly to other sectors a charter for service to the customer client has recently come out. The concept of quality in teaching and learning (and all HE matters) has become an important issue. Academics are being urged to create opportunities for improved business performance. In the increasingly competitive marketplace of HE expensive failures in the academic and educational sphere cannot be afforded. At the individual level, academic lecturers will be increasingly involved in the support, development and management of the learning environment (MacFarlane, 1992).

Academics, more than ever, are becoming involved in course management, course administration and course design. In addition to these administrative/ management tasks the academic needs to participate in research and innovative teaching. To many the introduction of IT to aid one or many of these activities has become a necessity. It seems worthwhile considering what lessons can be learnt from business which may be valuable to the 'hybrid' academic.

THE HYBRID MANAGER AND THE (BUSINESS) ORGANISATION

Significant research and literature emphasises that for IT to be successful the development of IT-based systems must be matched to the organisation's business development (Morton, 1991; Galliers, 1985; O' Connor, 1993; Computer Bulletin, 1991; BCS, 1990(b)). Case examples suggest that the involvement of hybrid managers in projects helps to achieve this. For example, Esso UK claim that the involvement of hybrid managers in projects increased the percentage of projects delivered **on time** and **within budget** from 60% to 90% (Meiklejohn, 1990). This is a significant achievement and reflects the thinking suggested by the BCS report that, (British Computer Society, 1990a) the presence of hybrids makes it easier for the organisation to understand and so control project delivery.

The role of the hybrid is context dependent: dependent on the size and nature of the organisation and dependent on the formation of the development team. Case studies do not appear to encompass the rapidly expanding organisational base of the UK - the Small to Medium-size Enterprise (SME) (Howell, 1994). This is an important contextual consideration in our systems understanding of the hybrid. More research work requires to be undertaken in the area of the SME.

Existing research highlights competencies and roles exercised by the hybrid. Primary examples of these include the hybrid manager as **an enabler of change** (Palmer and Ottley, 1990; BCS, 1990a) whose characteristics "should provide a presence which .... can smooth the process of managing change and improve the understanding of what will happen." (BCS, 1990(a)). The characteristics of the hybrid manager include business acumen (Robson, 1994); technical competence, business confidence and organisational skills (Earl, 1989). The hybrid manager is able to employ business judgement and business skills in a manner which improves project success and is able to identify and tenaciously pursue business ideas and opportunities.

If IT is to be employed to the organisation's best advantage then hybrids must be developed or recruited.

The hybrid manager is seen as helping 'bridge the gap' between IS and business (Earl and Skyrme, 1992) - to provide an interface between business systems development and IT developers.

There is a need not just for making the most senior IS executive a hybrid, but for hybridisation at all levels (Earl and Skyrme, 1992).

The concept of a career path to take hybrids between levels of the same or different business organisations may be considered outmoded, too inflexible and inappropriate as a career route (Keen, 1988). More appropriate is the concept of a trajectory or alternative trajectories. A scenario based approach may be more suitable for the generalist hybrid moving around jobs or projects.

THE HYBRID MANAGER AND THE ACADEMIC

New educational programmes are being created to suit the changing breed of managers within the business world. The academic world is changing and a new educational programme may result from this development. The hybrid academic is required to be a lecturer, a tutor, a researcher and a manager of resources including the teaching and learning environment. To take a holistic perspective on the new academic we should consider first the context of his/ her environment. The development of the hybrid has impacted upon academics in two ways,

1. the set of skills needed to teach/ train the hybrid manager;
2. the 'force' which drives the hybrid is acting upon the academic world in a similar way to the business world and demands that many academics require to be both academics and managers.

THE FUTURE

The hybrid manager may be a transient term associated to some degree with a transient phase of IT. Within a few years managers will, for example, be able to build spreadsheets vocally. Key IT concepts will continue to be required to be able to direct the building of spreadsheets and the ensuing use of these in decision-making. Perhaps then, job titles may appear which reflect the role of a hybrid manager within organisations.

CONCLUSION

The hybrid manager is someone who can enable change within organisations and is therefore able to draw from a broad portfolio of competencies. Employing these competencies purposefully requires the use of 'soft skills' as well as 'hard skills'. These soft skills include interpersonal skills, teamworking skills and leadership skills. In decision-making the hybrid requires to employ business knowledge and acumen. Emphasis is on softer skills gained from a business area, not necessarily IT. Hybrids require hard skills also, sufficient to allow them to make viable and competent decisions to support the progress of projects developing IT-based systems. They must therefore have IT technical knowledge and competence commensurate with the level of decision-making.

Academics work in a changing environment, an environment that also follows some of the trends of the business world. In the more competitive, collaborative or complementary world of changing higher education more academics may begin to think of themselves as hybrids. HE institutions might also consider providing the necessary training, education and experience relevant to this new academic.

REFERENCES

Adair, J., 1983, Does your organisation develop leaders ?, *in:* "Effective Leadership: A Self Development Manual", Gower, London.

British Computer Society; 1990(a), Hybrids - a critical force in the application of Information Technology in the 1990's, *in*: A report of the British Computer Society Task Group on Hybrids", BCS, London.

British Computer Society; 1990(b), Hybrid Managers': What Should You Do?,*in*: "A report of the British Computer Society", BCS, London.

Computer Bulletin, 1991, British computer Society, London.

Earl, M.J., ed, 1989, "Management Strategies for Information Technology," Prentice-Hall, Hemel Hempstead.

Earl, M.J., and Skyrme, D.J., 1992, Hybrid managers - what do we know about them ?, *JIS*: 3:169-187.

Earl, M.J., and Skyrme, D.J., 1992, Hybrid Managers: What do We Know About Them ?, *in*: "Report of the Oxford Institute of Management", Templeton College, Oxford.

Financial Times, 1993, "Businesses complain of bank pains," September 27th, London.

Galliers, R.D., 1985, An approach to information needs analysis, *in:* "Human Computer Interaction - INTERACT 84", B. Shackel, ed., North Holland, Amsterdam.

Howell, J.G., 1994, Strategic information systems: a new perspective, *in:* "Proceedings of the Fourth International Conference of Information Systems Development", Zupancic J. and Wrycza S. (eds.), Bled, Slovenia.

Howell, J.G., 1994, Interview notes of applicants seeking places on the PgD/MSc in Business Information Technology. Paisley.

Keen, P.G.W., 1988, Rebuilding the human resources of IS, *in:* "Information Management: The Strategic Dimension", M.J. Earl ed., Oxford University Press, Oxford.

O' Connor, A.D., 1993, Successful strategic information systems planning, *JIS*:3:71-38.

MacFarlane, A.G.J., 1992, Teaching and Learning in an Expanding Higher Education System, *in*: "Report of a Working Party of the Committee of Scottish University Principals", Polton House Press, Edinburgh.

Meiklejohn, I., 1990, Whole role for hybrid, *Management Today*, March:113-116.

Morton S., 1991, Corporations in the 1990s: Information Technology and Organisational Transformation, *in*: "Final report of the MIT/Management in the 1990s Programme," Oxford University Press, New York.

Palmer, C., and Ottley, S., 1990, From potential to reality : hybrid's - critical force in the application of information technology in the 1990's, *in*: "A report by the BCS Task Group in Hybrids", British Computer Society, London.

Robson, W., 1994, "Strategic Management and Information Systems - An Integrated Approach", Pitman, London.

Stewart, R., 1986, The manager and the specialist, *in:* "The Reality of Organisations," Pan Books, London.

Rockart, J.F., 1988, The line takes leadership - IS management in a wired society, *Sloan Management Review*, 29 (4):57-64.

102

OPENING UP SYSTEMS

A Review of Systems Teaching in the Open University

Graham Paton*

The Open University, South West Region
4 Portwall Lane, Bristol
BS1 6ND
United Kingdom

INTRODUCTION

For over a decade, the Open University's systems courses have provided a sound foundation for undergraduate students wishing to use systems concepts to help understand the complexity of the modern world of technology. Indeed, the justification for creating these courses is precisely that this complexity can only adequately be grasped through the employment of systems concepts and that their understanding facilitates effective action. At the start of any consideration of the pedagogic approach to systems concepts however there are two particularly difficult problems. The first is whether to teach systems concepts overtly or covertly. The overt approach would design a course around the major systemic themes of communication and control, emergence and hierarchy. These concepts could only be brought to life and illustrated effectively by extensive use of case studies and examples from the everyday world; nevertheless such enlivening illustrations would be secondary to the main structure of presentation. In contrast, the covert approach would reverse this relationship. The courses in this category would be structured around some issue or case study and the way in which this subject matter was explained would draw in systemic concepts, without necessarily making a big show about them.

At first sight making the decision about which approach to adopt may seem a simple matter, related mainly to questions like which approach best presents systems ideas in assimilable form and which is likely to be most efficacious in attracting students. But there are deeper issues too. These relate to the second problem: that is, is 'Systems' a true discipline, a meta-discipline or what? This is a big debate for which there is not space here. Insofar as systems can be viewed as a discipline in its own right, it is important to teach it clearly as such. This thinking tends to favour the overt approach. But if systems is really a meta-discipline then it can only be presented through the medium of other disciplines; this thinking favours the covert approach.

* Tel: 0117 9299641; Fax: 0117 9255215; E-mail: G.PATON@UK.AC.OPEN.

Over the years the Systems Group at the Open University have evolved a quite sophisticated response to this situation. A portfolio of Systems courses have emerged, other teaching vehicles have been provided and utilised and there has been a diasporadic spreading of systems people. Let us take a look at each of these elements of the response in turn. The Systems course portfolio comprises three courses. Two are at second level, (each at 30 CATS points); T245 “Managing in Organizations” and T247 “Working with Systems”. The other is at third level, (rated at 60 CATS points); T301 “Complexity, Management and Change: applying a systems approach”.

MANAGING IN ORGANIZATIONS

T245 is due for first presentation in 1995. However it is a revised version of its highly successful predecessor of the same name T244 and follows the same structure. This is described in the following terms:

> “If managing were something that could be easily taught it would not be such a problem. In fact a great deal is simply learned by experience, and one purpose of this course is to help you learn from and make sense of your own experiences in organizations. So the course is decidedly practical: it is appropriate for anyone who has to get things done with or through other people, and may be especially suitable for those moving into positions with greater managerial or administrative responsibilities, perhaps from a more technical background.

The course’s aim is not just to teach ‘tools for thought’ applicable to organizational matters but to ensure that you can apply them to your own organizational life. It will help you to see your own work in a wider context; to understand better your organizational relationships (whether interdepartmental or supervisory); to recognise when your initial interpretation of problems may be inadequate; and to know how to go about generating a more rounded understanding of and response to complicated issues. The course neither uses nor teaches mathematics." (Undergraduate Courses, 1995).

There is a set book, ‘Organizations: Cases, Issues, Concepts’ by Rob Paton et al (eds) (1984) and both the ‘Graphic Guide’ by Carter et al (1984) and ‘Writers on Organizations’ by Pugh et al (eds) (1994) are recommended preparatory reading. It is interesting that, throughout this description, the word ‘system’ appears nowhere! This course is a very good example of the covert approach to presentation of systems concepts, which are introduced by way of explanation of the complexity of issues inherent in work groups, organizations and inter-organizational relations. These three blocks comprise the bulk of the course. They are introduced by a block on ‘Problems about organizations’ which discusses the nature of organizations and their problems and then goes on to call into question

> “...some conventional ideas about organizations and organizational problem solving” (T245 Course, 1995).

The Work Group block “...explains practical concepts in individual and group psychology to account for the frustrating (and the creative) ways in which organizational relationships can develop. The ‘control model’ is introduced as a simple framework for thinking about the control of processes, tasks and work groups.”

The Organizations block “...deals with organizational structures and processes, with power and conflict, and with decision-making. Several more practical frameworks for thinking about problems are taught and a variety of organisational problems and practices are discussed along the way.” (T245 Course, 1995).

The Inter-organizational relations block “...shows that inter-organizational relationships often provide the wider context needed to make sense of events and difficulties. Market (self-regulation) and institutional (hierarchical) patterns of inter-organizational relations are contrasted, and some reasons why neither works quite as intended are explored." (T245 Course, 1995).

Finally, the course concludes with a block entitled 'Wider Perspectives' which "...draws together some of the systems ideas that have been built into the course and prepares for part of the examination by giving a structured exercise based on a case study. The last unit provides further integration of the course material, but from an entirely different point of view." (T245 Course, 1995). In this way the course addresses a full range of management issues at all levels of organizational hierarchy, providing a general systemic view of management within organizations which, whilst based on specific issues and cases, overall provides a generic view.

WORKING WITH SYSTEMS

The complementary second-level course T247 "Working with Systems" is now well established, in its fifth year of presentation. This adopts a more overt approach to the presentation of systems concepts. The course description highlights this:

> "We depend on health and financial systems and they depend on information systems. We call parts of our bodies systems and insecticides systemic. We all work in and with these complex systems and we don't really understand them. We haven't got the time or the information to think through all the interactions that made the systems as they are and that determines the success or failure of what we do. We probably haven't got the mental capacity either. So in order to be effective we need to adopt a different way of thinking about these systems. The purpose of this course is to teach this distinctive and practical way of thinking.

The course will be of interest to anyone who has had the experience of being intrigued, puzzled, frustrated or enraged by the behaviour of a system and wants to make it work better. Such people could be managers at any level in the public, private or voluntary sectors, technologists in engineering or computing, NHS medical staff or administrators - in fact, the course has relevance for most jobs in a highly developed and interdependent society. By the end of it you will be able to generate radical yet feasible ideas for bringing about change in a wide variety of systems." (T247 Course, 1991).

In this description the word 'system' appears frequently. Nevertheless, as we have noted, even this overt presentational approach needs vehicles for illustration of system concepts. This is achieved in T247 through the introduction of a range of topics, which "...include the processing of work through a factory or office, ethical investments, pests and pesticides on crops, managing change in organizations, medical diagnosis and how groups make decisions. There are examples of many different kinds of system so that you can see the very wide application of the course's concepts and skills." (T247 Course, 1991). These are utilised to teach three main skills:

> "...diagramming, modelling and applying concepts and techniques to problems you are facing. Diagramming will enable you to represent your own thoughts, and those of others, in a form that allows new perceptions and insights into complex problems. By the end of the course you will be able to build and use spreadsheet models on a computer to help you to generate and try out new ideas for changing and improving the functioning of a wide variety of systems. The last skill, putting what you have learnt to use, is taught through frequent exercises in the text and the assignments." (T247 Course, 1991).

MOVING BEYOND SECOND LEVEL

As many students wish to take both these second-level course (often in the same study year), they are designed to reinforce each other. The different presentational styles intermesh neatly to achieve this effect.

Although systems concepts are introduced very well in these two courses, there is no space to address the use of systems analysis methodologies. This is left to the third level course T301. However students taking one (or both) of the second-level courses attend a discipline-based Summer School at which both a Hard and Soft Approach to Systems analysis are presented and explored practically. This summer School also provides a link into T301 which explores systems analysis in depth.

COMPLEXITY, MANAGEMENT AND CHANGE: APPLYING A SYSTEMS APPROACH

T301 is described as a course which "enables you to apply systems ideas and methods to difficult issues in the real world:

> "All of us are involved in management in its widest sense - private or personal, play schools or prisons, people or products. Simple problems may require only simple skills, but a systems approach starts from the premise that the decisions people face cannot be nearly separated. they are much more likely to be what have been called 'messes', systems of interrelated problems. The course provides skills, techniques and experience to deal with 'messes' that arise in systems like companies, schools and governments.

The course will be of particular value (now or later in your career) when you have to deal with intransigent human and technical problems, in short-term tactical ways or through longer-term strategies. By the end of it you will be able to use systems methods to understand complicated situations, model them, consider the systems consequences of various actions and, with others, improve the original situation.

The course gives you three complementary systems approaches: the analysis of systems failures and catastrophes; a systems modelling approach to organisational decision-making; and a systems approach to organisational change. An extended project will give you experience of applying one of these. The main teaching takes place in the first twenty weeks of the course and is followed by ten weeks of project work. The last two weeks draw together the themes in the course." (T301 Course, 1993).

The course commences with an introductory block entitled 'Systems Ideas' which presents systems concepts and the features of a systems approach. Students need no prior need no knowledge of systems work; the introductory block should bring all students to the same level of understanding.

The 'Graphic Guide' used by T247 is also used for this section. This is followed by 'The analysis of Systems Failures' which "...develops a practical method for understanding failures ranging from small-scale incidents to large catastrophes, and from local problems to questions of regional or national policy. The particular approach used for understanding failure is presented and you gain competence by applying it." (T301 Course 1993). This block uses case studies from the reader 'Understanding Systems Failures' by Bignell and Fortune (1983). The course then moves on to cover the Hard Systems approach: "A systems modelling approach to organizational decision-making studies systems amenable to precise modelling. Its 'hard' systems approach solves management problems and aids decision-making when the systems objectives can be defined unequivocally. The introductory unit outlines the T301 approach and the role of modelling. Its use is then pursued in case study material on the agrochemical industry." (T301 Course 1993).

Then the course addresses the Soft Systems Methodology (SSM) of Peter Checkland: "For a systems approach to clarifying and facilitating organisational change when there is an agreement about what is wrong and what would be right, modelling is appropriate and powerful. But many problems are not like that: people have conflicting perceptions, different ideas of 'the problem' and different ideas about a solution or improvement. To cope with this a 'soft' systems approach has been developed. This established and tested method helps people to understand the 'messes' in which they are involved, and helps them to define beneficial policies and actions. You will acquire the steps of a soft systems approach and gain experience of using it." (T301 Course, 1993). Finally all three Approaches are compared and contrasted in a concluding block.

THE PRESENT AND ITS CHALLENGES

Taken as a set then, the three courses provide a rather covertly systemic view of management; a more overtly systemic view of the workplace and the processes underway in a range of organizational contents; and finally a set of analytical tools with which to both make sense of and take purposeful action in such settings. Since their inception, the component courses of this set have, variously, changed in size, emphasis and the type of illustrative scenarios employed. However, the basic relationship of focus on management, work process and analysis has been preserved.

Producing these courses, when no similar courses existed in the UK, involved a major creative effort on the part of the Systems Group. Once this task was complete it was natural for those involved to ask 'what next?' Since the systems courses themselves were intended to teach systems concepts, techniques and methodologies; precisely because they were efficacious it is entirely logical that the Group should 'take their own medicine' and go onto apply this systems thinking to substantial issues. This is indeed what happened. Some members of the group moved into the embryonic Open Business School where a number of courses were influenced by systems ideas. Others created the Third World Development Course U208. Most significantly, several members of the group helped to form the Course Team for the Technology Faculty Foundation Course, "Living with Technology" (Currently in presentation with Code T102). Through T102 systems ideas (albeit in simple form) are introduced strongly at Level 1 in the Faculty's Course profile and this is available to the majority of Technology students at the outset of their studies.

The need to achieve a high standard of maintenance, cyclic review and updating of all these courses is now a major commitment for the Systems Group and requires considerable resource. However, resource could also be deployed to address several other tasks. New courses within the Technology faculty would benefit from the injection of systems ideas. There are several topical academic areas of activity (eg 'the built environment' and 'telematics') where a strongly systemic approach would be very effective. There is also the ever-present opportunity to pass on understanding of systems thinking to other groups and faculties within the University. Although those involved in systems recognise that the fundamental ideas have been around for some years now, and have been widely incorporated into the current scientific metaphor, nevertheless there are many people who have studied within the confines of one disciplinary area and to these systems ideas are vital, new and enlivening.

Finally, as the University seeks to build up its provision of post-graduate courses there are exciting opportunities both to include an element of systems thinking generally and to produce appropriate specifically systemic courses to build on the undergraduate portfolio (in particular on T301). Already, in addition to work on MSc Courses, members of the Group are deeply involved in the Faculty's MBA (Technology) portfolio and in the emerging European Masters of Technology (EMOT) programme. It will not be easy either to decide between these conflicting requirements, or to address them all well, given present levels of resource.

THE FUTURE AND ITS POSSIBILITIES

So what options are open to the Systems Group as it seeks to plan a strategy for the next decade? It is possible to envisage an increasing dispersion as individuals seek to apply systems ideas in an ever wider range of subject areas. This would be highly beneficial to the strength of these subjects individually. However, such a future would render the Group no more than a socially supportive network. It could not survive long in this form. Indeed the gap between those working in the Faculty and those ex-Group members who have transferred to the Open Business School (OBS), is now quite wide and evidences the future if this dispersion continued.

What would be lost if the Systems Group disintegrated in this way? The briefest survey of where systems thinkers are located in other institutions shows a variety of resting places: Computing science departments, Engineering Faculties, Business Schools and many more. As a meta-discipline Systems finds it hard to establish an academic space of its own. It is best suited to an interdisciplinary environment and for this reason the Open University, as an Institution which focuses on interdisciplinarity, provides a very natural and supportive home. That the Systems Group is one of the most innovative and productive in the Open University is testament enough to the fact that the relationship is a symbiotic one. So preserving the integrity of the Group, it is reasonable to suppose, should preserve this powerful symbiosis to the benefit of the Open University generally.

It therefore seems appropriate to assert that the Systems Group must retain a coherent form and corporate identity, that this necessitates finding the resources to maintain and develop 'core' systems courses in the undergraduate and post-graduate programmes and that this work should have the highest priority. Operating at the forefront of current systems teaching in this way has two considerably advantageous effects beyond the Open University itself. Firstly it provides a focus for research and scholarship and establishes a place to which widely dispersed and often isolated systems thinkers in other institutions can look for support and collaboration. Given the nature of Systems as a discipline this is a significant provision and could be a powerful aid to the wider promulgation of systems ideas. Secondly, it should not be forgotten that it is tutors as well as students who benefit form participation in Open University teaching. The tutor network within the UK associated with the delivery of the Open University Systems courses is itself a powerful mechanism to facilitate the spread of systems knowledge. Preservation of strong core systems courses provides this enthusiastic network with excellent material which eventually works its way into other institutions' courses, business practices and organisational cultures. Course Teams at the Open University have influence beyond their immediate student audience!

As a very new member of the Open University's full-time staff I have tried to outline here something of the work of what I see as a uniquely powerful and effective intellectual group who are currently engaging in the process of defining a path into the future. I have tried to portray my own view of where the swamps and the solid ground lie in the environment ahead. I look forward with immense interest to see what route is established.

REFERENCES

Undergraduate Courses 1995, The Open University, Milton Keynes.

Paton, R., and Armson R., (eds), 1984, "Organizations: Cases, Issues, Concepts", Paul Chapman Publishing.

Carter, R., Martin, J., Mayblin, B., and Munday M., 1984, "Systems, Management and Change: a Graphic Guide", Paul Chapman Publishing.

Bignell, V., and Fortune, J., 1983, "Understanding Systems Failures", Manchester University Press, Manchester.

Pug,h D.S., and Hickson, D.J., (eds) 1994, "Great Writers on Organizations", *Omnibus Edition*, Dartmouth Publishing Company, Plymouth

T245 Course Material, 1995, The Open University, Milton Keynes

T247 Course Material, 1991, The Open University, Milton Keynes

T301 Course Material, 1993, The Open University, Milton Keynes

COMMUNITY AND ENVIRONMENTAL SYSTEMS

TOTAL SYSTEMS INTERVENTION IN STRATEGIC PLANNING

R. Keith Ellis[1] and Andrew J. Humphreys[2]

[1] University of Humberside
School of Computing and Information Systems
Cottingham Road
Hull, HU6 7RT
United Kingdom
[2] North Yorkshire Police
Newby Wiske Hall
Newby Wiske, Northallerton
North Yorkshire, DL7 9HA
United Kingdom

INTRODUCTION

This paper is based upon the use of systems thinking in a large public sector service organisation. Specifically, the work involved the application of Total Systems Intervention (TSI) (Flood and Jackson, 1991), to develop "...a strategic planning process within...North Yorkshire Police (NYP)." (Humphreys, 1994:1). Both authors' were involved in the project which commenced with the expressed need of NYP to 'think long-term' at the top level of the organisation, and ended with a strategic planning process which is now in use within NYP. During our application of TSI several issues became apparent which not only contributed to our own understanding but, which may assist with the future development of the methodology.

BACKGROUND

North Yorkshire Police is responsible for the policing of over 3200 square miles of a geographically and socially diverse area (Humphreys, 1994:3). This policing is achieved through seven divisions which are known as 'service deliverers'. The operational service deliverers are supported from the force headquarters at Newby Wiske in North Yorkshire. Located at Newby Wiske are the Chief Officers (Executive) and Senior Civilian Staff. Organisational operations within the force are related to the planning levels and meetings structure:

- Strategic Planning : Force Strategy Group; Chief Officers.
- Policy : Service Providers/Deliverers Policy Groups; Superintending Officers.

- Tactical : Divisional/Departmental Management Teams; Superintending and Inspecting Officers: Team Meetings; Sergeants and Constables.

NYP employs some 2000 people and has an annual budget in excess of £63 million. In recent years NYP has had imposed upon it significant change including Quality of Service; Measures of Performance; Costed Policing Plans; Police Authority Powers; Rank Structure; Devolved Budgeting; etc. All of these factors led to the recognition "...that for the first time policing is in the public domain for inspection and comment." (Humphreys, 1994:10). To cope with such demands NYP introduced local (i.e. Divisional) planning. However, no long term planning system existed which resulted in a lack of direction and purpose for the organisation as a whole. It was this lack of strategic planning that led to this project such that NYP would be able "...to meet the challenges of organisational change in an unstable environment." (Humphreys, 1994:11)

THE APPLICATION OF TSI

Our first difficulty was to gain an understanding of the problem situation in terms of 'top level' strategic planning in NYP. We took the view that it was necessary to obtain a wide range of views from internal and external stakeholders. (Humphreys, 1994:49). Our chosen information gathering method was that of structured interviews. We selected our stakeholders for interview using the criterion of those that "...could directly affect, or be affected by, 'top level' strategic planning in NYP should be consulted." (Humphreys, 1994:49). This resulted in 42 people being interviewed over a period of six weeks using a standard set of questions. All interviews were tape recorded and then transcribed for analysis.

Creativity Phase

This phase was, in fact, the interviews of stakeholders and the resulting analysis of the interview tapes. The process produced a 'metaphor picture' of the organisation which included:-

- The organisation as a 'machine';
- The organisation as an 'organism';
- The organisation as a 'brain';
- The organisation as a 'culture';
- The organisation as a 'coercive system'.

Detailed analysis of our 'metaphor picture' and discussion with the Chief Officers revealed that there was an urgent need to clarify Chief Officer functions relative to strategic planning; that the ownership and involvement of stakeholders in strategic planning needed to be established; and that there was a need to develop a communication forum and audit trail for strategic planning.

We concluded that the dominant metaphors were **organism/brain** and that the dependant metaphors were **culture/political**.

Choice Phase

This involved 'placing' NYP within the framework of Systems of Systems Methodologies (Flood and Jackson, 1991). Material from the stakeholder interviews was used to show that NYP operates as a complex system and that relationship amongst stakeholders was pluralist. hence, we 'placed' NYP as a **systemic-pluralist** organisation.

We were faced with a 'real world' problem situation and our investigation thus far suggested that two systems based methodologies would assist in the development of a strategic planning process. These methodologies were:

- Viable System Model (VSM), (Beer, 1985);
- Interactive Planning (IP), (Ackoff, 1981).

Implementation Phase

We used both methodologies in accordance with the complementarist principle of TSI. The VSM was used to analyse NYP as an organisation and then IP to develop the strategic planning process. (Humphreys, 1994:81).

The VSM was used as a diagnostic tool, based upon the stakeholder perceptions of NYP. From our diagnosis it became clear that:

- There was a lack of co-ordination and control;
- Audit information was restricted;
- Feedback was poor, even non existent in some areas;
- Executive 'interference' was evident at lower levels of the system;
- System Four was non existent and that strategy was missing.

The strategic planning process was then developed using IP. We should point out that the 'Mess Formulation' Stage of IP had been carried out during our Stakeholder Interviews. Therefore, we used the remaining stages of IP (Ends Planning; Means Planning; Resource Planning and Implementation Control) to develop the strategic planning process. The intention was always that Ackoff's (1981) principle of 'idealised design' should guide our work.

From this our proposed strategic planning process emerged, and it put forward aims to fulfil specific criteria :

- To involve...all those who can affect, or who will be affected by strategies relating to policing...North Yorkshire;
- To inform and communicate, through involvement, to avoid the pitfalls of 're-telling' the strategy;
- To increase the relevance of strategic planning to operational...policing by concentrating upon the core service areas of policing;
- To ensure that planning is a continuous process;
- To offer a clarity of purpose for consultative forums by providing a guiding method for working within them;
- To provide...innovative dynamism...required by organisations in a highly unstable environment." (Humphreys, 1994:104).

The proposed strategic planning process was submitted to critical examination in the form of a Workshop which involved many of the stakeholders interviewed during the Creativity Phase of TSI. The proposed process was modified as a result of the Workshop, but it is now being used and the first NYP 5 year Strategic Plan has been developed and is in operation.

REVIEW AND CONCLUSION

It is to be expected that "...during the application of any systems methodology in a 'real world' situation issues arise that may assist in creating learning about the principles of the methodology or its application." (Humphreys, 1994:109). Using TSI (Flood and Jackson,

1991) was not without its difficulties. Space does not permit a full discussion here, but we offer a sample for consideration.

Use of Metaphors

Police officers are highly pragmatic and mistrust 'jargon'. Acceptability of the overt use of concepts such as 'metaphor analysis' as an approach to the problem situation was linked "...with the credibility that the project would require through its life and subsequent operationalisation." (Humphreys, 1994:110). The question became that of should the approach be covertly used by the interventionist? It was decided, given the nature of the organisation, to take the covert approach. Inexperience with TSI was a contributory factor as was the 'feeling' that the client would be unwilling to accept an apparent extension of time through the use of some mystical metaphor analysis.

Methodology

Flood and Jackson (1991) argue that the System of Systems Methodologies (SOSM) is fundamental to TSI. We rigorously applied SOSM, and it was very useful to a 'new' systems practitioner in terms of learning about the complementary nature of TSI. However, it can also be argued that, as a practitioner becomes more experienced with and knowledgeable about systems methodologies, there is a danger that it can be subjected to short cuts. it can also be viewed as little more than a "...pigeon holing..." (Flood and Jackson, 1941:242) exercise.

Ideology

The commitment to human emancipation as a central principle of TSI (Flood and Jackson, 1991) may raise "...practical application issues unless 'fee free' projects can be tackled." (Humphreys, 1994:112). Information gained from the stakeholder interviews was gathered 'in confidence'. Much of this information was controversial, but it was fed back to the Chief Officers who were the project clients. Even though the information was accepted by the Chief Officers it was critically challenged during the stakeholder workshop. The result was that the 'messages' which came from the stakeholders during the workshop were, in some cases, very different from those received during the interview programme. We suggest that this 'discrepancy' could be associated with individual values and the courage of people "...to expose personal values in the presence of powerfully dominant people in the organisation." (Humphreys, 1994:112). This raised doubts in our mind concerning emancipation particularly if clients and stakeholders are brought together. We feel that these changes in views were associated with power, giving concern with respect to the claim that TSI is emancipatory in nature.

Utility of TSI

TSI permits a thorough examination of a problem situation and the associated issues. Complementarism creates a framework within which various perspectives can be surfaced. However, the language of TSI - as it currently stands - does not allow practising managers to understand and use it effectively as a problem solving tool. There is a real danger that if the development of TSI, which is now necessary, remains at the theoretical level, then it will remain in academia and fail to transfer to the world of management consultancy. This is a critical issue for systems theory and practice.

ACKNOWLEDGMENTS

We are indebted to the stakeholders of North Yorkshire Police who, by giving so freely of their time during the interview programme, made a significant contribution to our understanding and the development of the strategic planning process.

We are particularly grateful to the Chief Constable of North Yorkshire Police, Mr D. Burke, QPM., BA., for his encouragement, his participation and his permission to publish this paper.

Any errors or omissions are the responsibility of the authors.

REFERENCES

Ackoff, R.L., 1981, "Creating the Corporate Future", John Wiley & Sons, New York.

Beer, Stafford., 1985, "Diagnosing the System for Organisations", John Wiley & Sons, Chichester.

Flood, R.L., and Jackson, M.C., 1991, "Creative Problem Solving : Total Systems Intervention", John Wiley & Sons, Chichester.

Humphreys, A.J., 1994, "Total Systems Intervention : Developing a Strategic Planning Process in North Yorkshire Police", unpublished M.A. Dissertation, University of Hull, Hull, UK.

SYSTEMIC REFLECTIONS ON CORPORATE RESPONSIBILITY

Raul Espejo[1] and Neil Stewart[2]

[1] School of Computing and Information Systems
University of Humberside
Hull, HU6 7RT
United Kingdom
[2] Manchester Business School
Manchester
United Kingdom

INTRODUCTION

In this paper we discuss the contemporary issue of corporate responsibility. Our aim is firstly to offer a general introduction to current concerns and practices under this topic. Of these environmental care is the specific concern of this paper. Following this we use the Viable System Model (VSM) to reflect upon the relations between corporations and those stakeholders concerned with the environment. The VSM offers a systems approach to think about these relations and a strategy to manage their complexity. This reflection is at the core of the paper. We finally draw general conclusions.

CORPORATE RESPONSIBILITY

An increasing area of both academic and corporate interests in recent years has become the study of corporate (social) responsibility - that is, the study of the relationship between corporations and the society in which they operate. Corporate responsibility implies that a business acts to secure benefits not only for itself but also for society at large. This does not necessarily mean purely altruistic behaviour, because in many ways corporate responsibility is enlightened self-interest. The assumption which underlies it is that by taking a responsible approach to society, business improves its chances of long term viability. In the chemical industry, in which corporate responsibility is built into the Responsible Care initiative, the industry regards responsible behaviour as preserving its "licence to operate" (Simmons and Wynne, 1993).

A key factor in establishing an effective corporate responsibility policy is the identification of those parts of society which the company should address as significant; these might be thought of as those bodies which can influence the destiny of a company. The

Critical Issues in Systems Theory and Practice, Edited by K. Ellis et al.
Plenum Press, New York, 1995

notion of 'stakeholder' encapsulates this thinking. Stakeholders have been defined as "any group or individual who can affect or is affected by the achievement of the organisation's objectives" (Freeman, 1984).

Corporate responsibility consists of a number of different aspects of community support such as corporate partnership in the community, corporate governance, the relationship between personal and corporate values (ethics) and environmental care. This latter aspect is the main focus of this paper.

ENVIRONMENTAL CARE

Environmental legislation and codes of practice are a major part of governmental activity at several levels. An important feature within all these levels of debate is the behaviour of business. Business is perceived to be the cause of most environmental damage and therefore is expected to find solution pathways to a *sustainable* world.

Two concepts are emerging which capture the essentials of best practice; 'Producer Responsibility' and 'Corporate Dialogue'. We have been directly involved with these concepts within the chemical, nuclear and aluminium industries. These industries can certainly cause damage to the environment if they are managed irresponsibly; thus being confident of responsible behaviour on their part is of considerable social importance.

Producer Responsibility

The notion that the 'polluter pays' is built in to much governmental thinking on environmental issues in most countries. Where liability exists, it means that the company or organisation which has caused the damage is held accountable. Economists speak of taking the costs to repair environmental damage, which are regarded as externalities viewed from the perspective of the firm's activities, and assigning the cost to the industrial activity which creates the damage. This principle is encapsulated, for example in the 'Superfund' legislation in the USA, where a levy on industry is used to fund the clean-up of contaminated land, in many of the directives produced by the European Union, and the Environmental Protection Act in the UK.

Producer responsibility goes beyond this. For example, when the UK government needed to set targets for reducing the packaging waste which is produced in domestic households, it made the packaging industry as a whole responsible for setting up an infrastructure to do this. It set targets for the recovery and reuse of packaging waste and charged the industry to find ways to meet these targets. The Producer Responsibility Industry Group has made its proposals and is now in active dialogue with the UK government about how to realise them. Part of the proposal is that government itself should produce enabling legislation to support what industry is trying to achieve - in this case to provide the basis for a levy on producers to fund a recycling infrastructure (ENDS, 1993).

This is by no means an isolated example. The German DSD system for recovery and recycling domestic packaging waste (Corbett and van Wassenhove, 1993) has been established on the basis of a guarantor group for the industry which will oversee the achievement of the government's very ambitious targets. Similarly also, when the European Commission selected certain hazardous waste streams as priority for improvement - tyres, chlorinated solvents and medical waste amongst them - industry was called to create and define effective solutions to the problem (European Commission, 1990).

The essential features of *producer responsibility* are that it involves government setting standards for performance of an industry in a specific and perhaps specialised area. It also involves industry as a whole, mainly through its trade associations, finding ways to meet the government's targets. Once its plans have been accepted by government, it then requires industry to set out expectations of its member companies in the form of codes of practice and other guidelines. Finally government needs to support the industry's initiatives with measures of its own - most notably supporting regulations and laws.

Corporate Dialogue

A second aspect of best practice in corporate responsibility concerns the notion that companies should have an open and co-operative discussion with representatives of society (stakeholders) where issues of public concern are involved. These discussions should be seen as going beyond public relations and are an input to the strategic planning of the company or industry involved. The best examples of this concern technological activities where society perceives that a risk to health and safety of the public at large are involved. The stakeholders involved in the debate will come from different sections of society, some of them with the necessary skills and knowledge to provide a 'peer review' of the industry's specialised proposals.

At a different level, the public enquiries in the UK which have almost always accompanied evaluation of development plans of the nuclear industry, especially in the search for safe disposal of nuclear waste, also represent a form of dialogue between industry and society (Berkhout, 1991). Similarly, the work of the Royal Commission on Environmental Pollution taking evidence on the best means for disposing of waste (RCEP, 1993), and the House of Lords Select Committee (1993) investigating biotechnology, involve the same process.

Thus at both corporate and industry level, public debate on issues of concern, especially focusing on areas of knowledge not easily understood by laypeople, is becoming an increasing part of society.

SYSTEMIC REFLECTION ON CORPORATE RESPONSIBILITY

There is an increasing awareness of the local, regional and global effects of corporate activities. Manufacturing processes may not only have an effect on people's health, or cause depletion of scarce natural resources and cause damage to the local environment; they may also contribute with their emissions to ecological problems on a regional and global scale. Moreover, in the past 20-30 years we have become increasing aware that corporations need to take responsibility for the full life cycle of their products. This awareness, often triggered by pressure groups but also, increasingly, accepted by the corporations themselves, puts pressure on them. Socially, we are recognising that the viability of local communities as well as the viability of regions and the globe itself require urgent action to avoid the worst effects of corporate activities. Corporations themselves are increasingly aware that their own viability depends on dealing effectively with these issues. Similar to total quality, clean processes and concern for the environment is seen by better companies as good business. The more companies integrate environmental protection into their processes, the more likely they are to secure cost advantages in the longer term.

Systemically, the problem for corporations is to work out their relationship with society in order to align their own viability with the viability of the local, regional and global environments. This is a complex relationship which imposes on corporations the need to increase their appreciation of the full impacts of their processes and products. It is necessary for companies to understand the dynamics of a sustainable environment, and take account of this throughout the life-cycle of their products from cradle to grave. At the same time as society, through its political institutions, creates more stringent criteria for 'acceptable' products and processes, corporations need to create, through research and other means, the capacity to anticipate and exceed these criteria; to be *responsible producers* gives them a competitive advantage. For this purpose, they need to be aware of, and anticipate, people's concerns and this they can only realise through open and effective *corporate dialogue* with them.

It is apparent that the complexity of this relationship is large indeed and that society and corporations need to understand it better and to find effective strategies to manage it. This is not a matter of choice, but of survival. The longer we take to appreciate the negative side effects of production activities or their threats to a sustainable environment, the more expensive and perhaps the more intractable will it be to solve the related problems. However, a company cannot talk to the whole population and the whole population cannot talk to each company - the number of individual conversations which might be needed is simply too vast. How can systemic thinking and the Viable System Model (Beer 1979, 1981, 1985; Espejo and Harnden 1989; Espejo and Schwaninger 1993) help us in this endeavour? In the next section we use this model to offer criteria to design these strategies.

THE VIABLE SYSTEM MODEL ('VSM')

The model sets out conditions for viability in complex and unpredictable environments and must therefore be relevant to the debate within society about how sustainability is achieved. The VSM offers structural criteria for viability; it helps organisations to work out how best to use their resources in order to maintain an effective interaction with their environments. These criteria make apparent that the structure of viable organisations, whether they are aware of it or not, is one of autonomous units within autonomous units; this is imposed by the much larger complexity of their relevant environments. We say that viable organisations have *structural recursion*; the structural requirements for an organisation to maintain its viability will recur in each autonomous unit within it, and in each autonomous unit within each autonomous unit. Structural recursion is by and large an outcome of self-organisation; it is a natural way to increase the complexity of an organisation in order to adjust itself to environmental disturbances. The problem is to achieve this structural recursion at the least cost to people; indeed many organisations fail in this process. Effective structural recursion requires, on the one hand, the cohesion of the organisation's autonomous units at all structural levels and, on the other, the balance of all these autonomous units with their relevant environments at socially acceptable levels of performance. The VSM is a model of the mechanisms necessary for the organisation's cohesion and for its stability in the environment; we call these the mechanisms for monitoring control and adaptation (Espejo, 1989).

Matching Environmental Issues and Organisations

Producer responsibility should be the concern of all production units, from the shopfloor to the total corporation and the industry. It is clear that a sustainable environment depends on the responsible autonomous behaviour of everyone; unless everyone develops an awareness of the impact of their actions on the environment and cares for it, for their own long term benefit, no amount of exhortation or policing will be enough to save the environment. However, we recognise that this highly commendable proposition may not be enough. In addition it is necessary to take pro-active steps to align the interests of society and corporations.

Individual plants may only deal with environmental issues at the local level, but regional and global problems require regional and global responses. Industrial associations at these higher levels are in the correct position for this purpose. Equally, from a social perspective, local, regional and global authorities have the best overview of environmental issues at their respective levels.

A plant, as an autonomous production unit, is embedded in at least three larger organisations; the company, the community and the industry. The plant, as a responsible producer, concerned with its own long term viability, needs to understand its local environment. Indeed, there is no one better than the plant itself to deal with the local environmental issues generated by that plant. For this purpose it needs to develop *corporate dialogue* with the local community. This is a necessary 'intelligence' effort to create the internal policies to care for the local environment. However, the plant's environmental policies need, to different degrees, to achieve an alignment with:

- -the local policies as developed by any form of local authority,
- -the company's environmental policies, and
- -the industry's environmental policies as developed by 'industrial associations' in response to national/regional environmental requirements.

In its turn, the industry's environmental policies may need not only an alignment with regional environmental requirements but also with global environmental policies as developed by international/global authorities in response to global concerns. Similarly to the plant's *dialogue* with representatives of its local environment, we may expect parallel *dialogues* at the industry level to improve their chances for long term viability. Unfortunately, it is often the case that these associations pay lip service to their industries as they operate with the lukewarm commitment of their member corporations. They must adapt their roles and responsibilities however if sustainability is to be a realistic aim of society.

The authors have been involved directly or indirectly with cases of these dialogues. For instance, the aluminium industry in Norway set out specifically to consult members of the public, including the relevant scientific community, about the environmental questions it should answer in order to minimise the negative effects of its activities (Hydro Aluminium, 1994). A similar process was used in Switzerland in order to account for the effects of a toxic waste incinerator (Ciba-Geigy, 1994). In yet another study, a pharmaceutical company established a set of dialogues with external groups to direct its research into finding effective and acceptable treatments for the common cold (Steinmann, 1993). In all cases, active dialogue about the issues involved was developed and research agendas were developed as a result.

Public enquiries, a form of social dialogue, may play a similar systemic role but to support the local and regional viability rather than the producers' viability. In the UK, social dialogue is represented by such bodies as the "Royal Commission for Environmental Protection" (RCEP) which has carried in depth studies about such issues as the impact of traffic on the environment (RCEP, 1994), and the best means to cope with domestic waste (RCEP, 1993). Both of these studies have looked at present and future trends in society and given impetus to future directions.

Autonomy and Cohesion

Producer responsibility requires the designing of a web of communications in order to support autonomy and cohesion. There is a risk of stifling important entrepreneurial freedoms if regulations from local, national or global authorities on companies are unnecessarily constricting. Often this is the outcome of understanding control as 'control from the outside' rather than 'producer responsibility or control from the inside'. Designing effective communication networks (i.e. webs) which avoid control dilemmas between levels, is facilitated by creating the conditions for the producer self-regulation. A proper balance between 'extrinsic' and 'intrinsic' control is necessary; on the one hand inappropriate external restrictions may cripple producer performance and inhibit responsibility, but, on the other, total lack of external restrictions may produce unaligned social and corporate policies.

In areas of major concern in the UK and elsewhere, it is clear that government sees itself increasingly as providing the framework for control activities. Thus, as noted above, the Department of the Environment in the UK and its equivalent in Germany set the respective industries' targets for recovery of packaging waste and asked the institutions involved to devise a system to meet those targets. However, unrealistic targets for the German industry led to major difficulties and the need to adjust the legislation. Similar problems have been experienced by the energy industry in Norway.

Another revealing example of the relevance of the intersection between the responsibility of the producer and the corresponding role of the state has been the establishment of regulatory bodies for privatised industries in the UK. When the water industry was privatised, it was given tough targets by government for improved environmental performance. When government was freed of the direct ownership of the industry, it could separate more clearly the broader interests of society from those of the industry. Thus it is that government starts to represent the interests of all the stakeholders, and not just the one dominant aspect of ownership. It is also clear that the dilemmas faced by the industry in confronting the issue of water purity are more clearly in the public domain - the issue of price for water versus. levels of cleanliness is an issue aired to a large extent in public (OFWAT, 1992). However, the social control over these industries requires not only the definition of environmental targets but also the creation of the context for producer responsibility and the capacity to monitor their progress. These two additional requirements, highlighted by the VSM, are often under-resourced, something that may lead to abuse of stakeholders and poor industrial performance.

CONCLUSION

The Viable System Model provides structural criteria to improve current best practice for environmental care. Our discussion has been of a general nature. However it can be developed further to work out requirements in specific situations. For simplicity, we have considered only three levels of social structure; the local, regional and global levels. For the same reason we have considered only the plant, company and industry levels. In a particular situation we would need to refine this broad definition of levels and consider a much more complex web of interactions. Our hope is that this paper has helped the reader to see, albeit at a very general level, the value of systemic thinking in providing a framework to study aspects of corporate responsibility.

REFERENCES

Beer, S., 1979, "The Heart of Enterprise", John Wiley & Sons, Chichester.

Beer, S., 1981, "Brain of the Firm", John Wiley & Sons, Chichester.

Beer, S., 1985, "Diagnosing the System for Organizations", John Wiley & Sons, Chichester.

Berkhout F., 1991, "Radioactive Waste. Politics and Technology", Routledge, London.

Ciba-Geigy, 1994, "The Regional Incinerator", Case Study presented to the European Business Ethics Network Conference, St. Gallen, September, 1994.

Corbett,C. J. and van Wassenhove, L. N., 1993, The green fee: internalizing and operationalizing environmental issues, *California Management Review,* Vol 36, pps 116-135.

ENDS (Environmental Data Services), 1993, "Packaging Waste Plan", September No.236, p13 and supporting reports in ENDS 222,229, and 234.

Espejo, R., 1989, The VSM revisited, *in*: "The Viable System Model", R. Espejo and R.Harnden, John Wiley & Sons, Chichester.

Espejo, R. and Harnden R., 1989, "The Viable System Model", John Wiley & Sons, Chichester.

Espejo, R. and Schwaninger, M., 1993, "Organizational Fitness: Corporate Effectiveness through Management Cybernetics", Campus Verlag, Frankfurt/New York

European Commission, 1990, "Analysis of priority waste streams", Manual of the Services of the Commission of EC, September.

Freeman, R. E., 1984, "Strategic Management, a Stakeholder Approach", Pitman, London.

House of Lords Select Committee, 1993, "Regulation of the United Kingdom Biotechnology Industry and Global Competitiveness", 7th Report, Vol 1. HMSO, UK.

Hydro Aluminium, 1994, "The Effect Study", Hydro Aluminium, Norway.

OFWAT, 1992, "The Cost of Quality: a Strategic Assessment of the Prospect for future Water Bills", OFWAT, Birmingham.

RCEP, 1993, "Incineration of Waste", 17th Report, HMSO, UK.

RCEP, 1994, "Transport and the Environment", 18th Report, HMSO, UK.

Simmons, P. and Wynne, B., 1993, Responsible Care: Trust, Credibility and Environmental Management, *in* "Environmental Strategies for Industry", K. Fischer and J. Schot, eds., Island Press, Washington, USA.

Steinmann, H., 1993, Private business and public interest, "Proceedings of the Conference 'Economy and Environment', Düsseldorf, 1992", Schaffer-Poeschel, Stuttgart.

THE USE OF SYSTEMS TECHNIQUES IN INTEGRATED CATCHMENT MANAGEMENT (ICM)

Its Promise and Limitations

W. E. Hutchinson

Edith Cowan University
Churchlands Campus
Pearson Street
Perth
Western Australia 6024
Australia

INTRODUCTION

Integrated Catchment Management (ICM) has all the characteristics of a 'messy' problem. It involves numerous interests all with a different perception as to what ICM actually means. It involves technical, social, economic and environmental aspects as well as numerous government, industrial, agricultural and civic groups. Water itself is a multipurpose resource and has time, place and quality characteristics (Watson & Johnson, 1993). ICM is a 'complex' problem situation as classified by Flood & Jackson (1991). The other axis to their classification regards the problem situation as either unitary, pluralist, or, coercive. The sheer number of interests in ICM tends to make some aspects of power play appear, so it is legitimate to call the situation coercive. Whatever classification is used for management purposes, the problem is definitely a systemic one. It is surprising to find, however, that explicit systems techniques are rarely used when designing ICM systems and organisational structures although some attempts have been made on the periphery by academics, e.g. Levy & Synnott (1993).

ICM IN WESTERN AUSTRALIA - THE CONTEXT

Western Australia has a number of land and water problems. For instance, some 3.3% of agricultural land is salt affected and this is likely to rise to 15% and about 52% of potentially divertible water resources in the south west of the state are saline. The government agencies directly involved in these problems are numerous e.g. Water Authority, Department of Agriculture, Waterways Commission. Other agencies are also involved in a less direct

Critical Issues in Systems Theory and Practice, Edited by K. Ellis et al.
Plenum Press, New York, 1995

way e.g. Health, Planning, Local Authorities. These agencies are at three levels: Commonwealth, State and Local. There is a state coordinating body for ICM called the Office of Catchment Management which basically encourages other agencies to cooperate with each other and citizen groups. In a number of catchment areas, Catchment Coordinating Groups (CCGs) have been set up and consist of members of the public (local farmers, councillors, etc.) as well as staff from government agencies. The agency involved with each CCG depends on the location. In a catchment with large tracts of forests managed by the Department of Conservation & Land Management, this agency would be represented whereas they might not be in an area dominated by farmland where the Department of Agriculture would dominate. Staff from the Office of Catchment Management would also attend; it is the other agencies, however, who have all the statutory power. Almost all the CCGs are in rural areas although one successful group is in the metropolitan area.

The whole structure is relatively arbitrary and its success is yet to be determined. Even the keenest advocates of ICM state that it has not really been implemented yet in its complete form. The Office of Catchment Management is under-resourced and appears to be being downgraded by the present state government.

THE USE OF SYSTEMS TECHNIQUES IN ICM

All environmental management problems, like ICM, have particular characteristics (Burton & Dukes, 1990):

- complexity
- involves mutually exclusive and strongly held views
- data is complex and difficult to obtain
- numerous separate parties may be involved
- there is no common deadline
- goals may be quite different
- parties have greatly different resources and power

These factors often make the techniques practiced within a single organisation not suitable. For example, Soft Systems Methodology (SSM) can be successfully used within an institution and even between a limited number of institutions if they have some common goal. ICM involves a large number of stakeholders each with varying levels of technical knowledge, interest, and power. For involved government agencies, the main way around this problem is to hold meetings where members of the public can voice their opinions. There are benefits in this tactic, of course, but it is time consuming, involves many of the vocal and powerful interest groups and is usually based around a plan already formulated by some government agency. This gives the impression of public involvement but is really just a way allowing public 'tinkering' of a pre-ordained plan. It does not include disparate views in the plan creation. This approach to the system design is, therefore, not systemic but gives the impression it is.

It is not practical on many occasions to include large numbers from the public and government agencies in a SSM session, for example. How, therefore, can this constraint be overcome?

The initial problem for a systems investigation is to determine what the actual problem is. As ICM can be practiced in a coercive situation, the various concepts of power and influence also need to be examined. These imply such approaches as SSM and Ulrich's Critical Systems Heuristics (see Ulrich, 1993). Modified approaches of these are needed, however.

Checkland & Scholes (1990) tend to emphasise the learning process of SSM, whilst Wilson (1990) tends to emphasise the output. In the investigation into ICM, it was the output from the SSM process which was required. It was therefore decided to take the first two 'stages' - that of creating a Rich Picture and Root Definition - and develop these from interviews with selected stakeholders in the problem. Adding 'Implementors' to the groups suggested by White (1973) to be involved in planning, the following were interviewed:

- Management (e.g. Public Servants, Politicians)
- Academics (e.g. Environmental Scientists, Economists, Engineers)
- Implementors (e.g. Catchment Management Group members)
- Interested Parties (e.g. members of the public, 'Green' Groups)

Within each group, those with apparently opposing views were selected, e.g. conservative and radical politicians. This selection was then reinforced using O'Riordan's (1989) classification of viewpoints - Ecocentric and Technocentric, and selecting people from the above groups who it was thought fell into these groups. After the interview each participant was asked for a person who disagreed with their opinion to expand the views used in the study. This process is not democratic and is designed not to represent the dominant viewpoint but rather the span of opinions.

Each participant was interviewed using a standard set of questions which was sent to them a week before the interview. The questions were a combination of Ulrich's 12 points (see Jackson, 1991(a)) and open questions designed to find the information to create a Root Definition. They were classified as questions which examined:

- the underlying philosophy of the interviewee
- the perceived power structures of the present situation
- the power structure that should be in place
- the structure of the desired system e.g. who are the clients, actors?
- the required outputs of the system e.g. how to measure efficiency?

Just over 100 people were interviewed. The data collected from the interviews were supplemented by investigating the literature produced by government agencies, academics and pressure groups.

The results from these questionnaires were then combined to produce the Rich Pictures thereby illustrating the context of the problem. Root Definitions were created from answers to specific questions and then Conceptual Models. The reasoning behind this was to determine what the system should be before 'harder' systems techniques should be used to determine how they should be implemented.

Once the Conceptual Models were completed, specialist literature was examined as to possible solutions to the tasks needed. Using Influence and Systems Dynamics Diagrams, each factor (e.g. the theory of water permits) was modeled to produce a 'primitive' model. Each of these could be combined with other 'primitives' to examine the effect of relationships and feedback on different system options.

To increase the comprehensiveness of the models, they were matched against the criteria for planning defined by Winpenny (1994):

- economic efficiency
- efficacy
- equity
- environmental impact
- fiscal effects
- political and public acceptability
- sustainability
- administrative feasibility

The organisational side of the problem was modeled using the Viable Systems Model (VSM) developed by Beer (1979, 1981). Each of the various possibilities of organisational structure were then examined and designed using VSM.

The end product is therefore a set of organisational and implementation policies which would meet the needs determined by the interviews.

DRAWBACKS OF THE TECHNIQUE

Checkland & Scholes (1990) emphasise the learning aspect of SSM. The participants learn the different worldviews involved and the creation of the outputs becomes the vehicle for change. The actual output, e.g. Root Definitions, Conceptual Models, are secondary to the process. In the ICM situation, this would be an applicable method for each CCG where the number of members is about 20. At a state level, where the whole of the ICM programme is to be determined, the numbers of people involved and the associated logistics would make the task a formidable one. Just obtaining viewpoints by individual interview lacks the interaction of the participants. This not only deprives the process of organisational learning but also of creative solutions brought about by the interaction of the people involved. It can also lessen the acceptance of the solutions created. In 'interactive' SSM the solution has been developed by the participants and therefore should be accepted by them (although only the careful selection of participants would determine its general acceptance in an organisation). The technique used for the ICM investigation could also be very much affected by the analyst whose own worldview might inadvertently prejudice the outcome. Overtly expressing the analyst's views can help in this but it is difficult to see how to overcome this completely in any technique used.

Criticism could come from who believe that some views are of less value than others. 'Experts' could legitimately complain that those without appropriate training cannot understand the implications of their opinions. As O'Riordan (1976) states "...few people can conceptualise all the possible outcomes of an anticipated course of action". To counter this, of course, is the argument that if different perspectives are included the solution is going to be 'holistic' and hence cover much that is missed in a linear, technical solution.

Power is also a problem. In a sense, personal interviews aided in getting honest views about where the power really lay. People from all levels of the official power structure were interviewed. In a group some might have been intimidated by those with power whether it be sapiential, charismatic, role, or, contact (Wilson, 1989). Those who would be classified as emancipatory systems thinkers e.g. Flood (1990) might disapprove of the technique. Although it does attempt to cover the five commitments of critical systems thinking - critical and social awareness, complementarism at the levels of methodology and theory and human emancipation - as laid out by Jackson (1991(b)), it does not pretend to be a radical approach in terms of social power redistribution. It is, however, difficult to know how to get over this problem in a practical situation where policies have to be put in place which will affect numerous interests. To do something to alter the situation of the powerless, power is needed: the powerless therefore have power, the argument is circular. The above methodology recognises power structures and in a sense copes with them. Some of the models produced would have curbed the power of some government agencies but it needed a 'higher' power to implement them, i.e. a government minister or above. There would therefore be a 'loser' and a 'winner' i.e. one viewpoint would have overridden another. Using the above technique coped with power by making all opinions equal. The models produced therefore emancipated all those who participated in the design stage. Implementation is another problem. It goes someway to "...freeing knowledge that is suppressed by dominant institutions in society" (Flood, 1993).

Like all models there is a temporal side. The social, economic, psychological and environmental situation is dynamic and, hence, the opinions expressed as to the ideal system might change radically. For instance, straight after a drought the supply of water for irrigation or households might have greater priority than saving wetlands. Environmental values in society change over time (Lothian, 1994). Using this technique might , therefore, give less than optimal environmental, social, or, economic solutions because they are based on short term opinions. It is difficult to overcome this as even log termed focused, technical 'experts' are prone to popular, contemporary mindsets of their peers (Ortolano, 1984).

THE BENEFITS OF THE TECHNIQUE

Many hard systems techniques are used in environmental management to formulate specific technical solutions in economics and ecology but not in the general management itself. Many techniques such as SSM have been used for institutional problems (Wilson, 1989) rather than general social problems. Even the contemporary systems literature seems to concentrate on the philosophical basis for systems thinking rather than providing practical solutions for really complex social and environmental problems.

Techniques can be manipulated to suit the problem. For instance, Kartowisastro & Kijima (1994) altered SSM for use in an autocratic society (Indonesia). This exercise with SSM is an attempt to do something similar. Using the gamut of systems techniques available, a choice was made to use the most appropriate to meet a specific aim. The purity of the technique was compromised but this is better than abandoning systemic techniques altogether. The techniques used were practical and hopefully created solutions which were more comprehensive than traditional techniques.

The process of obtaining the deliverables normally associated with SSM gave focus to the study and forced the analyst to include all opinions - even if these were interpreted by the analyst.

CONCLUSION

ICM is a complex, social and environmental problem. To obtain generally acceptable and successful systems and organisational structures, it is feasible to use systems techniques even where large numbers of interests involved. This may mean compromising the purity of the particular techniques but does not necessarily lessen its usefulness. The management systems so created should meet Walker's (1994) requirements by being both resistant (having ability to maintain functioning under stress) and resilient (having a capacity to restore the system once stress has destabilised it). The time saved creating these systems models could make environmental management and decision making more appropriate to different factions within society and to the environment itself. If nothing else, the 'plans' presented to the public for comment by government agencies should be more likely to meet the requirements for a variety of social and ecological needs.

REFERENCES

Beer, S., 1979, "The Heart of the Enterprise", John Wiley and Sons, Chichester.
Beer, S., 1981, "The Brain of the Firm", John Wiley and Sons, Chichester.
Burton, J., & Dukes, F., 1990, "Conflict: Practices in Management, Settlement and Resolution", St Martins Press, New York.

Checkland, P.B., & Scholes, J., 1990, "Soft Systems Methodology in Action", Wiley, Chichester.

Flood, R.L., 1990, "Liberating Systems Theory'", Plenum Press, New York.

Flood, R.L., 1993, Practising freedom, *in:* "Systems Science - Addressing Global Issues", F. A. Stowell, D. West and J. Howell, eds., Plenum Press, New York.

Flood, R.L., and Jackson, M.C. 1991, "Creative Problem Solving: Total Systems Intervention," Wiley, Chichester.

Jackson, M.C., 1991(a), "Systems Methodology for the Management Sciences", Plenum Press, New York.

Jackson, M.C., 1991(b), Five Commitments of Critical Systems Thinking, *in:* " Systems Thinking in Europe", M.C. Jackson, G.J. Mansell, R.L. Flood, R.B Blackham and S.V.E. Probert, eds., Plenum Press, New York.

Kartowisastro, H., & Kijima, K., 1994, An enriched soft systems methodology (SSM) and its application to cultural conflict under a paternalistic value system, *Systems Practice.* 7(3).

Levy, M., & Synnott, M., 1993, A systems approach to environmental management, *in:* "Systems Science - Addressing Global Issues", F.A. Stowell, D. West and J. Howell, eds., Plenum Press, New York.

Lothian, J.A., 1994, Attitudes of Australians toward the environment, *Australian Journal of Environmental Management.* 1(2).

O'Riordan, T., 1976, Attitudes behaviour and environmental issues, *in:* " Human Behaviour and Environment, I. Altman and J.F. Wohlwill, eds., Plenum Press, New York.

O'Riordan, T., 1989, The challenge for environmentalism, *in:* " New Model for Geography, R Peet and N. Thrift, eds., Unwin Hyman, London.

Ortolano, L., 1984, "Environmental Planning and Decision Making, John Wiley and Sons, New York.

Ulrich, W., 1993, Some difficulties of ecological thinking, considered from a critical systems perspective: a plea for critical holism, *Systems Practice.* 6(6).

Walker, K. J., 1994, "The Political Economy of Environmental Policy: An Australian Introduction", UNSW Press, Kensington, NSW.

Watson, G. & Johnson, M., 1993, Pricing: Cheap Water or an Environmental Perspective, *in:* "Water in Australia", M. Johnson and S. Rix, eds., Pluto Press, Sydney.

White, G.B., 1973, Public Opinion in Planning Water Development, *in*: "Environmental Quality & Water Development", C. R. Goldman, J. McEvoy and P.J. Richerson, eds., W. H. Freeman, San Francisco.

Wilson, B., 1989, Soft systems research and application, *in:* "System Prospects", R. L. Flood, M.C Jackson and P. Keys, eds., Plenum Press, New York.

Wilson, B., 1990, "Systems: Concepts, Methodologies and Applications", Wiley, Chichester.

Winpenny, J., 1994, "Managing Water as an Economic Resource", Routledge, London.

106

PROPAGATION OF UNCERTAINTY IN SOCIO-ECONOMIC SYSTEMS

J. Korn and A. Takats

Middlesex University
Bounds Green Road
London, N11 2NQ
United Kingdom

INTRODUCTION

Whether a person acts on his own or part of a group, all human activities are directed towards the attainment of a goal. A goal is expressed in terms of specified and envisaged properties based on the corresponding initial properties attached to an object. The achievement of such a change referred to as outcome, requires a cause which is generated by the purposive action of a group of objects referred to as a 'system'. Thus, a proposed change of properties then engenders the appropriate system, an approach different from conventional view (Korn,1994).

When people are gathered into interacting groups performing specific functions in the course of trying to attain a goal, we have a social system. Since people need products for their existence created by manufacturing systems, money, or other means, is usually circulated. Thus, it may be more appropriate to use the term 'socio-economic system'. When a number of such systems operate side by side without a common goal being set, or interact with objects and activities in their environments, changes of properties, or events, appear to take place by chance. The activity of a system, or systems, in their environments, is called a scenario.

It has been difficult to create a suitable symbolism in the form of a predictive inferential structure, for the representation of purposive situations even in engineering.

In the field of socio-economic systems, the task is even more difficult :

1. People can create simple and complex ideas as goals in response to perceptions from their environments and, also, from within themselves.They can set systems in action in a number of ways for the achievement of goals.
2. People exhibit will, i.e. they can change their minds.
3. People can make choice.

Thus, the behaviour of a socio-economic system is not repeatable and qualitative properties are of predominant importance.

Critical Issues in Systems Theory and Practice, Edited by K. Ellis et al.
Plenum Press, New York, 1995

Accordingly, mathematical modelling is not generally applicable and we have adopted natural language as the basis for creating a predictive, inferential structure (Korn, Huss and Cumbers,1993). Simple and complex declarative sentences describing a scenario are converted by means of linguistic analysis into basic sentences and modifiers of nouns and verbs. These are expressed as a set of conditional statements which can then be transformed into zero-order fuzzy logic conditionals called inference rules. The various modifiers are represented as fuzzy sets. These inference rules are aggregated to form the logic structure of a scenario, called the situation-frame. In order to investigate various outcomes, it is necessary to instantiate the situation-frame with specific fuzzy statements. These statements will initiate the propagation of fuzzy properties within the situation-frame, producing a computed instance of the statement in the form of a fuzzy distribution which designates the outcome of the scenario (Campos and Gonzalez,1993).

The objective of this paper is to outline how fuzzy properties as adjectives and adverbs, are propagated and how various configurations of fuzzy sets affect the outcome in a socio-economic system.

A SCENARIO

The following scenario describing a system and its environment, expressed in natural language, is used as a demonstration : 'Swamps covering a large area, absorb heat from the Sun converting the warm, humid, atmospheric conditions with presence of larvae, into hot and wet conditions with the presence of more larvae, which are favourable for mosquito production.

Large, agricultural holdings exploit workers by making them work hard for low wages under severe conditions. Thus, the physical state of these workers deteriorate, some of whom are carriers of malaria parasite as a result of having been bitten by mosquitoes which carry the parasite. Other workers are also bitten by mosquitoes which have fed on human carriers and, since the resistance to infection of their bodies is low, they become ill with malaria leading to many uncultivated holdings'.

SITUATION-FRAME AND COMPUTATION

The analysis of a description is carried out in stages.

Sentence Analysis

The first sentence of the description is complex. It is divided into a number of basic sentences and modifiers. For example,

1. Swamps absorb

i. Swamps covering a large area, s1(1)
(noun phrase is modified by an adjectival -ing participle in a non-finite clause)

ii. Swamps absorb heat from the Sun, r1,1(1)
(one-place verb is modified by a noun phrase)

also showing the appropriate symbolism. The process is then carried on until the whole description will have been broken down into basic sentences and modifiers. Terms such as 'large area' and 'absorb heat' can be expressed as fuzzy sets over numerical referentials in units of km2, for instance.

Conversion into Conditionals

A dynamic verb describing a process, implies a causal statement and a statement that indicates how the process is enabled to take place. The example above, is expressed as a conditional 'If swamps cover a large area and swamps absorb heat from the Sun then swamps become warm'. Symbolically:

$$s1(1) \wedge r1,1(1) \rightarrow p2(1)$$

Semantic Diagram

Following the conditionals, the complete description can be converted into a diagram as shown in Fig. 1. The diagram yields a number of causal chains which are obtained by tracing the paths from the outcome, in this case, the property 'uncultivated' carried by o14 'holdings', and from prompting objects, o7 and o12.

The causal chains are :

1. 14,8,13,11,10,9,8.
2. 12,11,6.
3. 7,6,5,15,4,3,2,1.

Deduction Graph

Following the causal chains, the inference rules can be written. For example, for part of the first causal chain :

$$s8(1) \wedge s8(2) \rightarrow r8,9(1,2)$$

$$r8,9(1,2) \rightarrow p10(1)$$

$$p10(1) \rightarrow r10,10(1)$$

$$r10,10(1) \rightarrow p11(1)$$

and so on.

After eliminating the conditionals with single literals in their premises, the deduction graph containing the relevant data to be computing, is drawn as in Fig. 2.

Computation

This is carried out using the software developed on the basis of the (Campos and Gonzalez,(1993)) theory. Specific fuzzy statements are used to instantiate, in this case, the s-terms as inputs indicated in Fig. 2 and computing their effect on the outcome, p14(1) in Fig. 1.

The statement 'mosquitoes are with parasites',s5(1), is allowed to vary as shown in Fig. 3 with all other s-terms to remain unchanged. The effect of this variation on the property 'uncultivated' is depicted in Fig. 4.

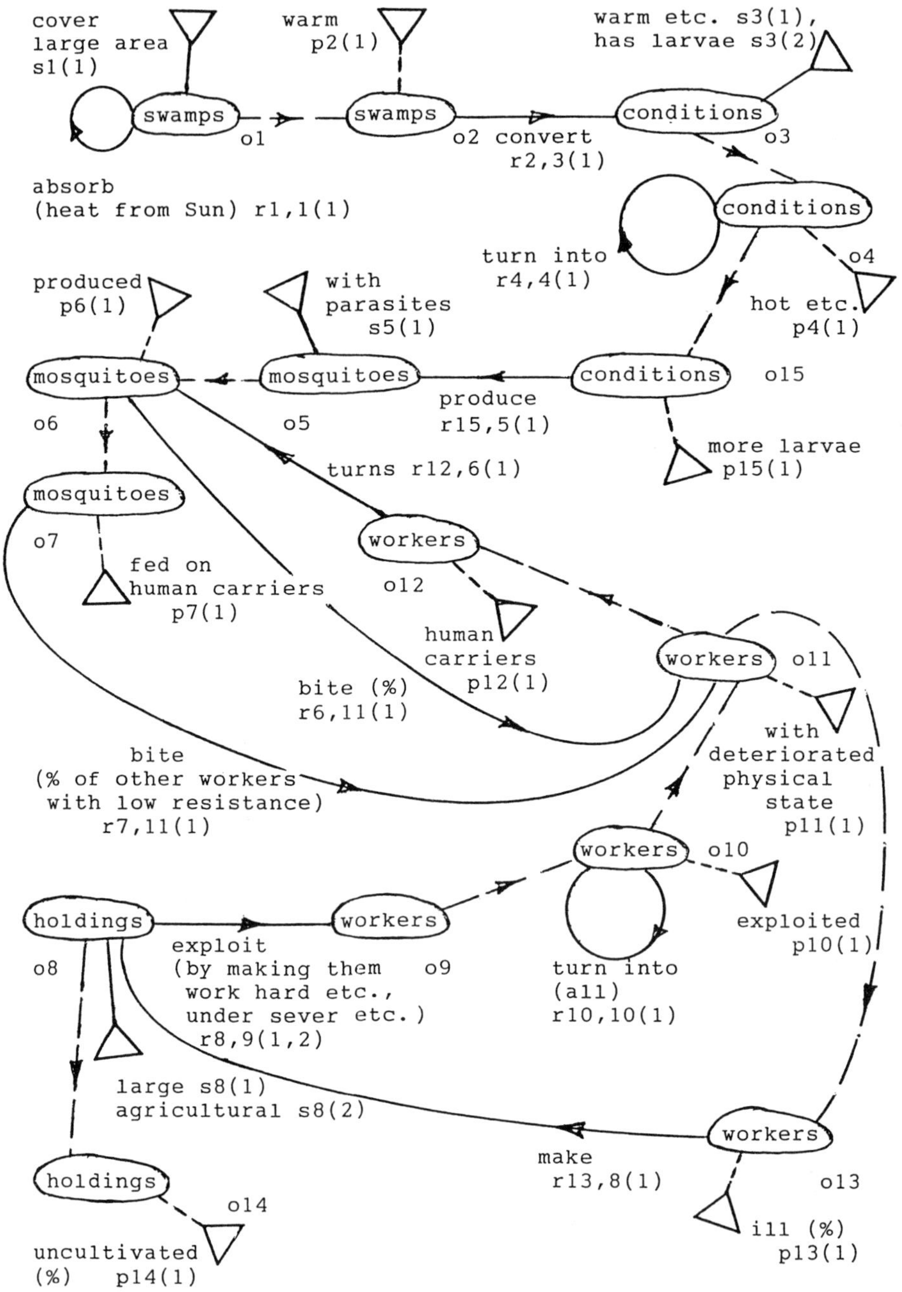

Figure 1. Semantic diagram of a situation.

The fuzzy distributions in Fig. 3 indicate a diminishing trend in the number of mosquitoes with parasites. This trend as shown in Fig. 4, results in the possibility of the amount of uncultivated holdings becoming less and less. This is so since fewer mosquitoes with parasites is bound to lead to fewer workers becoming ill. The result confirms expectation in a fuzzy, quantitative manner.

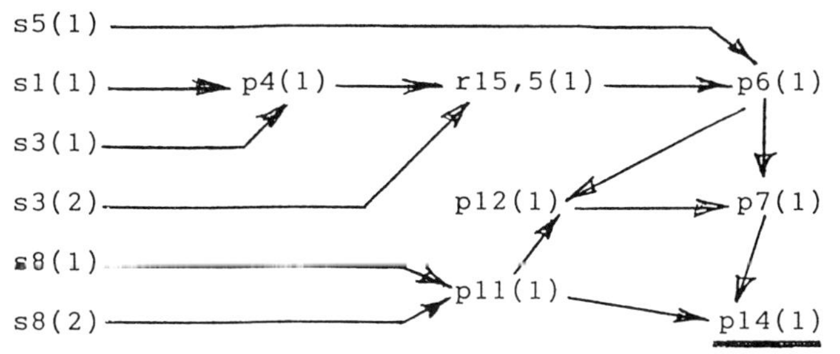

Figure 2. Deduction graph.

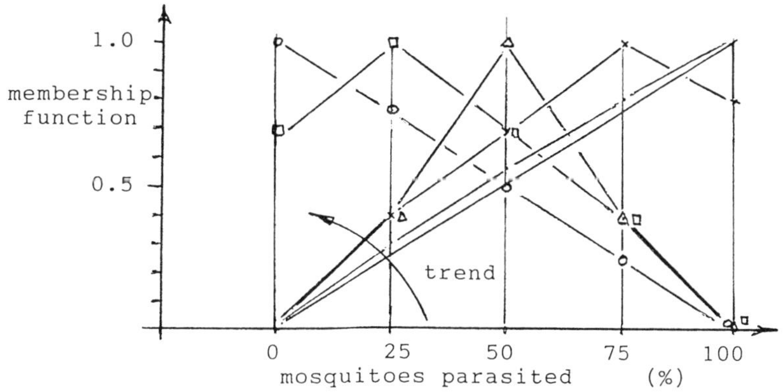

Figure 3. Instantiation of of 'mosquitoes are parasited' s5(1).

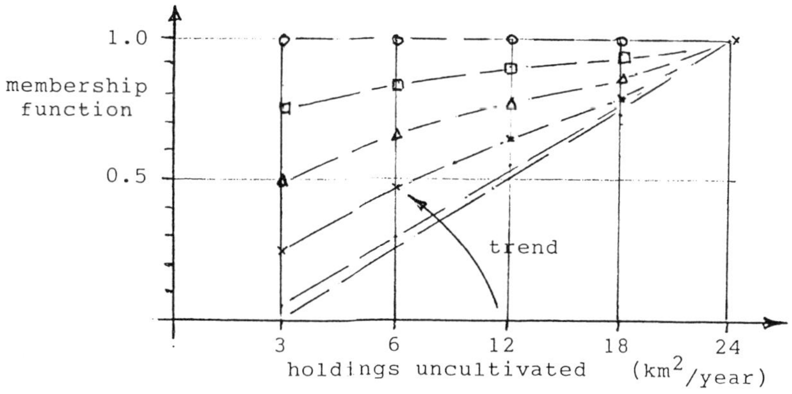

Figure 4. Distribution of outcome ' holding becomes uncultivated' p14(1).

ACKNOWLEDGMENT

The authors express their gratitude to colleagues, Dr F Huss and Mr J Cumbers who contributed a great deal to the development of linguistic modelling.

REFERENCES

Campos de, L.M., and Gonzalez, A.,1993, A fuzzy inference model based on an uncertainty forward propagation approach, *Int.J. of Approximate Reasoning*, 9:139-164.

Korn, J., 1994, A new direction in systems research,*Systemist*, v16(2).

Korn, J., Huss, F., and Cumbers, J., 1993, Systems modelling with natural language and fuzzy logic, *in*: "Systems science addressing global issues", F.A Stowell, D. West, and J.G. Howell, Plenum Press, NY.

THE AUTOPOIETIC VIEW OF PRISON ORGANIZATION AND OF CORRECTIONAL REFORMS

Effi Lambropoulou

Panteion University of Social and Political Science
136, Syngrou Avenue
176 71 Athens
Greece

According to the link of a system with the environment, closed and open systems are distinguished (Bertalanffy, 1956). For the closed systems external environment is of minor importance. As it is commonly known, the transfer of the idea of open systems into the social sciences is mainly due to Talcott Parsons (1977).

Parson's theory was a turning point in the sociology of prison during the 1960's and 1970's. Through research conducted in the light of the functional model, the effect of the inmates' groups (subcultures) on the internal equilibrium of the prison system and the importance of the balanced operation of the correctional system for the wider social context have been thoroughly examined. On the contrary, no attention has been paid to the effect of the external environment and the prisoners' social background. For this reason, it has been regarded by some that the functional model considers prison to be a closed system (Clemmer, 1940; Sykes and Messinger, 1960).

The importation model, conceived later, regards the external environment of great importance for the general operation of the organization, in the sense that the effect of the environment is expressed through the roles and the identities of the prisoners. The organized response of the prisoners, in respect to the problems they faced in prison, is associated with their social background (Schrag, 1961).

The changes in the social policy after the 'Big House' Era, along with a differentiation of criminality which influenced the role configurations, the prisoner values and the organized way of reacting to the problems generated by the deprivation of liberty, supported the importation model (Irwin, 1970). However, the inmate subculture and the alleged failure of the treatment programmes indicate that although the prison system is open to receiving information and stimuli (e.g. social and criminal background of the inmates, rehabilitation programmes, attitude of the staff and public opinion), it is closed as regards information and stimuli processing, and the inner control structure (reproduction of roles and relations, effectiveness of programmes, etc.).

That approach responds to the modern interpretations of social systems and, suggest that the existence of a system presupposes not only an external environment but an internal

Critical Issues in Systems Theory and Practice, Edited by K. Ellis et al.
Plenum Press, New York, 1995

one as well, the latter being quite different from the former (von Bertallanffy, 1956). The internal environment allows the system to change, when necessary to maintain its stability (self-reference) irrespective of problems that might appear in the external environment (Varela, 1979, Zeleny, 1981; Maturana and Varela, 1992). This new type of system has been termed 'self-referential system'.

On the basis of the theory of self-referential systems, the system itself defines and modifies its goals, in order to serve its self-reference. Self-reference constitutes a 'modus operandi' of the system, it is the relation of a system with itself. A more specific case of self-referential systems are the autopoietic ones. Autopoiesis is the mechanism which makes beings autonomous, able to formulate their rules and their special characteristics (Maturana and Varela, 1992, p. 71). The main characteristics of self-referential systems are self-reproduction and autonomy. A system aims at its self-reproduction, at its self-maintenance and it asserts itself not only when it attains a goal (Etzioni, 1973, pp. 33-37; Luhmann, 1988, pp. 165-166). The autonomy of the system is based on its closeness, namely its self-reproduction, while its openness is based on the internal processing of the external information which enrich and contribute to its self-reference (Maturana and Varela, 1992, pp. 98-103; Luhmann, 1984, pp. 191-192).

Since research about prisons questioned the closeness and the openness of the prison system, we will try to investigate if a prison is an autopoietic system. To answer the above query, we should highlight what the difference is between the prison system and its environment.

In the past, it was accepted that the identification of an object was the prerequisite for acquiring knowledge of the object. According to the theory of autopoiesis the aquisition of knowledge of a being, an object or a system is achieved through a clear-cut distinction from its environment (Maturana and Varela, 1992, p. 68; Luhmann, 1984, p. 61). The difference of the system from the environment re-enters the system processing the basic code of its self-reference (identity). All the proceedings taking place in the system are based on this code (Spencer Brown, 1972). The system understands only what is interpreted into this code. Therefore, a system exists only after it has defined the basic code of its self-reference. So an attempt will be made at the investigation of prison system's binary code (cf. Foucault, 1989, p. 239).

Penalty constitutes the 'apparatus' for the control of criminal behaviour. Until recently, the correctional laws of most countries, declared that the enforcement of the criminal penalty aims at the rehabilitation of prisoners. Rehabilitation composes the most important part of the special deterrence and the latter along with general deterrence constitute the instrumental function of the penalty.

By the surveys conducted, the success of the rehabilitation of offenders has been strongly questioned: neither was recidivism reduced nor were the prisoner's chances for social integration augmented. There was also little evidence that the deprivation of liberty or the low living standards have been able to prevent ex-convicts from committing new crimes (Rusche and Kirchheimer, 1939; Sutherland and Cressey, 1978, p. 348; Foucault, 1989, pp. 349-354, 365-369).

Since the instrumental function of prison is questioned, its contribution to the social system must be spotted. With the support of mass-media prison fabricates a hostile idol or a 'scapegoat'. According to psychoanalysis, this constitutes a mechanism, through which law-abiding citizens express their aggressiveness and revenge. Through projection and punishment of the offender, the fair image we have for ourselves asserts itself and our legal behaviour is fully justified (cf. Haffke, 1976, p. 164). Usually, the role of the enemy is played by some groups of persons which belong to the social margin, such as certain categories of offenders and prisoners.

Mass media offer a simplistic point of view to crime; deviance is identified with crimes which are committed in a violent way. By placing emphasis on these crimes, mass media create the impression that criminality is higher than it actually is (Voß, 1979). So, people feel threatened by dangerous perpetrators, thus regarding the taking of strict protection measures as indispensable (cf. Foucault, 1989). In this way, the mechanism of social control is justified in the eyes of the public. There is also the impression that the State protects its citizens: therefore, the reliability and prestige of the State are restored, the trust of the public in the dispensation of justice is reinforced and perhaps the violation of law is prevented. By placing emphasis on crime and the severity of sentences, public opinion is usually misled (Mathiesen, 1979, pp. 157-162), since there exist equally grave criminal acts, such as environmental pollution, gun trade, 'dirty' wars, or severe social, economic and political problems, which are not seriously dealt with. The disorientation of the citizens from serious problems, the justification and the gloss of efficiency of the mechanism of social control in the fight against crime constitute the symbolic function of prison.

The fabricated hostile idol mentioned above, which is often personified by offenders and prisoners, ensures the coherence amongst law-abiding members of society, since they feel that they need to protect themselves from the enemy that threatens the social rules in force, covers up existing class conflicts and reinforces the authority of the broken law (Mead, 1918; Parsons, 1961, pp. 309-321; Durkheim, 1977, pp. 147-149). This is the social function that prison fulfills.

Therefore, although the primary aim of the deprivation of liberty is the decrease of criminal behaviour through the rehabilitation or the intimidation of offenders, other functions are fulfilled. These functions contribute to the stabilization of legal behaviour in society through a 'sophisticated' intimidation, exercised especially on social strata which are at great risk to violate the law. This kind of deterrence might reduce criminal behaviour and support crime control. It seems that prison system manages the binary code "stabilization-destabilization of legal behaviour" (cf. Foucault, 1989, pp. 349-383).

As already mentioned, the primary interest of a system, and in our case of prison organization and corrections system, in general, is its maintenance and not the achievement of its publicly declared target. Society tolerates the ineffectiveness of prison concerning resocialization, because it fulfills other functions, equally significant with the instrumental one.

If the corrections system was interested in the rehabilitation of the prisoners, then it should try hard to increase the possibillities of its success. Of course, one could maintain that at times, rehabilitation programmes were implemented, but recidivism did not decrease, which caused the programmes to cease. However, the programmes were mainly implemented in organizations with a structure, suitable for 'just doing time' and not for rehabilitation. The majority of the wardens wished for the maintenance of the old hierarchy and did not contribute to the success of the new model. Correctional officers were inadequately trained and they did not approve of the treatment of the penal offenders, while the specialized personnel failed to produce a climate of cooperation with the correctional officers.

Public opinion was not positive as to the cost of correctional institutions, and for that reason the programmes were poorly funded, while politicians exploited this reaction and achieved the suspension of the programmes. Prisoners also questioned the model, because they experienced negative social reactions to their integration and understood that social factors influenced their future, irrespective of their participation in the programmes.

The rehabilitative ideal, was finally rejected because it was not compatible with the operation code of the corrections system. It was not the rehabilitative ideal which failed, but the corrections system could not absorb it. An autopoietic system accepts the information from the external environment and controls it through the basic self-reference code. If the external information cannot be decoded or absorbed by the system, then it is rejected as superfluous.

Therefore, we may conclude that reform efforts failed, among other reasons, because they overlooked the internal resistance of the prison, as its goal seems to be self-preservation through the maintenance of people who serve as an example to be avoided by certain social strata.

REFERENCES

Bertalanffy, L. von, 1956, General Systems Theory, *General Systems* 1:1-10.
Clemmer, D., 1940, "The Prison Community", Holt, Rinehart and Winston, New York.
Durkheim, E., 1977, "Über die Teilung der sozialen Arbeit", Suhrkamp, Frankfurt a.M.
Etzioni, A., 1973, "Soziologie der Organizationen", 4th ed., Juventa Verlag, München.
Foucault, M., 1989, "Epitirisi ke Timoria. I Genisi tis Filakis", Pappa, Athens ("Surveiller et Punir. Naissance de la Prison", Galimard, Paris 1976).
Irwin, J., 1970, "The Felon", Prentice-Hall, Englewood Cliffs, N.J.
Haffke, B., 1976, "Tiefenpsychologie und Generalprävention. Eine strafrechtstheoretische Untersuchung", Sauerländer, Aarau, Frankfurt a.M.
Luhmann, N., 1984, "Soziale Systeme", Suhrkamp, Frankfurt a.M.
Luhmann, N., 1988, Organisation, *in*: "Mikropolitik. Rationalität, Macht und Spiele in Organisationen", W. Küpper and G. Ortmann (eds.), Westdeutscher Verlag, Opladen, pp. 165-185.
Maturana, H. and Varela, F., 1992, "To Dendro tis Gnossis", Katoptro, Athens ("The Tree of Knowledge", Scherz Verlag, Bern, Munich, Vienna 1984).
Mathiesen, T., 1979, "Überwindet die Mauern! Die skandinavische Gefangenenbewegung als Modell politischer Randgruppenarbeit", Luchterhand, Darmstadt.
Mead, G.H., 1918, The Psychology of Punitive Justice, *AJS* 23:557-602.
Parsons, T., 1964, "The Social System", Free Press, New York.
Parsons, T., 1977, "Social Systems and the Evolution of Action Theory", Free Press, New York.
Rusche, G., and Kirchheimer, O., 1939, "Punishment and Social Structure", Columbia University Press, New York.
Schrag, C., 1961, Some Foundations for a Theory of Correction, *in*: "The Prison. Studies in Institutional Organization and Change", D.R. Cressey, (ed.), Rinehart and Winston, New York, pp. 309-357.
Spencer Brown, G., 1972, "Laws of Form", Julian Press, New York.
Sutherland, E. and Cressey, D., 1978, "Criminology", 10th ed., Lippincott, Philadelphia.
Sykes, G., and Messinger, S.L., 1960, The Inmate Social System, *in*: "Theoretical Studies in the Social Organization of the Prison", R. Cloward, et al (eds.), Social Science Research Council, New York, pp. 14-43.
Varela, F., 1979, "Principles of Biological Autonomy", Elsevier, New York, Oxford.
Voß, M., 1979, "Gefängnis für wen?", AJZ, Bielefeld.
Zeleny, M., 1981, "Autopoiesis. A Theory of Living Organization", Elsevier, New York, Oxford.

108

DETECTING ETHICS IN SOCIAL SYSTEMS

Gianfranco Minati

Associazione Italiana per la Ricerca sui Sistemi,
10 Via Jenner
20159-Milano
Italy

ABSTRACT

A list of practical rules for companies and banks is proposed in order to identify and measure ethics. A service named Ethical Quality Measurement (EQM) is proposed. Important references with some activities of the European Economic Community (EEC) in defining the ECOLABEL and ECOAUDIT programs are identified. A very important reference to the research activity about the concept of 'logical openness' in General Systems Theory (GST) is identified and a parallel with the research on semantic processing in artificial intelligence is proposed.

INTRODUCTION

This paper is based on the 1993 Asilomar Conversation on education. It consists of my reflections following the group discussion about ethics. After the Conversation, I designed a type of service, named Ethical Quality Measurement (EQM), that will be proposed for use in Italy. This service is based upon the following key points:

1. It is unethical to design a system for somebody else;
2. Designing is ahead of the process of producing knowledge (production of knowledge);
3. There is a dialectic between the production and the discovering processes;
4. We have responsibility for our freedom;
5. There are many levels of reality;
6. Ethics is a process
7. Ethics is a production and a discovery at the same time;
8. We need not only to solve but also to manage;
9. Ethics is always present, even when we do not declare it (implicit ethics);
10. How can we design a system able to produce ethical awarness? (explicit ethics);
11. Make the ethics of the system evident;
12. The system always has ethics, even when they are not evident;

Critical Issues in Systems Theory and Practice, Edited by K. Ellis et al.
Plenum Press, New York, 1995

13. Ethics is only for living systems;
14. It is unethical to be disinterested in the use of her/his own product/service;
15. The designer is not always a human being (self-organization process);
16. Cooperation and competition are complementary;
17. There are degrees of ethics; make them measurable;
18. There is a relationship between ethics and logical openness. (Collen and Minati, 1993; Collen, Minati, and Ciapessoni, 1994);
19. Examples of unethical systems are:

- social institutions designed for the workers inside more than for the users (specifically, hospitals for physicians rather than patients);
- schools that graduate incompetent students;
- when a person has an unreasonable expectations of another person.

20. Ethics is a silent (shadow) variable;
21. s is a construction of the mind;
22. The uses *and* the production of knowledge depends on the ethics of the social system

After discussion with people in companies and some students, I noted the additional points:

23. Ethics comes before the law;
24. Ethics is strategic; no strategy is possible without ethics;
25. Ethics should be as the 'operating system software' for social systems (as they were *computing* systems)

I consider these points very fundamental for the application and achievement of a service for the community and future generations.

PRINCIPLES FOR AN ETHICAL QUALITY MEASUREMENT (EQM) FOR COMPANIES

Following the 25 points above I developed a check-list for an (EQM) Ethical Quality Measurement to be offered to companies, banks and institutions. The list consists of 7 parts.

i. The Company is interested in its product from the beginning to the end of its life-cycle. The physical impact on the environment is examined, as the ecological impact of the production; the ecological impact of the product usage; the recyclability and the re-usability of the product and of the packaging; the relationships with other products; the necessity of other non-ecological products; the transportation, packaging, preservation problems;
ii. The Company is interested in the way workers produce (examine the social impact on the workers). The company examines the need to use the car for the workers, the social problems related to the need to be physically present in the company (e.g. dealing with the bureaucratic paperwork in the public institutions), the usage of cheaper labour (workers with no rights as in other non-developed countries);
iii. The Company examines the social impact of its products and of the usage of these products (the social impact on the people is examined). Related problems or needs are detected and examined (unexpected reactions are considered). The company examines its behaviour globally as part of the total social system and it collaborates with research institutions (such as Universities) for this purpose;

iv. The company *designs* ecological ways to use its products (especially in combination with others ones), advertises them, proposes them when the customer buys the product and keeps the customers continuously informed;
v. The company makes the committment to encourage some ecological usages, supporting and giving orientation to the customers with suitable accessories;
vi. The company publishes the costs of its activities for the global social system (taking into consideration the cost of the usage of services: sanitary, education, transportation, and pollution, ...);
vii. With the documentation about the product, the company gives all clients its 'Ethical Statement' as completion of the guarantee of the product itself.

This list is the base for the set of items annexed to the balance. This *ethical accounting statement* can be used:

- for advertising;
- for special credit services as the ones expected to be activated by the EEC with the program ECOAUDIT (Council Regulation, 1993; DG XI UK Department of the Environment, 1993); for the right to use the European ECOLABEL label, giving the companies trading advantages and assurance of quality to customers (EEC Eco-label Award Scheme, 1993; EEC-IP (93) 539, 1993).

A SHORT DESCRIPTION OF THE COMMUNITY'S ECO-LABEL AWARD SCHEME

"The eco-label will be awarded to products which have a reduced impact on the environment. Manufacturers who have obtained the appropriate authorisation will be permitted to use the official logo - a flower with the community's star symbolism- on their approved products. Thus consumers will be able to distinguish environment-friendly products from others.

The Community Eco-label Award Scheme is voluntary. The decision whether or not to apply for the label to be awarded to a product is up to the manufacturer to take. But it is clear that the Eco-label will gradually become a clear point of reference for the general public when buying goods.

The strength of the label is its European dimension. Once approved by one Community Member State, it can be used throughout the other 11 states. This avoids having to make an application in every country where a national label exists and thus avoids time-consuming and costly procedures.

In December 92, the Commission began the process of consultation of the principal interest groups in relation to eco-labelling proposals for washing machines and dishwashers (EEC-IP(93)539, 29 June 1993).

In January 93 the Consultation Forum gave its opinion to the proposals and in March 93 the Regulatory Committee agreed on them..." (ECC Eco-Label Award Scheme, 1993)

PRINCIPLES FOR AN ETHICAL QUALITY MEASUREMENT (EQM) FOR BANKING

1. The bank gives its customers the 'Ethical' contract signed by the bank itself. The bank gives in the same statement to its customers its 'mission' (e.g. to support with special credit ecological activities; to support employment for the younger

generation; to support artistic production or preservation; to support scientific research; to support tourism; to support trading; to support craft, ...);
2. The Bank encourages (banker's credit) and supplies credit to companies involved in the complete life-cycle of its product/service and publicly declares the usages. So the bank takes into consideration whether the products and their usages are ecological, how long-term the projects are and what their ability to generate value is (*global wealth*). The bank, by the usage of credit, is bound to support the growth of companies and of the market towards the production and/or the distribution of ecological and strategic products/services;
3. The bank undertakes to understand the needs of customers always suggesting more favourable solutions and keeping them informed of developments regarding their activity with the bank;
4. The bank tries to be as accessible, clear and easily understood as it can be (production of guiding-videotapes, for explanations, for didactics, for simulations and for particular needs);
5. The bank makes the committment to design a fixed procedure for every principle;
6. The bank makes provision for fines in the case of errors (as happens with products under guarantee);
7. The bank offers a 'maintenance service' (as with products). With the customer the bank examines the service at regular intervals and tries to fix problems by proposing solutions. The Bank accepts regular services and sporadic services;
8. Annually the bank examines the relationship with the global context in terms of: safety, accessibility, availability, reliability, understandability, timeliness, credibility, efficiency, effectivenness, and publishes this report;
9. The bank displays a style in its behaviour (rigour, kindness, availability, simplicity);
10. The bank undertakes to understand its market from a sociological point of view with reference to its particular needs.

This list is the base for the set of items annexed to the balance. This *Ethical Accounting Statement* can be used: to be different and identifiable in the market, in order to improve fidelity and to give workers pride in the Bank itself.

The bank offers the market a special (EQM) for companies through a special Division or Association with other banks, companies or institutions.

CONCLUSION

The system view allows us to link ecology and ethics. Is ethics the 'social software' for designing the ecology for human systems?

Ethics appears to be strongly related to the *arbitrariness* of the human beings (very different from the *casuality* emulable by a computing machine) and his ability to act as an open system (Collen and Minati, 1993) at the maximum logical level (the ability *to decide* to be open at certain levels and to switch continously from an open to a closed status and viceversa).

We need practical and theoretical tools for the manipulation of ethics, how to produce it, measure it, detect it , compare it, improve it, give it economical meaning.

Is ethics just a set of rules for the social systems? The theoretical centrality of the human being is foundamental in my view and is the source of openness, arbitrariness and ethics. It's because of it that the General Systems Theory is not just another tool. The concept of ethics is very close to the concept of 'social operating system'. The process of identifying ethics in the social systems and to decide it (by the usage of the EQM) is very similar to the process of designing (social) software for (human) systems as they were computing systems.

This is very close to the problems we have in human cognition for semantic processing (Minati, 1993). The need for theoretical tools in logical openness emerges. (Collen, Minati, and Ciapessoni, 1994).

REFERENCES

CEC, DG XI UK Department of the Environment, June 1993, Research into the Development of Codes of practice for Accredited Environmental Verifiers within the framework of the Proposed Eco-Management and Audit Regulation, Draft Guidelines.

Collen, A., and Minati, G. , 1993, Logical Openness and the ethics of globality, *in*: proceedings of the "Asilomar Conference", Asilimar, USA

Collen, A., Minati, G., and Ciapessoni, E., 1994, Logical Openness in Systems, *Systems Research*, 11:2

Council Regulation (EEC) No. 1836/93, 29 June 1993, Allowing voluntary partecipation by companies in the industrial sector in a Community eco-management and audit scheme.

Minati, G., 1993, The Project Interscience - a project for semantic processing based on general systems theory, *in*: Proceedings "Second European Congress on Systems Science, European Systems Science Union", Vol.1: 5-8, Prague.

Press release EEC-IP(93)539, 29 June 1993, The Commission adopts for the first time ecological criteria for the award of the eco-label to two product groups, Brussels.

The EEC brochure The Community's Eco-label , January 1993, Award Scheme and Europe Environment: Our Joint Future

THE APPLICATION OF SYSTEMS THINKING TO MODERN POLICING

Trevor Pearce[1] and Joyce Fortune[2]

[1] Kent County Constabulary
Police Headquarters
Sutton Road
Maidstone, Kent
ME15 9BZ
[2] Faculty of Technology
The Open University
East Midlands Region
143 Derby Road
Nottingham , NG7 1PH
United Kingdom

INTRODUCTION

Policing is a complex activity that is subject to growing and ever changing political, social and economic pressures. This paper argues that it is an area that is particularly suited to systems thinking. The paper presents the findings of a search of UK literature into the extent to which systems have already been applied to policing, and suggests further areas of activity where systems thinking could usefully be applied thus putting forward an agenda for further research.

COMPLEXITY IN POLICING

It has been said that systems research has come into being in response to the growing complexity of an over-technical world. However, it is not just technological developments that have confronted the police service over recent years, but also political, social and economic changes within society. The true impact of these latter features was perhaps being alluded to by Scarman (1981) in his report on the Brixton disorders of that year. He indicated that any inquiry into policing methods had to take account of the social context in which the police operated.

Building on the above, it is possible to identify further features which account for the increased complexity of policing as it exists in the 1990's. For example, the Police Service has had to assimilate advances in information technology, forensic science, changes in

Critical Issues in Systems Theory and Practice, Edited by K. Ellis et al.
Plenum Press, New York, 1995

legislation, new management practices and procedures, structural alterations, and a move to greater multi-agency co-operation. Taking all these features together, the high degree of complexity originates from: the number of elements which policing now comprises; the nature of the elements; the potential for greater interaction between them; and the degree of organisation that has had to develop in response external pressures. So, as Bradley, Weaver and Wilkie (1986) rightly point out: "Both Policing and its management are difficult and complex tasks, and ... there is little doubt they will become more difficult and more complex in future" p.vii.

SYSTEMS THINKING APPLIED TO POLICING

It should not be necessary in this paper to restate the value of systems thinking in examining complexity. Within the police service however, there has been little official recognition of the benefits which the use of systems concepts can achieve in the analysis of problems or issues affecting it. There have though been individual attempts, mainly by serving police officers, to develop the use of systems thinking and concepts within the policing context.

Hart (1982) used a systems approach to consider community-based police operations, looking in particular at contemporary policing problems. (Specifically increased demands on police resources.) He examined the objectives of the police service in relation to its structure and process, and conceptualised reactive and proactive systems. These allowed him to specify principles which led to the design of an operational policing system.

Beckett (1989) has also made use of systems concepts. Emphasising the need for local policing, he set out the need for clear analysis of how this should be achieved and the requirement for clearly defined systems of policing. Beckett also focused upon the reactive (public) demands upon the service and as a means of breaking what he termed the 'Extended Reactive Spiral', developed a geographic policing system.

Bond (1988) looked at decision making by 'police managers'. In a survey he examined the varying expectations held by operational officers, police managers and the public, and found that these groups had different expectations. Police managers were applying management models that often failed to take account of public demands. The study also identified that the police managers were not sufficiently aware of environmental complexity and Bond proposed an alternative, more complex, model based upon Beer's Viable Systems Model (VSM).

Organisational communication in the North Yorkshire Police was examined by Green (1992) through the Total Systems Intervention meta-methodology. Green's research revealed the differing perceptions of what would constitute effective communication within a hierarchical, bureaucratic organisation, and identified metaphors for the organisational structure. By intervening in the situation via use of the VSM and Soft Systems Methodology (SSM) a better understanding of local autonomy was achieved, and a shared understanding between actors of the consultation process in decision making was developed.

North Yorkshire Police is also the setting for a case study by Flood (1993) on public sector quality. He notes that in the course of developing quality of service, use was made of the VSM to produce a vision of the organisation. This then formed an essential background when structures for the implementation of Total Quality Management were considered.

Organisational structure was also considered, in systems terms by Pearce (1992(a)) who used the Formal System Model (FSM) to look at organisational change in the Kent County Constabulary. The analysis found, that a new model, based upon a two tier devolved organisational structure would, overall, be able to meet the requirements outlined for effective systems performance and behaviour. Based on this work he also suggested that the FSM was a useful tool for validating proposed change.

In a separate study Pearce (1992 (b)) used the Failures Method to analyse three perceived failures in policing activity. These were of three types: catastrophic, organisational and frequently repeated occurrences. A critique was also made of the conventional methods for conducting inquires into these types of failure in policing, and attention drawn to the benefits which the use of systems thinking might bring.

A case study focusing on the policing strategy towards the carrying of offensive weapons was used by Flood and Jackson (1991) in an examination of Critical Systems Heuristics (CSH). The case study itself was based on an earlier analysis of the problem situation by Brydges (1989) using SSM, but Flood and Jackson suggested that the use of SSM was not appropriate given the context of the problem situation because of the macro social difficulties and the wide range of views held by different actors. They argued that CSH was more appropriate because of its ability to expose the underlying agendas of those responsible for planning, especially in coercive situations when multi-agency activity is taking place.

Notwithstanding the comments above, SSM has certainly been used effectively in a police setting. When Gaisford (1989) (also reported in Flood and Gaisford (1989)) applied it to a contemporary social problem, namely vice in an inner city area and the response of the police to it.

A survey of the international literature is now being carried out, but where the UK is concerned even this brief summary of the literature should make clear that most applications to date have been concerned with the way police forces are structured and managed, or contemporary social issues affecting the police. Thus they have been organisation-based or issue-based and as such have neglected those areas of policing activity which can be categorised as function-based. Procedures and methods which the police service employs in complex operational activity have been ignored.

A RESEARCH AGENDA

In looking at the functions undertaken by the police, some stand out as providing clear scope for the application of systems thinking. The first area where it can be employed concerns the command structure applied to the management of major incidents and public disorder. Command and control of dynamic incidents involving the management of a large number and variety of resources is highly complex, and failures in such activity have in the past led to much criticism of the police service (see for instance Taylor, 1990). Systems thinking would enable an examination of relationships between the various command levels, and allow consideration of the appropriateness of the command structure to a variety of operational incidents. Such a study would also be able to focus on tensions which may exist between the command levels due to function, location and hierarchy. Pearce and Fortune (1995), have already commenced an investigation into this area, through a systems assessment of the Gold, Silver and Bronze command structure often employed by the police service in major incidents and events.

A second activity which might benefit from consideration using a system framework, is intelligence gathering and intelligence-based law enforcement activity. A recent report (Audit Commission, 1993) recommended that the police service should make greater use of intelligence in law enforcement, particularly in respect of targeting the known prolific offender rather than investigating historic reports of crime. Although the police service is taking a more pro-active approach to crime investigation, the proposals of the Audit Commission are still radical. They raise immense issues in respect of intelligence gathering such as public sensitivity, legal constraints, and cultural change within the police organisation. These, together with the requirement for robust operational procedures, increase the overall complexity still further. An investigation of this new approach through a systems framework may provide an interesting insight into its operation, potential emergent features, interaction of components and the environmental constraints.

Another vital area of policing activity into which systems thinking has the potential to provide further insights is the management of major crime investigations. Serious and series crime is subject to rigorous investigation by the police service, often making heavy demands on resources. A successful investigation requires the synthesis of many activities over and above the pure investigative process. These may include: the management of staff; victim care; financial and resource management; legal processes; scientific investigation; information and intelligence management; and many others. The complexity of the task is undeniable.

It is hoped that further research in these particular areas will have two outcomes. First, contribute to the debate about what constitutes effective and relevant policing and how it might best be accomplished; and secondly, assess the practical value, from a policing perspective, of adopting a systems approach.

REFERENCES

Audit Commission for Local Authorities in England and Wales, 1993, "Helping with Enquiries: Tackling Crime Effectively", HMSO, London.

Beckett, I., 1989, "Policing Systems", Ph.D. Thesis, The City University, London.

Beer, S., 1985, "Diagnosing the System for Organisations", John Wiley & Sons, Chichester.

Bond, K.P., 1988, "The Organisational Effectiveness of Policing in an Age of Complexity", Ph.D. Thesis, The University of Aston, Birmingham.

Bradley, D., Weaver, N., and Wilkie, R., 1986, "Managing the Police: Law Organisation and Democracy", Wheatsheaf Books Ltd, Brighton, Sussex.

Brydges, T.H., 1989, "A Systemic Analysis of the Rapid Rise in Crimes Associated with the Use of Offensive Weapons", M.Phil. Thesis, The City University, London.

Flood, R.L., 1993, "Beyond TQM", John Wiley & Sons, Chichester.

Flood, R.L., and Jackson, M.C., 1991, "Creative Problem Solving: Total Systems Intervention", John Wiley & Sons, Chichester.

Flood, R.L., and Gaisford, P., 1989, Policing vice in the west end, *OR Insight,* Vol. 2, No. 3, p 10-15.

Gaisford, P.J., 1989, "A Systemic Analysis of Female Street Prostitution Within the Environs of Central London", M.Phil. Thesis, The City University, London.

Green, S.M., 1992, Total systems intervention: organisational communication in North Yorkshire Police, *Systems Practice,* Vol. 5, No 6, p.585-599.

Hart, J.W, 1982, "A Systems Based Perspective of Police Operations", B.Sc. Degree Project, The City University, London.

Pearce, T.W., 1992(a), "The Formal Systems Paradigm used in the predictive mode to examine the structural reorganisation of a provincial police force", Unpublished Bramshill Paper Junior Command Course 4/92, Police Staff College, Bramshill.

Pearce, T.W., 1992(b), "Systems Failure Analysis and its Suitability as a Tool in Police Management", M.Phil. Thesis, The City University, London.

Pearce, T.W., and Fortune, J., 1995, "Command and Control in Policing - A Systems Assessment of the Gold, Silver, and Bronze Structure", in preparation.

Scarman, The Right Honourable Lord Justice, 1981, " The Brixton Disorders 10 -12 April 1981", HMSO, London.

Taylor, P., 1990, "The Hillsborough Stadium Disaster 15.4.1989", HMSO, London.

AUTHOR INDEX

INDEX